Underworld

UNDERWORLD

The Mysterious Origins of Civilization

GRAHAM HANCOCK

Photographs by SANTHA FAIIA

 THREE RIVERS PRESS · NEW YORK

Published by Three Rivers Press, New York, New York.
Member of the Crown Publishing Group, a division of Random House, Inc.
www.crownpublishing.com

THREE RIVERS PRESS is a registered trademark and the Three Rivers Press colophon
is a trademark of Random House, Inc.

Published in hardcover by Crown Publishers, a division of Random House, Inc., in 2002.

Originally published in Great Britain by Michael Joseph, an imprint of The Penguin Group, London.

Printed in the United States of America

Libary of Congress Cataloging-in-Publication Data is available upon request.

ISBN 1-4000-4951-2

10 9 8 7 6 5 4 3

First Paperback Edition

For Santha . . . for being there. Again

With all my love.

Contents

Acknowledgements

Underworld has been a huge, all-consuming quest spread out over a period of almost five years. I can only thank here a small number of the very many people who have contributed to it in one way or another.

First and foremost, thanks to my wife Santha, who travelled every step of the journey with me, took all the risks side by side with me, did every dive with me, faced up to every challenge with me and lived and breathed *Underworld* for five years just as I have done. Of course, all the photos in the book are Santha's but there has only been space here to reproduce a tiny fraction of them. Many more of her wonderful pictures from our adventures appear on the section of my website (*http://www.grahamhancock.com*) that is dedicated to *Underworld*.

Special thanks to Sharif Sakr, my brilliant researcher, who joined me straight out of Oxford University in the summer of 2000 when the writing phase of the book was just beginning. Sharif is, in every sense, exactly what a great researcher should be – an original thinker and an individualistic self-starter with huge intelligence, boundless energy and limitless initiative who never needs to be told what to do but who always just gets on and does it. Sharif's contribution to the strengths of *Underworld* has been enormous.

Thanks also to John Grigsby, my researcher for some years before Sharif joined me, and to Shanti Faiia for her excellent work on researching, planning and coordinating many of the diagrams in the book. Thanks to Sean Hancock for researching Ice Age chronologies and chasing rumours of underwater ruins for us at Pohnpei and Kosrae. Thanks to Leila Hancock for her research on the nature and attributes of Siva. Thanks to Shakira Bagwandeen for research notes on various issues of Indian religion and prehistory.

Dr Glenn Milne of Durham University's Department of Geology played a crucial role in generously providing all the inundation maps used in *Underworld*. Glenn's kindness in supplying these maps should not be taken as any sort of endorsement on his part of the broader theories and ideas presented here – which are entirely my own responsibility.

Thanks to Ashraf Bechai for showing us the mysterious underwater megalithic sites off Alexandria, particularly the gigantic blocks of Sidi Gaber, which orthodox archaeology has not yet come to terms with.

In what I have to say about Malta I drew heavily on the remarkable research of Dr Anton Mifsud and want to express my thanks to him for allowing me to report his findings so extensively here. If, as I believe to be the case, an entirely new chapter in the prehistory of Malta is about to open, then it is due to Anton's tireless search for the truth and the far-reaching investigation that he continues to conduct into the Maltese past. Thanks also to Anton's co-authors on his various books – Charles Savona Ventura, Simon Mifsud and Chris Agius Sultana.

In India I owe a debt of gratitude to all at the Archaeological Division of the National Institute of Oceanography (NIO), in particular to Kamlesh Vora, Sundaresh and Dr A. S. Gaur. Special thanks, too, to Dr Ehrlich Desa, head of the NIO, who did so much to clear the way for our dives at Dwarka and Poompuhur and showed such good will and kindness towards us when Santha and I first turned up at the NIO's headquarters in Donna Paula, Goa.

Thanks also to India's National Institute of Ocean Technology (NIOT), whose ground-breaking discoveries in the Gulf of Cambay are reported for the first time in this book. I mention in particular Dr S. Kathiroli, Project Director of the NIOT, Dr S. Badrinaryan, Geological Consultant to the NIOT, and G. Janaki Raman, Manager, Vessel Management Cell.

So many good people have helped us in Japan over the years that it is simply impossible to mention all of them here. I hope those whose names are left out will forgive me. Special mention must be made of our friend Shun Daichi, the Japanese translator of my books who accompanied me and Santha on our journeys in Japan – both above and below the water. Thanks too to Seamen's Club, Ishigaki, and to the staff and management there whose help made all our diving adventures in Japan possible. Outside Seamen's Club, Kiyoshi Nagaki, Isamu Tsukahara, Kihachiro Aratake, Yohachiro Yoshimaru, Mitsutoshi Taniguchi and Kuzanori Kawai all also dived with us and helped us in our underwater investigations.

Last but not least, Santha and I want to thank our children Ravi Faiia, Shanti Faiia, Sean Hancock, Leila Hancock, Luke Hancock and Gabrielle Hancock for putting up with our preoccupations and prolonged absences from home. All the children have played their own parts in the quest, have learned to dive and have joined us for some of our diving adventures. We're proud and happy to have such a bright and enterprising group of young people around us.

Graham Hancock,
London, January 2002

PART ONE

Initiation

1 / Relics

If you do not expect it, you will not find the unexpected, for it is hard to find and difficult.

Heraclitus

Five kilometres off the south-east coast of India, submerged at a depth of 23 metres beneath the murky, shark-infested waters of the Bay of Bengal, an ancient man-made structure sits on the bottom of the sea. The structure is U-shaped, like a huge horseshoe; its periphery measures 85 metres and its walls are about 1 metre thick and 2 metres high.[1]

The discovery was made by a team of marine archaeologists from India's National Institute of Oceanography (NIO) in March 1991, working off-shore of the Tranquebar-Poompuhur coast of Tamil Nadu near Nagapattinam. Their equipment included side-scan sonar, which transmits an acoustic signal up to 1000 metres wide and measures the strength of the returning echo. Towed behind a research vessel, side-scan sonar is capable of building accurate maps of sea-bed contours and of identifying any obvious anomalies such as shipwrecks.

On 7 March 1991 a shipwreck at a depth of 19 metres was pinpointed by the sonar. It was investigated by divers on 8 and 9 March, who found many scattered objects including lead ingots and iron cannon on the surrounding sea-bed. The official report of the project then states:

> Till 1.00 p.m. [on 9 March] the divers were working on the scattered objects. T. C. S. Rao who was carrying out sonar survey 5 km opposite Chinnavanagiri [not far from the wreck] reported another object 40 × 10 metres having the shape of a ship (?) recorded on sonograph. Shri Bandodkar was sent to the site (designated PMR2) and he placed two marker buoys there. By 2.00 p.m. Manavi and Chinni dived but as the buoys had drifted the object could not be explored.[2]

A second side-scan sonar survey later in the afternoon refined the measurements, suggesting that the object was oval and measured '30–35 metres east to west and 10 metres north to south with an apparent opening in one side'.[3]

On 16 and 19 March T. C. S. Rao continued the survey and now reported:

> There are actually three objects, the central one being oval-shaped with an opening on the northern side. Its longer axis is 20 metres. There is a clay deposit on the eastern

flank beyond which another semi-circular structure is seen. To the north-west of the central object one or more oval-shaped objects are found.[4]

Finally on 23 March 1991 three divers were able to go down but only had sufficient air to study the central structure. The official report describes what they saw as follows:

> a horseshoe-shaped object, its height being one to two metres. A few stone blocks were found in the one-metre-wide arm. The distance between the two arms is 20 metres. Whether the object is a shrine or some other man-made structure now at 23 metres depth remains to be examined in the next field season . . .[5]

Deep can mean very old

In the event no work could be done at the site in the next season, but in 1993 the structure was examined again by the NIO's diver archaeologists, who took careful measurements and eventually reported their findings as follows:

> The structure of U-shape was located at a water depth of 23 metres which is about 5 kilometres off shore. The total peripheral length of the object is 85 metres while the distance between the two arms is 13 metres and the maximum height is 2 metres. The height of the eastern arm is greater than that of the western arm. The centre of the object is covered with sediment but some patches of rock were noticed. Hand fanning showed that the central part of the object is rocky at a depth of 10–15 centimetres. Divers observed growth of thick marine organism on the structure, but in some sections a few courses of masonry were noted.[6]

Since 1993, for want of funding, no further marine archaeology has been conducted along the Poompuhur coast and the general impression has been disseminated in archaeological literature that the NIO has not found any submerged structures there that are older than the third century BC.[7] This is certainly true of numerous structures that were excavated very near to the shore, usually in depths of less than 2 metres of water and often half-exposed at low tide.[8] But the U-shaped structure at 23 metres – more than 70 feet – is another matter altogether and cannot by any means be automatically assigned to the third century BC. On the contrary, since we know that the sea-level has been continuously rising during the last 19,000 years,[9] common sense suggests that structures now submerged by 23 metres of water must be much older than structures in just 2 metres.

'Nobody has looked . . .'

In February 2000 I travelled to Bangalore to the home of the doyen of India's marine archaeologists, S. R. Rao, founder of the Marine Archaeology Centre at the NIO and the man who had led the Tranquebar-Poompuhur survey. Rao is a distinguished, lean-faced man in his mid-seventies, with boundless energy and enthusiasm for his subject. After the pleasantries were over I told him that I was intrigued by the U-shaped structure his team had found at Poompuhur: 'Twenty-three metres is deep. Doesn't that mean that it could be very old?'

'Correct, definitely,' Rao replied. 'That is what we are also thinking. In fact we took our ocean engineer also to see whether the structure had gone down as a result of erosion by the sea or by its own weight. I don't think that is the case, because it is a huge structure which has been built at that depth – at that time the sea was further out. This was built when it was above water. Then does the sea rise so much within such a short period was the question – 23 metres just within 2000 years or so?'

'Maybe the sea-level rise that covered this structure took place a lot earlier than that,' I offered. 'Maybe it belongs to a much earlier period than the 2000-year-old ruins of Poompuhur up in the intertidal zone? There have been sea-level rises that could have done something like this but they took place a long time ago – at the end of the Ice Age.'

'Correct. At that time it happened. You are correct.'

'There were three large floods at the end of the Ice Age – and even the most recent of these takes us back 8000 years. Is that a possible date for the U-shaped structure?'

'We don't know,' Rao replied, 'because you see from whatever we have got we are not able to decide its date at all.'

'Why is that?'

'Because amongst the samples we took we found no organic materials that could be dated by carbon 14 and no pottery that could be dated by thermoluminescence or by type. We have only stone which cannot be dated in any meaningful sense.'

'Except by one factor – which is that the structure is now under 23 metres of water. So the sea-level rise itself can be helpful in indicating a date.'

'Correct. I do know that for the Gulf of Kutch in north-western India an oceanographic study has been made and the oceanographers themselves have said that at 10,000 BC the sea-level was 60 metres lower than it is today. If that is true there it is also true here.'

'Which raises the possibility that we may be looking at remnants of a previously unknown ancient culture . . .'

'Ancient. Definitely!' Rao exclaimed. 'And, in fact, where really was the origin of India's earliest-known civilization – the Indus Valley civilization? Scholars guess, but nobody knows. The Indus Valley script itself is already a highly developed script when it first appears in the third millennium BC. The early architecture is already developed – you have got brick structures, you have got drains, everything is planned and all that – so there must be something before that. Where is the evolutionary phase? We don't know.'

Dr Rao was getting close to the real reason that I had come to see him. 'Maybe the evidence of the evolutionary phase is underwater?' I suggested.

'It's underwater. Quite possible.'

'If so, then this underwater structure at Poompuhur could be incredibly important – simply because of its depth . . .'

'Twenty-three metres . . .'

'Twenty-three metres. That's right. Now if we can rule out land subsidence, and further work must be done before we can rule that out, but if we can rule that out and if it's an issue exclusively of sea-level rise, then we have a discovery here that calls into question the accepted chronology of civilization.'

Rao pondered for a moment before replying: 'You see, some people, some traditions, do say that there was a continent in the Indian Ocean, a very long time ago, more than 10,000 years ago, that got submerged . . . Quite possible. You see, we are not doing thorough research. If we had taken more time and more funds and all that, perhaps we could find many more structures, not only that one, and then you could come to some kind of conclusion about the much earlier epoch.'

I told Rao that I was familiar with the south Indian traditions to which he was referring. These describe extensive lands, submerged about 11,000 years ago, that had once existed in the Indian Ocean to the south of the present Cape Comorin. The name of these lost lands was Kumari Kandam. At the time of their inundation, the traditions say, they had been the home of a high civilization that had even boasted an 'Academy' of advanced learning where philosophy and literature were cultivated.

'It must have existed,' Rao asserted. 'You can't rule that out at all. Particularly, as I have said, since we have found this structure at 23 metre depth. I mean, we have photographed it. It is there, anybody can go and see. I do not believe that it is an isolated structure; further exploration is likely to reveal others round about. And then you can go deeper, you see, and you may get more important things.'

I asked if there had been any further attempt since 1993 to find underwater structures off southern India.

'No,' Rao replied. 'Nobody has looked.'

Ken Shindo's story

In 1996, four years before my meeting with Rao, my book *Fingerprints of the Gods* became the number-one bestseller in Japan, a country that had fascinated me since childhood. The book's success gave me my first opportunity to travel there.

I visited Japan twice that year to give a series of public lectures about the issues I'd explored in *Fingerprints of the Gods*. On the second visit I was approached after a lecture by a photojournalist named Ken Shindo, who works for the influential Kyodo-Tsushin News Agency. He showed me striking underwater pictures that he had taken of a bizarre terraced structure, apparently a man-made monument of some kind, lying at depths of up to 30 metres off the south coast of the Japanese island of Yonaguni. My central research and writing interest, for years, has been the possibility of a lost civilization destroyed in the cataclysmic global floods that brought the last Ice Age to an end. So I was immediately fascinated by Shindo's story: 'An underwater ruin here in Japan!' I exclaimed. 'Is it definitely man-made?'

Shindo laughed: 'Some people say it's a freak of nature but they haven't spent time on it like I have. I'm absolutely certain it's man-made.'

'Does anyone know how old it is?'

Shindo told me that he had been working with Professor Masaaki Kimura, a marine seismologist at the University of the Rykyus (Okinawa), who had been studying Yonaguni's mysterious underwater structure since 1994. Kimura too was convinced it was man-made. His extensive survey, sampling and measurement had shown that it had been hewn out of solid bedrock when the site was still above water. If sea-level rise were the only factor to take into account, then provisional calculations would indicate a date of inundation of around 10,000 years ago.

That's approximately 5000 years older than the oldest known monumental buildings on earth – the ziggurats of ancient Sumer in Mesopotamia.

Davy Jones' Locker

I knew that I had to learn to dive and talked my wife Santha into doing lessons with me when we were on a visit to Los Angeles. We took our PADI Open-Water courses in the chill, kelpy waters off Catalina Island in November 1996.

My first reaction to diving was that it was a weird and scary experience, contrary to the laws of nature, and that I was unlikely to survive it. I was wrapped up like the Michelin Man in a full-body neoprene wetsuit, and there seemed to be a ludicrous amount of equipment strapped, velcroed or clipped on to me.

Let's start at the feet. Here the diver wears short rubber boots tucked inside the ankle-cuffs of his wetsuit. The wetsuit works by taking in a thin layer of water between the skin and the suit; this is rapidly warmed to body temperature and remains warm for some time because the neoprene of the suit is an excellent insulator. Over the boots are strapped the diver's fins, without which he would be almost as clumsy and immobile submerged as he is on land with all his gear on, and would unnecessarily waste a great deal of energy thrashing about. Strapped to his calf there should be a strong stainless-steel knife with a sharp blade – this can be life-saving if you get caught up in a drifting fishing net or some other equally uncompromising, usually man-made, hazard.

Around the diver's waist is a belt through which are threaded a number of lead weights to compensate for the natural buoyancy of the body and the additional buoyancy of the wetsuit. These days I can often get away with 2 kilos, but inexperienced divers need a lot more. On my first dives back in 1996 and into the first half of 1997, I remember having to use 12 and in one case even 14 kilos – a horrendous load.

Moving on up the body, the next item of equipment the diver wears is a partially inflatable sleeveless jacket called a Buoyancy Control Device – 'BCD', or just 'BC' for short. The scuba tank which provides the diver with air to breathe underwater is strapped on to the back of the BC and typically comes in 10, 12 and 15 litre sizes. A mid-sized tank weighs more than 15 kilos and for most dives is filled with nothing other than normal air under enormous compression. This is delivered to the diver through two transformers which step down the pressure of the air to a level where it can be breathed easily. The 'first-stage' is mounted immediately on top of the tank and removes most of the pressure; from here a rubber hose leads to the 'second-stage', or 'regulator', which is placed in the diver's mouth and provides air on demand. Three other rubber hoses also emerge from the first-stage. One of these connects to the BC, allowing the diver to power-inflate it direct from the tank. One leads to a dangling instrument-console usually containing a compass and gauges that tell you how much air you have left and how deep you are. The last, called the 'octopus', is a spare second-stage for use in emergencies – for example to provide air to another diver whose own tank is empty.

Sometimes divers wear a rubber hood, since heat loss from the unprotected head is very rapid. A glass-fronted mask, without which the human eye can only perceive blurred images under water, covers the eyes and nose. The final major pieces of equipment are a small wrist computer, which can save your life by warning you if you are ascending too fast from depth, and a pair of gloves to keep your hands warm and prevent grazing or accidental contact with unpleasant marine organisms like fire coral.

Wrapped up in all this stuff, with our total scuba experience at that time amounting to just three half-hour swimming-pool dives each, Santha and I contemplated the waters of the Pacific with certain misgivings. To be honest, we were afraid. It looked deep and dark and dangerous down there, down amongst the waving streamers of kelp, down in Davy Jones' Locker . . . But if we wanted to see that incredible underwater structure in Japan for ourselves then we were going to have to do this. On our instructor's command we jumped in and paddled out from shore.

Four days later we were licensed but definitely not yet experienced enough to dive at Yonaguni.

A generous offer

I did not know when we would be able to organize a diving trip to Japan but knew only that it would be expensive. Then a strange synchronicity occurred. Out of the blue some time in January 1997 I received a fax from an American company representing a Japanese businessman. The fax said that the businessman had read *Fingerprints of the Gods* and would like to invite Santha and me to fly first-class to Yonaguni at his expense to explore the island and to dive at the monument. He would ensure our safety by sending a group of top-flight diving instructors with us from the Seamen's Club, a hotel and dive school on the neighbouring island of Ishigaki. He would also provide us with a fully equipped dive boat and all other facilities.

There were no strings attached to this generous offer, which we accepted. In March 1997 we flew from London to Tokyo and then via Okinawa to Yonaguni to do our first dives there. This was the beginning of a long-term friendship with the businessman (whose privacy I protect) and of what began as an informal project to explore, document and try to understand the sequence of ancient and highly anomalous structures that have been found underwater at Yonaguni and at other islands in south-west Japan.

Yonaguni

The first anomalous structure that was discovered at Yonaguni lies below glowering cliffs of the southern shore of the island. Local divers call it Iseki Point ('Monument Point'). Into its south face, at a depth of about 18 metres, an area of terracing with conspicuous flat planes and right-angles has been cut. Two huge parallel blocks weighing approximately 30 tonnes each and separated by a gap of less than 10 centimetres, have been placed upright side by side at its north-west corner. In about 5 metres of water at the very top of the structure there is a kidney-shaped 'pool' and near by is a feature that many divers believe is a crude rock-carved image of a turtle. At the base of the monument, in 27 metres of water, there is a clearly defined stone-paved path oriented towards the east.

If the diver follows this path – a relatively easy task, since there is often a strong west-to-east current here – he will come in a few hundred metres to 'the megalith', a rounded, 2 tonne boulder that seems to have been purposely placed on a carved ledge at the centre of a huge stone platform.[10]

Two kilometres west of Iseki Point is the 'Palace'. Here an underwater passageway leads into the northern end of a spacious chamber with megalithic walls and ceiling. At the southern end of the chamber a tall, lintelled doorway leads into a second smaller chamber beyond. At the end of that chamber is a vertical, rock-hewn shaft that emerges outside on the roof of the 'Palace'. Near by a flat rock bears a pattern of strange, deep grooves. A little further east there is a second megalithic passage roofed by a gigantic slab that fits snugly against the tops of the supporting walls.

Two kilometres to the east of Iseki Point is Tategami Iwa, literally 'The Standing God Stone', a natural pinnacle of rugged black rock that soars up out of the ocean. At its base, 18 metres underwater, there is a horizontal tunnel, barely wide enough to fit a diver, that runs perfectly straight west to east and emerges amidst a scatter of large blocks with clean-cut edges.

A three-minute swim to the south-east brings the diver to what looks like an extensive ceremonial complex carved out of stone. Here at depths of 15 to 25 metres there are massive rectilinear structures with sheer walls separated by wide avenues.

At the centre is the monument that local divers refer to as 'the stone stage'. Into its south-facing corner either man or nature has carved an image that looks to some like a gigantic anthropoid face with two clearly marked eyes . . .

Kerama

At Aka Island in the Kerama group 40 kilometres west of Okinawa, local divers have been aware for some years of the existence of a series of underwater stone circles at depths of 30 metres. There are also associated rectilinear formations within the same general area that show some signs of having been cut and worked by human beings.

Diving conditions at Kerama are atrociously difficult (as indeed they often are at Yonaguni too). There is a killer current, but this drops away almost to nothing for approximately an hour between tides. Only in that lull is it possible to get any serious work done and to gain a perspective on the enigmatic structures without constantly having to fight against the sea.

Kerama's most spectacular feature is 'Centre Circle', which has a diameter of approximately 20 metres and a maximum depth of 27 metres. Here concentric rings of upright megaliths more than 3 metres tall have been hewn out of the bedrock surrounding a central menhir.

A second, similar circle, called 'Small Centre Circle' by local divers, stands immediately to the north-east. It is not noticeably smaller than the first.

A little to the south is 'Stone Circle', which is made up of much smaller, rounded stones. It has a huge diameter of about 150 metres. Within it are subsidiary stone circles sometimes touching one another at their edges like the links of a chain.

Aguni

Aguni Island, 60 kilometres north of Kerama, has steep and forbidding cliffs. On the south-west side of the island these cliffs overlook an area of turbulent water that local fishermen call the 'washing machine'. The turbulence is caused by the presence of a sea-mount that ascends from much greater depths to form a small plateau only 4 metres under the surface. This plateau, perpetually swept by strong currents, contains a series of circular holes that look initially like well-shafts.

As they are lined with small blocks, there is little doubt that these shafts are man-made. The largest and deepest has a diameter of 3 metres and reaches a maximum depth (below the summit of the sea-mount) of about 10 metres. Others are typically 2 to 3 metres in diameter with a depth of less than 7 metres. A few are narrower and shallower. One has a small subsidiary chamber cut sideways into the wall of the main shaft.

Chatan

The coastline around Okinawa has been the subject of intensive development during the past half-century. Thirty kilometres north of the capital Naha, on the west coast of the island, is the popular resort area of Chatan. Here, less than a kilometre off-shore, at depths of between 10 and 30 metres, is strewn a looming underwater fantasia of 'walls' and 'battlements' and 'step pyramids'. Are these weird submerged structures natural or man-made? And if they are man-made then when, and by whom?

One possibility suggested to me by local fishermen is that the 'structures' could be artefacts of relatively recent military dredging. Certainly, several large US Air Force bases are located very close to Chatan and the site is constantly overflown by all kinds of American warplanes doing manoeuvres. I still remain open to the possibility that dredging could have produced some of the features to be seen underwater, but against this I have received a report from Akira Suzuki, a Japanese historical researcher, who has carefully investigated both US and Japanese archives in Okinawa and has been unable to find any record of such operations in this area.[11]

The most striking of the Chatan structures is a wall with its base on the sandy bottom at a depth of 30 metres. It rises to a 'battlement' with a sunken 'walkway' about 10 metres above the sea-bed. At a certain point the walkway is broken by a vertical U-shaped shaft cut through the entire height of the wall.

To dive at Chatan is to be reminded of an episode in the *Nihongi*, one of Japan's most ancient texts, a chronicle of the earliest times. Here, in a long introductory section entitled 'The Age of the Gods', there is a passage that describes how a deity named Ho-ho-demi no Mikoto climbed into an upended waterproof basket and descended to the bottom of the sea. In this makeshift submarine 'he found himself at a pleasant strand . . . proceeding on his way he suddenly arrived at the palace of the Sea-God. This palace was provided with battlements and turrets and had stately towers.'

No doubt the many curious things that the *Nihongi* has to say about the Age of the Gods may all be explained as mythology and imagination. Still, I find it curious in Japan, where there are so many underwater 'anomalies', that such a venerable ancient text contains a clear tradition of submerged structures that can only be visited by divers.

15,000 years

Between 1996 and 2000, while I increased my practical diving experience of Japan's underwater ruins, I several times got caught up in the virulent debate about their provenance. Some scholars and journalists think they are entirely natural or 'mostly natural' (Robert Schoch of Boston University, for example). Others, such as Professor Kimura and Professor Teruaki Ishii of Tokyo University, remain convinced that they are man-made but are uncertain as to their antiquity (in addition to sea-level rise, complex factors such as possible land subsidence – through volcanism, plastic flow or isostatic rebound – must be taken into account when determining the date of submergence of any given site).[12] No early resolution of this debate can be expected, since we are dealing here as much with matters of opinion as with matters of generally agreed fact. Those who think the structures are natural are likely to go on thinking so no matter what the other side says – and vice-versa. It looks like a stalemate.

Yet there is a potentially fruitful line of inquiry, capable of shedding light on this problem, which neither side has yet considered. Whether they were flooded by rising sea-levels or because of some form of land subsidence into the sea (quite possible in an area of great seismic instability like Japan) all the underwater ruins were above water at some point between 17,000 years ago (the end of the Last Glacial Maximum) and 2000 years ago – the latest date that anyone has suggested for their submergence.

What happened in Japan during this 15,000-year period? Could it be that there is something concealed in the remote prehistory of these islands that would provide a context and perhaps even a completely rational explanation for the underwater ruins?

Alexandria

During 1998 and 1999 the Egyptian Mediterranean city of Alexandria was much in the news. French archaeologists, led by the melodiously named Dr Jean-Yves Empereur of the National Centre for Scientific Research, had announced the discovery of submerged ruins, complete with underwater columns, sphinxes and granite statues. In the same location they also claimed to have found the remains of the famed Pharos, or Lighthouse – 135 metres tall and one of the Seven Wonders of the ancient world[13] – that had overlooked Alexandria's Eastern Harbour from the point where the fort of the Mameluke sultan Qait Bey now stands. Though it was thought to have been built in the early third century BC, historical reports suggest that at least part of the giant lighthouse remained intact until 8 August AD 1303, when a tremendous earthquake struck the Egyptian coast.[14]

Researching my earlier books had given me little reason to go to Alexandria. During a decade of travels in Egypt my focus had always been on the oldest sites – those going back to the third millennium BC and perhaps further – sites like Giza, with the three Pyramids and the Great Sphinx, Saqqara, where the remarkable *Pyramid Texts* are inscribed inside the tombs of Fifth and Sixth Dynasty Pharaohs, and Abydos, with First Dynasty boat graves and the mysterious Osireion.[15]

Since it was common knowledge that Alexandria had not existed until 332 BC, the date of its foundation by Alexander the Great,[16] I had always felt that it was unlikely to hold much of interest to me. I was vaguely aware that it had been built upon the site of an earlier settlement named Rhakotis or Raqote, but since this was usually described as 'an obscure fishing village',[17] I never suspected for a moment that there might be significant traces of earlier monumental constructions in the area.

None of the underwater discoveries that were made public at the end of the 1990s did anything to change my view. They too belonged to what is called the Ptolemaic period of Egypt, named for the ruling dynasty – of which Cleopatra was the last monarch – established soon after Alexander's death by his general Ptolemy. I was at first intrigued to learn that inscriptions belonging to much earlier Pharaohs had been found amongst the underwater ruins – the cartouche of Rameses II (1290–1224 BC) on pink-granite 'papyriform' columns from Aswan, an obelisk of his father Seti (1306–1290 BC), a sphinx from the time of Senuseret III (1878–1841 BC) and numerous other artefacts and objects bearing ancient inscriptions.[18]

On good grounds, archaeologists did not regard such discoveries as evidence of any earlier monumental settlement in Alexandria but rather of a well-known Ptolemaic habit of borrowing pieces of religious art and architecture from temples that had been built throughout Egypt by earlier Pharaohs.[19] Jean-Yves Empereur was very clear on this point:

> The numerous products of the Pharaonic period – sphinxes, obelisks and papyrus columns [found underwater around Qait Bey] – do not make any significant difference to what we already know about the history of Alexandria and its foundation by Alexander the Great.[20]

Diving with Empereur

A research trip to Alexandria was easy to talk myself out of. Since what was known of its history was that it had no history before the end of the fourth century BC, there was obviously no good reason for me to go there. The ruins of the Pharos and of what looked like an extensive complex of buildings seaward of it had not been submerged in the period I was interested in – the end of the last Ice Age – but between the fourth century BC and the thirteenth century AD, most probably as a result of what geologists call 'vertical tectonic subsidence' caused by earthquakes.[21] Besides, there is a complicated permissions ordeal which one must undergo if one wishes to dive at Alexandria involving the Ministry of Information, the Ministry of National Security, the Supreme Council of Antiquities, the Police, Customs and the Navy. The whole process routinely takes a month . . .

So I'd pretty much quashed the idea before it took shape when I remembered that my good friend Robert Bauval was born in Alexandria and that several members of his large, globe-trotting family were still living there. On a whim I telephoned him – he lives just outside London – and asked him if he knew anything about Empereur and whether he thought it would be possible to fix up a day of unofficial diving with the French team.

Rob is reputed to have worked miracles in Alexandria, even from as far away as England. I therefore wasn't too surprised when he called me back the next day and informed me that he had spoken to his great-aunt Fedora, who knew Empereur well; she in turn had put in a good word with the archaeologist. The upshot was that we would be allowed to dive at Qait Bey without formality, any time that suited us in the next few weeks.

Sleep of years

On 30 September 1999 Santha and I, hefting our gear, met up with Robert at the gatehouse to Qait Bey fort. He ushered us inside its medieval limestone walls, soothing the guard in Arabic, and led us to a yard where scuba tanks were laid out and a group of young archaeologists, the men muscular, with stubbly chins, the women tanned and serious, were donning wetsuits and checking gear.

Empereur, in his late forties, was older than the rest of his team. He was wearing a tropical linen jacket and a Panama hat and carrying a briefcase. 'Excuse me,' he now said as we shook hands, 'but I have to rush off, so I won't be diving with you today.'

'No problem. I'm really very grateful to you for allowing us to do this at all at such short notice.'

Empereur shrugged: 'My pleasure. I hope you enjoy yourselves.' He introduced us to the other team members, then we shook hands again and he strode away.

Because it's hard to take notes underwater, I normally document my dives on video. It was my intention to do so now, but as we were getting ready I was told that this would not be permitted. Santha, likewise, was asked to leave her three Nikonos 5s behind. Apparently it was something to do with an exclusive deal that had been signed with the French photo agency Sygma. Robert protested vociferously on our behalf and as a compromise it was ultimately agreed that Santha could use her cameras but that my video would not be allowed under any circumstances.

Once that was settled we were led down through a series of dank stone corridors with arrow-slits overlooking the sea until we emerged at the edge of the island – long since connected to the mainland by a causeway – on which Qait Bey stands. Here we put on our gear and tanks, jumped into the water with one of the archaeologists as our guide and descended at once into a submarine wonderland less than a dozen metres below us.

It may be the most beautiful ancient site I have ever had the privilege to explore. The visibility was poor, which added a kind of foggy glamour to the scene, and we had to criss-cross the ruin-field many times, over three lengthy dives, before I began to appreciate how vast and how heterogeneous it was. There were huge numbers of columns, some broken, some virtually intact, but all tumbled and fallen. There were Doric column bases surrounded by tumbled debris. Here and there one or two courses of a wall could be seen, rising up out of the murk. There were dozens of metre-wide hemispherical stones, hollowed inside, of a type that I had never encountered before in Egypt. There were several small sphinxes, one broken jaggedly in half, and large segments of more than one granite obelisk seemed to have been tossed about like matchsticks. There were also quarried granite blocks scattered everywhere. Most were in the 2–3 square metre range but some were much larger – 70 tonnes or more. A notable group of these behemoths, some a staggering 11 metres in length, lay in a line running south-west to north-east in the open waters just outside Qait Bey. When I researched the matter later I learnt that they were amongst the blocks that Empereur had identified as coming from the Pharos:

> some of them are broken into two or even three pieces, which shows that they fell from quite a height. In view of the location the ancient writers give for the lighthouse, and taking into consideration the technical difficulty of moving such large objects, it is probable that these are parts of the Pharos itself which lie where they were flung by a particularly violent earthquake.[22]

There were exquisite moments when the sun broke through the clouds that lay over Alexandria that day and cast a beam of light down into some dark corner of the submerged ruins. Then the vanquished structures over which we were diving seemed to regather their former stature, like ghosts returning to flesh, before collapsing once again into their sleep of years.

Treasure of the sunken city

A few weeks later I still hadn't been able to get the images of what I'd seen underwater off Qait Bey out of my mind, or quite rid myself of the feeling that I might have missed something important there. Without any particular objective I began to buy books about Alexandria and to acquaint myself better with the story of its past. Visiting Amazon.com one evening in mid-October, I found that someone was offering a second-hand copy of *Alexandria – A History and a Guide* written during the First World War and published in 1922 by the British novelist E. M. Forster.[23] I bought it at once, for it is rumoured to be a fount of wisdom. Then I snapped up, in quick succession, *The Library of Alexandria – Centre of the Ancient World*, edited by Roy Macleod; *Life and Fate of the Ancient Library of Alexandria* by Mostafa El-Abbadi; *Philo's Alexandria* by Dorothy L. Sly; and *The Vanished Library* by Luciano Canfora.[24] Oddly enough, Amazon's search-engine couldn't immediately find me anything when I entered the keyword Pharos. While I was thinking about what to search for next – maybe Seven Wonders of the ancient world? – I called up Jean-Yves Empereur's name to see the complete list of his publications. I already owned his book *Alexandria Rediscovered*, which told the story of the underwater excavations at Qait Bey, but I hoped that he might have written other books about the region. He hadn't and I found myself looking at Amazon's sparse sales page for *Alexandria Rediscovered.*

There was one review, from a reader in Phoenix, Arizona. He wrote that he wished no disrespect to Dr Empereur; however, after seventeen years as an archaeological diver in Egypt, he could not agree that Empereur's team had found the Pharos. What they had found was interesting, yes, important, yes, but it was definitely not the Pharos.

What was someone who'd worked for seventeen years as an archaeological diver in Egypt now doing in Phoenix, Arizona? And what did he know – or think he knew – about the Pharos? My instincts told me that there could be a story here, and although the reviewer did not give his name, there was an e-mail address. I sent him a message at once, explaining my interest in the underwater ruins of Alexandria and asking him to elaborate on his views about the Pharos.

The next day, 17 October, I received this reply:

> Mr Graham,
> My name is Ashraf Bechai. I am the former leader of the Maritime Museum underwater team (1986/89). I am also a former diving engineer of the Institute of Nautical Archaeology. You can find a little more about me on the Institute web page. I will be glad to help you with any question you have.
> Sincerely, Ashraf Bechai,
> Phoenix AZ, USA.

Attached was an extraordinary 23-page report titled *Treasure of the Sunken City: The Truth About the Discovery of the Lighthouse.*

Ashraf Bechai's story

What came across in Ashraf's Bechai's angry and impassioned report was a sense, above all else, of intellectual outrage. In his view Jean-Yves Empereur and his team had been altogether too narrow-minded in their interpretation of what they had found underwater at Qait Bey:

> During the last three years there have been many claims that the French marine-archaeological team that has been working underwater in the area of Qait Bey Fort has found the remains of a great building, identified by French and Egyptian archaeologists as the remains of the Pharos lighthouse.
> But is it the Pharos?
> I don't see why we have to take it as they say without asking any questions. I don't see why we're expected to suspend our common sense just because this stuff is underwater and looks very spectacular on television.

Bechai pointed out that if the Pharos had indeed been more than 100 metres tall, as all historical sources maintain, then it must have been a truly enormous building. The Great Pyramid of Giza, for example, which is 150 metres tall, with a base area of more than 13 acres, weighs 6 million tonnes and consists of 2½ million individual stone blocks.[25] Since the building technology of the fourth century BC was, if anything, inferior to that of the third millennium BC, it is therefore unlikely that the lighthouse – with a reported height of 135 metres – could have had a base area of less than 12 acres or a weight of much less than 5 million tonnes. 'Imagine how big the pile of stones that should remain from a building like that,' suggested Bechai:

> Could this great amount of stone just disappear? Vanish in the water? The truth is that this much stone would have created an island in the sea and all the statues, sphinxes and other ancient Egyptian artefacts that the French team have found intermingled with the blocks would have been buried forever under a great pile of rock.

Even if one supposes – against the evidence – that a far superior building technology existed in Alexandrian times than in the times of the Great Pyramid, and even if one reduces the height of the Pharos from 135 metres to 100 metres, it is still extremely unlikely that it could have been built with less than half a million individual stone blocks (as against the Pyramid's 2½ million blocks). But let us reduce it still further – to just 100,000 blocks, or even 50,000.

Yet Empereur writes: 'As soon as one puts one's head under the water around Qait Bey one begins to feel dizzy at the sight of the 3000 or so architectural blocks which carpet the sea-bed.'[26] It was precisely this 'dizzying' spectacle of only 3000 blocks that bothered Bechai. If the ruins around Qait Bey were the remains of the lighthouse and associated structures, then 3000 blocks was nowhere near enough:

> Three thousand blocks wouldn't even build a large temple let alone a lighthouse 100 metres high! And many of the blocks in Empereur's survey are scattered very far from Qait Bey. Some are almost a kilometre away. There is even one 75-ton granite block half a kilometre out to sea and 1.5 kilometres distant from Qait Bey. Are we supposed to believe that the earthquake was powerful enough to throw a 75-ton block as far as that?

Bechai also made another valid point. Ancient texts referring to the Pharos concurred that it had been built of blocks of 'white stone' – limestone – which is plentifully available locally. Yet the underwater ruin-field outside Qait Bey consists primarily of scattered granite blocks and other architectural elements, such as columns, also made out of granite – a much more intractable material that had to be brought to Alexandria from quarries almost 1000 kilometres to the south. Whilst admitting that limestone does have a much faster rate of erosion than granite, Bechai did not believe that the vast amount of limestone that would have been required for the Pharos could possibly all have eroded away. He concluded:

> What we have at this site are scattered artefacts from different ages, different designs of blocks, columns and statues – not an indication of one thing but an indication of many things.

The giant blocks of Sidi Gaber

Before I was half-way through the report I realized that it pinpointed paradoxes and anomalies that I had completely missed during my dives with the French team. No doubt Empereur would have answers to all these questions but at this stage I had to admit that the questions themselves sounded reasonable.

As I read on I realized that Bechai was agitated about much more than just the problem of the Pharos. He wrote: 'I have seen things underwater in Alexandria during the last 17 years that challenge all our knowledge of the history of this area.' As an example he reported how in 1984 he had gone spear-fishing with some friends off-shore of Sidi Gaber, a district along Alexandria's crowded Corniche, some 3 kilometres to the east of Qait Bey:

> We were about two kilometres from shore, diving off a small boat. I remember that the visibility underwater was exceptionally good. We hadn't been expecting that because there had been a storm a few days before which moved around a lot of the sand and silt on the bottom. Suddenly I saw hundreds of huge sandstone or limestone blocks laid out in three rows, each two courses high, that had been exposed on the sea-bed at a depth of about six to eight metres. The blocks appeared to be of identical dimensions – four metres wide by four metres long by two metres high. They were stacked up on an underwater ridge of some sort, because there was deeper water between them and the shore. All around there were hundreds more blocks of similar size that were heavily eroded, or damaged, or had fallen out of line.
>
> This group of blocks has been seen on and off by fishermen and divers over at least 25 years and there is still no proper explanation for it. I have never been so lucky with the visibility there again, nor the same bottom conditions, and despite many subsequent attempts to relocate the site I have so far failed to do so.

Another interesting site, one that Bechai hadn't seen himself, was the so-called Kinessa, an Arabic word meaning 'church' or 'temple':

> If you have lived in the wonderful city of Alexandria long enough and had connection with fishermen who do commercial net fishing then you must have heard about 'Al Kinessa'. Some say that it is out in the open sea about one kilometre to the north of Qait Bey and that when an east wind blows and the waters are clear you can sometimes see what look like the remains of a building underwater. Others claim it

is much further north – perhaps as much as five kilometres out from shore. Three different people told me very specifically that it is five kilometres north to north-west of Qait Bey. Before reaching it the sea-bed slopes down to 40 metres where the bottom is sandy with a few patches of rock; then you pass an area of rocky pinnacles, some as much as 20 metres high jutting out of another sandy bed; then the bottom profile rises up sharply from 40 metres to just 18 metres in depth creating a smooth-sided, flat-topped hill five kilometres from shore in the middle of nowhere. That is where they say the *Kinessa* is.

Mystery of the sea

After I had read Ashraf Bechai's report I began to correspond with him about specific points by e-mail, and in due course we agreed that we would dive together to try to relocate the Sidi Gaber blocks and the Kinessa during the summer of 2000. Although his home was now in Phoenix, Arizona, where he ran a business, he told me that he still returned to Alexandria for at least three months every year and would be happy to work with me there so long as I could extract the necessary permits from the authorities.

There were other travels to do in the meantime. On one trip, I don't remember where, I took E. M. Forster's *Alexandria – A History and a Guide* with me as airplane reading. In it I was intrigued to learn that Forster had drawn attention to a report published in 1910 by the French archaeologist Gaston Jondet and entitled *Les Ports submergés de l'ancienne île de Pharos*.[27] According to Jondet, Forster said, someone had built a series of huge megalithic walls and causeways some distance off the coast of Alexandria beyond the island of Pharos that were now submerged to a depth of up to 8 metres beneath the sea. The character of these constructions, he judged, was 'prehistoric'.[28] Summarizing reactions to the discovery, Forster wrote:

> Theosophists, with more zeal than probability, have annexed it to the vanished civilization of Atlantis; M. Jondet inclines to the theory that it may be Minoan – built by the maritime power of Crete. If Egyptian in origin, perhaps the work of Rameses II (B.C. 1300) ... The construction ... gives no hint as to nationality or date. It cannot be as late as Alexander the Great or we should have records. It is the oldest work in the district and also the most romantic for to its antiquity is added the mystery of the sea.[29]

I wondered how many archaeologists today shared Forster's view about the antiquity and romance of the prehistoric harbour. I knew for sure that Jean-Yves Empereur did not. His on-the-record opinion, in full accord with the mainstream scholarly view, was that before Alexander's arrival 'the only inhabitants of the area must have been a few fishermen and perhaps also a garrison stationed here to guard the approaches to the Delta'.[30] But if so, then who had built the much older and now submerged harbour – if it was indeed a harbour? And how did it fit in, if at all, with the megalithic blocks underwater at Sidi Gaber, or the elusive Kinessa that fishermen said appeared and disappeared beneath the sparkling waves – now you see it, now you don't – like the Sea King's castle?

Rumours of the deluge

Descriptions of a killer global flood that inundated the inhabited lands of the world turn up everywhere amongst the myths of antiquity. In many cases these myths clearly hint that the deluge swept away an advanced civilization that had somehow angered the gods, sparing 'none but the unlettered and the uncultured'[31] and obliging the survivors to 'begin again like children in complete ignorance of what happened . . . in early times'.[32] Such stories turn up in Vedic India, in the pre-Columbian Americas, in ancient Egypt. They were told by the Sumerians, the Babylonians, the Greeks, the Arabs and the Jews. They were repeated in China and south-east Asia, in prehistoric northern Europe and across the Pacific. Almost universally, where truly ancient traditions have been preserved, even amongst mountain peoples and desert nomads, vivid descriptions have been passed down of global floods in which the majority of mankind perished.[33]

To take these myths seriously, and especially to countenance the possibility that they might be telling the truth, would be a risky posture for any modern scholar to adopt, inviting ridicule and rebuke from colleagues. The academic consensus today, and for a century, has been that the myths are either pure fantasy or the fantastic elaboration of local and limited deluges – caused for example by rivers overflowing, or tidal waves.[34] 'It has long been known,' commented the illustrious anthropologist Sir J. G. Frazer in 1923,

> that legends of a great flood in which almost all men perished are widely diffused over the world . . . Stories of such tremendous cataclysms are almost certainly fabulous; [but] it is possible and indeed probable that under a mythical husk many of them may hide a kernel of truth; that is they may contain reminiscences of inundations which really overtook particular districts, but which in passing through the medium of popular tradition have been magnified into worldwide catastrophes.[35]

Unquestioningly following Frazer's lead, scholars to this day still persist in seeing flood stories as

> recollections – vastly distorted and exaggerated . . . of real local disasters . . . There is not one deluge legend but rather a collection of traditions which are so diverse that they can be explained neither by one general catastrophe alone, nor by the dissemination of one local tradition alone . . . Flood traditions are nearly universal . . . mainly because floods *in the plural* are the most nearly universal of all geologic catastrophes.[36]

Not all mainstream academics toe this line. But amongst those who don't it seems to have been generally agreed that almost any explanation, however harebrained, is more acceptable than a simple literal interpretation of the myth of a global flood – i.e. that there actually was a global flood . . . or floods. For example, this from Alan Dundes, Professor of Anthropology and Folklore at the University of California, Berkeley, is regarded as a perfectly acceptable scholarly position on the problem: 'The myth is a metaphor – a cosmogenic projection of salient details of human birth insofar as every infant is delivered from a "flood" of amniotic fluid.'[37]

My guess is that such thinking will not much longer survive the steady accumulation of scientific evidence which suggests that a series of gigantic cataclysms, exactly like those described in the flood myths, changed the face of the earth completely between 17,000 years ago and 8000 years ago. At the beginning of this period of extraordinary climatic turbulence and extremes, fully evolved human beings of the modern type are thought to have been in existence for 100,000 years[38] – long enough in theory for at least some of them to have evolved a high civilization. While much of the land they formerly lived on is now submerged beneath the sea, and as unfamiliar to archaeologists as the dark side of the moon, how certain can we really be that some of them did not?

Dark zone

SCUBA is the acronym for the 'Self-Contained Underwater Breathing Apparatus' invented by the late Jacques Cousteau and Emile Gagnan in 1943.[39] At first thought likely to be expensive and of use only to specialists, the technology rapidly entered the mass market and, today, scuba-diving is the world's fastest-growing sport.[40]

Although it should be obvious, it is worth remembering that only since scuba-diving was introduced has any kind of systematic marine archaeology become possible. Moreover, funds for this kind of research are limited, and the oceans are extremely large – constituting, in fact, more than 70 per cent of the earth's surface.[41] Marine archaeologists have barely been able to begin the investigation of the millions upon millions of square kilometres of coastal shelf inundated since the end of the last Ice Age. As a result, the underwater world continues to constitute a gaping dark zone in human knowledge; it is entirely possible that archaeological surprises and upsets await us there.

> Question: Why has the first extensive evidence of large-scale prehistoric structures beneath the sea come from Japan?
> Answer: Japan has more scuba-divers than any other country and it follows that its coastal waters have been more thoroughly explored than those of any other country.
> Question: Why have the main underwater structures in Japan all been found south of the thirtieth parallel?

Answer: Because most sport divers prefer warm water. There may be structures further north as well which simply haven't been noticed yet because few divers are attracted to the cold or stormy seas in which they lie.

India is the opposite of Japan. It has almost no leisure-diving industry (just a couple of dive-shops in the whole subcontinent)[42] but it does have marine archaeologists like S. R. Rao whose minds are open to extraordinary possibilities. Rao's work around Poompuhur was guided by ancient Tamil traditions that speak of the submergence of large masses of land off southern India thousands of years ago.[43] And he himself admits that the 'U-shaped structure' found at 23 metres is hard to explain within the orthodox framework of history.

'11,000 years old, or older'

In August 2000 I took on a new research assistant, Sharif Sakr, who had just graduated in Human Sciences from Oxford University. One of the first tasks I gave him was to find me a top-flight academic, in Britain, who would be prepared to act as a kind of 'resident expert' on sea-level rise and who would be qualified to give an authoritative opinion on the date of submergence of almost any underwater structure in the world. Sharif came back to me with Dr Glenn Milne, a specialist in glacio-isostacy and glaciation-induced sea-level change at Durham University's Department of Geology. Milne and his colleagues have established a worldwide reputation predicting ancient sea-level changes and the corresponding changes in the earth's coastlines. Their predictions are based on a sophisticated computer model that has been under development since the 1970s and that takes into account many variables beyond changes caused solely by the melting of ice-sheets – the technical term is *eustacy*.[44]

In October 2000 Sharif approached Milne on my behalf and asked him to calculate the latest date that the large U-shaped structure and other nearby structures off the coast of Poompuhur could have been submerged.

Thursday 12 October 2000, Sharif Sakr to Glenn Milne:

Hi Glenn,

Hope everything's OK.

Just a quick question: I've got a series of structures 5 kilometres off the south-east coast of India (Tamil Nadu region, probably roughly around 11N, 80E as a rough guess).[45] The structures are 23 metres underwater – which is extremely deep. If we assume only eustatics, then the implication would be that the structures are older than around 7000 BC. But there is also isostatic subsidence to consider: what proportion of that 23 metres depth, as a rough off-the-record guess, could be explained away through subsidence?

Does the depth of the structures still suggest great antiquity, even when isostatics are brought into the equation?

Thursday 12 October 2000, Glenn Milne to Sharif Sakr:

Hi Sharif,

I did a quick model run for that site and the predicted sea-level curve shows that areas currently at 23m depth would have been submerged about 11,000 years before the present. This suggests that the structures you mention are 11 thousand years old or older!

No civilization known to history . . .

Although I could not be certain of anything until I was able to dive on it myself, the early descriptions of the U-shaped structure by the NIO's marine archaeologists left little doubt that it was man-made. The 'stone blocks' and 'courses of masonry' that had been reported by all these experienced witnesses seemed to exclude any possibility that it could be natural or recent – or indeed anything other than the ruins of a very old stone building, resting on bedrock, constructed here before the ocean rose to cover it.

Now, as I studied the e-mail from Glenn Milne, I knew just how ancient the U-shaped structure really might be – at least 11,000 years old. That's 6000 years older than the first monumental architecture of ancient Egypt or of ancient Sumer in Mesopotamia – traditionally thought of as the oldest civilizations of antiquity. Certainly, no civilization known to history existed in southern India – or anywhere else – 11,000 years ago. Yet the U-shaped structure off the Tranquebar-Poompuhur coast invites us to consider the possibility that it was the work of a civilization that archaeologists have as yet failed to identify – one whose primary ruins could have been missed because they are submerged so deep beneath the sea.

2 / The Riddle of the Antediluvian Cities

And the Lord planted a garden eastward in Eden . . . And out of the ground made
the Lord God to grow every tree that is pleasant to the sight, and good for food . . .
And a river went out of Eden to water the garden . . .

Genesis 2:8–10

I think we are going to get many surprises yet on land, and under the sea.

Thor Heyerdahl, June 2000

Millions of square kilometres of useful human habitat swallowed up by rising
sea-levels at the end of the Ice Age. Myths of an antediluvian civilization
destroyed by global floods. Sightings and rumours of inexplicable submerged
structures in many different parts of the world. Could there be a connection?

In order to investigate this problem systematically what I really needed was
some method of correlating the facts about land loss at the end of the Ice Age
with the localities suggested by the myths and with any eye-witness reports of
anomalous underwater structures. I needed, in other words, something like an
'antediluvian Encarta' – an electronic atlas of the world as it had looked before,
during and after the sea-level rise that accompanied the end of the Ice Age.
Ideally I should be able to see, on demand, any coastline, any island, any expanse
of ocean, as it had looked at millennium intervals throughout the entire period
of the meltdown.

Such a program, unfortunately, does not exist commercially, nor is information of the extremely specific kind I needed gathered together in any single
work of reference. Detailed studies of scattered areas are available but no
comprehensive, time-factored global picture. Yet, as I was to discover, cutting-
edge research into post-glacial sea-level rise is underway at many universities
and the information necessary to create a useful and reasonably reliable 'antediluvian atlas' does in fact exist – though not in published form. Glenn Milne
and his colleagues at the Geology Department of Durham University are the
leading UK specialists in the field and from September 2000 onwards it was
they who came to my rescue. As noted in chapter 1, the state-of-the-art computer model that they have developed calculates the relevant variables to the
extent that they are known and produces printable screen images of any location
at any epoch during the past 22,000 years. Since the model does not incorporate
tectonic motion and not all its variables are known with great certainty, it is

most accurate at predicting shoreline changes in tectonically inactive regions and over time intervals of several centuries or more – beyond that, its predictions are useful as approximate guides. The processing is not instantaneous and skilled man-hours are required to extract the required information from the program location by location. So Glenn was kind beyond measure in cheerfully and helpfully preparing all the inundation maps that are used in the later chapters of this book.

But I had made forays into antediluvian geography before I met Glenn Milne. This was feasible wherever sufficiently detailed sea-level data was accessible to build up a sense of how the inundation of a particular region had progressed over a period of several thousands of years. Thanks to the work of Kurt Lambeck, a geologist at the Research School of Earth Sciences of the Australian National University, such data has been on public record for the Persian Gulf since 1996. Lambeck's findings (which I was later able to confirm against Glenn Milne's modelling of the post glacial shorelines of the Gulf) were of enormous interest to me because the Persian Gulf was the home of a mysterious and extraordinary ancient culture – the Sumerians. Their flood myths seem to form the archetype for the much later Noah story in the Old Testament, and they are regarded by archaeologists as the founders of the oldest high civilization in the world.

Inundation data for the end of the last Ice Age has never before been thought likely to have a bearing, one way or another, on the problem of the origins of civilization and has therefore never been used as an investigative tool by archaeologists interested in this problem. But since the relevant data was available for the Persian Gulf, I decided to try to find out what it might show.

The five antediluvian cities of Sumer

Located immediately to the north-west of the present coastline of the Gulf between the Euphrates and Tigris rivers, ancient Sumer flourished during the fourth and third millennia BC and the earliest surviving written version of the global flood 'myth' was found during excavations of the Sumerian city of Nippur[1] (located on the Euphrates 200 kilometres south of the modern city of Baghdad). Inscribed on a tablet of baked clay, the Sumerian tradition is accepted by scholars as the source of the later Babylonian Epic of Gilgamesh[2] (which likewise speaks of a universal flood that destroyed mankind) and also bears a close relationship to the much-better-known flood account in the Old Testament.[3]

The Sumerian text is from a fragment – the lower third – of what was once a six-column tablet.[4] And while it is clear that it belongs to a very ancient and widely dispersed family of flood traditions, it nevertheless remains – in itself – a 'unique and unduplicated' document. 'Although scholars have been "all eyes and ears" for new [Sumerian] deluge tablets, not a single additional fragment has turned up in any museum, private collection or excavation.'[5]

What a rare and precious thing this little slab of baked mud is! And what a

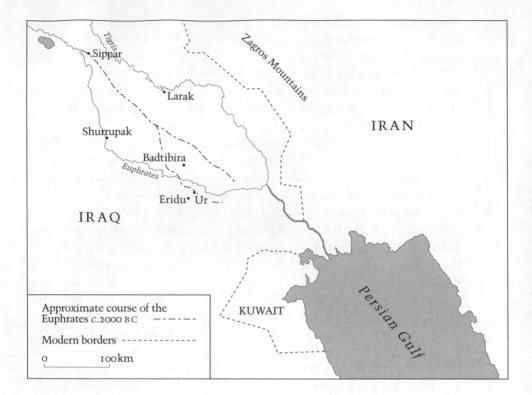

Approximate course of the
Euphrates *c.*2000 BC   — · — · —

Modern borders ----------------

0   100 km

tale it has to tell. When I first read it I was instantly intrigued, because it contains explicit references to the existence of five antediluvian cities which, we are informed, were swallowed up by the waters of the flood. If such cities ever existed, then where should we expect to find their ruins today?

The first thirty-seven lines of the Sumerian tablet are missing, so we do not know how the story begins, but at the point where we enter it the flood is still far in the future.[6] We hear about the creation of human beings, animals and plants.[7] Then another break of thirty-seven lines occurs after which we find that we have jumped forwards in time to an epoch of high civilization. We learn that in this epoch, before the flood, 'kingship was lowered from heaven',[8] a phrase that is eerily reminiscent of similar sky-ground symbolism contained in ancient Egyptian scriptures such as the *Pyramid Texts* (c.2300 BC), the *Book of what is in the Duat* (c.1400 BC) and the much later *Hermetica* (c.AD 300).[9]

Then comes the reference to the foundation of Sumer's antediluvian cities by an unnamed ruler or a god:

> After the lofty crown and the throne of kingship had been lowered from heaven,
> He perfected the rites and the exalted divine laws . . .
> Founded the five cities . . . in pure places,
> Called their names, apportioned them as cult centres.

The first of these cities, Eridu . . .
The second Badtibira . . .
The third Larak . . .
The fourth Sippar . . .
The fifth Shurrupak . . .[10]

'A flood will sweep over the cult centres . . .'

When we rejoin the narrative after a third 37-line lacuna the scene has changed bewilderingly. Although the flood is still in the future, the foundation of the five antediluvian cities is now far in the past. It is apparent from the context that in the intervening period the cities' inhabitants have behaved in such a way as to incur divine displeasure and that a convocation of the gods has been called to punish mankind with the terrible instrument of an earth-destroying flood. At the moment where we pick up the story again a few of the gods are dissenting from this decision and expressing their unhappiness and dissatisfaction with it[11]

Without preamble, a man called Zisudra is then introduced – the Sumerian archetype of the biblical patriarch Noah. The text describes him as 'a pious, god-fearing king'[12] and allows us to understand that one of the gods – unnamed – has taken pity on him. The god tells Zisudra:

> Take my word, give ear to my instructions:
> A flood will sweep over the cult centres.
> To destroy the seed of mankind,
> Is the decision, the word of the assembly of gods.[13]

A text break of forty lines follows, which scholars deduce, from the many later recensions of the same myth, 'must have continued with detailed instructions to Zisudra to build a giant boat and thus save himself from destruction'.[14] When the story resumes the cataclysm has already begun:

> All the windstorms, exceedingly powerful, attacked as one,
> At the same time the flood swept over the cult centres.
> For seven days and seven nights the flood swept over the land,
> And the huge boat was tossed about by the windstorms on the great waters.[15]

Throughout the cataclysm the skies remain dark. Then, on the eighth day, the sun breaks through the clouds, and the rains and raging storms cease. From the deck of his survival ship Zisudra looks out over a world that has changed for ever and sacrifices an ox and a sheep to the sun-god.[16]

An infuriating lacuna of thirty-nine lines follows, presumably telling us about the place where Zisudra makes landfall and the steps that he takes thereafter. When we pick up the story again, near the end of the text, we find him in the presence of the high gods of the Sumerian pantheon, An and Enlil, who have repented of their earlier decision to wipe mankind entirely from the face of the earth and are now so grateful to Zisudra for building his Ark and surviving the flood that they decide to make him immortal:

> Life like a god they gave him;
> Breath eternal like a god they brought down for him,
> . . . Zisudra the king.
> The preserver of the name of vegetation and of the seed of mankind.[17]

The final thirty-nine lines are missing.[18]

Picking and choosing

In his classic book *The Sumerians*, the late Professor Samuel Noah Kramer, one of the great authorities on ancient Sumer, observes that there are 'tantalizing obscurities and uncertainties' in this oldest surviving written version of the worldwide tradition of the flood.[19] What there can be no doubt about at all, however, is that the tablet speaks of an urban civilization that existed *before* the flood somewhere in the Persian Gulf area and provides us with the names of its sacred cities: Eridu, Badtibira, Larak, Sippar, Shurrupak. These cities, we are told quite specifically, were swallowed up in the deluge. Moreover, long after Sumerian civilization itself had ceased to exist, a rich tradition concerning the five cities, the antediluvian epoch and the flood survived in Mesopotamia almost down to Christian times.[20] Indeed it is fair to say that the traditional history of this region, as it was told in antiquity, is very clearly divided into two different periods – before and after the flood – and that both periods were regarded by the peoples of the region as absolutely factual and real.

It is only later scholars who have picked and chosen from the histories, accepting half of what they say as the basis for orthodox Sumerian chronology and rejecting the other half – concerning the antediluvian period – as myth and fantasy. Their logic is that there is no archaeological evidence for any high urban civilization in Sumer earlier than the fourth millennium BC and indeed their digs have revealed none.[21] Yet, as the cliché goes, absence of evidence is not necessarily to be taken as evidence of absence – and even Kramer obviously had his doubts. In *The Sumerians* he recounts how, before 1952, archaeologists were unanimous in their opinion that Sumer had been uninhabited (and uninhabitable) marshland until about 4500 to 4000 BC:

> This figure was obtained by starting with 2500 BC, an approximate and reasonably assured date obtained by dead reckoning with the help of written documents. To this was added from fifteen hundred to two thousand years, a span of time large enough to account for the stratigraphical accumulation of all the earlier cultural remains down to virgin soil, that is right down to the beginning of human habitation in Sumer.[22]

But then, continues Kramer, two geologists, Lees and Falcon, 'published a paper which carried revolutionary implications for the date of Sumer's first settlement'.[23] They demonstrated that Sumer had ceased to be uninhabitable marshland long before 4500–4000 BC. Now that this was understood:

> It was not impossible that man had settled there considerably earlier than had been generally assumed. The reason traces of these earliest settlements in Sumer have not as yet been unearthed, it was argued, may be because the land is sinking slowly at the same time that the water-table has been rising. The very lowest level of cultural remains in Sumer may, therefore, now be under water and may never have been reached by archaeologists, since they would have been misled by the higher water level into believing they had touched virgin soil. If that should prove to be true, Sumer's oldest cultural remains are still buried and untapped, and the date of Sumer's very first settlements may have to be pushed back a millennium or so.[24]

But why only a grudging millennium or so? Once we've admitted it is possible that archaeologists may never have reached the oldest layers of human habitation in Sumer, why should we assume that further digging might only push the horizon back by a thousand years? Why not five thousand years? Or ten thousand years? What is this worship of the recent that archaeologists indulge in?

The reason I ask these questions with a certain amount of exasperation is that Kramer, whose work has influenced several generations of students, does not for a moment consider the possibility that Sumer's antediluvian traditions might be based on anything real at all. Indeed, he devotes only three pages of his book to the prehistory of this ancient land before giving thirty pages to the historic period – as though all of the former is nothing more than a preamble to the latter.

I'm very struck by the extent to which Kramer relies on original Sumerian sources to build up his chronology of rulers, which begins, he says, with:

> The first dynasty of Sumer whose existence can be historically attested, the so-called First Dynasty of Kish, which according to the ancients themselves followed immediately upon the subsidence of the Flood . . . The first ruler of Sumer whose deeds are recorded is a king by the name of Etana of Kish, who may have come to the throne quite early in the third millennium BC[25]

It is in precisely this way that every Sumerian text about the period after the flood is treated as grist to the mill by historians constructing chronologies while every Sumerian text about the period before the flood is relegated to the realm of the mythologists . . .

So little to go on

Kramer's recognition, with the geologists Lees and Falcon, that people could have settled in the fertile valley between the Tigris and Euphrates rivers much earlier than had previously been assumed has been entirely vindicated by subsequent discoveries of the traces of 'primitive agricultural villages' dating back more than 8000 years.[26]

But the clues that have come down to us from this remote period are scanty and often ambiguous.

For example, with a tiny evidence base, are archaeologists absolutely certain that they could tell the difference between a small group of 'primitive' farmers and a small group of shattered and demoralized survivors from an urban civilization destroyed in a terrible flood?[27] Not a river-flood, no matter how big . . . but a real marine flood, deep and wild and sweeping in over the land, carrying all before it like the one described in the story of Zisudra.

Woolley's deluge

It is a river flood that has traditionally been suggested by scholars as the event described in the Zisudra text.[28] This goes back to the excavations of the renowned British archaeologist Sir Leonard Woolley at the Sumerian city of Ur in 1922–9. Digging inspection trenches through thousands of years of habitation layers, he suddenly reached a layer of silt almost 3 metres deep which he described as 'perfectly clean clay, uniform throughout, the texture of which showed that it had been laid there by water'.[29] The silt itself was void of habitation evidence, but there were further habitation layers below it that he dated to 3200 BC.[30]

Woolley declared that he had found the first concrete proof of the cataclysm described in the Zisudra story and the biblical flood of Noah and added:

> The discovery that there was a real deluge to which the Sumerian and the Hebrew stories of the Flood alike go back does not of course prove any single detail in either of those stories. This deluge was not universal, but a local disaster confined to the lower valley of the Tigris and Euphrates, affecting an area perhaps 400 miles long and 100 miles across; but for the occupants of the valley that was the whole world![31]

Woolley may not have been right that the inhabitants of the Tigris/Euphrates river valley thought of it as the 'whole world' but he needed to see them as geographically naive in order to explain why they had described his 'local disaster' as a 'universal' flood that threatened the survival of mankind as a whole. Neither was he necessarily right about the riverine nature of his silt layer; other, more recent, voices have suggested that it may have been laid down a few hundred years earlier than he suggested and that the agency is more likely to have been a massive transgression of the sea, followed by a gradual retreat of the waters with deposition of silt, than the work of the Tigris and Euphrates.[32]

Rising seas

In the 1990s Kurt Lambeck of the Australian National University carried out a detailed study of the Persian Gulf in order to map and simulate its 'palaeo-shorelines' from 18,000 ago – around the end of the Last Glacial Maximum – right up to today. He calculates that the modern shoreline of the region

> was reached shortly before 6000 years ago, *and exceeded* as relative sea-level rose 1–2 metres above its present level, inundating the low-lying areas of lower Meso-potamia.[33]

This marine transgression, which occurred between approximately 6000 and 5500 years ago, flooded the coastal plains of Sumer and extended the north-western shoreline of the Gulf to the doorsteps of Eridu and Ur – where the rising waters may have temporarily peaked as high as 3 metres above today's level before receding.[34] Geneticist Dr Stephen Oppenheimer, who has made a special study of floods and ancient migrations, suggests that this could have been the event that left behind the thick inundation deposit that Leonard Woolley excavated at Ur – not a river-flood at all as Woolley had believed, but a marine flood.[35]

In his important book *Eden in the East* Oppenheimer argues that what happened in the Gulf at this time, between approximately 6000 and 5500 years ago (4000–3500 BC), was the local effect of a worldwide episode of rapid, relatively short-term flooding known as the Flandrian transgression – which had a significant impact not only along the shores of the Gulf but in many other parts of Asia as well.[36] Noting that 'the destructive effect of the Flandrian transgression in wiping out coastal archaeological sites up to about 5500 years ago is now well recognized,' he launches the interesting speculation that in the case of Sumer:

> Eridu may be the oldest coastal city *not* destroyed by the invading sea. In other words it could have been the last old city to be built at the post-glacial high water point.[37]

Likewise, the distinguished Sumeriologist Georges Roux argues that between 6000 and 5000 years ago the shoreline of the Gulf was approximately 1 or 2 metres above its present level, so that its north-western coast lay 'in the vicinity of Ur and Eridu'. Thereafter, 'gradual regression, combined with silting from the rivers, brought it to where it is now'.[38]

Eridu

So I was back to the mystery of the antediluvian cities again and how they could possibly have been 'swept over by the flood', as the Zisudra story claimed, when Eridu had so obviously survived into historical times. In fact, as I was soon to learn, *all* the antediluvian cities had survived into historical times; none of them was presently underwater and at least one of them – Eridu – appeared *never* to have been underwater!

Between 1946 and 1949 Eridu's ruins, located in the south of Sumer near the Euphrates, a little to the north and west of the modern city of Basra,[39] were thoroughly excavated by a team from the Iraqi Directorate of Antiquities led by Fuad Safar.[40] The archaeologists paid particular attention to the temple of Enki, the Sumerian god of wisdom and Eridu's tutelary deity.[41] Here they dug a deep trench through many different layers of construction and reconstruction from about 2500 BC down until they finally reached the temple's very first building phase. Originally thought to have dated to about 4000 BC[42] – itself an epoch of fabulous antiquity – the excavators kept finding older and older material.

The central structure of the site is its principal ziggurat – step-pyramid – which was erected around 2030 BC by a Sumerian king named Amar Sin.[43] But it, too, turned out to stand on top of a series of earlier structures. Under one of its corners the archaeologists unearthed the ruins of no less than seventeen temples,

> built one above the other in proto-historic times. The lowest and earliest of these temples (Levels XVII–XV) were small, one roomed buildings which contained altars, offering tables and a fine-quality pottery decorated with elaborate, often elegant geometric designs.[44]

Judging by the pottery, these earliest shrines of Eridu go back much further than 4000 BC and probably as far as 5000 BC – i.e. 7000 years ago.[45] That, says Georges Roux, makes 'Eridu one of the most ancient settlements in southern Iraq' and a 'remarkable' choice in the mythology as the oldest of the antediluvian cities.[46]

There therefore seems to be no dispute that there was a settlement of some sort here *before* the region was flooded by the Flandrian transgression around 5500 years ago. Yet the excavations, which only stopped when the archaeologists reached 'virgin soil', 'yielded no trace of a flood'.[47] How could that be explained in an antediluvian city supposedly inundated not just by any old flood but by *the* flood? And how was I to make sense of the fact that ancient Ur, less than 20 kilometres away and on slightly higher ground,[48] was not even named in the flood tradition and yet *did* show evidence of a severe, silt-bearing inundation?

In 1992 Jules Zarins, a geologist at Southwest Missouri State University,

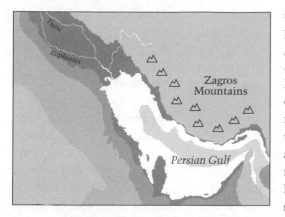

suggested a possible solution to this problem. In a paper published in the *Journal of the American Oriental Society*, he showed that in spite of Eridu's location in a low-lying depression south-west of Ur 'an eight-metre scarp of the Upper Fars formation (the Hazim) runs well to the north and south, possibly blocking any marine infilling into the depression'.[49] Now I was better equipped to understand what Oppenheimer and Roux were getting at. Looking at a map of the valley of the Tigris and Euphrates rivers, I could easily see how the relatively small and temporary increase in sea-level associated with the Flandrian transgression could have flooded low-lying areas of ancient Sumer – in fact up to about 180 kilometres inland[50] – in what are now Iran, Kuwait and Iraq. This would have brought the northern shoreline of the Persian Gulf very close to Eridu while quite conceivably carrying it just beyond Ur, thus leaving behind the flood deposit that Woolley had found.[51]

Shurrupak and Sippar

The archaeological results at the antediluvian city of Shurrupak, about 100 kilometres north of Eridu on the Euphrates river, also show evidence of a flood in the form of 'sizeable deposits of water-borne clay and sand due to a major and prolonged inundation'.[52] Since Shurrupak was renowned as the birthplace of Zisudra, the Sumerian Noah who had 'preserved the seed of mankind',[53] I thought at first that this might be a promising lead. But it fizzled out. The Shurrupak flood was securely dated to 4900 years ago – probably six or seven hundred years later than the flood recorded at Ur – and was almost certainly riverine.[54]

Dedicated to the sun-god Utu,[55] Sippar is the furthest inland of all the antediluvian cities and plays a special role in the Sumerian flood story. In fragment 4a of the few scattered remnants of the once widely renowned *History* of the Babylonian priest Berossos (who wrote in the third century BC but whose work is thought by scholars to convey authentic Sumerian traditions),[56] Sippar is remembered as the place where the knowledge of the antediluvian race was hidden away before the flood and preserved for use by the survivors of mankind.

The Noah figure in this version of the story is named Xisouthros (instead of Zisudra). A god visits him in a dream, warns him that humanity is about to be destroyed in a terrible deluge, and orders him to build a huge boat of the usual dimensions in the usual way.[57] So far this is all very familiar, but then comes a feature not found in the other versions of the tradition. The god tells Xisouthros that he is to gather up a collection of precious tablets inscribed with sacred wisdom and to bury these in a safe place deep underground in 'Sippar, the City of the Sun'.[58] These tablets contained 'all the knowledge that humans had been given by the gods' and Xisouthros was to preserve them so that those men and women who survived the flood would be able to 'relearn all that the gods had previously taught them'.[59]

The story of the flood itself is then given and of the journey of Xisouthros and his protégés in the Ark. Immediately after they make landfall Xisouthros steps down from the great ship, offers a sacrifice to the gods and then vanishes – having been transported immediately to immortal life. Those who remained on board are now leaderless and confused until a voice is heard from the heavens telling them to sail the ship back towards Babylon and to seek out the city of Sippar, which will have survived the flood. They are to 'dig up the tablets that were buried there and turn them over to mankind':[60]

> And those who had arrived in Babylonia dug up the tablets in the city of Sippar and brought them out. They built many cities and erected temples to the gods and renewed Babylon.[61]

An uncomfortable feeling

A quick inventory shows that we have so far identified three cities in the Persian Gulf area called Sippar, Shurrupak and Eridu in the historical period and three counterpart cities with exactly the same names which tradition says existed before the flood. We have Ur, very close to Eridu, which is not spoken of as an antediluvian city but which clearly suffered a major episode of flooding that laid down almost 3 metres of silt around the middle of the fourth millennium BC. We have Shurrupak, which was also inundated but not until about 700 years later. Meanwhile Sippar, the northernmost and farthest from today's Gulf coast of the five antediluvian cities, is named in the Berossos text as a place where it would have been practical for documents buried before the flood to be retrieved after the waters had subsided.

The remaining two antediluvian cities of the Sumerian tradition – Badtibira and Larak – have also been identified with archaeological sites in Iraq;[62] however (as indeed is the case with Sippar, Shurrupak and Eridu as well), these sites are not particularly large, splendid or significant as one might expect of such sanctified ground. As William Hallo of Yale University comments, 'The cities in question are not outstanding in importance . . . They are distinguished, rather, for their antiquity.'[63]

Since excavations at Eridu found the earliest occupation layers to have been laid down as much as 7000 years ago the city is indeed technically 'antediluvian' (by more than 1000 years) with respect to the Flandrian transgression – and the same is already known to be the case at Ur, where Woolley's excavators found habitation traces not only above the flood layer but also below it.

On the face of things, then, it seems reasonable to agree – and many scholars from Woolley onwards have agreed – that it was this flood at this time, or at any rate one of the frequent large-scale floods both riverine and marine to which the region was much prone in antiquity, that must have given rise to the Sumerian flood tradition. The new evidence revealing the extent of the flooding of southern Mesopotamia between approximately 4000 and 3500 BC – just on the edge of the historical period – should, if anything, have strengthened this hypothesis.

So why didn't I feel comfortable with it?

Heyerdahl on Sumer

The floods that had been archaeologically testified in the valley of the Lower Euphrates and Tigris took place too soon after the date for the foundation of Eridu and the other 'antediluvian' cities to fit in with the sense of grandeur and vast age that the traditions conveyed. When I looked again at the story of Zisudra, the story of the Babylonian flood hero Atrahasis,[64] the Epic of Gilgamesh,[65] the fragments of Berossos, and numerous other recensions and variants, I found that all of them set the antediluvian city-building period in the frame of vast expanses of time – frequently running into tens of thousands and even hundreds of thousands of years.[66] While I could understand why William Hallo felt that 'this chronology, measured in millennia, is obviously fantastic,'[67] I found his own proposed chronology equally absurd. 'Mesopotamian urbanism,' he argued in the prestigious *Journal of Cuneiform Studies*, 'was only some two centuries old at the time of the flood . . .'[68]

In June 2000 I met the explorer and adventurer Thor Heyerdahl, then eighty-six years old, at the excavation of a group of step-pyramids on Tenerife in the Canary Islands. We spent the afternoon together, under the blazing sun, exploring the site that he had brought to world attention.

Heyerdahl was everything I had expected him to be – impatient with protocol, a powerful presence, with piercing blue eyes, endearing vanities, a bawdy sense of humour, and an open, inquiring, restless mind. His *Tigris* expedition in 1977, which had begun in the Persian Gulf and culminated in Djibouti in the Horn of Africa, had proved that the reed boats of ancient Mesopotamia were sufficiently seaworthy and technically advanced to make long-distance marine voyages. Evidence of trans-oceanic trade at the very beginning of Sumerian history suggested very strongly that they had indeed made such voyages as early as the fourth millennium BC – and perhaps even earlier. Moreover, wherever archaeologists excavate they find amidst the ruins of Sumer's most ancient cities all the signs of a civilization that was already highly evolved, accomplished and sophisticated when those cities were founded more than 5500 years ago.

'Now we know that man is more than two million years old,' exclaimed Heyerdahl, 'it would be very strange if our ancestors lived like primitive food collectors for all that time until suddenly they started in the Nile valley, in Mesopotamia and even in the Indus valley, to build a civilization at peak level pretty much at the same time. And there's a question I ask that I never get an answer to. The tombs from the first kingdom of Sumer are full of beautiful ornaments and treasures made of gold, silver, platinum, and semi-precious stones – things you don't find in Mesopotamia. All you find there is mud and water – good for planting but not much else. How did they suddenly learn – in that one generation just about – where to go to find gold and all these other things? To do that they must have known the geography of wide areas, and that takes time. So there must have been something before.'

I pointed out that the First Dynasty of Sumer defined itself as the first dynasty *after* the flood. The historical Sumerians had always believed that their history was connected to an earlier episode of city-building and civilized life that had begun many thousands of years in the past and from which this deluge separated them. 'We're coming to a controversial idea,' I suggested, 'which is that the great civilizations of historical antiquity may have received some kind of legacy from an antediluvian culture – an idea orthodox archaeologists detest.'

'I know that,' Heyerdahl replied, 'but I mean they cannot give any answer to how could the Sumerians five thousand years ago know where to go and find these different kinds of raw material. They must have known the world. So, and I mean it, it is for me almost as fantastic as Erich Von Daniken who brought in people from space, to say as the archaeologists do – oh no, no, they sat in Egypt and Mesopotamia and the Indus valley, and they decided, bang, suddenly, just like that, we are going to build pyramids, we are going to go and find gold and we are going to do all this . . . It's ridiculous. I say it straight out – it could not be possible.'

'The idea of a lost civilization drives archaeologists mad and they seem to want to stop people thinking about it.'

'Well I understand why! Too many people have brought this up together with fairytale stories . . .'

'Which has put the historians off, so that they simply never explore this kind of question?'

'Yes, and this is a great pity. Because I mean even the sunken Atlantis story, which they all dismiss, is interesting – because why did the early Greeks write this story and why did they get it from the Egyptians, and for that matter why does every civilized and half-civilized nation in the world talk about the flood? Don't let us throw it away until we know that this is impossible. There has to be a possibility . . . and I think that we should look for it with the modern technical means we have. I think we are going to get many surprises yet on land, and under the sea.'

No surprises: what the archaeologists say about 'before'

Heyerdahl had arrived at his misgivings about the orthodox chronology of Sumer because he felt that it did not allow time for the evolution and development of the advanced urban civilization that archaeologists now knew had flourished there from the fourth millennium BC. 'There has to have been something before,' he reminded me when we parted. 'Look for whatever was before.'

Of course, there *had* been something before – a well-worked-out stratigraphical sequence that traced the development of human civilization in Mesopotamia back through 'proto-history' before the early dynastic period and thence into the Neolithic, Mesolithic and even the Palaeolithic epochs – a long, gradual, unsurprising process spread out over 30,000 years that Georges Roux sums up as 'from cave to farm and from village to city'.[69]

At risk of grossly abbreviating the painstaking archaeological work that has gradually uncovered this sequence, here are a few of the main mileposts:

Shanidar Cave in the Kurdish mountains of what is now northern Iraq: occupied by Neanderthal man *c*.50,000 years ago to 46,000 years ago; occupied by anatomically modern Upper Palaeolithic humans around 34,000 years ago; occupied by Mesolithic peoples around 11,000 years ago.[70]

Jarmo, also in northern Iraq – a Neolithic agricultural site which may perhaps date as early as 8750 years ago. It has a 7 metre high artificial mound resting on top of a very steep hill and is formed of sixteen layers of superimposed habitations.[71]

Hassuna, again in northern Iraq (35 kilometres south of Mosul). The first settlement here has the appearance of a more primitive Neolithic farming community living in huts or tents. Overlying this layer archaeologists found six layers of houses, progressively larger and better built.[72]

Umm Dabaghiya – about 8000 years old: more sophisticated features found, including beautiful murals and floors made out of large clay slabs 'carefully plastered with gypsum and frequently painted red'.[73]

The Samarra period – named after a widespread pottery style created by what Roux describes as 'a hitherto unsuspected culture which flourished in the Middle Tigris valley during the second half of the sixth millennium BC' – i.e. approximately 7500 years ago.[74] The geneticist Luca Cavalli-Sforza suggests that this date should be pushed back to 'about 8000 years ago'.[75] There is evidence that this culture used irrigation techniques, grew large surpluses of wheat, barley and linseed, and built spacious houses out of mud-brick[76] – later the favoured method of construction in the cities and temples of historical Sumer.

As well as Samarra several other 'proto-historical' cultural phases have been identified in which elements of Sumer's future civilization can be witnessed taking shape in increasingly organized and recognizable forms. Two of these phases stand out prominently in the archaeological record – the 'Ubaid' period (roughly 7200 to about 5500 years ago[77] and including the first temple at Eridu),[78] and the 'Uruk' period (6000 years ago[79] down to about 5200 years ago, showing further developments in the evolution of temple architecture).[80] The Uruk period, which some archaeologists prefer to see as a subdivision of the Ubaid,[81] then merges fairly seamlessly into the early dynastic period of Sumer.[82]

All of the above dates are of course approximate and are subject to processes of continuous revision and refinement by scholars. Nevertheless, they are thought likely to be accurate to within about 300 years.[83] In general the academics also agree that the direction of the 'flow' of the urban lifestyle in Mesopotamia is from north to south – with the first village-style settlements and large houses established in the north before being seen in the south. However, and paradoxically, Sumerian civilization as a distinctive entity, the origins of which archaeologists now trace back at least as far as the Ubaid period if not further, appears to be a phenomenon that had its origins in southern Mesopotamia. According to Georges Roux:

> During the fourth millennium BC the cultural development already perceptible during the Ubaid period proceeded at a quicker pace and the Sumerian civilization finally blossomed. This, however, took place only in the southern half of Iraq, the northern half following a different course and lagging behind in many respects.[84]

The word 'Sumerian' is derived from Shumer, the ancient name of southern Iraq.[85] Archaeologists believe that they have distinguished the presence of three distinct ethnic groups living in close contact in this region at the dawn of history around 5000 years ago. These were:

> the Sumerians, predominant in the extreme south from approximately Nippur [near modern Diwaniyah] to the Gulf, the Semites, predominant in central Mesopotamia (the region called *Akkad* after 2400 BC), and a small, diffuse minority of uncertain origin to which no definite label can be attached.[86]

Apparently, the only distinguishing features of these three groups are their languages.[87] Otherwise:

> All of them had the same institutions; all of them shared the way of life, the techniques, the artistic traditions, the religious beliefs, in a word the civilization

which had originated in the extreme south and is rightly attributed to the Sumerians.[88]

The Sumerian problem

With so much known about the evolution and development of the magnificent urban civilization of Sumer, it comes as a surprise to discover that there is such a thing as 'the Sumerian problem'.[89] I prefer to let the scholars speak for themselves:

> Who are these Sumerians? Do they represent a very ancient layer of population in prehistoric Mesopotamia, or did they come from some other country, and if so, when did they come and whence? This important point has been debated again and again ever since the first relics of the Sumerian civilization were brought to light more than a century ago. The most recent discoveries, far from offering a solution, have made it even more difficult to answer . . .[90]

And there is a mystery about the Sumerian language. It can be read and studied because later civilizations, such as the Babylonians, kept archives of Sumerian texts and also helpfully translated them into their own languages. However, Sumerian has a distinct peculiarity. It is unrelated to any of the known language families of the world.[91] So although there is a real sense in which Sumer and its precocious urban culture fit in very nicely with long-term developmental trends in ancient Mesopotamia – as I believe the scholars have successfully demonstrated – there is also a sense in which the Sumerians are definitely a bit different, a bit special . . . and conspicuously attached to the south . . .

I've been dealing with archaeologists long enough now to realize that they don't like myths or traditions very much ('can't weigh 'em, can't measure 'em, can't carbon-date 'em'). I was therefore not surprised to learn that they discounted what the Sumerians themselves had to say about their own origins:

> Sumerian literature presents us with the picture of a highly intelligent, industrious, argumentative and deeply religious people, *but offers no clue as to its origins* [emphasis added]. Sumerian myths and legends are almost invariably drawn against a background of rivers and marshes, of reeds, tamarisks and palm-trees – a typical southern Iraqi background – as though the Sumerians had always lived in that country, and there is nothing in them to indicate clearly an ancestral homeland different from Mesopotamia.[92]

But, as we have seen, the Sumerians had very clear ideas about their own origins . . . In their myths and legends they remembered a time, before the flood, when they had lived in five great cities. And they remembered a deluge so ferocious that it threatened the existence of all mankind . . .

The Seven Sages: what the Sumerians said about 'before...'

Sumerian myths and legends of the antediluvian world do much more than speak of the five cities. They also tell an extraordinary story of how their ancestors, who lived in the 'most ancient times', were visited by a brotherhood of semi-divine beings described as half men, half fish, who had been 'sent [by the gods] to teach the arts of civilization to mankind before the Flood' and who had themselves 'emerged from the sea'. The collective name by which these creatures were known was the 'Seven Sages' and the name of their leader was Oannes. Each of them was paired as a 'counsellor' to an antediluvian king and they were renowned for their wisdom in affairs of state and for their skills as architects, builders and engineers.[93]

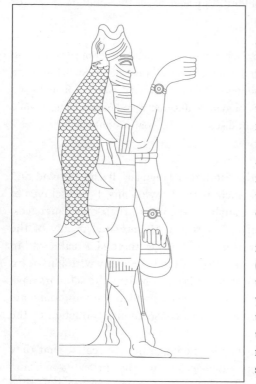

Fish-garbed figure taken from stone relief on Assyrian temple, possibly representing Oannes, leader of the Seven Sages.

The priest Berossos compiled his *History* from the temple archives of Babylon (reputed to have contained 'public records' that had been preserved for 'over 150,000 years').[94] He has passed on to us a description of Oannes as a 'monster', or a 'creature'. However, what Berossos has to say – ridiculous though this may sound – is surely more suggestive of a man wearing some sort of fish-costume. There is also a geographical anomaly in the text that may prove worthy of further consideration:

> There appeared from the Red Sea in an area bordering on Babylonia a frightening monster, named Oannes ... It had the whole body of a fish, but underneath and attached to the head of the fish there was another head, human, and joined to the tail of the fish, feet like those of a man, and it had a human voice. Its form has been preserved in sculpture to this day ...
>
> This monster spent its days with men, never eating anything, but teaching men the skills necessary for writing and for doing mathematics and for all sorts of knowledge: how to build cities, found temples, and make laws. It taught men how to determine borders and divide land, also how to plant seeds and then to harvest their fruits and vegetables. In short, it taught men all those things conducive to a civilized life. Since that time nothing further has been discovered. At the end of the

day, this monster, Oannes, went back to the sea and spent the night. It was amphibious, able to live both on land and in the sea ... Later, other monsters similar to Oannes appeared.[95]

Did they come from the east?

In 1944 Benno Landsberger, one of the great Sumerian scholars of the twentieth century, commented in an obscure essay that in his opinion:

> The legend of the Seven Sages who, emerging from the sea, imparted all technical skills and all knowledge to the Babylonians, may quite possibly have some historical basis.[96]

What he had in mind here was 'the Sumerian problem' – i.e. the as yet unanswered question: where did the Sumerians come from? Earlier than most archaeologists, he fully understood that 'the essential civilizing process on Mesopotamian soil must be ascribed to the pre-Sumerian population'. But at the same time the Sumerians were distinctively different and much more advanced than their immediate neighbours in terms of the level of development of their intellectual and philosophical ideas. 'In the area of intellectual culture,' he wrote, 'only the Sumerians possessed creative powers.'[97]

In fact they were *so* different in this respect that Landsberger was convinced they must have been migrants from somewhere else. He felt that only such a migration could account for the creation of the unique and idiosyncratic early dynastic culture

> which is considered to be so specifically Sumerian and which in its later manifestations indeed represented the Sumerian essence in its purest state. In all probability the Sumerians came from the East. Not only does the density of the settlement indicate a settling from south to north, but the absence of Sumerian elements in the mountain ranges north and east of Babylonia favors the thesis that the Sumerians came across the sea.[98]

In further support of his thesis Landsberger pointed out that the island of Bahrain, in the south of the Persian Gulf near Qatar

> possessed deities with authentic Sumerian names such as the chief god En-zak and his spouse Me-skil-ak. This circumstance supports an overseas origin for the Sumerians, since it is improbable that the island was colonized from southern Mesopotamia.[99]

Landsberger went on to speculate that the spark of Sumerian genius might have been imported from the Indus Valley civilization across the Arabian Sea to the east,[100] an interesting idea in itself. However, because he was writing in the 1940s he did not have access to modern knowledge about the astonishing changes that took place in the Persian Gulf at the end of the last Ice Age. He was thus unable to consider a far more radical possibility that the new science has revealed.

Explosive implications

Kurt Lambeck's work on the Persian Gulf initially drew my attention because it spoke of a marine flooding incident – the Flandrian transgression between about 6000 and 5500 years ago – that temporarily shifted the northern coast of the Gulf more than 150 kilometres inland and made Ur and Eridu beachfront property.

Lambeck's study was published in 1996 in the *Earth and Planetary Science Letters*, a specialist geological journal that probably does not cross the desks of a great many archaeologists.[101] He had focused on the period from 18,000 years ago – around the peak of the last glaciation – until today and had taken into account all the key variables including

> the response of the earth to glacial unloading of the distant ice sheets and to the meltwater loading of the Gulf itself and the adjacent ocean. Models for these glacio-isostatic effects have been compared with observations of sea-level change, and palaeoshoreline reconstructions of the Gulf have been made.[102]

Now, as I looked more closely into Lambeck's research, I realized that it could have unexplored and potentially explosive implications for the prehistory of Sumer:

> From the peak of the glaciation until about 14,000 yr BP [years before the present] the Gulf is free of marine influence out to the edge of the Biaban shelf. By 14,000 yr BP the Strait of Hormuz had opened up as a narrow waterway and by about 12,500 years ago the marine incursion into the Central Basin had started. The Western Basin flooded about 1000 years later. Momentary standstills may have occurred during the Gulf flooding phase at about 11,300 and 10,500 yr BP . . .

In other words the whole of the Persian Gulf – in fact to a point well beyond the Strait of Hormuz in what is now the Gulf of Oman – was dry land between 18,000 and 14,000 years ago. Only then did the sea begin to transgress into the Gulf itself, first as a narrow waterway, later as a recurrent cycle of powerful short-lived floods, each followed by a partial recession of the floodwaters, then a standstill, then renewed flooding at irregular intervals.

I knew from my first encounter with Lambeck's research that the present shoreline of the Gulf had been reached, and then temporarily exceeded, around 5500 years ago during the Flandrian transgression. But what I had not immediately understood was the extraordinary geological drama that had unfolded between 14,000 years ago, when the Gulf first began to flood, and 7000 years ago, when the city-state of Eridu was established at the north-western end of the Gulf and, along with it, the way of life that would soon flower as Sumerian civilization.

The floor of the Gulf

Lambeck himself was convinced that there must be some connection between the flooding of the Gulf and 'the Sumerian problem':

> The early record is incomplete and numerous questions have been raised. Who were the Sumerians, where did they come from? When did they arrive? Did they arrive from a mountainous region beyond Iran or did they arrive by sea? Were they descendants from earlier Neolithic settlers in the region, from the Ubaid culture at 4500–3500 BC or from the even earlier Eridu culture at about 5000 BC [archaeologists often refer to the Eridu culture as 'Ubaid I' – i.e. the earliest stage of the Ubaid culture].[103] Whatever directions the search for answers to such questions may take, a significant element in the puzzle must be the evolution of the physical environment of the Gulf itself.[104]

The last observation sounded particularly relevant to my concerns; however, Lambeck went on to qualify it by suggesting that the only epoch that historians and archaeologists really need to pay attention to is 'the latter period of the flooding of the Gulf and the subsequent flooding of the low-lying delta region [the Flandrian transgression] when sea-level rose perhaps a few metres above its present level between 6000 and 3000 yr BP'.[105] If archaeologists were interested in the earlier period between 18,000 years ago down to as recently as 7000 years ago – when a large part of the Gulf floor was still dry land – then they should focus on its role as a corridor of migration: 'a natural route for people moving westwards from east of Iran. Is this the route travelled by the ancestors of the Sumerians?'[106]

What Lambeck did not do, anywhere in his paper, was invite consideration of another possibility, even though it is suggested by some of his own data. This is the possibility that the dry floor of the Gulf could *itself* have been a place of permanent settlement at some point during the 11,000 years between 18,000 and 7000 years ago.

If it was, then why shouldn't an urban culture have evolved here, just as the myths of the antediluvian cities suggest?

After all, orthodox archaeology has already accepted the existence of very ancient cities elsewhere in the Middle East – such as Catal Huyuk in Turkey (at least 8500 years old), Jericho in Palestine (more than 10,000 years old)[107] and, indeed, Eridu in Mesopotamia (where, as we've seen, the oldest shrines are thought to be about 7000 years old). Knowing the inundation history of the Gulf as well as we now do, therefore, we cannot rule out the possibility that the ruins of cities that are literally 'antediluvian' could be concealed beneath its increasingly polluted, industrialized and militarized waters . . .

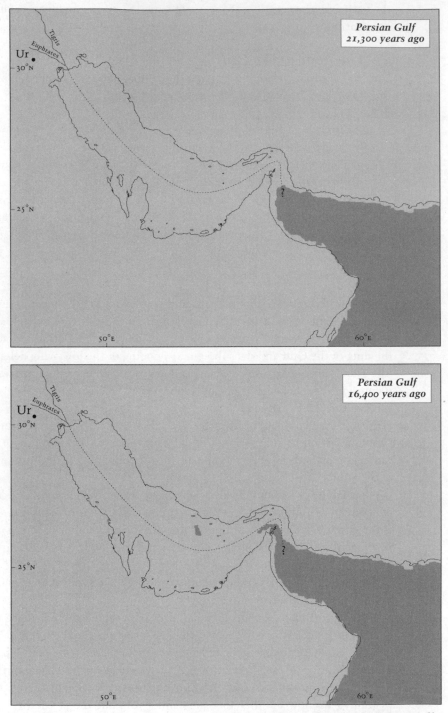

Dotted line represents projected course of Tigris-Euphrates through the Palaeo-Gulf.

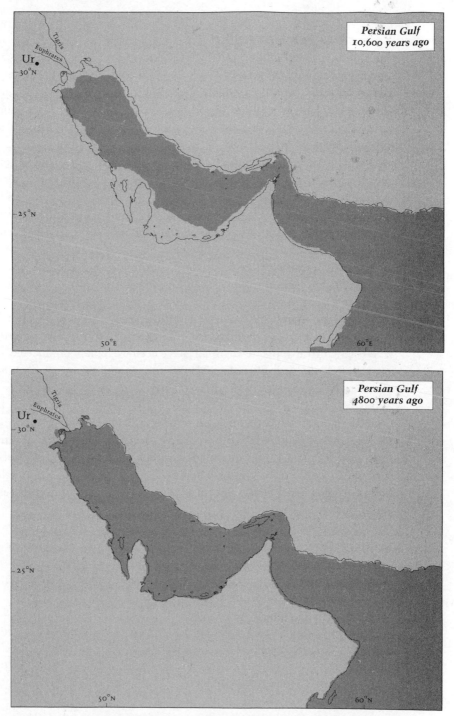

In these and all inundation maps in this book, the black lines represent modern coastlines, the light tint is land and the dark tint is sea.

A river ran through it

During the period from the Last Glacial Maximum until about 10,000 years ago the Ice Age world was generally colder and more arid than it is today, with average temperatures depressed by several degrees even in tropical and equatorial zones. However, these conditions are likely to have been much less severe within the micro-region of the antediluvian Gulf – essentially a large, well-protected, low-lying valley.[108]

Its notable feature, which undoubtedly would have been a magnet for life of all sorts including human beings – was that the Tigris and the Euphrates flowed through it, united as a single mighty river.[109] The river's course seems to have run along the northern side of the valley and at different periods appears to have passed through as many as three large, freshwater lakes in the Gulf's Western and Central and Eastern Basins.[110] It exited the Gulf through the narrows now known as the Strait of Hormuz and formed its delta on the Biaban Shelf to the east.[111] The delta was relatively small for such a large river, which suggests to scientists that it must have dumped most of its load of fertile alluvial silt in and around the shores of the lakes that it filled along the way.[112] Over thousands of years this would have created areas of great natural fertility within the valley where agriculture, if practised, might have been extremely productive.

For a while things could only get better and, despite the remorseless advance of the sea after the Strait of Hormuz was breached 14,000 years ago, conditions in the rest of the Gulf may for a long while have remained extremely pleasant. I was particularly interested to learn of a comprehensive study done in 1988 by the COHMAP group which showed that 'the Indian monsoon system penetrated into the southern and eastern portions of south-west Asia during the period of 12,000 to 9000 years ago, and then retreated'.[113] The implication was that throughout this period the Gulf, along with other parts of south-west Asia, would have

> enjoyed both winter rains and in some areas also summer rains or ephemeral summer storms. This rainfall would have increased grazing opportunities, particularly in semiarid areas, but would have had little effect on the growth of winter cereals that formed a principal base of early agriculture.[114]

A protected valley ... a great river ... lakes ... fertile soils ... bountiful rainfall ... The palaeo-climatological literature left me with the distinct impression that the Gulf around 10,000 or 12,000 years ago could have been a very unusual place ... indeed a secret garden blessed with an ideal climate, offering nearly optimum conditions for the emergence of a civilization.

A sea change

What changed everything was the sea. As Lambeck tells it·

> By 14,000 yr BP the Hormuz Strait has opened up as a narrow waterway and the
> flooding of the lowlands to the west begins, first with the flooding of the Eastern
> Basin by marine water soon after 13,000 BP. Marine influence is first experienced in
> the Central Basin before about 12,500 BP . . . The Western Basin lake remains free
> from marine incursion until about 11,500 BP. The northern part of the Gulf remains
> dry at this time, as does a vast area south of the palaeo-Gulf, although this plain
> contains numerous shallow topographic depressions. Until about 11,000 BP the
> northern part of the Persian Gulf floor would have been a relatively flat but narrow
> plain, hemmed in between the palaeo-Gulf and the southern foothills of the Zagros
> mountains forming the present coastline.
>   As the sea-level rises the Gulf continues to expand and the marine influence
> spreads into the northern region. By about 10,000 BP the north-east margin of the
> Gulf has approached its present position in several localities, particularly east of
> about 52 degrees longitude. Much of the southern part of the Gulf remains exposed
> until about 8000 BP and areas such as the Great Pearl Bank are not submerged until
> shortly after this time.[115]

I have deliberately chosen not to summarize Lambeck's blow-by-blow
account of the flooding of the Gulf, but to let him speak for himself. He
does not dramatize or interpret his data but presents it neutrally, without
speculation, as a good scientist should.

I am not a scientist and I have a different approach. What I see here is first
and foremost a mystery – the mystery of Sumerian origins – 'the Sumerian
problem' as archaeologists like to call it. When I look closer I find that not only
do we not know where the Sumerians came from but also that their language
is unique in the world – apparently unrelated to any other known language.
Closer up still and I learn that the Sumerians preserved traditions of a terrible
flood that had nearly obliterated mankind from the earth and that had inundated
the five antediluvian cities of their ancestral homeland. There had been survivors in a great ship who had been carried by the floodwaters to another land
and had settled there in order to renew the ruined earth, replenish the seed of
mankind, and preserve the ancient wisdom and the worship of the gods. For
this reason those who later traced their line and religion from these survivors
always remembered history as being divided into two periods – before and after
the flood – and recorded the dynasties of their rulers in exactly the same way –
with the list of the historical kings preceded by the list of the antediluvian
kings, the latter reigning for a very long period.[116]

I review the archaeological literature for rational explanations of the Sumerian
flood tradition and find that most of the experts agree it must have been rooted

in some kind of historical truth; they point to the temporary inundation of Ur around 5500 years ago, either by gigantic river floods or by the marine incursion known as the Flandrian transgression. But when I look further and try to match up the details of the flood tradition to the archaeological facts I find that nothing really fits; nevertheless there are strange resonances between the evidence and the myths.

For example, we've seen that Eridu, always named as the first and oldest of the antediluvian cities, was never flooded; yet the archaeological evidence does make it a strong contender, with its 7000-year-old shrines to the water-god Enki, for the title of 'oldest' Sumerian city.

Conversely, Ur, which is not mentioned in the flood tradition at all, was most definitely flooded around 5500 years ago. Shurrupak, which is named as one of the antediluvian cities, was likewise flooded, but not until 700 years later.

So, for me, the theory that connects the Sumerian flood tradition with whatever event it was that flooded Ur is a 'dog that don't hunt'. I would honestly sooner conclude that the Sumerians had made the whole thing up than agree that they were so geographically ignorant and historically naive that they were incapable of distinguishing between a universal flood capable of wiping out humanity and a local flood – however large. Since we respect them so highly in other departments – as the builders of the world's first schools, for example, the inventors of the world's first bicameral congress, the compilers of the world's first law codes,[117] etc. – shouldn't we also respect the Sumerians' own evaluation of the great deluge that they say swallowed up the cities of their ancestors so long ago in the past?

A new hypothesis

Then I come across Kurt Lambeck's data. What it tells me is that the floor of the Persian Gulf was entirely exposed until as recently as 14,000 years ago, that between 12,000 and 9000 years ago it would have been a veritable Garden of Eden, and that, despite continuous flooding, large areas of the Gulf floor remained above the waves until somewhere between 8000 and 7000 years ago. Since these included the Great Pearl Bank – between modern Dubai and Qatar near Bahrain – I find it difficult to believe it is a coincidence that deities with authentic Sumerian names were worshiped in ancient Bahrain or that the first definite evidence of an identifiably 'Sumerian' presence in Iraq is at Eridu around 7000 years ago – so soon after the Great Pearl Bank was inundated.

In short, although I stress again that I'm no scientist, I believe that Kurt Lambeck's data is strong enough to justify an entirely new hypothesis on the subject of 'the Sumerian problem'. I think it's time to consider seriously the possibility that the true story of Sumerian origins may have proved so elusive because it is veiled beneath the waters of the Persian Gulf. In that case, Eridu and the other four 'antediluvian' cities of Mesopotamia might well bear the same relationship to the original antediluvian cities of the Gulf floor as Halifax,

Nova Scotia bears to Halifax, England or as Perth, Australia bears to Perth, Scotland. They could, in other words, have been named in memory of other, older cities somewhere else – normal, well-testified behaviour by migrants of almost all cultures in every epoch. Moreover, in this case we are not even required to imagine that the migration came from very far away but merely from the flooded lowlands of the Gulf towards the nearest higher and productive ground that was blessed by the same Tigris/Euphrates river system as the floor of the Gulf had once been.

At this point I find that the hypothesis and the existing archaeological evidence begin to converge nicely. Yes, it seems to be true that Eridu stands out as one of the earliest 'nascent' cities of Sumer, yes, the date of submersion of the Great Pearl Bank coincides quite closely with the date of foundation of the first shrines to Enki at Eridu, and yes, the Sumerians did have distinct memories of an advanced antediluvian culture that had been destroyed by a great flood.

But still, the flooding of the Gulf was a long-term event, wasn't it, spread out over more than 6000 years? Surely something that gradual, that predictable, is no more likely than the localized flooding around Ur 5500 years ago to have inspired the Sumerian tradition of the sudden world-destroying flood that threatened the survival of mankind?

Before I attempted to test my Sumerian hypothesis further by trying to set up a proper diving expedition in the Gulf (written authorization required in triplicate from Saddam Hussein, the US Navy, the CIA, Texaco, the President of Iran, the King of Saudi Arabia, and the Emirs of Kuwait, Bahrain, Qatar, Sharjah, Abu Dhabi and Dubai) I decided that I had to learn more about the behaviour of the world's oceans in the key 7000-year period from roughly 14,000 to 7000 years ago.

I knew already that this had been the peak period of the meltdown of the last Ice Age. I knew already that it had been a period of great turbulence and instability. It was therefore by no means impossible that something had happened at the global level during these millennia that could have projected a truly cataclysmic flood into the sheltered valley of the Gulf.

In fact, as I was to discover, it could have happened more than once . . .

3 / *Meltdown*

Athenian: Do you consider that there is any truth in the ancient tales?
Clinias: What tales?
Athenian: That the world of men has often been destroyed by floods . . . in such a way that only a small portion of the human race survived.
Clinias: Everyone would regard such accounts as perfectly credible.

Plato, Laws, vol. I, book III

It is clear that the [Beverley Lake] drumlins . . . must have been submerged in the formative flow . . . minimum depths of about 20 metres were required . . . On a helicopter traverse along the north shore of Georgian Bay, a single field of bedrock erosional marks was noted that had a width of at least 50 kilometres . . . [These] drumlins and erosional marks indicate meltwater floods that were competent to remove the largest boulders . . . Flow widths, equal to the widths of drumlin and erosional-mark fields, were in the range of 60 to 150 kilometres . . . Volumes of water required to sustain such floods would have been of the order of one million cubic kilometres, equivalent to a rise of several metres in sea level over a matter of weeks.

John Shaw, Professor of Earth Sciences, University of Alberta

As recently as 20,000 years ago, North America had an array of large animals to rival the spectacular wildlife of modern Africa. Mammoths bigger than African elephants, as well as smaller, pointy-toothed mastodons, ranged from Alaska to Central America. Herds of horses and camels roamed the grasslands while ground sloths the size of oxen lived in the forests and bear-sized beavers built dams in the streams. By about 10,000 years ago, all of these animals – and others such as American lions, cheetahs, sabertooth cats and giant bears – were gone. Some 70 North American species disappeared, three-quarters of them large mammals. Why?

Washington Post, 21 November 2001

If you study the literature and talk to the experts on the last Ice Age, you will find that there are wide differences of opinion over such fundamental matters as the main sequence of events, the chronology and consequences of these events, and even the terminology used to describe them.

The very idea of 'the last Ice Age' is poorly defined and is used differently by different authorities. For some it refers to the period from roughly 125,000 years ago, when the ice-caps of the northern hemisphere began their most recent advance, down to about 21,000 years ago, when they reached their maximum extent (LGM – 'the Last Glacial Maximum') and then began to melt. Even here, though, there seems to be variation in the scientific literature, as I have seen the LGM dated as early as 25,000 years ago and as late as 18,000 years ago.[1]

Another school of semantics takes a longer view, pointing out that the 'last Ice Age' was merely the most recent surge in a boom-and-bust cycle of glaciations and deglaciations going back some 2.6 million years. To them it is this longer cycle that is the Ice Age – and it is not 'the last Ice Age' because we are still in it. They point out that the process of deglaciation after 17,000 years ago was extremely rapid – being largely over within 10,000 years – but not far beyond the norm set by previous deglaciations. Likewise, the relatively congenial conditions that we have enjoyed during the 7000 years since then are perhaps a little better than those in some previous interglacials, but not spectacularly so.

Although I am not concerned in this inquiry with epochs millions of years in the past, I note in passing how curiously the fortunes of the creature called man seem to be intertwined with the long chronology of the Ice Age:

- The traces of our earliest, upright-walking ancestors of the genus *Homo* first begin to appear in the fossil record about 2.6 million years ago, when the great cycle of the current Ice Age began.
- Another coincidence occurs approximately 125,000 years ago, the onset of the most recent surge of the ice-sheets. It is at about this time, or a little after, that the earliest remains of *possible* anatomically modern humans are found.
- The earliest *undisputed* remains of anatomically modern humans are much more recent – perhaps 40,000 years old. This is around the same time that the first traces of classic European 'cave art' begin to appear – already mature and fully formed – in such locations as the Chauvet Cave in France.
- The earliest undisputed remains of large-scale permanent settlements with monumental stone architecture are found around 10,000 years ago – Jericho for example, which stands in the Jordan valley in Palestine. Other impressive sites include Catal Huyuk in Turkey, dating to perhaps 8500 years ago. The whole idea of permanent settlement, however, does not seem to take very wide root until after about 7500 years ago. This is the time when the world's climate begins to stabilize again after 10,000 years of unbelievable turbulence, melting ice and rising sea-levels.
- The same chronology, more or less, and the same loose correlation to the end of the last glaciation, applies to accepted scientific models of the spread of agriculture.

But does it? Or is it possible that important parts of the story of our past could have been veiled from us by the upheavals of the glacial cycle?

Although I know that it was just the most recent of many glaciations, I use the term 'the last Ice Age' to refer to the latest glacial expansion between 125,000 and 17,000 years ago. When I use the term Last Glacial Maximum (LGM) I refer not to a specific moment but to a *period* of approximately 5000 years between 22,000 years and 17,000 years ago during which the ice-sheets remained at or near their maximum extent. There was some melting and sea-level rise after around 19,000 years ago but the volume was relatively small and there was little impact on coastlines. What may truly be described as the epoch of the 'meltdown' began immediately afterwards – say 16,500 years ago – with the mass of ice-sheet wasting and associated sea-level rise complete by 7000 years ago.

Before the flood

Imagine the world before the flood. Seventeen thousand years ago, at the end of the Last Glacial Maximum, most of northern Europe and North America were buried under ice several kilometres thick. So much water was tied up in these continental ice-caps that global sea-level was between 115 and 120 metres lower than it is today. The antediluvian world, therefore, looked very different from the world we are familiar with.

- A land-bridge joined Alaska and Siberia across what is now the Bering Strait.
- It was possible to walk from southern England to northern France across the dry valley that would later become the English Channel.
- Many more islands were exposed in the Mediterranean than are visible today and existing islands were much larger. Malta, for example, was certainly joined on to Sicily. Corsica and Sardinia formed a single huge island.
- Further east, we've already seen that the whole of the Persian Gulf as far as the Strait of Hormuz was dry 17,000 years ago but for its great alluvium-rich river and its life-giving lakes . . .
- Further east still, India's coastlines were much more extensive at the end of the last Ice Age than they are today and the shape of the subcontinent was strikingly different. Sri Lanka was joined to the mainland and south of Sri Lanka, sprawling across the equator, the Maldive islands were far larger than they are today.
- Around modern Malaysia, Indonesia and the Philippines, and stretching as far north as Japan, lay the endless plains of 'Sunda Land', a fully fledged antediluvian continent. It was submerged very rapidly some time between 14,000 and 11,000 years ago.

- Up until about 12,000 years ago, the three main islands of Japan formed a continuous landmass.
- In the southern seas lay the gigantic Ice Age continent of Sahul, formed out of the united landmasses of Australia, Tasmania and New Guinea.
- Across the Pacific the thousands of small, remote islands of today were integrated into much larger archipelagos 17,000 years ago.
- In the western Atlantic, in the same epoch, the Grand Bahama Banks, now shallowly submerged, formed a huge plateau 120 metres above sea-level, and all of the Florida, Yucatan and Nicaragua shelves were exposed.[2]

In short, the habitable landmasses that modern civilizations have inherited from the meltdown of the last Ice Age only began to take their present form in the ten millennia between 17,000 and 7000 years ago.

Before that, areas that are densely populated today, Chicago, New York, Manchester, Amsterdam, Hamburg, Berlin, Moscow – in fact most of North America and northern Europe – were absolutely *uninhabitable* due to the fact that they were covered by ice-caps several kilometres thick. Conversely, many areas that are uninhabitable today – on account of being on the bottom of the sea, or in the middle of hostile deserts such as the Sahara (which bloomed for about 4000 years at the end of the last Ice Age) – were once (and relatively recently) desirable places to live that were capable of supporting dense populations.

Geologists calculate that nearly 5 per cent of the earth's surface – an area of around 25 million square kilometres or 10 million square miles – has been swallowed by rising sea-levels since the end of the Ice Age.[3] That is roughly equivalent to the combined areas of the United States (9.6 million square kilometres) and the whole of South America (17 million square kilometres). It is an area almost three times as large as Canada and much larger than China and Europe combined.[4]

What adds greatly to the significance of these lost lands of the last Ice Age is not only their enormous area but also – because they were coastal and in predominantly warm latitudes – that they would have been among the very best lands available to humanity anywhere in the world at that time. Moreover, although they represent 5 per cent of the earth's surface today, it is worth reminding ourselves that humanity during the Ice Age was denied useful access to much of northern Europe and North America because of the ice-sheets. So the 25 million square kilometres that were lost to the rising seas add up to a great deal more than 5 per cent of the earth's useful and habitable landspace at that time.

Now, imagine if you were to discover a hidden secret: the entire orthodox account of world prehistory as it is presented in the classroom, at university, through books and in the media has been created by archaeologists with no

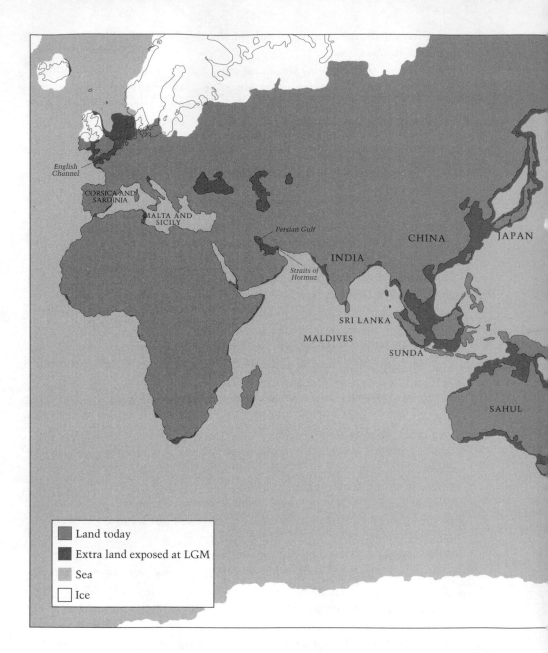

English
Channel

CORSICA AND
SARDINIA

MALTA AND
SICILY

Persian Gulf

CHINA   JAPAN

INDIA

Straits of
Hormuz

SRI LANKA

MALDIVES

SUNDA

SAHUL

■ Land today
■ Extra land exposed at LGM
■ Sea
□ Ice

reference whatsoever to China and Europe, or to South America and the land-mass of the USA. Having missed out entirely such large areas from their excavations and research wouldn't you feel that their conclusions about world prehistory and the story of the origins of civilization were likely to be – to say the least – flawed? Well, it is a similar story with the 25 million square kilometres lost at the end of the Ice Age. Marine archaeologists have barely even begun a systematic survey for possible submerged sites on these

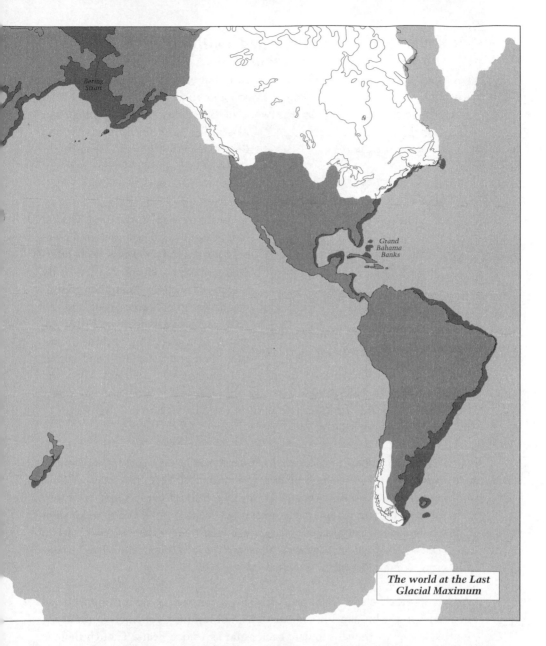

Bering
Strait

Grand
Bahama
Banks

**The world at the Last
Glacial Maximum**

flooded lands. Most would regard it as a waste of time even to look. In consequence, whether in Australia or Europe, the Middle East, India or south-east Asia, the enormous implications of the changes in land-use and rising sea-levels between 17,000 and 7000 years ago do not appear *ever* to have been seriously considered by historians and archaeologists seeking the origins of civilization.

A case history: the drowned 3 million square kilometres of Sahul

Let's look more closely at what happened to Sahul – also known as 'Greater Australia' – between approximately 17,000 and 7000 years ago. Much of the story has been unravelled by the work of Jim Allen, an archaeologist at Australia's La Trobe University, and Peter Kershaw from the Department of Geography and Environmental Science at Monash University, Melbourne.[5]

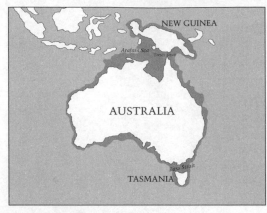

Sahul at the Last Glacial Maximum.

Until the end of the Last Glacial Maximum 17,000 years ago, and probably for several thousand years afterwards, New Guinea was fully integrated with the Australian continent across the Torres Strait and the Arafura Sea, Tasmania was fully integrated in the south – the Bass Strait then being dry land – and 'other smaller, now offshore, islands were also incorporated'.[6] In total Allen and Kershaw estimate that Sahul of 17,000 years ago extended 'from almost exactly the Equator to nearly 44 degrees S and from 112 degrees E to 154 degrees E'.[7]

Then came the meltdown:

> Between circa 16,000 BP and 7000 BP Greater Australia was reduced in area by *more than three million square kilometres* – an area much larger than Mexico. Three major landmasses existed where previously there had been one . . . Coastal sites were either submerged or preserved on islands, while sites of the former arid interior became coastal . . . In places *the postglacial marine transgression reduced the width of the coastal plain by up to several hundred kilometres*, thus presumably drowning many terminal Pleistocene sites in the process . . .'[8]

And how much else? There are, after all, a number of discontinuities and mysteries in the human story in Australia, not least the venerable antiquity of its first settlers – thought to date back as far as 50,000 years. Though there is no archaeological evidence whatsoever that a high civilization in the technical, material or urban senses ever flourished here before the modern era, there are certain aspects of Aboriginal culture that are frankly puzzling and do not fit in. These include evidence of sophisticated astronomical ideas from a very early date and the use of an 'astronomical terminology' that is also found in other very distant regions of the world. Thanks to the research of the Russian prehistorian Boris Frolov, for example, we must now ask ourselves whether it is a coincidence that indigenous tribal peoples as far afield as North America, Siberia and Australia all called the Pleiades star-group 'the Seven Sisters'.[9] Frolov's own view is that coincidence is not a satisfactory explanation and that only an extremely ancient shared heritage can account for this and many other thought-provoking parallels that he has uncovered.[10] But if Frolov is right, as the Cambridge anthropologist Richard Rudgley observes in his groundbreaking *Lost Civilizations of the Stone Age*, then the implication is:

> a tradition of communicable knowledge of the heavens that has existed for over 40,000 years, since a time roughly coinciding with the beginning of the Upper Palaeolithic. This is something that is extremely awkward for most widely accepted views of the history of knowledge and science – in short it is far, far too early for most people to accept.[11]

Of course, it is true that archaeologists excavating Australian terrestrial sites have not turned up any evidence there of the kind of social infrastructure that would normally be associated with the spread of a global astronomical tradition. But with more than 3 million square kilometres of Greater Australia submerged between 16,000 and 7000 years ago, and almost entirely unexplored by archaeologists, who can be sure what yet might be found?

Floods and civilization

Were the post-glacial 'floods' really floods at all? It doesn't take a mathematical genius to work out that 120 metres of sea-level rise spread out over 10,000 years amounts to an average of not much more than a metre a century. Inconvenient, certainly . . . But surely not enough to submerge and sweep away all traces of a great civilization? Surely not enough to inspire the global myth of the flood – so often accompanied, as it was in Sumer, by the unshakeable conviction that the gods had resolved to obliterate mankind?

In previous books I have discussed the cycle of the Ice Ages. Over the past 2.6 million years, this cycle shows strong correlations with the (slowly changing) obliquity and precession of the earth's axis and the varying degree of eccentricity of its orbit around the sun. Some scientists feel that these large-scale astronomical influences are sufficient, on their own, to explain the recurrent glaciations and deglaciations of our planet. Others feel that trigger factors must also be involved – extreme episodes of volcanism, asteroidal or cometary impacts, a realignment of the earth's crust or mantle, and so on and so forth.

Irrespective of the cause, however, there is no dispute about the biggest consequence of the meltdown of the last Ice Age: sea-level is now 120 metres higher than it was 17,000 years ago. This, by any standards, represents a dramatic change in the distribution of habitats for human settlement and should, one might expect, be a matter of great interest to archaeologists. When I began to research this subject I was therefore surprised to learn that this is not at all the case:

- only an infinitesimal amount of marine archaeology has been done along continental shelves (infinitesimal in relation to the total area of land submerged worldwide);
- of the marine archaeology that has been done, the largest part has been focused upon the discovery and excavation of shipwrecks and of sites submerged in historical times;[12]
- with the exception of Robert Ballard's exciting underwater survey of the Black Sea for the National Geographic Society, which got underway in 2000 and has been oriented directly towards an investigation of a colossal incursion of the Mediterranean through the Bosporus narrows 7500 years ago, marine archaeology has simply not concerned itself with the possibility that the post-glacial floods might in any way be connected to the problem of the rise of civilizations.

I am aware that there is a new mood of political correctness amongst archaeologists and a willingness to accept, and state publicly, that the peoples of the Stone Age were neither ignorant savages nor lowbrow 'cave men' – although one need only spend a moment glancing at the transcendental art of Lascaux to realize that! But still it seems to me true to say that the great majority of archaeologists see no particular trend or connection that obviously links the 'Palaeolithic' way of life, 17,000 or even 12,000 years ago, to the urban way of life that first appears at Jericho, Catal Huyuk and a handful of other sites between 10,000 and 7000 years ago. This is why, although they are certainly more open than they were before to the spirituality and high artistic culture of the ancients, archaeologists – almost without exception – do still assume that the population of the earth was at a uniformly hunter-gatherer level of social and economic development 17,000 years ago, and still about 7000 years away from founding the first cities. They therefore have no particular reason to be interested in the fact that millions of square kilometres of continental shelf were flooded in the intervening years, changing the face of the habitable earth completely.

If, on the other hand, the level of development of different cultures in that period was *not* uniform (as is the case in the world today) and if one or several cultures had concentrated along the ancient sea-shores – or in any other areas which might have been rapidly and cataclysmically inundated – then it is possible that the post-glacial floods could have had *enormous* significance for the story of civilization.

Moreover, the rise in sea-level of 120 metres over those 10,000 years between 17,000 and 7000 years ago is large enough to have engulfed entire cities for ever and either demolished or covered up with millennial deposits of silt and muck all evidence of their former existence. If the waves rose slowly, such hypothetical cities would have been pounded for centuries in the high-energy intertidal zone which makes short work even of granite structures. But if the sea-level rise was due to some cataclysmic surge, then walls of water would have borne down on and crushed beyond recognition much that stood in their path.

Many things happening at once

It is hard to know where to begin to tell the story of the meltdown of the last Ice Age, because it is really many different stories woven together into a single fabric.

- Part of it concerns large-scale climate flips, sudden radical thaws and equally radical freezes, volcanism on a planetary scale, earthquakes of unparalleled ferocity and mass extinctions of animal species.
- Part of it, which I've already touched on, is the huge loss of habitable land, of low-lying coastal plains and fertile river deltas that occurred as the sea-level rose – a 'lost continent' scattered around the world like the pieces of a jigsaw puzzle with a combined land area of 25 million square kilometres.
- Part of it concerns the speed and the sheer magnitude of the post-glacial flooding.
- Part of it is the need to understand the processes that led the earth into this devastating cycle of inundations.
- Part of it is a complexity: *yes*, global sea-level did rise by about 120 metres between 17,000 and 7000 years ago; *no* this 'eustatic' rise (i.e. pertaining to sea-level alone) has not been uniformly reflected in changing shorelines through time. Thus, in some parts of the world sea-level relative to ancient shorelines has remained quite stable for millennia; in others, submersion of a particular locality may be deeper than expected from eustatic changes; in yet others submersion may be shallower than expected from eustatic changes. Such variations can be caused by local land subsidence or land rise following earthquakes or volcanic activity; however, a much more potent and extensive agent of changing land-levels is known to geologists as isostacy.

Kicking the gel-filled football

The earth's surface, which seems solid beneath out feet, can yield and deform when subjected to sufficiently large pressures. It behaves a bit like a football that has been loosely filled with a thick, heavy gel: pressure at one point on the gel-filled ball will result in an indentation in that area, a displacement of the fluid mass within and a corresponding rise in a roughly circular area surrounding the indentation. Geologists call this process isostacy, and it plays an important role not only during Ice Ages but also for thousands of years after all the ice has melted away. The reason it does so is that the vast weight of the ice-caps is sufficient to force down the earth's crust into great basin-like depressions beneath them. When the ice melts, that pressure is suddenly removed and the

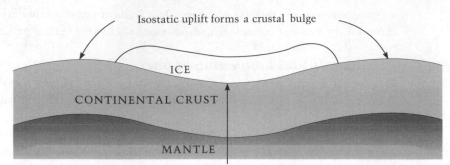

Isostatic uplift forms a crustal bulge

ICE

CONTINENTAL CRUST

MANTLE

Isostatic subsidence forms a crustal depression

Ice-loading causes a depression in crust under ice, and an isostatic bulge effect beyond it. Based on Wilson and Drury (2000).

floors of the basins begin to rebound; they will, if sufficient time is allowed, rise again to their original levels.

At the LGM 17,000 years ago, the ice-caps over large parts of North America and northern Europe were *between 2 and 4 kilometres thick* and applied loads of thousands of billions of tonnes to the continental landmasses on which they had formed.[13] Thomas Crowley and Gerald North, both oceanographers at Texas A&M University, observe that North America's Laurentide ice-sheet

> extended from the Rocky Mountains to the Atlantic shore and from the Arctic Ocean southward to about the present positions of the Missouri and Ohio rivers. In Europe the Fennoscandian Ice Sheet reached northern Germany and the Netherlands. The

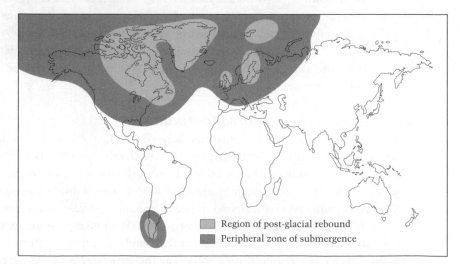

Region of post-glacial rebound
Peripheral zone of submergence

The post-glacial world showing regions of isostatic rebound (light shade) and submergence (dark shade). Based on Wilson and Drury (2000).

weight of the massive ice sheets depressed the crust by as much as 700–800 metres, resulting in gravity anomalies that are still detectable.[14]

On average it has been found that 100 metres of ice-loading depresses continental crust by 27 metres.[15] But this is only part of the story. The water of the world's oceans also has weight; indeed it is denser than ice. Thus, 100 metres of water-loading depresses the sea-bed beneath it by 30 metres.[16] Since all the ice formed on land during the last Ice Age was made out of water extracted from the sea, it follows that while the crust was pressed down beneath the continents, it actually rose up beneath the oceans (as the water-burden above it lightened). Conversely, after all the ice had melted and returned to the oceans as water, the burden on the sea-bed would have again increased. R. C. L. Wilson, Professor of Earth Sciences at Britain's Open University, calculates that a layer of water 165 metres deep was subtracted from the oceans to make the great ice-caps of the last glaciation. This, however, only produced a net drop in relative sea-level of around 115 metres between the onset of glaciation 125,000 years ago and the onset of LGM 104,000 years later – the reason for the discrepancy being that reduced water-loading in the oceans during the Ice Age allowed the sea-bed to rise by 50 metres through the process of isostatic compensation.[17]

Let's stop for a moment and take another look at this see-saw system swing by swing:

1. 125,000 years ago the most recent glacial surge begins, turning a worldwide layer of ocean 165 metres deep into ice-caps thousands of metres high piled up (for the most part) in North America, Greenland, northern Europe, South America and the Himalayas.

2. The maximum extent of ice formation is reached 21,000 years ago and largely maintained until 17,000 years ago; by this time the continental crust beneath the big ice-caps has been depressed into huge basins nearly a kilometre deep.

3. Simultaneously, as the ice-burden on the land increases, the water-burden on the sea-bed decreases; by the Last Glacial Maximum this had allowed the ocean-floor around the world to rise by 50 metres.

4. Soon after the LGM the ice begins to melt and to flow back as water to the oceans, a process that is substantially over within 10,000 years.

5. Since a layer of water 165 metres deep was taken out of the oceans to begin with to make up the ice-caps, it follows that a layer of water 165 metres deep is returned to the oceans with the complete melting of the ice-sheets.

6. Professor Wilson observes that the rate at which the crust and mantle respond to loading and unloading is 'much slower than the build-up or melting of ice caps. This is why areas that were buried beneath

several kilometres of ice 18,000 years ago are still rising today, thousands of years after the ice sheet melted away.'[18]

7. It also follows that the average 50 metre rebound of the ocean floor between 125,000 and 17,000 years ago would take thousands of years to be forced down again by isostatic subsidence to its original level.

8. Measured at a warm point in a long interglacial, and after 17,000 years of isostatic subsidence, today's sea-level is probably quite close to the final balance in the equation of rising seas and sinking sea-beds. But there must have been many times during the meltdown of the Ice Age when the speed of the former far outstripped any compensating effects of the latter.

Is it not possible, perhaps even probable, that this combination of a higher sea-floor than today's and rapid influxes of meltwater from the decaying ice-caps could have produced relative temporary rises in sea-level much greater than the average annual rate projected over the full period of the meltdown?

See-saw

Examples of segments of continental crust that continue to rise through isostatic rebound since the removal of the ice-sheets include the highlands of Scotland[19] (where the ice-cap that once covered most of Britain was at its thickest), the floor of the Gulf of Bothnia in what is now the Baltic Sea (reported to be rising at a rate of a metre per century),[20] large parts of the coasts and mainland of Sweden, Denmark and Norway, the north-east coast of Canada,[21] and parts of southern Chile.[22]

Complicating the picture is the fact that around each zone of 'post-glacial rebound', there lies what geologists call a 'peripheral zone of submergence' – which is always larger than the zone of rebound.[23] Thus, while it is not uncommon to find such phenomena as raised beaches in the highlands of Scotland[24] (demonstrating graphically that areas that were once at sea-level, and formed an ancient coastline, have now been lifted well above it), other areas of the British Isles are visibly sinking into the sea. This is because the downward pressure of the Fennoscandian ice-sheet on the northern European continental crust at the LGM was transformed by the mechanism of isostatic compensation into a huge 'forebulge' several hundred kilometres beyond the ice-margin – literally as though one end of a see-saw had been forced down, pushing the other end up. As the ice melted the weight that was holding the end of the 'see-saw' down was released, allowing it to rise again and causing the other end – the 'forebulge' – to fall.

This is exactly what is happening in the English Channel today, which we've seen was entirely dry at the LGM. The Isle of Wight stood on the forebulge of the Fennoscandian ice-sheet, forced upwards by isostatic compensation. Then when the ice-sheet melted, the dynamics of isostacy again came into play and the forebulge began to subside – taking the Isle of Wight (and much of southern England) down with it.

Isostatic Atlantis

An ingenious theory of the lost land of Atlantis, the first that I am aware of that is explicitly based upon the relationship between isostacy and rising sea-levels, was put forward in the late 1990s by Vitacheslav Koudriavtsev, a member of the Russian Geographical Society of the Russian Academy of Sciences.

It is well known that the story of Atlantis was set in writing in the fourth century BC by the Greek philosopher Plato – in his dialogues *Critias* and *Timaeus*. But before that, Plato tells us, it had been an oral tradition passed down within his family from his ancestor Solon, the revered Athenian law-maker. Solon had been told it during a visit that he had made to Egypt at around 600 BC. His informant, in turn, had been an elderly Egyptian priest at the Temple of Sais in the Delta, who said that he had drawn the information from written records, then more than 8000 years old, lodged in the temple's archives.

There are four essential ingredients in Plato's story:

- Atlantis was a relatively advanced, well-organized and prosperous civilization.
- It flourished and was destroyed 9000 years before Solon's time – in other words, approximately 11,600 years before our time.
- It was located on a large island 'opposite the Pillars of Hercules' – presumed to be the modern Straits of Gibraltar.
- Its destruction was the result of a global cataclysm: 'There were earthquakes and floods of extraordinary violence, and in a single dreadful day and a night . . . the island of Atlantis was . . . swallowed up by the sea and vanished.'[25]

There have been a thousand theories about the location of lost Atlantis, moving it around in time according to individual researchers' whims and placing it everywhere from the Mid-Atlantic Ridge to Indonesia and from the Andes mountains to Crete. What Koudriavtsev is suggesting is just another theory. Nevertheless, it has the great merit of requiring no liberties to be taken with Plato's text either in respect of the location of 'Atlantis' (beyond the Straits of Gibraltar in the Atlantic Ocean) or of the date of its submergence – 11,600 years ago.

Koudriavtsev's location is an area known to fishermen as the Little Sole Bank, situated on a vast underwater plateau called the Celtic Shelf, 200 kilometres to the south-west of the British Isles and Ireland. Although the shallowest part of Little Sole Bank is now 57 metres beneath the waves, and thus might be expected to have been about 60 metres above sea-level just before the end of the last Ice Age, Koudriavtsev's research shows that it and a large area of the surrounding shelf may have been tilted dramatically upwards during the build-up to the Last Glacial Maximum by the see-saw effect of isostatic forces emanating from the continental ice-mass. In brief, his theory is that there was an unusually rapid collapse of the forebulge in this area around 11,600 years ago, coinciding with a ferocious episode of ice-melting and global flooding – the sudden inundation of Atlantis described by Plato.

'In my opinion,' states Koudriavtsev,

> the most serious argument in favour of the assumption that Atlantis was not invented by Plato is that the time when it vanished, as indicated by Plato – about 11,600 years ago – and the circumstances of its vanishing described by him (the sinking into the deep of the sea), coincide with the findings of modern science about the end of the last Ice Age and the substantial rise of the level of the World Ocean that accompanied it.[26]

Three global superfloods

Anyone who has read the *Timaeus* and *Critias* carefully knows that what Plato describes in his account of the destruction of Atlantis is indeed a global flood that took place approximately 11,600 years ago and that swallowed up huge landmasses as far apart as the eastern Mediterranean and the Atlantic Ocean. I would have thought that a first line of approach for scholars investigating Plato's claims would be to find out whether anything on this scale might actually have happened in the world 11,600 years ago. So far as I can discover, however, not a single historian or prehistorian has ever made the effort to do so – although many of them have put forward theories, usually widely applauded by their peers, locating Atlantis anywhere but in the Atlantic, where Plato says it was, and any time within the epoch of recorded history, rather than considering the prehistoric date of 9600 BC given by Plato. One of the ludicrous (but positively peer-reviewed) claims put forward to divert the debate endlessly into trivia is that Plato meant 9000 *months* before Solon's time, not 9000 years, when he spoke of the submergence of Atlantis.

In my experience historians and archaeologists will go through Houdini-like contortions of reason and common sense rather than consider the possibility that their paradigm of prehistory might be wrong – so I am not surprised that they have never attempted to investigate at face value the Atlantis tradition of a devastating global flood 11,600 years ago. However, there are scholars – trained in other disciplines and not hobbled by the same preconceptions – who are more open to the possibility that the flood tradition in general, and the Atlantis story specifically, might be rooted in the real events of the meltdown of the last Ice Age. This view has been entertained positively by the late Cesare Emiliani, for example, former Professor in the Department of Geological Sciences at the University of Miami[27] – one of the pioneers of the isotopic analysis of deep-sea sediments as a way to study the earth's past climates.[28] Moreover, Emiliani's fieldwork in the Gulf of Mexico has produced striking evidence of cataclysmic global flooding 'between 12,000 and 11,000 years ago'.[29] Robert Schoch, Professor in the Department of Geology at Boston University, observes that there was also a dramatic warming of the earth's climate in the same period[30] – the 'Preboreal' – and that overall there is a

> stunning line-up in time between the sudden warming of 9645 BC, Emiliani's scenario of a massive freshwater flood pouring into the Gulf of Mexico, and the date Plato ascribed to the sinking of Atlantis. Whatever the accuracy of specific details, this curious coincidence points to the effect sudden climatic changes can have – and no doubt have had – on civilization.[31]

Science writer Paul LaViolette likewise argues that 'there may be much truth to the many flood cataclysm stories that have been handed down to modern times in virtually every culture of the world. In particular, the 9600 BC date that Plato's *Timaeus* gives for the time of the deluge happens to fall at the beginning of the Preboreal at the time of the upsurge of meltwater discharge.'[32]

Before rejecting the possibility of a lost civilization of the last Ice Age, therefore, I urge historians and archaeologists to take a close look at the mass of data that now exists about the sequence of cataclysmic floods that swept the earth between 17,000 and 7000 years ago.

Yet this too is a contentious area of debate. For while scientists now agree on the approximate figure of 120 metres for sea-level rise during the 10,000 years of post-glacial flooding, many do not accept that these were 'floods' at all – and certainly not in the cataclysmic sense. Averaging the rise over the time span as we did earlier, they see a fairly gradual and distinctly non-cataclysmic process in the range of a metre a century. This remains the majority view. But since Emiliani's findings first began to undermine it in the 1970s there has been more and more research to show how *very* cataclysmic the meltdown of the Ice Age could in fact have been.

In brief what is being suggested is that during the long span of the meltdown – in addition to countless episodes of smaller-scale flooding – there were *three global superfloods* which have been dated within the following approximate time-bands: 15,000–14,000 years ago, 12,000–11,000 years ago and 8000–7000 years ago. I have found that estimates of these dates vary by more than a thousand years either way, depending upon which authority you consult, but the general point is clear enough: there now exists a strong case that nearly half the total meltwater release at the end of the last Ice Age was concentrated into these three relatively short episodes, creating conditions of concentrated damage after long periods of stability – precisely the combination of circumstances and bad luck that could have led ultimately to the destruction of an antediluvian culture.[33]

Professor Emiliani's ice dams

Cesare Emiliani made many original contributions to scientific understanding of the meltdown of the last Ice Age. He was also among the first to work out the precise mechanism behind the characteristic 'rhythm' of this 10,000-year period – millennia of slow melting and gradual sea-level rises interrupted, apparently randomly, by much shorter episodes of extremely severe global flooding and rapid, destructive oceanic transgressions:

> During the last Ice Age, ice reached its maximum extension 20,000 years ago. Deglaciation started almost immediately and progressed rapidly. Sometimes ice meltwater would pile up behind an ice dam and when the dam collapsed a huge flow would follow. One such great flood occurred in the American northwest 13,500 years ago when an ice dam holding back about 2000 cubic kilometres of ice meltwater (Lake Missoula) collapsed. A huge mass of muddy water and debris rushed across the area into the Columbia River, cutting broad channels called coulees and forming the so-called Channelled Scabland . . . As a result of the flood that formed the Scabland, the sea-level rose very rapidly, from minus 100 to minus 80 metres [vis-à-vis today's level]. By 12,000 years ago more than 50 per cent of the ice had returned to the ocean, and the sea-level had risen to minus 60 metres. At that point other giant floods occurred, down the Mississippi River valley into the Gulf of Mexico and down the Siberian river valleys into the Arctic Ocean. The Mississippi flood carried pebbles, which are now confined to the upper reaches of the Missouri-Mississippi system, all the way down to the delta. Sea-level rose very rapidly from minus 60 metres to minus 40 metres.[34]

The key phrase that caught my attention when I first read this passage was 'ice dam'. It was very simple, and yet it explained so much. Averaged out over 10,000 years it was true that the total global sea-level rise of 120 metres at the end of the last Ice Age only amounted to a little more than a metre a century. But what Emiliani was now suggesting was the intriguing possibility that enormous quantities of the glacial meltwater could have been detained for thousands of years behind ice dams on continental Europe and continental North America – *and then released into the open ocean all at once.*

The ice-caps that formerly covered these areas were up to 4 kilometres thick, as we've seen, and larger than present-day Antarctica in both cases.[35] Emiliani reminds us how:

> The weight of the ice on the land surface below created bowl-shaped depressions about 1 km deep. Heat from the interior of the earth was trapped under the ice sheets, the bottom ice melted, and great freshwater lakes formed. Twice in North America and western Siberia these lakes busted through the ice margins and created huge floods. Sea-level rose abruptly around 13,000 years ago and again 11,000 years ago

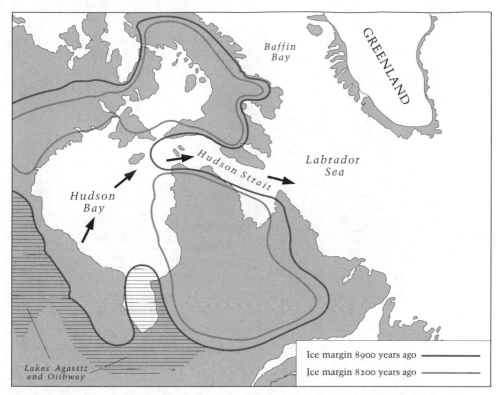

Between 8900 and 8200 years ago, the Laurentide ice-sheet disintegrated in the Hudson Bay, facilitating catastrophic drainage of the massive Agassiz/Ojibway glacial lakes into the Labrador Sea. Based on Barber et al. (1999).

and then more slowly as the residual ice continued melting. Some have hypothesized that these prehistoric floods generated the flood legends common to many civilizations.[36]

Professor Shaw's abrupt steps

John Shaw, Professor of Earth Sciences at the University of Alberta, is one of the world's leading experts on the last Ice Age and on its catastrophic meltdown. The author of an impressive list of peer-reviewed scientific papers, his research is at the forefront of inquiry in this field and has focused on the reasons for the superfloods. This is the graphic account that he gave us:

The big ice-sheets that covered Canada, most of Scandinavia and much of northern Russia – instead of them being pure ice and rock – it seems that at a late stage there was rock at the bottom and then a sub-glacial lake or reservoir of water, then the ice. And it's possible that when warming occurred, the top of the ice started to melt, and the ablation zone and the sub-glacial water got bigger and bigger and bigger. And yet

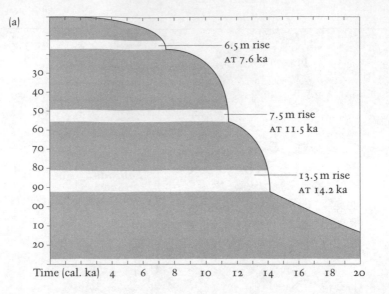

(a)

30
40
50
60
70
80
90
00
10
20

6.5 m rise
AT 7.6 ka

7.5 m rise
AT 11.5 ka

13.5 m rise
AT 14.2 ka

Time (cal. ka)   4   6   8   10   12   14   16   18   20

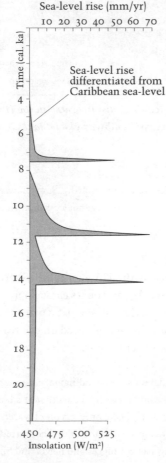

(b)   Sea-level rise (mm/yr)

10 20 30 40 50 60 70

Sea-level rise
differentiated from
Caribbean sea-level

450   475   500   525
Insolation (W/m²)

Graph of sea-level in the Caribbean against time since the LGM showing three abrupt steps around 14,000, 11,000 and 8000 years ago. Based on Blanchon and Shaw (1994).

Graph of rate of sea-level rise against time since the LGM. Based on Blanchon and Shaw (1994).

for good reason the ice-sheet seals around the edges. And then one time the big system on top connects – it's a little bit like a toilet bowl, you sort of open the valve and the water comes surging through.

[In Canada on one occasion] the water literally came spewing out all over, except to the east of the Hudson Strait, because there was a big ice barrier there. So it came out southwards and through the St Lawrence, through the finger lakes, down through the Red river, South Winnipeg and the Winnipeg Lakes, and out through parts of Saskatchewan and out over the Milk river – which is the continental divide south of Alberta. The Milk river water flowed north to the Arctic, to the east to Hudson Bay, south to the Gulf of Mexico. And a huge amount of water went north into the Arctic Ocean. So you were suddenly introducing a vast amount of water to the oceans. And the duration of the flows was probably measured in weeks. And the kind of flow that we're talking about, just for a small filament in Alberta, would have been 10 million cubic metres per second – that would drain Lake Ontario in about four days. And sea-level would have risen instantly, and somewhere in the region of 10 metres. This is about 15,000 years ago, when there were people living in many places. And the sea-level would have suddenly risen, and if you had lived by the sea-shore collecting jellyfish or something like that, and your house was suddenly underwater, you'd notice it. I imagine that it had quite an impression on the oral tradition and myths.

So the big event came from under the ice about 15,000 years ago. And then about 11,000 years ago there was a big lake in the southern part of the ice-sheet called Lake Agassiz that covered a big part of Canada. There was an equally big lake called the Baltic Ice Lake in Scandinavia. And then recent evidence suggests that there were big lakes across northern Asia and the north of the Soviet Union. These lakes were dammed by ice and tended to drain very suddenly. And as a result you get a similar effect, with a sudden rise in sea-level. Then last of all, about 8,000 years ago, there was the last lake in North America associated with the Laurentide ice-sheet which is called Lake Ojibway, and it lay just south of the Hudson Strait. And that lake drained catastrophically.

So originally it was thought that the rise of sea-level was steady at the end of the Ice Age, but now we are able to see that it rose abruptly in steps.[37]

Floods, volcanoes, earthquakes

Professor Shaw's 'abrupt steps' were, arguably, the most traumatic experiences of global cataclysm that our species has ever undergone. To those alive then, the end of the last Ice Age with its sudden global floods must have seemed like the end of the world. Continental plates were shifting upwards relieved of the weight of the ice they'd supported for 100,000 years. Huge earthquakes and outbreaks of volcanism accompanied this extensive crustal rebalancing. The earth would have rung like a bell with tremendous sounds and vibrations. The sky would have been heavy with volcanic dust and black, bituminous rain. And at the same time the oceans were remorselessly, apparently unstoppably, rising.

One of the geo-climatological mysteries of the last Ice Age is that the period of the meltdown – roughly from 17,000 to 7000 years ago – was also a period of dramatically enhanced volcanic activity. A paper published in *Nature* in October 1997 draws particular attention to what at first sight seems like a bizarre correlation between the rate of global sea-level change and the frequency of explosive volcanism in the Mediterranean area – with a distinct episode of enhanced volcanic activity registered in the geological and palaeo-climatological records between 17,000 and 6000 years ago:[38]

> In areas where active volcanism and glaciation coincide, the correlation between the events can be explained by the effect of changing ice volumes on crustal stress. In contrast the effect of ice-sheet volume changes on unglaciated volcanic areas remains problematical. Several authors have proposed that meltwater loading and unloading could influence volcanic activity at sites distant from areas of ice-accumulation through the global redistribution of water, although this hypothesis has never been tested.[39]

The international team of scholars behind the *Nature* article counted *tephra* layers in deep-sea cores from the bottom of the Mediterranean (*tephra* is a general term for solid matter ejected during volcanic eruptions) and conclude that:

> The frequency of *tephra*-producing events and, by proxy, notable explosive eruptions at Mediterranean volcanoes, can be related to rapid variations in sea-level change. In particular we draw attention to the quiescent phase centred at 22,000 years ago and corresponding to the last low sea-level stand, and to the most intense period of *tephra* layer formation between 15,000 and 8000 years ago which accompanied the very rapid rise in post-glacial sea-levels.[40]

The authors think that 'the existence of a single causal link between the rate of sea-level change and the level of explosive activity is unlikely' and point out that 'the unique response of individual volcanoes to large changes in sea-levels requires detailed study of each eruption record'.[41] Where this has been done, however, 'The level of explosive eruptions is seen to fall to a marked low between 22,000 years ago and 15,000 years ago, coincident with the last low sea-level stand.'[42]

I find it intriguing that the end of a 7000-year period of volcanic quiescence 15,000 years ago, and the beginning of the period of violent eruptions, both overlap with the first of John Shaw's global superfloods; likewise the end of the period of enhanced volcanic activity around 8000 years ago follows Shaw's third and last superflood.

Addressing this point, the scientists writing in *Nature* argue for broad-scale influences operating, for example, through stress changes in continental margins and at island arcs. These may promote the ascent of fresh batches of magma into volcanoes, while increased levels of regional seismicity related to load distribution may play a role in destabilizing already weakened volcanoes.

On a global scale the number of volcanoes susceptible to the above-mentioned effects is large. Current spatial distributions of active volcanoes show that 57 per cent form islands or occupy coastal sites while a further 38 per cent are located within 250 kilometres from a coastline. Assuming a similar distribution for around 1500 volcanoes active during [the last Ice Age], then 1400 are likely to have been subject to the more direct effects of rapid sea-level change . . . Furthermore, the rapidity of these sea-level changes, and consequently their potential to trigger responses in active volcanic structures, are only now becoming apparent.[43]

Despite its authors' caution about identifying a single cause, the evidence set out in the *Nature* paper does suggest that the earth's own isostatic rebalancing process, sparked off by the sudden meltdown of the ice-sheets and rapidly rising sea-levels at the end of the last Ice Age, must have been what awakened the volcanoes. The implication is that isostatic adjustment does not always proceed at a constant, steady rate – otherwise volcanism would presumably be constant as well – but must at times involve large, rapid shifts transmitting shock-waves through the earth's crust powerful enough to set the volcanoes raging around the globe.

It is precisely a shift of this speed and magnitude that Koudriavtsev envisages with his hypothesized 'overnight' collapse of the Celtic Shelf on the forebulge of the Fennoscandian ice-sheet 11,600 years ago. Moreover, researchers have found evidence that the meltdown of the same ice-sheet also unleashed tremendous forces during other periods of rapid worldwide flooding. At the time of Shaw's third great flood around 8000 years ago, for example, the stresses and earthquakes became so severe that immense waves were formed *in the ground*. One of these, in northern Sweden, is 150 kilometres long and 10 metres high and has been described as a 'rock *tsunami*'[44] that can only have been caused by 'earthquakes of unbelievable magnitude'.[45]

Descent of hell

Snaking across a bleak landscape, Sweden's Parvie ('wave in the ground') as it is known locally, is a remarkable and somewhat disturbing feature, exactly resembling a three-storey-high *tsunami* made of solid rock caught forever in freeze-frame as it rears up just before breaking. The most remarkable – and disturbing – thing about it, however, is that this part of northern Sweden is a zone of extremely low seismicity and stands on what geologists define as a 'stable continental region' (SCR) of the tectonic plate.[46] There should be no reason for catastrophic earthquakes *ever* to happen in an SCR. Yet the evidence unambiguously demonstrates that a catastrophic earthquake – indeed 'the largest earthquake ever known within the stable continental regions'[47] – did throw up the Parvie:

> Studies over the last two decades show that it formed suddenly by earthquake faulting in the late glacial to early postglacial times of the great Fennoscandian ice sheet (approximately 8000 to 8500 years ago), suggesting a genetic relationship between the two.[48]

The precise nature of this relationship and the true magnitude of 'post-glacial faults' (PGFs) such as the Parvie have been studied by Ronald Arvidsson of the Seismological Department of Uppsala University. He has shown that such faults – of which there are a whole series in northern Sweden – *frequently cut as far as 40 kilometres deep into the earth's crust.* All were caused by *different* gigantic earthquakes and all these earthquakes occurred within the same thousand-year period between 9000 and 8000 years ago.[49]

Arvidsson's widely agreed estimate is that the Parvie quake measured 8.2 on the Richter scale.[50] Another scholar, Arch C. Johnston of the Centre for Earthquake Research at the University of Memphis, points out that quakes of this magnitude only occur today along the edges of tectonic plates. The force that formed the Parvie ground-wave must, therefore, have been enormous:

> The Fennoscandian PGF's are . . . a remarkable consequence of rapid crustal unloading as the ice-sheets of the last Ice Age melted. The *Parvie* and other PGF's . . . represent the faults of *induced* earthquakes, events that would not have happened without externally-imposed . . . conditions.[51]

Johnston then goes on to note that, although 'induced seismicity' is known today,

> the post-glacial earthquakes are easily the largest known examples of this class. Surface quarrying can generate earthquakes of 2 to 4 [on the Richter scale];[52] deep mining and deep-well waste disposal 5 to 6 events; and large hydro-reservoirs mid 6 events. Excluding PGF's there are no earthquakes exceeding 7 confidently considered induced. The earthquake magnitude seems to scale with the agent of change of crustal stresses: great ice-sheets can induce great earthquakes.[53]

Now a characteristic of the Richter scale, not widely understood by those who live outside earthquake zones, is that it is calibrated so that each increase of one unit represents a tenfold increase in the magnitude of the quake.[54] So a 2 is ten times bigger than a 1, a 3 is ten times bigger than a 2, a 4 is 10 times bigger than a 3, and so on. The earthquake that hit Kobe in Japan on 17 January 1995, killing more than 5000 people in twenty seconds, measured 7.2.[55] With a Richter scale value of 8.2, the Parvie quake was ten times bigger than Kobe. The largest earthquakes ever recorded on the scale – rare events in subduction zones under oceans or between continental plates – have not exceeded the value of 9.[56]

The clear implication of Arvidsson's and Johnston's research, therefore, is that crustal rebound and isostatic rebalancing did at times take place *very rapidly* as the ice-caps melted down into cascading floods – rapidly enough to trigger extremely violent earthquakes and sudden massive faulting (penetrating to hitherto unheard-of depths of 40 kilometres and radiating laterally for up to 160 kilometres).[57] Writing up his findings in *Science* magazine, Arvidsson concludes:

> I interpret the earthquakes as signs of a progressive rapid rise of the land from the centre of postglacial rebound . . . to the outer reaches of the ice-sheet . . . More than 9000 years ago a nearly isostatic equilibrium was reached due to the depression of the lithosphere by the ice. After a quick removal of the ice-sheet a non-isostatic condition caused compressional stresses within the crust which triggered the earthquakes.[58]

Since the Parvie is only one of many giant post-glacial faults associated with the collapse of the Fennoscandian ice-sheet, what Arvidsson is really talking about – I think – is the descent of hell in northern Europe for a reign of 1000 years centred on 8000 years ago. As we follow his evidence, we must envisage extraordinary scenes of geological turmoil in which continuous deep tremors vibrate all the way through the Baltic Shield crust and the earth repeatedly roils, fractures, rears up and collapses – seemingly about to tear itself apart . . . While this is happening the ancient ice-cap over Fennoscandia is in a state of runaway meltdown, close now to the point of total collapse, and huge chunks of decaying ice the size of islands are falling into the sea, generating cataclysmic displacement waves. The ice-cap over North America is behaving in much the same way . . .

And let's not forget that the earth by this time – 8000 years ago – has *already* suffered the consequences of 7000 years of intense volcanism, 7000 years of rising sea-levels and sudden and unpredictable marine floods, 7000 years of continental shelves, land-bridges and islands vanishing beneath the waves, and 7000 years of spectacular climatic instability. Indeed, the palaeo-climatological record testifies to all of the following – and much more – between 15,000 and 8000 years ago: cold oceans, high winds, mountains of dust in the atmosphere[59] and wildly unpredictable temperature shifts.[60]

To give an example of the latter, Romuald Schild of the Polish Academy of Sciences cites an abrupt warming that took place in the northern Atlantic at around 12,700 years ago, stopped and equally abruptly went into reverse 10,800 years ago – when there was a sudden 800-year plunge to almost full glacial temperatures – then turned again to another episode of abrupt warming about 10,000 years ago.[61] Robert Schoch reports that the bulk of the first warming –

'approximately 27 degrees Farenheit, a massive increase' – occurred after 11,700 years ago:

> Remarkably, the ice-core data suggests that half of the temperature change, in the neighbourhood of 14 degrees Farenheit, occurred in less than 15 years centring around 9645 BC. That's a bigger temperature increase, and faster, than the scariest doomsday scenario about global warming in the twenty-first century.[62]

It also happens to coincide, almost exactly, with Plato's date of around 11,600 years ago for the sinking of Atlantis, when, the reader will recall, 'There were earthquakes and floods of extraordinary violence, and in a single dreadful day and night ... the island of Atlantis was ... swallowed up by the sea and vanished.'[63]

'You remember only one deluge . . .'

I'm not trying to 'find' Atlantis, or even to guess where it might have been located – if it ever existed at all – since it is well known that such inquiries lead to madness. I prefer to treat it like any other archaic flood account, whether in the form of myth or purporting to be history, and to consider it solely in terms of its general level of plausibility – a task made easier by its unusual detail and precision. What it tells me at that level is at least the following:

1. *A devastating global flood occurred around 11,600 years ago.* This is interesting, the date coincides with the second of John Shaw's super-floods and with Cesare Emiliani's data from the Gulf of Mexico.
2. *The flood was accompanied by enormous earthquakes.* This is plausible because of the close correlation between huge earthquakes, enhanced volcanism, rapid ice melting, and fast post-glacial flooding.
3. *The island of Atlantis was swallowed up by the sea and vanished in a day and a night.* We have seen how isostatic rebalancing sometimes occurred very rapidly and cataclysmically at the end of the last Ice Age and how it is theoretically possible that intense isostatic subsidence in a suitably weakened area of the earth's crust could have brought about just such a sudden collapse as Plato describes.

There is one further element of the story that also resonates with scientific evidence, and this is that the flood that destroyed Atlantis 11,600 years ago was but one of many floods . . .

Remember that the source of the Atlantis tradition is supposed to have been an ancient Egyptian priest, in conversation with Plato's ancestor Solon. Here's how Plato reports the exchange in the *Timaeus*:

Egyptian priest: Oh Solon, Solon, you Greeks are all children, and there's no such thing as an old Greek.

Solon: What do you mean by that?

Egyptian priest: You are all young in mind, you have no belief rooted in old tradition, and no knowledge hoary with age. And the reason is this . . . With you, and others, writing and the other necessities of civilization have only just been developed when the periodic scourge of the deluge descends and spares none but the unlettered and the uncultured – so that you have to begin again like children, in complete ignorance of what happened in early times . . . You remember only one deluge, though there have been many . . .[64]

As a general synopsis, I have to say that the priest's comments fit reasonably well with the three global superfloods and countless lesser deluges that we now know did occur at approximately 15,000, 11,000 and 8,000 years ago. Moreover, his placing of the Atlantis flood *anywhere* in this period (the only period in the last 125,000 years when there actually were floods of the kind described) is – if you stop to think about it – quite an achievement in itself.

An aggressive little bugger from Yorkshire . . .

We've seen that it was Cesare Emiliani who first drew serious attention to the possibility of post-glacial superfloods. In a paper published in *Science* magazine in 1975, he and a group of colleagues presented startling evidence from deep-sea cores from the north-eastern part of the Gulf of Mexico. The evidence revealed 'a 2.4 per cent isotopic anomaly between 12,000 and 11,000 years ago', which the authors correctly interpreted as having been caused by 'the occurrence of major flooding of ice meltwater into the Gulf of Mexico . . . centring at about 11,600 years before the present'.[65]

At the time Emiliani's ideas were not well received. As Isaac Asimov was later to comment: 'The suggestion was largely ignored because it was difficult to imagine the ice melting that fast, but in 1989, John Shaw . . . made a suggestion as to just how such floods might come about . . .'[66] I thought that I had already fully understood Professor Shaw's catastrophic scenario of how the three great deluges were caused by sudden releases into the world ocean of pent-up meltwater from behind ice dams. But as I looked more closely at his research, and at the transcript of the lengthy interview he had given us in February 1999, I began to realize that his story had hidden complexities and that the cataclysms he described could have been far more severe than I had initially supposed. For it was not just a matter of very rapidly rising seas submerging and washing away low-lying coastal areas – although there was an immense amount of that! – but also of the true character and extent of the run-off floods *on land* as the ice-caps melted down and the glacial lakes burst their ice barriers.

Shaw's interest in this problem does not begin with floods but with drumlins:

Drumlin: elliptical, streamlined hill composed of till [unstratified glacial deposit consisting of boulder clay and rock fragments of various kinds] deposited beneath moving glacial ice. Drumlins commonly are found in clusters with their long axes roughly parallel to the direction of the ice movement. They slope steeply in the direction from which the glacier came and gently in the direction in which it moved. They vary in height from 6 to 60 metres and in length up to several miles . . . Drumlin fields may contain as many as 10,000 drumlins; one of the largest fields is in the north-western plains of Canada.[67]

Based at the University of Alberta, Professor Shaw has Canada's drumlins at his doorstep, at least in a manner of speaking, so it's not surprising – as a geologist – that he should have views about them. But the reactions that his views have elicited amongst other geologists are harder to understand:

When I go to conferences, people yell at me, people get angry and they yell and scream, and are constantly bringing in diversions because they don't want the story to be told. And being an aggressive little bugger from Yorkshire anyway, I tend to fight back.[68]

At a recent conference in Sweden a senior Quaternary geologist instructed Shaw: 'Don't bring your ideas here':

So I looked at him and grinned, and next day I gave the paper. And then it was rejected and not published in the conference proceedings so I put it on the Net, and that's where it is now . . . If I were a young assistant professor I wouldn't be kept and I wouldn't have published either and people would say my ideas were barmy.[69]

What, one might ask, is all the fuss about? It seems hard to believe that geologists could come close to excommunicating such a senior and widely respected colleague as Professor Shaw simply for expressing an original scientific *opinion* on the matter of elliptical, streamlined hills. I mean, who cares?

In fact, we should care, says Shaw, because the drumlins and other 'hummocky' landforms strewn across Canada are evidence of continental floods of biblical proportions – floods of water in some cases hundreds of metres high – that roared out from beneath the ice-caps during the last deglaciation, destroying or mangling everything in their path. Shaw explicitly suggests that many elements of the universal myth of the deluge may be explained by such floods pouring down off the land – intimately linked, as they were, to the episodes of sudden and ferocious sea-level rise that took place between 15,000 and 8000 years ago.[70]

Slow and gentle or fast and furious?

Although there is no single explanation for the formation of drumlins to which all geologists subscribe, most see them as the result of a relatively slow subglacial process involving first the lodgement of a huge mass of 'till' on the bedrock beneath the glacier and subsequently its moulding into the classic 'streamlined-hill' shape by the flow of the ice itself.[71] Such gradualistic theories have dominated the earth sciences and archaeology since the end of the nineteenth century, creating an exceptionally difficult environment in both disciplines for the exploration of alternative hypotheses requiring any kind of sudden change or catastrophic agency. Because John Shaw's theory requires both, it was inevitable that it would face stern opposition. Nevertheless, he has stuck to his guns since first putting his ideas forward in the 1980s and has gradually seen a convergence of evidence building up in his favour, including 'subglacial landforms, surface water isotopic composition of the Gulf of Mexico, and the sedimentology of cores from the Gulf'.[72]

At risk of reducing a massively documented and complex argument to statements of ludicrous simplicity, I think it is fair to say that Shaw himself does not claim to have found any definitive, all-inclusive explanation for the formation of drumlins but believes them to be features that are caused in *different* ways by *different* kinds of cataclysmic *floods* and not, as has traditionally been thought, by ice moulding. For example, 'on the evidence of form and structure', his interpretation of the Livingston Lake drumlins in northern Saskatchewan is that they are 'infills of inverted erosional marks scoured in the ice-bed by subglacial meltwater'.[73] In other words, forget about the old notions of 'lodgement' and 'moulding' that generations of geologists have had hard-wired into their logic-circuits. Consider the possibility, instead, that the end of the Ice Age was much less genteel – as, indeed, we already know that it was in almost every other measurable characteristic that we have encountered – and that the vast drumlin-fields at Livingston Lake were created by apocalyptic meltwater floods.

This is precisely Shaw's scenario and he believes that the 'subglacial landforms' – i.e. the drumlins themselves – are his most powerful evidence:

> When I first looked at drumlins – this is how it all started for me really – I thought, My, they look just like erosional forms on the sea-bed – which are negative forms of course – but these ones are positive. How can that be? Then the idea came to me, OK, if you erode upwards into the ice and then fill in the cavities with sediment that's what you would get. And so we went and dug holes and found out that the sediment corresponded to filling in from below and very catastrophically.[74]

In brief, Shaw's argument is that at certain stages during the collapse of the Laurentide ice-sheet between 15,000 and 8000 years ago, parts of the slowly

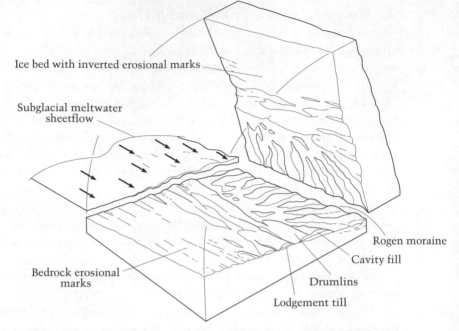

Ice bed with inverted erosional marks

Subglacial meltwater sheetflow

Rogen moraine

Cavity fill

Bedrock erosional marks

Drumlins

Lodgement till

Different kinds of landforms created by subglacial meltwater floods of varying depths. Based on Shaw (1998).

moving ice-mass – more than 3 kilometres thick and weighing as much as a giant mountain range – must have rested not on bedrock but on *a deep layer of meltwater moving at high speed and under enormous pressure*. These 'turbulent-flows' would have carried with them tremendous volumes of sediment ranging from finely grained clays to huge stones and boulders, and it is easy to see how a cavity eroded into the base of the ice-mass – where it rested on the running water – would quickly have become filled up and densely packed with sediment forced in from below. The result, like any object created in a mould, would have taken on the characteristic shape of the mould – which in the case of this kind of erosion is streamlined, elliptical and hill-shaped – and might then have been sealed within the ice, and carried further by it, until it was ultimately released by generalized melting.[75]

Take a few thousand such objects of varying sizes, dump them in northern Saskatchewan, and you have the Livingston Lake drumlin-field.

Shaw believes that other drumlin-fields in Canada have been created in a different way – again involving glacial meltwater rather than ice, but this time as a direct erosional agent on bedrock or depositional landforms:

> Drumlins around Peterborough and Trenton, Ontario, are mainly erosional; their
> internal stratigraphy is relatively undisturbed . . . Drumlins in Ireland contain com-

plex glacigenic sequences . . . The form of these Irish drumlins . . . is almost entirely erosional.[76]

Returning again to his notion of powerful floods running under immense pressure at the base of the ice-sheets, Shaw draws attention to the drumlins of Beverley Lake field in Canada's Northwest Territory, which he suggests were sculpted by these floods, and to erosional marks – also caused by floodwater – in the bedrock near Kingston, Ontario:

> Concerning the depth of the flow, it is clear that the [Beverley Lake] drumlins . . . must have been submerged in the formative flow . . . minimum depths of about 20 metres were required . . . Erosion marks in the bedrock in the Kingston area, Ontario, indicate subglacial meltwater flows that have widths of more than 60 kilometres. Spectacular erosional marks along the north shore of Georgian Bay, Ontario, also indicate broad subglacial meltwater flows. On a helicopter traverse along the north shore of Georgian Bay, a single field of bedrock erosional marks was noted that had a width of at least 50 kilometres . . . [These] drumlins and erosional marks indicate meltwater floods that were competent to remove the largest boulders . . . Flow widths, equal to the widths of drumlin and erosional-mark fields, were in the range of 60 to 150 kilometres.[77]

I think it is worth re-emphasizing Shaw's figures, and their implications. He is talking about turbulent, energetic floods 20 metres deep flowing in vortices at high speed and pressure, *under* the main ice-sheets, across fronts up to 160 kilometres wide. Only floods on such a scale and of such violence could have sculpted the drumlin-fields and hummocky terrain and tortured pitted scablands of Canada and the United States and carved out other remarkable features such as the extremely large through valleys – including those containing the Finger Lakes – that lie to the south of drumlin-fields in northern New York State.[78] 'Volumes of water required to sustain such floods', observes Shaw, 'would have been of the order of *one million cubic kilometres* equivalent to a rise of several metres in sea-level over a matter of weeks.'[79]

Drowned coral and floating ice

Of course, when water flows under ice, severing its connection to bedrock, the ice begins to move – 'surge' is the technical term:

> Subglacial meltwater sheets with thicknesses of several tens of metres occurred over vast areas of the Laurentide Ice Sheet. The decoupling of glaciers from their beds as a consequence of increased water pressure is used increasingly to explain their rates of sliding. The scale of this process implied here is much larger than that considered for modern glaciers. Nevertheless, the effects should be similar . . . In short, the glacier is expected to surge.[80]

There is indeed compelling evidence of a series of massive glacial surges at the end of the last Ice Age. These correlate with meltwater pulses and peaks of sea-level rise, recorded, for example, in 'drowned' reefs of *Acropora palmata* from the Caribbean-Atlantic region near the island of Barbados. *Acropora* is an efficient tracker of rising sea-level because it is a light-loving coral that dies at depths greater than about 10 metres. The Barbados reefs were drowned three times at the end of the last Ice Age – at approximately 14,000, 11,000 and 8000 years ago[81] – and so suddenly and deeply on each occasion that they now form three distinct steps, one for each flooding peak (rather than having crept towards shallower water as would have been the case with more gradual sea-level rises). Shaw and his colleague Paul Blanchon at the University of Alberta conclude in a 1995 paper in *Geology* that the reef data confirm:

> three catastrophic, metre-scale sea-level rises during deglaciation. By converting radiocarbon-dated marine and ice-sheet events to a sidereal chronology we show that the timing of these catastrophic rises is coincident with ice-sheet collapse, ocean-atmosphere reorganization and large-scale releases of meltwater.[82]

There is also evidence that a cataclysmic feedback mechanism may have been at work between even relatively small eustatic sea-level rises due to meltwater alone and much larger and more sudden events caused by the destablization of entire ice-sheets extending over continental shelves.[83] Indeed, in an article in *Nature*, geologists D. R. Lindstrom and D. R. Macayeal go so far as to identify ice-sheet mechanics 'as a controlling factor in meltwater production'.[84] They then make the very radical and original suggestion that:

> sudden and significant changes in sea-level due to the floating of formerly grounded ice-sheets and attendant ice-dome drawdown might have accompanied the meltwater pulses and *these 'jumps' in sea-level might not have been recorded in the reef accretion data.* Thus a logical mechanism exists by which sea-level may have risen *faster and to higher levels* than represented by the reef-accretion histories at Barbados.[85]

In other words global floods that *already* appear to have been extremely sudden and severe on the basis of the coral-reef data alone – and each 'drowning' event required a minimum instantaneous sea-level rise of 5 metres before it would take effect[86] – may temporarily have been several magnitudes *more severe* than the coral-reef record shows. Shaw and Blanchon suggest that a global eustatic hike in sea-level of between just two-tenths and four-tenths of a metre in a period of a few weeks would have been 'sufficient to free grounded ice and stimulate further ice-sheet wasting, additionally elevating sea-level on the order to 5 to 10 metres or more'.[87]

Armadas of icebergs

Induced by sudden sea-level rises, such sudden wasting at the sea-margins of the ice-sheets would have manifested in equally sudden launchings of fleets of gigantic icebergs. In 1988 the German oceanographer Hartmut Heinrich was the first to come up with the firm geological evidence for such a cataclysmic 'iceberg-calving' process during the last Ice Age. By examining deep-sea drill cores sampled at various points across the North Atlantic he demonstrated the existence of widely dispersed layers of 'ice-rafted detritus' – millions of tonnes of rocks and rocky debris that had once stood on land, that had been clawed up by the ice-sheets and that had ultimately been carried out to sea frozen into huge icebergs:

> As they melted they released rock debris that was dropped into the fine-grained sediments of the ocean floor. Much of this ice-rafted debris consists of limestones similar to those exposed over large areas of eastern Canada today. The Heinrich layers as they have become known, extend 3000 kilometres across the North Atlantic, almost reaching Ireland.[88]

The Heinrich layers record at least six separate discharges of 'stupendous flotillas of ice-bergs'[89] into the North Atlantic – discharges that are now known, obviously enough, as 'Heinrich Events' and that are thought to have unfolded in concentrated bursts of activity that may, in each case, have lasted less than a century.[90] Because of the progressive thickening of the Heinrich layers towards the western side of the Atlantic and the continuation of this trend into the Labrador Sea in the direction of Hudson Bay, it is obvious to geologists that 'much of the floating ice was sourced from the Laurentide ice-sheet'.[91]

However, other debris has been found intermingled in some Heinrich layers that 'could only have come from separate ice-sheets covering not only Canada, but Greenland, Iceland, the British Isles and Scandinavia'.[92] Likewise, research into southern hemisphere ice-caps in the Andes and New Zealand shows that these too 'grew and then collapsed synchronously with the ice-rafting pulses recorded in the North Atlantic'.[93] The implication, admits Professor R. C. L. Wilson of Britain's Open University, is that some 'global rather than regional forcing of climate change' must have been at work.[94]

With this reminder of the interconnectedness of all the great ice-sheets of the last glaciation – and the broad similarities of all their biographies – let's take a closer look at one of them. What happened to it also happened, to a very similar degree, to all of the others. Its apocalypse is therefore the end of the Ice Age in cameo.

Laurentide

Thomas Crowley and Gerald North, oceanographers at Texas A&M University, describe the melting of the great ice-sheets at the end of the last Ice Age as 'one of the most rapid and extreme examples of climate change recorded in the geologic record'.[95] As we have seen, most of the changes were concentrated into a period of just 7000 years between 15,000 and 8000 years ago. Like the other ice-sheets, the Laurentide did not really go into meltdown until after 15,000 years ago, and like the others it experienced three primary episodes of collapse correlating closely with Professor Shaw's three global superfloods (at approximately 15,000, 11,000 and 8000 years ago).

It is known that an immense meltwater reservoir in the Laurentide ice-sheet was catastrophically released between 15,000 and 14,000 years ago:

> The volume of water discharged produced regional-scale fields of drumlins, giant-flutings and extensive tracts of scoured bedrock. Such large amounts of meltwater could potentially destabilize ice sheets grounded below sea-level.[96]

The period between 13,000 years ago down to about 10,000 years ago saw recurrent outburst-flooding from a series of glacial lakes and lake complexes in the Laurentide – notably glacial Lake Agassiz which 'periodically emptied into the Gulf of Mexico via the Minnesota spillway and the Mississippi drainage basin'.[97] The reader will recall Emiliani's evidence for a peak flooding event of Laurentide meltwater into the Gulf at around 11,600 years ago. Within a thousand years of that date glacial Lake Missoula (in Montana in the western United States) also underwent one of its periodic outbursts, sending what Crowley and North calculate to have been 'a wall of water 600 metres high on to the Columbia plateau of eastern Washington'.[98]

Another series of large outburst floods occurred around 9400 years ago. According to Charles Fletcher and Clark Sherman of the Department of Geology and Geophysics at the University of Hawaii, each event added an estimated 4000 cubic kilometres of water to the world ocean.[99] By 8400 years ago yet more calamitous melting had allowed Lake Agassiz to merge with its formerly separate (and almost equally massive) eastern neighbour, Lake Ojibway. This confluence created a titanic inland sea, with a surface area of more than 700,000 square kilometres, poised behind an ice dam over Hudson Bay at elevations of between 450 and 600 metres above sea-level.[100]

At some point between 8400 and 8000 years ago the dam broke and the almost unimaginable mass of water burst through and emptied almost instantaneously into the North Atlantic:

> The breakout occurred into the Hudson Bay lowland, lowering lake level by at least 250 metres and resulting in a total discharge of between 75,000 and 150,000 cubic kilometres, possibly the single largest flood of the Quaternary Period.[101]

This outburst may have single-handedly raised global sea-level by half a metre or so. But this is a good place to remind ourselves that the spiralling decay and collapse of the Laurentide ice-sheet was not an isolated event but was part of a global pattern and feedback system – and that floods of almost equal magnitude poured in tandem off the Fennoscandian ice-sheet on the other side of the Atlantic Ocean. This is why, at around the same time as the collapse of the Laurentide, the north-eastern side of Britain close to the Fennoscandian margins also experienced severe flooding. Here there was a very rapid rise in sea-level which

> submerged an area in the North Sea the size of modern Britain . . . Most of this 100,000 square mile British 'Atlantis' [not to be confused with Koudriavtsev's suggested site of Atlantis on the Celtic Shelf] was there in 8000 BC and gone by 6500 BC. By then only a 140 mile long, 5000 square mile island, where the Dogger Bank is now, survived.[102]

The separate meltwater floods originating in different ice-caps would, of course, have mingled in the world ocean and multiplied their effects by floating and breaking up grounded ice on the continental shelves. Stephen Oppenheimer calculates that the ice 'flushed out through the Hudson Strait' from what had once been the centre of the Laurentide ice-dome between 8400 and 8000 years ago may have been as much as '1.6 kilometres thick and a third the size of Canada'.[103]

Such statistics beggar the imagination and require common sense to rebel against what is still very much the establishment view – namely that the sea-level rises at the end of the last Ice Age – though large overall – were too small on a year-by-year basis to have caused cataclysmic flooding, and thus to have inspired global flood myths, or to be of any relevance at all to traditions of lost civilizations and antediluvian cities.

Although very few historians are presently taking any interest, the geological and oceanographic evidence has begun to turn against this 'gradualist' and 'uniformitarian' view of the meltdown, and there are more and more reasons to suspect that 'the world of men', as Plato's Athenian comments in the passage from the *Laws* quoted at the beginning of this chapter, might indeed have often been 'destroyed by floods . . . in such a way that only a small portion of the human race survived'.

Entering the realm of the unknown

At any of the three nodes of peak flooding around 15,000, 11,000 and 8000 years ago the convergence of evidence suggests *very fast* global sea-level hikes of the order of 5–10 metres – and sometimes far more – in each case complicated and exacerbated by induced ice-sheet break-up and other factors. In particular, as we have seen, experts believe that there may have been several temporary rises in sea-level during these periods – caused by the sudden floating of vast masses of ice – that far exceeded the margins recorded in the oceanographic record.[104]

Moreover, rising sea-levels – bringing floods from sea to land – are only part of the story of the end of the last Ice Age. Of at least equal, perhaps greater, importance are the terrible walls of water hundreds of metres high that again and again rolled out from the monstrous ice-domes – and thence over low-lying land, and from land to sea – when ice dams ruptured and glacial lakes spilled, or when pressurized subglacial meltwater burst from under the ice-sheet.

We know that relatively minor sea-level rises could set off major ice-sheet break-ups, and it has been suggested by Stephen Oppenheimer that the tremendous earthquakes caused by isostatic rebalancing at the end of the Ice Age could have stirred up 'mountain-topping superwaves' in the northern regions of the Atlantic and Pacific Oceans.[105] Other than Oppenheimer's own investigations, however, my impression is that while many brilliant individual scientists have studied individual post-glacial phenomena in great depth, very little has yet been done to investigate all these phenomena together as part of a complex system or to consider the effects on the earth and its human population of multiple, interacting cataclysms – floods, lands subsiding into the sea, earthquakes, volcanic eruptions – all occurring at the same time.

We are entering the realm of the unknown here – because science has only recently begun to consider the end of the last Ice Age as a cataclysm at all and the evidence is still coming in about just how devastating and extensive that cataclysm might have been. Nevertheless, some observations that I believe deserve special attention have been made by the researcher Paul LaViolette in his 1997 book *Earth Under Fire*:

(1) At peak moments of the meltdown any hypothetical civilizations living around the edges of partially enclosed seas that served as drainage areas for the great ice-sheets could have suffered disproportionately large and rapid changes in sea-level. In a sophisticated and original argument, LaViolette draws particular attention to the Mediterranean:

> Glacial meltwater [from the nearby European ice-sheets] would have entered the Mediterranean much more rapidly than it could escape through the Straits of Gibraltar, and, as a result, the temporary rise in Mediterranean sea-level would have been much greater than in the surrounding oceans ... [Such meltwater surges] could have temporarily raised the Mediterranean by some 60 meters, flooding all coastal civilizations.[106]

(2) Mega-avalanches of rock and ice must have repeatedly thundered into the world's oceans during the epoch of the meltdown because of the effects of isostacy on continental margins and the breakaway collapse of the gigantic ice-sheets. From an example in recent history we know how severe avalanche-induced floods can be. In July 1958 in Alaska's Lituya Bay '40 million

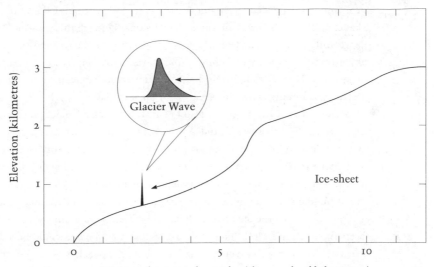

Glacier wave sweeps down side of ice-sheet, growing in height as it descends.
Based on LaViolette (1997).

cubic metres of ice and rock weighing 90 million tons, avalanched from the glaciated slopes of the Fairweather Range and fell 900 metres into one of the bay's inlets.' The resulting wave roared inland up the bay's steep opposite shore for a distance of more than a kilometre at a speed of 200 kilometres per hour and 'destroyed ten square kilometres of forests to a height of 540 metres'.[107]

What then would have been the effects of the fall into the North Atlantic of a sheet of ice a third of the size of Canada?

(3) The phenomenon of outburst floods from catastrophically released glacial lakes, already understood to have had apocalyptic regional-scale effects, may have been far more severe that previously thought:

> Ponds and lakes on a glacier's surface, as well as natural caverns within the glacier, are known to store large quantities of glacial meltwater. From time to time the contents of such reservoirs can discharge suddenly to create potentially destructive floods called glacier bursts or glacier floods . . .
>
> During periods of intense climatic warning, the Earth's ice-sheets were melting extremely rapidly, with most of the melting taking place on their upper surfaces. Consequently large quantities of meltwater would have collected on the ice-sheet surface to form numerous supraglacial lakes perched at elevations of up to 3.5 kilometres. In cases where the impounded waters were restrained by ice jams and where mounting pressures caused these jams to give way, large floods of glacial meltwater would have poured out over the ice-sheet surface. As one such glacier burst swept forward, gradually descending the ice-sheet's surface, it would have

incorporated any ponded meltwater that lay in its path, triggering these supraglacial lakes to discharge their contents and add to its size. Through this snowballing effect a single initial glacier burst would have progressively grown in size and kinetic energy during the course of its downhill journey, eventually becoming of mountainous proportions. This so called continental *glacier wave* would have produced catastrophic floods unlike anything seen on our planet today . . .

Waves of greater height travel faster. Accordingly, as a glacier wave proceeded across an ice-sheet to lower altitudes, gaining in height and kinetic energy, it would have accelerated to higher speeds. By the time it had journeyed thousands of kilometres to the edge of the ice-sheet, it could have attained heights of 600 metres or more, a cross-sectional breadth of as much as 40 kilometres, and a forward speed of several hundred kilometres per hour. Such a wave could have extended thousands of kilometres along the ice-sheet . . . Glacier waves issuing from the surface of ice-sheets in North America, Europe, Siberia and South America would have had sufficient kinetic energy to travel thousands of kilometres over land to devastate regions far removed from the ice-sheet's boundary. Upon entering the ocean, the wave would have continued forward as a *tsunami* to cause considerable damage on the shores of distant continents. Because of its immense energy, a glacier-wave *tsunami* would be far more destructive than any tidal wave observed in modern times.[108]

Yesterday . . .

There is much that we do not know about what happened to the earth, and to mankind, between 17,000 and 7000 years ago. And though science has made great strides towards a fuller understanding of that epoch, there is much that we may never know. Yet it is to this precise period of unrecorded prehistoric darkness set amidst epic climatic and environmental turmoil that archaeologists trace the origins of civilization: the first settlements, the first signs of structured hierarchical communities, the domestication of plants, the invention of agriculture, building with bricks and stone, etc. – in other words the whole suite of economic and social attributes that set mankind on the road to science and reason and the technological achievements of the modern world.

Proper 'history' doesn't begin until after 5000 years ago when we have written records to go on and thus the basis to build up a reasonably accurate picture of past events – although even then there are huge gaps. Before 5000 years ago, in the absence of written records, all we have to light up our collective yesterdays are the conjectures of archaeologists based upon their interpretations of extremely scanty material evidence elevated from tiny areas of archaeological sites that become more and more scarce the further we go back in time. And almost all of these sites, of course, are on land. Thus far the contribution of marine archaeology to the debate has been risible. So this is the flimsy, hopelessly incomplete, and wholly inadequate basis on which we rest our understanding of the unwritten past and passively accept, as though we are drugged or senseless, that there is no mystery in it.

PART TWO

India (1)

4 / Forgotten Cities, Ancient Texts and an Indian Atlantis

The lasting gift bequeathed by the Aryans to the conquered peoples was neither material culture nor a superior physique, but a more excellent language and the mentality it generated . . . At the same time the fact that the first Aryans were Nordics was not without importance. The physical qualities of that stock did enable them by bare fact of superior strength to conquer even more advanced peoples and so to impose their language on areas from which their bodily type was almost completely vanished. This is the truth underlying the panegyrics of the Germanists; the Nordics' superiority in physique fitted them to be the vehicles of a superior language.

Vere Gordon Childe, Professor of Prehistoric Archaeology,
University of Edinburgh, 1926

In the end there is no reason to believe today that there ever was an Aryan race that spoke Indo-European languages and was possessed of a coherent or well-defined set of Aryan or Indo-European cultural features.

Gregory Possehl, Professor of Anthropology,
University of Pennsylvania, 1999

The word 'city' is etymologically linked to the word 'civilization'. It is therefore of interest that mankind's first cities have been traced by historians to the following regions and dates: (1) Mesopotamia, late fourth and early third millennia BC; (2) Egypt, late fourth and early third millennia BC; (3) India, late fourth and early third millennia BC; (4) China, mid-second millennium BC; (5) Central and South America, mid-second millennium BC.

In four of the five regions – Mesopotamia, Egypt, China and the Americas – nothing remains of these ancient civilizations except their extraordinary stone monuments together with more or less incomplete collections of their inscriptions, legends and traditions. These, by good fortune, have come down to us and have proved amenable to translation. But the cultures that created the monuments and the scriptures are long gone and thus inaccessible to study – except through inference and deduction from the material remains they left behind.

In the fifth region, the Indian subcontinent, matters are very different. Here the oldest cities are ascribed to the 'Indus Valley civilization'. It was forgotten by history and unknown to archaeologists until the 1920s, when the first two

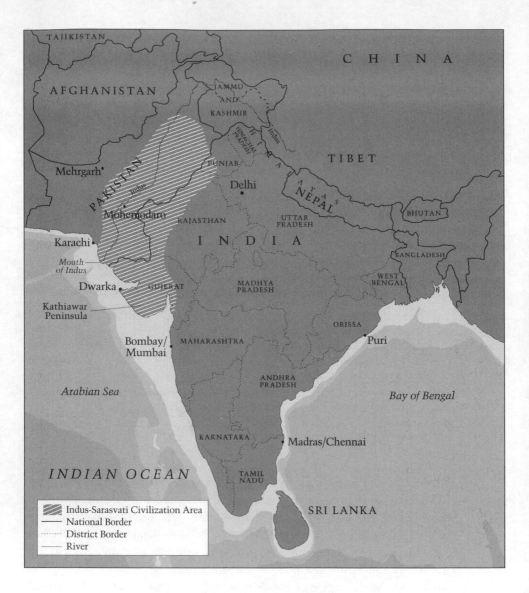

Indus-Sarasvati Civilization Area
National Border
District Border
River

sites to be discovered – Harappa and Mohenjodaro on the Indus river in what is now Pakistan – were excavated and found to be about 5000 years old. It is because of the location of these two 'type sites' that the name 'Indus Valley civilization' was coined – while at the same time the characteristic 'Bronze Age' archaeological assemblage of this civilization was referred to as 'Harappan' – since Harappa was the first site to be explored. Subsequent excavations, continuing today, have led to the realization that the majority of the approximately 2600 'Harappan' sites so far discovered in fact lie outside the Indus valley, particularly to the east along the course of the ancient Sarasvati, a river that has been dry for almost 4000 years. This wide distribution of sites has been recognized by scholars, many of whom now prefer to speak of the 'Indus-Sarasvati civilization' – the term that I shall generally use here, since it more accurately describes the very large geographical catchment area in which this mysterious culture flourished.

It was an architectural culture, carrying out prodigious feats of civil engineering and building its gigantic cities out of bricks so strong, so uniform and so well made that even after thousands of years they could safely be reused on modern construction projects (something that happened frequently in the nineteenth and early twentieth centuries before Harappa and Mohenjodaro were recognized as archaeological sites). The Indus-Sarasvati civilization was also the first in the world to experiment with 'New Towns' – towns literally planned and built from scratch, according to a blueprint – the first to institute scientifically designed urban sanitation systems and the first to build an efficient tidal dock.

It was a literate culture. Altogether, some 4200 objects – mainly pottery and seals made from steatite and terracotta – have been found bearing the Indus-Sarasvati script. Many of the seals are inscribed in 'mirror image' (so as to produce a positive impression when stamped, for example, into damp clay) and are thought to have been used by merchants to brand-mark their goods. The earliest inscribed seal (excavated in Harappa) dates to 2600 BC while the pottery is a little older. The average inscription contains five signs, the longest twenty-six, and there are many with just one sign. Despite the best efforts of the world's leading linguists, it has not proved possible to translate any of inscriptions (although quite a number of translations have been attempted and then rejected by the academic community). There is, however, a general consensus that the script, as presently known 'emerged as a fully-formed system of abstract signs called graphemes . . . After careful comparison of all the signs, most scholars agree that there are between 400 and 450 different signs or graphemes.'[1] The mature form of the script, in other words, appears suddenly in the archaeological record some time before 2600 BC. There are no indications of evolution or development. One day it wasn't there, next day it was.

How is this to be explained?

A missing literature

It could simply be because the traces of the script's evolution exist but have not yet been found by archaeologists, or that such traces once existed but have now all been destroyed. It could be that the script did not 'evolve'. Perhaps it was invented and introduced all at once, a bit like the script for the previously unwritten Somali language that was invented in the 1960s and introduced in the Horn of Africa in 1972.[2] Or it could be that the Indus-Sarasvati civilization did not regard written documents as a suitable medium in which to preserve its great literary and religious compositions. What I mean to suggest by this is the possibility that the Indus-Sarasvati script might have been devised to serve strictly limited *commercial* and *bureaucratic* functions such as labelling merchandise, naming the owners of goods, naming the contents of pots, etc. It could be that the nature of the society was such that it would have been regarded as a desecration to use the script to write down anything that was revered or sacred like a wonderful story from antiquity or the prayers, hymns and recitations used in religious services.

To live in the twenty-first century is to live in a world in which it is increasingly difficult to imagine how any kind of civilization could exist without large-scale written communications. We regard writing as an essential intellectual skill, as well as the only way to preserve proper long-term records. In our society to call someone 'illiterate' is therefore an insult; people who do not read and write fluently often have feelings of inadequacy; and there is widespread unstated agreement that the written word is in itself a virtue, that's its absence is a curse, and that no high civilization could possibly develop without it. This great, universally accepted 'given', as unimpeachable as motherhood, is one of the reasons why historians and archaeologists regard evidence of the introduction and extensive use of writing as amongst the defining characteristics of a 'civilization' – to such an extent that 'preliterate' cultures are automatically regarded as much less civilized than literate ones.

But isn't this exactly the perspective that one would expect of a highly literate technical society looking at the past? Wouldn't it tend to seek out its own image there, in however early a form, and define that as 'civilization'?

I believe that this may be what has happened with the vexed issue of the indecipherable Indus-Sarasvati script. The very brevity of the inscriptions (which is part of what makes them so difficult to decipher) means that they cannot have been used to tell complex stories containing numerous details and large quantities of information – and I do not think any scholar would disagree with me on that point.[3] Yet, to my mind, it is inconceivable that a society so large, so complex, so well ordered and so intelligently run as the Indus-Sarasvati civilization did not possess a literature, did not possess religious and spiritual compositions, did not have vital sacred records that it wanted to preserve. I am certain that it had all these things, and since I know that this society understood the principle of writing, and indeed had developed a writing system with more than 400 different signs, I am not at all inclined to conclude that it did not possess any information of great cultural importance but rather that it must have made a deliberate *choice* not to use its script to convey such information.

A potion for remembering . . .

A clue as to why a civilization might not regard writing as an automatic virtue, and why its leaders might even take an ethical decision to restrict the use of writing, has been passed down to us by Plato. In *Phaedrus* he has Socrates pose a rhetorical question: 'What feature makes writing good, and what inept?' He then declines to give an immediate answer to this question but instead continues:

I can tell you what I've heard the ancients said . . . Among the ancient gods . . . in Egypt there was one to whom the bird called the ibis is sacred. The name of that divinity was *Theuth* [*Thoth*, the ancient Egyptian god of wisdom], and it was he who first discovered number and calculation, geometry and astronomy, as well as the games of checkers and dice, and, above all else, writing.[4]

What the ancients said about Thoth, Socrates reports, was that having invented writing he had gone to the god Amon, 'the King of all Egypt at that time', and urged him to introduce it amongst the populace, with these words: 'O King, here is something that, once learned, will make the Egyptians wiser and will improve their memory; I have discovered a potion for memory or wisdom.' But Amon replied:

O most expert Theuth, one man can give birth to the elements of an art, but only another can judge how they can benefit or harm those who use them. And now, since you are the father of writing, your affection for it has made you describe its effects as the opposite of what they really are. In fact it will introduce forgetfulness into the soul of those who learn it: they will not practise using their memory because they will put their trust in writing, which is external and depends on signs that belong to others, instead of trying to remember from the inside, completely on their own. You have not discovered a potion for remembering but for reminding; you provide your students with the appearance of wisdom, not its reality. Your invention will enable them to hear many things without being properly taught, and they will imagine that they have come to know much while for the most part they will know nothing.

Later in the discussion Socrates makes it clear that he feels there are certain matters and certain kinds of information that should not be available to all but should be limited to 'those with understanding':

Once it has been written down, every discourse roams about everywhere, reaching indiscriminately those with understanding no less than those who have no business with it, and it doesn't know to whom it should speak and to whom it should not . . .[5]

These passages in the *Phaedrus* may be interpreted in many different ways, but one of the things they definitely are is a sturdy defence of the oral tradition and a clear statement that scripts may not, ultimately, be the best way to preserve precious cultural documents. Because a script depends on signs there is always the possibility that a time will come when those signs will no longer be understood (as has indeed happened with the Indus-Sarasvati script today). In such a case any knowledge consigned to the future exclusively in the 'ark' of that script will have been utterly and irredeemably lost. Because a script is accessible to anybody who reads it means there is no guarantee that compositions expressed in it will be delivered only to those whom they are intended for. If the compositions contain sacred material that is aimed exclusively at initiates within a cult, for example, and cannot be properly understood without specific information possessed by those initiates, then it is probable that such compositions – even if they can be 'translated' in a literal sense – will appear meaningless, nonsensical or absurd to outsiders. Last but not least, because a script eliminates much of the need for memory its introduction in any society will inevitably lead to a reduction in the value of the science of memory and in due course that science will be forgotten. Memorization is not a highly regarded skill in our society today (and increasingly less so as the years go by), yet it is possible that a powerful memory, developed through discipline and training, could operate as a catalyst to other intellectual and perhaps even spiritual skills which would otherwise lie dormant.

By keeping communications within a strictly oral tradition all of these problems can be avoided. From generation to generation, from initiate to initiate directly, the sacred archives (or hymns, or utterances) can be passed down and their obscurities explained where necessary, no doubt evolving to some extent as the language in which they are carried evolves, perhaps even being translated into other languages – but always strictly through the medium of the spoken and memorized word, with suitable interpretation and explanation by a wise practitioner at hand, never, never, never through the medium of the written word.

Consider sacred texts that are valuable to 'advanced' technological societies such as Japan and the United States. In Japan the *Nihongi* and the *Kojiki* are revered for the antiquity and wisdom of the traditions they contain. In the United States the Old Testament and the New Testament of the Bible are equally revered amongst Christians. But in neither country does more than a tiny handful of people (if indeed any at all) have these enormous and complicated texts off by heart. In consequence, although they may be found in many household libraries, they are not often discussed or consulted by the majority of Japanese or Americans today.

Now, by contrast, consider the case of India with its population of one billion.

Almost supernatural feats of memory

Unlike in other big modern industrial nations that have long ago lost all sense of the sacred and all respect for 'what the ancients said', the sacred life still permeates India through and through to such an extent that an appeal to the authority of scripture can still settle all disputes. And unlike the cultures of ancient Egypt, Mesopotamia, China and the Americas, where only spectacular fossils of architecture and language remain, the culture of ancient India is still vibrantly *alive* today in the subcontinent and offers as its gift to the present a vast library of archaic rituals, dances, games, ceremonies, festivals and customs as well as an immense oral literature that has not only been preserved and continuously passed on in the memory of *sadhus* and *rishis* (sages, wise men) for thousands of years but that is also celebrated, rehearsed, admired and relished in hundreds of thousands of Hindu villages from the Himalayas to the sea.

The oldest elements of India's oral tradition are the *Vedas* (the word *veda* means 'knowledge'), consisting of four major *samhitas* (compilations of hymns): the *Rig Veda* (the most ancient and the most revered), the *Sama Veda*, the *Yajur Veda* and the *Atharva Veda*. The language used is a very archaic form of Sanskrit, and there is a great deal of it! The *Rig* alone has an extent of around 450,000 words (about twice as long as this book) expressed in 1028 hymns made up of 10,589 verses.[6] The total compendium of the four *samhitas* probably runs to almost double that. But what is most amazing about these hymnodies is not so much their overall length, which is awesome, but that for most of their history it is probable that no written versions of them ever existed – and not because they *could* not be written down but because the priests of the Vedic religion that evolved into Hinduism believed that they *should* not be written down but should be kept alive instead in human memory.[7]

> The Vedic texts were originally part of an oral literature. They are *sruti*, or 'Heard', and Brahmins [the priestly caste in Hinduism] were expected to memorize all four books, some parts of which were clearly composed and arranged to assist in this learning process. It can be surmised then that there was a period of composition, when new material was added and older verses were edited and changed. But at some point this flexibility in composition stopped and the priests defined their text as immutable, not to be changed by one word or even one syllable, and the slightest mispronunciation or deviation from the canon was believed to be a sacrilege.[8]

Significantly there is no mention of writing in the *Rig Veda*. Moreover, even when writing had become widespread in ancient Indian society for other purposes, strict proscriptions continued to be enforced against writing the *Vedas* down. This ban was respected until about 1000 years ago, from which period the earliest surviving written versions have reached us.[9]

Gregory Possehl, Professor of Anthropology at the University of Pennsylvania, and one of the world's leading experts on ancient India and the Indus-Sarasvati civilization, comments:

> The Indian Brahmins took the memorization of the *Vedas* very seriously, and developed means to ensure accuracy and the careful reproduction of the same words and sounds from generation to generation. Careful, even exact oral replication of the *Vedas* was part of the Hindu faith, institutionalized during the learning process and maintained through peer observation and pressure through the life of a Brahmin. This community of faithful Brahmins was large and they all went through the same learning process, which was standardized to some degree. Deviation from the . . . path of exact replication would have brought powerful forces of censure to bear on the offender . . .

There is also good agreement between the written *Vedas* that exist from Medieval times on, and the oral versions. It is thought that the oral tradition may not have been contaminated by the literate, but we cannot really know for sure. Still, the writing down of the *Vedas* was not favoured, nor widespread . . .

The noted Sanskritist J. A. B van Buitenen told me that in the eighteenth and nineteenth centuries the Europeans who were learning Sanskrit were impressed by the fact that no matter where they went in the subcontinent, when they heard Brahmins recite the *Vedas* they heard the exact same thing. From Peshawar to Pondicherry, or Calcutta to Cape Comorin, hundreds of thousands, even millions, of Brahmins who had no direct contact knew these texts in precisely the same way . . .

[There are therefore] some reasons to believe that this oral tradition is different from most, and that what we have today as texts may be remarkably close to those of deep antiquity.[10]

The problem of the Aryas

How deep? How ancient is the content of the *Vedas* really? And from what wellspring of philosophy, insight and religious speculation do they flow?

Scholars like Gregory Possehl, with the (almost) unanimous backing of non-Indian Indologists and Sanskritists, believe that the Vedic hymns were 'codified' at around 1200 BC. They admit that the actual compositions must be older than that but it is clear that they would be unlikely to accept a date – even for composition – that is earlier than about 1500 BC, perhaps begrudgingly 1800 BC in some rare cases.[11] Why should this be so when the archaeological record makes it is clear that the second millennium BC in India, if not a time of total decay and collapse as it has sometimes been painted, was certainly not a time that was magnificently fruitful intellectually and does not look like the sort of epoch that would have produced a sublime intellectual creation like the *Rig Veda*? On commonsense grounds alone, isn't this enigmatic text, which we will explore in chapter 6, at least as likely to have been the work of the equally enigmatic Indus-Sarasvati civilization? And why is it only now that such a possibility is beginning to be tentatively explored by some scholars while the majority still won't even consider it?

The answer is that the Vedic peoples are referred to repeatedly in the *Rig* as the 'Aryas' and that from this a great and sustained error of orthodox historical scholarship was spawned. Even though the adjective 'Aryan' in ancient Sanskrit actually means 'noble' or 'cultured' – and therefore the Aryas are essentially 'the "noble" or "cultured" folk', and thus as easily a religious cult as an ethnic group – it was assumed by historians and archaeologists that they were a race and that they had invaded India around 1500 BC. Known as the 'Aryan invasion theory', this error was only brought to light and dropped from official curricula during the last quarter of the twentieth century. Because it has far-reaching implications, and requires the wholesale rewriting of canonical academic texts and standard works of reference, it is the kind of error that historians are not normally eager to admit. Yet in this case, to their credit, it is the orthodox scholars themselves who have exposed it.

It is not an error that has ever made the headlines. But since the early 1990s it has been increasingly widely discussed in academic journals and books and taken into account, more or less completely, in all new thinking and teaching on the subject. So there is no question at all of a cover-up or even of significant denial by those whose specialisms have been most directly affected or whose publications in scientific journals are now out of date.

The Aryan invasion of India

The attribution of the *Vedas* to 'Aryan invaders', the date of 1200 BC for the codification of the *Vedas*, and the Aryan invasion theory itself can all be traced back to an idea that had already planted roots by the beginning of the nineteenth century. It was then that a number of Western scholars began to notice that Sanskrit, the classical language in which the *Vedas* are written, and its modern relatives in north India such as Hindi, Bengali, Punjabi, Gujerati and Sindhi, have extremely close affinities with modern and ancient European languages such as Latin, Greek, English, Norwegian and German. How, the scholars asked themselves, had this amazingly widespread distribution of what are now known as the 'Indo-European' family of languages come about?

Fairly soon a predictable doctrine began to take form. 'This', explains Gregory Possehl,

> had to do with the Aryan race, proposed to be the people who spoke the languages of the Indo-European family. European intellectual and moral superiority was a fore-gone conclusion to most savants of the nineteenth and early twentieth century. The success of European colonialism, Christianity and the Industrial Revolution proved that. This condition of innate superiority was seen in the Classical Greeks and to have been carried forward by Rome. With the discovery of the Indo-European family of languages there was evidence for an even earlier history, one set within a pre-historic past that only archaeology could uncover. The Aryans, or Indo-Europeans, must have been blessed with this 'superiority' since they too were successful conquerors of vast lands, from the Bay of Bengal to the outer islands of Scandinavia and the United Kingdom.[12]

It was against this ideological background of inevitable European superiority, combined with misunderstood references to the Aryas in the *Rig*, that the doctrine of the 'the Aryan invasion of India' arose and gained universal accept-ance amongst scientists as an event that had taken place at a specific moment in history and that had involved a mass movement of peoples from a European 'homeland' into India.

Indeed, the earliest version of this scenario remained widely accepted until the twentieth century was quite far advanced. It held that India – which before had been inhabited exclusively by dark-skinned aboriginal and Dravidian tribes – was invaded from the north-west through the passes of Afghanistan by a light-skinned and perhaps even blue-eyed European race at some time during the second millennium BC. The pale nomadic invaders, mounted on horses, armed with iron weapons and driving fast war chariots, called themselves the 'Aryas'. They rapidly overwhelmed and subjugated the indigenous inhabitants, whose civilization was at a lower level than their own. At the same time they imported their own naturalistic religion – expressed in the *Rig Veda* – which they imposed on the 'inferior' conquered races of India.

The second scenario began to take shape after the discovery and excavation of the Indus valley sites of Harappa and Mohenjodaro during the 1920s and 1930s. It rapidly became clear that these sophisticated, centrally planned cities were much older than the supposed 1500 BC date for the Aryan invasion of India and that they belonged to a previously unidentified high civilization of remote antiquity, perhaps almost as old, it was speculated, as Sumer or Egypt – in other words, dating back to 3000 BC or earlier.

Like other resilient bad ideas, the Aryan invasion theory survived what should have been critical evidence against it by adapting. Although the chronology had to be increasingly stretched to fit in with the new archaeological discoveries, historians were for a long while able to cling on to the notion of an invasion by 'Aryan' hordes in the second millennium BC.

What changed was the background. Previously, the pale Aryas had overrun primitive tribes of dark-skinned hunter-gatherers. Now it had to be admitted that they had overrun a sophisticated urban civilization that had flourished in India for at least a thousand years before their arrival and that had been far ahead of them in culture but no match for their superior military prowess and technology. Previously the Aryas had been the bringers of civilization to a benighted and barbaric India; now they were the destroyers of a far older civiliz-ation than their own – a literate civilization, moreover, and one that had clearly been prosperous for a very long time.

It was generally agreed that this earlier race of city dwellers had been Dravidi-ans – an ethno-linguistic group, principally represented by Tamil-speakers, that is now almost entirely confined to southern India. With no more evidence than the authoritative (and in this case incorrect) opinion of the revered British archaeologist Sir Mortimer Wheeler concerning a few dozen skeletons thought to display wound marks that had been found at Mohenjodaro, scholars adopted the theory that the invading Aryas had 'massacred' the Dravidian inhabitants of the Indus-Sarasvati cities, forcibly taken over their lands and driven the survivors towards the south.

Although the massacre theory was later discredited (the skeletons came from different epochs, showed no signs of fatal wounds, and were not the result of any one event),[13] the idea of a violent invasion of India by a non-Indian people calling themselves the Aryas survived in at least some enclaves of mainstream scholarship into the early 1990s – when even its most ardent supporters began to distance themselves from it. By 1999 the standard texts on the subject had caught up and Gregory Possehl was able to write the definitive obituary of the Aryan invasion hypothesis in his massive tome *Indus Age*:

> In the end there is no reason to believe today that there ever was an Aryan race that spoke Indo-European languages and was possessed of a coherent or well-defined set of Aryan or Indo-European cultural features.[14]

1500 BC or 15,000 BC?

So it is not controversial to state that the top scholars in this field now accept, absolutely, that there was no Aryan race and no Aryan invasion. Strangely, however, very few of them seem to have noticed that these conclusions must have implications for the history that we ascribe to the *Vedas* – hitherto assumed to have been *composed* by the Aryan invaders, and codified by them into the form that is with us now, during the first few centuries after their arrival in India around 1500 BC.

It turns out that this assumption, which in all logic cannot stand now that the core idea of an Aryan invasion has been abandoned, is one of the pillars of the orthodox chronology of the *Vedas*. This dates the codification of the four principal books – the *Rig Veda*, the *Atharva Veda*, the *Yajur Veda* and the *Sama Veda* – to between 1200 and 800 BC (with the three centuries between 1500 BC and 1200 BC allocated to the actual composition of the hymns).

The second pillar has to do with metals and the supposed date of the 'Iron Age' in India. The *Rig Veda*, which is thought to be the oldest Vedic text, uses a general term, *ayas*, for metal. By the time of the codification of the slightly later *Atharva Veda*, however, a new term has been introduced: *krsna ayas*, meaning 'black metal'. Scholars have taken this to be a reference to iron, and have drawn very large chronological conclusions from it. Gregory Possehl:

> There is some content of the *Rig Veda* that hints at its age. There are references made to metals . . . but not iron. However, by the time of the *Atharva Veda* iron is known. This can be used to suggest that the *Rig Veda* was codified prior to the widespread use of iron in northern India and Pakistan and that the *Atharva Veda* is on the other side of this timeline; nominally 1000 BC or slightly earlier.[15]

Possehl describes this as nothing more than a 'reasonable or interesting observation, not a hard and fast historical point'.[16] This is certainly a wise caution. For example, the metal *krsna ayas* might have been known in Rig Vedic times but simply not mentioned in the *Rig* itself. Or, as a number of authorities have argued, it may be that *krsna ayas* has been mistranslated as iron and that some other dark-coloured metal was intended. Or again, with no indication given in the texts as to how the *krsna ayas* was acquired or manufactured, it is also possible – even if 'iron' was intended – that the references are to meteoritic iron (as opposed to man-made smelted or forged iron). This is widely understood to be the case, for example, with the many references to 'iron' – *bja* – in the ancient Egyptian *Pyramid Texts* (c.2300 BC, long before the Egyptian 'Iron Age') and there is no reason from the context why it should not also be so in the *Atharva Veda*.

The third pillar supporting the orthodox chronology of the *Vedas*, and the one most relied upon for dating the *Vedas* today, is a linguistic argument extrapolated from a 'feeling' certain specialized scholars have about the pace at which Sanskrit might have evolved. Gregory Possehl again, setting out the orthodox view as it stood in 1999:

> Based on the language of the *Rig Veda*, its vocabulary and grammar, Vedic Sanskrit can be thought of as the archaic form of this language. The Sanskritists on whose judgement I rely, feel that the date for the codification of the *Rig Veda* is not likely to be earlier than 1200 BC nor later than 800 BC. There is some bias toward the later date. These dates are not based on a process of reasoning rich in data and cross-checks. They emerge instead from a sense of how rapidly Sanskrit might have changed, using the grammar of Panini (c.5th century BC) as a baseline and working backward from this point. There are few chronological checkpoints in this process and the period between 1200 BC and 800 BC emerges as a scholarly judgement; a kind of ballpark guess . . .[17]

Possehl then goes on to warn that since 'this date for the *Rig Veda* is based primarily on language', it gives at best 'the approximate date for the codification of the text, but not for the history that may be represented there, which is certainly earlier; how much earlier is simply not known'.[18]

Likewise, it is surely significant that Max Muller, perhaps the most eminent Indologist of all, and in fact the first Sanskritist to propose a codification date of 1200 BC for the *Rig Veda*, was himself much more hesitant than the generations of scholars following uncritically after him, who have allowed the date of 1200 BC to crystallize into received wisdom. It is clear that Muller became aware during his own lifetime that such a 'crystallization' process was underway – and that he resisted it. 'I have repeatedly dwelt on the *entirely hypothetical* nature of the dates which I ventured to assign to the first three periods of Vedic literature,' he protested at one point.[19] Again, in his Gifford Lectures in 1890, Muller warned his students that 1200 BC was a purely arbitrary date based on unproven assumptions about the rate of evolution of Sanskrit: 'Whether the Vedic hymns were composed in 1000 or 1500 or 2000 or 3000 BC no power on earth could ever fix.'[20] And in his book *The Six Systems of Indian Philosophy*, which describes the *Vedas* as 'tombs of thought richer in relics than the royal tombs of Egypt', Muller cautions:

If we grant that they belonged to the second millennium before our era, we are probably on safe ground, though we should not forget that this is a constructive date only, and that such a date does not become positive by mere repetition . . . Whatever may be the date of the Vedic hymns, whether 1500 or 15,000 BC, they have their own unique place and stand by themselves in the literature of the world . . .[21]

Alchemy

Despite Muller's insistent and repeated caveats, the date of around 1200 BC that he had once 'ventured to assign' to the codification of the *Rig Veda* was the date that stuck. The master himself never saw it as anything more than a hypothesis, but the alchemy of his own prestige and authority transformed it after his death into a 'fact'.

Such cults of the personalities of great men have converted opinions into facts before – usually only for short periods of time until common sense reasserts itself. But Muller's nineteenth-century hypothesis about Vedic chronology is still treated as a fact virtually universally in the twenty-first century even, as we have seen, amongst such wise and insightful scholars as Gregory Possehl. To give just one further example out of many that are available to make the point, Professor Jonathan Mark Kennoyer of the University of Wisconsin, another leading authority on the Indus age, states as fact in his 1998 book *Ancient Cities of the Indus Valley Civilization* that:

> The *Rig Veda* is a compilation of sacred hymns that was codified in its present form during the mid-second to first millennium BC at around the same time as the Indus cities were declining . . .[22]

As anyone who knows their work can attest, Kennoyer and Possehl are far from being dogmatic about the interpretation of the past. On the contrary, they are amongst a number of really fine thinkers and brilliant field-researchers in universities all around the world – not least in India itself and in Pakistan – who are today confronting the enduring riddle of Indian antiquity with a formidable combination of open minds and scientific method. It is important also to remind ourselves that they are only proposing *codification* dates for the *Rig Veda* and fully endorse Muller's earlier recognition that many of the compositions within the standardized collections may have had an *extremely long* prior existence in India's ancient and fantastically elaborate oral tradition. So while their approach does recognize a date of approximately 1200 BC for codification, Possehl, Kennoyer and others are advocates of much earlier dates of composition. Kennoyer in particular seems willing to explore the possibility of continuity between Indus-Sarasvati motifs and the Rig Vedic hymns[23] – when not so long ago such a line of thought would have been inconceivable for mainstream scholars.

Yet so far neither Possehl nor Kennoyer, nor any other Western Indologist of whom I am aware, nor any Western historian, archaeologist, linguist or any other academic from any other discipline working in a university outside India itself, has ever seriously considered the possibility that the Indus Valley civilization, hitherto believed 'mute' because its script cannot be deciphered, could in reality have been speaking to us all along through the medium of Vedic Sanskrit.

Having taken two big steps towards such a conclusion – dumping the Aryan invasion theory, and accepting that the *Vedas* are likely to be significantly older than their date of codification – it is, I think, rather strange that scholars outside India have not yet been prepared to take the third obvious step, which would involve giving proper consideration to the possibility that the true parent of these orphaned scriptures could be the Indus-Sarasvati civilization itself rather than the evaporated 'Aryan invaders' of the second millennium BC.

Could it be that the reason for this reluctance is the same as the reason that the Aryan invasion theory was allowed to flourish during the colonial era in the first place?

How to have your Aryan invasion and not admit it

There can be little serious doubt that the evolution and lengthy survival of the Aryan invasion theory was underpinned by an ingrained conviction on the part of European scholars that the presence in India of a 'superior' language such as Sanskrit that was related to European languages *must* imply a movement of that language from Europe to India in remote prehistory rather than from India to Europe.

Vere Gordon Childe, Professor of Prehistoric Archaeology at the University of Edinburgh and later Director of the Institute of Archaeology, University of London, was one of the most influential exponents of this gross scholarly racism. In 1926, while Harappa and Mohenjodaro were actually under excavation, Childe eulogized the 'gift' that he believed had been given to India by brawny 'Nordic' Aryans:

> The lasting gift bequeathed by the Aryans to the conquered peoples was neither material culture nor a superior physique, but a more excellent language and the mentality it generated ... At the same time the fact that the first Aryans were Nordics was not without importance. The physical qualities of that stock did enable them by bare fact of superior strength to conquer even more advanced peoples and so to impose their language on areas from which their bodily type was almost completely vanished. This is the truth underlying the panegyrics of the Germanists; the Nordics' superiority in physique fitted them to be the vehicles of a superior language.[24]

Such ideas, endorsed and propagated by the leading archaeologists and ethnologists of the time, played a crucial role in the growth of the Nazi cult of 'Aryan' racial superiority during the 1930s and 1940s and led, ultimately, to the abomination of the Holocaust. One would expect, therefore, that archaeologists of today would take an entirely different line. This is what Colin Renfrew, Professor of Archaeology at Cambridge University, has to say on the subject:

> As far as I can see there is nothing in the *Rig Veda* which demonstrates that the Vedic-speaking population were intrusive [to India]; this comes rather from a historical assumption about the 'coming' of the Indo-Europeans . . .[25]

Renfrew blames Vere Gordon Childe's contemporary Sir Mortimer Wheeler for the widespread diffusion and rapid uptake of the 'invasion' idea, which

> is rooted entirely in assumptions . . . When Wheeler speaks of 'the Aryan invasion of the Land of the Seven Rivers in the Punjab', he has no warranty at all, so far as I can see. If one checks the dozen references in the *Rig Veda* to the Seven Rivers, there is nothing in any of them that to me implies invasion: the Land of the Seven Rivers is the land of the *Rig Veda*, the scene of the action. Nothing implies that the Aryas were strangers there.[26]

Finally Renfrew makes the significant observation that despite Wheeler's attempt to hold the Aryas responsible for massacres they never committed in the Indus-Sarasvati cities, and to blame them for those cities' collapse in the second millennium BC:

> It is difficult to see what is particularly non-Aryan about the Indus Valley civilization, which on this hypothesis would be speaking the Indo-European ancestor of Vedic Sanskrit.[27]

But ultimately Renfrew too turns out to be proposing an Aryan invasion of India – only in a freshly scrubbed, politically correct incarnation. Renfrew's scenario enables him to keep a non-Indian origin for Sanskrit while abandoning the now untenable theory of an invasion in the second millennium BC. His argument, in the simplest terms, is that the 'invasion' was actually a peaceful agricultural 'migration' or 'dispersal' and that it took place much earlier than the second millennium BC – indeed he prefers a date at the beginning of the Neolithic perhaps as much as 9000 years ago.[28] In his important study *Archaeology and Language* he makes the case that Anatolia (in modern Turkey, occupying the peninsula between the Black Sea, the Mediterranean and the Aegean) was

> a key area where an early form of the Indo-European language was spoken before 6500 BC. From there the distribution of the language and its successors into Europe was associated with the spread of farming . . . The zone of early farmers speaking Proto-Indo-European extended east to northern Iran and even to Turkmenia at the

outset. The spread of Indo-European speech to the south, to the Iranian plateau and to north India and Pakistan, can then be seen as part of an analogous dispersal, related to demographic changes associated with the adoption of farming.[29]

After their forefathers had arrived in India, Renfrew's hypothesis has it that the descendants of the original Neolithic migrants remained there and developed their society and religious ideas *in situ* for thousands of years. In his view they continued to speak an evolving form of the language brought with them from Anatolia that was to become Sanskrit – in which the *Vedas* would ultimately be composed. And although he has not explored the implications further, he clearly has no objection in principle to the idea that it was also they who founded the Indus-Sarasvati civilization.

Two sides of the same coin

Outside the cosy Pall Mall club of Western scholarship, Indian academics have been forthright in contemplating direct links between the Indus-Sarasvati civilization and the Vedic texts. Like Renfrew, Dr S. R. Rao, famous as the founder of marine archaeology in India, believes that the language of the Indus-Sarasvati cities was an early form of Vedic Sanskrit – and has even gone so far as to propose a full interpretation on this basis of all known examples of the Indus-Sarasvati script.[30] A number of other leading scholars, such as Dr R. S. Bisht, Director of the Archaeological Survey of India, and S. P. Gupta, Professor of the History of Art in the National Museum Institute, New Delhi, also have similar ideas.

Bisht, for example, has argued that the hierarchical layout of Harappan towns was organized according to the Rig Vedic *trimeshthin* system which advocates three distinct sectors of settlement: Parama-Veshthina (Upper Township), Madhyama-Veshthina (Middle Township) and (Avama-Veshthina) (Lower Township). He also points out that the Harappan city of Dholavira in Gujerat, which dates back to the third millennium BC, measured 771 metres from east to west at its maximum extent and 616.8 metres from north to south, the ratio being 5:4. The Citadel, or Upper Township, measured 114 metres from east to west while from north to south it measured 92.5 metres, the ratio being again 5:4. Bisht does not think it is a coincidence that the same ratio is specifically mentioned in ancient texts setting out the proper construction of Vedic fire-altars.[31]

S. P. Gupta likewise points out that all the key characteristics ascribed to Rig Vedic religion and culture are already found in the mysterious ancient cities along the Indus and Sarasvati rivers. First and foremost amongst these characteristics are the cities themselves – since, contrary to the old view that the *Vedas* portray only a pastoral or nomadic lifestyle, all scholars now acknowledge that cities are frequently mentioned in the *Rig* and other Vedic texts as the homes of Aryans. Additional archetypally 'Vedic' characteristics that have been confirmed by excavation of the Indus-Sarasvati sites include the presence of cattle and of the domesticated horse, the use of fire-altars, and evidence of widespread international trade and deep-sea navigation. Gupta concludes:

> Once it becomes reasonably clear that the *Vedas* do contain enough material which shows that the authors of the hymns were fully aware of the cities, city life, long-distance overseas and overland trade, etc., which characterized the Indus-Sarasvati urban gamut of cultural elements, it becomes easier for us to appreciate the theory that the Indus-Sarasvati and Vedic civilizations may have been just two complementary elements of one and the same civilization.[32]

Unlike Renfrew and other Western experts, however, the Indian scholars are not inclined to support any kind of European or central Asian origin for Vedic civilization. Instead, with good reason, they prefer to see it as a wholly indigenous development of their subcontinent – Indian through and through like the Indus-Sarasvati cities.

In this way they have begun the long-overdue process of bringing together one of the greatest and most profound spiritual literatures of antiquity with what is arguably the greatest and most remarkable urban civilization of antiquity. As well as resolving the paradox of a sophisticated urban culture with a script but no literature, and of a sophisticated literature with no urban culture evident behind it, this process has the potential to link the *Vedas* to known history and prehistory and to definite archaeological remains rather than to vapid speculations about an 'Aryan invasion'.

Perhaps we are coming to a time when ancient India will speak for herself again after millennia of silence . . .

My Indian childhood

On a bright morning in July 1954, when I was three years and eleven months old, I got off a ship in the port of Bombay with my mother and father. We then made an immense journey across India by rail that I remember very little of (although I remember the ship very well), and eventually arrived in Vellore in the state of Tamil Nadu in the far south. There my father took up the post of general surgeon at the Christian Medical College Hospital.

We lived in a flat on the campus of the CMC with other doctors' families and medical staff. We had a verandah to the rear of the flat that overlooked some distant palm trees at the edge of a field. During the monsoon season, if I plugged the drains of the verandah, it would fill up with rainwater like a swimming pool. The view of the palm trees bent double in the big winds of the monsoon used to make my heart race and my chest feel tight and I still remember it now as though it were yesterday.

Our flat was on the first floor. There was a dust-patch below in which I once found a lizard's soft-shelled eggs. There was a lily-pond with enormous frogs. And there were trees to climb, including one with a tree-house.

I remember often being in Vellore, 5 kilometres away from the campus. Sometimes I would be at the CMC Hospital following my dad around. Or I would be at the Tamil school I attended at around the age of six where a fellow pupil once stabbed me in the left forearm with a pencil; I still bear the scar.

My father was on a missionary salary in India, so we thought we were as poor as church mice. Still, we employed a servant, who must have been a lot poorer than us. His name was Manikam. I remember he used to bring me my lunch every day in a skyscraper of circular aluminium tiffin tins and take me for rides on rickshaws through narrow streets jammed with tremendous crowds of people.

We had holidays too – Kodai, up in the mountains, where Trixie, my dog, was bitten by something rabid and had to be put down, and Mahabalipuram, on the coast just south of Madras, where I learned to swim. Imprinted on my memory for years afterwards – until I returned there, in fact, and was able to overlay old memories with new ones – were images of the eerie rock-hewn temples of Mahabalipuram, overlooking the Bay of Bengal.

My childhood encounter with India was formative and I am grateful that I was introduced at such an impressionable age to its aura of intriguing and impenetrable mystery, its velvety warmth and depth, its intense colours, sights, sounds, tastes and smells, its joyous, erotic beauty, its cruelty, its love, its passion and its never-ending drama of stark contrasts – past and present, sun and storm, desert and meadow, wealth and poverty, life and death . . .

My baby sister Susan was born in India and died less than a year later of some nameless disease. Then my brother Jimmy was born with an immune system so weak that he could not even fight off the most minor infections. Soon he too was teetering on the edge of death, his lungs ravaged by *pneumocystis carinii* pneumonia – known today as one of the most awful opportunistic infections of AIDS. So, on a dark night in March 1958, when I was about seven and a half years old, I climbed on board an aeroplane with my mother and father and tiny, sad, sickly Jimmy almost invisible inside his portable oxygen tent.

And that was it. That was the end of my Indian childhood.

We flew back through the darkness. We stopped in Egypt, where I saw an ocean of sand from the air. We stopped in Zurich. It was snowing and I was bought my first-ever bar of Toblerone, a truly unforgettable experience. For a while I somehow became briefly separated from my father while we were on the ground and had terrible fears that the plane would leave without me. Finally we landed in London, where my parents rushed to Great Ormond Street Children's Hospital in a desperate but ultimately hopeless attempt to save Jimmy. Meanwhile, I was taken to Edinburgh by my grandmother. There I became entranced by snow, got soaked and frozen playing in it and promptly went down with a life-threatening case of pneumonia.

Indian Atlantis

Many years later, in the summer of 1992, a letter was forwarded to me by my publishers from an Indian lady resident in Canada. She had just read my then newly published book *The Sign and the Seal* and had noticed that it contains a few pages on the subject of Atlantis and considers the possibility of a lost civilization destroyed in a flood cataclysm. The reason for her letter was to tell me of an Indian tradition, which she rightly thought I might not have heard of, that spoke of something quite similar – a great city that had been swallowed up by the sea thousands of years previously. The name of the city, she said, had been 'Dwarka' or 'Dvaraka' and it was referred to in India's sacred texts. More interestingly, a team of Indian marine archaeologists had been to the site where Dwarka was said to have been submerged and had found the remains of gigantic walls and fortifications underwater.

At the time I received the letter I was already deeply embroiled with research for my next book, *Fingerprints of the Gods*, (eventually published in 1995) and half considering a trip to India anyway. By then I was married to Santha, who is of Tamil origin (although she was born and brought up in Malaysia), and she too was keen on the idea. But it was the synchronicity and obvious potential relevance of the letter from Canada that focused our minds. We agreed that we would go if the Dwarka story checked out.

First I confirmed that there are indeed scriptural references to antediluvian Dwarka in ancient Indian texts. There are many. They speak very clearly of Dwarka's foundation in a bygone age by the god Krishna in human form and of its submergence soon after Krishna's death.

Next I looked to see if Dwarka, which the texts clearly locate in north-western India, had any counterpart on land in historical antiquity. I found that not only did it have such a counterpart but that there is still, today, a sacred city called Dwarka, which is one of India's major sites of pilgrimage. It is located just where it should be, in the state of Gujerat on the north-western corner of the Kathiawar peninsula overlooking the Arabian Sea. And as my informant had correctly indicated, Indian marine archaeologists (led by S. R. Rao) had been diving about a kilometre off-shore and had discovered a very large submerged site. Although no datable artefacts had been found, the ruins had been assigned to the 'late period' of the Indus-Sarasvati civilization, perhaps as late as 1700 to 1500 BC.

Santha and I didn't dive in those days but it still seemed worth going to Dwarka just to get the flavour of the place and see if we could learn anything. So we began to plan a journey of about five weeks for November and December of 1992. We would go to Pakistan first to visit the world-famous Indus valley cities of Mohenjodaro and Harappa – cities that had traded with Sumer, cities as old as the Great Pyramid of Egypt. Then we would fly north to Nepal to visit Shanti and Ravi, Santha's two children from her first marriage, who were attending the American School in Kathmandu. From Nepal we would travel to Delhi, the Indian federal capital, and then east to the state of Orissa to the sacred solar temples of Puri and Konarak on the Bay of Bengal. The next stop would be Tamil Nadu, so that we could visit Vellore, my childhood home, and explore Santha's connections with southern India. From there we would fly to Gujerat and spend a week in Dwarka.

Well it didn't quite work out that way. The best-laid plans in India almost never do. Riots and demonic hate-killings between Hindus and Muslims had led to a partial imposition of martial law. At the same time, for entirely unrelated reasons, the main domestic carrier, Indian Airlines, had gone on strike and was stranding passengers all over the subcontinent.

So although we did in the end reach Dwarka on that trip it was not by air but by road.

The flooding of Dwarka and the descent of the Kali Age

Indian thought has traditionally regarded history and prehistory in cyclical rather than linear terms. In the West time is an arrow – we are born, we live, we die. But in India we die only to be reborn. Indeed, it is a deeply rooted idea in Indian spiritual traditions that the earth itself and all living creatures upon it are locked into an immense cosmic cycle of birth, growth, fruition, death, rebirth and renewal. Even temples are reborn after they grow too old to be used safely – through the simple expedient of reconstruction on the same site.

Within this pattern of spiralling cycles, where everything that goes around comes around, India conceives of four great epochs or 'world ages' of varying but enormous lengths: the Krita Yuga, the Treta Yuga, the Davapara Yuga and the Kali Yuga. At the end of each *yuga* a cataclysm, known as *pralaya*, engulfs the globe in fire or flood. Then from the ruins of the former age, like the Phoenix emerging from the ashes, the new age begins.

And so it goes on – birth, growth, fruition, death, rebirth – endlessly across time. At the end of each cycle of four ages there is a super-cataclysm and then a new cycle of *yugas* begins.

Each cycle and each *yuga* within a cycle is believed in India to possess its own special character: the Krita Yuga is a golden age 'in which righteousness abounds'. The Treta Yuga that follows sees a decline and 'virtue falls short'. In the Davapara Yuga 'lying and quarrelling expand, mind lessens, truth declines'. In the Kali Yuga 'men turn to wickedness and value what is degraded, decay flourishes and the human race approaches annihilation'.

The story of Dwarka is tightly intertwined with this scheme of things. Reported in the ancient Indian epic known as the *Mahabaratha* (thought to have been composed a few hundred years after the *Rig Veda*) and in later sacred texts such as the *Bhagvata Purana* and the *Vishnu Purana*, it straddles two of the great world ages.

Towards the end of the most recent Davapara Yuga, the texts tell us, Dwarka was a fabulous city founded on the north-west coast of India. Established and ruled over by Krishna (a human avatar of the god Vishnu), it was built on the site of an even earlier sacred city, Kususthali, on land that had been reclaimed from the sea: 'Krishna solicited a space of twelve furlongs from the ocean, and there he built the city of Dwarka, defended by high ramparts.'[33] The gardens and the amenities of the city are praised, and we understand that it was a place of ritual and splendour.

Years later, however, as the Davapara Yuga comes to an end, Krishna is killed. The *Vishnu Purana* reports: 'On the same day that Krishna departed from the earth the powerful dark-bodied Kali Age descended. The ocean rose and submerged the whole of Dwarka.'[34] The Age of Kali thus ushered in turns out to be none other than the present epoch of the earth – our own. According to the Hindu sages it began just over 5000 years ago at a date in the Indian calendar corresponding to 3102 BC.[35] It is an age, warns the *Bhagvata Purana*, in which 'people will be greedy, take to wicked behaviour, will be merciless, indulge in hostilities without any cause, unfortunate, extremely covetous for wealth and wordly desires . . .'[36]

5 / *Pilgrimage to India*

Mahabalipuram became soon celebrated beyond all the cities of the earth; and an account of its magnificence having been brought to the gods assembled at the court of Indra, their jealousy was so much excited at it that they sent orders to the God of the Sea to let loose his billows and overflow a place which impiously pretended to vie in splendour with their celestial mansions. This command he obeyed, and the city was at once overflowed by that furious element, nor has it ever since been able to rear its head.

William Chambers, *The Asiatic Researches*, vol. 1, 1788

On the same day that Krishna departed from the earth the powerful dark-bodied Kali Age descended. The ocean rose and submerged the whole of Dwarka.

Vishnu Purana

It is a curious thing that if one wishes to select a date that truly does seem to mark the beginning of some kind of 'new age' in the Indian subcontinent, then it would have to be around about 3100 BC – the epoch traditionally signalled as the beginning of the Kali Yuga. It was at this time, at any rate, along the river valleys extending down from the Karakoram and Himalayan mountain ranges, that the largest urban civilization of antiquity began to stir. As we have seen it would later be called the Indus Valley civilization, or the Indus-Sarasvati civilization.

At its peak around 2500 BC this mysterious prehistoric culture boasted at least six large inland cities – others may yet await discovery – with populations in excess of 30,000. These urban hubs were linked to hundreds of smaller towns and villages and to several key ports like Lothal and Dholavira at strategic locations along its coastline and up its navigable rivers. Its borders enclosed an area larger than western Europe – 1.5 million square kilometres, extending from Iran in the west and Turkmenia and Kashmir in the north to the Godavari valley in the south and beyond Delhi in the east.[1] It also had outposts overseas, including a once thriving colony in the Persian Gulf, and it had an extensive trading network supported by a large merchant navy.[2]

In November 1992, when Santha and I boarded the PIA flight from London to Karachi, I had heard enough about the 'Indus Valley civilization' (the only name by which I knew it then) to be intrigued by it, but was ignorant about the details. Like most people who know of it at all I identified it only with the first

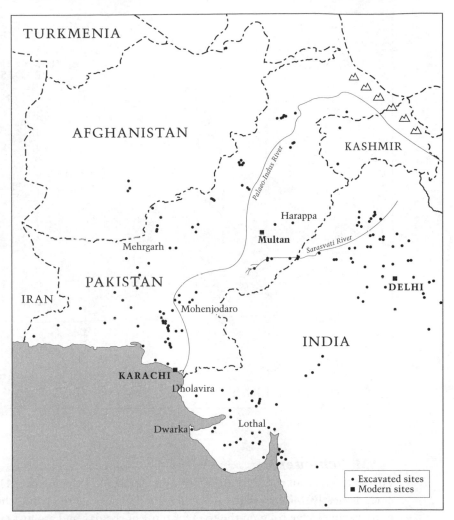

Based on Possehl (1999).

two sites to be excavated – Harappa and Mohenjodaro – which had attracted worldwide headlines and won everlasting renown when they were discovered in the 1920s.

After spending a day sleeping off jet-lag in a seedy hotel in Karachi we flew north to the city of Multan, itself the shrine of a famous Islamic saint. There we found an English-speaking taxi-driver who was willing to drive us first north to Harappa, then south to Mohenjodaro, and finally to drop us off in Karachi – a total journey of about 1000 kilometres.

Mohenjodaro

I'll pick up a bit of the story from my 1992 notebook, skipping over Harappa since, honestly, Mohenjodaro can stand for both places. At the point where the entry begins we've been on the road for most of the day and are just entering the province of Sind:

Monday 16 November 1992

Cross from Punjab into Sind quite late – 9.30 or 10 p.m. Checkpoints fairly thorough. Atmosphere of increased security in Sind. Finally arrive in Sukkur, crossing the Sukkur Barrage, around 10.50 p.m. and check into hotel in some dusty suburb around 11.50 p.m.

Hotel receptionist, who also cooks us dinner around midnight, inquires what time we will be leaving in the morning. I ask why he wants to know. He says because there is a big security problem in Sind – *dacoits* (bandits). Recently one Japanese and one Taiwanese traveller were kidnapped on the road with a total ransom required of six million rupees – their families paid half; Pakistan government paid half. Foreigners very much in demand by kidnappers as all are believed to be enormously rich.

It turns out we must have an escort to drive between Sukkur and Hyderabad via Mohenjodaro. Mohenjodaro itself, in Larkana district, is 'very dangerous' apparently.

It also turns out that a police guard will be required at the hotel all night, because we are there, to prevent us from being snatched from the room!

Leave hotel at 9 a.m. next morning accompanied by four armed police escorts in the back of a Toyota pick-up. They have an array of weapons – one G3, one AK47 and two much older carbines.

We follow and discover that we are part of a well-coordinated escort operation that will see us 'passed', like the baton in a relay race, from police vehicle to police vehicle – a total of fourteen in all between Sukkur and Hyderabad. Often the escort cars drive very fast, headlights flashing, sirens sounding, pushing through traffic with us behind. In general we are treated like VIPs and the police coordination is impressive with the next vehicle already pulling out ahead of us as the previous vehicle pulls in at the end of its jurisdiction. They're all in touch with each other by radio and the whole province of Sind, it seems, is under martial law, controlled by the army, with the police subordinate to the army.

We arrive at Mohenjodaro around 11.30 complete with our police escort – at this point four guards in a lorry with two up front. En route we have broken down once and spent an hour at the side of the road with the four armed policemen standing in a cordon around us, presumably to prevent us from being snatched by the twenty or so Sindhi villagers who milled curiously and unthreateningly around us in their little Sindhi hats.

At any rate, we go straight into the site, still closely followed and guarded by our armed escorts, who politely refuse to leave us alone, even for a second, advising that there would be a real risk of our being snatched if they did. We therefore progress through the dusty ruins with an entourage of armed men. It all feels slightly surreal and peculiar.

Because the Harappan culture only very rarely decorated the bricks used in the construction of its massive buildings, Sir Mortimer Wheeler [*The Indus Civilization*, 3rd edition, 1968] describes the vast remnants of Mohenjodaro as 'impressive quantitatively and significant sociologically' but 'aesthetically miles of monotony'.[3] Surveying the very extensive brick ruins through the heat-haze of midday, I found little to disagree with in Wheeler's words. There is a certain monotony and sameness about the acres of red brick under the red dust that lies everywhere. At the same time, paradoxically, this strange place manages to be overwhelming: dense, solid, truly impenetrable.

We approach the main area of ruins up some steep steps and around the western edge of the eroded Buddhist stupa built here 2000 years ago [long after the Indus-Sarasvati civilization had ceased to exist]. From here there is a view down in a westerly direction over the structure that the archaeologists call the 'Great Bath' and Mohenjodaro's geometry of neat orderly streets organized into a strict north–south/east–west grid with rows of brick houses and covered drains. Beyond the

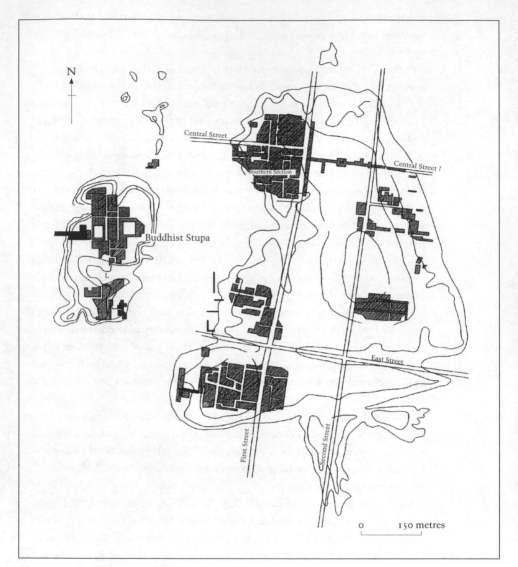

City plan of Mohenjodaro. Based on Possehl (1999).

Bath, again towards the west, what's left of the 'Granary'. And beyond that the old course of the Indus.

The Great Bath – presumed to have been for ritual bathing and purification – looks exactly like a medium-sized rectangular swimming pool and measures 11.89 metres in length (north to south) and 7.01 metres wide (east to west), the depth being 2.44 metres.[4] The close-jointed brickwork and the use of bitumen damp-courses and gypsum mortar to waterproof it all bespeak a high culture with much experience of architecture – experience that could not have evolved overnight . . . Particularly impressive is the drainage system, whereby water was released from the Great Bath, passing through a deep channel covered by a high brick corbel vault.

Moving on from the Great Bath area we then walked half a mile or so to the east of the stupa to the 'DK' residential area of probably wealthy or noble families. It's called DK after its unfortunately named excavator, a certain D. K. Dikshitar, who worked here in the 1920s.

DK would have been an imposing residential suburb. Many of its buildings had two, sometimes even three, storeys and some walls still stand up to four metres high. Evidence that wooden beams, long since rotted away, once supported floorboards and ceilings. Also evidence of municipal street-lighting (lanterns in wall-sockets – one such lantern in museum) and municipal refuse collection – with public rubbish-bin enclosures. Even more impressive is the obvious concern with sanitation evidenced by the miles of covered drains and by the fact that many of the houses had private toilets, somewhat of the modern Western type, which vented down carefully made angled brick slipways into the sewers or into refuse pots that stood outside in the street under the vents and that are thought to have been cleared away at regular intervals by municipal sewage squads. Inside the main sewage drains themselves, spaced at regular intervals and again regularly cleaned out, were rectangular sump-pits that trapped solid waste while allowing liquid waste to flow away.

These people, in short, knew a great deal about urban life and urban architecture. And that knowledge, I'm sure, was already old and evolved, handed down, a legacy, when they first began to build Mohenjodaro . . .

Science

At its peak in the mid-third millennium BC the total inhabited area of Mohenjodaro exceeded 250 hectares and it is possible that its population may have risen as high as 150,000.[5] By then it was part of a vast network of other cities, towns and villages within the Indus-Sarasvati civilization, the majority of them built out of baked mud bricks produced from moulds with standard proportions. One size of brick (measuring $7 \times 14 \times 28$ centimetres) was used in house construction, and a different size ($10 \times 20 \times 40$ centimetres) was used in the building of city walls. But both sizes of brick have identical proportions: thickness=1, width = 2×1, length = 4×1.[6]

Like Mohenjodaro, some of the other Indus-Sarasvati settlements (though by no means all) were laid out according to a strict grid with the major thoroughfares and buildings accurately aligned to the cardinal directions – north–south and east–west. This suggests a high degree of planning and deliberation – after all, in most cultures settlements grow up haphazardly, a bit at a time, but apparently that didn't happen here: in the case of many Indus-Sarasvati sites the template was set out right at the beginning. Moreover, the precision of the alignments of major structures leaves little doubt that the planners employed the services of astronomers in their architectural teams. Several scholars have reasonably deduced that astronomy may have been a highly regarded science in the Indus-Sarasvati cities and was perhaps linked to whatever religion was practised there.[7]

It has also been noted that weights and measures found at Mohenjodaro, Harappa and many other widely separated Indus-Sarasvati sites are not only extremely accurate and consistent but demonstrate a high level of mathematical development. The weights appear to have been designed according to a binary scale: 1, 2, 4, 8, 16, 32, etc., up to 12,800 units (with one unit being equivalent to 0.85 grams).[8] Measures, on the other hand, made use of a decimal system: 'In Mohenjodaro a scale was found that is divided into precise units of 0.264 inches. The "foot" measured 13.2 inches (equalling 50 × 0.264).'[9] Likewise in the Indus-Sarasvati port of Lothal, S. R. Rao excavated a scale with tiny divisions of just over 1.7 mm:

> Ten such divisions . . . (are equal to . . . 17.78 mm. The width of the wall of Lothal dock is 1.78 metres, which is a multiple of the smallest division of the Lothal scale marked in decimal ratio. The length of the east–west wall of the dock is 20 times its width. Obviously the Harappan engineers followed the decimal division of measurement . . .[10]

In Rao's opinion the material remains of the Indus-Sarasvati civilization – whether in terms of the alignments of its city blocks, the design and civil engineering of its efficient public sewerage systems, or the use of standardized weights and measures in precise mathematical relationships – provide ample proof of 'the scientific approach of the Harappans'.[11] In some cases this approach was so scientific that 'even today', as Jonathan Kennoyer admits,

> many aspects of Indus technology are not fully understood as scholars attempt to replicate stoneware ceramics from ordinary terracotta clay and to reproduce bronze that was as hard as steel.[12]

'Almost everything that was ever written about this civilization before five years ago is wrong . . .'

It is inconceivable that a civilization as developed and well organized as the one that boomed 4500 years ago along the banks of the Indus and Sarasvati rivers in northern India and Pakistan could have simply appeared from nowhere, fully formed, with all its principal accomplishments already in place. Common sense suggests that there must have been a very long developmental phase – somewhere – before such a civilization could have reached maturity. Yet for most of the twentieth century the archaeological record refused to reveal evidence of a sufficiently long period of development anywhere in the subcontinent.

The result was a vacuum in which European scholars felt free to conclude that the Indus Valley civilization might, in its origins, have been alien to India. Many seem to have been attracted to this convenient explanation of the advanced state of Indus-Sarasvati culture. For example, as S. P. Gupta points out, not only did Sir Mortimer Wheeler teach that Mohenjodaro and Harappa had been destroyed by invading Aryans; also he never quite brought himself to accept that cities as advanced as these could originally have been the creation of India herself and argued that at least 'the "idea" of "city" as a way of life' must have come to India 'from Mesopotamia'.[13] He even tells us, Gupta notes with annoyance,

> that at least some Mesopotamian masons must have been working in Mohenjodaro directing the method of construction involved in brick masonry. All this simply means that at the operational level not only the 'idea' but also the 'men' came from Mesopotamia to India to give the latter her first cities.[14]

When Wheeler died in 1976 his theory of the Mesopotamian origin of the Indus Valley civilization died with him. But the reason it did so had less to do with his passing than with the start of excavations in 1974 by the French archaeologist Jean-François Jarrige at a previously unexplored site named Mehrgarh overlooking the western edge of the Indus valley from the rugged Bolan pass.

What Jarrige and his team have unearthed since then is the archaeological equivalent of the Holy Grail – an intact sequence of occupation layers at Mehrgarh extending uninterrupted from approximately 6800 BC, 4000 years before the urban boom at Harappa and Mohenjodaro, until the decline of these cities in the second millennium BC.[15] The excavations are still actively underway and the pace of analysis at Mehrgarh, and other nearby sites such as Naushoro that are equally ancient, has quickened since the mid-1990s with results that have a dramatic bearing on the origins of the Indus-Sarasvati civilization. Indeed, these results are so dramatic that when we spoke with Gregory Possehl by telephone in October 2000 he had this to say: 'You want to know something? I'm teaching a class and I told them that almost everything that was ever written about this civilization before five years ago is wrong.'[16]

In chapter 8 we will return to the mystery of Mehrgarh, but in 1992, when Santha and I visited Harappa and Mohenjodaro, I was ignorant of the place and knew nothing of its extraordinary implications.

From the Himalayas to the sea

After leaving Pakistan on 19 November 1992 we travelled first to Nepal, where the bookshops in the narrow streets of Kathmandu's cosmopolitan Thamel market are stocked with interesting and unusual reference works on ancient Indian religious thought – including many of the hard-to-find primary texts. At Pilgrims Bookshop I was able to buy the entire unabridged six-volume set of Ralph Griffith's 1881 translations of the *Rig Veda*, the *Atharva Veda*, the *Yajur Veda* and the *Sama Veda*. But, because at that point I had no reason to disagree with the 1200–800 BC time-span that scholars assigned to the *Vedas*, I again and again postponed studying these huge, daunting books over the next several years and gave my attention instead to texts from Sumer and Old Kingdom Egypt, which I supposed to be much more ancient.

I was about to learn in due course that a new generation of scholars both from within and beyond India are beginning to be convinced that the opposite may be true and that the Vedic hymns could be, by a margin of several thousand years, the most ancient surviving scriptures on earth. In 1992, however, this was just another one of the many possibilities about India's mysterious past that I was ignorant of.

From Nepal we flew on to northern and eastern India – Delhi, Khajuraho, Puri, Konarak – and then south to Tamil Nadu:

Sunday 6 December 1992

Arrive Madras around 10 a.m. – with a migraine. Dr Ramni Pulimood, who worked with my father in the 50s at the Christian Medical College, has sent a taxi to pick us up. We motor the 150 kms to Vellore, passing the spot where Rajiv Gandhi was assassinated. There is a small memorial to him which we visit.

I'm in a coma with my migraine for most of the journey, but rouse myself when we are about 50 kms outside Vellore. Is the countryside familiar? I don't know really. Don't seem to recognize anything. Then we cross a bridge over a very wide dried-out river bed – and I'm sure I remember that from my dreams of childhood, just as I'm sure I remember a dried-out river bed suddenly filled to overflowing with the roiling, rearing waters of a flash flood. And I remember, too, palm trees bent double in the monsoons, the warm splash of fat drops of rain on my bare back, red spider-mites teeming across the earth, and the smell of distant thunder.

We reach Vellore – a medium-sized, dirty, bustling south Indian town full of garish modern signs and vegetarian restaurants. I still remember very little, even when we pull up for a moment right outside the CMC Hospital.

Then we drive through the town and out again towards the CMC compound. I do seem to remember an old school that we pass. Finally I see to my left College Hill rising greenly to a rocky summit and, far away to my right, Toad Hill – so named after the toad-shaped boulder that squats on its peak. I *do* remember both of these landmarks quite vividly, and remember climbing them as a child with my dad and our dog Trixie, but the college buildings into which we now pull ring no immediate bells. I realize later that this is so because they now stand to either side of a busy main road. In the 50s there was no road like this.

We go to 'the big bungalow' and meet Ramni Pulimood, who accommodates us there as previously agreed. Inside, I remember the ancient green cloth blinds which were also standard fitments in the Men's Hostel where we lived and in which I once found a trapped bat.

Half an hour later Ramni and her son drive us out to the Protestant cemetery, where we hope to find my sister Susan's grave. Santha brings flowers, but despite

1. On the waterfront, Alexandria. The author (right) and Ashraf Bechai (second right) discussing locations of underwater sites with fishermen.

2. Megalithic blocks of Sidi Gaber, Alexandria – a site unrecognized by orthodox archaeologists.

3. *Megalithic blocks of Sidi Gaber, Alexandria.*

4. *The 'Great Bath', Mohenjodaro.*

5. Brick foundations, Mohenjodaro.

6. Street with intact drainage, Mohenjodaro.

7. *Exposed well-shaft, Mohenjodaro.*

8. The fairytale city of Dwarka.

9. Sadhu reading the Vedas, Dwarka.

10. *The Dwarkadish temple, dedicated to Lord Krishna, Dwarka.*

11. *Vedic school, south India.*

12. *An Indian ascetic seeking spiritual enlightenment through detachment from the material world.*

13. *Image from 'the penance of Arjuna', Mahabalipuram: ascetic performing austerities.*

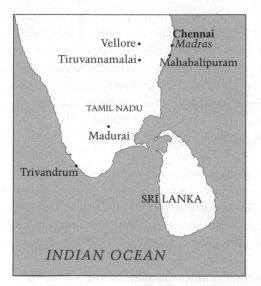

pacing up and down in the peaceful late-afternoon sun we find nothing. We ask the caretakers to check the records, but they too fail to find the grave.

That evening Santha and I climb College Hill, beautiful as the sun goes down with commanding views over a green, half-remembered landscape.

Monday 7 December 1992

Things are coming back a bit more to me now. We visit the CMC Hospital in the morning. Then take a rickshaw to Vellore Fort and then back to the CMC compound via the Protestant cemetery once again. Still we can't find the grave. It's strange to reflect that my sister lies buried and forgotten somewhere here. I dreamed of her a few nights ago, dreamed that she spoke to me. I would like to have known her and – really for the first time – am acutely aware of a missing presence in my life. It's all years ago now, and far away, but I do miss you, Susan. I wish I could just pick up the phone and call you sometimes. Instead I'm an only child, wandering in a graveyard, feeling sorry for myself.

Santha and I complete our visit to Vellore by exploring the CMC compound. I remember the lily-pond – still there – with its frogs. And I do remember the great old tamarind tree and the general outline of the two wings of the Men's Hostel.

Finally we climb up College Hill again for a last look around and then set off on the four-hour drive to Madras on the coast of the Bay of Bengal.

The mystery of the Seven Pagodas

The next day our target was Mahabalipuram, 50 kilometres south of Madras, where I planned to indulge some more childhood memories – this time of a rock-hewn temple standing by the sea. As in Vellore, I didn't really feel that I was there to do research, more on a journey of personal reminiscence. Since the temples were thought to be less than 1500 years old, and had been made on the orders of known historical kings, I had no reason to expect they might be relevant to my primary interest in the possibility of a lost civilization of the last Ice Age more than 12,000 years ago.

Perhaps it was because I went to Mahabalipuram in this frame of mind in 1992 that it gave me back exactly what I expected – i.e. nothing of interest. And yet all along, as I was to discover much later, there was something that I needed to know there. It was hidden away in an anthology of travellers' journals and reports edited by a certain Captain M. W. Carr in 1869 under the title *Descriptive and Historical Papers Relating to the Seven Pagodas of the Coromandel Coast.*[17]

I found the anthology in a second-hand bookshop in Madras after visiting Mahabalipuram in 1992 but did not read it until the year 2000. It was then I discovered for the first time that 'Seven Pagodas' is the old mariners' name for Mahabalipuram – and that the Coromandel coast is the coast of the Bay of Bengal from Point Calimere in the south to the mouth of the Krishna river in the north.

In one paper J. Goldingham, Esq., writing in 1798, spoke of the part of Mahabalipuram that I remembered best from my childhood – the 'Shore Temple', carved out of solid granite, lashed by waves:

> The surf here breaks far out over, as the Brahmins inform you, the ruins of a city which was incredibly large and magnificent . . . A Brahmin, about 50 years of age, a native of the place, whom I have had an opportunity of conversing with since my arrival in Madras, informed me his grandfather had frequently mentioned having seen the gilt tops of five pagodas in the surf, no longer visible.[18]

An earlier traveller's report, from 1784, describes the main feature of Mahabalipuram as a 'rock, or rather hill of stone', out of which many of the monuments are carved. This outcropping, he says:

> is one of the principal marks for mariners as they approach the coast and to them the place is known by the name of 'Seven Pagodas', possibly because the summits of the rock have presented them with that idea as they passed: but it must be confessed that no aspect which the hill assumes seems at all to authorize this notion; and there are circumstances that would lead one to suspect that this name has arisen from some such number of Pagodas that formerly stood here and in time have been buried in the waves . . .[19]

The same author, William Chambers, then goes on to relate the more detailed oral tradition of Mahabalipuram – given to him by Brahmins of the town during visits that he made there in 1772 and 1776[20] – that prompted his suspicion of submerged structures.

According to this tradition, which is supported by certain passages in ancient Hindu scriptures,[21] the god Vishnu had deposed a corrupt and wicked Raja of these parts at some unknown date in the remote past and had replaced him on the throne with the gentle Prahlada, whose reign 'was a mild and virtuous one'.[22] Prahlada was succeeded by his son and then by his grandson Bali, said to have been the founder of the once magnificent city of Mahabalipuram (which, translated literally, means 'the city of the great Bali' or more likely 'the city of the giant Bali').[23] Bali's dynasty continued with his son Banasura – also portrayed as a giant[24] but during his reign disaster struck:

> Aniruddha, the [grand]son of Krishna, came to his [Banasura's] court in disguise and seduced his daughter, which produced a war in the course of which Aniruddha was

taken prisoner and brought to Mahabalipuram; upon which Krishna came in person from his capital Dwarka and laid siege to the place.[25]

Although the god Siva himself fought on the side of Banasura, they could not prevail. Krishna found a way to overthrow Siva, captured the city and forced Banasura into submission and lifelong fealty.[26]

An interval followed, after which another Raja – whose name was Malecheren – took the throne at Mahabalipuram. He encountered a being from the heavenly realms who became his friend and agreed 'to carry him in disguise to see the court of the divine Indra' – a favour that had never before been granted to any mortal:[27]

> The Raja returned from thence with new ideas of splendour and magnificence, which he immediately adopted in regulating his court and his retinue, and in beautifying his seat of government. By this means Mahabalipuram became soon celebrated beyond all the cities of the earth; and an account of its magnificence having been brought to the gods assembled at the court of Indra, their jealousy was so much excited at it that they sent orders to the God of the Sea to let loose his billows and overflow a place which impiously pretended to vie in splendour with their celestial mansions. This command he obeyed, and the city was at once overflowed by that furious element, nor has it ever since been able to rear its head.[28]

There are puzzles about this myth.

First, it was collected, written down and published in the eighteenth century. This was long before any of the ancient inscriptions of Mesopotamia could be read, yet the story of Mahabalipuram bears some striking resemblances to the flood myths of Mesopotamia. In the original Sumerian flood text cited in chapter 2, and in all later variants of it – including the Babylonian versions, the Old Testament account of the flood of Noah and, for that matter, Plato's (supposedly unrelated) story of Atlantis[29] – the gods are angry with or jealous of mankind, exactly as they are said to have been in the Mahabalipuram myth. In all the other myths (with the exception of the Noah story) the gods meet in assembly – again as they are said to have done at Mahabalipuram – before resolving to destroy upstart mankind by sending a flood. And in all the other myths cities and cult centres are submerged by the flood:

> *Sumer*: 'All the windstorms, exceedingly powerful, attacked as one; at the same time the flood swept over the cult centres.'

> *Mahabalipuram*: 'The God of the Sea . . . let loose his billows and . . . the city was at once overflowed by that furious element . . .'

It is also obvious that there are resonances between the Mahabalipuram flood tradition from south-eastern India and the Dwarka flood tradition from the north-west. It is not just that Dwarka is specifically mentioned in the Mahabali-puram story (somewhat surprising in itself) but also that Mahabalipuram and Dwarka, like lost Atlantis and the five antediluvian cities of Sumer, all suffer the same fate – which is to be swallowed up by the sea.

In the case of Dwarka there is also another matter to consider – the end of the former age of the earth and the dawn of the Kali Yuga.

Travels in the Kali Yuga

Our journey from Mahabalipuram to Dwarka in December 1992 was fraught with reminders that we live in the Kali Yuga today – an age of spiritual darkness that the Vedic sages always knew would be filled with the worst kinds of human cruelty and evil. On 6 December 1992 Hindu *kar savaks* (volunteers) violently attacked and pulled down the mosque at Ayodhya in Uttar Pradesh intent on building a new temple for Ram (Rama, another incarnation of Vishnu), whose birthplace is believed to have been on the site of the mosque. This act of 'reclamation' sparked off a wave of violence and mass murder between Hindus and Muslims throughout India which reached a peak in the city of Surat on the Gulf of Cambay in south-western Gujerat. There whole families were roasted alive on fires made up of heaps of their own possessions and in one grisly incident a woman was subjected to multiple rape by a crowd of frenzied males, then burned, and finally beheaded with a sword.

With martial law declared in most cities and go-slows and strikes being staged by Indian Airlines, it took us two days to fly via Madras and Trivandrum to Bombay. From there we arranged to travel the remaining 1000 kilometres or so to Dwarka by road and hired a Maruti van (a motorized roller-skate) and a stalwart Gujerati driver named Vinhod to get us there.

Saturday 12 December 1992
Set off north from Bombay in our little Maruti van. The country towards Gujerat is surprisingly lush, jungly and hilly. The roads are completely crazy and this is an interminable day of driving. It becomes clear that we cannot reach Dwarka in less than two full days like this, and that we may require three – so we set our sights for the first night on Lothal, the Indus-Sarasvati port of the third millennium BC that lies in central Gujerat near the northern end of the Gulf of Cambay. Unfortunately, Vinhod and most people along the way don't seem to know where or what Lothal is and the maps we have are not clear. But part by luck, part by trial and error, we arrive at a truck stop called Pakota late, late at night which turns out to be just 18 kms from Lothal. One of the truckers directs us to a rundown hotel.

Lothal and the ships from Meluha

Lothal turned out to be a quiet sleepy mound in the midst of flat, productive countryside, but in the third millennium BC it was the greatest port of the Indus Valley civilization, connected to the sea by a tidal river channel that has long since dried up. Its dominant architectural feature still surviving today is its great trapezoidal dock.

A major problem with river ports in general is that they can quickly become choked by silt and useless. At Lothal a scientific solution was found to this problem 4500 years ago. First a huge artificial basin was cut into the ground on the eastern side of the town. Then a walled structure measuring 219 metres in length (north–south), 38 metres in width (east–west) and 4.15 metres high, was built into it. The walls were almost 1.78 metres thick at the base, narrowing to just over 1 metre thick at the top, and millions of the best quality kiln-fired bricks were used in their construction.[30] According to the report of S. R. Rao, Lothal's excavator, the inner faces of the dock walls are plumb and 'no steps or ramps are provided anywhere as the primary purpose was to see that the edge of the boat should touch the wall-top to facilitate easy landing and handling of cargo'.[31] At the same time 'three offsets were provided on the outer face of the western wall and two in the case of other walls to resist the overturning movement due to water thrust'.[32]

The dock has a major inlet in its north wall, a second inlet at the southern end of its east wall and a spillway, fitted with an efficient water-locking device, in its south:[33]

> Ships entering the Gulf of Cambay had to be moored along the river quay on the western side of the town and sluiced into the basin at high tide through the first inlet (12 metres wide) provided in the northern arm. A spillway with 1.5 metre thick walls was built at right angles to the southern arm for escape of excess water at high tide. The water-locking device provided in the spillway ensured a minimum draught at low water [2 metres as against 3.5 metres at high water]. Easy manoeuvrability of large ships of 60 to 75 tons capacity and measuring 18 to 20 metres long was possible as they could enter from the shorter side and move along the longer side. The easy flow of water at high tide through the basin ensured automatic desilting. The scouring effect of the tidal waters was arrested by constructing a buttress wall on either margin of the inlet, traces of which can be seen in the case of the northern inlet and more clearly in the second-stage inlet. When the river changed its course and started flowing 2 kms away from the town, a new inlet 2 metres deep, was dug to connect the river with the eastern arm of the dock, but large ships could not enter the basin after 2000 BC.[34]

Archaeologists and engineers are in little doubt that the design of the dock testifies to a long-accumulated experience within the Indus-Sarasvati civilization of the particular problems and challenges posed by such structures. According to N. K. Panikkar and T. M. Srinivasan:

> The Lothal dock being purely a tidal one, the Lothal engineers must have possessed adequate knowledge of tidal effects, the amplitude, erosion and thrust. From this

knowledge they developed competence at Lothal for receiving ships at high tide and ensuring flotation of ships inside the dock at low tide. This is perhaps the earliest example of knowledge of tidal phenomena being put to a highly practical purpose both in the selection of site having the highest tidal amplitude and in adopting a method of operation for entry and exit of ships.[35]

The builders of Lothal lived in the same epoch of early history as the builders of the wonderful Great Pyramid of Egypt and – though obviously on a smaller scale – the dock is a reminder that the peoples of the Indus-Sarasvati civilization possessed a scientific approach, design skills, and hands-on experience of construction problems comparable to those that were evident amongst the ancient Egyptians.

Moreover, it is thought likely that there were both direct and indirect contacts between the Nile and Indus valleys, and between Asia and Africa in general, going back to very ancient times. In the on-site museum at Lothal we were able to see certain items excavated by Rao's team that are indicative of this. These include a terracotta figurine of a gorilla, a species that is found only in sub-Saharan Africa, and a second terracotta figure reminiscent of an Egyptian mummy.[36]

Finds in Egypt also suggest contact. Of special interest, because it dates back to the predynastic 'Gerzean' period (roughly 3500–3300 BC),[37] is a ripple-flaked flint-bladed knife with a beautifully carved ivory handle that was excavated at Gebel-el-Arak in Upper Egypt. In one of the reliefs that decorate the handle a bearded man in fine robes is depicted gripping two powerful male lions by the throat. According to the Egyptologist and art historian Cyril Aldred, this scene 'shows, subduing two lions, a hero who resembles the Mesopotamian Gilgamesh, "Lord of the Beasts" '.[38] Aldred notes that 'this same unusual theme appears on a wall-painting in a Gerzean tomb at Hierakonopolis'[39] – which is indeed the case. He seems unaware, however, that, with minor variations, the scene also appears in the art of the Indus-Sarasvati civilization – for example, on ornate terracotta and steatite seals, excavated in many sites, and on a particularly striking moulded tablet from Harappa that Jonathan Mark Kennoyer describes as:

> a figure strangling two tigers with bare hands [which] may represent a female, as a pronounced breast can be seen in profile. Early discoveries of this motif on seals from Mohenjodaro definitely show a male figure, and most scholars have assumed some connection with the carved seals from Mesopotamia that illustrate episodes from the famous Gilgamesh epic. The Mesopotamian motifs show lions being strangled by a hero, whereas the Indus narratives render tigers being strangled by a figure, sometimes clearly male, sometimes ambiguous or possibly female. This motif of a hero or heroine grappling with two wild animals could have been created independently for similar events that may have occurred in Mesopotamia as well as the Indus Valley.[40]

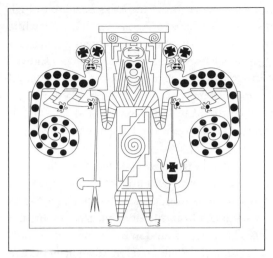

Gilgamesh-like figure between two felines from a bronze breastplate, Tiahuanaco.

Gilgamesh-like figure between two felines from a Gerzean period knife, Egypt.

Perhaps. But I wonder if Kennoyer's conclusion is not a little hasty, and whether it is strengthened or weakened by the fact that almost identical figures of a 'man between two felines' have also been found amongst the art of the prehistoric megalithic city of Tiahuanaco in South America.[41] Such similarities *may* depict similar events that occurred by coincidence in different places, but other explanations might also fruitfully be sought for why the same – 'unusual' – symbolic device is found in ancient Egypt, ancient Mesopotamia, the ancient Indus-Sarasvati civilization and ancient South America.

On the other side of the Gebel-el-Arak knife handle is a second scene that suggests contact between Indus-Sarasvati peoples, predynastic Egyptians and the ancient civilizations between the Tigris and the Euphrates rivers in Mesopotamia. In Aldred's description it shows a water battle in progress: 'In the upper row, the boats have vertical prows and sterns rather like the *belems* of the Tigris, in the lower they have the normal appearance of Egyptian boats of Gerzean date.'[42] But the archaeologist Ernest J. H. Mackay, who carried out extensive excavations in both Egypt and the Indus valley during the first half of the twentieth century, noticed something else. Describing a representation of a boat carved on a seal that he found at Mohenjodaro, he commented:

> The bindings of its hull suggest that this boat was made of bundles of reeds, as were so many contemporary craft of ancient Egypt. It is mastless, which perhaps indicates that it is a river boat. The uprights at either end of the cabin carry flags or emblems and a seated steersman holds a pair of rudders, as on the modern Indus craft. This vessel, it is interesting to note, is singularly the one portrayed on the well known Gebel-el-Arak ivory knife handle.[43]

The specific comparison being made here is to the mastless boats with high prows and sterns, which Aldred separately likens to Tigris river craft, and while the similarities cannot be taken as conclusive evidence of contact amongst all three regions in prehistory they are at least suggestive. Thor Heyerdahl showed long ago with his Tigris and Ra expeditions that reed-boats are capable of trans-oceanic journeys.[44] Besides many representations and terracotta models of masted sea-going boats have been found in Indus-Sarasvati sites – and at Lothal itself trade goods and inscribed seals from the Persian Gulf have been excavated.[45]

The indications are that the bulk of this trade was carried on ships of the Indus-Sarasvati civilization – a civilization that was known to its neighbours in the Persian Gulf as Meluha.[46] Inscriptions from ancient Babylon and Akkad speak proudly of the number of great boats from Meluha that have moored in their harbours. Five such references have been found in the cuneiform records of the time before Hammurabi (1792 BC).[47] One concerns Sargon of Akkad (2334–2279 BC) and tells us: 'the ships from Meluha . . . he made tie up alongside the quay of Akkad'.[48]

Interestingly, a terracotta seal from Mohenjodaro shows a large high-prowed ship with a spacious on-deck cabin. Fore and aft of the cabin perch two birds which archaeologists believe are 'land-finding birds (*diskakas*)'.[49] As the reader will undoubtedly be aware, many ancient traditions of the global flood, not least the biblical story of Noah, make prominent mention of the role played in the navigation of the survival ship by birds just such as these.[50]

The city of Krishna

After leaving Lothal in the late afternoon we spent another night on the road at Jamnagar, the regional capital, and completed our journey to Dwarka the next morning. This final two-hour leg was across the barren, sun-baked flatlands of Gujerat's Kathiawar peninsula, uninhabited and for the most part overgrown with thorn trees and scrub vegetation. Through the open windows of the van we began to sense first the humidity, then the salty tang, of the approaching Arabian Sea. Next, a glimpse of distant water came into view and, rising above it through the heat haze, a shimmering pyramidal mound, topped by the spectacular Dwarkadish temple, sacred to Lord Krishna, soaring skywards on its 72 granite columns.[51] At the apex of the mirage fluttered a colourful flag decorated with astronomical symbols, while around its base the medieval labyrinth of Dwarka's streets and houses clustered tightly packed, as though seeking protection.

We asked Vinhod to bring us closer and we eventually pulled up in a crowded market area directly in front of the temple. From this vantage point I could make out weird figures like the gargoyles of a Gothic cathedral carved into the corners of the roof and walls – here an elephant, there a swan, there a winged sphinx with a woman's face . . . It was easy to imagine the temple as an avatar's palace magically brought into being in the midst of the sea, charged with the mantric energy of pilgrims' prayers and surrounded by a force-field of divine grace.

In Book X of the *Bhagvata Purana* we read how Krishna used 'his supernatural yogic powers',[52] in a crisis of battle, to transfer all his people to Dwarka where he could protect them from the enemy in 'a fortress inaccessible to human beings' [literally 'bipeds']:

> the Lord caused a fortress constructed in the western sea. In the fortress he got built a city twelve yojanas (96 miles) in area and wonderful in every respect.
>
> The building of the city exhibited the expertise in architecture and the skill in masonry of Tvastr, the architect of the gods. The roads, quadrangles, streets and residential areas were constructed in conformity to the prescribed tenets of the science of architecture pertaining to city building.
>
> In that city, gardens planted with celestial trees and creepers and wonderful parks were laid out. It was built with sky-scraping, gold-towered buildings and balconies of crystals. It had barns built of silver and brass which were adorned with gold pitchers. The houses therein were of gold and big emeralds.[53]

But that was the first Dwarka, the original Dwarka – India's lost Atlantis swallowed up by the sea long ago at the beginning of the Kali Age. This Dwarka of today, whatever it was, and this Krishna temple, were much more recent – built to commemorate the inundated city perhaps, but not to be confused with it.

Santha and I checked into a mosquito-infested hotel with the bonus of several dozen hornets drowsing irritably in the curtains of our room and then took a stroll through the town in the late afternoon. It was dusty, of course, dirty, of course. There were people, everywhere, of both sexes and all age-groups – selling to one another, buying from one another. Nobody seemed to be miserable or angry or in a grouchy mood. A whole menagerie of animals roamed the streets, grunting and squawking, barking and mewing, bleating and mooing. There were cows everywhere – a normal sight in Hindu India but here the sacred animals seemed to be more than usually serene and unhurried. I suppose it helps that just about everybody in Gujerat, and definitely everybody in Dwarka, is a strict vegetarian – so strict that not only are animals safe from them but also eggs, onions and garlic as well.

Through the maze of narrow lanes and cobbled alleyways lined with tiny, garish one-roomed shops and makeshift stalls we worked our way down to the bank of the Gomati river where it runs along the edge of the town and enters the Arabian Sea. Here, a large group of giggling children fed breadcrumbs to small fish, and orange-robed *sadhus*, their faces smeared with ash, sat with their backs to an ancient brick wall, reciting verses from the *Rig Veda*. The air was filled with frankincense and ganja and the sound of chanting, and the

December sun, setting out over the sea to the south-west, had infused the vast horizon with an otherworldly glow.

Continuing the remaining few hundred metres along the embankment in the gathering dusk we came to the small circular temple of Samudranarayana – the temple of Samudra, God of the Ocean – perched directly above the point where the Gomati flows into the sea. A breeze was picking up, stirring the waves into white caps, and I walked to the edge of the jetty and looked out.

I had read the reports of the marine archaeologists and I knew that a city of gigantic proportions lay underwater less than a kilometre in front of me. I reminded myself that a conjectural date of approximately 1700–1500 BC had been assigned to the site by S. R. Rao and that he believed it to be one of the late works of the Indus-Sarasvati civilization – much too late to have had anything to do with any hypothetical lost civilization of the last Ice Age.

But there were areas of doubt. Although it seemed astonishing, and was perhaps just the result of incomplete research on my part, I had not been able to find evidence in the scientific literature that any Indus-Sarasvati artefacts – though reasonably plentiful in the countryside round about – had ever been recovered from the submerged ruins of Dwarka (or, for that matter, evidence that any datable objects of any kind had ever been found there). All that the archaeologists had discovered underwater were the looming remains of huge stone walls built of undatable megalithic blocks often interlocked with one another by means of L-shaped dovetails. Since there were thick silt deposits around the site, it was possible that many further structures remained as yet unexcavated beneath those that had already been mapped. Moreover, no thorough survey had been done further out from the shore in water that was deeper than 20 metres.

All in all it seemed to me that the chronology that the archaeologists had proposed here might be right or might be wrong, but was far from settled. And what complicated the picture even more was the opaque history of relative sea-level rise in this part of India which had included several intense episodes of tectonic activity to do with mountain-building in the Himalayas during the past 20,000 years. It had therefore proved difficult to establish the date of Dwarka's submergence from geological clues alone.

The sun was now half-sunk in the ocean and the light was fading fast as the waves piled up against the jetty.

It would be another four years before I learned to dive and four more after that before I could return to Dwarka to explore the underwater city.

6 / The Place of the Ship's Descent

Sages who searched with their heart's thought discovered the existent's kinship in the non-existent . . . Who verily knows and who can here declare it, whence it was born and whence comes this creation? The Gods are later than this world's production. Who knows then whence it first came into being? He, the first origin of this creation, whether he formed it all or did not form it, whose eye controls the world in highest heaven, he verily knows it, or perhaps he knows not . . .

Rig Veda (Book 10, Hymn 129, Verses 4–7, Griffith translation)

'Scientific progress in historical, genetic, linguistic and archaeological research has proved during the past decade that the Hebrew Torah – which is the fundamental scripture of Judaism and which also serves Christians as the Old Testament of the Bible – is not the work of the Jewish people, and in fact that there is no reason to believe today that there ever was a Jewish race that spoke the Hebrew language and was possessed of a coherent or well-defined set of Jewish or Hebraic cultural features.'

Suppose that this statement is supported by powerful evidence and, moreover, that it comes from a distinguished academic source – a Professor at the University of Pennsylvania for example – regarded as a world authority on Jewish culture. Having just read the statement, and knowing the authority of its source are you: Shocked? Surprised (that you have not seen any headlines on this)? Sceptical? Disbelieving? Disoriented (if the Jews didn't write the Old Testament, then who did)? Angry? All of the above? None of the above? Or do you know enough about the *Torah* and about Jewish culture to have realized at once that the statement is a complete fabrication? No such scientific evidence has ever been produced and the identification of the *Torah* with the Jewish people and the Hebrew language remains unassailable today. This is so because the sacred book is comprehensively rooted and grounded in a known cultural background of great antiquity and fits perfectly into its historical and archaeological context.

The same cannot be said of the *Rig Veda*, the fundamental scripture of Hinduism. The abandonment by scholars of the theory that India was invaded around 1500 BC by a people calling themselves the Aryas, and the recognition that there never was any such thing as an Aryan race that spoke Indo-European languages, have had the unfortunate side-effect of orphaning the *Rig* – because it was hitherto believed that these very same Aryas had been its authors. We've also seen how it has been claimed by Renfrew and others – probably correctly – that Indo-European languages have been present in north India for at least 8000 years. Logically, therefore, the fact that the *Rig Veda* is expressed in Sanskrit – an Indo-European language – can no longer be used to substantiate a chronology for the *Rig Veda* that brings the culture that is supposed to have composed it into India (via the non-existent Aryan 'invasion') as late as 1500 BC.

In other words, the ship of the *Vedas* presently has no one at the helm. These sublime hymns, these cleverly coded riddles from antiquity, which form the core scripture of a thousand million Hindus in the twenty-first century, now stand in the astonishing position of having no known authors, no known cultural background and no known historical or archaeological context into which they fit. Moreover, although their moorings to an 'Aryan' race in 1500 BC have been severed, most orthodox historians and archaeologists living outside the Indian subcontinent seem content to leave the *Vedas* drifting and unassigned – the scriptures of no known people composed at no known time.

In such a situation where history has little to offer and a huge blunder to retract, it becomes reasonable to inquire: what do the *Vedas* have to say on the subject of their own origins?

Some points of terminology, some basic information

In ancient Sanskrit the word *veda* means 'knowledge', '*gnosis*', 'insight' (deriving from the root *vid*, meaning 'to see, to know'),[1] and the word *rig* (*rc* or *rik*) means 'verses' or 'hymns'.[2] So *Rig Veda* means 'Verses' or 'Hymns' of 'Knowledge'. We've seen that there are three other *Vedas*. These are, respectively, the *Sama Veda* – the *Veda* of song or chanted hymn (a reordering for liturgical purposes of certain verses of the *Rig* with new verses added);[3] the *Yajur Veda* – an annotated text of the instructions and sacrificial formulae required at Vedic rituals;[4] and the *Atharva Veda*, which Gregory Possehl describes as the 'least understood of the *Vedas* . . . a book of magic, spells and incantations in verse'[5] and Griffith as 'the *Veda* of Prayers, Charms and Spells'.[6]

As well as these, many Indian scholars also list the following massive and venerable bodies of text within the Vedic corpus:[7] the *Brahmanas* (very ancient prose commentaries on the *Vedas*),[8] the *Arankyas* (a later development of the *Brahmanas*, given over to 'secret explanations of the allegorical meaning of the *Vedas*')[9] and the *Upanishads* (philosophical speculations arising out of the *Vedas*).[10]

The *Upanishads* are often referred to in Sanskrit as the *Vedanta*, meaning 'conclusion of the *Veda*', since they are thought to represent the final stage in the tradition of the *Vedas*.[11] However, there are other important later texts of Hinduism which unerringly continue the same essential teaching and cosmology rooted and grounded in the *Vedas*, and which will therefore also be cited in this inquiry from time to time. These include the *Mahabaratha* (which is about eight times as long as Homer's *Odyssey* and *Iliad* put together!),[12] the

Ramayana, and the *Puranas*. The *Mahabaratha* and the *Ramayana* are both epics consisting of a mass of legendary and instructive material worked around a central heroic narrative.[13] Embedded within the vast text of the *Mahabaratha* is the famous *Bhagvad Gita*, ('Song of the Lord'), described as 'the single most important text of Hinduism'.[14] The *Ramayana*, which tells of the deeds of the hero Rama, an incarnation of Krishna, is traditionally ascribed to the semi-legendary poet Valmiki.[15] Last but not least, the *Puranas* (Sanskrit for 'Ancient Lore') are collections of myth, legend and genealogy.[16]

A generally agreed chronology for all these texts (with arguments usually about periods of hundreds rather than thousands of years) is in use amongst scholars. We saw earlier that the *Rig Veda* tends to be dated anywhere in a broad range from 1500 BC (the supposed date of the non-existent Aryan invasion of India) down to 800 BC. Dr John E. Mitchiner, a great authority on the ancient Sanskrit texts, prefers a narrower range of 1400–1100 BC for the *Rig*, with the *Sama* and *Yajur Vedas* dated 1200–1000 BC, the *Atharva Veda* 1300–900 BC, the *Brahmanas* 900–600 BC, the *Aranyakas* 700–500 BC, the *Upanishads* 600–400 BC, the *Mahabaratha* 350 BC – AD 350, the *Ramayana* 250 BC – AD 200, and the *Puranas* AD 200–1500.[17]

While this is convenient as a summary of what is still, amazingly, the accepted scholarly chronology, I feel it is essential to bear in mind that these dates are a house of cards founded on the redundant hypothesis of an Aryan invasion of India in the second millennium BC. Whether starting in 1500 BC, 1400 BC or 1200 BC, the timelines that have been suggested for the compilation and codification of the *Rig Veda* all rest on this now thoroughly falsified and bankrupt idea. And since the chronology that scholars have 'established' for the *Rig* is the foundation of the entire literary history of India, it follows that if the previously accepted dates for it are proved by further research to be badly in error then the dates for much of what comes after it are also likely to be wrong. In this connection, Mitchiner himself concedes that 'the dating of Sanskrit texts is a notoriously difficult problem'[18] – one that is further complicated by many texts 'which may be relatively late in their overall or final composition yet contain passages of considerable antiquity alongside much later additions'.[19]

Amidst this tangled maze of texts, all of which once lived as memorized recitations within an oral tradition before they were written down, only one story is offered – the same story repeated again and again with minor variations and additions – as an explanation and account of the origins of the *Vedas*. This is the story of Manu, the father of mankind – India's Noah – and of a mysterious brotherhood of ascetics called the 'Seven Sages', said in many of the recensions to have accompanied Manu in the Ark when the great flood overtook the world.

The father of mankind

Manu (whose name has the same root as the English word man) was the first and greatest patriarch and legislator of the Vedic peoples and is unambiguously described throughout the ancient texts as the preserver and father of mankind and of all living things.[20] Ralph Griffith, the translator of the *Vedas*, describes him as 'the representative man and father of the human race and the first institutor of religious ceremonies'.[21] And in the *Rig Veda* the people who called themselves the 'Aryas' – an epithet meaning literally the 'noble', or 'pure', or 'good' or 'enlightened' folk (a puzzle that we shall return to in another chapter) – are also referred to as 'Manu's progeny',[22] while Manu is known as 'Father Manu'[23] and even the gods are named as 'Manu's Holy Ones'.[24] At the same time the *Rig* does not take the trouble, anywhere, to tell us exactly what it was that Manu supposedly did to earn these honorifics; only that the events took place 'long ago'.[25]

Manu's literary predicament much resembles that of Osiris in ancient Egypt. Nowhere in the entire corpus of ancient Egyptian scripture, from the *Pyramid Texts* to the last versions of the *Book of the Dead*, is the full story of Osiris ever told. We get fragments of it, bits and pieces here and there, records of his titles and honorifics, many axioms ('the truth is great and mighty and it has never been broken since the time of Osiris', etc. but never a connected, continuous narrative which states clearly what it was that Osiris did to deserve all this honour and prominence. Only in a later, non-Egyptian, text – Plutarch's *Isis and Osiris* – does the whole story come out. Plutarch states that his sources were Egyptian priests and the details that he provides are so convincingly identical to the much more fragmentary details contained in the much earlier ancient Egyptian material that each, in a way, provides corroboration for the other. Scholars therefore believe that Plutarch got the story just about right and that it was never spelled out in detail in the earlier scriptures because it was simply too well known for this to be necessary.[26]

It looks as though the same sort of process must have been at work within the *Rig Veda*. Here, like Osiris for the Egyptians, Manu is a household name even incorporated into aphorisms such as 'may we speak like Manu'[27] – which, Griffith says, was universally understood to mean 'with the wisdom and authority of Manu who was instructed directly by the Gods'.[28] Yet nowhere in the *Rig* is there anything even remotely resembling a continuous Manu narrative which would explain the awe within which he was held and the fundamental role assigned to him as the saviour and the progenitor of Vedic civilization. As with the case of Osiris in Egypt, it is probably safe to assume the full story of Manu was simply so well known amongst the practitioners of the *Vedas* that the composers and compilers saw no need to spell it out in detail.

A flood to carry away all creatures

The earliest surviving glimpse of a more complete version of the story of Manu is provided by the *Satpatha Brahmana*. The setting is antediluvian India some years before it is to be destroyed by the flood and Manu is a king and leader of men (specifically identified in the later *Bhagvata Purana* with 'a South Indian or Dravidian king named Satyavrata'):[29]

> In the morning they brought to Manu water for washing the hands. When he was washing himself a fish came into his hands. It spake to him the word 'Rear me, I will save thee!' 'Wherefrom wilt thou save me?' 'A flood will carry away these creatures: from that I will save thee.' 'How am I to rear thee?' It said, 'As long as we are small, there is great destruction for us: fish devours fish. Thou wilt first keep me in a jar. When I outgrow that, thou wilt dig a pit and keep me in it. When I outgrow that, thou wilt take me down to the sea for then I shall be beyond destruction.' It soon became a large fish . . . Thereupon it said, 'In such and such a year that flood will come. Thou shalt attend to me [i.e. to my advice] by preparing a ship; and when the flood has risen thou shalt enter into the ship and I shall save thee from it.' After he had reared it in this way, he took it down to the sea. And in the same year which the fish had indicated to him, he attended to the advice of the fish by preparing a ship; and when the flood had risen he entered into the ship. The fish then swam up to him, and to its horn he tied the rope of the ship and by that means he passed swiftly up to yonder northern mountain. It then said, 'I have saved thee. Fasten the ship to a tree; but let not the water wash thee away[30] whilst thou art on the mountain. As the water subsides, thou mayest gradually descend!' Accordingly he gradually descended, and hence that slope of the northern mountain is called 'Manu's descent'.[31]

In this version, Manu survives the deluge alone, with no mention of the 'Seven Sages' and with no other human companions. How then does he qualify for his Vedic role as the father of mankind?

According to the *Satpatha Brahmana*:

> Being desirous of offspring, he engaged in worshipping and austerities. During this time he also . . . offered up in the waters clarified butter, sour milk, whey and curds. Thence a woman was produced in a year . . . With her he went on worshipping and performing austerities, wishing for offspring. Through her he generated this race, which is this race of Manu . . .[32]

'The ship whirled like a reeling and intoxicated woman . . .'

Maintaining the sequence of the established chronology, the next properly connected version of the Manu story comes to us in the *Mahabaratha*. In this recension of the old tale Manu is not a king but a powerful *rishi* (sage, seer) who spends a supernaturally long time practising yogic austerities:

> standing with uplifted arm, on one foot, he practised intense, austere fervour. This direful exercise he performed with his head downwards, and with unwinking eyes, for 10,000 years. Once, when clad in dripping rags with matted hair, he was so engaged, a fish came to him on the banks [of a river] and spake, 'Lord I am a small fish; I dread the stronger ones, and from them you must save me.'[33]

With a few more details the tale then proceeds in the same manner as in the *Satpatha Brahmana* with the fish being cared for and attended to by the kindly Manu, outgrowing various habitats and finally being placed by him in the ocean:

> When he had been thrown into the ocean he said to Manu: 'Great lord, thou hast in every way preserved me: now hear from me what thou must do when the time arrives. Soon shall all these terrestrial objects . . . be dissolved. The time for the purification of the worlds has now arrived. I therefore inform thee what is for thy greatest good. The period dreadful for the universe has come. Make for thyself a strong ship, with a cable attached; embark in it with the Seven Sages and stow in it, carefully preserved and assorted, all the seeds which have been described of old . . . When embarked in the ship, look out for me: I shall come recognizable by my horn . . . These great waters cannot be crossed over without me.[34]

When the deluge came:

> Manu, as enjoined, taking with him the seeds, floated on the billowy ocean in the beautiful ship. [The arrival of the enormous fish is then announced.] When Manu saw the horned leviathan, lofty as a mountain, he fastened the ship's cable to the horn. Being thus attached the fish dragged the ship with great rapidity, transporting it across the briny ocean which seemed to dance with its waves and thunder with its waters. Tossed by the tempests the ship whirled like a reeling and intoxicated woman. Neither the earth, nor the quarters of the world appeared; there was nothing but water, air and sky. In the world thus confounded, the Seven Sages, Manu and the fish were beheld. So, for very many years, the fish unwearied drew the ship over the waters; and brought it at length to the highest peak of *Himavat* [the Himalayas]. He then smiling gently, said to the Sages, 'Bind this ship without delay to this peak.' They did so accordingly. And the highest peak of Himavat is still known by the name of Naubandhana ('the Binding of the Ship').[35]

Thereafter, through his advanced yogic powers Manu, the father, 'began visibly to create all living beings'.[36]

'The sea was seen overflowing its shores . . .'

A third example – amongst so many more that it is invidious to chose – comes from the *Bhagvata Purana* where Manu first bears the name of Satyaravrata, 'the lord of Dravida'[37] [south India]. In the usual way this Manu encounters a small fish, it grows big and he eventually throws it into the sea. It then reveals itself to him as none other than an incarnation of the god Vishnu, who warns him of the impending flood – which here, as the *Mahabaratha* also hints, acquires the cosmic and universal dimension of the great *pralaya* that brings each *yuga*, or age of the earth, to an end:

> On the seventh day after this the three worlds shall sink beneath the ocean of the dissolution. When the universe is dissolved in that ocean, a large ship, sent by me, shall come to thee. Taking with thee the plants and various seeds, surrounded by the Seven Sages . . . thou shalt embark on the great ship and shalt move without alarm over the one dark ocean . . .[38]

The fish incarnation of Vishnu then vanishes, promising to return at the right moment. Seven days later: 'The sea, augmenting as the great clouds poured down their waters, was seen overflowing its shores and everywhere inundating the earth.'[39]

Next, the ship of Vishnu appears and Manu and the Seven Sages embark in it – with Manu not failing in his duty to bring on board 'the various kinds of plants'.[40]

Last but not least the great fish returns. Manu's Ark is moored to its horn and towed safely across the flood and storm waves.[41]

Fleshing out the Vedic flood myth

Is this ancient tradition entirely mythical and symbolic, or could it be anchored at some level in geological reality and historical time?

My impression, perhaps quite wrong, is that the later texts of the tradition *deliberately* begin to fill in and clarify the details of the Manu narrative missing from the numerous 'customary' allusions to him in the *Vedas* that seem to take a widespread and detailed knowledge of his story for granted.

Perhaps this setting down in writing of the ancient tradition in its late days arose from a recognition that such widespread knowledge could no longer be relied upon and a fear that the oral compositions might eventually be completely lost. The result, at any rate, is that we can now guess exactly why the *Rig* speaks of Manu as the father of mankind. It is because in the ancient traditions of the Vedic peoples – so well known to all in the early days that no written elaboration was thought necessary – he was remembered as the survivor of the universal flood through whose virility and yogic powers the human race and all living beings were propagated again after the cataclysm.

We now also have the following other pieces of information at our disposal:

1. Manu made a special point of bringing something very precious and significant with him from the world before the flood – a cache of 'plants and various seeds' by means of which agriculture could be restored in post-diluvian times.

2. Also with Manu in the ship were the Seven Sages.

3. The character of the flood was that 'the sea . . . was seen overflowing its shores and everywhere inundating the earth'.

4. Borne up on the waters of the flood, and towed by a god, Manu's survival ship travelled towards the north.

5. Manu and the Sages made landfall on the slopes of the 'Northern Mountain' in Himavat – the Himalayas.

6. They were to descend from the mountain 'gradually', and only as the flood subsided, making sure never to put themselves in a position where they could be 'washed away'.

7. Manu was believed to have practised yoga.

8. Manu was believed to have been, in antediluvian times, a king of the Dravidian people of south India.

A ship in the Himalayas?

Despite the formidable reputation of India's oral tradition for preserving and transmitting extremely ancient information, I realize that some linguists and historians are likely to be sceptical of any attempt to connect what may be relatively late texts about Manu's survival of the flood to his earlier more fleeting appearances as a 'household name' in the *Vedas*. Nevertheless there is a strange, isolated passage in the *Atharva Veda* (AV), and another in the *Rig* itself, which add further merit to the view that the Vedic peoples at the dawn of their civilization were already fully conversant with all the details of the flood myth as they are given in the much later texts – and even used similar symbols, imagery and language.

Of course, it is possible that the later compositions simply echo the older ones, but if that were so I would expect them not just to be similar but to be much *more* similar than they in fact are. In my opinion a sufficient degree of difference is evident in the terminology to make it quite unlikely that the *Satpatha Brahmana*, the *Mahabaratha* and the *Bhagvata Purana*, etc., are simply copying the AV and the *Rig* and much more likely that the earlier and the later written texts both descended *separately* from a common, extremely archaic, oral source. My view on this is buttressed by the fact that the relevant passages in the AV and the *Rig* are opaque and meaningless if left to stand alone but begin to make sense to any reader – or listener – who *already* has knowledge of the broader tradition of Manu and the flood. This creates a knotty logical paradox for those who wish to believe that the connected Manu/flood story is an invention of the later texts and was not in circulation at the time of the AV and the *Rig*. The knot can be untangled very simply, however, if we accept that the full connected Manu/flood story must indeed have been in circulation (perhaps even very wide circulation) in the earliest Vedic times *but was simply not written down then* and remained for much longer in the exclusive domain of the oral tradition.

As far as I am aware, the peculiarity of the passage in the *Atharva Veda* was first commented on in the nineteenth century by Professor Albrecht Weber, a well-known German Indologist.[42] The passage can be found in Book 19, Hymn 39, Verse 8, and a modern translation has recently been provided by Sanskrit scholar David Frawley: 'At the place of the ship's descent at the top of the Himalayas, there resides the vision of immortality.'[43] Griffith's (1895) translation of the same verse reads as follows: 'Where is the Sinking of the Ship, the summit of the Hill of Snow, there is the embodiment of life that dies not.'[44] In a footnote Griffith then adds:

> *The Sinking of the Ship*: or the place where the ship sank or glided down; probably the Naubandhana of the later Epos [i.e. the *Mahabaratha*], the highest known peak of the Himalayas, to which in the great flood Manu fastened his ship.[45]

Weber's 1882 comment on the passage had made essentially the same comparison of the *Rig Veda* and the *Mahabaratha*. In the latter, the peak of the Himalayas to which the ship was tied was afterwards called Naubandhana (meaning 'the binding or tying of the ship'). Weber pointed out the curious imperfect similarity of this concept to the central idea of AV, 19, 39, 8, 'where the term *Navaprabhramsana* or "Gliding down of the Ship" is used in connection with the summit of Himavat'.[46]

Since one would not normally expect to see a ship either moored to a mountain or gliding down one, I submit that the presence of such imagery in the AV without an accompanying explanation only makes sense if we assume that the singers of the Vedic hymns *were already very well acquainted with a story of how a ship got itself into the Himalayas*. There are also extremely good reasons to assume that the story in oral circulation then was an early version of the compositions that were much later written down in the *Satpatha Brahmana*, *Mahabaratha*, etc.

The passage in the *Rig Veda* is, if anything, even more indicative of the long pre-existence of this story, with all its essential ingredients. In Book 2, Hymn 23, Verse 13 there is suddenly a reference to 'pure medicines ... those that are wholesomest and health-bestowing, those which our father Manu hath selected ...'[47] In the mid-nineteenth century the Vedic scholar Horace Haymann Wilson was the first to conclude that 'this alludes to the vegetable seeds which Manu, according to the *Mahabaratha*, was directed to take with him into the vessel in which he was preserved at the time of the deluge'.[48]

Finally, to return to the *Atharva Veda*, there is one other unexplained matter raised in AV, 19,39,8. This concerns the association of immortality – 'life that dies not' – with the 'Place of the Ship's Descent' in the Himalayas (or the 'Place of Manu's Descent', as the *Satpatha Brahmana* calls it). Once again, later texts provide the background story that is presupposed in the *Vedas* by telling us that as his reward for saving mankind and the seed of all living creatures the gods granted Manu insight into 'the mystery of the soul',[49] mastery over 'all knowledge'[50] and more than human powers with a lifetime of millions of years so that he might reign for 'one *manvantara*'.[51] A *manvantara* is a period of time which the Vedic sages (with uncharacteristic vagueness) describe as 'about 71' complete cycles of four *yugas*,[52] equivalent to 64,800,000 years[53] – effective immortality.

As readers may already have noticed, there is something familiar about this tradition that Manu was rewarded by the gods with immortality – or at any rate an extremely long life! The same gift was also bestowed (by a supposedly different group of gods) upon Zisudra, the Sumerian flood survivor whose travails are described in chapter 2:

> Life like a god they gave him;
> Breath eternal like a god they brought down for him,
> . . . Zisudra the king,
> The preserver of the name of vegetation and of the seed of mankind.[54]

Two times seven

Another extraordinary similarity concerns the presence of Seven Sages in both the Sumerian and Vedic traditions. Most ancient societies, I concede, had their sages or seers or wise men – in India they were, and still are, called *rishis*. But it seems to me to be stretching coincidence too far to find a group specifically named the 'Seven Sages' prominently associated with two separate ancient cultures and to imagine that this did not come about through some sort of connection.

In the case of Sumer the Seven Sages were depicted as amphibian, 'fish-garbed' beings who emerged from the sea in antediluvian times to teach wisdom to mankind.

In the case of the *Vedas* the focus is not on the antediluvian period but on the flood itself and those antediluvians who are claimed to have survived it, namely Manu and the Seven Sages.

What do we have so far?

- Two groups of seven antediluvian sages, one in ancient Sumer, one in ancient India.

- Both groups are associated with fish symbolism of some sort – the Seven Sages of Sumer are themselves half men, half fish, and the Vedic Seven Sages take refuge on Manu's survival ship, which is towed by a gigantic fish through the raging waters of the deluge.
- Both groups of sages perform an identical function – which is to preserve the gifts of civilization and bring them to mankind in their respective areas.
- Both groups of sages set an example of asceticism and teach and promote the spiritual life.
- Paradoxically, both groups of sages also play an absolutely fundamental and extremely distinctive earthly role as king-makers and as advisers to kings.

Perhaps the similarities result from direct cultural exchange and transfer of ideas between ancient India and ancient Sumer? This option is at least worth considering, because we already know that the Indus-Sarasvati civilization – which has been proposed as the likely mother of the orphaned *Vedas* – and the civilizations of ancient Mesopotamia were contemporary and did have contact with one another. The problem as before, however, is that the similarities are not similar enough – or, to put it another way, that there are too many differences between the traditions – for them to have resulted from the direct transmission of the 'Seven Sages' idea from one society to the other. Besides, although the Indus-Sarasvati people and the Sumerians undoubtedly traded with and knew one another and have left proof of this, the archaeological record also shows that they simply did not exchange cultural ideas, themes and motifs – even at the most basic level such as jewellery design, let alone so fundamental a religious and historical concept as the Seven Sages.

The only explanation left then is coincidence.

Or the possibility that the two traditions are after all related – not directly, but through a shared legacy from a more ancient and perhaps even forgotten common ancestor . . .

An institution for saving the Vedas

What is particularly striking about the Indian tradition is the way that the story of Manu and the Seven Sages is bound up with the ancient *yuga* theory of the cyclical destruction and rebirth of worlds. To this extent it is reminiscent of the story of the inundation of Dwarka; however, in Dwarka's case we hear of only a single city being destroyed while in the case of the flood of Manu – a true *pralaya* – the waters overtake the whole earth and (improbably!) reach high enough to maroon a ship in the Himalayas.

The Sanskrit texts make it clear that a cataclysm on this scale, though a relatively rare event, is expected to wash away *all traces* of the former world and that the slate will be wiped clean again for the new age of the earth to begin. In order to ensure that the *Vedas* can be repromulgated for future mankind after each *pralaya* the gods have therefore designed an institution to preserve them – the institution of the Seven Sages, a brotherhood of adepts possessed of unerring memories and supernatural powers,[55] practitioners of yoga, performers of the ancient rituals and sacrifices, ascetics, spiritual visionaries, vigilant in the battle against evil, great teachers, knowledgeable beyond all imagining, who reincarnate from age to age[56] as the guides of civilization and the guardians of cosmic justice.

But I'm getting ahead of myself. Let's start with first principles.

The Seven Godlike Sages

The earliest surviving written references to the Seven Sages are in the *Rig Veda*. But as with Manu it is apparent from the nature of the compositions that an initiated audience has been assumed and that no attempt has been made to render a full connected narrative (quotations from the Griffith translation);[57]

> Our fathers then were these, the Seven Sages . . . (4, 42, 8)

> They value One, only One, beyond the Seven Sages . . . (10, 82, 2)

> Those Gods of old, Seven Sages who sat them down to their austere devotion . . . (10, 109, 4)

> So by this knowledge men were raised to Sages, when ancient sacrifice sprang up, our Fathers. With the mind's eye I think that I behold them who first performed this sacrificial worship. They who were versed in ritual and meter, in hymns and rules, were the Seven Godlike Sages. Viewing the path of those of old, the [later] sages have taken up the reins like chariot-drivers. (10, 130, 6 and 7)

There are many additional accounts of individual *rishis* – and of their deeds, their knowledge, their powers, etc., but the four passages cited above contain the only direct and explicit references to the Seven Sages (*Sapta Rishis*) in the entire half-million-word corpus of the *Rig Veda*. The references are tantalizingly brief. Yet they are at the same time surprisingly rich in information – rich enough, I think, to allow us to make a few tentative deductions about Vedic beliefs on this subject:

1. The Seven Sages were considered in some way as the 'fathers' of those *rishis* who controlled the rituals and recited the *Vedas* in later times.
2. The Seven Sages were held in enormously high esteem, second only to 'the One, the only One' – the supreme divine power in the universe.
3. The Seven Sages had formerly been mortal men and had been elevated, through their possession of 'knowledge', at the time 'when ancient sacrifice sprang up' – presumably at the dawn of the Vedic religion.

4. The Seven Sages were in some way 'Gods' or at any rate 'Godlike'.
5. The Seven Sages performed austerities.
6. The Seven Sages were ritual specialists who knew the ancient rules of metre and memorization that made it possible to preserve and transmit the 'verses of knowledge' for the benefit of future mankind.
7. Later generations of sages who continued to perform the ritual functions and to memorize and recite the verses of knowledge – i.e. the *Vedas* – were (in the words of one nineteenth-century commentator) 'only imitators of those who preceded them'.[58] It appears that one of the techniques used by subsequent generations to follow 'the path of those of old' may have involved yogic visualization (in the 'mind's eye') of the primal gathering of 'the Seven Godlike Sages ... who first performed this sacrificial worship'.

Makers of the Vedas

As with the story of Manu and the flood, the overlapping story of the brotherhood of Seven Sages who survive the deluge in the Ark with Manu is a difficult jigsaw puzzle scattered across thousands of pages of ancient Sanskrit texts. The leading expert on the subject is Dr John Mitchiner, whose Ph.D. thesis at London University's School of Oriental and African Studies was on the Sanskrit traditions of the Seven Sages and who later published the definitive book, *Traditions of the Seven Rsis*[59] (he uses the Sanskrit term throughout, being a stickler for detail, and not satisfied that the English words 'sage' or 'seer' perfectly translate all the nuances of the Sanskrit *rsi* or *rishi*).[60]

Mitchiner points out that a fundamental connection exists in Indian thought between the Sages and the origins of the *Vedas* – so fundamental that an inquiry into the latter inevitably ends up being an inquiry into the former as well:

> The Seven Rsis are ... frequently described as being those who composed, are most conversant with and supremely knowledgeable in the *Vedas* – as makers of the *Vedas*, knowers of the *Vedas* and masters of the *Vedas* ... [They are] thought to be composers of Vedic hymns, and ... to come to the earth periodically in order to renew Vedic knowledge among men; they are further depicted as teaching the *Vedas* and other sacred works to various individuals and pupils, and as praising the learning, study and recitation of the *Vedas*.[61]

Despite the apparent clarity of the statement, the relationship between the Seven Sages and the composition of the *Vedas* is and always has been difficult to unravel. According to the doctrine of India's *yuga* system as set out by the great nineteenth-century Hindu savant Bal Ganghadar Tilak:

> The *Vedas* were destroyed in the deluge, at the end of the last age. At the beginning of the present age the Sages, through *tapas* [meditation and yogic austerities], repro-

duced in substance, if not in form, the antediluvian *Vedas*, which they carried in their memory by the favour of god.[62]

So on the one hand we are to understand that it is the role of the Seven Sages to 'reproduce' and repromulgate the 'antediluvian' *Vedas* (which themselves were believed to have been the result of an earlier such process of reproduction and repromulgation). On the other hand, and confusingly, there are other hymns in which the Sages are referred to as 'making', or 'generating' or 'fashioning' – i.e. *composing* – the *Vedas*.[63] Last but not least there are passages which leave no doubt that the hymns were believed originally, in some remote epoch, to have been 'inspired', 'given', or 'generated' by the gods and are thus, in essence, *revealed* knowledge.[64]

Secret communication

During the long journeys both intellectual and physical that I have made in India I have learned to live with a certain level of ambiguity. Remember that the Hindu religion is the child of the *Vedas* and that in this religion what we think of as 'reality' (i.e. 'the world of form', the material universe) is held to be *maya* – an illusion or mass hallucination sustained by ignorance and dispersable only by the special knowledge, insight or *gnosis* that is concealed within the *Vedas*.[65] Since this knowledge was intended to be *earned* through individual study and personal asceticism, and yet was conveyed in publicly recited hymns, it was necessary for it to be coded in some way, or for it to make use of cues, images or ideas that might have one set of meanings for the laity and a totally different set of meanings and associations for those on the path to *gnosis*. That such a system of coding or secret communication was in use is confirmed by the *Rig Veda* itself in Book 1, Hymn 164, Verse 45 (Griffith translation):

> Speech hath been measured out in four divisions, the Brahmans who have understanding [*gnosis*] know them. Three keep in close concealment cause no motion; of speech men speak only the fourth division.[66]

Wilson translates the same passage this way:

> Four are the definite grades of speech: those Brahmans who are wise know them: three deposited in secret indicate no meaning; men speak the fourth grade of speech.[67]

The new and the old

There are enough similar hints[68] scattered here and there throughout the ancient Sanskrit texts to justify a cautious approach to the ambiguities about the Seven Sages and their role in either merely 'reproducing', or actually 'composing', Vedic hymns – while these hymns are at the same time understood to consist of revelations from the gods.

Bal Gangadhar Tilak, who devoted his scholarly life to unravelling the *Vedas* and who approached the subject with an extremely lucid and open mind, suggests that there is a way to reconcile these seemingly conflicting utterances. This involves making a distinction

> between the expression, language, or form on the one hand and the contents, substance or subject matter of the hymns on the other; and to hold that while the *expression* was human, the *subject matter* was believed to be ancient or superhuman. There are numerous passages in the *Rig Veda* where the bards speak of ancient poets (*purve rishayah*), or ancient hymns (1.1.2; 6.44.13;.7.29.4; 8.40.12; 10.14.15, etc.) . . . [or where a hymn is said to be] new (*navyasi*), yet the god or the deity to whom it is addressed is old (*pratna*) or ancient (6.22.7; 62.4; 10.91.13, etc.). This shows that the deities whose exploits were sung in the hymns were considered to be ancient deities. Nay, we have express passages where not only the deities but their exploits are said to be ancient, evidently meaning that the achievements spoken of in the hymns were traditional and not witnessed by the poet himself.[69]

The *Rig Veda* is therefore best understood as a multi-layered construct containing some extremely ancient information (which is either repeated verbatim, as handed down from antiquity, or in various ways spoken of, or referred to, in later compositions) and also a fair amount of much less ancient information associated, perhaps, with the various stages and locales of repromulgation and dissemination of the *Vedas*. Moreover, while linguists and historians can debate endlessly about the origins, authorship and antiquity of these amazing compositions and of the later bodies of texts that descend from them, the compositions themselves are absolutely clear on all these points.

The Vedic palimpsests

The *Vedas* describe themselves as being in essence primordial, having been revealed to mankind by the gods. After that initial revelation, when the *Vedas* entered human space and time, a mechanism had to be found to protect the path to *gnosis* enshrined within them from the vicissitudes of the material world – of which the greatest and most deadly of all is the *pralaya*, the cataclysm, that separates one age of the earth from the next. The function of the Seven Sages is to ensure that the *Vedas* are not lost during these periodic episodes of destruction; instead, they are to preserve the hymns in their memories, survive the flood, and repromulgate the entire corpus again to the new age of men.

It is important to note, in the *Vedas*, and the later explanatory hymns as we know them today, that this was already understood to have happened *many times* before[70] – in other words these *Vedas* were not believed, even by those who recited them in antiquity, to be the first *Vedas* but rather a younger recension separated by countless aeons from the original, salvaged from the most recent *pralaya* by the Seven Sages in the Ark of Manu, brought to 'the

Place of the Ship's Descent' in the Himalayas, and from there repromulgated to the present race of men. Moreover, further study of the texts makes it perfectly clear that even these events are cast far in the past in the Vedic scenario – that the time of the flood, Manu and the Seven Sages was itself perceived as having occurred long, long ago by those who said they were the descendants of Manu and by those later sages who spoke of the Seven Sages as their 'Fathers'. Tilak summarizes the issue in the following way:

> The Vedic Rishis were themselves conscious of the fact that the subject-matter of the hymns sung by them was ancient or antediluvian in character, though the expressions used were their own productions.[71]

The hymns are therefore 'oral palimpsests', each imposed on top of an earlier hymn which itself has been 'reproduced' from an earlier hymn, which was reproduced from an even earlier hymn – and so on, back into the night of prehistory. Often the older layers of the palimpsest show through in the younger compositions so that everything is jumbled – like archaeological strata that have been turned over with earth-moving machinery indiscriminately mixing older and more recent artefacts.

As we will see in a later chapter, progress has been made in separating the truly ancient from the more recent information tangled up in the Vedic hymns – and the results have been surprising.

Meanwhile, in summary, it is at least clear that the essential task of the Seven Sages, whose own story is set in the remotest antiquity, was that having learned the *Vedas* from the Sages of an even earlier age they should survive the cataclysm and go forth at the beginning of the new age[72] to 'repromulgate the knowledge inherited by them, as a sacred trust, from their forefathers'.[73] According to the *Matsya Purana*: 'What the Seven Sages heard from the Sages of the preceding age, that they narrated in the next age.'[74]

Connections hidden in the stars?

There are repeated hints in the Sanskrit texts concerning something that sounds very much like a *lineage* of Sages – or perhaps a monastic order or a cult known as 'the Seven Sages' which was believed to have replenished its ranks in each generation. Indeed, in some of the texts detailed lists are provided of many groups of Seven Sages and of the past ages of the earth in which they lived.[75] The *Mahabaratha* makes explicit mention of 'the many Seven Sages'.[76] There are even different groups of Seven Sages assigned to different regions – particularly to northern and southern India[77] – which apparently were believed to have coexisted in different areas at the same time. Out of all this confusion, however, the names Visvamitra, Jamadagni, Bharadvaja, Gotama, Atri, Vasistha and Kasyapa are most frequently mentioned in the early literature as comprising the 'main' group of Seven Sages,[78] with Agastaya sometimes cited as an eighth.[79]

But another group of seven 'Great Sages' (with Atri and Vasistha overlapping), is given at least equal prominence: Marici, Atri, Angiras, Pulastya, Pulaha, Kratu and Vasistha.[80]

It is this latter group that is assigned most often to southern India. But at the same time, curiously and strikingly, there are traditions that associate its members very firmly and vividly with seven stars in the *northern* sky – specifically the stars that form the prominent 'Big Dipper', or 'Plough', within the larger circumpolar constellation of the Great Bear.[81] The identification of this constellation with a bear is extremely ancient and found in many supposedly unconnected cultures.[82] This may shed light on an otherwise peculiar passage in the *Satpatha Brahmana* which informs us: 'The Seven Rishis were in former times called the *Rikshas* [bears].'[83] Mitchiner comments:

> In later times the term *rksa* came to be given a more general meaning, denoting . . . any star . . . This more general meaning is, however, in all probability derivative of the original and more specific meaning denoting the shining stars of the Bear or Ursa Major.[84]

The identification of the Seven Sages with this particular group of stars, so apparent in the Indian tradition, is peculiarly resonant of the well-known ancient Egyptian belief in the stellar destiny of the soul.[85] I cannot help but be reminded of the Pharaoh's wish, repeated countless times in the *Pyramid Texts*, that if in this lifetime his spirit has been 'perfected' then it should upon his death be transformed into a star in the sky.[86]

Two areas of the sky were favoured for stellar rebirth by the ancient Egyptians – the region of the constellation of Orion in the southern sky and the region of the circumpolar, never-setting, 'Imperishable' stars – particularly Kochab[87] in the Big Dipper – in the northern sky. Regarding a circumpolar destiny we read in Utterance 419 of the *Pyramid Texts*: 'Arise . . . raise yourself that you may travel in company with the spirits . . . Cross the sky . . . Make your abode among the imperishable stars . . .'[88] Regarding a destiny in Orion we read in Utterance 466: 'O King, you are this great star, the companion of Orion, who traverses the sky with Orion.'[89]

I do therefore find it odd, to say the least, that ancient India's Seven Sages are given a stellar 'manifestation' as the Big Dipper at the heart of the circumpolar region of the sky, just where the Egyptian Pharaohs wanted to go. Even odder, however, as Mitchiner reports, is that one of the Sages, Visvamitra, is said in both the *Ramayana* and the *Mahabaratha* to have transferred a king of ancient India named Trisanku to the sky in bodily form 'where he now shines as the constellation of Orion'.[90]

Knowledge and balance

Just like the Heliopolitan priesthood who oversaw the construction of the Great Pyramid of Egypt, what the Sanskrit texts suggest to me is the possibility that the 'Seven Sages' of ancient India were not a small group of remarkable individuals but an *institution* that persevered through time – perhaps for many thousands of years – that recruited new members in each generation, and that was dedicated to the preservation and transmission to the future of a body of spiritual knowledge from the remote past.

The highly initiated Sages of India were understood to be ascetics who shunned material pleasures and material things. They are said to have worn simple clothes made out of natural products such as bark-cloth and to have smeared their bodies with ashes. They did not cut their hair but allowed it to grow long and matted. They were strict vegetarians who gathered fruits and roots to live on, praised abstention from meat[91] and spent the greater part of their time in the snow-covered mountain fastnesses of the Himalayas. There it was said that they withdrew to perform the *tapas* – or yogic austerities – by means of which they were able to strengthen their spiritual power.[92]

But the ancient texts also tell us that the Sages did intervene and involve themselves extensively in mundane affairs – in particular as king-makers and as advisers to kings who influenced and shaped state policy.[93] Their role in this respect again parallels the role of the Heliopolitan priesthood of ancient Egypt, the king-makers of the Pyramid Age.[94] In both cases the purpose of secular involvement was the same: to guide, shape, form, and maintain indefinitely a society in perfect balance with itself and with the universe – a society constructed in accordance with what the ancient Egyptians called *maat* (earthly and cosmic harmony, truth, balance, the 'right way') and what the Hindus still call *dharma*, a concept that has exactly the same meanings.[95]

Thus we discover that the Seven Sages would from time to time take over as the rulers of kingdoms during an interregnum or in the prolonged absence of the legitimate ruler.[96] They would instruct rulers on the duties of kings.[97] They would also 'obtain sons for kings' (if necessary by impregnating the king's wives themselves!) thus ensuring the longevity of royal dynasties[98] – since it was felt (in both ancient India and in ancient Egypt) that the presence of a king or pharaoh was an essential aspect of cosmic balance. When through some mishap there was no king, then it was the task of the Seven Sages to seek out and appoint a new one. In this regard the *Mahabaratha* tells how, after the destruction of the kingly caste, 'the earth – being without kings – started to sink in distress, whereupon Kasyapa supported the earth and found new kings for her'.[99]

Amongst many other roles related to rulers and the secular order it is interesting to note that the Seven Sages also frequently *cursed* kings if they abused their powers (and it was a very dangerous thing – often fatal – to be cursed by a Sage). 'In such contexts,' observes Mitchiner:

The Rsi comes to be seen not merely as an upholder or teacher of *dharma* who strives to maintain righteousness and proper conduct among men, but as the very embodiment of *dharma* itself, manifesting *dharma* in his words and deeds, and purging with his curse the *adharmic* actions of others.[100]

A spiritual basis to history?

In conclusion, the more I learned about 'the Seven Sages' on my journey through the ancient texts and commentaries, the more they began to sound to me like a religious cult armed with powerful spiritual ideas, fired by yogic asceticism and the quest for *gnosis*, manipulating the development of 'kingdoms' in India from retreats in the Himalayas. And maybe not only kingdoms in India, but elsewhere in the archaic world as well?

We've seen that the Sanskrit texts speak of two groups of Seven Sages, one for south India, one for north India – regions that are widely separated geographically. But beyond India it's worth reminding ourselves again that it was Seven Sages – also associated with the dissemination of a system of knowledge – who served as the advisers to kings in ancient Sumer. Is it not a coincidence too far to discover that Seven Sages fulfilled exactly the same function in Egypt? According to the remarkable *Edfu Building Texts*, which I examined at length in an earlier book,[101] these Seven Sages and other gods came originally from an island, 'the Homeland of the Primeval Ones', said to have been destroyed suddenly in a great flood during which the majority of its 'divine inhabitants' were drowned.[102] Arriving in Egypt, those few who survived became 'the Builder Gods, who fashioned in the primaeval time, the Lords of Light . . . the Ghosts, the Ancestors . . . who raised the seed for gods and men . . .'[103]

Most historians and archaeologists today more or less automatically project the 'materialist' basis and structure of modern society (whether in its 'capitalist' or 'socialist' form) back on to societies of the remote past. This belief – that civilization is simply a function of economic forces – has in turn dictated research and excavation strategies in the field and profoundly influenced the way that scholars look at ancient texts such as the *Vedas*. In recent years, however, a thought-provoking counterview has begun to emerge. 'Our political and economic interpretations of history', argues the Sanskritist David Frawley, 'cannot be true if enlightenment or spiritual realization is the real goal of humanity.'[104]

Frawley draws attention to the ancient science of yoga in India – how ancient it may really be is one of the subjects we will consider in the later chapters – and points out:

The modern view of the development of human civilization is far removed from the evolution of man according to the system of Yoga. The modern idea of civilization developing gradually through the growth of technology and scientific thinking contradicts the yogic point of view which rather sees culture as having been originally formulated and passed down by sages . . . If the essence of civilization is technology then the modern view may be right, but if it is the culture of spirit, it is quite wrong. By my interpretation civilization was founded by yogis, seers and sages.[105]

Is it conceivable that the Indus-Sarasvati civilization of ancient India could have sprung up exactly in the way that the Vedic traditions tell us? Could it have been the outcome of a programme or even a 'policy' instituted by religious ascetics to protect a precious system of knowledge – knowledge from before the flood that was said to have reached India in the Ark of Manu, preserved in the memories of the Seven Sages?

7 / *Lost India*

When Varuna and I embark together and urge our boat into the midst of the ocean, we, when we ride o'er the ridges of the waters, will swing within that swing and there be happy.

Rig Veda (8, 88, 3)

The Vedic flood story, which is also the story of Father Manu and the Seven Sages, contains seemingly absurd elements: a gigantic fish towing the survival ship; no women on board, so Manu must create a wife and progeny by magical means; and a flood so huge and so high that the ship is carried to the Himalayas. There it is ultimately moored to the peak of the 'northern mountain', also referred to as 'the mountain of snow', in a legendary spot known in the *Mahabaratha* as Naubandhana ('the Binding of the Ship') and in the *Atharva Veda* as Navaprabhramsana, 'the Place of the Ship's Descent' (or 'the Place of the Sinking of the Ship').

Although it is true that the Himalayas are young mountains in geological terms – mountains that were indeed once under the sea and that are still rising as India pushes up against the mass of Asia – I know that I am on absolutely safe ground to state that no oceanic flood in the entire evolutionary history of mankind has ever reached into or even anywhere near these 9000 metre high snow-covered ranges. It is, in other words, a geophysical impossibility for Manu's Ark to have been marooned in the Himalayas as the sacred texts of India claim.

Yet it is also true that large areas of the Indian subcontinent *did* experience severe oceanic flooding at the end of the Ice Age – particularly between 15,000 years ago and 8000 years ago. The floods of that epoch were global phenomena, as we saw in chapter 3. In the Arabian Sea and the Bay of Bengal, however, they were fuelled and amplified locally by the spectacular meltdown of the Himalayan ice-cap, which was much deeper and more extensive in the Ice Age than it is today.

So, although I remained puzzled by the references to a ship in the Himalayas, I was not yet prepared to join the scholars in their opinion that all of this was complete fantasy with no historical value. It was time to get more detail on exactly what did happen to India in the crucial epoch of post-glacial flooding from 15,000 to 8000 years ago.

Two anomalous sites ... and counting

In chapter 1 I reported a baffling discovery that was made in the early 1990s by marine archaeologists working in the Bay of Bengal along the Tranquebar-Poompuhur coast of southern India near Nagapattinam. Although they did

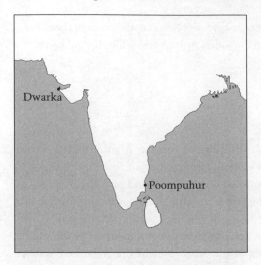

not at the time have sufficient funding to undertake more than a cursory examination, they were able to identify, and recommend for future investigation, a large, man-made 'U-shaped structure' flanked by a 'semi-circular structure' and an 'oval-shaped' mound. What is baffling about these submerged ruins, on which 'a few courses of masonry' can be made out under thick encrustations of marine growth, is the fact that they were found 5 kilometres off the present shoreline and at a depth of 23 metres.[1]

I had discussed the Poompuhur structure with S. R. Rao some months previously (see again chapter 1) and had for a long time regarded them as being of great potential significance. Nevertheless, local sea-levels in many parts of the world can (and do) rise and fall for all sorts of reasons independent of global sea-level rise – so, while tempting, I knew that it would be a mistake to jump to conclusions about the age of the Poompuhur ruins just because they are deeply submerged. This was why I put the problem to Dr Glenn Milne of Durham University, one of the world's leading experts in the cutting-edge science of 'inundation mapping' – which uses a powerful computer program to calculate the complex variables and to produce accurate models of ancient shorelines at chosen dates in chosen locales.

Milne ran the programme for the coordinates of the Poompuhur site and e-mailed the result on 12 October 2000:

> areas currently at 23 m depth would have been submerged about 11,000 years before the present. This suggests that the structures you mention are 11 thousand years old or older![2]

The possibility that the traces of a forgotten episode of global prehistory might indeed lie underwater off the shores of the Indian subcontinent suddenly looked a good deal more plausible. Previously I had focused on only one anomalous submerged site – in the north-west off the coast of Gujerat at Dwarka – and it was of uncertain date. But now I had confirmation of a second strong candidate located at the opposite end of India – in the south-east off the coast of Tamil Nadu – with a provisional dating to the end of the last Ice Age.

The next step was to ask Milne and his colleagues in the Department of Geology at Durham to prepare detailed inundation maps of the whole coastline of greater India as far to the south as the Maldive islands – which straddle the equator – as far to the north and west as Pakistan's Makran coast half-way to the Persian Gulf, and as far to the north and east as the Ganges delta at the top of the Bay of Bengal.

Milne e-mailed the results of this new inquiry in mid-December 2000.

India 21,300 years ago

He had prepared four high-resolution maps. The earliest of these (see page 152) shows the subcontinent as it would have looked 21,300 years ago – around the time of the Last Glacial Maximum (LGM) when the world ocean had sunk to its lowest level.

In that epoch India's coastal plains were everywhere more extensive than they are today, in some areas they were much more extensive, and in two areas in particular – around Gujerat in the north-west and around Tamil Nadu in the south-east – they were so much more extensive as to make ancient India virtually unrecognizable. Is it by chance that it is in these two areas exactly – where marine encroachment during the Ice Age meltdown was more dramatic than anywhere else in the subcontinent – that anomalous underwater ruins have been found?

At the LGM a strip of territory *at least* 100 kilometres wide that is now entirely submerged was exposed along almost the whole of the west coast of India – a linear distance of 2000 kilometres from the far south, beyond present Cape Comorin, to as far north as the Indus delta. However, at about latitude 15 degrees north this strip began to widen rapidly. Off modern Goa it was 120 kilometres wide, four degrees further north it was close to 500 kilometres wide and at 21 degrees north the Gulf of Cambay was a pleasant valley and the site on which the city of Surat now stands would have been as much as 700 kilometres from the sea.

But as I studied Milne's inundation map in December 2000 I was most struck by what it revealed about Gujerat's distinctive Kathiarwar peninsula. Today surrounded on three sides by the sea (with the Gulf of Cambay to the south, the Gulf of Kutch to the north and the Arabian Sea to the west), it was completely landlocked 21,300 years ago. Even Dwarka with its mysterious submerged ruins – now poised on the extreme north-western 'horn' of the peninsula – would then have been about 100 kilometres from the sea.

All in all, I realized that what western India had lost to the global floods that followed the Last Glacial Maximum amounted to a vast coastal domain, nearly the same size and roughly the same shape as modern California and Baja California put together, with an area of close to half a million square kilometres.

The second part of the map that was almost unrecognizable was in the south-east, where the underwater structures had been found off Poompuhur.

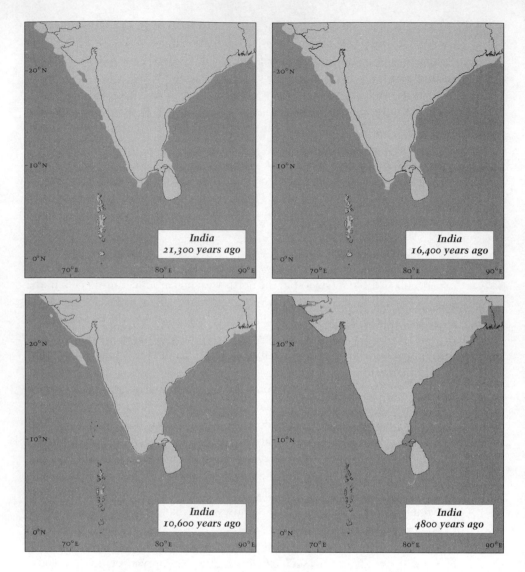

India
21,300 years ago

India
16,400 years ago

India
10,600 years ago

India
4800 years ago

Milne's calculations demonstrated that the Poompuhur site would have been almost 100 metres *above* sea-level at the Last Glacial Maximum, and would have stood towards the northern edge of a great peninsula roughly the same size and shape as the modern Koreas. Enclosing the Palk Strait, which was then a valley, and grafting a much-enlarged Sri Lanka firmly to the mainland, this lost Ice Age realm extended from a little below Dondra Head, at about 6 degrees north, as far as modern Pondicherry at around 12 degrees north. Mahabalipuram, with its neglected legends of the Seven Pagodas and the flooded city of Bali, lies at 12.37 degrees north and would have been at least 50 kilometres from the sea at the Last Glacial Maximum. Meanwhile, to the west of the Sri Lankan peninsula, forming the other side of the Gulf of Mannar – a large enclosed bay at the LGM – a snout of land extended into the Indian Ocean more than 150 kilometres beyond modern Cape Comorin. Finally, off-shore to the south-west, the 'necklace' of tiny atolls that make up the Maldives in the twenty-first century appeared as an imposing archipelago on Milne's map. Greatly enlarged and increased in number because of the lowered sea-level, they included thousands of square kilometres of continuous landmasses at the Last Glacial Maximum that have long since completely vanished.

So here again what the inundation map revealed was a substantial, integrated area – an entire sub-region of India – that had been above water 21,300 years ago and that is submerged today.

16,400 years ago

Milne's second map did not look very different from the first, although it showed India almost 5000 years later – at 16,400 years ago.

To my eye the south-eastern portion was to all extents and purposes identical in both maps. In the south the snout-shaped peninsula below Cape Comorin was slightly reduced in width, but still about the same length, and some of the larger Maldives had begun to break up.

In the south-western sector of the mainland (northwards from the Cape) the 100 kilometre wide strip of coastline up as far as latitude 15 degrees north was thinner – generally between 20 and 50 kilometres thinner – than it had been at the LGM. But beyond 15 degrees north, where the strip began to widen, the loss of land had been much less severe, indeed negligible. The Gulfs of Cambay and Kutch were still filled in, the Kathiarwar peninsula was still landlocked, and Dwarka was still about 100 kilometres from the sea.

In the light of what I'd learned so far about the chronology of the post-glacial cataclysms, the general lack of dramatic change during this period made perfect sense: 16,400 years ago the meltdown of the last Ice Age had only just begun and the first of the three global superfloods identified by Professor John Shaw and discussed in chapter 3 was still more than a thousand years away.

The reader will remember the approximate chronology of those floods, which were actually prolonged *episodes* of flooding in all cases – 15,000–14,000 years ago; 12,000–11,000 years ago; and 8000–7000 years ago.

10,600 years ago

Glenn Milne's third map showed India as it had looked 10,600 years ago, after the first two of the three episodes of flooding had done their work. In the far south the 'snout' that had protruded beneath Cape Comorin was now almost completely inundated, leaving only a lonely island anchored in the Indian Ocean about 80 kilometres off-shore.

To the south-west the Maldives archipelago was much reduced, although the residual islands were larger than their modern counterparts.

In the south-east, I was surprised to see Sri Lanka still attached to India –

albeit by a diminished land-bridge – as late as 10,600 years ago. On the mainland the coast of Tamil Nadu had in general been reduced almost to today's levels. Five kilometres off-shore the Poompuhur structures had been inundated. At Mahabalipuram the coastal plain still extended 2 or 3 kilometres further into the Bay of Bengal 10,600 years ago than it does today – far enough, in theory, for the legendary city of Bali to have been built there as late as that date.

On the south-west side of the Indian mainland the strip of coast running from Cape Comorin at 8 degrees north up as far as 15 degrees north now extended less than 5 kilometres beyond today's level. At about 17 degrees north it began to widen as before, but much more gradually. A very large part of the landmass directly below the Gulf of Cambay was now flooded by the sea and it was possible to make out the emergence of the modern shape of the Kathiarwar peninsula. Nevertheless the Gulf of Cambay was still entirely above water 10,600 years ago, so too was the Gulf of Kutch, and the present coastline of the peninsula was still surrounded by a healthy margin of dry land. Dwarka was at least 40 kilometres from the sea. Off-shore of Dwarka to the south-west there was an island about 50 kilometres in length – a remnant of the formerly much extended coastline in these parts. A second much larger island – 400 kilometres long and almost 100 kilometres wide – lay a little further to the south and extended down to well beyond modern Bombay.

4800 years ago

When I turned to study the final map of the four received from Milne it showed that sea-level was slightly *higher* 4800 years ago than it is today, marking the post-glacial high-stand of the sea. In the far south the Maldive islands had almost completely vanished and Sri Lanka was fully isolated from the mainland and in its modern form. On the mainland itself most parts of the coast were indistinguishable from those on a modern map, although the eminence on which Dwarka stands today would have been an island at that date. Much more significant marine incursions into areas that are now mostly dry land were shown into the Rann of Kuch and the Gulf of Cambay in the north-west and around Poompuhur-Tranquebar in the south-east.

But this made sense. I remembered that in the Persian Gulf too the sea-level had been a metre or two higher around 5000 years ago – as a result of a worldwide episode of rapid, relatively short-term flooding known as the Flandrian transgression.[3] Presumably in India, as in the Gulf, the land had later been recovered thanks to the subsequent regression of sea-level to the modern value, combined with the local effects of silting. Indeed the salt-flats of the Rann remain susceptible to marine transgressions to this day and by 4800 years ago had become, temporarily, a large navigable extension of the Gulf of Kutch, scattered with numerous islands, that would not dry up for another thousand years. Into that gulf as far as Dholavira, the trade of the Indus-Sarasvati civilization was soon to be brought in great high-prowed ocean-going ships – the ships depicted on the terracotta seals of the mid-third millennium BC, the ships that also sailed further south, through the extended Gulf of Cambay, to the now landlocked port of Lothal.

The amount of time that Glenn Milne was able to spend making inundation maps for me was strictly limited, but there was a period within the range of 21,300 to 4800 years ago that I particularly wanted him to do some more modelling on. I already knew by comparing his map for 21,300 years ago with his map for 10,600 years ago what lands had been surrendered to the sea during the first two global floods (15,000–14,000 years ago and 12,000–11,000 years ago). Now I wanted more fine detail on what had happened between 8000 and 7000 years ago, when the third episode of global superfloods had been unleashed. Just to be on the safe side I asked Milne to give me a complete sequence of maps covering the period from 13,500 years ago down to the present.

What if?

India is so big that I sometimes find it difficult to conceive of it all at once. Now after my first session with the inundation maps it seemed to be dividing itself conveniently into the two great cultural, linguistic and geographical regions into which it has always divided itself – at least since the time of the *Rig Veda* – namely the Dravidian-speaking south and the Indo-European-speaking north.

In both these areas there had been extensive post-glacial flooding, and I was determined to dive in both if I could. But the south was far from the Himalayas, with which the *Vedas* associate the escape of the Seven Sages and Manu from the flood, while the north-west coast around modern Gujerat was not only much closer but also had lost more land more rapidly than any other part of India.

The conjunction begged an obvious speculation. What if by extraordinary bad luck some kind of civilization had been based in precisely this area, on land that had been inundated 11,000 or 8000 years ago at the end of the Ice Age?

If so, then it was by no means inconceivable that the survivors might have fled to the Himalayas, pretty much as the Vedic traditions state. They could not have got there by boat, of course. But if a boat had played an essential part in their survival of the flood then it was easy to see how the whole adventure might have been dramatized and remembered in later times as a boat journey.

I could think of several good arguments against this scenario. In no particular order: (1) What right had I to assume that there had been any civilization at all, anywhere, 11,000 or 8000 years ago? (2) Even in the unlikely event that a culture that was a little out of the ordinary had existed at that time, and had so far escaped discovery by archaeologists, why should it have chosen to concentrate itself in the very part of India that would suffer the most extensive post-glacial inundations – when there were so many other parts of India to choose from? (3) Even if both the prior improbabilities are granted and we accept that a civilization was there and was flooded, why did its survivors retreat all the way to the

Himalayas? There was perfectly safe land in between that would have been much more congenial for settlement and for agriculture (presumably an important priority to Manu, who made such a point of 'saving the name of vegetation' and of bringing with him 'all the seeds which have been described of old').[4]

Yet history is full of examples of improbable things that have happened. It was thought improbable in the nineteenth century that a European army could ever be defeated in battle by an African army – until the Abyssinians routed the Italians at Adowa in 1896. It was thought improbable that the *Titanic* would sink on her maiden voyage, but she did. The residents of Pompeii obviously thought it was improbable that their city would be smothered by an eruption of Vesuvius, but it was.

So let's just ask the question and be damned: what if a prehistoric people, with more sophisticated spiritual ideas and a more developed culture than is known to have existed elsewhere in India at this time, had evolved on the California-sized coastal domain between Goa and the Indus delta before it was inundated at the end of the Ice Age? What would have happened to that culture when the deluge came? What sort of story might its survivors have told? And – the heart of the matter really – could it be that story that is expressed in the *Vedas*?

The hypothesis that no one has tested

Even in the twenty-first century, long after it supposedly relinquished its grip, the dead hand of the 'Aryan invasion of India' theory still moulds our perceptions of the *Vedas*. The assumption that there ever was such a thing as an invasion, or even a distinct ethnic group called the Aryas, may have been abandoned, but we've seen in previous chapters how scholars have retained the closely related assumption (albeit within a much wider time-scale) of an overland *migration* of semi-nomadic or transhumant tribes *towards* India *from* somewhere in the general direction of Europe.

Underlying this assumption are other assumptions about the state of development of the migrants (in the early days of 'the transition to agriculture'); about the kind of land that they might have inhabited before coming to India (plains, valleys, mountains); and about the various 'environmental challenges' (desertification, drastic changes in rainfall and temperature regimes, etc.) or 'economic pressures' (overpopulation, competition for scarce resources) that might have compelled them to migrate in the first place.

Because assumptions are free and everybody is entitled to one, the quest for the 'Indo-European homeland' has become something of a scholarly equivalent of the quest for Atlantis. By various highly ranked authorities at various times it has been placed as far afield as the North Pole, Scandinavia, central Europe, southern Russia, central Asia and eastern Anatolia.[5] The suggestion that it might have been within India itself has only very rarely been made and then not by European scholars. Indeed in a survey of 'Recently proposed homelands of the original Indo-Europeans' the Sanskritist David Frawley, along with historian of religion George Feuerstein and Professor Subash Kak of Louisiana State University, found that only one out of ten of the homelands that had been proposed was in India (and that by an Indian academic) while the other nine were all set much further to the north and west.'[6]

Never, so far as I am aware, has a reputable scholar – Indian or otherwise – ever suggested a Vedic homeland located not only *within* India but also *exclusively* on the subcontinent's coastal margins inundated at the end of the Ice Age. Nor for that matter do I know of any reputable scholar who has ever considered oceanic flooding in *any* shape or form amongst the 'environmental' challenges that might have compelled a migration of the 'proto-agricultural' Vedic peoples out of their 'homeland' (wherever that was) and into a wider theatre.

This seems like an oversight, since the origins of settled agriculture and 'civilization' in India – indeed of the very urban lineage that culminated millennia later in the Indus-Sarasvati civilization itself – are now known by scholars to go back *at least as far as 8500 years before the present*. That is the approximate date – 6500 BC – of the first habitation strata at the extraordinary prehistoric town of Mehrgarh in Pakistan's Bolan pass,[7] an archaeological site of great mystery, as we shall see. It is also an early enough date to lie firmly within the time-frame of the three episodes of global superfloods at the end of the last Ice Age.

A maritime culture?

What sort of ancient culture would have chosen to locate itself exclusively in a region so close to the sea that the recurrent cycles of post-glacial floods might have seriously endangered it?

In my opinion only a maritime, sea-going culture – indeed a culture that was dependent on the sea – fits the bill. Moreover, there can be no objection in principle to the existence of such a culture in India 8000 or even 15,000 years ago – since scholars accept that early humans may well have been seafarers as much as 40,000 years ago and that by 10,000 years ago lengthy oceanic journeys and difficult navigational feats were being accomplished by supposedly 'Stone Age' peoples in many different parts of the world.[8]

Yet the assumption continues to be that the founders of the Vedic religion – the forefathers of those who sang the Vedic hymns that have come down to us – were hunter-gatherers or nomads or farmers who only reached India after a long overland journey (itself thought to have been motivated by the demand for more land). Most Western Indologists studying the *Rig Veda* have therefore never seen the need to analyse the many references that its ancient hymns contain to 'seas' and 'oceans'. Indeed, only David Frawley, who is far from the mainstream but whose knowledge of the *Vedas* cannot be faulted, has attempted a serious investigation of this problem:

The modern, generally Western idea is that the *Rig Veda* is the product of a nomadic people invading India from the northwest, who, therefore, could not have known anything of the sea . . . However this idea does not come from the *Veda* itself. It is a preconception used to interpret it. We can only discountenance the many references to the ocean in the *Rig Veda* by redefining the regular Sanskrit terms for ocean presented in it to have meant nothing more than any large body of water, river or lake. If we take them as they appear . . . they fairly clearly show a maritime culture.[9]

Frawley argues that although forests and deserts are also mentioned in the *Vedas*, familiarity with these does not prove non-familiarity with the ocean:

> The scope of Vedic geography is quite large, with mountains, plains, rivers and seas. This allowed scholars to focus on one side of it and become caught up in that one aspect. Yet the oceanic symbolism appears to be the most common.[10]

So much so, Frawley points out, that Ralph Griffith, the translator of the *Vedas* – who did not accept that the Vedic peoples had any experience of oceans – was compelled almost 100 times to translate various Vedic terms as 'ocean' or 'sea', because this is exactly what those terms mean and no alternative translation is possible.[11] Other more ambiguous maritime references, in Frawley's view, were mistranslated or treated simply as metaphors. And while he admits that the word 'ocean' in the *Vedas* is sometimes used as a metaphor (the 'ocean of heaven' for example), he argues persuasively that

> such images do not reflect a lack of contact with the earthly ocean . . . They show great intimacy with the sea, not just as a practical fact but as a poetic image impressed on them by life in proximity to it.[12]

Nor are the maritime images in the *Vedas* confined to seas and oceans. They also include descriptions of sailing, of ships and of ship-borne trade. According to Professor S. P. Gupta:

> There are . . . references to sea, i.e. *samudra*, and traders, i.e. *panis*, engaged in seaborne trade; *navah*, *samudriah*, *sata-aritra*, etc. are such terms which clearly indicate it. Even piracy is mentioned. Attack by unscrupulous people on boats laden with goods in order to capture them finds clear mention in terms like *duseva*, *tamovridha*.[13]

If you listen to the Vedas *you can hear the ocean*

Scholars have long regarded it as legitimate to make firm deducations about the biblical world – its economy, its history, its environment, its sense of geography, its social organization, etc. – by studying the Old Testament.[14] When the same approach is applied open-mindedly to the *Rig Veda*, you can hear the ocean:

All sacred songs have magnified Indra, expansive as the sea. (1, 11, 1)
He [the god Varuna] knows the path of birds that fly through heaven, and . . . of the sea, He knows the ships that are thereon . . . (1, 25, 7)

Like as a watery ocean so doth he [Indra] receive the rivers spread on all sides in their ample width . . . (1, 55, 2)

The Seven mighty Rivers seek the ocean. (1, 71, 7)

O thou whose face looks every way, bear us past foes as in a ship . . . As in a ship convey thou us for our advantage o'er the flood. (1, 97, 7 8)

Come in the ship of these our hymns to bear you to the hither shore. (1, 46, 7)

Yea Asvins [two 'divine intermediaries' or 'guardian angels' frequently referred to in the *Vedas*], as a dead man leaves his riches, Tugra left Bhujyu in the cloud of waters . . . Ye brought him back in animated vessels . . . Bhujyu ye bore . . . to the sea's farther shore, the strand of ocean . . . Ye wrought that hero exploit in the ocean which giveth no support, or hold, or station, what time ye carried Bhyjyu to his dwelling borne in a ship with hundred oars, O Asvins. (1, 116, 3–5)

Ye ever-youthful Ones . . . ye brought back Bhujyu from the sea of billows . . . uninjured through the ocean . . . (1, 118, 14–15)

O Asvins . . . Ye made for Tugra's son [Bhujyu], amid the water floods, that animated ship with wings [sails?] to fly withal, whereon . . . ye brought him forth. And fled with easy flight from out the mighty surge. Four ships, most welcome in the midst of ocean, urged by the Asvins, saved the son of Tugra, him who was cast down headlong in the waters . . . (1, 182, 5–6)

O Maruts [sky and storm gods], from the Ocean ye uplift the rain, and fraught with vaporous moisture pour the torrents down. (5, 55, 5)

Earth shakes and reels in terror at their [the Maruts'] onward rush, like a full ship which, quivering, lets the water in. (5, 59, 2)

May Aja-Ekapad, the God, be gracious, gracious the Dragon of the Deep, and Ocean ... (7, 36, 13)

Let not the sinful tyranny of any fiercely-hating foe smite us as billows smite a ship. (8, 64, 9)

As rivers swell the ocean, so, Hero, our prayers increase thy might. (8, 88, 8)

Ye furtherers of holy Law, transport us safe o'er many woes as over water-floods in ships. (8, 72, 3)

When Varuna and I embark together and urge our boat into the midst of the ocean, we, when we ride o'er the ridges of the waters, will swing within that swing and there be happy. (8, 88, 3)

In both the oceans hath his home, in eastern and in western seas. (10, 136, 5)

Well knoweth Savitar [the personification of the Sun as a life-giving force] where ocean, firmly fixt, o'erflowed its limit. (10, 149, 2)

Although the Vedas are eloquent on their own behalf, the passages above (quoted from the Griffith translation and representative of many other passages not reproduced here) do seem to raise a number of queries.

For example, as well as confirming a knowledge of the relationship between rivers and oceans – with references to rivers seeking the ocean, pouring into it, etc. – we are also presented with the concept of rivers *filling up* the ocean, quite a different matter. When was the last time that human beings are likely to have seen rivers literally filling up the ocean (rather than just flowing into it and making no difference to its level as they do today)? Could it have been the time when the ocean, previously thought to have been firmly fixed in its place, 'o'erflowed its limit' and when only those on board ships were safe from its floods?

And what about the Maruts, the storm gods, who 'from the Ocean . . . uplift the rain, and fraught with vaporous moisture pour the torrents down'? Knowledge of the workings of our planet's great ocean-evaporation-cloud-rainfall cycle is not something that we normally ascribe to proto-agricultural nomads who have never been near an ocean in their lives. But the idea should occur naturally to anyone who lives in sight of a coast – where, at times, the clouds do seem visibly to be drawing up moisture from the sea.[15]

Also amongst the quoted passages are references to the 'eastern and the western seas', and to 'both the oceans'. These references suggest a rather wide-spread maritime experience (at the very least, presumably, of the Arabian Sea to the west of India and of the Bay of Bengal to the east).

Then we must consider the question of all those references to ships – hardly a subject of great interest or relevance to landlubbers but something that we would naturally expect to encounter in the discourse of mariners. And what ships! Ships in which to ride out the 'water-flood' as we have seen . . . ships so formidable and so secure that they are used as a metaphors for safety, security and protection . . . ships, with great sails and banks of oars, that fly across the waves so fast they hardly seem to get wet . . . ships that can brave the billows and pull off the spectacular rescue 'from out the mighty surge' of a man lost overboard and then return him safe to his dwelling on 'the strand of ocean'.

Last but not least, and again as we would expect with a maritime people, there is knowledge both of the dreads and dangers of the sea and of its joys and pleasures. Thus, on the one hand, there is the delightful hymn to Varuna which could only have been composed by someone completely at ease with the motions of the sea and the way that a sailing ship behaves as it skips the ridges of gentle waves or lies at anchor rocking on the swell. On the other hand, these ancient compositions also offer an insight into the awful predicament of the human lost alone in the ocean 'which giveth no support, or hold, or station'. In a few simple words and images they allow us to know the fear and victimization felt by those on board a ship that is being mercilessly pummelled by storm waves 'smiting' it 'like a fiercely-hating foe'. With the same minimal but effective description we learn of the 'terror' experienced by its sailors when an injured ship 'quivers' and begins to 'let the water in'. And then there are such creatures to appease as the 'Dragon of the Deep' – aquatic monsters that would be out of place in fields or mountains but seem quite at home in the fantasies and experiences of a maritime people.

I therefore find much in the *Rig Veda* to recommend the hypothesis that its original composers must have lived close to the sea and been familiar with the ways of the sea over a long period of time. This, at the very least, improves the odds in favour of a possibility briefly raised in previous chapters – namely that the *Vedas* (a superb religious literature with no known parent) might in fact have been the work of the undeniably maritime Indus-Sarasvati civilization (which was long known to have possessed a script but apparently had no religious literature).

In that case the mystery of the origins of the *Vedas* would converge with the mystery of the origins of the Indus-Sarasvati civilization – origins that are receding further and further back into the past with each new turn of the archaeologists' spade at sites such as Mehrgarh and Nausharo in Baluchistan that are already confirmed to be more than 8000 years old.

I remind the reader again that 8000 years before the present is within the time-frame of the great post-glacial floods.

Hidden treasures

We've seen that the scholarly chronology really has no bearing, one way or another, on the ultimate age of the *Rig Veda*. Even the date of 1200 BC that is generally used turns out to be for codification only, with all concerned ready to admit that many of the actual compositions must be older – although exactly how much older nobody knows.

It's also obvious that the *Rig* is a composite work, recension after recension, layer upon layer, and that part of the difficulty of interpreting it probably comes from a jumbling of earlier with later material. Similarly, as Gregory Possehl argues, it looks like a work that underwent a long period of composition, 'when new material was added and older verses were edited and changed'. Then at some point 'this flexibility in composition stopped and the priests defined their text as immutable, not to be changed by one word or even one syllable, and the slightest mispronunciation or deviation from the canon was believed to be a sacrilege'.[16]

So in a sense what the *Rig* presents us with is a dynamic body of scripture and oral history that kept on changing and growing, retaining its dynamism – conceivably even for thousands of years – before being frozen in amber and then preserved eternally in its interrupted form for later study and reflection.

I see no need to get into the argument about when, precisely, that 'freezing in amber' might have occurred, or join with the scholars in bickering about a few hundred years here or there. I'm much more interested in the possibility that layers of extremely ancient oral history and tradition could be concealed alongside the much more recent material that the *Rig* also undoubtedly contains.

The case of the vanishing river

There is a river, spoken of repeatedly in the *Rig Veda*, that vanished into the earth – though not from human memory – thousands of years ago and that was only revealed again by satellite imaging and remote-sensing technology in the latter half of the twentieth century. It is the Sarasvati – the very same ancient river which now gives its name to the Indus-Sarasvati civilization, because large numbers of 'Harappan' and 'pre-Harappan' archaeological sites, dating back at least to the fourth millennium BC, have been discovered close to its former course. The Sarasvati began to dry out towards the end of the third millennium BC and to all extents and purposes had ceased to flow by the early second millennium BC. Even now, however, notes Gregory Possehl,

> there is a river bed, kilometres wide in some places and heavily cultivated, that the people of Haryana refer to as 'Sarasvati'. During the monsoon, parts of this channel carry small amounts of water, most of which is quickly captured for irrigation. Thus the river that today's people call Sarasvati is not entirely dead . . .[17]

There is a bigger question to ask, however: when was it entirely alive? When, for example, was the Sarasvati alive enough to merit these descriptions of it in the *Rig Veda*?

> Sarasvati, the mighty flood . . .[18]

> Coming together, glorious, loudly roaring – Sarasvati, Mother of Floods . . . with fair streams strongly flowing, full swelling with the volume of their water . . .[19]

> She with her might . . . hath burst with strong waves the ridges of the hills . . . Yea, this divine Sarasvati, terrible with her golden path, foe-slayer . . . whose limitless unbroken flood, swift-moving with a rapid rush, comes onward with tempestuous roar . . . Yea, she most dear amidst dear streams . . . graciously inclined, Sarasvati hath earned our praise.[20]

In the footnotes to his 1889 translation, long before the era of satellites and remote sensing, Griffith commented on the use of the word 'she' in the above verse and expressed a certain geographical puzzlement:

> *She*: Sarasvati as a river. The description given in the text can hardly apply to the small stream generally known under that name; and from this and other passages which will be noticed as they occur it seems probable that Sarasvati is also another name of *Sindhu*, or the Indus.[21]

Griffith did not for a moment consider the possibility that the Sarasvati of the *Vedas* might have been a much greater 'stream' in the distant past than it is today (thus justifying the *Rig's* description), and even translated without comment another passage that negates his own hypothesis by speaking of *both* rivers in the same verse:

> Let the great Streams come hither with their mighty help, Sindhu [Indus], Sarasvati, and Sarayu with waves. Ye Goddess Floods, ye Mothers, animating all, promise us water rich in fatness and in balm . . .[22]

Because the *Rig* is in fact clear on the matter, scholars have long since given up the attempt to brush off the anomalous descriptions of the Sarasvati by trying to pretend that the Indus was meant. Nor – because of the perfect conformity between the ancient descriptions of a massive Sarasvati and the latest scientific evidence of a formerly massive Sarasvati – does there seem to be much mileage in writing it all off as hyperbole or poetic licence. Thus Possehl is prepared to concede:

> The image created in the *Rig Veda* for the Sarasvati River is one of a powerful, full-flowing river, not easily reconciled with the literal meaning of the name 'Chain

of Pools'. The discrepancy cannot simply be dismissed; swept under the carpet. It is a good example of how difficult it can be to use the *Rig Veda*, and the Vedic texts generally, as historical sources.

It could be that when the composers of the *Vedas* first came to the Sarasvati it was a river of great magnitude, and these recollections are what we read in their texts. But over time the stream was robbed of its headwaters and dried up, becoming a chain of pools. For whatever reason, the name was changed and Sarasvati is the name that was preserved in the texts; awkward to be sure, but probably not insurmountable. This carries an interesting chronological implication: the composers of the *Rig Veda* were in the Sarasvati region prior to the drying up of the river and this would be closer to 2000 BC than it is to 1000 BC, somewhat earlier than most of the conventional chronologies for the presence of Vedic Aryans in the Punjab.[23]

Possehl understates his case. The 'chronological implications' of Vedic Aryans in the Punjab by 2000 BC are much more than 'interesting'. They are potentially devastating for the academic edifice of Indian literary history founded on a date for the *Rig Veda* of around 1200 BC – and thus for every assumption about Indian prehistory that has ever been based on such a date for the *Rig*. At the very least, if this is what the references to a full and powerful Sarasvati mean, then the possibility of a connection between the Indus-Sarasvati civilization and the Vedic religion must be greatly enhanced.

But the plot thickens . . .

From mountain to ocean

As well as presenting us with images of a powerful, fast-flowing, roaring river (that would seem to be have been historically accurate for the Sarasvati at any time up until the end of the third millennium BC) the *Rig Veda* tells us something else, very, very clearly, that at first sight does not appear to be historically accurate at all. It tells us that the Sarasvati known to the Vedic priests and sages ran unbroken from the mountains to the ocean:

> This stream Sarasvati with fostering current comes forth, our sure defence . . . the flood flows on, surpassing in majesty and might all other waters. Pure in her course from the mountains to the ocean . . .[24]

The problem, in a nutshell, is this: the satellite studies indicate that the last time the Sarasvati flowed into any ocean may have been more than 10,000 years ago – in other words during the final millennia of the post-glacial meltdown. In a paper in the specialist journal *Remote Sensing*, S. M. Ramaswamy, P. C. Bakliwal and R. P. Verma make the following observations about the satellite data from which they draw this very important conclusion about the 'palaeo-Sarasvati':

The occurrence of well-developed tentacles of palaeo-channels in the vast Indian Desert [north-east of the Rann of Kutch] and the final arm of the palaeo-channel as the Ghaggar . . . show that River Sarasvati flowed close to the Aravalli hill ranges [and] met the Arabian Sea in the Rann of Kutch.[25]

The exact epoch in which the Sarasvati stopped flowing 'pure in her course' to the Arabian Sea and began to lose her way instead in the thirsty sands of the Indian Desert is not yet known with any certainty. Nevertheless, Ramaswamy, Bakliwal and Verma are quite satisfied that it was not in the 'Holocene' (the most recent geological age) but in the 'late Pleistocene' – about 12,000 years ago.[26] The same approximate date has also been suggested by Bhimal Ghose, Anil Kar and Zahrid Jussain in a study for the Central Arid Zone Research Institute, Jodhpur,[27] and by Ghose et al. in the *Geographical Journal*.[28] B. P. Radhakrishna of the Geological Society of India similarly indicates the period between 8000 and 6000 BC as the time when melting ice-sheets in the Himalayas, accompanied by a great increase in precipitation, allowed 'Sarasvati and all its tributaries [to flow] in full majestic splendour'.[29] If all these scientists are interpreting the data correctly, then it is only to follow Possehl's own logic to observe that the combination of the remote-sensing evidence and the textual evidence carries an interesting chronological implication: the composers of the *Rig Veda* were in the Sarasvati region at a time when that river still ran all the way to the sea, and this would be closer to 8000 BC than it is to 1000 BC.

It goes without saying that such a date is not just 'somewhat earlier' but dramatically, startlingly, inexplicably earlier than any of the conventional chronologies for the presence of Vedic Aryans in the Punjab. So has the modern science of remote sensing revealed one of the deeper layers of the Vedic palimpsest? Or is it just a fluke that what appears to be an accurate geographical account of the Sarasvati river as it last looked 10,000 or even 12,000 years ago seems to have been preserved in the *Rig*?

Since leading mainstream scholars like Gregory Possehl have already all but accepted the heretical possibility that Vedic civilization was present in the Punjab by 2000 BC (on the basis of the colourful description of a full and turbulent Sarasvati) it seems invidious of them to ignore or sidestep the *Rig*'s equally colourful description of the Sarasvati flowing to the sea. However, this is exactly what Possehl does. Quoting the relevant passage ('pure in her course from the mountains to the ocean'), he admits that 'the Vedic pundits thought that the Sarasvati went to the sea' but explicitly advises students to treat this observation 'critically, not literally'[30] – presumably because to take the observation literally would imply an 'impossibly' early date for Vedic civilization.

Under Vedic skies

There are other passages within the *Rig* – not to do with rivers at all – which also appear to contain material of very great antiquity. These particularly concern astronomical observations of various stars and groups of stars at set seasons – the spring and autumn equinoxes and the summer and winter solstices. Because of a phenomenon known as the precession of the equinoxes, the technical details of which need not detain us here,[31] the constellations seen at these seasons slowly and magisterially trade places, as though revolving on a great belt in the heavens, at the rate of one degree every seventy-two years with a full cycle of just under 26,000 years.[32] Thus, if an ancient text says 'we saw the star such-and-such or the constellation such-and-such rising at dawn at midsummer', then it is possible with modern astronomical formulae to calculate approximately when that observation must have been made.

There are numerous statements of this sort about stars and the seasons in the *Rig Veda* which, if taken at face value, suggest that the Vedic sages made observations of the sky for thousands of years and from time to time added verses or hymns incorporating new astronomical data to the pre-existing compilation. The problem is that the range of dates, going back to the same epoch as the Sarasvati material, has always been thought of as too outlandish to be taken seriously by the majority of scholars.

This is, however, not quite a uniform view. Two of the highly respected Vedic scholars of the late nineteenth century, Professor H. Jacobi and Bal Ganghadar Tilak, were in no doubt that very ancient celestial observations are embedded within the *Rig*. On the basis of astronomical references Jacobi dated most of the hymns to the epoch of 4500–2500 BC.[33] And although Tilak's more comprehensive study found the greatest concentrations of references pointing to approximately the same period, he noted that earlier dates were also flagged.[34] Tilak thought that the most prolific epoch of Vedic composition had been between 4000 and 2500 BC – the 'Orion period' as he called it – in which references are found 'from the time that the vernal equinox was in the asterism of *Ardra* to the time when it receded to the asterism of the *Kritikas* [the Pleiades]'.[35] But he also identified an older sub-layer of Vedic hymns with what he called 'the *Aditi* or the pre-Orion period', stating: 'we may roughly assign 6000–4000 BC as its limit'.[36]

More recently David Frawley has pointed to other references which may carry the *Rig Veda*'s astronomical testimony back even earlier than 6000 BC, indeed 'possibly as early as 7000 BC when the [winter] solstice first entered [the constellation of] *Ashwini*'[37] (i.e., when the winter solstice was at or very near the beginning of the constellation of Aries).[38] Frawley concludes:

> The *Vedas* look back to a time when the winter solstice, the Path of the Gods or northern course of the Sun, began near the beginning of the sign Aries . . . This does

not mean that the hymns which use such symbolism were all composed during this era ... It means that the *Rig Veda* looks back in its mythology to this era as determining much of the symbolism of its Gods and the order of its rituals . . .'[39]

The Era of the Seven Sages

Why should the *Rig* look back in time towards such a distant epoch, roughly between 7000 and 6000 BC, if it does not have some very real and significant connection with that epoch?

Oddly enough, exactly the same question can be asked of a system of calendrical reckoning still in use in some remote highland parts of India today, notably Kashmir.[40] Described at length in the *Puranas*, it is called, suggestively, 'the Era of the Seven Rishis'.[41] Although it operates completely independently of the *yuga* system it does intersect with it at certain points and, indeed, it is this very Saptarishi calendar which provides the referents that pundits have used to calculate the onset of the Kali Yuga to a date of 3102 BC.[42]

To state a complicated matter briefly, the Saptarishi calendar envisages a series of revolving cycles, each of 2800 years duration (much shorter than those of the *yuga* system). And while the *yuga* system has no real beginning or end, the Saptarishi calendar has a definite start date – a very first 'Era of the Seven Rishis'. This start date is 6676 BC.[43] According to John Mitchiner's detailed study:

> The complete cycle wherein occurs the start of the Kali Yuga will commence with Krittika in 3876 BC ... while the preceding complete cycle will commence with Krittika some 2800 years earlier, namely in 6676 BC ... and the following complete cycle will commence with Krittika in 1076 BC ... The date of 6676 BC was in some sense regarded as being a starting point for Indian chronology.[44]

Mitchiner points out that that there is historical corroboration for a seventh-millennium BC start-point for Indian chronology in the works of Greek and Roman authors. Notable examples are Solinus and Pliny (AD 23–79), who said of the Indians that from the time of the founding-father of their civilization to the time of Alexander the Great: 'they reckon the number of their kings to have been 154 and they reckon the time as 6451 years and 3 months'.[45] Alexander entered the Punjab in 326 BC and left in the same year. The implication is that the 'Father' figure (associated with Bacchus in the Roman texts) 'was thought to have reigned in India in 6451¼ + 326 = 6777 BC'.[46]

Since Pliny and Solinus drew on reports sent back by Rome's ambassadors to India's Maurya court,[47] their chronology is regarded as first-hand information and is thought to transmit an accurate representation of ancient Indian beliefs about the past. Mitchiner is therefore intrigued by the fact

> that the date of 6777 BC which is given ... by Pliny and Solinus is only a single century in advance of the date of 6676 BC which is suggested in the Indian texts to

represent the starting point of Indian chronology, as based upon the Era of the Seven Rsis. We may therefore conclude that such a date was indeed regarded – from at least the 4th century BC – as being a starting point of Indian chronology.[48]

Connections

I already knew that it was the ancient function of Rishis – Sages, Seers – to sustain the institution of kingship on earth. It was to this end, and in order to preserve and repromulgate the *Vedas*, that the Seven Sages were said to have travelled to the Himalayas with Father Manu in the time of the great flood.

Now I also knew that an Indian calendar system identified with the Seven Sages, with a father figure and with a line of kings, had a start date of around 6700 BC – a date that fell well within the time-frame of the greatest floods the earth has known in the past 125,000 years.

Last but not least, I could not forget that 6700 BC is extremely close to the date at which the first settlement of the remarkable site of Mehrgarh in Baluchistan took place – a site where the systematic planting and cultivation of cereals and vegetables, as well as systematic animal husbandry, was apparently introduced into India for the first time.

Inevitably I began to wonder if all these things might not in some way be connected.

8 / The Demon on the Mountain and the Rebirth of Civilization

Why humans came to domesticate plants and animals at some particular point in history remains somewhat of a mystery. It seems to be a phenomenon that developed just after the opening of the Holocene in several regions of both the Old and New Worlds. Why it did not occur earlier is not known.

Professor Gregory Posschl, University of Pennsylvania, 1999

Geological record indicates that during Late Pleistocene glaciation, waters of the Himalaya were frozen and that in place of rivers there were only glaciers, masses of solid ice . . . When the climate became warmer, the glaciers began to break up and the frozen water held by them surged forth in great floods, inundating the alluvial plain in front of the mountains . . . No wonder the early inhabitants of the plains burst into song praising Lord Indra for breaking up the glaciers and releasing waters which flowed out in seven mighty channels (*Sapta Sindhu*). The analogy of a slowly-moving serpent (*Ahi*) for describing the Himalayan glacier is most appropriate . . . With the hindsight we possess as geologists, we at once see that the phenomenon described in the *Rig Veda* was no idle fancy but a real natural event of great significance connected with the break-up of Himalayan glaciers and the release of pent-up waters in great floods.

B. P. Radhakrishna, Geological Society of India, 1999

In its study and interpretation of the past, archaeology depends heavily on material evidence produced at excavations. The dependence becomes total when the culture being investigated has left no documents or inscriptions to tell us about itself.

The Indus-Sarasvati civilization was a literate culture, but the archaeological interpretation of it has been strictly limited to excavated material remains and has never been able to draw upon the civilization's own texts. This is because all attempts to decipher the enigmatic 'Harappan' script have failed, and because (at least until very recently) the Sanskrit *Vedas* were regarded as the work of another, later culture and were assumed to have had nothing to do with the Indus-Sarasvati civilization. Well into the twentieth century, this approach simply meant that there was no Indus-Sarasvati civilization. It was not part of the archaeological picture of India's past and was never even contemplated. It was, in other words, as 'lost' as Plato's Atlantis until the material evidence that proved its existence began to surface when excavations were started at Harappa and Mohenjodaro in the 1920s.

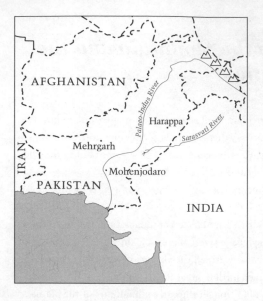

Many more characteristically 'Harappan' sites were discovered during the next half-century of excavations in Pakistan and India but, as luck would have it, none of these were significantly older than Harappa and Mohenjodaro themselves. For a long while, therefore, the prevailing view amongst scholars was that these great cities had sprung up suddenly, with none of the long-term local development, evolution and growth that would normally be expected to underlie such a huge leap forward into organized urban life. For some archaeologists this was proof that the Indus-Sarasvati civilization was an offshoot of what was assumed to be the much older civilization of Sumer in Mesopotamia. Others just took it as an enigma and preferred to get on with the more practical business of understanding the evidence in hand.

The breakthrough came with the start of excavations at 'the village farming community' of Mehrgarh in Baluchistan in 1974. Now joined by Nausharo and a number of other equally ancient sites, its earliest settlement layers are dated to around 7000 BC. Two things are particularly striking about Mehrgarh: (1) from the very beginning its people were efficient and productive farmers; and (2) invaluably for archaeology, the site remained *continuously inhabited* until as late as the first millennium BC.

Moreover, many sites of intermediate age, between Mehrgarh in 7000 BC and Harappa around 2500 BC, have also subsequently been found in the ever-widening Indus-Sarasvati catchment area – and all of them are now regarded by archaeologists as the direct antecedents, represented at various stages of an entirely normal and reassuringly *gradual* process of evolutionary development, of the Indus-Sarasvati civilization itself.

This is often lauded as an example of how archaeologists are open to new facts and at the same time as proof that if you dig deep enough and far enough afield you will sooner or later expose a lengthy phase of evolution behind *any* highly developed civilization. In other words, great cities with a mature and efficient agricultural base *don't* spring out of nowhere, *ever*. They may *seem* to, for a while; but in the end they always turn out to have a background.

Professor S. P. Gupta of the National Museum Institute in New Delhi provides a useful summary of current archaeological thinking on the origins of the Indus-Sarasvati civilization:

It is common knowledge that the history of Indian civilization begins in the Neolithic cultures of the north-western hills and the piedmont regions dating back to the late eighth millennium BC at sites like Mehrgarh on the Bolan River in Baluchistan. Unfortunately . . . Mehrgarh . . . was not put to excavation [until] 1974 . . . However, after the excavations conducted at Mehrgarh our entire perspective of the hill cultures of Baluchistan, hence about the beginning of the Indus-Sarasvati Civilization, has undergone a sea-change.

We now no longer talk of Baluchistan either in terms of a 'corridor' through which Iranian or Turanian[1] cultures passed on their way to the Indus Valley and caused the Indus-Sarasvati Civilization, or in terms of a rugged mountainous region with 'as many cultures as there are now hills'. Instead, we now see the hills and sub-mountainous regions of Baluchistan as the 'nuclear zone' which gave birth to a very long succession of cultures starting from the aceramic Neolithic, datable to the 8th millennium BC, to the beginning of the Indus-Sarasvati Civilization in the mid 4th millennium BC. In other words, what was once thought . . . to be a loose chain of autonomous Neolithic and Chalcolithic cultures inspired by Iranian cultures can now be seen as parts of well-integrated cultural systems operating on an inter-regional basis all along the sub-mountainous regions, skirted by the Kirthar and Suleiman mountains, and the basins of the Indus, Ravi, Chenab, Satluj and the Sarasvati along with their tributaries. It is this system which eventually gave birth to the Indus-Sarasvati Civilization in the plains of the Indus and the Sarasvati.[2]

What archaeology knows

So let's be clear about the mainstream archaeological position today:

1. The 'nuclear zone' out of which the Indus-Sarasvati civilization emerged was the 'submontane' or 'piedmont' region in the foothills of the Hindu Kush, Karakoram and Himalayan mountain ranges.
2. This 'first stirring' of what was ultimately to become the largest urban culture of the ancient world, took place around the end of the eighth millennium BC and the beginning of the seventh.
3. The earliest surviving and most complete site so far found that bears witness to it is Mehrgarh in the Bolan pass, which dates to around 7000 BC.
4. Since Mehrgarh, the story of the evolution and development of the Indus-Sarasvati civilization is well known, with close to 3000 sites excavated. It is therefore extremely unlikely that any more major surprises await archaeologists researching the 5000-year period from 7000 BC down to 2000 BC.

I feel it is important to stress that all these points represent entirely reasonable deductions from the evidence now to hand and that the orthodox scholarly picture of the origins and development of civilization in India since the time of

Mehrgarh is likely to be correct – not only in broad outline but also in most of its finer details. In the absence of texts there will certainly be some aspects of the process that have been misunderstood or not even recognized – particularly matters to do with religious or symbolic expression – but there is no doubt that the archaeologists (these days mostly indigenous teams from India and Pakistan) have done diligent and extensive work and that by and large they have got the chronology and the connections right.

What archaeology doesn't know

The same cannot be said of the period *before* Mehrgarh, as the scrupulously honest Gregory Possehl informs us:

> Almost nothing is known of the time between the late Glacial Age at circa 15,000 BC and the beginnings of Mehrgarh at circa 7000 BC ... The first period at Mehrgarh has fully-developed domestic architecture based on mud brick ... So while Mehrgarh ... is undoubtedly an early village farming community, there is also a sense that the excavations there have not documented the beginnings of this tradition or the beginnings of food production and domestication in the region. It is certainly nothing like a terminal hunting-gathering site with the intensive collection of cereals, pulses and sophisticated hunting. These people were already farmers.[3]

Quite a mystery, in my view!

Possehl explains the 'sudden' appearance of this strangely sophisticated village farming community at Mehrgarh as an artefact of incomplete excavations and is confident that 'the beginnings of food production and domestication in the region' will eventually be traced – within the region itself.[4] Also he relates the level of development that archaeologists have exposed in the first period of Mehrgarh, c.7000 BC, to that of so-called PPNB ('Pre-Pottery Neolithic "B"') sites in the Levant. The PPNB represents the period between 8600BC and 7000 BC, when farming economies first came to dominate the Levant and southeast Anatolia (though there is highly localized evidence of agriculture in the Levant a thousand years before that).[5] Possehl is careful, however, not to imply any causal connection or influence in one direction or the other and admits:

> Why humans came to domesticate plants and animals at some particular point in history remains somewhat of a mystery. It seems to be a phenomenon that developed just after the opening of the Holocene in several regions of both the Old and New Worlds. Why it did not occur earlier is not known.[6]

Why, in other words, did the shift to food production and domestication happen suddenly and specifically *then* – after 12,000 years ago (the date that geologists have set as the end of the 'Pleistocene' glacial age and the beginning of the modern 'Holocene') rather than at some other time? This is precisely the moment, Possehl observes, 'near the beginning of the Holocene, following the retreat of the last great continental glaciers' that the 'origins of settled life in the northwestern sector of southern Asia can be documented'.[7]

We are entering here one of the truly great riddles of prehistory: not just why did humans begin to domesticate plants and animals at a particular moment in the Indian subcontinent, but why did they do so in the first place *anywhere* in the world – and when and where (if anywhere) did this process really begin?

There have been many attempts to understand the driving forces behind the food-producing and domestication revolution in human history:[8]

> Propinquity, overpopulation, cultural readiness, systems feedback, climatic change and stress, population pressure, even a kind of historical inevitability have all been offered, acting alone or in concert with other forces, to explain this revolution.[9]

By the mid-1990s the abrupt climate changes at the Pleistocene/Holocene boundary that accompanied the end of the Ice Age were becoming a focus of special interest to quite a number of researchers interested in the origins of agriculture.[10] McCorriston and Hole (1991) and Bar-Yoseph and Meadow (1995) were amongst many to argue that:

> The origins of agriculture must be viewed in the context of a fluctuating climatic regime that broadened and then constricted areas suitable for productive hunting and gathering and later for cultivation and pastoralism . . . abrupt climate shifts are seen as triggers.[11]

The counter-argument to this position offered by Gregory Possehl in 1999 is persuasive and worth hearing in detail:

> Those who use the 'short-term climatic trigger' hypothesis are essentially proposing that . . . when the climate reduced resources, there was only room for one response: food production with domestication. That may have been a possibility, but there must have been other conceivable reactions to such climatic stress: e.g. migration (probably only partial) to other environments, broadening the adaptation to include plants and animals not already part of the subsistence regime, population reduction, some combination or partial implementation of these solutions.
>
> The San !Kung bushmen seem to have lived through a three-year drought in Botswana and hardly noticed it. Neighbouring Bantu-speaking pastoralist-farmers lost 100,000 cattle, and food for 200,000 farmers and herders had to be brought in as relief. In fact, the hunter-gatherers are reported to have helped the Bantus who came into their area to gather. We learn from this that the human response to drought and natural adversity is difficult to predict. The hunting-gathering adaptation can be extraordinarily resilient and provide very deep, very reliable insulation against adversities of nature.

We should not imagine that the relationship between humans and the natural world involves such unsophisticated responses as those proposed by the climatic and environmental stress models. The notion that early Holocene hunting and gathering populations . . . were just fine until the weather turned bad and that this caused them to domesticate plants and animals is just too simple . . . Moreover, placing the burden of the final shift to food production on a deteriorating climate relies on the notion that the people who 'invented agriculture' were under stress and impoverished.[12]

What the Vedic sages knew (1): flood survivors

In summary, isn't it much more likely that 'the people who invented agriculture' would have been part of a society with the means and time to undertake what scholars have described as 'the leisurely process of domestication', rather than people on the brink of starvation?[13] Such a scenario, at the very least, seems to offer an alternative explanation for why the inhabitants of Mehrgarh were *already* farmers when the first bricks were laid there 9000 years ago: *either*, as Possehl suggests, they evolved their food-producing skills in the submontane belt around the foothills of the Karakorams and the Himalayas earlier than 9000 years ago. In this case we must suppose, as he does, that the traces of this vital evolutionary phase – between sophisticated hunter-gathering on the one hand, and full-scale agriculture and livestock management on the other – still await discovery (despite the admittedly intense archaeological investigation of these areas during the past fifty years); *or*, they evolved their skills somewhere else, in the Levant or another place where archaeologists have not looked, and migrated into the submontane regions of north-west India from there.

Oddly enough, it is the second possibility, not the first, that is favoured by the ancient traditions of India itself. We've seen how these explain that Manu and the Seven Sages retreated to the Himalayas from a place that was not the Himalayas at the time of a terrible oceanic flood, and that they brought with them from their antediluvian homeland not only the *Vedas* but also all the 'seeds' that would be necessary to re-establish permanent food-producing settlements.

The sacred texts also tell us that Vedic society was guided by a brotherhood of these Seven Sages – Rishis, wise men – who oversaw its evolution, established the institution of kingship within it for the general benefit of mankind, and ensured that those kings ruled justly. The fundamental ethic taught by the sages was asceticism – which is indeed the eternal ethic of ancient India for as far back as the memory of man extends – and while recognizing the necessity of a society that could meet all the basic material needs of human beings, it is unlikely that the 'economic policies' of such sages would ever have encouraged overproduction or the growth of luxury.

A relatively simple lifestyle, with few material preoccupations and a focus on spirituality and yogic self-discipline would be more along the lines of what would be expected – a lifestyle very much like that of Mehrgarh 9000 years ago at the end of the Ice Age.

Mehrgarh's story

The Bolan pass connects the western side of the Indus valley with the highlands of Baluchistan and beyond. Mehrgarh nestles at the foot of the pass on the alluvial Kachi plains beside the Bolan river. It is a well-chosen spot: sheltered location; plenty of water; good for agriculture; and good as a transit point for any trade or travel that is going on between the mountains on one side and the lowlands and the Arabian Sea on the other. Mehrgarh is far enough from the coast – about 500 kilometres – to have been safe from oceanic inundation (still an issue 9000 years ago with one further major episode of global superfloods yet to come). Moreover, although rugged, Baluchistan is not high enough to have supported an ice-cap during the last glaciation. Other than occasional unavoidable flooding of the Bolan river, we may therefore speculate that Mehrgarh would have enjoyed a moderate climate threatened by no obvious environmental or geological hazards when it was founded around 9000 years ago.

So it's easy to see why those first inhabitants – who were already farmers and clearly knew a thing or two about agricultural land – chose to settle at Mehrgarh rather than somewhere else. What is not so clear is whether there was any special motive or purpose or plan or inspiration behind the settlement or whether it is just to be seen the way scholars usually portray it – i.e. as part of some general, haphazard 'trend' towards sedentarization and intensified food production in north-west India that had in some vague way been prompted by climate change.

Mehrgarh is extensive, running north to south along the west bank of the Bolan river in a strip up to a kilometre wide and more than two kilometres long – although not all sectors were occupied at the same time. The Period 1 material is clustered towards the northern end of the site, where it is estimated to cover an area of approximately 3–4 hectares. Of this only a very small proportion (75 square metres) has as yet been excavated.[14]

One of the several things about Mehrgarh that I find puzzling, given the generally high level of development and discipline shown by its people from the beginning, is that the first settlers either did not know how to make pottery, or for some inexplicable reason chose not to use it. At any rate no pottery has been found in the earliest occupation layer (Period 1A) dated to around 9000 years ago; it begins to show up in Period 1B, about a thousand years later.[15]

This 'aceramic' phase suggests that Mehrgarh's first inhabitants must have been relatively unsophisticated; however, other evidence – notably concerning their competence as builders – contradicts this view. From the outset, for example, they built with well-made mud bricks of regular size (33 × 14.5 × 7 centimetres)[16] and oriented certain structures to the cardinal directions.[17] Many of the structures are simple dwellings with relatively strong walls made out of two courses of bricks laid side by side and with floors on which the ancient impressions of reeds can sometimes still be made out. The average size of these

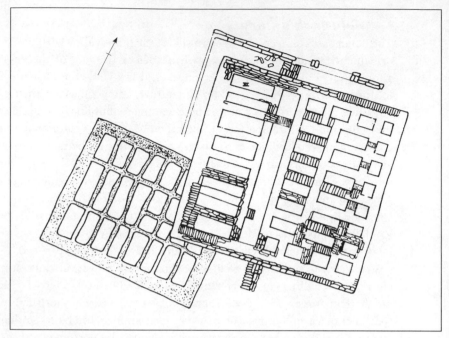

Plan of Compartmented buildings at Mehrgarh. Based on Rao (1991)

dwellings is small, just 5 by 4 metres, and yet they are frequently subdivided into several small rooms:[18]

> Ovens and hearths . . . were usually found in the corners of rooms and signs of their use can be seen as traces of smoke on the plastered walls. One circular oven was lined with bricks and had a dome [like the tandoor ovens of Pakistan and northern India today] which was traced in its collapsed condition.[19]

Some of the Mehrgarh structures bear a striking family resemblance to much later buildings of the Indus-Sarasvati civilization – notably the so-called 'Granary' of Mohenjodaro, which has numerous narrow, cell-like compartments and has been interpreted as a storage facility.[20] The same interpretation has been given by the French archaeological team to 'Structure B' at Mehrgarh, which measures:

> 6.3 metres by 6.7 metres, is oriented north–south, and is made up of six rectangular rooms. Three rooms measure 2.25 metres by 1.5 metres and the other three 3.3 metres by 1.5 metres. No doorways between rooms were found even though there are two, three or four preserved courses of bricks. The walls were made of two rows of bricks . . . The floors of five of the rooms were covered with pebbles (three rooms were completely covered with them).[21]

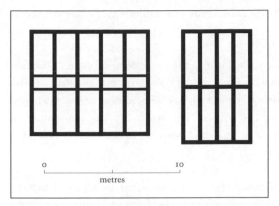

Diagram of cell units at Mehrgarh. Based on Quivron (1991).

There are traces of many other compartmented structures at Mehrgarh from several successive periods in the life of the town. Some of them are preserved up to a height of more than 15 courses of bricks and in none of them have doors or windows been found. The cell units are often no larger than 1 square metre and it is presumed that they must have been entered through their roofs.[22]

So, although they did not make pottery, the very first inhabitants of Mehrgarh did make a range of brick buildings – and these look like the work of people who knew what they were doing. The compartmented structures may not have been 'granaries' – there's no definite evidence – but, whatever they were, they clearly had a function and were built according to some sort of protocol. Such a protocol must, logically, have antedated the foundation of Mehrgarh in order to feature in an already developed form in the oldest habitation layers there.

The first people of Mehrgarh were accomplished farmers, from the beginning, as Gregory Possehl has pointed out. They grew domesticated wheat and barley, still two of the principal food grains of northern India today.[23] In their suite of crops they also included other carefully chosen domesticates: lentils, peas and chickpeas:

> The pulses, annual legumes cultivated for their seed, are an especially interesting group of plants because they are able to fix atmospheric nitrogen in symbiosis with the bacterium Rhizobium found on their roots. They add nitrogen to the soil, rather than consume it, and if these plants are rotated and mixed with the food grains, higher yields are achieved through increased soil fertility.[24]

Because agricultural knowledge like this ought to take centuries, maybe millennia, to build up, Gregory Possehl is not alone amongst archaeologists in his conviction that Mehrgarh does not represent the beginnings of the food-producing tradition in north India but an already developed stage of it.

There is also evidence that the domestication of wild species of goats, sheep and cattle was undertaken by Mehrgarh's first settlers, with great success, as though this was something else that they already understood how to do from experience that they had acquired in another location. Moreover, they seem to have arrived at Mehrgarh with this animal-domestication programme already in mind and in the initial years supplemented their diet with hunting on the Kachi plains (gazelle, swamp-deer, blackbuck, wild pig, elephant, etc.) while the development of their domesticated herds was underway. 'What we see at Mehrgarh,' concludes Possehl,

> is a sequence of events that seems to document the local domestication of animals. The sheep, goats and cattle start out looking wild, and were manipulated ... Over time the potential domesticates came to look like domesticated animals (smaller, with the osteological hallmarks of domesticated beasts) ... The contribution of domestic or 'pro-domestic' stock to the faunal assemblages came to surpass that of other animals early in the aceramic.[25]

I note in passing that the food-production sequences that archaeologists have been able to piece together at Mehrgarh show a good level of fit with the Manu story – which, unlike the Noah story, says nothing about animals on the Ark, but which does tell us that the archetypal Indian flood survivor brought on board, 'carefully preserved and assorted, all the seeds which have been described of old'.[26]

Other materials excavated at Mehrgarh add to our understanding of its first settlers: they used small amounts of copper 'thought to be of the native variety, not smelted'; their primary tools, fashioned from flint, include sickle blades bearing the characteristic sheen imparted when such blades are used to harvest crops; they wove textiles; they made baskets, sometimes waterproofing them with bitumen; they fashioned awls, spatulas and needles from bone; they also possessed a well-developed bead-making industry producing tiny disc-shaped beads in black steatite, barrel-shaped beads in calcite and bangles of polished conch shell;[27] Dentalium shells – long, hollow tubes that form natural beads – have likewise been found in Mehrgarh. These shells are endemic to the Gulf of Cambay.[28] There is also evidence of contact with coastal areas 'and long distance trade networks as attested by the presence of marine shells, lapis lazuli, and turquoise in even the earliest graves'.[29]

Mention of these earliest graves raises another mystery that surrounds the first inhabitants and founders of Mehrgarh. Unlike later occupants of the site, they buried their dead with great care and ceremony. The bodies were carefully arranged in a 'flexed' or embryonic posture, oriented with the head towards the east and the feet towards the west,[30] surrounded by personal effects and sometimes by offerings of food and drink for sustenance on what was clearly believed to be some form of afterlife journey of the soul.[31] Such burials – 166 graves in total – began right at the start of aceramic Period IA and were sustained over more than a thousand years down to Period IIA before gradually being abandoned.[32] A particularly interesting 'side-wall' grave from Period IB contained the remains of an adult male or female

> alongside a very eroded wall. At the feet were a polished stone axe, a large flint core, a piece of a red ochre lump, a bovine bone, and two fragments of a double-pointed

bone tool, a third fragment of which lay in front of the thorax and provides evidence for the intentional breaking of the tool before burial. Also associated were two turquoise beads (as a belt) and other bovine bone fragments.[33]

Ritual burials of this nature, with more or less elaborate grave goods, were conducted again and again in the early years of Mehrgarh. The practice is firmly established at the beginning, with a number of distinct conventions in place concerning the style and orientation of the grave and the types of objects and ornaments interred with the deceased. All of this suggests a complex religious and funerary culture – one that must already have been in use by Mehrgarh's first inhabitants when they established the site.

But in use for how long? And where? Where did the mature religion with afterlife beliefs that we get a glimpse of at Mehrgarh 9000 years ago have its origins?

Although most archaeologists consider the origins of Indian agriculture to lie either in the Near East or in the sub-Himalayan piedmont region, there is one discordant observation about the first settlers which raises doubt. Although the observation was published in 1983 in the peer-reviewed journal *Current Anthropology*, and although its validity has not been challenged by any of the archaeologists working at Mehrgarh, it seems that no scholar has yet got fully to grips with what it could mean.

The observation, arising from research conducted by dental morphology specialist John Luckacs, concerns 'the high frequency of shovel-shaped incisors among the inhabitants of Mehrgarh Period I. *This is a distinctive feature of populations of eastern and southeastern Asia.*'[34] According to Luckacs, the teeth of the Period I inhabitants of Mehrgarh

> contrast strongly with the European dental complex [generally found in India and in the neighbourhood of Mehrgarh from antiquity] and share several dental features common with the Sundadont pattern ... The Neolithic people of Mehrgarh may represent the western margin of South-Southeast Asian phenotypic dental pattern known as Sundadont.[35]

Though passed off in a low-key manner, the implications of this discovery are actually quite extraordinary – since the way overland from south-east Asia to north-west India is very long indeed and since the Sundadont characteristics found at Mehrgarh have never been observed anywhere else in the sub-continent.[36] Moreover, south-east Asia's extensive Sunda Shelf – the home of Sundadont teeth and a continent-sized landmass above water at the Last Glacial Maximum – was submerged in several rapid stages between 16,000 and 11,000 years ago.

The implications seem obvious at first, i.e. that forced out of their original homes (where they had established agriculture, religion, etc.) by the flooding of the Sunda Shelf, the first settlers somehow sailed all the way from south-east Asia to the north-west coast of India then sailed up the Indus and then finally crossed overland to the foot of the Bolan pass, where they founded Mehrgarh. Yet the teeth don't warrant such a large conclusion. They are not pure Sundadont but rather 'share several dental features in common with the Sundadont pattern' and are more likely to have come from some intermediate place – though where that might have been cannot be guessed from the dental evidence alone.

Besides, if flooding is to be cited as the reason why settlers – hypothetically – would have left the Sunda Shelf and sailed to India, then why do we need to look so far afield when we have half a million square kilometres of good land to the north, south and east of Gujerat that was inundated during the same period? Aren't hypothetical flood refugees much more likely to have reached Mehrgarh from there, less than a thousand kilometres away, than from distant Indonesia or Malaysia on the Sunda Shelf?

At the very least, the similarities to the Sundadont pattern seen in the teeth of Mehrgarh's Period I people do seem to rule out any possibility that they had migrated to Mehrgarh overland from the west. As Jonathan Kennoyer confirms:

> They do not have strong morphological relationships to known Neolithic populations of West Asia. On the contrary their dental morphology associates them with a distinctively Asian gene pool.[37]

The mystery of who exactly it was who founded Mehrgarh therefore remains unsolved to this day, and the whole issue has been somewhat neglected – perhaps because of its potential to cause controversy. Scholars also continue to have no idea as to what it was that brought the settlers to Mehrgarh in the first place, though they seem to have arrived with a definite plan and purpose in mind. Last but not least, we should not draw conclusions about the state of mental and intellectual development of the first inhabitants from the rather simple and austere nature of their homes, their tools and their lifestyle. This 'archaeological assemblage' is consistent with the orthodox historical model of how people at the threshold of sedentarized food production should have looked and behaved when they set up their first permanent settlements.[38] But Mehrgarh is also consistent with another model – the model that is suggested in the *Rig Veda* of a society established by yogic sages to meet simple needs with great efficiency, but showing no interest in material luxuries or excesses that might lure humans away from the pursuit of spiritual enlightenment and the immortal destiny of the soul.

Rising seas and melting ice-caps

Mehrgarh Period I takes us back to about 9000 years ago, but the radiocarbon results are frequently confusing,[39] 'the stratigraphy at the site is extremely complex',[40] and because of the margins of inaccuracy that apply to any attempt to date sites as old as this one it is by no means inconceivable that Mehrgarh may in fact be closer to 10,000 than to 9000 years old.[41]

I decided to find out more about what had been happening in the north-western Himalayas in the millennia leading up to the foundation of Mehrgarh, during the catastrophic meltdown at the end of the last Ice Age. It was at this time, immediately following 'the retreat of the last great continental glaciers', as Possehl puts it, that the food-producing explosion began in north-west India. But strangely neither he nor any other major scholar looking at the revolutionary cultural developments of that epoch has considered the possibility that the melting glaciers and rising sea-levels were more than just symptoms of generalized climate change and might in some way have been *directly* connected to the introduction at Mehrgarh of a settled agricultural way of life that was apparently new to the subcontinent.

We've already seen how dramatically India's coasts were inundated after 15,000 years ago. But what about the 'supply' end of the rising sea-level equation? What about the ice-caps, in runaway meltdown as glaciers collapsed, that sent huge floods roaring down from the mountains to fill up the oceans? If there were cataclysmic outburst floods from glacial lakes in North America and in Europe, then why not in the Himalayas too?

Double meanings

The language of the *Rig Veda*, even after its passage from a spoken, oral tradition to a written Sankrit tradition, and after its more recent transformation from ancient Sanskrit into modern and often prosaic English, remains intensely mysterious – filled with symbols, metaphors and riddles that sometimes seem to have been designed to blur the borderline between image and reality, between the symbol and the thing symbolized.

A small but possibly significant example of this concerns the use of certain Sanskrit words in the Manu story with what can only have been the deliberate intention of exploiting ambiguities and innuendoes in their meaning. This is surely the case, argues David Frawley, with the Vedic word for 'boat' – *nau* – which also means 'word' or 'Divine Word', while the word for 'thought', *dhi*, also means 'vessel'.[42] Such puns could offer a rational explanation for the improbable image of a ship marooned in the Himalayas that the Manu story leaves us with. For example, although the words used speak literally of a ship attached to the peak of a high snow-covered mountain, the relevant passages could very easily have been intended to suggest that the 'word' – the revealed 'Divine Word', i.e., the *Vedas* themselves – had been brought to the Himalayas for safekeeping in the memories of the Seven Sages. That would make sense of the caution supposedly given to the refugees by Vishnu that the 'ship/word' should not be allowed to descend from the mountains too fast lest the waters sweep it away. Perhaps the community of Sages that is hinted at in the texts decided to stay for a long time in retreat in the Himalayas, perhaps even for many generations, storing and preserving the seeds of already domesticated varieties of cereals and pulses that they had brought from their homeland until such a moment as they felt it was safe for the 'Word' once again to be promulgated amongst men. In this case we should read the term Naubandhana in the *Mahabaratha* (see chapter 6) not as so much as 'the place of the binding of the ship' but as 'the place of the protection of the Word'.

Another interesting area of ambiguity concerns the many shades of meaning that have been found in the name of the Sarasvati river. Possehl renders it 'Chain of Pools', Frawley reads it as 'She who flows'.[43] Griffith's authoritative translation, on the other hand, is 'The Watery'.[44]

What therefore are we to make of one of the most ambiguous and symbolic ideas that the *Vedas* have to offer: the great myth known as 'the Freeing of the Seven Rivers' that seems to speak of a flood cataclysm in the Himalayas?

What the Vedic sages knew (2): the meltdown in the Himalayas

The *Rig Veda* conjures up a compelling image of a demon in the form of a great dragon, or serpent, that has wrapped itself around the ice-covered mountain ranges that hem in northern India and strangled seven great rivers. The name of the demon is sometimes Ahi but more often Vrtra and the story of how he is slain by the god Indra and of how the seven rivers are freed, is repeated again and again in the hymns of the *Rig Veda*:

> I will declare the manly deeds of Indra, the first that he achieved, the Thunder-wielder. He slew the Dragon, then disclosed the waters, and cleft the channels of the mountain torrents. He slew the Dragon lying on the mountain; his heavenly bolt of thunder Tvastr [the artificer of the gods] fashioned. Like lowing kine in rapid flow descending, the waters glided downward to the ocean ... Indra with his own great and deadly thunder smote into pieces Vrtra ... There he lies like a bank-bursting river, the waters taking courage flow above him. The Dragon lies beneath the feet of torrents which Vrtra with his greatness had encompassed ... Rolled in the midst of never-ceasing currents flowing without a rest for ever onward, the waters bear off Vrtra's nameless body ... O Indra ... thou hast let loose to flow the Seven Rivers. (1, 32, 1–12)

> Indra hath hurled down the magician Vrtra who lay beleaguering the mighty river. Then both the heaven and earth trembled in terror at the strong Hero's thunder when he bellowed. (2, 11, 9)

Thou, slaying Ahi, settest free the river's path. (2, 13, 5)

Indra, whose hand wields thunder, rent piecemeal Ahi who barred up the waters, So that the quickening currents of the rivers flowed . . . Indra, this Mighty One, the Dragon's Slayer, sent forth the flood of waters to the ocean. (2, 19, 2–3)

Thou in thy vigour having slaughtered Vrtra didst free the floods arrested by the Dragon. Heaven trembled at the birth of thine effulgence; Earth trembled at the fear of thy displeasure. The steadfast mountains shook in agitation: the waters flowed and desert spots were flooded. (4, 17, 1–3)

Thou slewest Ahi who besieged the waters . . . the insatiate one, extended, hard to waken, who slumbered in perpetual sleep, O Indra. The Dragon stretched against the seven prone rivers, where no joint was, thou rentest with thy thunder. (4, 19, 2–3)

Indra for man made waters flow together, slew Ahi and sent forth the Seven Rivers, and opened as it were obstructed fountains. (4, 28, 1)

E'en now endures thine exploit of the Rivers, when, Indra, for their floods thou clavest passage. Like men who sit at meat the mountains settled. (6, 30, 3)

Indra . . . ye slew the flood-obstructing serpent Vrtra . . . Heaven approved thine exploit. Ye urged to speed the currents of the rivers, and many seas have ye filled full with waters. (6, 72, 3)

A common explanation that is offered for this myth, both by foreign scholars and by the Indian commentators, sees Vrtra as a symbol for large, dark rainclouds which Indra bursts open with his thunderbolt. The rivers in this scenario are said to symbolize 'streams of rain'.[45] Thus Horace Wilson writes:

the original purpose of the legend of Indra's slaying of Vrtra . . . is merely an allegorical narrative of the production of rain. Vrtra . . . is nothing more than the accumulation of vapour condensed or figuratively shut up in, or obstructed by a cloud. Indra, with his thunderbolt, or atmospheric or electrical influence, divides the aggregated mass, and vent is given to the rain which then descends upon the earth.[46]

It is true that some descriptions of Vrtra in the *Rig Veda* do unambiguously depict the demon as a withholder of rain ('the rain obstructor', 1, 52, 6) and equally clearly associate his destruction with the onset of 'floods of rain' (1, 56, 5) – so any attempt to assess Vrtra's character must take such descriptions into account. Nevertheless, I do not feel that Wilson's elegant allegory satisfactorily explains certain key features of the myth outlined in the passages cited above:

the constant references to the 'freeing of the Seven Rivers' (if 'rivers' are really 'streams of rain', then why are there just seven of them?); the description of pieces of Vrtra's body being carried away in the waters, 'rolled in the midst of never-ceasing currents' (surely more consistent with what is seen during powerful floods than it is with rainstorms?); the clear statement that the released waters cut channels in the mountains and descend in rapid flow to the oceans; the way that the flooding of 'desert spots' is connected to this downrush of waters from the mountains; and most of all the way that the released waters are said to flow 'above' the Dragon Vrtra as he lies abased 'beneath the feet of torrents' (whereas, if he were merely a rain-cloud dispersed by Indra's thunderbolt, one would have expected what was left of his 'body' – the remaining wisps of cloud? – to have been *above* the freed waters, not beneath them).

Uncomfortable with Wilson's pure symbolism for precisely these reasons, other scholars have offered a more literal interpretation of the myth in which the rivers are the seven physical rivers of ancient north-west India – an area that is indeed referred to as early as the *Rig Veda* as the 'Land of the Seven Rivers'.[47] The rivers concerned are generally presumed to be the Indus, the Sarasvati and the five rivers of the Punjab[48] which 'often entirely dried up in the summer'.[49] According to this variant, Indra is 'the god of the rainy season' who calls the rivers back to life and Vrtra is the demon of summer drought.[50]

But there are problems here too. Most significantly, Indra's 'exploit of the Rivers' is not portrayed in the *Rig Veda* as an annually or seasonally recurring event but as a one-off, unrepeatable event of awe-inspiring proportions that took place a long time ago (so long ago it is described as Indra's first manly deed and the poet remarks with wonder that 'e'en now' its fame endures). When I read the accounts in the *Rig* I find it impossible to convince myself that the sages of remote antiquity who composed these hymns were talking about something that happened every year when they described this epic conflict that took place in the snow-covered northern mountain ranges. On the contrary, the texts leave no doubt that when Vrtra was slain he was slain for ever: 'When Indra and the Dragon strove in battle, Maghavan ["Lord of Bounty", an epithet for Indra] gained the victory for ever' (1, 32, 13).

So I think there's room for a third scenario – one that the scholars haven't looked at.

Ice dragon

Suppose that Vrtra symbolizes glaciation – more specifically the Himalayan ice-cap, which would have been greatly extended at the Last Glacial Maximum and might indeed at times have choked off the headwaters of the Seven Rivers. If so, then it can be seen that the myth is quite consistent with the tumultuous collapse of ice-caps all around the world at the end of the Ice Age – and with what one might have expected to witness in the Himalayan and Karakoram mountains at this time:

- Before the heroic intervention of Indra, the demon Ahi in his lair high in the mountains is explicitly described as being 'extended' and 'stretched' against the seven prone rivers' and also as being locked in a 'perpetual slumber' – a suitable metaphor for an ice-cap in deep-freeze.
- Indra's slaying of Ahi/Vrtra is compared to the sudden opening of obstructed fountains.
- The floods pouring down off the mountains are incredibly strong – strong enough to cleave rocks and ridges asunder as they carve out their paths.
- Large chunks of the central dome of the ice-cap get flushed out with the powerful onrushing floods ('Rolled in the midst of never-ceasing currents flowing without a rest for ever onward, the waters bear off Vrtra's nameless body').
- Filled with jostling icebergs, the waters are turbulent and noisy, like stampeding herds of cattle, as they foam out of the rocky gorges and rush towards the ocean.
- The dramatic effects of the meltdown include tremendous descending waves ('glacier waves', see chapter 3) that form in the vast pools of meltwater on the surfaces of large glaciers ('There he lies like a bank-bursting river, the waters taking courage flow above him. The Dragon lies beneath the feet of torrents').
- Gigantic earthquakes are unleashed as the burden imposed by the ice-cap on the land beneath is suddenly reduced; in the Himalayas and Kara-korams, which are anyway amongst the fastest-rising regions on earth, such isostatic rebound might have been amplified by normal mountain-building processes ('the steadfast mountains shook in agitation').
- Distant desert areas far downstream are flooded.
- The floods are of a nature to fill 'many seas'.
- After the catastrophic events that denuded the Himalayas and the Kara-korams of much of their Pleistocene ice-cover and that perhaps left them looking very much as they do today, the Seven Rivers that previously had been dammed up or frozen at their headwaters by the expansion of the ice-cap were set free and began to flow again in their normal courses.

Plausible? Some of it, perhaps. But this is one of the problems with the game of interpreting myth: the meaning ascribed may be more in the eye of the beholder than anywhere else . . .

Still, after reviewing the whole Vrtra mystery, I thought it made sense to look more closely into the scientific literature about the Himalayas. What did the palaeoclimatologists say had been happening there during the 10,000 years after the end of the Last Glacial Maximum when every other ice-covered area in the world, as far afield as New Guinea, the Andes, North America and northern Europe, was simultaneously experiencing the danger and the drama – but also the promise for a better future for mankind – of a ferocious meltdown?

Flying through ELA Land

Scientists studying ice-caps and glaciers make much use of the acronym ELA, which stands for Equilibrium Line Altitude, 'the altitude on a glacier at which annual accumulation [of ice] is exactly matched by annual ablation [melting], so that the net mass balance is zero'.[49] As one might expect, numerous studies have confirmed that ELAs across the Himalayan and Karakoram mountains were significantly lower at the Last Glacial Maximum than they are today (i.e. the ice-coverage descended further into the valleys and the ice-cap was therefore deeper – although opinions differ somewhat as to exactly how much deeper). A few examples from the literature are sufficient to illustrate the consensus on this matter:

> It is evident that there is still considerable room for disagreement on the glacial succession in the north-west Himalaya and Karakoram, and even on the details of the events during the last Pleistocene glaciation. This is illustrated by the continuing divergence of opinion on the ELA depression during the Last Glacial Maximum, the maximum (of Haserodt) being 1250 metres and the minimum (of Scott) being 720 metres ... Despite the apparent diversity in the estimates of ELA-depression-values for the Last Glacial Maximum, values for the north-west Himalaya, Greater Karakoram and Swat Kohistan tend to cluster in the range 800–1000 metres.[52]

> For the Dunde ice cap on the northern flank of Tibet ... we have interpreted a temperature decrease of four to six degrees Centigrade and consequent lowering of equilibrium line altitude (ELA) in the range of 700–850 metres during the last glacial stage.[53]

> Estimated maximum depressions of ELAs range from approximately 1100 metres below present values (Swat Kohistan and the Hunza Valley in the Karakoram range) to 600 metres (southern side of the Zanskar range).[54]

> Depressions of ELA were calculated from glacial geological mapping of the former extent of the glaciers. Maximum ELA depressions were 700 metres below present values in the Ningle Valley, 750 metres in the Liddar Valley, and 800 metres in the Sind Valley.[55]

> ELAs were reconstructed for the Last Glacial Maximum advance ... The results show an ELA depression of approximately 1000 metres below present values in the Ladakh range.[56]

One would not go far wrong by saying that the average lowering of ELA over the Himalayan/Karakoram ice-cap at the Last Glacial Maximum was probably of the order of 750 metres – i.e. about three-quarters of a kilometre.

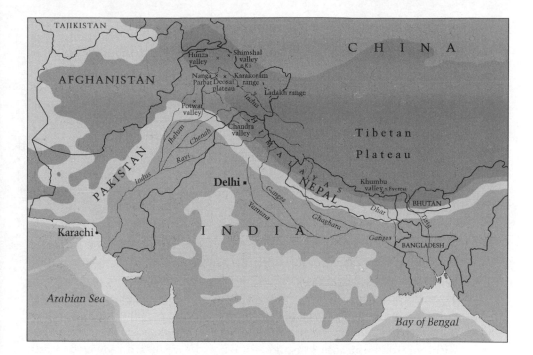

Now what does this mean in practical terms? Writing in *Science*, Nicholas Borozovic, Douglas Burbank and Andrew Meigs helpfully provide an answer to this question with special reference to the north-western Himalayas and the Karakorams at the Last Glacial Maximum:

> small changes in ELA significantly increase the percent surface area covered by glaciers when the region lies at an altitude similar to the ELA . . . For deeply incised mountainous regions (Nanga Parbat, the Karakoram, and Haramosh and Rakaposhi) there is an approximately linear relation between ELA lowering and the area above the snowline. Modern-day glaciers in the Karakoram are extensive; conditions at the LGM would have nearly doubled the area above the snowline available for their accumulation areas. For the Nanga Parbat and Haramosh and Rakaposhi regions, LGM conditions could have nearly quadrupled the area above the snowline . . . For the plateaus and dissected plateaus, the effect of lowering ELAs is even greater on the landscape. The Deosai Plateau is unglaciated today but would have been blanketed by an ice sheet during LGM conditions.[57]

Years ago, so long ago it seems like a former incarnation, I flew in a five-seater Alouette helicopter over the bleak high plains of the Deosai plateau above Skardu. At one edge of the plains, which, if not permanently glaciated, were certainly deeply blanketed in snow, there is a lake, frozen most of the year round, called Shershar. Hovering over it in the thin air, we could see the distant peaks of the surrounding mountains, ice-bound, marching away in all directions.

It was March or April of 1981, I was still thirty and I was working with Mohamed Amin – a great friend and a great photographer who much later tragically lost his life in the Ethiopian Airlines hijack of 1996. We spent an exhilarating, nerve-racking fortnight flying around the Karakorams in the Alouette, which was owned by the Pakistan Army and piloted by a lieutenant colonel and a major with impressive handlebar moustaches. We were based in Gilgit, in the shadow of the 7,788 metre shark-tooth peak of Rakaposhi, and every day we went out and flew at ridiculous altitudes through the mountains – sometimes plunging down below the snowline into secret, verdant valleys – so that Mo could get the spectacular photographs that would later feature in our book *Journey Through Pakistan*.[58] On the third morning, in all seriousness, I wrote out a will and left it with my passport in my hotel room.

The Alouette had a service ceiling of around 3300 metres, but we frequently struggled and clattered up to over 5200 metres – the pilots said it was a training exercise for them – and then just hung there suspended amidst the glaring white wilderness under the bright blue sky. It was a very macho thing to do with no oxygen on board and the machine wasn't really built for it, but it brought home to me, more clearly than any other experience could possibly have done, how immense these mountains are. When we flew by Rakaposhi at 5000 metres, with our rotors almost brushing its flank, its peak still towered nearly 3000 metres above us. And within a 160 kilometre radius of Gilgit there are 100 peaks over 5486 metres high, including K2 which, at 8610 metres, is the world's second-highest mountain.[59]

In an area of such superlatives it is hardly surprising that the north-west Himalayas and the Karakorams contain some of the longest valley glaciers in the world outside of the polar regions[60] – and these huge glaciers coil through the ranges like ancient serpents of myth, their backs ridged with serried ranks of ice-scales.

At the Last Glacial Maximum they may have been up to four times as massive and the whole landscape surrounding them would have been locked and frozen in deep ice-cover extending to altitudes of 4000 metres – as much as a kilometre further down than today.[61]

Imagine what must have happened when all that ice melted down.

So, what did happen?

The scientific literature covering various effects and phenomena of the Ice Age in the Himalaya/Karakoram area is growing fast – as is interest in this subject amongst palaeoclimatologists and geologists.

One important issue that has been much debated concerns the glaciation and deglaciation of the Tibetan plateau at various periods during the past 2.5 million years. It has even been controversially suggested that the geologically recent uplift of Tibet as a result of mountain-building forces in the Himalayas between 3 and 2.5 million years ago may have been the specific trigger that set the

Pleistocene Ice Age in motion 'through the effects this had on the Earth's rotation as well as on the circulation of ocean and atmosphere'.[62]

A related area of active debate concerns the overall extent of the Himalayan ice-cap. Here, explains Edward Derbyshire of the University of London's Quaternary Research Centre, the broad measure of agreement that exists on the magnitude of the ELA depression at the Last Glacial Maximum:

> is not matched by agreement on the regional extent of the last glaciation which has been described, at one extreme, as an ice sheet of continental scale and, at the other, as an Alpine glaciation in the Karakoram-northwest Himalayan region with some trunk valleys remaining unglacierized.[63]

How is it possible for serious and respected scientists, reporting their studies in peer-reviewed journals and working from essentially the same evidence base, to have come up with such divergent views about the extent of the Himalayan glaciation? 'The explanation of the apparent paradox,' suggests Derbyshire, lies in the difficulty of interpreting the chaotic geological record in this extremely mountainous region:

> The world's greatest relief is a locus of enormous geodynamic energy consisting of a complex interplay between tectonics and glacial and fluvial erosion associated with widespread and frequently catastrophic mass wasting. One obvious product of such a situation is the problem of reliably discriminating between diamictons deposited by glacier ice and those laid down by other processes. The two suites of processes are frequently intimately related, posing a recurrent challenge to those attempting to establish limits of past glaciations.[64]

'Diamicton' is a general term used to describe a mixture of sand, clay, silt or gravel that is laid down by various geological processes – notably the forces of flowing rivers, or moving glaciers, or lakes draining catastrophically. Derbyshire's point is that where ongoing geological activity results in a continuous mixing up and redeposition of the materials being studied – as is most definitely the case in the Himalayas – then there is obviously going to be uncertainty over the extent of glaciation in the region at any particular moment in the past.

The *range* of the uncertainty surrounding the extent of the ice-cap at the LGM is, however, surprisingly large – since there is all the difference in the world between 'an ice-cap of continental scale' on the one hand, and a regional 'Alpine glaciation' on the other. Moreover, this uncertainty seems even greater when it comes to immediately post-glacial events. Indeed, although a great deal is known about the cataclysmic meltdown of other ice-caps in this period, I was surprised to discover that the literature has relatively little to say about what happened in the Himalayas after the LGM.[65]

Before and after

Scientists have been able to pick up traces of at least one cataclysmic melting event that took place in the area *before* the LGM. It is another measure of the uncertainty of the data available for study that the date-range offered for this flood is very wide – it could have happened any time between 28,000 years ago and 43,000 years ago.[66] Fortunately, its imprint on the landscape has not been as badly obliterated and jumbled as those of earlier and later floods and geologists have narrowed its location to the Upper Chandra valley in the Lahul Himalaya. Using landforms and sediment data, Peter Coxon, Lewis Owen and Wishart Mitchell, writing in the *Journal of Quaternary Science*, conclude that former glacial Lake Batal – which had backed up the Chandra valley for about 14 kilometres – suddenly burst through its ice dam. When it did so it released almost one and a half cubic kilometres of water into the valley in less than a day: 'This cataclysmic flood was responsible for major resedimentation and landscape modification within the Chandra valley.'[67]

Further striking but unfortunately undatable evidence of colossal ancient outburst floods is provided by the presence of numbers of large boulders scattered across the Potwar plateau – so-called 'Punjab erratics' – which geologists now believe were 'carried down the Indus valley by catastrophic flooding, probably in iceberg rafts'[68] The traces of violent outburst floods long *after* the post-glacial meltdown was over, have also been widely recognized and there are a number of eye-witness accounts. In 1959, for example, there was

> a sudden outburst from an ice-dammed lake in the Shimsal valley which caused a flood wave of approximately 30 metres to be produced, destroying the village of Pasu at the confluence with the Hunza River, 40 kilometres down-valley.[69]

Similarly, when a moraine-dammed glacial lake in the Khumbu area of eastern Nepal called Dig Tsho burst on 4 August 1985, the consequences for the region were catastrophic:

> The destruction of a newly-built hydroelectric power plant, 14 bridges, about 30 houses, and many hectares of valuable arable land, as well as a heavily damaged trail network resulted from 5 million cubic metres of water plummeting down the Bhote Kosi and Dudh Kosi valleys. The breaching of the moraine was triggered by wave action following an ice avalanche of 150,000 cubic metres into the lake. The surge had a peak discharge of 1600 cubic metres per second; 3 million cubic metres of debris were moved within a distance of less than 40 kilometres.[70]

The most spectacular event, however, was undoubtedly the great Indus flood of 1841 – a deluge of near biblical proportions which, like the return of the waters of the Red Sea after the Hebrews had passed safely through to the other side, destroyed a vast army.

The first step was an earthquake in late 1840 or early 1841. The earthquake caused the collapse of the Lichar Spur, part of the flank of Nanga Parbat, which blocked the Indus valley to a depth of 300 metres, strangled the downstream flow of the Indus to a trickle for six months and caused a lake 60 kilometres long and 300 metres deep to back up behind it. When the blockage was breached in June 1841 a gigantic flood wave was released. The wave raced downstream along the (by then almost dry) course of the Indus at a terrifying pace and fell upon a Sikh army that was camped on the Chach plain near Attock, 400 kilometres downstream.[71] Eye-witnesses later reported that:

> A wall of mud, many tens of metres high, rushed down the watercourses. Those people not fast enough to reach the high ground, numbering several thousand troops and camp followers, were lost. Trees were uprooted, buildings destroyed, artillery guns scattered, and farmland washed away. Large areas of the Vale of Peshawar were flooded as the various tributaries banked up against the Indus floodwaters.[72]

Today there is increasing awareness of the dangers posed by outburst floods specifically related to glaciation. It has been pointed out, for example, that more than thirty glaciers in the Karakoram mountains are presently in a position to 'form substantial dams on the Upper Indus and Yarkand river systems. Many more interfere with the flow of rivers in a potentially dangerous way.'[73] According to Kenneth Hewitt of Wilfred Laurier University, Canada:

> A particularly large and dangerous dam occurs where a glacier enters and blocks a major river valley of which it is a tributary . . . In one region of the world, . . . the Karakoram Himalaya and neighbouring ranges, there has been a substantial number of these main valley glacier lakes in modern times. Outbursts from a series of dams . . . between 1926 and 1932 brought devastating floods along more than 1200 kilometres of the Indus. Some even larger landslide dams and outburst floods occurred here in the nineteenth century and an exceptional concentration of surging glaciers has been found. Some of the latter have formed main valley ice dams . . . Thirty-five destructive outburst floods have been recorded in the past two hundred years.[74]

Stocktaking

There are a few details that are worth holding on to.

The Equilibrium Line Altitude of glaciation in the Himalayas at the LGM was about three-quarters of a kilometre or more lower than it is today.

The ice-cap at the LGM was much more extensive than it is today – although there is no agreement over exactly how much more extensive.

There have been catastrophic outburst floods from the Karakorams and the

Himalayas in the past, floods that reshaped landscapes, floods that carried icebergs full of huge impacted rocks all the way down to the Potwar plateau.

Such outbursts continue to occur and even in the much reduced conditions of today's glacial cover they can produce floodwaves 30 metres high capable of smashing whole villages to smithereens and destroying armies.

The region is uniquely plagued by the particularly dangerous and rare phenomenon of its main river valleys being dammed by gigantic landslides or by the encroachment of glaciers – a sure recipe for catastrophic outburst flooding.

Paradoxically, despite the evidence for catastrophic outburst floods before the Last Glacial Maximum, as well as in much more recent times, the literature pays scant attention to the issue of outburst flooding in the Himalayas during the 10,000 years after the LGM.[75]

But this shouldn't prevent us from asking a few common-sense questions:

1. If main river valleys are threatened by glaciers today, and if even a giant river like the Indus can be blocked for six months, then isn't there every probability that the threat would have been much bigger and much worse under LGM conditions?
2. Is it unreasonable to speculate – as the *Rig Veda* has been telling us all along – that there could have been a time, within the memory of man, when some of the great rivers of north India were indeed choked off, most likely by giant glaciers entering and blocking their main valleys up in the Karakoram and Himalayan ranges? If so, then those glacial dams would eventually have burst asunder and the rivers chained up within them would have been set free once again . . .
3. Last but not least, is it so far-fetched to wonder if such a sequence of events might have inspired the great Vedic myth of Indra's slaying of Vrtra with its specific symbolism of the freeing of the Seven Rivers?

Probably no more far-fetched than the more orthodox 'cloud-demon' and 'drought demon' ideas, but hardly foolproof as a theory. For example, there's the absence of evidence of flooding in the Himalayas after the LGM – but that means very little given the state of the geological record (and the level of disagreement amongst geologists on the actual extent of the maximum glaciation).

More seriously there is the other 'face' of the Vrtra myth – the clear association that some of the hymns make between the presence of the Dragon and the withholding of rain on the one hand, and between the slaying of the Dragon by Indra and the return of the rain on the other.

How is that to be explained if Vrtra is a symbol for glaciation?

The dry and the wet

Sediments in ocean-bottom cores taken in the Arabian Sea off the south-west coast of India contain pollen traces that tell us about the types of vegetation that grew on the subcontinent at different periods going back to the Last Glacial Maximum – and since vegetation cover is determined by climate, reliable deductions can be made from these pollen records about India's climate in past epochs.

The Arabian Sea cores demonstrate that there was a period of extreme cold and aridity in India between 25,500 years ago and 21,500 years ago.[76] This period is described by Elise Van Campo of the *Université des Sciences et Techniques du Languedoc* as 'the LGM interval'[77] and coincides exactly with other indications from around the world of the duration of LGM conditions (i.e., the Last Glacial Maximum was not a peak reached for a very short time, but rather a *plateau* of extreme glaciation that was sustained, in India at any rate, for 4000 years). When warming did set in, it set in quickly and between 21,500 years ago and 13,000 years ago the Indian climate did a 180-degree flip from cold and arid to warm and wet:

> The major fluctuations of the Indian monsoon climate are characterized by two extreme periods, a very arid period around [25,500 to 21,500 years ago] and a very humid period culminating at [13,000 years ago] ... The climate conditions of the LGM interval were greatly different from modern conditions. The southwest monsoon, which produces a strong asymmetry between the western and the eastern coasts of the Arabian Sea, was considerably reduced and arid conditions were very similar on both sides) ...[78] (Carbon-14 dates in original text replaced with approximate equivalents in calendar years.)

What this would have meant in the Himalayas between 25,500 and 21,500 years ago was 4000 years of deep freeze as the ice tightened its grip on the valleys and the headwaters of the rivers in the mountains.

Then at the peak of the LGM interval, some time soon after 21,500 years ago, the phase of warm, wet climate in India abruptly kicked in. Back to the Arabian Sea cores, which demonstrate:

> an increase of monsoonal rainfall as early as about [19,700 years ago] at 10 degrees north and at [18,500 years ago] at 15 degrees north. This period ... culminates synchronously at [13,500 years ago] at 10 degrees and at 15 degrees north and is considered as the period of the greatest abundance of monsoonal rains.[79]

Worldwide, we know that the period of 14,000 to 13,000 years ago, which coincides with the peak of abundant monsoonal rains over India, was marked by violent oceanic flooding – in fact, the first of the three great episodes of global superfloods that dominated the meltdown of the Ice Age. The flooding was fed not merely by rain but by the cataclysmic synchronous collapse of large ice-masses on several different continents and by gigantic inundations of meltwater pouring down river systems into the oceans.[80]

If this was happening in other glaciated regions such as North America and northern Europe between 14,000 and 13,000 years ago, then things are unlikely to have been very different in the Himalayas, and it seems safe to assume that there must have been episodes of exceptionally powerful outburst flooding and that all the great rivers from the Indus to the Ganges would at that time have been in full flow.

So is 14,000 to 13,000 years ago a candidate epoch for the events recounted in the *Rig Veda* as the slaying of Vrtra and the freeing of the Seven Rivers?

The answer has to be no – simply because the previous 7000 years had witnessed a continuous worldwide increase in temperature and because 14,000 to 13,000 years ago was the peak and the climax of this long, humid warming phase in India. As such, it is most unlikely that the glaciers in the Karakorams and the Himalayas would have been surging or advancing so as to block or 'enchain' rivers in the way that the *Rig* seems to describe. On the contrary, everything suggests that the flow of the rivers should have been uninterrupted from the end of the cold, dry LGM interval 21,000 years ago until the clear end of the humid phase that shows up in the cores at around 13,000 years ago.

Moreover, the Vedic myth portrays the slaying of Vrtra as being followed by the release of the waters – both rivers and rain. This is very clear and, in a way, the point of the whole thing. But that was *not* what happened.

A Dragon called the Younger Dryas

What happened, at around 13,000 years ago, was that the long period of uninterrupted warming that the world had just passed through (and that had greatly intensified, according to some studies, between 15,000 years ago and 13,000 years ago)[81] was instantly brought to a halt – all at once, everywhere – by a global cold event known to palaeoclimatologists as the 'Younger Dryas' or 'Dryas III'.[82] In many ways mysterious and unexplained, this was an almost unbelievably fast climatic reversion – from conditions that are calculated to have been warmer and wetter than today's 13,000 years ago,[83] to conditions that were colder and drier than those at the Last Glacial Maximum, not much more than a thousand years later.[84]

From that moment, around 12,800 years ago, it was as though an enchantment of ice had gripped the earth. In many areas that had been approaching terminal meltdown full glacial conditions were restored with breathtaking rapidity and all the gains that had been made since the LGM were simply stripped away:

> Temperatures . . . fell back on the order of 8–15 degrees centigrade . . . with half this brutal decline possibly occurring within decades. The Polar Front in the North

Atlantic redescended to the level of Cabo Finisterre in northwest Spain and glaciers readvanced in the high mountain chains. With respect to temperature the setback to full glacial conditions was nearly complete . . .[85]

For human populations at the time, in many except the most accidentally favoured parts of the world, the sudden and inexplicable plunge into severe cold and aridity must have been devastating. And in the Karakoram-Himalayan region, as in other glaciated areas, it is very likely that it was accompanied by a significant readvance of the ice-cap that previously had been in recession for some 7000 years.

Is it possible that that this hypothetical readvance of the Himalayan ice-cap between 12,800 years ago and 11,400[86] years ago could be the event personified in the *Rig Veda* as Vrtra the Dragon, the enchanter, the great magician, 'who barred up the waters'?

Since the slaying of Vrtra resulted in the release of the waters to flow to the sea, it obviously made sense to find out if there was evidence of sudden large-scale meltwater floods off the mountains shortly after 11,400 years ago when the 'climate switched back to warm, moist Holocene conditions, over only a few decades'.[87]

Salt and freshwater

I did find evidence of floods. It was in another set of cores taken off the Indian coast. According to a report in *Nature* by a team of Australian scientists:

> Microfossil, sediment and oxygen-isotope studies of deep-sea cores from the Bay of Bengal and northern Arabian Sea have revealed strong contrasts between high late Pleistocene and low early Holocene salinity values, indicative of major changes in runoff from the large rivers of southern Asia.[88]

Some definitions: salinity values measure the 'saltiness' of the sea, so 'high salinity values' mean a saltier sea and low salinity values mean a less salty sea – i.e, a sea with more freshwater in it. The Pleistocene–Holocene boundary is set, arbitrarily, at 12,000 years before the present. 'Late Pleistocene' is loose language but generally means the few thousand years before 12,000 years ago. 'Early Holocene' is loose language too but generally means anywhere between 12,000 years ago and 10,000 years ago.

Why were India's seas so salty just before 12,000 years ago? The most likely explanation is that the flow of the great rivers draining the Karakoram-Himalayan region had virtually ceased because of the advance of glaciers into their main valleys during Dryas III – pretty much as the *Rig Veda* tells us ('Ahi who besieged the waters . . . the insatiate one, extended, hard to waken, who slumbered in perpetual sleep'). Likewise, the explanation for the low salinity values that suddenly appear soon after 10,000 years ago is a sudden gigantic inrush of freshwater to the Arabian Sea and the Bay of Bengal on a scale that could have been caused by the breaching of ice dams in the Himalayas, the freeing of rivers pent up behind them, and the flushing out of parts of the ice-cap. ('The Dragon stretched against the seven prone rivers, where no joint was, thou rentest with thy thunder.' 'Like lowing kine in rapid flow descending, the waters glided downward to the ocean.')

All in all, therefore, even in the absence of direct evidence of flooding of the type described in the *Rig Veda*, the indirect evidence from the ocean cores does suggest that such floods must have occurred and that they could have followed a period, however brief, when the main rivers of northern India had in fact dried up. So the hypothesis that the Vrtra story in the *Rig Veda* might be describing glacial outburst floods remains a reasonable one.[89]

Conveniently, the ambiguity over Vrtra's character is also removed. Now he is at one and the same time an ice dragon blocking the flow of the mighty rivers and a rain-withholding demon whose period of grim enchantment over the Himalayas is brought to an end not only by the freeing of the rivers but also by the abrupt return to heavy rains and warm, wet conditions that we know followed the Younger Dryas.[90]

All this is speculation, of course, and implicit in it is a deeply heretical assumption – the assumption that the sages who composed at least some of the verses of the *Vedas* could have been in the Himalayas 12,000 years ago to witness the end of the Younger Dryas cold advance and to commemorate it as Indra's victory over Vrtra. This does not fit at all with the much later date that scholars habitually assign to composition of the *Rig Veda* – but then neither do the accounts of a full and turbulent Sarasvati that the *Rig* provides us with and that also seem to sketch out the archaic geography of 10,000 or more years ago.

Mehrgarh's yogic ethic

Growing up in the industrialized and now the electronic world, dominated as it has been by the rival material philosophies of capitalism and communism, we automatically imbibe from schools, peers and parents the idea that civilization is something that man invented in order to meet his material and economic needs. This is why, when archaeologists look for the origins of civilization, they look for the material and economic forces that might have driven hunter-gatherers to become farmers and to create the first permanent village communities.

But India, with its vibrant spiritual culture, its armies of ragged pilgrims and its remarkable *Vedas* raises the possibility that the real origins of civilization could be very different – not driven by economics but by the spiritual quest that all true ascetics of India still pursue with the utmost dedication. Such a quest does not deny that the basic material requirements of the human creature must be met but seeks to limit our attachment to material things and in general to subordinate material needs to mental and spiritual self-discipline.

In the sparseness, understatement and efficiency of Mehrgarh's most ancient period could it be that we are seeing the imprint of this essentially yogic ethic – which the *Vedas* anyway tell us was the ethic of most ancient India?

And since archaeologists are now in universal agreement that there is an unbroken continuity of culture from Mehrgarh I around 9000 years ago all the way down to the great cities of the Indus-Sarasvati civilization around 4500 years ago, shouldn't we expect signs of the same yogic ethic to turn up there?

PART THREE

India (2)

9 / Fairytale Kingdom

If Dwarka could be located and identified, well the personality of Krishna is not a myth but a fact.

       S. R. Rao, discoverer of the Dwarka underwater ruins, 29 February 2000

I stood in the Harappan Gallery of the National Museum in New Delhi peering through security glass at a small steatite seal from Mohenjodaro. Dated to approximately 2700 BC,[1] the seal depicts an ascetic seated in difficult posture of highly advanced yoga known as *mulubandhasana*.[2] Lean-waisted, bearded,

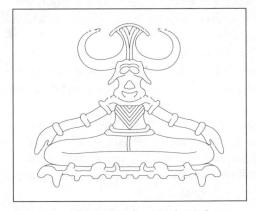

Pasupati seal (2700 BC) from Mohenjodaro, showing a god in a yogic posture.

half-naked, phallus erect, the figure wears a head-dress of buffalo horns over long, unkempt hair. His face might be a mask. It is powerful, almost hypnotic, and there is the suggestion of two further faces (or masks?) in profile looking to either side. He is surrounded, but clearly unthreatened, by dangerous big-game animals – wild buffalo, rhinoceros, elephant, tiger. His arms are covered with bangles and stretched out so that his hands rest loosely on his knees – the traditional signal of a state of profound meditation.

It is often said that we can never hope to learn much about the religious beliefs or the guiding philosophy of the Indus-Sarasvati civilization because we cannot read its script – a line of which appears above the meditating figure. Yet even though the inscription is opaque to us this enigmatic seal from Mohenjodaro does provide some definite and indeed rather intriguing information.

It tells us that at least the outward appearances of the ascetic mind-body disciplines of meditation and physical self-control which still lie at the heart of the spiritual lifestyle in Hindu India in the twenty-first century were being practised 4700 years ago in the Indus-Sarasvati cities.

It tells us specifically that yoga, one of the six orthodox schools of Vedic philosophy,[3] was already known 4700 years ago as a fully evolved system – since *mulubandhasana* cannot be achieved by beginners but requires the prior mastery of numerous intermediate postures.[4] Unless we are to imagine that yoga was miraculously conjured into being all at once as a complete system 4700 years ago, it tells us that the origins of the system must be much older even than that. And since variants of the lean, unkempt yogic figure performing *mulubandhasana* are 'amongst the most common motifs in Indus ritual art',[5] it tells us that the classic image of the *rishi*, the yogic sage or seer, that is summoned up again and again in the Vedas, was also ubiquitous amongst the Indus-Sarasvati people in the third millennium BC.

Moreover, if scholars are right in their universal consensus that the Mohenjodaro seal 'depicts the figure of a god seated in yogic posture'[6] then we are witness to an amazing continuity in religious iconography – for to this day the Hindu god Siva is 'the Lord of Yoga' and is to be seen depicted on temple walls throughout India as a lean, almost naked, meditating ascetic with shaggy hair and sometimes even with a similarly erect penis (the latter feature not meant to imply unconstrained lust but rather its opposite; in Tantric Hinduism Siva's erection symbolizes complete yogic control of bodily desires).[7] Siva, too, is called Pasupati, the 'beastmaster' or 'Lord of animals', because of his ability to tame ferocious beasts with his yogic powers – exactly in the manner in which the figure on the Mohenjodaro seal seems to be portrayed.[8] Even the phallic *lingam* symbol (the butter-smeared stone column erected in the inner sanctum of every Siva temple in India and regarded by worshippers as an embodiment of the god himself) is prefigured in the Indus-Sarasvati cities by conical sacred stones or 'proto-linga'.[9]

For all these reasons the yogic god on the steatite seal has been known as 'proto-Siva', and also routinely spoken of by archaeologists as 'the Pasupati figure', since its discovery during excavations in the DK area of Mohenjodaro in 1928/9.[10] Yet Western scholars like Jonathan Kennoyer attach little significance to the comparisons that invoke such epithets:

> The figure has been referred to as 'proto-Siva' because of its similarity to later iconography of the deity Siva from the Hindu pantheon. Whereas many later Hindu deities may have had their roots in earlier beliefs of the Indus Valley or other indigenous communities living in the subcontinent, we cannot confirm specific connections between the horned figure on the Indus seals and later Hindu deities. There are similarities in the iconography but the meaning relayed may have been significantly different.[11]

The Vedas *and archaeology*

I left the Harappan Gallery deep in thought and walked across the corridor into the Museum's circular central garden. I realized that I felt irritated by Kennoyer's caution. And it wasn't just because he was downplaying the many interesting iconographic links between Siva and the Mohenjodaro figure. Unspoken behind this was the larger problem of the Vedas, which also describe a Siva-like or 'proto-Siva' deity – the Vedic god Rudra[12] – and which bestow the utmost respect, even awe, upon seven *rishis* with yogic powers.

I found a shady spot to sit down, opened my notebook and scrawled the words *Summary of Vedic traditions about the origins of civilization in India* at the top of a blank page:

Summary of Vedic traditions about the origins of civilization in India:

1. An earlier civilization, which knew the *Vedas* and practised yoga, existed before the great flood and was destroyed by it.
2. Manu and the Seven Rishis (Saptarishi) were yogic adepts who survived the flood.
3. The role of the Seven Rishis was to preserve the *Vedas* through memorization and to repromulgate them amongst post-diluvial humanity.
4. The role of Manu was to re-establish agriculture after the flood, using a cache of seeds and plants that he had brought with him for this purpose, and to become the progenitor of future civilized humanity by fathering a dynasty of kings.
5. The *Vedas* and the traditions that descend from them depict the Saptarishi as a lineage of ascetics. After the flood their primary abode was in the Himalayas, where they would retreat to meditate and perform austerities, but they also played decisive roles in running and ordering secular affairs, and in the making and guidance of kings.
6. The so-called Saptarishi calendar of ancient India, which of course cannot be separated from the traditions of the Seven Rishis, has a start date around 6700 BC – almost 9000 years ago.

Summary of archaeological evidence about the origins of civilization in India:

1. Fully functional 'village farming communities' like Mehrgarh in the foothills of the Himalayas appear suddenly in the archaeological record somewhere around 9000 years ago. It's a bit of a mystery. No clear antecedents have yet been found. The original settlers came with seeds and already knew how to farm.
2. This happened in the midst of an epoch of cataclysmic global floods that saw huge areas of India's continental shelf inundated. The possibility, therefore, cannot be ruled out that the founders of Mehrgarh had previously lived on lands swallowed up by the rising seas.
3. There is an unbroken archaeological continuum between Mehrgarh 1 A around 7000 BC and the upsurge of Mohenjodaro and Harappa as great cities after 3000 BC. For some reason the rate of growth and development became particularly rapid between 2600 and 2500 BC – the mature phase of incredibly vigorous urban expansion – but you can see the roots

even of this phase in many small and large details more than 4000 years older exposed in the excavations of the first habitation layers at Mehrgarh.

4. The paramount ritual image to have come down to us from Mohenjodaro and Harappa, and therefore likely to be connected in some way to this ancient heritage, recognizably portrays a *rishi* seated in an advanced yogic posture and seemingly deep in meditation.

Question:

Why should the people of the largest and most sophisticated urban civilization of antiquity have specially venerated the figure of a half-naked ascetic meditating in a rural setting surrounded by ferocious animals?

If the *Vedas* were the scriptures of Mohenjodaro and Harappa, then an answer immediately suggests itself.

They would have venerated the image because they would have been taught from childhood that their civilization had been founded, and that it continued to be guided, by *rishis* looking exactly like this.

I closed my notebook and returned to the Harappan Gallery for another look at the cross-legged, three-faced, buffalo-horned *rishi* of Mohenjodaro. Well, not exactly cross-legged, in fact – because to perform *mulubandhasana* you first have to sit down and bring your heels together with your feet pointing forward whilst placing your knees flat on the ground. Next, with your feet still pointing forward, you tuck your heels in under your perineum. Then you turn your feet a full 180 degrees under your body so that they now point excruciatingly backwards – a manoeuvre that will disclocate the ankles of an inexperienced practitioner. Then you meditate.

How long, I wondered again, does it take to perfect a system like yoga? And if it was already perfect 4700 years ago, then how many thousands of years before that must its roots go back, what are we to conclude about the level of development of the supposedly Stone Age people who created it, and why is there no archaeological trace of them?

Return to the diving quest
February 2000

From Delhi I flew to Goa to meet marine archaeologists at India's National Institute of Oceanography, whose research, I hoped, might provide me with some answers. I had already been in contact with them by e-mail and telephone for more than a year, trying to arrange to dive at Dwarka – which still fascinated me, as it had since 1992, with its ancient legends of a flood at the end of a world age and its mysterious underwater ruins. The archaeologists seemed friendly enough, even enthusiastic, but answered to higher authorities in the Indian government whose blessing they needed before they could agree to let me dive with them.

By this stage, early February 2000, I still didn't have a clear chronology in which to place the underwater structures at Dwarka. Nor, it seemed to me, did the NIO. As I've reported in previous chapters, there was a general assumption that the ruins had been submerged by relatively recent land subsidence (not rising sea-levels) and that they belonged to a very late period of the Indus-Sarasvati civilization – 1700–1500 BC. But the marine archaeologists had not recovered any datable artefacts that could confirm or deny this theory.

All the more I wanted to look for myself and form my own opinion.

Legacy of a lost civilization
February 2000

On the flights to Goa, and the long stopover in Mumbai, I went back over some of the evidence on the origins of civilization in India I'd been considering in recent months, reread the notes I had made in the National Museum in Delhi, and then, in large letters, wrote the word *Hypothesis* at the top of an empty page:

Hypothesis:

The Indus-Sarasvati civilization, the development of which archaeologists have already traced back 9000 years, has an earlier episode of hidden prehistory. It was founded by the survivors of a lost Indian coastal civilization destroyed by the great global floods at the end of the Ice Age.

Such floods occurred many times between 15,000 and 7000 years ago, but a particularly bad episode is attested in high salinity levels in the Arabian Sea and the Bay of Bengal between 12,000 and 10,000 years ago.[13]

The convergence of archaeological evidence is that the first food-producing villages like Mehrgarh were established immediately after the worst flooding between 10,000 and 9000 years ago. For example, Gregory Possehl: 'There is no entirely satisfactory chronology for the Indus Age, especially for the internal stages and phases of prehistoric life. Present estimates, based on radiocarbon dates, suggest that it arises at 7000 or 8000 BC with the earliest villages, the domestication of plants and animals and the beginnings of farming and herding societies.'[14]

The survivors who established the early villages practised a 'proto-Vedic' religion that they had brought with them from their inundated homeland and probably spoke an early form of Sanskrit.

The survivors were experienced farmers, as the archaeological record confirms, and their cultural level was high, but religious and philosophical considerations (perhaps even a reaction to the supposed 'judgement' of the flood on their former lifestyle?) led them to create a sparse, utilitarian and ascetic new world – even as they moved gradually towards ever larger and more complex urban communities.

There were secular rulers but the real leadership of the new communities remained vested down the generations in the brotherhood of sages whose forefathers had escaped the deluge – the lineage of Vedic masters whose task it was to preserve and transmit a precious body of antediluvian knowledge. For thousands of years, from Mehrgarh to Mohenjodaro, it was the policies set by these great *rishis* in pursuit of that objective – rather than in response to economic or other material forces – that shaped the steady, peaceful, modest material development of the Indus-Sarasvati civilization.

It was a hypothesis – just that, nothing more. But I'd already been playing around with it in my mind for months as my research on India had progressed and it was time to set it down on paper. Nothing in it contradicted the archaeological evidence. It made sense of the sudden and fully formed appearance of village-farming communities like Mehrgarh between 10,000 and 9000 years ago. It took proper account, as other theories did not, of the latest science on the end of the Ice Age. It provided a rational basis in real events for the Indian flood myth. And it explained the phenomenal longevity and continuity of the Indus-Sarasvati civilization from the simplicity of its sudden beginnings at the end of the Ice Age until its equally sudden boom and collapse in the third millennium BC.

There was one way to prove the hypothesis very quickly. All I had to do was find ruins more than 9000 years old underwater on India's continental shelf. And that was the private hope I had for Dwarka.

Gatekeepers of the fairytale kingdom

The headquarters of the National Institute of Oceanography are in Dona Paula, Goa, in a pleasant university-style campus of trees and lawns. As well as occupying a modern block on the highest point of the campus, the Institute's many divisions, sub-divisions and laboratories sprawl outwards into a suburb of old-fashioned bungalows set beneath the trees. The Marine Archaeology Centre is in one of these, identifiable by a display of stone anchors and other stone objects mostly retrieved from depths of 5–10 metres amongst the underwater ruins at Dwarka.

My appointment was with Kamlesh Vora, the NIO's head of archaeology, with whom I had been corresponding. I appreciated that he had taken the trouble to process my proposal at all, since he could perfectly easily have dismissed it out of hand or just ignored it, but the fact was that many months had passed and there was still no sign either of approval or disapproval from the higher authorities – in Delhi as it happened – to whom he had submitted it.

'Now that you are here,' he said, 'perhaps it will galvanize them into action.'

He picked up the telephone and placed a call to the offices of the Scientific Research Council, the NIO's parent organization and an important spoke in the wheel of central government. A lengthy conversation then followed in Hindi. Finally, Kamlesh hung up: 'There is a certain lady within the SRC who

I need to talk to about your case.' He gave me a gloomy look: 'Unfortunately she is not at her desk today' A smile: 'But I'll find her tomorrow.'

'What do you expect the answer will be?'

Kamlesh became gloomy again and explained that never before had the NIO had to deal with a request from an author to dive with them at Dwarka. If I was an academic or governmental institution seeking to send an observer to the site there would be set procedures to follow and the permission process would go along according to a well-ordered routine. But since I was a private individual, non-governmental, non-academic, and non-Indian into the bargain (raising issues about what sort of visa I should be travelling on), no one knew what to do with me.

And here was the problem. The NIO's annual campaign in Dwarka, which I was hoping to join, was scheduled to go ahead in mid-February (less than two weeks hence) but would continue only until mid-March. So my permission had to come through before then. If it didn't I'd miss the campaign and therefore would lose my chance to dive at Dwarka until the following year.

'You mean you only dive there for one month every year?'

'If we're lucky. Our funds are very limited, but we do what we can.'

'What if I make my own arrangements? If the permission comes through after the NIO has gone is there any way that I can arrange to dive privately at Dwarka?'

Kamlesh was horrified: 'No, not at all. It is a protected national archaeological site, so our people have to be with you. Besides, there's no private diving at Dwarka. There are no facilities there. It's a very out of the way place. We bring our own compressor and tanks with us from Goa every year and take them away again when we leave . . .'

My heart sank. Since I'd first learned of it in 1992 as a non-diver, the underwater city of Dwarka had beckoned to me like a fairytale kingdom that seemed far beyond my reach. Eight years later I'd acquired the skills, but not yet the permission, to dive at it. And I felt helpless to influence the matter in any way.

'Come and see me mid-morning tomorrow,' Kamlesh said. 'I will try again with the SRC. Maybe I will have good news for you.'

Write a letter

I was back with Kamlesh by eleven the next morning, but there was no news, good or bad. The lady at the SRC was still not at her desk. He called her again. Still nothing. Finally, half an hour later, she answered her phone. Yes, she had received the paperwork concerning my proposed visit. Yes, it was being considered. No, there was no decision as yet. Kamlesh asked if anything could be done to speed things up. It might be a good idea, she told him, if I were to write a letter explaining in greater detail than in my original proposal exactly why I wanted to dive at Dwarka.

Suppressing a mood of rising irritation and bad temper, I took a taxi back to the Ciudad de Goa hotel, fired up my portable computer and began to draft the letter – which Kamlesh suggested I should address in the first instance to Dr Ehrlich Desa, the Director of the NIO. 'If he intervenes with the SRC on behalf of your case it will make a great difference.'

When I met Kamlesh later in the afternoon to review the text of the letter, he told me that he had spoken to Dr Desa who had agreed to see me at ten the next morning.

Two days later I left Goa. Permission had still not been given. But my meeting with Ehrlich Desa had been encouraging and he had promised his support in fast-tracking my application through the SRC. I felt confident that he and Kamlesh would do their best for me, and vaguely optimistic that somehow the necessary strings would be pulled to allow me to dive at Dwarka. We agreed to stay in touch by e-mail.

Interlude: the quest for Kumari Kandam

My trip to India in February 2000 had multiple objectives and I had intended from the beginning to be on the road until the middle of March. So although the hold-ups and uncertainties about Dwarka were worrying, they hadn't yet really inconvenienced me. It was perfectly possible that permission could still be granted . . .

Meanwhile Santha and I had long planned another journey in southern India and flew first to Madras, now called Chennai, to pick up where we had left off in 1992.

Then it had been a journey of personal reminiscence – Vellore and the shore temples of Mahabalipuram on the Coromandel coast. Now we would start in Mahabalipuram, travel inland from there to Tiruvannamalai, a temple sacred to Siva since time immemorial, and thence to Madurai, an ancient centre of Tamil learning linked again to the yogic god Siva. To the north-east of Madurai we planned to visit Poompuhur, and to the south-east Rameswaram on the thin spit of mainland that reaches out towards Sri Lanka, dividing the Palk Strait from the Gulf of Mannar. Then we would go on to Kaniya Kumari – Cape Comorin – on the southernmost tip of India.

During 1999 I had begun background research on southern India and had been intrigued by what I had found.

One source of information that had lain unopened in my library for far too long was Captain M. W. Carr's *Descriptive and Historical Papers Relating to the Seven Pagodas of the Coromandel Coast*.[15] As I reported in chapter 5, Carr's anthology preserves strong local traditions of a fabulous antediluvian city at Mahabalipuram swallowed up by the waters of a great flood. Those traditions had certainly been in wide circulation in the eighteenth and nineteenth centuries when the papers in Carr's anthology were written. I wanted to find out if they were still in circulation today and if there could be any substance to them.

I had also come across the work of David Shulman, Professor of Indian

Studies and Comparative Religion at the Hebrew University in Jerusalem. His wide-ranging investigation of Tamil flood myths had helped to put places like Poompuhur, Madurai and Kaniya Kumari on the map for me. In the Tamil epic known as the *Manimekalai* it was said that the ancient port-city of Kaveripum-pattinam had been flooded by the sea off the Poompuhur shore. Other traditions spoke of prehistoric wisdom schools or academies (*sangam*) established 'in an antediluvian Tamil land stretching far to the south of the present southern border at Cape Comorin'.[16] The name of this lost land, which had been swallowed up by the sea in two distinct inundations separated by thousands of years, was Kumari Kandam, and its last survivors were said to have fled to Madurai.[17]

As usual when I'm on the road I was carrying a shoulder bag full of books and reference materials with me, some brought from England, some picked up along the way. Following my few days in Goa, I had added substantially to my stack with a pile of bulky annual conference reports and back numbers of the NIO's *Journal of Marine Archaeology* that Kamlesh had given me.

Serendipitously the very first of these that I browsed through on the flight from Goa to Chennai (volume 5–6 of 1995–6) opened with a lengthy paper entitled 'Underwater Explorations off Poompuhur 1993.'[18] Much of the paper concentrated on an archaeological validation of the *Manimekalai* myth, connecting it to the submerged ruins of Kaveripumpattinam 'an ancient port town of 3rd century BC to 4th century BC' that the NIO's marine archaeologists had identified very close to the shore in water generally less than 3 metres deep.[19] But the paper also reported the anomalous U-shaped structure that the divers had found at a depth of 23 metres more than 5 kilometres out to sea.[20]

I immediately realized that this obscure and neglected reference to a 1993 exploration that the NIO had never had the funds to follow up was potentially significant. I did not then have access, as I would later, to Glenn Milne's computerized inundation maps. But at that depth and that distance from the shore, common sense alone suggested that the U-shaped structure must be extremely old.[21]

The main author of the report and team-leader of the Poompuhur exploration had been S. R. Rao, Kamlesh Vora's predecessor at the NIO and the original discoverer of the underwater ruins of Dwarka. Since he was now retired and living in Bangalore, only a short hop from Chennai, I decided on impulse that at some point on our journey in the south I would try to meet him.

'It must have existed . . .'
February 2000

My encounter with Rao, which I've already reported in chapter 1, took place on 29 February. To my amazement the doyen of Indian marine archaeology proved open to the notion that an antediluvian civilization could have existed on the Indian coastal lands flooded at the end of the Ice Age:

It must have existed. You can't rule that out at all. Particularly, as I have said, since we have found this structure at 23 metre depth. I mean we have photographed it. It is there, anybody can go and see it. I do not believe it is an isolated structure; further exploration is likely to reveal others round about. And then you can go deeper, you see, and you may get more important things.[22]

We'll return to the quest for Kumari Kandam in chapter 11. For me a big part of it unfolded there and then in the year 2000 and an even bigger part – the diving part – in 2001.

Meanwhile, a couple of days before my encounter with Rao, something suddenly shifted in the turgid backlog of Indian bureaucracy and Kamlesh e-mailed me with the good news that the permission had come through – 'at the eleventh hour' as he put it – and that I would be allowed to dive at Dwarka with the NIO team. Much was owed, apparently, to the robust support given to our adventure by Dr Desa. At any rate there would be no further obstacles and Santha and I should plan to reach Dwarka on 2 March.

The problem of Dwarka's age
March 2000

It felt good to be back in Dwarka again after so many years away and to have the opportunity at last to look into the mystery of its underwater ruins.

When I'd interviewed him in Bangalore, Rao had reaffirmed his longstanding view that the ruins are those of an Indus-Sarasvati port probably built between 1700 BC and 1500 BC during the final years of the civilization's decline and then flooded by an incursion of the sea. However, he admitted that the dates were a supposition not an empirical fact. Radiocarbon or thermoluminescence tests, which might settle the matter, had not been possible, since the latter requires pottery contemporary with the ruins and the former organic materials contemporary with the ruins – neither of which had yet been found in submerged Dwarka itself:

> Rao: I mean to be frank, you see, we did some thermoluminescence dating for the pottery extracted from the wall which is just on the shore – and of course it also partially gets submerged at some times. All right, that gives 1528 BC. But that is at a slightly higher terrace than the submerged one. So the submerged one must be earlier.
>
> GH: Would it be fair to say, concerning the underwater structures, that the minimum age would be about 1500 BC but that it is possible that they may be older?
>
> Rao: Oh yes, definitely, that you can definitely say. Minimum age would be about even 1500, 1600 BC, but an earlier date can't be ruled out. I mean there is every possibility of getting earlier dates.
>
> GH: My understanding is that underwater structures that have been identified so far go down to about 12 metres under the sea?

Rao: These structures go to about 10 metres depth. Of course, the ridge which was converted into a sort of wharf, that is at 12 metres depth. Beyond that we have gone, but not much.[23]

GH: Do you think there's any chance of further ruins being found further out into the sea?

Rao: Maybe. Maybe. I won't rule that out at all. Because, you see, what we did [beyond the 12 metre depth contour] was only side-scan sonar survey. I mean, a little diving as well we have done here, but not much, to be frank. I mean, if you dive for three days or four days only then you cannot expect to find much . . .[24]

Expecting the best

We were to dive at Dwarka off a small wooden sea-going trawler, a rough-and-ready working ship crewed by local fishermen that the NIO had chartered. Since its draft was too deep to approach the shore it was moored in the bay about half a kilometre to the south-west in front of the Gomati river mouth. We were ferried out to it in an inflatable dinghy that picked us up from the steps of Gomati Ghat, and as we chugged across the bay I found myself looking down impatiently at the water, hoping to get some glimpse of whatever lay below.

The ruins had been thoroughly mapped by the NIO across a large area between the mouth of the Gomati – which now lay behind our dinghy to the north-east – and a submerged rock ridge about a kilometre out to sea to the south-west that had been cut and modified as a wharf when it was above water in ancient times. This was the wharf that Rao had mentioned as the site's deepest known structure at 12 metres and which he suspected to have been part of its harbour.

All the other remains, revealing the outlines of a series of spacious rectilinear buildings, lay much closer to shore between just 3 and 10 metres with the majority concentrated between 5 and 7 metres.[25] These included twelve so-called 'citadels', protected by massive bastions, six on each bank of a now submerged section of the Gomati channel, where Rao told me he thought that 'not only the King but also the army chief, other officials or his ministers used to live'.[26] The ancient harbour city itself was divided into six blocks:

> All six sectors have protective walls built of large well-dressed blocks of sandstone, some as large as 1.5 to 2 m long, 0.5 to 0.75 m wide and 0.3 to 0.5 m thick. L-shaped joints in the masonry suggest that a proper grip was provided so as to withstand the battering of waves and currents. At close intervals semi-circular or circular bastions were built along the fort walls in order to divert the current and to have a proper overview of the incoming and outgoing ships . . . There are entrance gateways in all sectors as surmised on the basis of the sill of the openings. The fort walls and bastions, built from large blocks which are too heavy to be moved by waves and

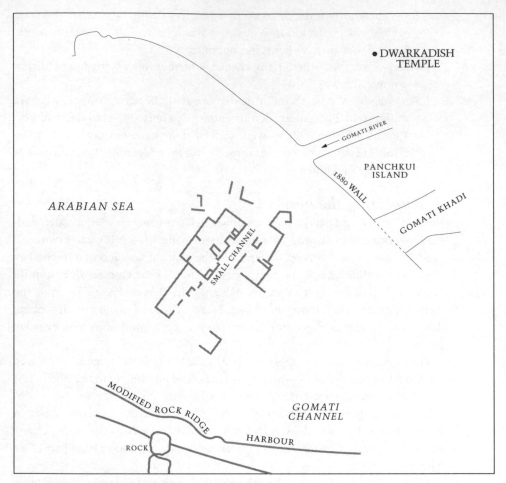

Map of submerged ruins off Dwarka. Based on Rao (1999).

currents, are *in situ* up to one or two metres height above the boulder foundation in the sea. In a few places as many as five courses of masonry are visible but in others the wall and bastion have collapsed.[27]

Prepped by such imagery of a fairytale underwater city, and the beautiful reconstructions of antediluvian Dwarka that feature in Rao's books, I confess I was expecting the best as I clambered out of the dinghy and up the side of the NIO's chartered fishing boat on the morning of 3 March 2000.

Fog, weed and sludge
In the relentless war of heat-exchange that goes on between a diver and the sea, it is the sea that always wins in the end. The process is faster in cold water, slower in warm water, and can be delayed further by an insulating wetsuit;

however, the end result is always the same. If the sea is colder than the diver's body temperature then the diver's body temperature will begin to fall.

I think of myself as a reasonably experienced diver but I'm fifty years old, way past my peak fitness, and I make mistakes. The mistake I made at Dwarka, though I'd been warned that the water was only 23 degrees centigrade (and thus 14 degrees below body temperature), was not to wear a wetsuit. This would have been fine if I'd been going down for just one or two short dives. But we did three dives that day, running to an hour or more each.

The first two dives were on the big concentration of ruins that the NIO had mapped between the 5 and 7 metre contour lines. Gone were the lofty turrets, battlements and bastions of Rao's reconstructions and of my imagination. All seemed to have been reduced to a ruin-field of haphazardly strewn stone blocks, the angles and edges of which poked here and there out of the thick sludge of sediment and slimy green weed that carpeted everything. And although the sea was calm that morning, allowing some settlement of silts carried down into the bay by the Gomati river, millions of tiny particles hung suspended in the water, scattering light like a fog.

Through the fog I was just able to make out beneath me several dozen large limestone blocks that seemed to have come from a collapsed section of wall, not quite megalithic in the strict sense of the term, but very close to it, tumbled on top of one another. The wall had been dry-stone – no mortar in the joints to keep the courses together. But I could see how the masons had dealt with the problem. Many of the bigger blocks had been designed to lock into each other with dovetails and, as Rao had commented, with carefully chiselled L-shaped joints which would have given extra structural stability.

The same architectural principle had been used in the massive curved bastions that had stood at the corners of the citadels. Although I found none intact, I several times came across huge curved monoliths, dressed and polished to very high standards and in one case still jointed to a second block.

Also protruding out of the slime and ooze on the sea-bed were carved hemi-spherical stones, some up to a metre across, with circular holes drilled through their centres. These were thought to have been door sockets.

And trapped amongst the rubble of ancient Dwarka there were still a number of three-holed triangular stone anchors that the NIO had not yet salvaged for the display outside their offices. Identical anchors, Rao had told me, were known to have been used in the Mediterranean by the merchant ships of Cyprus and Syria at around 1400 BC and also in the Persian Gulf and at the nearby Indus-Sarasvati port of Lothal.[28] Assuming the 1400 BC date for this type of anchor to be generally valid, he regarded their presence at Dwarka as good circumstantial evidence in favour of his 1600 BC date for the city. Certainly, they could only have been dropped here after the ruins had been submerged deeply enough for boats to sail over them.

But one mystery which began to nag at me on those first two dives, since we were supposedly in the heart of the ancient city, was that there didn't seem to be *enough* stone ruins here. This had nothing to do with the stark contrast between Rao's archaeological reconstructions of the antediluvian city and its actual appearance underwater today. What bothered me more was the almost equally stark contrast with photographs that Rao had shown to me from his personal collection that tracked the NIO's underwater excavations at the site from 1983 to 1994.[29] Although some of the features in those photographs were instantly recognizable on the sea-bed, many others were nowhere to be seen. Most notable by their absence were several partially intact walls of large stone blocks, in some cases up to five courses high, in some cases showing right-angled corners where two walls joined, in some cases extending in straight lines away from the camera as far as the eye could see – and the visibility was far better in those early shots than the fog that I was finning around in now.

So where were the missing walls?

Storms

After I'd surfaced from the second dive and clambered back on board the boat I asked Kamlesh this question and he signalled for Sundaresh and Anuruddh Gaur, two of the NIO's senior marine archaeologists, to join us. A geologist by training, Kamlesh himself was not then a diver. Gaur and Sundaresh, on the other hand, had been diving at Dwarka since the 1980s.

Their answer was that the majority of the intact walls that had been photographed before 1994 either no longer existed or could not be relocated. Apparently, a series of severe monsoon storms during the past six years had loosened and dislodged the great blocks and tumbled the walls over. Since then sedimentation and weed had covered up the debris which had been scattered over a wide area by the monsoon swells.

I remembered the section of fallen wall that I'd seen early on the first dive and thought no more about it. It was only much later that it struck me how odd it was that a site which had supposedly been submerged for more than three millennia, and at which so many intact structural features had been documented as recently as 1994, could have deteriorated so dramatically in just the last six years.

The rock-cut wharf

Slightly dodgy-looking curries were available for lunch, cooked by the crew on a kerosene stove in the cabin of the fishing boat and served out on a mixture of plastic and tin plates. The wind had come up since the morning and wavelets were freshening in the bay – not enough to stop us diving but potentially enough to stir up the bottom and worsen the visibility.

I wasn't feeling particularly well – headache, stiff-neck, nausea – and was aware that I had been cold on the last dive, but I didn't put the two together. I thought that what was making me ill was the exhaust gas from the diesel pump that the NIO had on board to provide air from the surface via long tubes to technical divers working down below. A powerful air-lift system was also operating, sifting silt around the foundations of the ruins in the still unsuccessful search for artefacts that could positively identify their period of construction. All the vibrations and the fumes were a bit much for me but I thought that I'd probably feel better when I got back in the water and could breathe the clean air from my tank.

At this point the voice of reason told me it was time to put my wetsuit on for the afternoon's work and the voice of stupidity urged me not to bother. The voice of stupidity won.

The dive we did that afternoon was with Gaur on the rock-hewn wharf at a depth of 12 metres about a kilometre out in the bay. Athough this was technically still a shallow dive, there was an oppressive darkness and gloom in the dirty green water and I began to feel more and more cold, weak and exhausted.

We swam east on the seaward side of the ridge. As well as its rock-cut features, including what were presumed to have been holes for mooring-ropes drilled through it at several points, there were a number of hulking megaliths scattered round about it on the sea-bed down to depths of about 18 metres. The official view was that these were natural slabs that had become detached from the rock ridge due to wave action when sea level had been much lower – and perhaps even before the wharf had been fashioned – but to my eye they looked in places as though they had been dressed and cut.

Quarter of an hour later, still heading east along the ridge, I saw a pattern of other smaller blocks, like large tiles, laid out in a square grid amidst a tangle of boulders. I went down to investigate and found that the regular pattern seemed to continue under the boulders. That was exciting. On the other hand, up close, the little blocks and the joints between looked less regular, less man-made, than I had thought . . .

I couldn't make up my mind. And other ambiguous features along the ridge left me, if anything, even more in doubt.

Whitecaps and lentil soup

I spent the next four days in bed in our dingy hotel room paying the price for being a fifty-year-old with no sense and mild hypothermia. A blinding, thudding headache was by far the worst of it and continued without any let-up for more than seventy-two hours. I felt weak, shaky and couldn't keep down anything I tried to eat.

But I wasn't missing much diving. The wind that had begun to pick up on that first afternoon grew steadily stronger during the night, whipping the waves in the bay into whitecaps, reducing the visibility to zero and making further diving impossible. The NIO's chartered boat headed back for the shelter of a nearby fishing port and everyone waited to see if the weather would improve.

By the time I dragged myself out of bed the wind had died down and the boat was anchored over the ruins again. But the underwater conditions, with the transparency of lentil soup, made it impossible to do any serious work. I tried a couple of dives at different locations on the site but could see nothing.

Then the wind came up once more; this time with a forecast that it would continue to blow for more than a week, and it became obvious to all that there would be no further diving that season.

Layer upon layer

How old is the city beneath the waves?

Sitting on the edge of the Gomati Ghat by the Temple of the Sea God on the last evening of our stay in Dwarka, I looked over the agitated waters of the darkening bay and tried to figure out the mystery.

When I'd interviewed Rao at his home in Bangalore, I remembered that he'd told me how he had first become involved with excavations at Dwarka more than twenty years before. In his work for the Archaeological Survey of India he had arranged the demolition of a modern building that stood beside the main Dwarkadish (Krishna) temple, blocking the view:

> *Rao:* It was demolished. When we removed this structure we were surprised to find a temple below it – a temple of Vishnu. [Krishna is considered to be an avatar, or manifestation in human form, of the Vedic god Vishnu][30] . . . It has beautiful sculptures and all that. We were surprised. You see this is a thirteenth- to fifteenth-century temple, the present one that we visit, but here is a ninth-century temple. How is it? When we dug for that we got two more temples below – below that there are two more temples.
>
> *GH:* So it's as though the existing Dwarkadish temple was built on top of an older temple?
>
> *Rao:* Not the existing one. The one just by the side of it. You see, actually, this temple, I mean the existing one, must have been built on top of an ancient one, because what we got is a small shrine, and the other shrine must be below the present temple.
>
> *GH:* But your excavation was beside the existing temple and there underneath you found earlier layers?
>
> *Rao:* Earlier layers. And further when we dug we came across a clear section showing erosion by sea, with pottery and other datable objects of about 1500 BC. So between 1500 BC and 1500 AD there must have been continuous occupation here of which we hardly know anything. But again sometime there is divine help for us. One professor by name of B. R. Rao, a geologist, had come to Dwarka to inspect the site for a proposed university. I showed him the section and he said yes, this is clear evidence for erosion by the sea. I showed him the pottery and he said there must have been a township near by. He said, what will you do? I said we have to excavate in the sea – that's marine archaeology.[31]

Rao then successfully arranged government funding for his proposed venture at Dwarka:

> But we did not know how to start the work. We had hardly any experience of marine archaeology. Then I thought what we should do now is take a bold step . . . Where to look for the structures was the question. Fortunately, there is the temple of Samudra Narayana, the sea god. So I said people have been making some offerings here. Maybe ancient times also there may have been some structure there and offerings might have been made. So we straight away started looking there. And then within a few days we got evidence of the structural remains there, underwater.[32]

An earlier town

Looking over the bay from the Samudra Narayana temple I reflected on Rao's dating of the underwater ruins to the second millennium BC and the 'late Harappan' period. I could see no reasons why the scattered structural remains that I had dived on should be any older than that – and even some to suspect that they might be younger, perhaps much younger. Except for the rock-hewn wharf, which itself was not particularly deep, most of the structures were in shallow water of 7 metres or less and might easily have been submerged relatively recently in land-subsidence caused by the immense earthquakes that periodically afflict Gujerat.[33] Besides, what I'd seen of the underwater ruins looked nothing like any of the 'late Harappan' settlements I knew of, on the contrary, the distinctive curved bastions and general style of the architectural blocks on the sea-bed looked much more like medieval Indian construction work than anything to do with the Indus-Sarasvati civilization.

But what intrigued me, and what Rao had been entirely open to, was the possibility that there might be other ruins further out to sea which the NIO had not yet found – indeed had not yet even looked for. Rao also reminded me that the ancient texts that seemed to have correctly predicted the presence of the underwater ruins that he had discovered also predict that other older ruins should exist in the vicinity – for Krishna was said to have built Dwarka on the site of an even earlier city called Kususthali:

> In fact I used to read the *Mahabaratha* and also other *Puranas* like *Vishnu Purana* and others, where it is clearly states that Dwarka was built at Kususthali in such a way that it was surrounded by the sea . . . So Krishna comes to Kususthali and then builds a town and calls it Dwarka and there existed an earlier town before Dwarka was built . . .

What is striking about the story of Krishna's city being built above an earlier city is the way it resonates with the firm evidence we already have from Rao's excavations around the Dwarkadish temple – revealing layers and layers of earlier constructions beneath it and around it, going back to a stratum at around

1500 BC that is roughly parallel to modern sea-level. The ruins that Rao then found underwater should, as he reasons, belong to the time-period immediately before 1500 BC – say 1700 to 1800 BC at the earliest – suggesting that the city that today clusters around the Dwarkadish temple and down to Gomati Ghat is where it is because it replaces the earlier city that lies submerged in the bay beneath it.

And that city in turn – the city of Krishna – is where it is, the legends say, because of the earlier city of Kususthali:

> GH: Is there a sense in the ancient texts that there had been a sacred centre at Dwarka in the remote past, a long time ago? Or was it absolutely newly established by Krishna?
>
> Rao: Well, you see, it says that [an ancestor of Krishna] had built that town Kususthali and he went to Brahamaloka [a higher world]. So some connection with mythology and all that is already there when Krishna comes to that place. So the earlier township had some sanctity about it . . .

In an epoch of rising sea-levels the obvious place to rebuild and reconsecrate a submerged shrine or sacred centre would be on the nearest area of coast still above water. When the new shrine was inundated in its turn it would have to be re-established on higher ground – and so on. So maybe this is what we're seeing at Dwarka: Krishna's Dwarka was built to replace the antediluvian sacred centre that the texts call Kususthali – and when Krishna's Dwarka was inundated, modern Dwarka was built to replace it. By inference, if we keep looking further out to sea, beyond what's left of Krishna's Dwarka – if it really is Krishna's Dwarka, as Rao believes – then we should find older, more deeply submerged ruins.

3102 BC

But are the underwater ruins that Rao discovered at Dwarka the remains of 'Krishna's city' – or of something else?

As I sat there overlooking the darkening waves, with the heady aroma of sacred ganja being exhaled all around me by the orange-robed *sadhus* who'd gathered to watch the sunset from Samudra Narayana, I remembered feeling that Rao couldn't have it both ways. He couldn't have his underwater ruins dating archaeologically to around 1800 or 1700 BC on the one hand and claim on the other that they were the ruins of Krishna's city – since, apart from one minor variant tradition, Krishna is universally believed in India to have died at a date equivalent to 3102 BC.[34] This date (see chapter 4) also marks the onset of the Kali Yuga.

But Rao wasn't trying to have it both ways:

GH: Another question concerning Krishna. The departure, or death, of Krishna's incarnation, if I understand correctly, is taken as the end of a previous age, of a *yuga*, and the beginning of the Kali Yuga. Now in many calculations that I've seen – numerous calculations – they all seem to point the beginning of the Kali Yuga to 3100 BC approximately.

Rao: Correct.

GH: Do you regard that as an impossible date? Because you seem to focus on a much later date, in the second millennium BC, for the submerged Dwarka.

Rao: Well, I wouldn't call it an impossible date. But what evidence we have got so far shows that about 1700 or 1800 BC, by that time this township that is now underwater must have been built. Now if so, how is that date wrong? I mean, the 3100 BC date. We have discussed this matter in a journal where we said that maybe we are yet to find some more antiquities of the same township . . . So we can't discard the earlier date totally.

But if the underwater ruins already excavated do really date back to 1700 or 1800 BC, then where is the logical place to search for ruins even older than that – the ruins of the city said to have been engulfed by a great flood at the beginning of the Kali Yuga in 3102 BC?

Further out, in deeper water

The connection of the death of Krishna and the submergence of Dwarka to the onset of the Kali Yuga is a powerful and widespread tradition in India, as is the connection of the Kali Yuga to a start date of 3102 BC.

We know that the city called Dwarka today is built on a mound made up of continuous occupation strata going down to present sea-level at 1500 BC and with 'a clear section showing erosion by sea' in the lowest stratum – indicative of a marine incursion (perhaps a tidal wave?) at that date.

We know that ruins have been found under that level beneath the sea and provisionally dated to 1800–1600 BC, though a more recent date is also possible. These ruins extend up to approximately 1 kilometre from the shore.

Therefore, it follows, if we wish to search for the ruins of 3100 BC and earlier that are hinted at in the traditions, that we are going to have to look further out, in deeper water.

In March 2000 I still didn't have Glenn Milne's inundation maps and imagined that Gujerat's Ice Age coastline might have extended 5 or at the most 10 kilometres beyond the modern shoreline of Dwarka. In fact, as the maps show, Dwarka was almost 100 kilometres from the sea 16,400 years ago when it was part of a vast antediluvian landmass around Gujerat that filled in the Gulfs of Kutch and Cambay – and was still 20 kilometres inland as late as 10,600 years ago, just after the rapid rise in sea-level attested in the deep-sea cores between

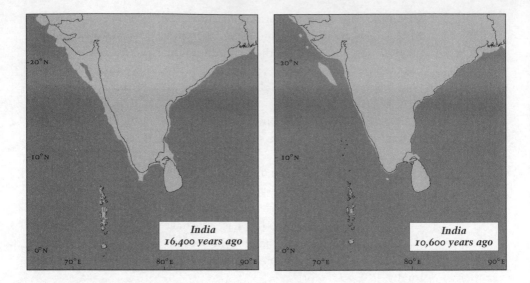

India
16,400 years ago

India
10,600 years ago

10,000 and 9000 BC and the sudden appearance of village farming communities along the piedmont of the Himalayas.

If anywhere in the world looks like a potential 'nucleus region', or 'Ice Age refugium', out of which the first settlers of Mehrgarh and the other 'aceramic Neolithic' food-producing settlements in north-west India might have sprung, then, surely, this is it? And doesn't it make sense that the descendants of those first settlers, who went on, in time, to create the Indus-Sarasvati civilization, might have continued to revere sacred coastal sites and to rebuild them further inland whenever the sea-level rose?

The mystery of the U-shaped structure

That night, over a farewell dinner with the NIO team, I produced the *Journal of Marine Archaeology* given to me by Kamlesh and opened it at the report on the underwater explorations off Poompuhur in the south-east – about as far away from Dwarka as it is possible to get and still remain in India. Both Sundaresh and Gaur had participated in the 1993 Poompuhur expedition and had co-authored the report with S. R. Rao. Now was my chance to quiz them about the anomalous U-shaped structure that they had found 5 kilometres from the shore and 23 metres deep and to launch the idea of mounting a further expedition with them to Poompuhur at some time in the future.

We began by discussing the less controversial – and for me less interesting – ruins of Kaveripumpattinam in the intertidal zone and the shallows down to 3 metres. These, Sundaresh and Gaur concurred, were in the range of 2000 years old, and I had no reason to doubt them.

'OK,' I said, 'so let's accept that dating for the inshore structures. Then what do you find as the water gets deeper?'

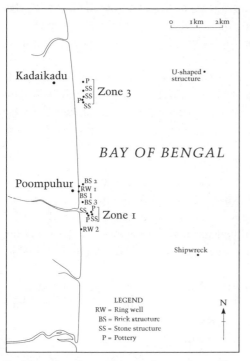

They told me that their survey had identified fairly extensive structural remains in the form of heavily eroded and scattered dressed sandstone blocks down to a depth of about 7 metres. At the same depth they had also located several curious circular cairns, some 10 metres in diameter, made up of rounded stones and some small upright stones. Nothing was seen deeper than 8 metres until the U-shaped structure and its neighbouring mounds suddenly appeared at 23 metres.

'Don't you think that's odd?' I asked.

Sundaresh and Gaur agreed that it was indeed odd since it suggested that the date of submergence of the U-shaped structure must be much earlier than the date of submergence of the structures inshore.

'How much earlier?'

'Maybe 8000 years earlier,' said Gaur after a moment's thought.

Position of various submerged structures off Poompuhur coast. Based on Rao et al. (1993).

'So if the Kaveripumpattinam structures in 1–3 metres are 2000 years old then what you're saying is that the U-shaped structure might be 10,000 years old?'

'I'm saying it would have been submerged by the rising sea-level about 10,000 years ago – maybe even before that. But I think it must be some sort of natural outcrop.'

I was genuinely puzzled. 'Everyone else who has dived on it seems convinced it's man-made. Courses of masonry were seen on it. That's in this report' – I pointed at the *Journal of Marine Archaeology* – 'which you co-authored by the way.'

Gaur laughed: 'Yes, but I have my own view and the more I think about it the more I am convinced it must be natural.'

'But why? What are your reasons?'

'Because it is a huge structure and we know that there was no culture anywhere in India at that time capable of mobilizing the necessary resources and organizing the necessary labour to build something so big.'

'That's just classic old-school historical chauvinism,' I complained. 'It's as though you're saying, "We archaeologists know everything about the past and we won't let a few contradictory facts get in our way."'

'It is a fact! We don't know of any culture 10,000 years ago that could have built this structure.'

'But maybe it was the work of a culture that you *don't* know about yet. Maybe this U-shaped structure, whatever it is, is the first concrete evidence for the existence of that culture. Maybe if you look you'll find even more structures, even further out, in deeper water.'

Sundaresh chipped in at this point that he did not agree with Gaur. In his opinion, he said, the U-shaped structure was not a natural outcrop: 'It is definitely man-made. And I have seen a second structure, a mound, about 45 metres away at the same depth where there are perfect cut blocks scattered on the sea-bed . . .'

'But what about the 10,000-year-old date?'

'Maybe the structures are not that old at all. Maybe there has been some great land subsidence here that we do not know of, or erosion of the coast by the sea.'

It was obvious that the only way to find out, and to settle the mystery, was by doing more diving and by careful measurement, observation and excavation of the site. But the problem was that since 1993 no funding had been available for a further expedition.

'So you have no plans to dive at Poompuhur in the coming year?' I asked.

'Rather you should say no budget,' Kamlesh intervened dolefully. 'If somebody will finance us to go – only then can we go.'

I bit the bullet. 'So what would it take to finance your team to go back there and dive on the site with me later this year or early next year – a sort of special charter, so to speak? Is it even possible to do something like that within the NIO's regulations?'

'Now that the SRC already know of you it should be possible,' said Kamlesh. 'I don't see why not.'

He spent the next three minutes doing calculations on the back of a napkin and finally quoted me a sum equivalent to the gross national product of a small European country.

I gulped but steadied my nerves. It was going to be a long negotiation.

10 / The Mystery of the Red Hill

The ground near it is not at all touched by the four oceans that become agitated at the close of the Yuga and that have the extremities of the worlds submerged in them . . . When the annihilation of all living beings takes place, when all created things are reabsorbed . . . all the future seeds are certainly deposited there . . . All the lores, arts, wealth of scriptures, and the *Vedas* are truthfully well-arranged there . . . Brahmanas who resort to the foot of that mountain are called by me after the deluge and I make them study the *Vedas* and make collections thereof . . .

Skanda Purana

February 2000, south India

Since 5 a.m., Santha and I had been climbing the winding track towards the rocky 800 metre summit of Arunachela, the sacred mountain of Tamil Nadu. It was now just after 6 a.m. and dawn had not yet broken. Except for the sound of our footfalls and distant cockcrows, everything was silent, everything still. Then we rounded a corner and the streetlights of Tiruvannamalai, the burgeoning town that clusters at the foot of the mountain, came suddenly into view beneath us. In its midst, due east of us, there lay a huge geometrical pool of deep darkness and shadow, like a giant doorway to another world. This place, where no lights yet burned, marked the precincts of Arunacheleswar, one of the five most important temples of Siva in all of India.[1] We found a ledge of rock to sit on and waited for the sun to rise . . .

After being drawn in by the charisma and magnetism of the 'proto-Siva' figure on the Mohenjodaro seal I began to realize that Siva is everywhere in India. Even in Dwarka, with its all-pervasive cult of Krishna, there is also a beautiful Siva temple. Yet the devotees of the yogic god are most numerous and most demonstrative in the south, amongst the Dravidian-speaking peoples of Tamil Nadu, and Tiruvannamalai is one of the true centres of his cult.

Very little in Hinduism is straightforward or exactly what it seems: identities change and merge, contradictions abound, one thing stands for another, gods may manifest in different ways at the same time, ambiguity is everywhere. All this is there in the ancient story of Siva's great temple at Arunachela:

The Supreme Being, the Ocean of Grace, Lord Siva once had a desire – 'Let me become many.' In accordance with this desire, Brahma and Vishnu came into existence spontaneously. They were delegated the duty of creating the worlds and protecting

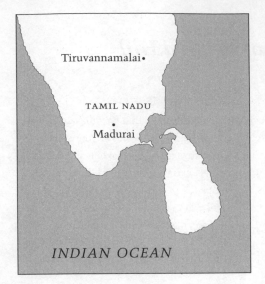

them. However, instead of merely carrying out the duty ordained by the Lord, they were caught up in an argument out of egoism which resulted in a major conflict. Seeing the terrible rage they had fallen into while battling with each other, the Lord of Compassion deemed it fit to reveal Himself in a form that would put an end to their fighting.[2]

To cut a long story short, Siva revealed himself on the spot where Arunachela now stands in the form of a limitless column of blazing light and scorching fire piercing the sky and pervading the universe. On seeing this dazzling and fearsome vision Brahma and Vishnu were not humbled but entered into a new competition to discover 'either the beginning or the end' of the column.[3] Only when both had proved themselves incapable of doing so did Siva at last emerge from the effulgence.[4]

There are a few other twists and turns in the story, but the upshot is that Siva forgives Brahma and Vishnu for their contentions, telling them: 'Carry on vigilantly with your work of creation and sustenance without forgetting me.'[5] He then announces that the effulgent column will remain eternally manifest on this spot in the form of a mountain of fire:

My Effulgent form will shine here forever as eternal, immutable Arunachela. Oceans will not submerge it even at the time of the great Deluge. The winds will not shake it and the world-destroying fire will not burn it.

On hearing these words Brahma and Vishnu humbly bowed down to Siva and prayed: 'Sustainer of the Universe! Let this Hill be the mainstay of the world as stated by you. But moderate its Effulgence, O Rudra, so that it becomes bearable, yet retains its boundless glory and remains a repository of everything auspicious.'[6]

In answer to Brahma and Vishnu's prayers, 'Siva reduced the blinding effulgence of his shining appearance in the column by transforming himself into this lacklustre mountain'[7] – the 'Red Hill' of Arunachela, of which it is said: 'Just as we identify ourselves with our body, Lord Siva identifies himself with this Hill where the reddish colour of the rocks suggests the primeval fire.'[8] In addition Brahma and Vishnu beseeched Siva:

Although this Red Hill exists for the welfare of all, none could worship it without your grace . . . [Therefore] we request you also to take the form of a Lingam on the East side of the Mountain so that we may worship you . . .'[9]

Again Shiva complied and a miraculous column of stone – the Sivalinga, or phallic symbol of Siva – appeared at the foot of the mountain on its eastern side. As a token of gratitude Brahma and Vishnu commanded Visvakarma, the architect of the gods, to erect a temple around it – the primordial temple of Arunacheleswar.

The temple that now stands on the site is of more recent origin. But believers maintain that it is the original natural stone lingam, 'self-generated' at the beginning of time, that still resides in the Holy of Holies and continues to be venerated by millions of pilgrims as the sign and the seal of Siva's presence on earth.[10]

Austerities

We watched the sun come up in the south-east, illuminating first the nine pyramidal *gopurams* that surround the temple complex and then the deeper shadows in the interested rectangles of its plazas, ambulatories and shrines. As the town's streetlights faded out in the rising glare of the day we could see that beyond the temple was a plain extending to the horizon in a great arc beneath us, its flatness broken here and there by isolated conical hills.

We resumed our climb of Arunachela. Although it is not particularly high, the way is steep and the winding path is long. After another hour had passed we still seemed to be nowhere near the summit, the sun was much hotter, and I was beginning to regret bringing only one bottle of water. Santha and I paused to take a swig each, looking down the way we had come to the distant towers of Arunacheleswar. Rising out of the morning haze the temple possessed an epic, otherworldly quality and it was not difficult to imagine it in the way that the ancient traditions describe it – as the work of the gods themselves, built at the dawn of the present cycle of time.

We started climbing again and when we next looked up we saw a lean but muscular young man, with the long tangled hair, ash-smeared forehead and orange loin-cloth of a *sadhu*, sitting cross-legged on a rock on the slope above us. He seemed oblivious to our presence but when Santha said good morning to him in Tamil his reply was friendly enough.

We passed him and continued to climb. When I glanced back a few moments later I saw that he was no longer seated on the rock but following immediately behind, barefooted and silent. Now, effortlessly, he increased his pace and overtook us and soon he had disappeared round a twist in the path ahead, shielded from us by piles of fallen boulders.

I guessed that he must be one of the devotees of Narayana Swami, the almost legendary figure I was hoping to encounter, who was reputed to have remained near the summit of Arunachela for the previous ten years, consuming no solid food of any kind and subsisting exclusively on small quantities of milk and tea brought to him by his acolytes.

By the performance of such austerities (*tapas*), which may range from relatively pleasant tasks like prolonged sexual intercourse without ejaculation to relatively unpleasant ones like holding one's arm permanently above one's head for decades, great yogis like Narayana Swami are believed to build up a special power of supernatural 'heat':

> The basic transformation brought about by the Rishi in his performance of *tapas* is the production of heat in the body. The fire of his *tapas* becomes such that it is transformed into Fire itself, burning the worlds with his heat and illuminating them with the light that radiates from his body . . . Powers of becoming invisible, walking on water and flying through the air are among those most frequently said to be obtained by performing *tapas*; while in the *Yoga-sutras*, a large number of such powers are listed as being attained through the practice of yoga – including, in addition to such 'physical' powers, various types of mental knowledge such as of previous existences and of the thoughts of others . . .[11]

The intense physical and mental discipline that *tapas* requires is also an essential step on the road to liberation from death. Thus, through their fearsome austerities, the Seven Rishis of the *Vedas* were said to have possessed

> powers of rejuvenation, of curing illnesses, and of restoring the dead to life . . . One of the aims of the Rsis in performing *tapas* was to attain to the realm of the immortals and to obtain immortality – even as it is said that the gods and demons themselves performed *tapas* in order to escape death.[12]

John E. Mitchiner, the expert on the traditions of the Seven Rishis, admits that 'such powers are indeed attributed to the Rsis throughout Indian literature'.[13]

But the question is why? Why the consistent association, throughout history, of great *rishis* with these extraordinary powers, and why do they always use the same means – yoga, austerity, meditation – in order to develop them? Is it all just imagination and fantasy on the part of the ancient storytellers? Or is it possible that something substantial lies behind these traditions?

I did not expect Narayana Swami to tell me the answer but I was nevertheless curious to set eyes on anyone who could live on tea and milk at the top of a mountain for ten years. I was also intrigued by the way that his presence there appeared to symbolize or bear out another tradition, recorded in a Tamil text known as the *Arunachela Mahamatmyam* ('Glory of Arunachela') to the effect that Siva himself always sits beneath a tree near the summit of the mountain in the guise of a *siddha*:

> Siva abides here forever as a *siddha* known as Arunagiri Yogi, wearing only a loin-cloth and with matted locks and forehead shining with marks of *vibhuti*. [sacred ash].[14]

Because I had gradually acclimatized myself to such material over many months, I now had no difficulty in understanding how Siva could, at one and the same instant, be a phallic stone column in the Holy of Holies of the temple at the foot of the mountain, an ascetic meditating under a tree at the top of the mountain, and the mountain itself – for it is said that 'unlike other mountains, which have become holy because the Lord dwells in them [e.g. Kailas in the Himalayas], Arunachela is Lord Siva himself'.[15]

Mainstay of the world

In the north Indian tradition of the flood we hear that Manu and the Seven Sages took refuge in the Himalayas and that it was from there that they spread out to re-establish agriculture and to repromulgate the *Vedas* in the 'Land of the Seven Rivers' between the Indus and the Ganges. For south India, a Tamil tradition recorded in the *Skanda Purana* assigns the same role – as a place of refuge from the flood and as a centre of subsequent teaching – to Arunachela, forever protected by Siva's guarantee that 'oceans will not submerge it even at the time of the great deluge':[16]

> The ground near it is not at all touched by the four oceans that become agitated at the close of the Yuga and that have the extremities of the worlds submerged in them . . . When the annihilation of all living beings takes place, when all created things are reabsorbed . . . all the future seeds are certainly deposited there . . . All the lores, arts, wealth of scriptures, and the *Vedas* are truthfully well-arranged there . . . Brahmanas who resort to the foot of that mountain are called by me after the deluge and I make them study the *Vedas* and make collections thereof . . . Sages of well-praised holy observances and rites, who abide within the caves of that mountain, shine with their matted hair. They have the refulgence of 100,000 suns and fires . . .[17]

The *Puranas* also tell us that the Seven Sages (normally associated with the post-diluvial preservation of the *Vedas* in the Himalayas) were amongst those who visited Arunachela.[18] And it was undoubtedly the case, I reflected, as Santha and I continued our climb – passing now through a zone of cooling mist and then entering a dark defile – that this red-granite mountain, which in fact belongs to one of the oldest exposed rock formations on earth,[19] would never have been flooded during the post-glacial meltdown. Even during the worst events, the great tidal waves would not have reached this far inland or this high.

So Arunachela might well have been perceived as a solid and reliable 'mainstay of the world' in a time of rapidly and unpredictably rising sea-levels around the coasts of southern India. How interesting, therefore, that it was remembered, like the Himalayas in the distant north, as a place where 'all the future seeds' were deposited for the later benefit of mankind, and as a refuge for sages where the ancient wisdoms of the *Vedas* were kept safe and from whence they were later repromulgated.

Master of all wisdom

Siva is a god of many dimensions and he has been present in India – all of India – for a very long while. We've seen that his form as a meditating *sadhu*, lean, naked, powerful, the Lord of Yoga, goes back at the very least to the Pasupati seals of Indus-Sarasvati times, 4700 years ago. The same is true of his manifestation as a phallic cone or column of stone – many examples of which have been excavated in Indus-Sarasvati sites.[20] He is also one of the primeval gods of the *Rig Veda*, where he is known as Rudra. It is in recognition of this ancestry that the names Siva and Rudra are used interchangeably (or jointly as 'Rudra-Siva') in many ancient Indian scriptures.[21] And Rudra is addressed as follows in the *Yajur Veda*: 'Thou art *Siva* [gracious, kind] by name.'[22]

Like Siva, Rudra is both terrifying and reassuring.[23] He is said to have 'two natures or two "names": the one, cruel and wild (*rudra*), the other kind (*siva*) and tranquil (*santa*)'.[24]

Like Siva, Rudra is the 'dweller in the mountain',[25] 'the blue throated one',[26] and 'Tryambaka' ('the three-eyed').[27] Like Siva, Rudra of the *Vedas* has a fair or white complexion[28] (but is also sometimes described as 'red'[29]), and is a great Yogi and the Lord of Animals.[30] Like Siva, Rudra has long, braided and/or matted hair, and healing powers.[31] Like Siva, Rudra is associated with fire.[32] And like Siva, Rudra's symbol in later Vedic tradition is *sthanu*, 'a post' or 'a pillar' signifying 'the timeless, motionless state of *samadhi* in which the Lord of Yoga dwells'.[33]

But above and beyond any of this, the true defining characteristic of Rudra-Siva is as the God of all Knowledge and of insight and inner wisdom (*jnana* – *gnosis*). This is why we read, in Book VIII of the *Rig Veda*: 'That mind of Rudra, fresh and strong, moves conscious in the ancient ways.'[34]

This is why Siva is frequently portrayed in Hindu religious art as Jnana-Dakshinamurti, Master of all Wisdom, 'sitting under a tree on Mount Kailasa with his foot on a dwarf who symbolizes human ignorance'.[35]

The highest knowledge to the most humble

The particular nature of Rudra-Siva as the God of Knowledge in the form of a powerful *rishi* with unkempt hair who lives in mountains and wild places is connected to a subtle and complex system of ideas which, even if one does not agree with it, must be admitted to be extremely well thought-out and (in view of the Mohenjodaro seal) extremely ancient. Ultimately, it seems to state that enlightenment, and true knowledge, cannot be attained without becoming the master of one's impulses and renouncing the lures of the material world – or at any rate one's 'attachment' to it. Conversely, a person's material wealth and physical beauty can tell us nothing useful about that person's mind and soul. It is to drive this point home, perhaps, that when the gods come to seek advice from Siva they find him 'accompanied by myriads of devoted followers, all of them naked, all deformed, with tangled curly hair'.[36]

Likewise the Orientalist Alain Danielou observes that:

> Already the *Vedas* picture Rudra as living in the forests and mountains, ruling over animals tamed and wild. The Saiva mythology shows him as the divinity of life, the guardian of the earth, who wanders naked through rich forests, lustful and strong. He teaches the highest and most secret knowledge to the most humble.[37]

The idea that true wisdom does not clothe itself in finery is also conveyed in another story of Brahma, Vishnu and Siva, where Brahma and Vishnu are once again contending as to which one of them is the supreme being:

> Thus Vishnu and Brahma disputed, and at length they agreed to allow the matter to be decided by the *Vedas*. The *Vedas* declared that Siva was the creator, preserver, destroyer. Having heard these words, Vishnu and Brahma, still bewildered by the darkness of delusion, said, 'How can the lord of goblins, the delighter in graveyards, the naked devotee covered with ashes, haggard in appearance, wearing twisted locks ornamented with snakes, be the supreme being?'[38]

The answer, since Rudra-Siva is in fact the supreme being, is that he can take any form he chooses. And it is his choice that leads him to smear himself with ashes and consort with the poor and humble who are pure in spirit. According to Professor Stella Kamrisch:

> He stood apart and was an outsider to other Vedic gods. He could be recognized by his weird, mad looks. He seemed poor and uncared for, neglectful of his appearance; the gods despised him, but he intentionally courted dishonour, he rejoiced in contempt and disregard, for 'he who is despised lies happy, freed of all attachment'. The fierce, self-humiliated Lord was a yogi ... He provoked contempt as a test of his detachment.[39]

So there is an *idea* here, a fairly consistent idea – perhaps it is better to say a system of ideas – behind the conception of Rudra-Siva as the God of Knowledge. Whatever knowledge and powers he possesses have been acquired through meditation, austerity and self-sacrifice – practices that are likely to have been part of a wider curriculum. And the same is true, unconditionally, of the Seven Rishis of the *Vedas*. They also, John Mitchiner observes,

> smother their bodies with ashes, and have their hair uncut, matted and tied in a knot: in other words they are depicted as being in appearance much as many other – especially Saiva – ascetics.[40]

There is even a tradition in the *Bhagvata Purana* that the greatest sages 'range over the world in the guise of mad persons' whilst imparting wisdom.[41]

At the very least the lesson of this is that it is worth showing respect and listening carefully to the words of any person. Appearances can be deceptive and you never know who you're dealing with.

In such a spirit I hoisted my weary body up the last few metres of Arunachela's crumbling granite scree and on to the muddy path overlooked by sloping rocks that led to Narayana Swami's mountain-top lair.

Tea and prayers

The *rishi* did not occupy the summit of the mountain – he would have been roasted by the sacred fire that is lit there every December to mark the apotheosis of Siva as a column of flame – but had set up his hermitage in a tree-lined bower that lay off to one side a few minutes' walk below the summit. He was attended by the young man who had passed us earlier on our climb, and four other Siva ascetics (Sivachariars) clad in orange rags, who now peered down from the rocks and greeted us from either side of the muddy path.

Suddenly, as soon as we'd arrived, we found ourselves in the middle of some sort of ceremony or routine. The young acolytes indicated that we should take off our shoes – because we were now approaching holy ground – and beckoned to us to accompany them down a little incline to the edge of the bower where Narayana Swami had presumably been sitting for the past ten years. In the shady gloom, buzzing with enormous hornets, we could just make out a little half-tent, like a refugee lean-to covered in plastic, underneath the overgrowing branches of the trees.

We never actually did get to see the *rishi*, this embodiment of Siva, face to face, let alone speak to him. He didn't speak to anyone, at least not in any known language, although he did mumble and grunt incoherently to his followers from time to time and they seemed to understand. The most we saw was a thin but strong arm with leathery skin reaching out sometimes, and a bony finger making patterns in the mud in front of the little plastic tent – and there was a great deal of mud around the *rishi*'s bower and pools of water lying in the hollows of the rocks.

Next we had to sit down in the mud and the acolytes brought us dirty half-coconut shells of what they announced to be tea that had been blessed by the *rishi*. Into this tea, which was lukewarm, they melted finger-sized dollops of butter and asked us to drink. We did so, with some trepidation (I was thinking amoebas, right from the start). Then there were prayers, reminding us that the tea had been blessed and that it would make us well in our bodies. Then more tea and more prayers. Then we were brought a cold, but somehow greasy, herbal drink with leaves floating in it – also blessed by the *rishi*. We drank it. More prayers followed, and more tea with butter and intestinal parasites.

After that one of the acolytes beckoned to us to line up behind him and led us in a clockwise direction on a brisk walking circuit (with each circuit requiring only twenty or thirty seconds to complete) of the path that runs around the inside of the bower and in front of Narayana Swami's shelter. There we knelt down in the mud and sacred ash was placed on our foreheads. Then we completed a few more circuits chanting as we went 'Siva, Siva, Siva, Raga Ra, Raga Ra' – or something like that.

It was very strange. We didn't ask for the ceremony and – most unusually in India – no money was required of us for participating in it.

Arunachela and Kumari Kandam

Was Narayana Swami genuinely mad, I wondered, as we made our way down Arunachela that afternoon. Or was he one of those great *rishis*, lit with the inner fire of *tapas*, said to roam the world disguised as a madman whilst imparting knowledge? To believe him to be wise if he was in fact mad would be the height of gullibility, but to believe him to be mad if he was in fact wise might be an even bigger mistake. Besides, whatever he was, his presence testified to the continuing vitality of the pan-Indian tradition that mountains such as this one had served as centres for the collection and repromulgation of the *Vedas* after the flood and as places where a brotherhood of ascetics preserved antediluvian knowledge that would be used to plant 'the seeds of the future'.

Setting aside for a moment its connection with Rudra-Siva, the Yogic god of wisdom, I felt that I needed more information on this 'flood' aspect of the Arunachela story. Specifically, I wanted to find out if was connected in any way to the mysterious lost land called Kumari Kandam that was said to have been swallowed up by the sea around south India thousands of years before. By the time Santha and I reached Tiruvannamalai in February 2000 I was already familiar with some details of this tradition – which is widely known amongst India's 200 million Tamils but almost unheard-of outside India. I now hoped to learn more from a Tamil pundit whom I had arranged to meet after our climb. A retired ship's captain who had given himself over to the life of contemplation, he now resided permanently at the Ashram of Sri Ramana Maharishi, which is positioned at the foot of Arunachela about 2 kilometres from the Arunacheleswar temple.

A loin-cloth, a water-pot and a walking stick

Maharishi means 'great *rishi*' and Sri Ramana seems in every way to qualify for this title. Like Naryana Swami, he had at one stage of his life exposed himself for several years on the slopes of Arunachela after first arriving there in 1896. At the time, it is recorded, Sri Ramana

> was completely oblivious to his body and the world; insects chewed away portions of his legs, his body wasted away because he was rarely conscious enough to eat, and his hair and fingernails grew to unmanageable lengths.[42]

This fugue had been brought on by a flash of spiritual insight that the real nature of the human creature is 'formless, immanent consciousness'.[43] After two or three years in this state Sri Ramana 'began a slow return to physical normality, a process that was not finally completed for several years'.[44] During this period followers began to gather about him and by the time of his death in 1950

> he was widely regarded as India's most popular and revered holy man . . . He made himself available to visitors twenty-four hours a day by living and sleeping in a communal hall which was always accessible to everyone, and his only private possessions were a loin-cloth, a water-pot and a walking stick.[45]

Since Sri Ramana's death his Ashram has continued to attract devotees and is a thriving, busy place today with a good library, extensive offices, private and communal accommodation, a canteen and a beautiful prayer hall. The pundit I had come to meet, Captain A. Naryan (no relation to Naryana Swami), was a tall, heavy-set moustachioed man in his early seventies, who explained to me that he was no great scholar, but that he had a personal interest in Tamil traditions which he had been able to pursue since his retirement, and that he hoped his small knowledge might provide me with a few clues for my search. 'Everyone calls me Captain,' he said, when I asked how I should address him.

As old as the hills

We began by talking through the story of Arunachela and how it was said that the mountain would never be submerged or swept away – even by the waters of the great deluge at the end of a world age. 'So we may assume that this has been the case in the past?' I was half asking, half affirming 'because there is a destruction at the end of each cycle of *yugas*, so somehow Arunachela has remained constant throughout all of this?'

The Captain nodded sagely.

'So it is the centre of everything,' I continued. 'Now the area which I'm trying to explore is the borderland between history and what comes before history. And we know that, historically, the temple here at Arunachela, there are documents which speak of its construction, and probably the temple as we see it now, most of it is less than 1000 years old and some parts may go back closer to 2000 years old, but at the heart of it is the Sivalingam, which is said to be much older. Can you tell me a bit about that lingam – which is supposed to be "self-created"? What does this mean?'

'"Self-created",' replied Narayan, 'means it is not chiselled by man in the way that other lingas are chiselled by man. But there are certain other lingas which come out of the earth, not made by man, but which conform to all the characteristics – like the proportion, the width, the circumference and the height. So just like a man-made Sivalinga it conforms to the correct proportions.'

'So it would look like a man-made one, but it's not?'

'It is not!' affirmed the Captain. 'It is more perfect. And it must be as old as Arunachela itself. Because as the *Purana* says, when the primal gods were beseeching the supreme being: "Since the mortals cannot see you in your effulgence form, you should take the form of a lacklustre hill. Even if you assume the form of a lacklustre hill, only the clouds can anoint you and only the sun and the moon can be the lamps lit for you. But we have to do *puja* [prayers, offerings] before you so you should assume the form of a smaller lingam." So Arunachela granted their wish and he told them I will appear in the form of a lingam and you may worship me . . .'

'And that is the lingam that's in the temple?'

'That is the lingam.'

'OK, fair enough. A naturally formed lingam that's literally as old as the hills. But at some point human beings must have found it, begun to treat it as a cult object, and built some sort of structure around it. What I'm trying to get at is when did the anointing and worship of this naturally formed lingam begin? It's presumably much earlier than the date of construction of the temple that's standing on the site today?'

'Yes. Yes, naturally. What the *Puranas* say is that gods came here and they were the first to build a temple around the self-generated lingam of the Lord. That's what the *Puranas* say. The primal gods Brahma and Vishnu built the temple, and cities were created by the heavenly builder Visvakarma around this place, around Arunachela.'

Cities of the gods

I was already familiar with the origin myth of Arunachela as it is told in the Tamil *Puranas*[46] and knew that it was like many other tales from around the world of cities and temples built by gods.[47] Frequently – as in the case of the *Edfu Building Texts* of ancient Egypt, for example – such traditions tell us that the gods embarked on these works of construction at carefully chosen locations on earth in the aftermath of a global cataclysm, typically a flood.[48] This is not what the *Puranas* say about the temples and cities supposedly built around Arunachela by the gods; nevertheless the central motif of the story is the eternal endurance of the Red Hill through the cataclysms that accompany the end of world ages, and it is specifically stated: 'Oceans will not submerge it, even at the time of the great deluge.'[49] So it was here that I wondered if there might be some crossover with the Kumari Kandam myth.

'This memory of gods building the first temple and cities at Arunachela,' I now asked, 'what period do you think it originates in? If those cities are supposed to have been built at the same time as the formation of the mountain and the self-generated lingam, then that's surely an awfully long time ago.

'Geology says it must have been 3.5 billion or 2.5 billion years ago that Arunachela first took its form as a mountain. But such a time-span seems outside any reasonable scale for the construction of cities and temples, since we know that the human race only came into being, what is it, 100,000 or 200,000 years ago? No "memory" of ours can be older than that.

'But if they're to be placed in the human scale, if they're not just something that's been made up by storytellers, then shouldn't archaeologists be able to find at least some traces of these former cities of the gods?'

The Captain shrugged. 'Probably during the previous destructions of the world their remains have been hidden from us and if we could search sufficiently widely probably we could find many cities below the surface of the earth.'

He seemed to reflect for a moment. 'You see,' he said at last, 'Arunachela is in the land of the Dravidians, where our language goes back more than 10,000 years.'

He then told me that the Red Hill was referred to in the most ancient surviving work of Tamil literature, the *Tolkappiyam*,[50] which itself makes reference to an even earlier work now lost to history which in turn had supposedly been part of a library of archaic texts, all now also vanished, the compilation of which was said to have begun more than 10,000 years previously. This had been the library of the legendary First Sangam – or 'Academy' – of the lost Tamil civilization of Kumari Kandam, swallowed up, as Captain Narayan put it, 'by a major eruption of the sea'.

And one of the members of the First Sangam, he added, finally making the direct connection that I suspected to the Arunachela story, had been Siva himself,[51] the god in the mountain, the god of yoga performing *tapas* beneath a tree at the top of the mountain, the god of cosmic knowledge compressed into the lingam at the foot of the mountain.

Academies of the gods

As Captain Naryan walked us to the gate of the Sri Ramana Ashram later that afternoon, he gave me the name and telephone number of a friend who he hoped might be useful to me in the city of Madurai, the next great centre of the cult of Siva that we intended to visit in south India. There, he told me, there were knowledgeable professors at many colleges and universities – for Madurai has been always been a place of scholarship and learning – who would certainly be able to tell me much more about Kumari Kandam and the Sangam tradition. Nor could there be any more appropriate place to mount such an inquiry, since Madurai itself was an important part of the Sangam tradition – having served as the headquarters of the Third Sangam . . .

'So let me see if I've got this right,' I asked in parting. 'We have a First Sangam thousands of years ago and it gets flooded – the city which it's in gets flooded?'

'You are right. Permanently flooded. It was overwhelmed by the sea.'

'And that city was?'

'It was called Tenmadurai – which means "Southern Madurai". It was in the southern part of Kumari Kandam. After it was gone, a city called Kapatapuram that lay further to the north was chosen as the headquarters of the Second Sangam. It endured for some thousands of years but ultimately it too was flooded. Our oldest surviving text, the *Tolkappiyam*, is a work of the Second Sangam.'

'And then?'

'Finally, when Kumari Kandam had entirely gone beneath the sea, the Third Sangam was established in the city of Madurai. Then it was called Uttara Madurai, "Northern Madurai".'

Lingam or omphalos?

Before we left Tiruvannamalai we visited the Arunachelswar temple in order to see Lord Siva in his lingam form.

Walking barefoot through the ambulatories and open stone-paved plazas, we passed rows of poor, homeless and hungry people, for the most part dressed in rags – here a mother with sunken breasts trying to suckle her child, there an old blind man, here a cripple, there a leper – waiting patiently for the charity soup kitchen to feed them.

If we looked up we could see the rugged red peak of Arunachela looming above us, framed by the tall towers of the *gopurams* that marked the main entrances of each of the temple's internested rectangular zones. Their steep pyramidal form, and their general arrangement in opposing pairs around a geometrical central plaza, as well as the scale of the whole enterprise, reminded me forcefully of the Mayan city of Tikal in Guatemala, and of Angkor Thom and Angkor Wat in Cambodia. Indeed, in general, it has for a long while struck me as worthy of note that so many of the world's ancient places of worship – in Europe, Egypt, Israel, Mesopotamia, India, south-east Asia, China, Japan, Central America and the Andes, for example – have assertively geometrical designs and architecture. What is this recurrent association of geometry with the religious quest? Certainly, it seems that there were many great thinkers in antiquity who, if asked 'What is God?', might well have replied, as St Bernard of Clairvaux did to the same question, 'He is length, width, height and depth.'[52]

Because all Hindu temples are part circus, we encountered a painted elephant surveying the world through a jaundiced eye, chained up in a stone pillared pavilion, and when we descended the steps to the sacred pool, known as Siva-ganga Teertham we were followed by a persistent fortune-teller who could only with great difficulty be persuaded to relinquish what he clearly felt was a fair claim on us.

Soon after we had shaken him off (not until Santha had relented and agreed to have her fortune told for 100 rupees) we were appropriated by a beautiful doe-eyed young man in flowing white robes who floated up to us declaring himself to be a Brahmin and the son of a senior priest of the temple. As though reading our thoughts he then led us towards the sanctuary where the 'self-generated' lingam of Siva resides, explaining as he did so that it was normally out of bounds to non-Hindus, but that we had happily chanced, in his person, upon just the man to get us inside. The only thing we would not be allowed to do, he said, was touch the lingam – a privilege that was reserved for initiated Sivachariars.

I have been offered illegal access to inaccessible areas in many temples around the world, and the young Brahmin's patter was so familiar that I could already almost count the 100 rupee notes changing hands. Still, we followed him through a maze of crowded rooms and hallways, visited various subsidiary shrines where we were fed puffed rice and sugar, had our foreheads liberally smeared with ash, and jumped a queue of worshippers at the entrance to the principal sanctuary. Then suddenly, for just a few moments, we were in the presence of the natural pillar or cylinder of stone that is venerated by the faithful as the eternal manifestation of Siva himself. The pillar, however, was so decked out with finery, robes, jewellery and an elaborate head-dress in the form of a rearing golden cobra hood that it was impossible to get a clear glimpse of any part of it. All that I can say is that it seemed to be less than half a metre thick and approximately 1.5 metres high and was rounded like a cigar-tube at the tip – very much, in other words, like 'unclothed' Sivalinga that can be seen in temples and shrines all around India.

So what was special about this one?

As he took my money, the Brahmin could only repeat the old mantras – that it is a wonder of nature wrought by the power of Siva, that it is ancient and nobody knows how old it is, and that the first temple to be built around it was the work of the gods.

The numbers of time and the world grid

In previous books I have grappled several times with the hypothesis that the earth and all its oceans may have been explored, mapped and accurately measured with lines of latitude and longitude – a pre-eminently 'civilized' and sophisticated activity – thousands of years before what we now think of as history began.[53] I want to avoid the tedious repetition of evidence and arguments that I have already presented in *Fingerprints of the Gods* and *Heaven's Mirror*, but, in summary, the problem is this: certain medieval and Renaissance maps seem to express sophisticated geographical and cartographic knowledge far ahead of the science of their age. A number of researchers attribute this knowledge to older source documents that have not come down to us. In his *Maps of the Ancient Sea Kings*, for example, Charles Hapgood draws attention to the accurate longitudes on the so-called 'portolano' charts of the fourteenth century (400 years before the invention of Harrison's Chronometer supposedly made the accurate measurement of longitude at sea feasible for the first time). Hapgood believes that the anachronism may be explained by the survival of ancient cartographical knowledge (either in the form of maps copied and recopied again and again down the generations, or in the form of oral traditions retained and passed on amongst mariners) that originated with a highly advanced, sophisticated and as yet unidentified seafaring civilization of prehistory. He makes the same argument for the appearance of Antarctica on the Oronteus Finnaeus map of 1539 (some 300 years before Antarctica is believed to have been 'discovered').[54]

Evidence that provides some tangential support for the general thrust of Hapgood's theory comes from a large sequence of numbers – including 18, 36, 72, 144, 2160, 4320, 25,920, etc. – that appears repeatedly and prominently in ancient myths, scriptures and traditions from all around the world.[55] According to the late Professor Giorgio de Santillana of the Massachusetts Institute of Technology and Professor Hertha von Dechend of Frankfurt University, these ubiquitous numbers derive from an archaic astronomical tradition which used shared, globally diffused conventions to record its observations of the stars. The central symbol of the system depicts a great wheel that rotates in heaven, 'churning' or 'milling' for thousands of years. The entire axis, spokes and bands that bind this wheel are said to be periodically broken by recurrent cataclysms – often flood and fire – at which point a new wheel is forged and the cycle begins again.

Santillana and von Dechend's explanation for this symbolism and for the numbers associated with it is that it is a metaphor for the celestial phenomenon that astronomers today call 'precession'. This is a slow, cyclical wobble of the earth's axis in space so that, if the tip of the north (or south) pole were imaginarily extended it would be seen to transcribe a great circle amongst the polar stars over a period of 25,920 years. Though it was not thought to have been detected until the time of the Greeks, it is Santillana and von Dechend's radical contention that precession was observed, and measured, thousands of years earlier than that by what they describe as 'some almost unbelievable ancestor civilization'.[56] They further claim that it is these same ancient measurements (all *time* measurements) that generate the mysterious numbers in the myths.

The most notable effect of precession is that it causes a slow, relentless drift of the background of stars against which the sun is seen to rise on the spring equinox (21 March, when night and day are of equal length). This is called 'the precession of the equinoxes'. Although it can be detected by relatively simple observations, these must be sustained over several generations before the sequence begins to emerge.

The ruling number in the sequence, Santillana and von Dechend suggest, is 72 – the round number of years required to observe one degree of the precession of the equinoxes.[57] This, they say, is why the tally of significant numbers in the myths includes 72 and multiples of 72 (e.g. 144, 720, 2160, 4320, etc.); 36 (half of 72) and multiples of 36; 24 (one-third of 72) and multiples of 24, etc. The system also uses other ways of combining these numbers – e.g., 72 + 36 = 108, a sacred number in many cultures, while half of 108 is 54, also a sacred number, as is 540 or 540,000, or 5,400,000, etc. and as are 108,000, 1,800,000, and so on.[58]

It may be that this powerful number system is *not* based on the observation of the precession of the equinoxes at all and that some explanation other than a lost civilization will ultimately be found for it. But what cannot be denied is the simple, well-evidenced fact that the system exists – whatever its source – and that it occurs in known texts of all the great archaic mythological and religious systems, amongst them ancient Sumer and Babylon, Vedic India, ancient Egypt, ancient Greece, ancient China, the Maya of Central America, the Old Testament Hebrews and many other cultures.[59]

It was only while I was writing *Heaven's Mirror* that I began to look into another and much more controversial possibility – that a network of sacred sites might have been established all around the globe according to a longitude grid based on precessional numbers. Thus, the massive sacred complexes on which stand the Great Pyramids of Giza in Egypt and the fabulous temples of Angkor in Cambodia are on meridians 72 degrees of longitude apart; Pohnpei is 54 degrees of longitude east of Angkor; Easter Island is today the closest dry land to 144 degrees of longitude east of Angkor; the Bay of Paracas in Peru, dominated by the massive cliff drawing of unknown origin known as the 'Candelabra of the Andes', lies 180 degrees east of Angkor. Frequently these sites are linked to flood myths, spoken of in ancient traditions as 'Navels of the Earth' (*omphalos* in Greek), and are rich in symbolism of obelisks, stone pillars, pyramids and other stone monuments.[60]

All this I was already well aware of during my travels in India in February and March 2000. Yet I honestly did not expect when I came to Arunachela, despite its obvious and prevalent *omphalos*/lingam symbolism, that it too would prove to be located at a meaningful point on the same hypothetical 'precessional grid'. I only looked it up in the longitude tables as a matter of routine. As soon as I did so, however, it was immediately obvious that a relationship based on significant precessional numbers does in fact exist between Arunachela and other grid sites – for it lies 24 degrees west of Angkor and 48 degrees east of Giza (respectively one-third and two-thirds of the 72 degrees of longitude separating the former from the latter).[61]

Apparent longitudinal 'correlations' linking sacred sites according to a sequence of numbers thought to have been derived from astronomical observations that occur in ancient myths and scriptures could, of course, arise by chance. I don't deny that possibility. But I wish to pursue what I believe to be the more interesting explanation – namely that such sites may originally have been established on specific longitudes to act as permanent markers and reference points for an archaic worldwide grid of earth measurements and to safeguard precious geodetic and navigational knowledge for the long-term benefit of mankind.

This, indeed, is little more than is already claimed in the ancient Indian accounts of the deluge, and the survival of it by a remnant of wise men, and their preservation and repromulgation of antediluvian knowledge in the new age of the earth. Moreover, it can hardly be an accident that the *yuga* system that lies at the heart of the Dwarka story, of the story of the flood of Manu, and of the Hindu concept of recurrent cycles of cataclysm and rebirth, is also denominated in terms of precessional numbers. According to the *Puranas*, for example, the duration of the Kali Yuga is set at 1200 'divine years', equivalent to 432,000 mortal years. The durations for the preceding Krita, Treta and Davapara Yugas are set respectively at 4800 divine years, 3600 divine years and 2400 divine years, such that one *mahayuga* – made up of the total of 12,000 divine years contained in the four lesser *yugas* – is equivalent to 4,320,000 years of mortals.[62]

Whatever the explanation ultimately turns out to be, and whether Santillana and von Dechend are basically right or basically wrong, the worldwide distribution of such an intricate sequence of numbers, not only in myths but also in architecture (e.g., the 72 pillars of the Dwarkadish temple), represents a serious problem that orthodox historians have so far failed to address.

If it is not 'coincidence', then what is it?

The riddle of Vishnu's three steps

Santha and I treated ourselves to a luxury in south India in February 2000, which we would never have dreamed of affording back in 1992. This was a comfortable, crème-white, air-conditioned Ambassador limousine (for what was going to turn out to be a journey of almost 3000 kilometres) with Palani, a small, wiry ex-army driver from Chennai, at the wheel. With his steady nerves and encyclopedic knowledge of the highways and byways of Tamil Nadu, he was the best possible guide and friend we could have had on such a journey. When I needed a beer in a 'dry' town he always knew (although never imbibing himself) where to obtain bottles of cold, illicit Kingfisher wrapped up in brown paper sacks. And more to the point he put us through no collisions, no nerve-jangling skids, no horrific misjudgements of the proximity of a pedestrian, no death-defying overtaking manoeuvres, and no falling asleep at the wheel.

From Tiruvannamalai we drove south all day towards Madurai through a rich, green, predominantly flat dreamscape of paddy fields and palm trees dotted here and there with the weird outcroppings of ancient red granite that are the characteristic feature of this region. There were people everywhere, Tamil peasant farmers at work in the fields in brightly coloured clothes, or strolling along the road, sometimes drying cattle fodder on the road itself, doing hard labour on building sites and eighteen-hour days in wayside shops and stalls – a tremendous mass of individual human lives surviving in many cases on the very edge of absolute penury yet somehow making do and getting by. It was fascinating to realize, and impossible to ignore, that the religion of all these industrious people was a peculiarly Saivite brand of Hinduism:

- Siva 'the embodiment of knowledge'.[63]
- Siva, the god of wisdom, who rules in 'the city of knowledge' (*jnana-puri*, literally '*gnosis* city').[64]

- Siva who takes the form of Arunachela, 'the mountain of knowledge'.[65]
- Siva who, through initiation into *gnosis*, has the power to inflict or to withhold death and to grant immortality.[66]

In some texts, I had been interested to learn, Siva is identified with Vishnu. In the *Mahabaratha*, for example, there is an episode in which the warrior Arjuna experiences a revelation after being wrestled to the ground by a huge stalwart being:

> Arjuna's limbs were bruised and he was deprived of his senses. When he recovered he hailed the god, saying: 'Thou art Siva in the form of Vishnu and Vishnu in the form of Siva . . . O Hari, O Rudra, I bow to thee.'[67]

In the *Rig Veda*, Vishnu's principal exploit, recounted and celebrated again and again, is the taking of 'three steps'.[68] Although it is agreed that these steps must symbolize something of profound importance, scholars have as yet reached no consensus as to their underlying meaning.[69]

I pulled the Griffith translation of the *Rig Veda* from the half-open satchel that lay perched between Santha and myself on the middle of the back seat and opened it at Book I, Hymn 104:

> I will declare the mighty deeds of Vishnu, of him who measured out the earthly regions . . . thrice setting down his footstep, widely striding. For this mighty deed is Vishnu lauded . . . He within whose three wide-extended paces all living creatures have their habitation . . . Him who alone with triple step hath measured this common dwelling place, long, far extended . . .[70]

All kinds of symbolism might indeed be intended in such a passage, but if we take the hymn at face value, then isn't it rather clearly saying that Vishnu measured out the earth by taking three footsteps? We might speculate on what precisely the footsteps represent, but the involvement of the whole enterprise in earth-measuring – i.e., geography – cannot reasonably be denied.

Other passages reinforce the same conclusion, describing Vishnu, for example, as 'He who strode, widely pacing, with three steppings forth over the realms of earth for freedom and for life . . .'[71] Two verses later we read that 'He, like a rounded wheel, hath set in swift motion his 90 racing steeds together with the four . . .'[72] What could the function of this latter verse possibly be if it is not to invite us to multiply 90 by 4, giving us the 360 degrees of the circle (or 'rounded wheel')? Remember, we have been told just beforehand that such an approach to measuring out 'the realms of earth' is a contribution to the cause of freedom and life – a clear incentive to its preservation!

In Book 6, Hymn 49 of the *Rig* we find Vishnu described as 'He who for man's behoof in his affliction thrice measured out the earthly regions.'[73] Again, the idea seems to be that Vishnu's earth-measuring endeavours were of great value and benefit to mankind and were, moreover, delivered in a time of 'affliction'.

Last but not least, in Book I, Hymn 164, we encounter the following riddle:

> Formed with 12 spokes, by length of time, unweakened, rolls round the heaven this wheel of during Order. Herein established, joined in pairs together, 720 sons stand . . .[74]

So here, represented by a multiple of its 'ruling' number 72, pops up Santillana and von Dechend's ancient precessional code combined in the same passage with the familiar 'wheel of heaven' metaphor of the precession of the equinoxes. The passage also provides further evidence that the convention still in use by modern geographers of dividing the circle into 360 degrees (or 720 half-degrees) was already in existence in Vedic times and is directly alluded to in this hymn. Likewise, the 12 spokes of the wheel are anachronistically suggestive of the 12 'houses' of the (supposedly Graeco-Babylonian) zodiac in which the sun rests for 30 'days' of each precessional month – each such month being equivalent to 2160 human years with the entire precessional cycle thus amounting to 12 × 2160 = 25,920 human years.[75]

Surviving the null hypothesis

Could there really be 'science', in the hard, empirical, modern sense, in the ancient Indian scriptures?

According to Dr Richard L. Thompson, who received his Ph.D. in mathematics from Cornell University, where he specialized in probability theory and statistical mechanics, the answer to this question is 'yes . . . probably'! In his impressively researched and thoroughly documented study *Mysteries of the Sacred Universe* Thompson takes a particularly close look at the *Bhagvata Purana* (a later compilation of oral traditions than the *Rig Veda* but one that nevertheless belongs, as we have seen, to the same body of knowledge).[76] In it he draws attention to a curious word picture called Bhu Mandala that the *Purana* conjures up and that consists of circles and internested spheres of precise, very large, dimensions. He argues that Bhu Mandala is a complex and cleverly designed cosmological model serving at one and the same time as an accurate map of the solar system and as a planar projection map of the earth.[77]

Thompson's arguments must be considered on their own merits backed up by the detailed evidence that he sets out in his book. But the centrepiece of his case is the electrifying correlation, to which he is the first to draw serious attention, between the dimensions given for the various circles of Bhu Mandala in the *Bhagvata Purana* and the actual dimensions of the planetary orbits within the solar system as determined by modern science.[78] Since the correlations turn out to be extremely close, Thompson concludes:

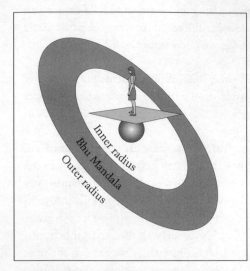

The Bhu Mandala shown as a tilted ring in relation to a local horizon on Earth. Based on Thompson (2000).

Orbits of Saturn and Uranus around Earth.

Comparison of Bhu Mandala's parameters with orbital radiuses of Saturn and Uranus		
	7928 miles	Earth's diameter
compare {	890,000,000 miles	radius of Saturn's orbit
	1,000,000,000 miles	inner radius of Bhu Mandala
compare {	1,790,000,000 miles	radius of orbit of Uranus
	2,000,000,000 miles	outer radius of Bhu Mandala

It is clear that Bhu Mandala, as described in the *Bhagvatam*, can be interpreted as a geocentric map of the solar system out to Saturn. But an obvious and important question is: Did some real knowledge of planetary distances enter into the construction of the Bhu Mandala system, or are the correlations between Bhu Mandala features and planetary orbits simply coincidental?[79]

Being a mathematician interested in probability theory, Thompson is better equipped than most to answer this question and does so through computer modelling of a proposed 'null hypothesis' – i.e.,

that the author of the *Bhagvatam* had no access to correct planetary distances and therefore all apparent correlations between Bhu Mandala features and planetary distances are simply coincidental.[80]

However, the Bhu Mandala/solar system correlations proved resilient enough to survive the null hypothesis. 'Analysis shows that the observed correlations are in fact highly improbable.'[81] Thompson concludes:

> If the dimensions given in the *Bhagvatam* do, in fact, represent realistic planetary distances based on human observation, then we must postulate that *Bhagvata* astronomy preserves material from an earlier and presently unknown period of scientific development . . . [and that] some people in the past must have had accurate values for the dimensions of the planetary orbits. In modern history, this information has only become available since the development of high-quality telescopes in the last 200 years. Accurate values of planetary distances were not known by Hellenistic astronomers such as Claudius Ptolemy, nor are they found in the medieval *Jyotisa Sutras* of India. If this information was known it must have been acquired by some unknown civilization that flourished in the distant past.[82]

Needless to say, a civilization that could make accurate maps of planetary distances, a hypothetical civilization of the distant past that had approached to within 200 years of our own level of development in astronomy, would have had no great difficulty in observing and measuring the precession of the equinoxes, or in dividing up the earthly and celestial spheres into degrees of longitude and latitude, or in consecrating a series of sacred sites at specific longitudes, and, in the process, exploring and mapping the globe.

Neither do I find it at all difficult to imagine how the geodetic and cartographic works of such an elder culture might have been remembered in much later and more superstitious times as gifts that had been handed down by the gods.

Had some stone pillar, now venerated as the self-generated lingam of Siva, been set up by prehistoric geodecists at Arunachela, for example, to mark the auspicious longitude of the Red Hill? The same symbolism of the lingam is, of course, found all over the temples of Angkor in Cambodia. And in ancient Egypt the conical Ben Ben stone, perched atop a stone pillar, was the symbol of the Heliopolitan priesthood that built the Pyramids of Giza.

Same symbolism in all three places.

Same gnostic quest for immortality.

Same use of precessional numbers in their architecture and their myths.

And there are 48 degrees of longitude between Giza and Arunachela, 24 degrees between Arunachela and Angkor, and 72 degrees between Giza and Angkor.

Coincidence?

Design?

Take your pick.

Madurai

A few hours later, well after dark, the Ambassador rolled smoothly across the thin membrane that separates rural from urban life in India, and we found ourselves in Madurai. As the reader will recall, Captain Naryan had told me that this city, with the great Meenakshi temple residing at its heart, was the site of the third and last Sangam, or Academy, of Tamil poets and philosophers – an institution that traced its origins back to the antediluvian civilization of Kumari Kandam.

While we drove through the crowded streets blaring with sound and lights I remembered that the First Sangam was said to have been established many thousands of years ago in an earlier 'Madurai' – Tenmadurai – that lay far to the south on lands subsequently swallowed up by the sea.

It is astonishing how little attention has been paid to these Tamil myths, and how little has been written about them outside the subcontinent. Even David Schulman, who has done more than most to fill this gap in knowledge, is dismissive of the significance of the traditions:

> The story of the three Cankam [Sangams] as it appears in our sources is suspect on many counts, and there is no geological evidence of any deluge affecting the area in historical times.[83]

Though I respect Dr Schulman's work, which offers a lucid exposition in English of the Tamil flood myths, he is dead wrong to consider only whether deluges have affected the area in historical times when massive geological corroboration exists for multiple deluges at the end of the last Ice Age – well within the time-frame of more than 10,000 years that is set out in the Sangam tradition itself.

Could it be the ruins of Kumari Kandam that are lying in 23 metres of water 5 kilometres off-shore of Poompuhur? And could those mythical antediluvians remembered by the ancient Tamils have been the source of the fragments of high cartographical and astronomical knowledge that seem to have been fossilized in the ancient Indian texts?

11 / *The Quest for Kumari Kandam*

The river Prahuli, and the mountain Kumari, surrounded by many hills, were submerged by the raging sea.

Silipathikaram xx: 17–20

With reference to the first two Sangams I may say that the account is too mythical and fabulous to be entitled to any credit and I do not think that any scholar who has studied the histories of the world will be bold enough to admit such tales within the pale of real history.

Professor Sesagiri Sastri, *Essay on Tamil Literature*, Madras 1897

February 2000–January 2001, south India

Madurai is an ancient city but it has little to show, other than a few texts of disputed antiquity,[1] to back up its claim to have been the headquarters of the third and last of the great Tamil Sangams ('Academies') It can produce no evidence in support of its further claim that the Third Sangam was the direct-line descendant of two earlier Sangams, *dating back thousands of years into prehistory*, located in antediluvian Tamil cities that had once existed far to the south of Madurai but that had been swallowed up by the sea. The very word 'Sangam' turns out not even to be derived from the Tamil language (it is Sanskrit) and does not appear in any of the texts that tradition attributes to the Third Sangam period.[2] Last but not least, the earliest surviving written account of the so-called 'Sangam Age' is not thought by scholars to be older than the sixth century AD.[3]

By his use of such arguments the late K. N. Shivaraja Pillai – whose highly regarded but rare *Chronology of the Early Tamils* I was able to consult at a research library in Madurai – stands out as the most persuasive opponent of the alluring notion of lost Tamil lands and a lost Tamil civilization in the Indian Ocean. He wags an admonishing finger at those tempted to wonder if there might be even a drop of the truth anywhere in the story of Kumari Kandam and the first two Sangams, and proclaims the whole thing to be

one of the most daring literary forgeries ever perpetrated. The incredibly high antiquity with which Tamil literature comes to be invested by this legend, and the high connection with divinity it brings about, were more than enough to secure for it a ready acceptance by a credulous public.[4]

The historical annals of most cultures contain examples of this kind of manipulation of the past in order to annex some dignity or aura of the divine to a fledgling royal dynasty, or to dress up a new cult in a cloak of antique venerability – or, for that matter, to render *arriviste* philosophies or literary works more acceptable to traditionalists by attaching them to existing or imagined traditions.[5] It is therefore easy to see the force of Pillai's arguments, and, since he published his *Chronology* in 1932, his view that Kumari Kandam is nothing more than a 'preposterous story'[6] has been the dominant one amongst serious scholars of Tamil history.

This, of course, by no means guarantees that his view is correct. On the contrary, as I continued my research in Madurai, the potential significance and implications of what the NIO had found in 1993 off the south-east coast of Tamil Nadu at Poompuhur began to weigh more and more heavily on my mind.

Lost lands and flooded cities

From the photographs and descriptions that I had by this time seen and read, everything about the U-shaped structure appeared to be strikingly anomalous. Yet equally striking was the way in which it had thus far attracted zero attention or interest outside the rather closed world of the NIO (which had been unable to do anything further about it because of insufficient funding). I found this lack of interest and knowledge to be almost unbelievable.

After all, the fully qualified Indian marine archaeologists who had dived on the structure in 1993 had not hesitated in their official report to pronounce it to be man-made with 'courses of masonry' plainly visible – surely a momentous finding 5 kilometres from the shore at a depth of 23 metres? But far from exciting attention, or ruffling any academic feathers, or attracting funds for an extension of the diving survey to the other apparently man-made mounds that had been spotted near by on the sea-bed – and very far indeed from inspiring any Tamil expert to re-evaluate the derided possibility of a factual basis to the Kumari Kandam myth – the NIO's discovery at Poompuhur had simply been ignored by scholarship, not even reacted to or dismissed, but just widely and generally ignored.

All the more I felt it was my role to be proactive and to stir things up around this matter. Because if the U-shaped structure was indeed man-made and more than 10,000 years old (remember at this stage I still did not have Glenn Milne's inundation maps that would later push the age of the ruins back to 11,000 years old or older) then things were going to have to change in south Indian history. Despite all the question marks that had been raised over it on literary and philological grounds, the myth of Kumari Kandam and of the two antediluvian Sangams would suddenly clamour to be taken seriously.

After all, it is one thing for scholars like Shivaraja Pillai, David Schulman and others, to belittle the historical significance of a myth for which there seems to be no substantiating evidence, but it is quite another to try to sustain such a

14. *The Temple of the Sea Lord, Dwarka, overlooking the underwater ruins.*

15. *View of Dwarka from the sea. The ruins are directly beneath the small boat.*

16. *Marine archaeologists of the NIO at Dwarka.*

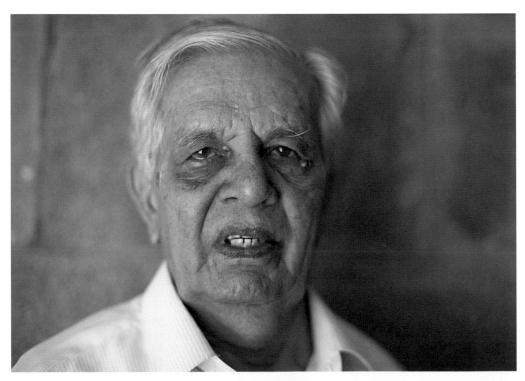

17. *S. R. Rao, the founder of marine archaeology in India.*

18. *Technical divers of the NIO entering the water at Dwarka.*

19. Underwater Dwarka, large blocks scattered on the sea-bed.

20. Circular stone anchor amidst underwater structures, Dwarka.

21. *Part of a curved bastion, underwater Dwarka.*

22. *Treasure trove of man-made artefacts brought up from two mysterious submerged cities discovered in 2001 in India's Gulf of Cambay.*

23. *Detail of artefacts and human remains from the lost cities in the Gulf of Cambay.*

24. *The author with NIO experts, examining plans of the two deeply submerged cities in the Gulf of Cambay thought to be more than 8000 years old.*

25. *Pilgrims flocking to a Siva temple on the seashore at Dwarka.*

26. *Siva temple, Dwarka. Although Dwarka is sacred to Krishna, the cult of Siva is also celebrated there.*

posture among a growing community of scholars and interested members of the public with access to inundation data like Milne's.

Reproduced here and in chapter 7, the Durham geologist's maps of south India between 17,000 and 7000 years ago have an eerie effect on me. Incorporating Sri Lanka in the south-east, extending southward, below Cape Comorin, and enhanced off-shore by the enlarged Lacadives/Maldives archipelago running all the way to the equator and into the southern hemisphere, the maps portray the region as no culture of the historical period is supposed to have known it: yet when I look at them through half-closed eyes I can almost imagine that someone has tried to *draw*, at various stages of its supposedly mythical inundation, the much bigger Dravidian homeland of thousands of years ago that is described in the Kumari Kandam tradition.

Coincidence? Or mystery?

- With its dominant motif of a once much larger Dravidian homeland, the opening of the Kumari Kandam flood myth is set in remote prehistory between 12,000 and 10,000 years ago.
- The work of Glenn Milne and other inundation specialists confirms that between 12,000 and 10,000 years ago India's Dravidian peninsula and its outlying islands would indeed have been far larger than they are today – but were in the process of being swallowed up by the rising seas at the end of the Ice Age.
- With its descriptions of flooded cities and lost lands, the Kumari Kandam myth 'predicts' that prehistoric ruins more than 10,000 years old should lie underwater at various depths and locations off the Tamil Nadu coast.
- The NIO's discovery of a large and apparently man-made structure at a depth of 23 metres off Poompuhur seems to confirm the accuracy of this prediction.

If the myth is right about the flooded cities, then what else might it be right about?

If there is anything at all to the story of the First and Second Sangams orchestrating a golden age of literary, artistic and musical creativity amongst the Tamils of 10,000 years ago and maintaining an archive of written records, then it means not only that an as yet unidentified culture of the last Ice Age may have flourished in the lost lands of the Indian Ocean, but also that we seem to be dealing with a *civilization* here that had reached a high level of development, organization and self-awareness.

The teachings of illustrious men

The sources for all that is known today about Kumari Kandam are limited and it is true, as the detractors of the myth point out, that the oldest written version dates from no earlier than the sixth century AD – some would even make it as young a document as the tenth century AD. Supposedly the work of the renowned medieval commentator Nakirar, this version appears in a learned gloss to the *Iriyanar Agapporul*, a grammar of classic Tamil love poetry in sixty *sutras*.[7] Our concern here is not with the *Agapporul*, but strictly and exclusively with Nakirar's gloss, which is itself said to have been 'handed down orally for ten generations before it was put into writing'.[8]

Other medieval commentators who support Nakirar by speaking of Kumari Kandam and of the first two Sangams not as myths but as historical entities are Nachinarkkiniyar, in his gloss to the *Tolkappiyam Poruladikaram*, the distinguished Per-Asiriyar in his commentary upon the *Tolkappiyam*, and Adiyarkkunelar, in his commentary on the *Silipathikaram*.[9]

As my research continued in Madurai, therefore, I was not surprised to learn that long before something looking very much like underwater ruins had been found off the south-east coast of India in exactly the depth/age-range that is predicted by the Kumari Kandam myth, the credibility lent to the flood and Sangam tradition by the illustrious men who passed it down to us had clearly begun to worry some otherwise sceptical modern historians:

> Three commentators of no mean scholarship and repute have unreservedly accepted the version of the commentator of the *Iriyanar Agapporul*. Though it is easy to dismiss these valuable works as unhistorical and uncritical and hence worthless to students of history, still we cannot afford to credit commentators with such ignorance of the subject which they were handling. When they quote with approval it means they were satisfied of the veracity of the tradition behind the account.[10]

The Kumari Kandam tradition (1)

Although I am (of course!) writing *Underworld* with the benefit of hindsight, I have sought to unfold the key information that it contains in something of the gradual and fragmentary manner in which it reached me. Thus I didn't learn about Kumari Kandam and the Sangam tradition all at once – but rather in dribs and drabs over a period of many months – and this is reflected in the details that I have already given about Kumari Kandam in earlier chapters.

Now, with all the resources of Madurai at my disposal, I was able to compile a more extensive and accurate summary of what the tradition actually says (as opposed to what others say about it):

- Over a period of just under 10,000 years, the Pandyans (a part-historical, part-legendary dynasty of Tamil kings) formed three Sangams or Academies in order to foster among their subjects the love of knowledge, literature and poetry: 'These Assemblies were the fountainhead of Tamil culture, and their principal concern was the perfection of Tamil language and literature.'[11]
- The first two Sangams were not located in what is now peninsular India

but in the antediluvian Dravidian land to the south 'which in ancient times bore the name Kumari Kandam'[12] (literally 'the Land of the Virgin' – or perhaps 'the Virgin Continent').[13]

- The First Sangam was headquartered in a city named Tenmadurai ('Southern Madurai'). It had 549 members 'beginning, with Agattiyanar (the sage Agastaya) . . . Among others were God Siva of braided hair . . . Murugan the hill god, and Kubera the Lord of Treasure.'[14]

- Patronized by a succession of eighty-nine kings, the First Sangam survived as an institution over an unbroken period of 4440 years, during which time it approved and codified an immense library of poems and literature. These classic texts, all now lost and known only by their titles, are said to have included works such as the *Agattiyam, Paripadal, Mudunarai, Mudukurgu* and *Kalariyavirai* – still well known and revered among Tamils today.[15]

- At the end of this golden age the First Sangam was destroyed when the deluge arose and Tenmadurai was 'swallowed by the sea' along with large parts of the land area of Kumari Kandam.[16]

- However, survivors of the antediluvian civilization were able to relocate further north, saving some of the First Sangam books, and the Second Sangam, said to have been patronized by fifty-nine kings, was established in another city – Kavatapuram. 'The *Agattiyam* and *Tolkappiyam*, the *Mapuranam, Isainunukkam,* and *Budapuranam* were their grammars. The duration of the period of this Sangam was 3700 years.'[17] Then, like its predecessor, the Second Sangam was 'swallowed by the sea' and lost for ever with all its works (with the possible exception, some claim, of the *Tolkappiyam*, which has survived to this day).[18]

- Following the inundation of Kavatapuram the survivors of the Kumari Kandam civilization again relocated northward, this time into peninsular India, where the headquarters of the Third Sangam was established in a city identified with modern Madurai – then known as Uttara Madurai or Vadamadurai ('Northern Madurai', presumably to distinguish it from its antediluvian predecessor 'Southern Madurai').[19]

- The Third Sangam survived for a further 1850 years: 'Forty-nine were the kings who patronized this Academy.'[20]

Choosing the right slot

A matter that I found hard to reconcile while I talked to the experts and read up the literature in Madurai was the way in which the very same Tamil authorities who brush off the First and Second Sangams as 'preposterous stories',[21] accept without demur the existence of the Third Sangam – or anyway some sort of genuinely Tamil institution of letters that might retrospectively have been referred to by the Sanskrit term Sangam. Most, moreover, agree upon dates of between AD 350 and 550 for the *termination* of this Third Sangam's activities.[22]

For example, Ramachandra Dikshitar proposes that 'the end of the fifth century AD marked the extinction of the Academy'.[23] He adds:

> Though the origin of the Sangam as an institution is shrouded in deep mystery, still the fact remains that there was something like an organized Academy ... and it continued to exist for several centuries. A definite stage was reached by the beginning of the sixth century AD [after the extinction of the Academy] when the Tamil language underwent some transformation in regard to style, metre, etc.[24]

According to Shivaraja Pillai – as ever pursuing his 'forgery' case against the scheme of things set out in the commentary on the *Agapporul*.

> The fabricator appears to have started from some authentic data before him. They were the so-called 'Third Sangam' works, which in all probability must have by that time assumed a collected form. These collections furnished the basis on which he proceeded to raise his imaginary structure of the Three Sangams.[25]

If we accept the generally agreed date of between AD 350 and 550 for the end of the – at least semi-historical – 'Third Sangam', then this gives us a fixed reference point on which to anchor the chronology of the myth:

- AD 350 minus the 1850 years given as the duration of the Third Sangam takes us back to 1500 BC (i.e., about 3500 years ago);
- 1500 BC minus the 3700 years given as the duration of the Second Sangam takes us back to 5200 BC (7200 years ago);
- 5200 BC minus the 4440 years given as the duration of the First Sangam takes us back to 9600 BC (11,600 years ago).

The date of 9600 BC for the formation of the First Sangam (or 9800 BC or 9400 BC for that matter) coincides closely enough with Plato's date for the inundation of Atlantis – also 9600 BC – to raise the hairs on the back of my neck.

And the question continues to be this: how could Plato less than 2500 years ago, or Nakirar less than 1500 years ago, have managed *by chance* to select the epoch of 9600 BC in which to set, on the one hand, the sinking under the waves of the Atlantic Ocean of the great antediluvian civilization of Atlantis and, on the other, the foundation of the First Sangam in Kumari Kandam – a doomed Indian Ocean landmass that was itself destined to be swallowed up by the sea?

If Plato and Nakirar were pure 'fabulists' working independently of any real tradition or real events, then isn't it much more likely that they would have chosen *different* imaginary epochs in which to set their flood stories?

Why didn't they chose 20,000 or 30,000 years ago – or even 300,000 years ago, or three million years ago – instead of the tenth millennium BC?

And was it just luck that this slot turns out to have been in the midst of the meltdown of the last Ice Age – the only episode of truly global flooding to have hit the earth in the last 125,000 years?

The Kumari Kandam tradition (2)

More information than I have already reported remains to be gleaned within the medieval commentaries. And outside the commentaries there are several allusions in Tamil literature that can also fairly safely be said to be part of 'the tradition behind the account' – even if they do not always refer to Kumari Kandam or to the first two Sangams by name. Some are in works of considerable antiquity and high renown, others are in less well-known sources, but all in one way or another add to our picture of the lost Tamil lands and of the floods that ancient peoples believed had swallowed them up.

According to V. Kanakasabhai, a specialist in south Indian history, the Tamils of the early first millennium AD preserved a tradition, already ancient in their time,

> that in former days the land had extended further south and that a mountain called Kumarikoddu, and a large tract of country watered by the river Prahuli had existed south of Cape Kumari. During a violent irruption of the sea, the mountain Kumari-koddu and the whole of the country through which flowed the Prahuli had disappeared.[26]

Kanakasabhai's sources include the *Kalittogai* (stanza 104:1–4) and the *Silipathikaram* (xx:17–20): 'The river Prahuli, and the mountain Kumari, surrounded by many hills, were submerged by the raging sea.'[27] Adiyarkkunelar fills in some of the detail when he tells us that in the time before the flood these forested and populated lands between the Prahuli and Kumari rivers were divided into 49 counties that stretched for '700 Kavathams' – about 1000 miles.[28]

The historian P. Ramanathan also draws attention to 'ancient Tamil poems and authentic traditions [that] refer to successive submersions of land to the south of India in the Indian Ocean and the consequent reduction of the extent of the Tamil land':[29]

> *Purunanuru* 6 by Karikishar and *Purunanuru* 9 by Nettimaiyar . . . refer to Kumari and Prahuli rivers both placed by ancient commentators in the submerged lands to the south of Cape Comorin [modern Kaniya Kumari]. *Kalittogai* 104 specifically refers to [a Pandyan king] losing his territories to the sea and compensating the loss by conquering new territories from the Chera and Chola rulers (to the north). *Silapathikaram – Kadukankathai* (lines 18–23) refers to the sea swallowing up the Prahuli river along with Kumarikoddu tract comprising many hill areas. The *Venirkathai* of *Silipathikaram* refers to the ocean as the southernmost frontier of Tamilaham *and commentator Adiyarkkunelar explains that the reference there is to the topography after the deluge.* The *Payiram* to the *Tolkappiyam* refers to

Venkatam as the northern boundary and [Kaniya] Kumari as the southern boundary of Tamilaham. In his commentary thereon Illampuranar states that the southern boundary (viz Kumari) was mentioned because, *before submersion by the sea there were lands to the south of Kumari* ... In his commentary on the *Tolkappiyam*, Nachinarkkiniyar mentions that the sea submerged 49 Nadus (counties) south of Kumari river . . .[30]

Ramanathan further reminds us that, according to tradition, the Pandyans are:

the oldest of the three ancient Tamil dynasties. Perhaps the oldest ruling dynasty in the world ... Some accounts ... say that Cheras and Cholas were mere branches of the Pandyan dynasty which separated long ago.[31]

He then repeats essentially the assertion of the *Kalittogai* cited above that:

One of the earliest Pandyan kings, Nediyon ('the tall one') is said to have organized the worship of the sea. Portions of his land to the south of Cape Comorin [Kaniya Kumari] were submerged by the sea and to compensate for the loss he conquered vast territories to the north of the Pandyan kingdom.[32]

Likewise, T. R. Sesha Iyenagar refers to Tamil traditions which suggest that, although Kumari Kandam may have included islands, a large part of it was mainland

connected with South India ... which was overwhelmed and submerged by a huge deluge. There are unmistakable indications in the Tamil traditions that the land affected by the deluge was contiguous with Tamilaham, and that, after the subsidence, the Tamils naturally betook themselves to their northern provinces.[33]

What secrets lie concealed in such fragments of folklore and tradition? In his paper 'The Cultural Heritage of the Ancient Tamils', Dr M. Sundaram, Chief Professor and Head of the Department of Tamil, Presidency College, Madras, sums up the evidence to conclude that:

The tradition of the loss of a vast continent by a deluge of the sea is too strong in the ancient Tamil classics to be ignored by any serious type of enquiry. In fact the first Tamil Sangam was said to have been functioning from South Madurai, in the lost continent. Ancient grammatical texts in Tamil and their latter day commentators testify that River Prahuli and Kumari Mountain ranges were lost by a deluge, a *Purunaruli* verse refers to the River Prahuli and *Silipathikaram* mentions the deluge in which the Kumari continent was lost ... There were 49 divisions between River Prahuli and mountain Kumari. The erudite commentator of *Tolkappiyam*, Per-Asiriyar, has stated that the Kumari river was left as Cape Kumari after a deluge.[34]

Last but by no means least, the Tamil epic *Manimekalai* speaks of the flooding of a city off-shore of Poompuhur as divine retribution upon a king who had failed to celebrate the festival of Indra.[35] Most archaeologists believe that the reference here is to the shallowly submerged ruins of the historical city of Kaveripumpattinam found just south of Poompuhur in the intertidal zone mainly at 3 metres or less and dated to between 300 BC and AD 300 (see chapter 9). However, the U-shaped structure that is now known to lie much further out from shore and in deeper water raises the possibility that what is remembered in the *Manimekalai* could be a far earlier event.

Ravana's antediluvian domain

If the Kumari Kandam tradition is in any way a true guide then we should expect to find underwater ruins not only in south Indian waters, but also in the waters of the island of Sri Lanka – ancient Ceylon. And because Sri Lanka was joined to the mainland during the Ice Age by a land-bridge close to Poompuhur (indeed, would have been an integral part of 'Kumari Kandam') logic suggests that Sri Lankan myths and legends should also have something to say on the subject of floods.

It is therefore reassuring to discover that the *Mahavamsa*, *Dipavamsa* and *Rajavali*, Ceylonese chronicles based on archaic oral sources that first began to be set down in writing by Buddhist monks around the fourth century AD,[36] 'speak of three deluges which destroyed a large land area that lay beyond Ceylon'.[37] For example the *Rajavali* remembers a time, long before its own compilation as a text, when

> the gods who were charged with the conservation of Ceylon became enraged and caused the sea to deluge the land . . . In this time . . . 100,000 large towns, 970 fishers' villages and 400 villages inhabited by pearl fishers . . . were swallowed up by the sea . . .[38] Twenty miles of the coast, extending inland [were] washed away.[39]

The same source also refers to a flood that affected Sri Lanka even earlier – indeed 'in a former age'[40] – during the time of the giant Ravana (the 'demon king' whose exploits feature, separately, in the Indian Sanskrit epic, the *Ramayana*). Ravana, it seems, had angered the gods with his 'impiety' and was punished in the usual way:

> The citadel of Ravana, 25 palaces and 400,000 streets, were swallowed up by the sea . . . The submerged land was between Tuticorin [south-east coast of modern Tamil Nadu] and Mannar [north-west coast of modern Sri Lanka] and the island of Mannar is all that is now left of what was once a large territory.[41]

I was later to realize that there is something remarkable about this. In December 2000 when I was first able to study Glenn Milne's inundation maps of the

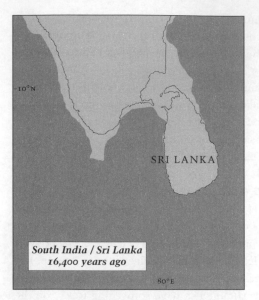

**South India / Sri Lanka
16,400 years ago**

Poompuhur region, I noticed that a large tract of land would indeed have been exposed between Tuticorin and Mannar – just as the chronicle said – at around 16,000 years ago. This was soon after the end of the Last Glacial Maximum, shortly before global sea-level began to rise steeply, and Milne's maps go on to show the flooding of Ravana's antediluvian domain by the post-glacial floods. Interestingly, the maps also show an area of higher relief that was never submerged and that is today, as the *Rajavali* correctly reports, the island of Mannar.[42]

Sir J. E. Tennant, among others who wrote long before the era of inundation mapping, disregarded 'the traditions of the former extent of Ceylon and submersion of vast regions by the sea' on the grounds that 'evidence is wanting to corroborate the assertion, at least within the historic period'.[43] But once again, as we now know, there is abundant evidence that *before* the historic period, at the end of the Ice Age, Sri Lanka was indeed much larger than it is today with the greatest extent of antediluvian land in the north-west bridging the Gulf of Mannar exactly where, 'in a former age', Ravana's citadel is supposed to have stood.

16,000 BC to 9600 BC

This notion of earlier flood epochs – with the parallel thought of layer upon layer of forgotten history receding deep into a past beyond remembrance – is reinforced in certain Ceylonese traditions about the ancient Tamils. Amongst these an intriguing statement is made that the total number of Sangams was not three, as most other accounts maintain, but seven[44] – implying the existence at unknown locations of four previous Sangams *before* the First Sangam set up its headquarters at Tenmadurai on the banks of the Prahuli river.[45]

In this connection I note that N. Mahalingam, Chairman of the International Association of Tamil Studies, refers in the *Proceedings of the Fifth International Conference of Tamil Studies* to Tamil traditions that speak of three episodes of flooding in the millennia *preceding* the supposed foundation date of the First Sangam:

> The first great deluge took place in 16,000 BC ... The second one occurred in 14,058 BC when parts of Kumari Kandam went under the sea. The third one happened in 9564 BC when a large part of Kumari Kandam was submerged.[46]

The date for the third of these archaic floods, as readers will note, overlaps, give or take forty years, with the date of 9600 BC for the foundation of the First Sangam (and thus also with Plato's date for the submersion of Atlantis). It is only a hint, but if there is any substance to it, then it raises the possibility that the First Sangam too, like its successors, might have been founded by flood survivors – perhaps even survivors of the very same episode of global floods that in another ocean gave rise to the Atlantis myth.

Cults of knowledge

At the heart of the Sangam story, whether it concerns three or seven ancient Academies, is a theme of entropy and degeneration, spiralling downwards through a series of stages from a golden age, powered by vast cosmic cycles of destruction and rebirth. There are curious echoes here of the *yuga* system at the heart of the Dwarka story, on the one hand, and of the Vedic notion of the *pralaya* – the global cataclysm that recurs at the end of each world age – on the other:

- In both cases we must envisage an antediluvian civilization of high spiritual and artistic achievement and a group of sages – the Seven Rishis in the case of the *Vedas*, the members of the 'Academy' in the case of the Tamil texts – who gather to serve the interests of knowledge and to provide an archive or repository for poetic and religious compositions.
- In both cases a cataclysm in the form of a global flood intervenes, swallowing up huge areas of land and destroying the antediluvian civilization.
- In both cases survivors repromulgate the ancient knowledge in the new age – which is portrayed as a decline from the age before – forming a new group of Seven Rishis or a new Sangam suitable to that age.

Needless to say there are many differences between the two traditions – too many for either to be the result of direct influence from the other. Nevertheless, the underlying idea is essentially the same – that recurrent cataclysms afflict the earth, threatening the obliteration of human knowledge and a return to ignorance, but that an institution or 'brotherhood' (the Seven Rishis, the Sangam) survives 'the periodic scourge of the deluge' and rises again after the recession of the waters to carry the cause of knowledge forwards into the new age and to 'bring glory and light to ignorant lands and peoples'.[47]

There are also prominent crossover figures suggestive of an unseen link. For example, the Sage Agastya, frequently listed amongst or alongside the Vedic Seven Sages, appears in Tamil traditions as a member of the First Sangam. Likewise, listed amongst the 549 members of the First Sangam is the Vedic god Rudra-Siva, master of animals, Lord of Yoga, 'he of the braided hair'. And while his presence there may well, as Pillai argues, be just an outcome of Tamil 'fabulists' seeking to concoct a divine heritage for their work, it is worth remembering that Siva's primary attribute is *gnosis* – or knowledge – and that whether in south India or the Himalayas he is associated with a cult of esoteric knowledge that is said to have been carried down from before the flood.

The tank and the pillar

Siva is everywhere in Madurai and stories of his deeds and miracles abound here. Even the Meenakshi temple is in fact two temples within a single walled complex – one, the smaller of the two, for the goddess Meenakshi, a wife of Siva, and one for Siva himself in his manifestation as Sundareshwar. The temple sits at the ancient geometrical centre of Madurai, occupying an area measuring approximately 220 × 260 metres[48] – as large as the footprint of the Great Pyramid of Egypt.[49] Its perimeter is embellished with eleven spectacular *gopurams* (entrance towers – the highest, in the south, rising to more than 50 metres), all of them luridly carved and painted with sensational three-dimensional scenes from Hindu mythology. Such scenes, made up of an estimated total of 33 million carvings,[50] crowd in everywhere upon the visitor who approaches this vast complex of buildings – from the walls of its medieval stone gateways to the columns of its Thousand Pillar Hall.

The temple is not aloof from the great city that surrounds it, but rather the life of the city continues within its walls at a different pace. Sometimes it has the atmosphere of a market with colourful, noisy crowds bustling from shrine to shrine, beggars seeking alms, hawkers selling souvenirs and long-horned cows wandering about as though they own the place. It is surprising how often you will see a businessman slip off his shoes to stroll inside, smear sacred ash upon his forehead and offer prayers amongst the cool shadows and garlanded statues. Lean pilgrims and wild-haired *sadhus* gather from all parts of India seeking alms and enlightenment, couples and families come here on outings, and classes of schoolchildren march bright-eyed through the corridors, adding their shrill laughter to the non-stop hubbub of conversation and chanting.

I entered through the southern *gopuram* and made my way across a sunlit ambulatory to the nearby Citra Mandapa, an elegant cloistered colonnade with painted walls and ceilings surrounding the Golden Lotus Tank – perhaps the Meenakshi temple's most spectacular feature. Legend has it that this very large tank, which measures 52 metres long by 36.5 metres wide, was 'used to judge the merits of Tamil literary works' during the Third Sangam period.[51] 'The manuscripts that floated were considered great works of literature, and if they sank they were dismissed.'[52]

In terms of general appearance and design the tank strikingly resembles the Great Bath at Mohenjodaro – only there the rectangular ritual bathing pool has been empty and dry for thousands of years; here it is filled with green water and is still used by pilgrims for purification ceremonies. Much of the temple as we see it today dates from the thirteen century AD or later – while the Indus-Sarasvati cities had fallen into ruin by the second millennium BC – but I knew

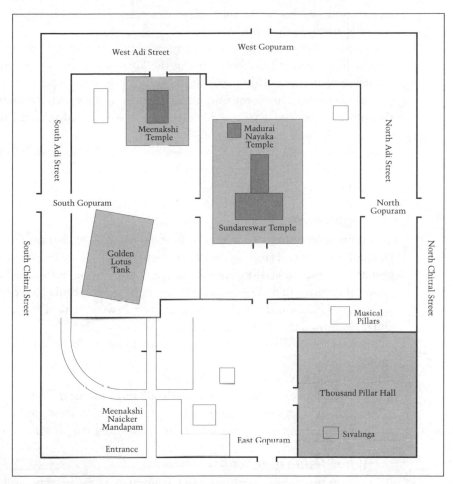

Floorplan of Madurai temple. Based on Howley and Dasa (1996).

that the tank 'prominently figures in legends connected with the origin of the shrine'.[53] As at Tiruvannamalai these legends also state that the temple stands where it does because of the prior existence there of a *sthala* or pillar of natural stone – a Sivalingam – that had manifested in primordial times. In the case of Madurai, however, the pillar did not appear at the foot of a sacred mountain but was found standing upright in a forest 'beneath a Kadamba tree' where the Vedic god Indra was said to have built the first prehistoric shrine around it.[54]

I was reminded of the cylindrical and conical stone pillars (of officially 'unknown', but I would have thought obvious, function) that have been excavated by archaeologists along the valleys of the Indus and the Sarasvati rivers at numerous Harappan and pre-Harappan sites.[55] These 'proto-Sivalinga' are antedated by even earlier stone pillars of the same sort excavated from Neolithic settlements in India[56] – so many of them that T. R. Sesha Iyenagar can write:

'the worship of Siva in the form of a linga existed in the Stone Age, which certainly preceded the Vedic Age'.[57]

The truth is that nobody really knows when the 'Vedic Age' began just as nobody has yet found the beginning of the Siva cult in India. Powerful and omnipresent from the Himalayas to the deep south, it always seems to have existed – in the worship of the lingam, in the worship of the sacred mountain, in the worship of the god of yoga and knowledge, cross-legged, deep in meditation, surrounded by wild beasts.

This enigmatic figure, and the complex system of ideas and symbols that he evokes, must have come from somewhere.

Perhaps Kumari Kandam?

Look south

'It was the most ancient continent in the whole world,' exclaimed Dr T. N. P. Haran, Professor of Tamil Studies at the American College in Madurai. 'The best and the ancient civilization existed there. And it belongs to Tamils.'

'And if I wanted to find it – whatever's left of it – where would I have to look?'

'Kumari Kandam was a big land. So many people were there. The sea came in and it swallowed the whole thing.'

'If I were to go diving off modern Kaniya Kumari, do you think I'd find ruins?'

'I've no idea! But I wish you all the best!'

I persisted: 'Should I look directly south of Kaniya Kumari?'

Haran thought for a while before replying: 'Yes, I think at least 300 kilometres south of Kaniya Kumari. If you go there you will be able to get something.'

What fishermen know

Before returning to dive with the NIO at Dwarka at the beginning of March 2000 (reported in chapter 9) Santha and I completed the rest of our long overland journey in Tamil Nadu with visits to four coastal towns: Kaniya Kumari in the south, Rameswaram in the south-east, where India reaches out towards Sri Lanka across the Palk Strait, and Poompuhur and Mahabalipuram along the Coromandel coast facing the Bay of Bengal.

- Mahabalipuram commands attention on account of the old myths of the Seven Pagodas and the sunken city of Bali (see chapter 5).
- Kaniya Kumari is explicitly referenced in the Kumari Kandam tradition as the new southern border of India after the hilly and well-watered land that formerly lay to the south of it had been swept away in the deluge.
- Rameswaram is identified in the *Ramayana* with what sounds like a land-bridge to Sri Lanka: 'To build a bridge across the sea, the bears and monkeys hurled trees and rocks into the water which by the power of Rama remained afloat. The Gods looked down enthralled as the monkey

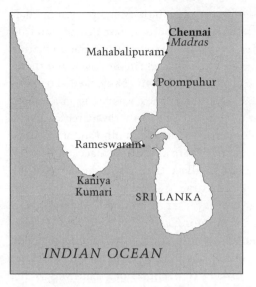

armies moved across the sea on Rama's bridge.'[58] (The 'monkey armies' – don't ask, it's a long story! – are on their way to Lanka to rescue Rama's wife Sita from Ravana, the same demon king of a 'former age' whose antediluvian domain is said in the Ceylonese Chronicles to have stretched between Tuticorin and Mannar. So much land-bridge imagery, from two different traditions, and in just the right places!)

- Poompuhur speaks for itself as the site of the submerged U-shaped structure. When I went there in February 2000 I knew that diving would be out of the question without going through a long

permissions and money rigmarole with the NIO first. But I wanted to get a sense of the land side of the story and at least dip my toes in the water.

As we explored and talked to more and more local people it began to dawn on me that the ubiquitous south Indian traditions of lost lands and flooded cities – which so many scholars simply ignore in their evaluation of history – are well known and almost universally believed to be true accounts by the general public of the region.

This in itself does not necessarily mean anything. Superstitions and follies abound amongst the public in every country. But many of my informants were hard-bitten professional fishermen who for the most part were clearly *not* relaying half-remembered folklore that they had heard from their grandfathers, but were speaking from direct personal experience. Indeed, in Poompuhur and again in Mahabalipuram I met fishermen, who had nothing whatsoever to gain by deceiving me, who claimed to have seen with their own eyes what they described as 'palaces', or 'temples', or 'walls' or 'roads' underwater when diving down to free trapped anchors or nets.

An underwater ruin, if it is of any size, will function as an artificial reef, attracting many different species of fish to the shelter and security that it provides – particularly in areas like south-east India, where the sea bottom is largely flat and featureless. And since fishermen are in the business of catching fish, they naturally look out for places in the ocean where fish congregate for any reason. In this way they are often the first to find unsuspected underwater sites – and frequently may know of sites that archaeologists are unaware of.

My instinct is that this may well turn out to be the case along extensive stretches of the south Indian continental shelf which, except off Poompuhur, has never been the subject of a marine archaeological survey. My travels from Kaniya

Kumari to Mahabalipuram have convinced me that the local sightings of anomalous submerged structures in these areas are too numerous, too consistent and too widespread to be safely ignored. Moreover, were it not for the NIO, no marine archaeology at all would have been attempted anywhere in the region. It is therefore surely significant that in the one place where the NIO has looked – Poompuhur – something as unusual as the U-shaped structure was found in a project lasting just a few days. It makes sense to suppose that if further systematic surveys and marine archaeology can be done underwater – at Poompuhur and at the other south Indian locations – then more discoveries are likely to be made . . .

At Mahabalipuram, in the little fishing village that lies in the curve of the bay a mile or so to the north of the Shore temple, Santha and I sat on the beach on a pile of drying nets with a large crowd gathering around us. Everybody in the village who might have an opinion or information to contribute was there, including all the fishermen – some of whom had been drinking palm toddy most of the afternoon and were in a boisterous and argumentative mood. What they were arguing about were their answers to the questions that I was asking and precisely who had seen what, where underwater – so I was happy to listen to their animated conversations and disagreements.

An elder with wrinkled, nut-brown eyes and grey hair bleached white by long exposure to the sun and sea spoke at length about a structure with columns which he had seen one day from his boat when the water had been exceptionally clear. 'There was a big fish,' he told me. 'A red fish. I watched it swimming towards some rocks. Then I realized that they were not rocks but a temple. The fish disappeared into the temple, then it appeared again, and I saw that it was swimming in and out of a row of columns.'

'Are you certain it was a temple?' I asked.

'Of course it was a temple,' my informant replied. He pointed to the pyramidal granite pagoda of the Shore temple: 'it looked like that.'

Several of the younger men had the usual stories to tell about heroic scary dives – lasting minutes, hearts thudding, their breath bursting in their lungs – to free fishing gear snagged on dark and treacherous underwater buildings. In one case, it seemed, a huge net had become so thoroughly entrapped on such a structure that the trawler that was towing it had been stopped in its tracks. In the case of another underwater ruin divers had seen a doorway leading into an internal room but had been afraid to enter it.

One strange report was that certain of the ruins close to Mahabalipuram emit 'clanging' or 'booming' or musical sounds if the sea conditions are right: 'It is like the sound of a great sheet of metal being struck.'

'And what about further away,' I asked. 'If I were to take a boat south following the coast what would I find? Are the underwater structures mainly just here around Mahabalipuram or are they spread out?'

'As far south as Rameswaram you may find ruins underwater,' said one of the elders. 'I have fished there. I have seen them.'

Others had not travelled so far but all agreed that within their experience there were submerged structures everywhere along the coast: 'If you just go where the fish are then you will find them.'

Which site to dive on?

If I had unlimited funds and complete freedom of action then I would long ago have organized full-scale marine archaeological expeditions at Kaniya Kumari, Rameswaram, Poompuhur and Mahabalipuram in the south and south-east of India, and all along the coast of the Gujerat peninsula and the Gulfs of Kutch and Cambay in the north-west. But I don't have unlimited funds – or time – and India, for all her magnetism, is a vast challenge and energy drain best approached with a flexible schedule and a spirit of compromise.

Besides, India is one facet of 'Underworld', not the whole mystery. After returning to England in March 2000, with the Dwarka dives behind me, I could not afford to forget that other research was also crying out to be completed and that other journeys had to be made – at the very least to the Maldives, the Persian Gulf, the Mediterranean, the Atlantic and Japan. Although I had no intention of abandoning the wider investigation in India I therefore decided that for the immediate future I would focus my energies on getting to dive at Poompuhur – which I had already begun to negotiate with Kamlesh Vora before leaving Dwarka – and that all the other potential Indian dive sites would have to wait their turn.

Poompuhur was the obvious first choice, head and shoulders above the other contenders. Here alone advance work had been done by the NIO, who, quite extraordinarily and with absolutely no fanfare, appeared to have found precisely what I was looking for – viz. a large, well-organized and apparently man-made structure that had been inundated more than 10,000 years ago at a time when there was no known civilization in the vicinity that could have built it.

While keeping the money and permissions process going with the NIO by e-mail, I used the next several months to complete an intensive series of research and diving trips to Malta, Alexandria, the Balearic islands, the Canary islands and twice to Japan (once in April/May for seven weeks and again in September for a further two weeks).

By October 2000 my attention was very much back on Poompuhur again, when Glenn Milne's calculations arrived showing that the U-shaped structure was in fact '11,000 years old or older' – putting its inundation squarely in the same time-frame as the supposedly mythical foundation of the First Sangam at Tenmadurai, and as the supposedly mythical submersion of Plato's Atlantis.

The next development came in December 2000 when Milne supplied me with a series of high-resolution inundation maps of India, spanning the period between 21,300 years ago and 4800 years ago, which tracked the changes in the subcontinent's coastline caused by rising sea-levels during the meltdown of the Ice Age (see chapter 7). The maps show not only the huge amounts of land that

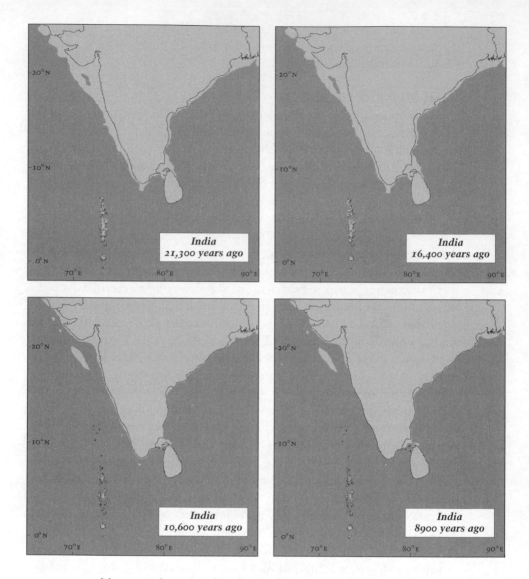

India
21,300 years ago

India
16,400 years ago

India
10,600 years ago

India
8900 years ago

antediluvian India surrendered to the rising seas but how practical a proposition it is that an unidentified high culture – or *cultures* – of Indian antiquity could have been lost to archaeology during this period.

In December 2000 I also received confirmation from the NIO that permission had at last been granted for me to dive at Poompuhur. The trip could take place in February 2001 – exactly a year after my previous visit. Mercifully, the final arrangements and negotiations (and the money that had to be paid to the NIO) had been taken over on my behalf by a film crew from Channel 4 TV in Britain who were now covering my story. I welcomed the fact that whatever the NIO had to show me at Poompuhur would be documented properly for television. I was convinced that only by allowing the greatest number of people to see the

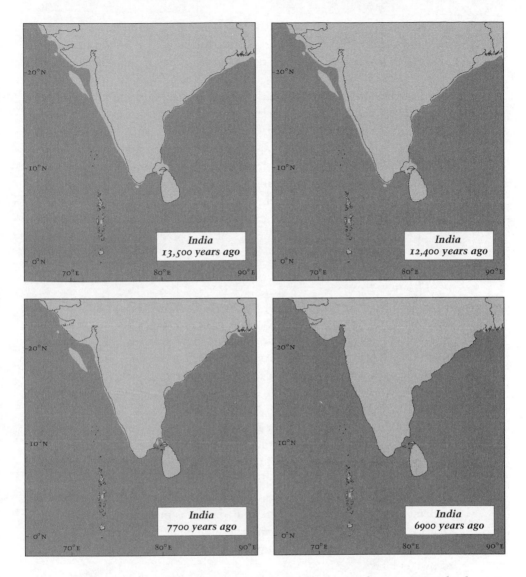

U-shaped structure for themselves and to make up their own minds about it did it stand a chance of getting the attention it deserved from the archaeologists who had hitherto ignored it.

Unfolding the Indian floods

In January 2001 Glenn Milne, who had been working overtime, sent me more Indian maps – a complete sequence of high-resolution inundation simulations for 21,300 years ago; 16,400 years ago; 13,500 years ago; 12,400 years ago; 10,600 years ago; 8900 years ago; 7700 years ago; 6900 years ago.

Although I had a rough idea of what to expect, it was still a revelation to flip rapidly through these maps from the oldest to the youngest and watch the entire process of the post-glacial inundation of India unfold before my eyes. What I found most striking of all, however, was the way in which the two areas rich in flood myths where underwater ruins had already been found – off the coast of Gujerat in the north-west and off the coast of Tamil Nadu in the south-east – were also the two areas most clearly flagged by Glenn Milne's maps as large and continuous antediluvian habitats in which it was conceivable that Ice Age civilizations could have flourished.

Moreover, now that I had the maps at virtually millennium intervals, it was possible to pinpoint periods when the extent of the ongoing loss of land to the sea had been particularly rapid and to note any correlation between these and (1) John Shaw's cataclysmic chronology for the post-glacial floods; (2) the relevant mythology; and (3) the accepted dates for the so-called 'Neolithic revolution' in India (i.e., the beginnings of food production at Mehrgarh and other sites).

The north-west

In the north-west, around Gujerat, the maps show that a huge land area was inundated between approximately 17,000 and 7000 years ago – an area contiguous to the domain in which archaeologists believe that the first recognizable roots of the Indus-Sarasvati civilization were planted during the last three millennia of the same period. As we saw in chapter 7, the submerged lands are at their most extensive around the modern Gulf of Cambay – south of which the map for 16,400 years ago shows an extensive depression, very likely to have been filled with a large freshwater lake, bounded by a further tract of land at least 100 kilometres wide and beyond that the Arabian Sea.

The next map in the sequence – 13,500 years ago – reveals that major changes occurred during the intervening 2900 years. The landmass around the Gulf of Cambay was much reduced in area and a large island, almost 500 kilometres long and 100 kilometres wide at its midpoint, was marooned off-shore in the Arabian Sea. Between the island and the mainland a marine strait, also 100 kilometres wide in some places, opened up through the basin of the former freshwater lake.

These rather dramatic land-losses between 16,400 and 13,500 years ago correlate well with the first of John Shaw's proposed episodes of global superfloods, which falls midway through the period at around 15,000 years ago.

Over the next 6000 years – between 13,500 years ago and 7700 years ago – the maps show that the large off-shore island and the coastal strip masking the outline of the Gujerat peninsula were continually nibbled away at by the rising seas, but that these events were gradual, extended over many lifetimes, and would have been unlikely to have been perceived as cataclysmic. As late as 7700 years ago the Gulf of Cambay was still the 'pleasant valley' that it had been, uninterrupted, since at least the Last Glacial Maximum and the island lying off-shore, though reduced, was still of formidable size – perhaps 300 kilometres in length and close to 80 kilometres wide.

This pattern for the Gujerat area, therefore, does not correlate well with the second of John Shaw's proposed episodes of global superfloods around 11,000 years ago. Nor does it suggest a motive for any memorable panic-migration of flood refugees out of this area at any point during this period – which straddles the supposed date of around 9000 years ago for the first settlement of Mehrgarh.

What happens next, however, provides a close match to Shaw's chronology of around 8000 years ago for the third flood. The maps for 7700 years ago and 6900 years ago show that in this relatively short period of 800 years the large remnant island below the Gulf of Cambay was completely wiped off the map and the Gulf itself was fully and permanently inundated to its modern extent. For any hypothetical coastal culture that had been forced to retreat and compact into the Gulf's pleasant valley over the previous 6000 years, or that had lived on the island, it goes without saying that these events would have been more than cataclysmic.

They would have looked like the end of the world.

The south

As we would expect, the inundation maps for 21,300 years ago and 16,400 years ago show that few significant coastline changes took place in the south during the five millennia or so of the Last Glacial Maximum. At that time Sri Lanka was joined to the mainland, as we have seen, and 'a substantial integrated area – an entire sub-region of India' that is today submerged[59] – was above water in the south and the south east (and indeed all along the Malabar coast in the west also). This lost antediluvian realm accords extremely well in a general sense with the central claim of the Kumari Kandam tradition that a large landmass did exist around the south of India in ancient times and that it was swallowed up by the sea in a series of floods.

The maps of 21,300 and 16,400 years ago reveal the full extent of the continental shelf that was exposed during the Ice Age, but a specific feature of great interest is the snout-shaped peninsula shown to have extended approximately 150 kilometres southwards into the Indian Ocean below modern Kaniya Kumari. As the reader will recall, such a peninsula in exactly this location is spoken of in the Kumari Kandam tradition:

> In former days the land . . . extended further south and . . . a mountain called Kumari-koddu, and a large tract of country watered by the river Prahuli had existed south of Cape Kumari. During a violent irruption of the sea the mountain Kumarikoddu and the whole of the country through which flowed the Prahuli . . . disappeared.[60]

The peninsula that Glenn Milne's calculations place on the inundation maps is not as large as the one described in the tradition (which was said to have been

'700 Kavathams', about 1500 kilometres, in length). Still it is there – precisely where the Kumari Kandam tradition says it should be, and in the correct time-frame. Moreover, the maps show another antediluvian landmass that has also for the most part disappeared beneath the waves standing in the open ocean to the south-west – the greatly enlarged Maldive islands as they looked at the Last Glacial Maximum.

What if the civilization of Kumari Kandam had been partially based along the coastal margins of southern India and Sri Lanka and partially on the antediluvian Maldives archipelago? If so, then the idea that Kumari Kandam once extended 1500 kilometres to the south of Kanya Kumari does not seem so far-fetched. Nor does the notion that a civilization that had once existed in this area could have been destroyed by recurrent cycles of catastrophic floods.

The tradition says that the last of these floods occurred 3500 years ago (supposedly the flood that destroyed the Second Sangam at Kavatapuram), and the one preceding it 7200 years ago (supposedly the flood that destroyed the First Sangam at Tenmadurai). In addition N. Mahalingam has cited further Tamil sources that speak of earlier floods: one around the date of foundation of the First Sangam, approximately 9600 years ago, one just over 16,000 years ago and the earliest 18,000 years ago.[61]

Once again there is a good general correlation between what scientists now know about the meltdown of the Ice Age (particularly the episodic and recurrent nature of the post-glacial floods) and what the Kumari Kandam tradition claims was happening in the world in precisely the same period (episodic and recurrent floods). There is by no means one-to-one agreement on the dates at which particularly severe inundations occurred – as is to be expected given the margins of inaccuracy that surround the estimating processes used by both Shaw and Milne, not to mention the scope for error and exaggeration in the tradition itself. Still, there is more than enough agreement on the general course of events to give us pause for thought. After all, how many times can we reasonably cry 'coincidence' when the medieval Tamil 'fabulists' keep on getting their palaeogeography right? Or did they in fact – as Shivaraja Pillai asks sarcastically – 'come upon some secret archive which had escaped the deluge'?[62]

Glenn Milne's inundation map for 13,500 years ago shows a dramatic change in the south Indian landscape since the previous map of 16,400 years ago: the coastal margins have been greatly reduced and the peninsula below Kaniya Kumari has been severed by the sea, leaving an island off-shore. In the Indian Ocean to the south-west the land area of the antediluvian Maldives archipelago has been reduced almost by half.

The map for 12,400 years ago shows little significant change, but in the map for 10,600 years ago the island to the south of Kaniya Kumari has been reduced to a dot, the Maldives have been further ravaged, and, for the first time, a neck of sea is shown separating Tuticorin on the mainland and Mannar in what is now Sri Lanka. This incursion seems very close to what is described in the Sri

Lankan myth of the flooding of Ravana's kingdom (said to have extended between Tuticorin and Mannar 'in a former age').[63] Moreover, the timing – between 12,400 and 10,600 years ago – coincides with Glenn Milne's date for the submersion of the U-shaped structure at Poompuhur and accords well with the second of John Shaw's episodes of post-glacial flooding around 11,000 years ago.

The map of 8900 years ago shows further minor erosion all around the south Indian coastal strip and a deepening of the marine incursion beyond Tuticorin and Mannar into what is now a bay beneath the war-torn Jaffna peninsula. However, the Palk Strait was still dry land 8900 years ago and, though much diminished in size, the land-bridge connecting Jaffna to the mainland was still in place at that date (and indeed was to remain there for another thousand years).

On John Shaw's estimates, the third of the three great episodes of post-glacial flooding was unleashed on the world's oceans around 8000 years ago – and we have seen how this correlates well with what happened at around that time when the Gulf of Cambay and neighbouring areas of the north-west of India were rapidly inundated. In the south-east the inundation maps show that in the same period between 7700 and 6900 years ago there was also significant further inundation of the Maldives, while the land-bridge between Sri Lanka and Tamil Nadu, which had clung on for so long, was at last swallowed up by the sea – leaving India looking very much as it does today.

Occam's razor

What are we to conclude about the Kumari Kandam myth?

In some respects there is no doubt that it has proved eerily, stunningly accurate. On the other hand much of it sounds wildly improbable and in places obviously 'manufactured'. For example, when one studies the way numbers are used in the myth (something that I have not sought to tax the reader with here) certain obvious patterns emerge that are more suggestive of a mathematical game, or code, than of true reports of the number of members of, or the number of royal patrons of, or the duration of this or that Sangam.

It will be recalled that the durations of the three Sangams were said to be 4440 years for the First Sangam, 3700 years for the Second Sangam and 1850 years for the Third Sangam.[64] It is obviously not an accident that each of these numbers is a multiple of 37 ($120 \times 37 = 4440$; $100 \times 37 = 3700$; $50 \times 37 = 1850$).[65] What the significance or purpose of this pattern is I cannot begin to guess, but it means that the chronology of the myth is suspect and cannot be treated as a reliable historical record.

Still, it does not follow from this and other criticisms that the whole myth must be tossed in the dustbin of history and forgotten – as it has been by most scholars. Although wildly out of line on some of the details and dates, the myth is right in the broad sweep. It is right that India's Dravidian peninsula was formerly much bigger than it is today. It is right that a series of huge deluges occurred over a period of several thousand years and that these swallowed up the antediluvian lands in stages. And the myth selects the correct epoch – smack in the middle of the post-glacial floods around 11,600 years ago – in which to set its flood story.

Besides, whatever one thinks of myths (and most historians and archaeologists regard them as useless to scientific inquiry)[66] there is the awkward and inescapable archaeological fact of the U-shaped structure 23 metres underwater and 5 kilometres off-shore of Poompuhur – a structure that is '11,000 years old or older'.[67] Isn't the most parsimonious way to explain its presence there the very one that the myth itself provides – namely, that a civilization of former times once flourished in this region but was swallowed up by the sea?

I could only learn more by diving.

12 / *The Hidden Years*

The period dreadful for the universe has come. Make for thyself a strong ship, with a cable attached; embark in it with the Seven Sages and stow in it, carefully preserved and assorted, all the seeds which have been described of old . . .

Satpatha Brahmana

An epoch of spectacular geological turmoil occurred at the end of the last Ice Age, with the most dramatic effects registered in a series of cataclysmic floods that took place at intervals between roughly 15,000 and 7000 years ago. Is it an accident that this same 8000-year period has been pinpointed by archaeologists as the very one in which our supposedly primitive forefathers made the transition (in different places at somewhat different times) from their age-old hunter-gatherer lifestyle to settled agriculture? Or could there be more to 'the food-producing revolution' than meets the eye? After all, most scientists already recognize a causative connection between the end of the Ice Age and the supposed beginning of farming – indeed an unproven hypothesis that rapid climate changes forced hunter-gatherers to invent agriculture presently serves as pretty much the sum of conventional wisdom on this subject.[1]

But there is another possibility. Nobody seems to have noticed that in the general vicinity of each of the places in the world where the food-producing revolution is supposed to have begun between 15,000 and 7000 years ago there is also a large area of land that was submerged by the post-glacial floods between 15,000 and 7000 years ago:

- We have seen that this is true for India, one of the world's ancient agricultural 'hearths',[2] which lost more than a million square kilometres in the south and the west and, most conspicuously in the north-west, at the end of the Ice Age.
- It is true for China and for south-east Asia, both important centres of palaeo-agriculture. Immediately adjacent to them, but now under as much as 100 metres of water, lies the Ice Age continent of Sundaland. Prior to its final inundation of about 8000 years ago, this consisted of more than 3 million square kilometres of prime antediluvian real estate extending from the Malaysian peninsula through what are now the Indonesian islands and the Philippines. Taiwan was incorporated with the Chinese mainland and northwards from there the coast expanded almost

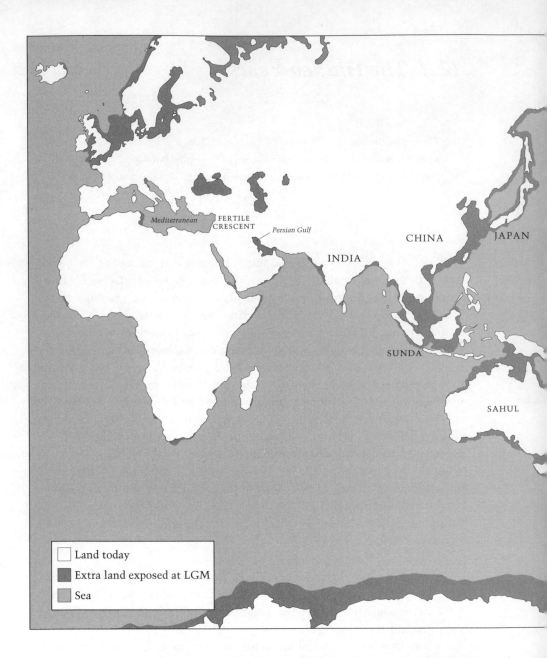

Land today

Extra land exposed at LGM

Sea

1000 kilometres to the east to fill what is now the Yellow Sea and incorporate the Korean peninsula fully with the mainland.

- It is true for the so-called Fertile Crescent – the prime agricultural 'hearth' of the Middle East, centred around lands watered by the Tigris and Euphrates rivers, that forms a rough semi-circle through parts of modern Israel, the Lebanon, Syria, Turkey, Iraq and Iran and ends up near the Persian Gulf. For not only was the Gulf previously dry – and flooded at

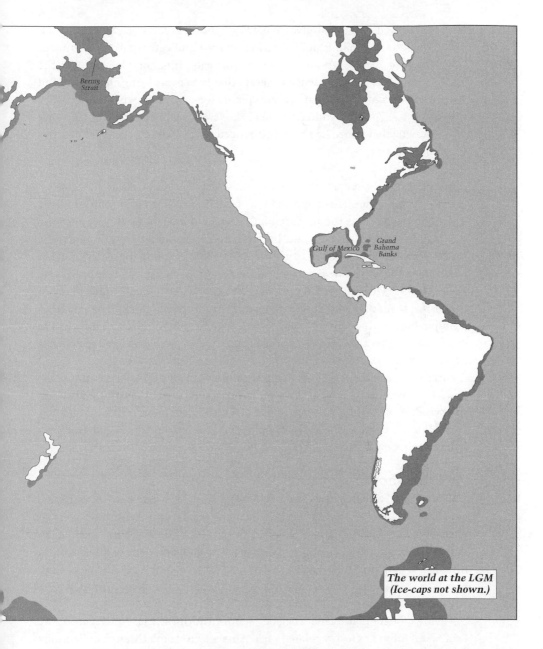

Bering
Strait

Gulf of Mexico

Grand
Bahama
Banks

**The world at the LGM
(Ice-caps not shown.)**

the end of the Ice Age, as we saw in chapter 2 – but a glance at the wider map also shows several other inundated areas near by in the Indian Ocean, the Red Sea and the eastern Mediterranean.

- And it is true for Central America, where agriculture is thought to have sprung up spontaneously, independent of developments in the Old World. Off the Gulf of Mexico, the Yucatan, Nicaragua, Florida and Grand Bahama Banks were imposing landmasses during the Ice Age that were swallowed by the post-glacial floods around 7000 years ago. Evidence from Mexico and Panama, published in July 2001, indicates that 'agriculture in the Americas began around 7000 years ago'. It is notable that: 'On the Gulf coast pollen evidence suggests that forest was being cleared around 5100 BC and domesticated maize plants were being grown only a century later . . . The San Andres site near the famous Olmec centre of La Venta showed that maize had been introduced and grown in a region of beaches and lagoons.'[3]

My curiosity about coincidences like these developed as I researched *Underworld* – because the sudden appearance of village farming communities at the end of the Ice Age was the first step on the road to modern civilization (so the stakes in this inquiry are high), and because the Ice Age lands that went under the sea cover an area of more than 25 million square kilometres of the earth's surface where, for obvious practical reasons, almost no archaeology has ever been done (so important evidence could very easily have gone undetected). Since many of the coastal lands that were inundated would have offered desirable refugia from inhospitable and unpredictable Ice Age conditions, the possibility surely has to be considered that the real story of the origins of food production and of civilization may yet await discovery because the evidence is underwater.

I decided to explore this neglected possibility with all the resources at my disposal, knowing when I did so that it would commit me to an exhausting and expensive schedule of travel and diving – much of which might prove fruitless – and that I would have to enter arcane areas of inquiry, ransack obscure libraries and rack my brains on uncompromising sciences if I was to have any hope of success.

Long shot

I needed a good research assistant and in August 2000 I found one – Sharif Sakr, who has proved to be the very best of the many good researchers I have worked with over the years. Right at the beginning, I asked Sharif to find me an authoritative scientist at a major university who could produce high-resolution inundation maps for us, virtually on demand, for any point on earth at any time during the meltdown of the Ice Age. This was the start of our long and productive working relationship with Glenn Milne.

Then, as the inundation data began to pour in during the last quarter of 2000, I set Sharif another closely related task. This was to comb through collections of ancient maps from the sixteenth century or earlier – i.e. before the world had been fully explored – to see if he could find any that showed correlations with Glenn Milne's reconstructions of Ice Age coastlines.

This touches on a problem – and a mystery – that I have long had an interest in and to which I devoted three chapters in my 1995 book *Fingerprints of the Gods*. To put matters at their simplest, it has been claimed by Charles Hapgood and others that certain maps dating roughly between the fourteenth and sixteenth centuries show Antarctica and other areas of the world not as they look today, but as they may have looked during the Ice Age when sea-levels were 120 metres lower. Moreover, many of the areas in question had not even been discovered when the maps were drawn (Antarctica was not discovered until the nineteenth century).

Hapgood explains such anomalies with the suggestion that a high civilization, which was subsequently destroyed, may have existed and mapped the world to near-modern levels of precision during the Ice Age. He further proposes that after the destruction of that hypothetical civilization some of the maps survived and were handed down from generation to generation, being copied and recopied many times as the original materials on which they were drawn perished. Perhaps facsimiles preserved and passed on in this manner eventually ended up lodged in the great libraries of late antiquity – notably at Alexandria in Egypt, which was for a long while a world centre of navigational and astronomical science. Perhaps some of the facsimiles were amongst other salvaged documents rescued from the fire that is said to have destroyed the Alexandria library in the early centuries of the Christian era. Perhaps a handful found shelter in other archives in the Middle East. Perhaps from there, after a few more centuries had passed, they were looted by Crusaders and redistributed around the Mediterranean where their value as navigational charts was recognized by mariners. And perhaps then, in the late thirteenth or early fourteenth century, a new era of copying began in which information from the highly revered and generally accurate ancient maps was integrated with the observations and measurements of contemporary sailors to create navigational charts of astounding accuracy. Since the Mediterranean was at that time conceived of by its inhabitants as the centre of the world, it would have been quite natural for the copyists to focus most of their work on reproductions of the Mediterranean and neighbouring coastal regions – even if their source documents showed a far wider area . . .

All speculation of course. Except the part about the sudden appearance at around the end of the thirteenth century, of uncannily good maps of the Mediterranean and immediately neighbouring parts of the Atlantic. That is completely true. They are called portolans or portolanos and several hundred have come down to us – all of which, eminent cartographers are agreed, show the influence of a single source map, now lost, that the great map historian A. E. Nordenskiold called 'the normal portalano'. Rarer, but fortunately still also surviving, are a handful of world maps and portions of world maps in recognizable portolan style – and it is mainly amongst these that the alleged similarities to Ice Age coastlines and topography are observed.

Many years have passed since Hapgood published his famous *Maps of the Ancient Sea Kings* in 1966 and there have been huge improvements in the technology for calculating post-glacial sea-levels. Moreover, although he has been repeatedly attacked and vilified by scholars who claim to have 'debunked' his work, the essential mystery upon which he touched remains unsolved to this day.

I'm not interested in reviewing Hapgood again – read *Fingerprints*, or better still read Hapgood! But in the light of the good inundation data we now had from Glenn Milne I asked Sharif to cast a fresh eye over some of the more intriguing ancient maps that Hapgood had drawn attention to and to look for others that might have a bearing on the problem. I suggested he exclude Antarctica from the search, since I had paid enough attention to it in 1995. And on the same grounds of redundancy I told him to ignore any correlations that Hapgood himself had already written up. I only wanted material that hadn't been observed and argued about before, that correlated well with the inundation maps, and that was substantial enough to withstand the rigours of hostile academic scrutiny.

It seemed a lot to ask for – a real long shot – but then in February 2001 Sharif e-mailed me about a map of India that he had been investigating. What was remarkable about this 1510 Portuguese map was the fidelity and degree of detail with which it which it portrayed areas of the Indian coast as they had last looked 15,000 years ago.

I was already in India when I read the e-mail on my laptop on 23 February 2001. I had just flown into Tamil Nadu from the Republic of Maldives, where I had spent four days working with the Channel 4 film crew.

The same night, after we had checked into the Fisherman's Cove hotel in Mahabalipuram, where we would be filming the next morning, we received confirmation from the NIO that their team had relocated the U-shaped structure at Poompuhur and would be ready to dive with us on the 26th.

13 / Pyramid Islands

The Redin came long before any other Maldivians. Between them and the present population other people had also come, but none were as potent as the Redin, and there were many of them. They not only used sail but also oars, and therefore moved with great speed at sea . . .

Thor Heyerdahl

Republic of Maldives 18–23 February 2001

This is the Maldives. Imagine you are flying in a specially equipped plane, under an endless blue sky over endless blue ocean . . . The plane is very fast and manoeuvrable, you can go where you want in it, and yet all you see is blue – just blue above and blue below.

Suddenly, in the distance, far away where the sky meets the water, your eye catches a glint of . . . something on the horizon. You turn the plane towards it, skimming at 200 metres over the ocean with little waves breaking into white horses below you.

Soon land comes into view – just a curving feather of sand no more than a kilometre wide and three kilometres long, adorned with plumes of lush green palm leaves seeming to float in a sea that is now not merely blue but that grades into incredible shades of azure and turquoise. Passing directly overhead you see an area cleared of jungle packed with tiny houses built out of white coralline limestone blocks and separated from one another by an orderly network of streets brushed with white coralline limestone sand – so that the whole Lilliputian village glares like a mirror in the morning sun.

You take the plane higher to get a better view (remember this is an imaginary journey and you can go as high as a satellite if you want), and you see that the stunningly beautiful but tiny inhabited island over which you have just flown is part of an even more stunningly beautiful ring of even tinier uninhabited islands and sandbars also shaped as rings and crescents and ellipses. This ring in its turn reveals itself to be just one of countless other rings and crescents and ellipses lying side by side to form a much larger ellipse in the ocean – the outer rim of a great Maldivian atoll 50 kilometres wide and more than 100 kilometres long. The atoll encloses a lagoon of hardly smaller dimensions (since the rim islands themselves are narrow), and within the lagoon are scattered dozens more small coral islands and sandbars in which the essential patterns of the entire Maldives chain – circles, ellipses, crescents – repeat themselves again and again.

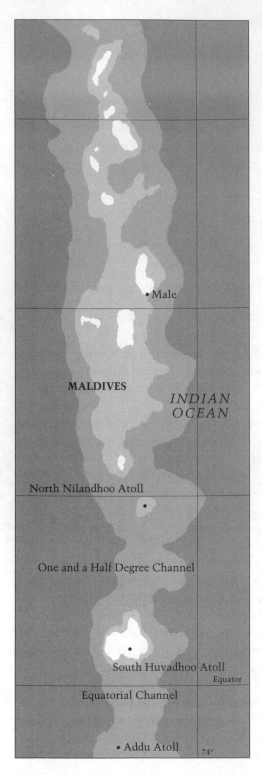

You urge the plane higher still, look down at last on the entire archipelago stretched out below you around the curve of the earth and discover that it consists of an assembly of similar atolls, twenty-six of them in all, strung together like the pearls in a necklace and draped in the form of an elongated ellipse 754 kilometres long from north to south and 118 kilometres wide from east to west.

Each atoll is the product of coral growth around the edges of a submerged volcanic mountain peak:

> In a scenario played out over hundreds of thousands of years, coral first builds up around the shores of a volcanic landmass producing a fringing reef. Then when the island, often simply the exposed peak of a submarine mountain, begins slowly to sink, the coral continues to grow upwards at about the same rate. This forms a barrier reef which is separated from the shore of the sinking island by a lagoon. By the time the island is completely submerged, the coral growth has become the base for an atoll, circling the place where the volcanic landmass or island used to be. The enclosed lagoon accumulates sand and rubble formed by broken coral, and the level of this lagoon floor also builds up over the subsiding landmass ... Coral growth can also create reefs and islands within the lagoon ...[1]

The lagoon floors are all submerged today, but at the Last Glacial Maximum, when sea-level was lower by about 120 metres, the huge basins within each and every one of the Maldives atolls were all dry land ...

You fly the plane lower again, spiralling downwards towards the sea, zoom-

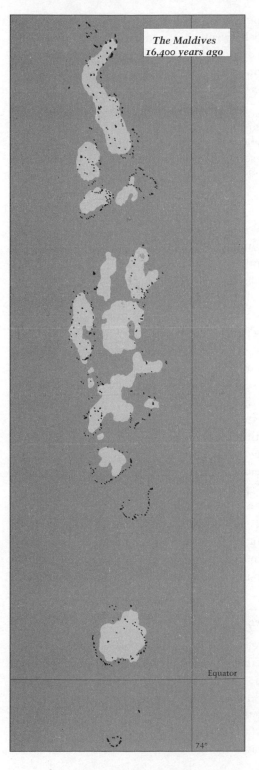

The Maldives
16,400 years ago

Equator

74°

ing in on one atoll, one emerald-green island. Within a beach perimeter of start-lingly white sand it seems at first to be just thick palm jungle from one side to another and apparently uninhabited.

Then you spot a clearing in the jungle less than half a kilometre from the sea. You fly closer. In the heart of the clearing, with a tree growing on its summit, is what looks like a conical hill. Closer still and you discover that the hill is not a hill at all, and it is not quite conical either.

It is a ruined and partially collapsed pyramid about the height of a two-storey building.

The necklace

The four-day trip that we made to the Maldives immediately before returning to India on 23 February 2001 was not intended to be an expedition to search for underwater ruins – hardly practicable in such a short time in an archipelago of almost 1200 tiny islands extending through eight degrees of latitude across 90,000 square kilometres of ocean. In all that mass of blue water the total area of dry land is presently less than 300 square kilometres and many scientists are of the opinion that even this remnant may be submerged before the end of the twenty-first century by rising sea-levels linked to global warming.[2]

The threat of extinction that hangs over the Maldives and its unique culture serves as a reminder that the world's oceans can and do rise, and that when they do they can swallow up low-lying countries – and all their history – with not a trace left visible above the water. And if that is true today, deep in what has so far been the most placid interglacial of the past 2.5 million years, then it doesn't take much imagination to work out how things must have been in the world when sea-levels were rising crazily between 15,000 and 7000 years ago.

Besides, thanks to the ingenuity of modern science, we have inundation maps to tell us the story – perhaps still not with 100 per cent accuracy (although that is being refined all the time) but based on the best data presently available.

And what the maps tell us about the Maldives is that the necklace of scattered coral atolls of which the archipelago now consists was almost continuous land at the Last Glacial Maximum, broken only by intermittent channels, bays and inlets, occupying perhaps 50,000 square kilometres out of the total of 90,000 square kilometres that the Republic presently encloses within its territorial waters. In other words, some 49,700 square kilometres of the Maldives that was above water between 21,000 and 16,000 years ago is underwater today.

In my investigation of the riddle of Kumari Kandam I could hardly ignore this lost antediluvian landmass in the Indian Ocean that had stretched towards the equator from a point roughly parallel to the extended southern tip of Tamil Nadu during the Ice Age. Even today the much reduced Maldives are a barrier to shipping, but 16,000 years ago, had anyone been sailing in these parts, they would have been confronted by an 800 kilometre long line of cliffs running north to south effectively blocking the east–west passage. Hypothetical Ice Age seafarers wanting to sail east or west would have been more or less obliged to make their way through one of two deep-water channels – the 'One and a Half Degree Channel' (so named because it slices across the Maldives one and a half degrees north of the equator) and the 'Equatorial Channel', then as now about 50 kilometres wide, which separates South Huvadhoo Atoll (in the northern hemisphere) from Addu Atoll (in the southern hemisphere).

So rather than the dots in the ocean that they are today, the Maldives 16,000 years ago would have been formidable. If such a thing as 'Kumari Kandam' ever did exist, centred as the myths suggest on the antediluvian coastal margins of southern India and Sri Lanka, then might it not also have included the great barrier islands of the Maldives just a few hundred kilometres to the south-west? As I noted in chapter 11, such a hypothesis would explain the old Tamil traditions which tell us that Kumari Kandam once extended into the Indian Ocean some '700 Kavathams' (about 1500 kilometres) beyond modern Cape Comorin.

The disappearance of prehistory

The ancient history of the Maldive islands is almost completely unknown[3] and their inundation profile suggests that their prehistory, if any, may have been lost beneath the rising seas at the end of the Ice Age. The matter is further complicated by the presence of an alarming 'gravity anomaly' centred here. In layman's terms what this means is that the archipelago is situated at the bottom of an enormous trough in the surface of the Indian Ocean itself – this trough being created by a strong local gravitational field which some believe may be linked to the mass of sunken mountains on top of which the Maldives atolls have grown. Like other gravity anomalies (several similar troughs have been measured in the world's oceans by satellites) it is not certain that this one has always remained in exactly the same location, or that its depth has always remained the same, or that it always will do so in the future.[4]

Very little archaeology of any kind has ever been done in the Maldives, but the view of most orthodox scholars is that 'the first settlers probably arrived from Ceylon not later than AD 500 and were Buddhists'.[5] Other authorities argue for an earlier date – back to about 500 BC – and note some south Indian, specifically Tamil, Hindu religious influence.[6] Thor Heyerdahl, who is one of the few to have conducted archaeological expeditions in the Maldives and whose book *The Maldives Mystery* is the only serious attempt to get to grips with the problems of the islands' ancient history, believes that they were settled much earlier than that – perhaps by 2000 BC or even 3000 BC – and that they may have played a part in an archaic Indian Ocean trading network involving ancient Egypt and the Mesopotamian and Indus-Sarasvati civilizations.[7] So far Heyerdahl has not been supported by the few carbon-dates obtained from the Maldives – none older than AD 540[8] – but in this as other matters he may yet be proved right. What we do not know about these islands far exceeds what we know:

> Usually the history of a nation begins with a potent king founding a dynasty. The Maldives is a definite exception. A long dynasty of kings was already there before known Maldive history started. This kingdom ended when Maldive history began. The last king was made a sultan by a pious foreigner who came by sea and started local history. He caused all the kings to disappear into oblivion, except one, the one he himself converted. With neither arms, nor with any Maldive blood in his veins, he introduced a new faith, new laws, and founded the present Moslem Maldive state.[9]

In other words, not only has the Maldives suffered the incursions of the sea and the usual depredations of time but also it was converted, in the year AD 1153 (the year 583 of the Holy Prophet), to the Islamic faith,[10] which led to further attrition of ancient structures, artefacts and inscriptions. As my old friend Peter Marshall, author of *Journey Through the Maldives*, explains:

> Recorded history only begins about the time of the conversion of Maldives to Islam ... As Christians in Europe begin their calendar from the birth of Christ and tend to dismiss all earlier religions as pagan, so Maldivians follow the Islamic calendar. Until recently they had very little interest in what happened before. Not only was Maldivian pre-Islamic history suppressed but most pre-Muslim artefacts were destroyed.[11]

So what archaeologists are left to work with in the Maldives, above the water at least (and nobody has yet looked underwater), is almost certainly just a fraction – and perhaps an extremely unrepresentative fraction – of what was once there.

Even so, buried deep in the jungle of islands up and down the archipelago – some uninhabited and all off-limits to tourists – there are several dozen partially collapsed and heavily overgrown pyramids, up to ten metres high, with their sides oriented to the cardinal directions. Although in a state of ruin today, these mounds of compacted earth and stone, in some cases with stepped courses of closely jointed megalithic masonry to be seen exposed under the earth fill, have a sombre and looming presence as they emerge out of the jungle. Called *hawitta* by the local people, the precise function and origin of these mounds have not been confirmed – though the carbon-dates put their construction between roughly AD 500 and 700.[12]

Most scholars think they are Buddhist *stupas* (relic mounds), which probably they are. Unimpeachably Buddhist sculptures, reliefs on stone and artefacts have been found amongst the ruins and some of the pieces are recognizably similar to other Buddhist work of the same period from India and Sri Lanka – so there is no doubt that Buddhism was extensively present on these islands in the centuries before the coming of Islam.[13] Indeed, a Sanskrit text of Vajrayana Buddhism dating back to the ninth or tenth century AD is the earliest surviving legible inscription thus far found in the Maldives.[14]

Still, as a number of observers have noted, there seems to be something strange about this Maldivian Buddhism. Could it be some other religious influence showing through – maybe a form of Hinduism that had preceded the Buddhist faith to the Maldives? Certain striking sculptures of grotesque human faces with bulging eyes, twirled mustachios and curved cat-like fangs 'may recall Hindu deities',[15] admits Arne Skjolsvold, an archaeologist with the Kon-Tiki Museum – who nevertheless prefers to explain such images as expressions of a localized subculture of Tantric Buddhism.[16]

There may be clues in Dhivehi, the Maldivian language. It belongs to the Indo-European family and is related to Sanskrit and thus also to Sinhalese, one of the two languages of Sri Lanka (the other being Tamil). Sinhalese has been heavily influenced and modified by its contact with Tamil,[17] and, according to Clarence Maloney, a Tamil/Dravidian sublayer exists in Dhivehi also, which suggests that 'Hinduism was present in the Maldives before the Buddhist period.'[18]

Interestingly, large numbers of 'phallic' sculptures have been recovered in archaeological excavations in the Maldives – for example amid the ruins of a vast temple complex in North Nilandhoo Atoll.[19] I was able to study a collection of such objects from different parts of the archipelago and in my opinion, despite some idiosyncrasies, they are nothing more nor less than Sivalinga.

That Siva's characteristic emblem should be found here in these remote islands on the edge of the southern hemisphere is in a way not surprising – since he was ever Daksinamurti, 'the God of the South'.[20] But Siva is an ancient and widely revered god whom the *Vedas* associate with the high peaks of the Himalayas far to the north and whose image as the ascetic Lord of Yoga and as Pasupati, Master of Beasts, goes back nearly 5000 years in the Indus valley cities of Harappa and Mohenjodaro.

Moreover, as we saw in chapter 11, so many lingam-like objects have been found in much older pre-Harappan sites that T. R. Sesha Iyenagar can exclaim: 'the worship of Siva in the form of a linga existed in the Stone Age'.[21] In this regard, therefore, the Kumari Kandam tradition once again proves itself to be in accord with the archaeological facts when it proclaims Siva's membership of the First Sangam, supposedly founded in the antediluvian city of Tenmadurai 11,600 years ago – a date deep in the Stone Age.

The riddle of the hawittas

Let's set to one side for a moment the intimations of vast antiquity for the religion and religious ideas that became Hinduism and Buddhism (for Buddhism is merely a 'protestant' offshoot of Hinduism and both trace their origins and authority back to the *Vedas*).

Let's accept the range of dates around the middle of the first millennium AD proposed by archaeologists for the construction of the pyramidal *hawittas* of the Maldives (or, strictly speaking, for the construction of the few that have thus far been excavated).

And let's accept the same date range for the religious sculptures, artefacts, etc. that have been found round about them. There seems no good reason not to do so; on the contrary, it looks as though the archaeologists have done their jobs well and that these dates are likely to be accurate within a reasonable margin of two or three hundred years either way.

But then the question arises, where did the distinctive religious art and architecture of the Maldives come from? Yes, its sculptures and its pyramids – or *stupas* – are similar to those of the Buddhists of Sri Lanka, but there are differences . . . And yes, they are similar to those of the Hindus of south India, but again there are differences. So where and when did these differences and unique characteristics incubate and take shape? There is no archaeological trace of any evolution of architectural and symbolic ideas behind the oldest structures in the Maldives. The *hawittas* just suddenly appear – we must assume around 1500 years ago from the carbon-dating – in an already fully designed, fully worked-out form and with all the required building skills already in place.

Were they the work of immigrants importing a pre-existing architectural canon from elsewhere? Perhaps – but if so, then where? No other trace of the distinct Maldives style has been found in India or Sri Lanka. Or is it possible that the ever-encroaching seas have simply swept away and covered up the earlier stages of the Maldives story – just as they will sweep away and cover up the little that is left of the archipelago before the end of this century?

Bill Allison's antediluvian tour

I dived a couple of times in the blue waters of the Maldives with Bill Allison, a tough, crew-cut, steely-eyed, flat-bellied 54-year-old Canadian who is conducting a long-term scientific survey of the islands' coral reefs. I've already noted that our rushed filming schedule and the vast area that would have to be covered ruled out any structured or useful exploratory diving during our short stay – for the same reasons that there is no point in looking for a needle in a haystack. So the producers' objective for these two dives was simply to film what they call 'pretties' – beautiful fish, beautiful coral, lush tropical waters with infinite visibility, sun effects, surge effects, etc., and generic shots of me finning around *in situ*. The 'motive' for our dives here in storytelling terms (as if anyone needs a motive to dive in the Maldives!) would be provided by Bill Allison – the coral reef expert – showing me – the eager historical detective – notches and caves at various depths that had been cut in the coral formations by waves during the lowered sea-levels of thousands of years ago.

After we had completed our dives we sat talking on the deck of the boat in the afternoon sun, moored in the open sea just on the outside edge of North Male Atoll. I asked Bill: 'How come the Maldives are here? We see coral under us, but what's the story of how it got there?'

> *Bill:* Well, it seems that as India drifted over towards Asia [continental drift hundreds of millions of years ago] the Maldives or what became the Maldives were left as a string of volcanoes behind it, and as these volcanoes sank into the earth's crust, coral grew on them and just kept growing. Right now there's over maybe 2000 metres of coral.
>
> *GH:* 2000 metres of coral on top of the original volcanoes?
>
> *Bill:* That's right.
>
> *GH:* Wow . . . (*pauses for thought*) – Now if we . . . if we go back to the period that I'm interested in, which is the period from the Last Glacial Maximum, through until about the beginning of historical times, about 5000 years ago or so – so say from 17,000 years ago down to 5000 years ago – what would we be seeing around us here, if we could be here 17,000 years ago?
>
> *Bill:* Well, we'd be right now where we are with respect to these islands, looking up about 130 metres to see those trees . . . Like the cliffs of Dover or something. It'd be a plateau with notches cut where the channels are, so the cliffs might be 130 metres high –
>
> *GH:* Wow.
>
> *Bill:* – and the channels –
>
> *GH:* So that would be towering above us?
>
> *Bill:* That's right. And the channels might be, oh, 80 metres, 90 metres high.

GH: Wow. And then once we're inside that area there (*pointing towards atoll*) presumably it would all be land?

Bill: Yeah.

GH: Or would there be some water too?

Bill: Well, it'd be depressed and it just depends. This is very porous material. Coral doesn't grow as a solid mass, just a lot of crevices and so on, so any water falling would drain rapidly. There might be temporary lakes, there'd be streams. They would probably develop into underground rivers and they'd probably empty into the sea through the ground or maybe through the channels.

GH: Would there have been rivers above ground?

Bill: Rivers? Probably. But probably not big rivers and probably disappearing into the ground pretty quickly, and we can imagine waterfalls cascading out of this plateau we're looking at, into the sea.

GH: So . . . so the land would be rearing above us. Does that mean we would or wouldn't be on the sea where we are now?

Bill: Well, we might be . . . we might be on part of the shelf, or on the island too, depending how far out from shore we are. (*Looks around and over side of boat.*)

GH: But in general, from island to island, what would the situation have been? Would they have been islands?

Bill: (*figuring out location of boat in relation to reef*) Oh, right, OK. We're on the outside of the atoll now so we'd still be on the sea . . . We'd be looking at this big plateau and the islands, what we now think of as sea bottom between the islands, would all be dry – unless it was raining and there were lakes forming – and there'd be vegetative jungle. It'd look a lot like the cockpit country in Jamaica in the present time.

GH: Right. So it would be –

Bill: That's how I imagine it.

GH: So it would be kind of lush, jungly country?

Bill: Yeah. On limestone, what's called karst topography, very rugged, with sink holes.

GH: And then what happens? That's 17,000 years ago. We're outside the atoll. We look inside. We see a huge amount of land – jungle – between what are now scattered individual islands. Then we know that after the Last Glacial Maximum, sea-level begins to rise. So if you could just talk me through what happens after that. And I understand it's a complicated problem, because at the same time the sea is rising, the volcanoes are very, very slowly sinking and the coral is growing.

Bill: Well, as the sea-level rose, we'd see all that vegetation and land inundated. A lot of the soil would become sediment suspended in the water. It would probably inhibit coral growth for a while, so some of the reefs would grow and others would not grow, and that probably accounts for some of the variation we see. We see reefs that are maybe at 50 metres . . . their tops are at 50 metres, yet now there's no obvious reason why they didn't grow, we can only assume that for some reason they drowned, whereas other reefs kept up and they're the ones we see on the surface today.

GH: I know from the studies that we've done that there were still substantial amounts of land exposed here down to 10,000, even as late as 8000 years ago. There was more land above water than there is now. Would there be any reason why these islands should be uninhabited at that time? Would they have been the kind of place where people could have lived?

Bill: I would have thought that they'd be relatively easy to find given how far they were out of the water, and presumably how far west the shelf around India might have extended, so given how much we're finding out about how our ancestors used to get around, I wouldn't be at all surprised if they'd made it here.

GH: Because it seems that this sea-level rise – I don't know if your studies underwater have given any indication of this, but what we've found out so far is that the sea-level rise seems not to have been gradual, but to have occurred in episodes and peaks when there were sudden flooding events and then a plateau and then another flooding event. Do you see signs of that underwater here?

Bill: Well, in fact probably not only was it intermittent, but there were also declines at certain times, and provided that the sea-level stood still for a long enough time – and I don't really know how long that was but probably centuries to a millennium – then you would get notches cut in the reef slope for example, and in some places the substantial notches that dissolved in water became grottoes or caves, like those we swam through this afternoon – and some of those collapsed, and you can see these collapsed structures here and there.

Bill Allison's tantalizing glimpse

I had what I thought was a final question for Bill – the obvious one: 'In all your years of diving around the Maldives,' I asked, 'have you ever seen anything underwater that looks man-made – and I don't mean something modern that's been dropped down there, but something old?'

There was a pause, then he replied rather hesitantly: 'Well, I did once when I was down where I shouldn't have been, and . . . I wouldn't trust what I saw.'

GH: How deep were you?

Bill: I was about 40 metres doing some work, and it was down below me and I can only estimate that it might have been at 70 metres, and it looked a lot like a stairway.

GH: Wow.

Bill: But given the distance between it and me, and the fact you can't resolve anything very clearly at that distance, and because your mind plays a few tricks on you at that depth . . . Well, I wouldn't want to bet the farm on it.

GH: But it looked like a regular cut stairway?

Bill: Yeah. And it was narrow, that's what made me think about it – that it wasn't an undefined width. It was clearly defined.

GH: With sort of side edges?

Bill: And had a step-like structure, yeah, as far as I could tell from that distance.

GH: So what was your feeling when you saw that? Hallucination?

Bill: No. I thought, 'That's interesting – I'd like to get back and have a closer look some time.' But I'd prefer to do it on Trimix and with proper surface support.

GH: How far is the site from here?

Bill: It's in the Vadhoo Channel – about an hour by boat, but I'm not at all sure that I could find it again.

GH: And is it close to islands? I guess everywhere around here is.

Bill: Yeah, it's right on the edge of an atoll rim. So if sea-level was 130 metres lower, or anything less down to about 70 metres, then to access the water or the land, you'd need something like that.

GH: You'd need something like a jetty or a wharf, something with steps, yeah.

Bill: But I mean I really . . .

GH: You can't guarantee it?

Bill: I'd give it a probability of about 20 per cent or less.

Even if Bill had rated the probability of relocating his steps at 2 per cent or less, I think I would still have wanted to go and see if we could find them.

But if we could find them – itself probably requiring several days of searching – I would have to do a lengthy, complicated and highly technical course in diving with Trimix (special mixed gases instead of compressed air) before I could safely descend to work at 70 metres (about 220 feet). So the most we would be able to do – and then only if the visibility was very good – would be to hover at 40 metres and look down at the steps as Bill had done before.

However, none of this was an option, because our filming schedule required us to fly to India the next day. Steps or no steps, we were going to have to pack up and leave . . .

The secret of the Redin

There are ancient oral traditions, still repeated by the elders of some of the more remote islands, which provide an explanation for the Maldives' atmosphere of lost prehistoric grandeur and for its strange ruins. These traditions speak of a mysterious people called the Redin, said to have built the *hawittas*, who were described to me by Naseema Mohamed, a scholar at the Maldives National Institute for Linguistic and Historical Research, as:

> Very tall. They were fair-skinned, and they had brown hair, blue eyes sometimes. And they were very, very good at sailing. So this story has been around in Maldives for many, many years, and there are certain places where they say the Redin camped here, and certain places which they say here the Redin were buried. But we don't really know how old or how long ago it happened.[22]

During his series of research visits to the Maldives, Thor Heyerdahl collected and compiled Redin legends from all parts of the archipelago. He concludes that in the memory of the islanders the Redin were 'a former people with more than ordinary human capacities':[23]

> The Redin came long before any other Maldivians. Between them and the present population other people had also come, but none were as potent as the Redin, and there were many of them. They not only used sail but also oars, and therefore moved with great speed at sea . . .[24]

Likewise, Peter Marshall reports a Maldivian tradition about the phenomenal maritime abilities of the Redin which tells of how on one occasion they cooked their food in the north of the archipelago then sailed so fast to the far south that they were able to eat the meal there still warm.[25]

Such notions of humans with supernatural or even god-like powers flying swiftly across the sea in their boats with sails and oars is strangely reminiscent of the imagery of the *Rig Veda* cited in chapter 7 concerning the Asvins – who are several times praised for having conducted a daring rescue in the deeps of the Indian Ocean:

> Yea *Asvins*, as a dead man leaves his riches, Tugra left Bhujyu in the cloud of waters . . . Ye brought him back in animated vessels . . . Bhujyu ye bore . . . to the sea's farther shore, the strand of ocean . . . Ye wrought that hero exploit in the ocean which giveth no support, or hold, or station, what time ye carried Bhyjyu to his dwelling borne in a ship with hundred oars, O Asvins.[26]

> O Asvins . . . Ye made for Tugra's son [Bhujyu], amid the water floods, that animated ship with wings [sails?] to fly withal, whereon . . . ye brought him forth. And fled

with easy flight from out the mighty surge. Four ships, most welcome in the midst of ocean, urged by the Asvins, saved the son of Tugra, him who was cast down headlong in the waters . . .[27]

A connection with the Gulf of Cambay?

Any connection with the Vedic Asvins is purely speculative. Nevertheless, Thor Heyerdahl makes a case that there is real history behind the Redin myth, that it is older than the date now confirmed by radiocarbon for the construction

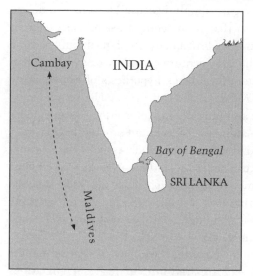

Possible prehistoric maritime connection between the Maldives and northern India.

of the *hawittas* – which tradition nevertheless attributes to the Redin – and that the people it refers to probably originated in north-west India, the primary setting of the *Rig Veda*. After visiting Gujerat and the great marine dockyard of the Indus-Sarasvati civilization at Lothal – where cowrie shells from the Maldives (*Cyprea Moneta*) have been excavated amongst the ruins and are to be seen in the site museum[28] – he comments:

I was convinced that at least the Hindu element in the Maldives had come from the north-western corner of India. And probably the Hindus were not even the first to have made the journey straight south from the Gulf of Cambay to the Maldives. Perhaps earlier sailors in the days of Mesopotamian and Indus Valley seafaring had been led by the sun to the Equatorial Channel, and survived in legend as the Redin.[29]

But if this could be so, then it is also possible that the real people upon whom the Redin myth is based could have arrived in the Maldives even earlier than that. Of particular interest is the fact that the Gulf of Cambay was not a gulf until it was suddenly inundated by the last of the three great episodes of post-glacial floods some time around 7700 years ago (see chapter 11). Prior to that, the further back you go in time the further the coast extends to the south of the Gulf, with another episode of tremendous land-loss registered at around 15,000 years ago.

More than one lost civilization?

Then there is the whole complicated question of the obvious but ancient role of Dravidian and south Indian culture in the prehistory of the Maldives and the way in which the enlarged Ice Age footprint of the Maldives dovetails with the Kumari Kandam myth of the Tamils.

On the other hand, there is the obvious Sanskrit and north Indian influence that is also present in the Maldives and that dominates its language, Dhivehi.

It is too easy, in my view, to argue, simply because Dhivehi belongs to the Indo-European language family, that it therefore must be derived from Sinhalese, the Indo-European language of Sri Lanka – which itself only became entrenched in that island around the sixth century BC following an invasion of settlers from northern India.[30] Thor Heyerdahl's hypothesis of a prehistoric maritime connection between the Maldives and Gujerat – and let us not be too hasty to put an upper limit on the antiquity of that connection – is an equally effective means of supplying the Maldives with an Indo-European language.

Behind all of these questions and problems is the wider issue of the relationship between the Dravidian culture of south India, the traditions and religious ideas of north India and the distinctive manner in which the Vedic and the Tamil flood myths intertwine, sharing gods, sharing sages, and sharing the same underlying story-line built up around the theme of recurrent cataclysms and the preservation of antediluvian knowledge.

Not for the first time I found myself wondering if we could be dealing in India with not one, but two different and *yet intimately interrelated* lost civilizations of the Ice Age – one predominant but not exclusive in the antediluvian northwest, with its own individual character, style and language, the other predominant but not exclusive in the antediluvian south, again with its own individual character, style and language.

Because of the spectacular land-losses that India had suffered at the end of the Ice Age, it was not difficult to imagine how both could have flourished along the subcontinent's coastal margins and outlying island chains at roughly the same time, both could have been swallowed up by the sea over roughly the same period, and both could have left survivors to repromulgate the antique system of knowledge that they shared – which claimed, through self-discipline, meditation and the asceticism of yogic austerities, to have marked out the straight and narrow path of spiritual transcendence in the material world.

14 / Ghosts in the Water

The great deluge took place in 16,000 BC . . . The second one in 14,058 BC, when parts of Kumari Kandam went under the Sea. The third one happened in 9564 BC, when a large part of Kumari Kandam was submerged.

N. Mahalingam, Chairman, International Association of Tamil Studies

Poompuhur coast, south India, 26 February 2001

The ancient religious teachings of India may be directed towards spiritual transcendence but the morning that we were going out to dive at Poompuhur I felt no inner peace. Instead, I was up brooding long before dawn, my head swirling with fears and anxieties, hopes and possibilities. I could feel the first leaden numbness and uneasy visual aura of an oncoming migraine – a perverse affliction with which I must deal whenever I am under great stress and am most in need of a clear head. I immediately treated myself with an injection in the thigh of the powerful drug Immigran, which will normally stop even a severe full-blown migraine in its tracks, but this time it only reduced and did not entirely eliminate the symptoms, leaving me feeling weak, drained and on edge.

I knew that these were going to be big dives for me, that there was a lot riding on them, and that the mysterious U-shaped structure that I had come to see would be filmed for the first time so that people everywhere, archaeologists and non-archaeologists alike, could make up their own minds about it.

What this meant was that I was being given the chance – the incredible opportunity funded by Channel 4's money and prestige – to test the basic proposition of the Underworld hypothesis, i.e. that evidence which might shed significant new light on the mystery of the origins of civilization could be lying under the sea. I realized that if Glenn Milne's inundation dating of '11,000 years old or older' for the U-shaped structure was correct, and if the earlier NIO marine archaeologists' reports that it was man-made rather than some natural outcrop of rock were also correct, then what was awaiting me on the sea-bed off Poompuhur was, quite possibly, the vindication of my quest.

It didn't matter much what the structure turned out to look like. For example a ruined pyramid, or a corbelled archway, or broken columns – though archetypal antediluvian images in popular culture – were not in the least required. Irrespective of how dilapidated it might be, irrespective of how covered in marine growth and sediment it might be, even should it prove dull and unexceptional to the eye, all that I needed to prove my case were the remains of a structure that was monumental in scope, man-made and more than 11,000 years old, sitting on the sea-bed off the south-east coast of Tamil Nadu.

If the U-shaped structure was all these things, then it could not be explained by the orthodox model of history. And if it was all of these things, then the hitherto discredited Tamil myths of a great antediluvian civilization called Kumari Kandam that had once existed around the southern coasts and islands of India might very well be true. So in a way, I reflected, if the U-shaped structure really was what the NIO said it was, then I was about to come face to face with my own personal Holy Grail.

How very annoying, therefore, that my film producers had scheduled just one day for the diving. Having gone to great lengths to get the NIO to cooperate with us over filming the U-shaped structure at Poompuhur, and having paid out a very large sum of money to hire the NIO diving team and marine archaeologists full-time for six days, here we were making use of them for just one day!

It struck me as a crazy, misguided, self-contradictory policy which on the one hand had moved heaven and earth to make it possible for me to dive at Poompuhur at all and on the other would only allow me two or at the most three dives at the site – thus making it almost inevitable that I would not be able to do a proper job there. I felt like Moses being told that he could see the Promised Land but would not be allowed to enter.

No wonder I had a headache.

That gentleman is not well . . .

The coastal plains around Poompuhur are exceptionally flat with a gentle seaward slope – a characteristic of topography that continues unbroken underwater for a very great distance out from shore and that would have multiplied the effects of even relatively small sea-level rises into rapid and catastrophic floods capable of inundating very large areas.

We met up with our NIO friends on the beach – Kamlesh Vora, Gaur, Sundaresh, Gudigar, Bandodkar and others – and a scene was shot of me greeting them and walking with them. The scene required three takes.

Then we all piled into a small open launch to make the run through the big breakers that were lashing the shallows to the point about a kilometre off-shore where the fishing trawler that the NIO had chartered for the diving was moored.

Another hour or so passed while we did the launch-to-trawler run twice more so that it could be shot from different angles.

Then finally we all climbed on board the trawler – not so easy since its sides towered more than 2 metres above the bottom of the launch – stowed our equipment, and headed out into the open sea.

I was irritable, withdrawn – certainly not very conversational – and felt like lying on my back and closing my eyes to ease the ominous symptoms of my returning migraine. Instead, for the next half hour as we chugged the remaining 4 kilometres towards the dive site, basic good manners required that I stay on

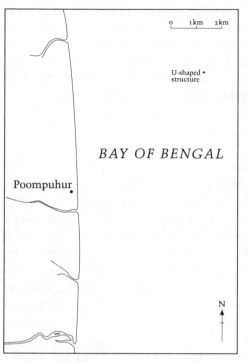

Position of the submerged U-shaped structure off Poompuhur coast. Based on Rao et al. (1993).

my feet, catch up on gossip with everyone from the NIO, and look cheerful, optimistic and positive. After all, I was a man being given an incredible opportunity. Shouldn't that put a smile on my long Scottish face?

Sundaresh and Bandodkar had already buoyed the site some days previously and while the trawler manoeuvred into position to anchor next to the buoy I wandered off to an unoccupied corner of the deck and surreptitiously gave myself another shot of Immigran. That made two, the maximum permitted dosage in twenty-four hours. Praying that this horrible, increasingly blinding and ghastly headache would now please go away, I lay down with a towel over my eyes for the next ten minutes, only sitting up again when it was clear that the anchoring operation had been completed.

'Feeling any better?' asked Kamlesh with genuine concern.

'Not sure,' I replied.

'That gentleman also is not well.'

I looked over to where Martin, our underwater cameraman, was indeed very definitely unwell, sprawled on the deck retching miserably . . .

It seemed unlikely that he would be going underwater any time soon.

Cornucopia

In the end it was decided that Stefan Wickham, the producer, would film the first dive. Hopefully, Martin would be well enough to shoot the second for us. There probably wouldn't be time for a third, because we still had to interview Gaur, Sundaresh and Kamlesh on the boat and had already used up most of the morning shooting the scenes from the beach and just getting to the dive site.

As I was rigging my tank I noticed that half a dozen small local fishing craft had arrived here ahead of us and that the fishermen, oblivious to our presence in their midst, were cheerfully casting out their lines and hauling them in again with big silver fish attached. It seemed that here, as elsewhere along the Coromandel coast, the location of underwater ruins was part of the essential survival knowledge and folklore of fishing communities – just as they knew the tides and the monsoons – because an underwater ruin meant one thing for sure and that was a cornucopia of fish . . .

Stefan jumped in the water first – intending to have the camera rolling before I jumped in. Instead, he was carried off in a brisk surface current and began rapidly to recede from view. Fortunately, the trawler had a motorized rubber dinghy in tow which was dispatched to retrieve him and fifteen minutes later he was back on board. The trick, explained the NIO divers, was not to try to fight the current but to grab hold of the buoy-line the minute you hit the water and then use it to pull yourself down the 23 metres to the ruin.

Dive 1: descent

Although the sky is now overcast the water isn't cold, not like Dwarka the year before. But compared to the life-giving iridescent blues of the Maldives, its sickly and unnatural green hue, through which light penetrates only dimly after the first few metres of the descent, has all the allure of radioactive fog after a nuclear disaster. Like blighted snowflakes, a blizzard of grey particles blows through the water on the current and I soon lose sight of the other divers on the line. I know that Sundaresh, my dive buddy today, is just a few metres below me, but I can't see him. In conditions like this there is really nothing much to be done except check your gauges, relax, trust in your own competence and head for the bottom.

Five metres deeper and the visibility suddenly begins to clear – not dramatically, but still much better than before. The current seems to have slackened too, as sometimes happens at greater depth. Visibility continues to improve and at one point looking down the line I can see all three of the NIO's divers spaced out at metre intervals below me, their yellow and blue tanks bright through the haze.

At about 18 metres I begin to get the first sense of something large standing out from the flat and sandy bottom. At this moment it's just a looming mass of darkness contrasted to lighter surroundings and my eyes can't resolve it into a definite shape.

The other divers above and below me leave the line, fan out and disappear from view. Gaur is working with Santha, who will be shooting stills. Gudigar is working with Stef on the video camera. Gaur, Gudigar and Sundaresh were all part of the team of marine archaeologists who first dived on the structure during the NIO's initial surveys in 1991 and 1993.

Sleeping with the fishes

Sundaresh, who is waiting for me at the bottom, wants to show me courses of masonry that he has noticed on his previous dives – but before I join him for the guided tour I let go of the line, establish neutral buoyancy and just drift about 2 metres above and 2 metres to the side of the structure. There's no current now at all, the visibility has gone very foggy again – probably sediment kicked up by some of the other divers – and I rest completely still in mid-water, adjusting my eyes to the gloom, trying to understand what I'm looking at.

The only thing I can tell immediately is that it's a big, squat, powerful-looking structure. In order to get any useful idea of its shape, extent and general situation, and even to form a first opinion of whether it might be man-made or natural, I need to be quite a bit further away from it than 2 metres. But if I do that, in these conditions, it rapidly fades from view, becoming just a vague, undefined darkness on the sea-bed again, and then disappearing entirely into the fog.

I swim around a bit, now closer, now further away, trying to get perspective, looking for an angle. And then unexpectedly the whole scene in front of me brightens – the sun must have broken through the clouds – and for thirty seconds I am confronted by a massive wall of deeply eroded and pitted stone. Although much broken and ruined, and incorporating a number of jagged vertical protrusions and step-like changes in level, I can see that the wall in general rises about 2 metres above the sea-bed to form the outside edge of an extensive platform.

It comes home to me, in this moment of illumination, that the structure has its own character – as many buildings do. It seems menacing but also forlorn, eerie but also sad. For as well as thick growths of unusually leprous marine organisms all over it, the shaft of sunlight shows it to be draped and tangled across its entire length in a strangling web of fishermen's nets – some made of old rope, ancient and rotting away, others in the sinister colours of indestructible modern synthetics – which seem to tie it down like the body of a Mafia victim sleeping with the fishes.

I find myself suppressing an involuntary shiver, as though reacting to an apparition, or a ghost, and swim back to find Sundaresh still patiently waiting for me at the bottom of the line.

Walls . . . passages . . . entrances

We begin by swimming slowly south along the upper outside edge of the platform wall – if indeed it is a platform, which I'm now beginning to doubt. Rather than flat as I'd initially assumed, its surface at this point seems to be slightly concave – or dish-like – and to be paved with a mosaic of small stones. I find myself wondering if it's possible that I'm looking at the retaining wall of an enclosure – I know its supposed to be U-shaped – filled up almost to the rim with some kind of sandy, stony aggregate.

The wall at this point is aligned north–south but soon begins to bend to the east to form the base of the 'U'. In another one of those little flashes of illumination as the sun breaks through the clouds I can see that we must have started our swim at the open end of the 'U' – the end spoken of in some of the NIO reports as 'the entrance' – and that the length of the structure along this axis is therefore roughly the distance we have just travelled, about 30 metres.

Not far before the bend begins I pass an opening to my left which I pause to investigate. It is a deep, narrow cleft with parallel sides a little wider than my shoulders slicing vertically through the whole height of the outer wall to penetrate the platform (or the stony fill, or whatever it is) that lies beyond. And for the first few metres at least, this gully, or unroofed passage (or whatever it is!) follows a curving path that seems to duplicate, from within the structure, the distinctive outer curve of the 'U'. Swathed everywhere with snagged and rotting nets, it is rough and broken in places, flat-floored and clean-edged with an almost quarried look in others.

Making a mental note to spend more time here before the end of the dive I turn back and resume my original course along the outside wall where it bends to the east, trying to catch up with Sundaresh. Looking for me, he meanwhile has swum all the way round and made his way back to the entrance where I eventually join him.

But is it really an entrance?

As though understanding my perplexity, Sundaresh points to a gap in the wall about a metre and a half wide to one side of which I can now see that the buoy-line is tied. Holding his hands up he reassuringly signals 'this is the entrance'.

I take a closer look.

What's confusing things once again is the stony aggregate that fills most of the structure – although I've noticed that it does so quite unevenly. Its presence here makes it hard to see the gap as an entrance because it doesn't seem to lead anywhere much. At the same time the thick retaining wall, generally in the range of 2 metres high, is at least a metre higher than that on either side of the gap – resembling a pair of gateposts. It also has a pronounced lip standing proud of the aggregate infill by almost half a metre – weighting the scales ever more in favour of the idea that the U-shaped structure must originally have been designed not as a platform but an enclosure, and that it certainly cannot be a natural formation.

But is the enclosure wall hewn out of living rock, like the great carved shore temples of Mahabalipuram, or is it a built structure made of bricks or stone blocks?

We use up the rest of the first dive searching for the courses of masonry that Sundaresh is convinced he saw in 1993. Yet how are we to find them under the thick and tenacious armour of marine organisms that coats the wall? Several times reaching into shadowy eroded hollows to see what's inside we must work our hands carefully around resident scorpion fish which flutter their poisonous spines as though to taunt: 'Go on, touch me – make my day.'

But we don't find any evidence of masonry.

Not on the first dive.

Disturbing

During the surface interval I fought down waves of nausea and the pounding in my temples, took another shot of Immigran, and felt sufficiently restored after half an hour to fall into an argument with Gaur about the U-shaped structure.

The reader will recall from chapter 9 that Gaur's position had been rather stark when he and I had first discussed the matter a year before: the structure was large; its depth meant that it was more than 10,000 years old; archaeology knew of no culture anywhere in India capable of building such a structure 10,000 years ago; therefore *either* the structure was not man-made *or* it was not 10,000 years old.

I asked him if he'd changed his mind in any way over the intervening year and told him of the findings of Glenn Milne and his team at Durham: 'We've had some geologists working with us on this project in Britain who are specializing in sea-level rise. And their computer model is quite sophisticated. It takes account of many, many different factors, including land subsidence. And they're very confident that for these bearings, for this location, that this site would have been submerged about eleven thousand years ago. What do you make of that?'

If anything, Gaur replied, this made his chronological problems with the data even worse: '11,000 years ago whatever settlements there may have been here were at the Mesolithic level. And we don't expect, we don't have any data to suggest, that such people, Mesolithic people, can build this kind of structure.'

'Such a large structure as this?' I prompted.

'Yes.'

'And you're saying that – presumably – on the basis of what you already know about the level of culture and civilization in this area in different periods?'

'Yes,' said Gaur: 'So I think – if it is man-made – it should be around 2500 years old, maximum date. Not earlier than that, particularly in this area.'

'And I think you're putting the cart before the horse,' I interjected. 'See, obviously I'm not an archaeologist and I come at this really from the point of view of a reporter or a journalist. So my response to this structure is first of all the facts. A structure is there. It's at 23 metres. Is it or is it not man-made? I feel the structure has to answer that question itself instead of us simply replacing what it has to say with our preconceptions about the nature of development of culture in India at this or that period. The structure should speak to us through archaeology. We should excavate it and find out really is it man-made or not. Although I must say that I personally find it very difficult to believe that nature could have deposited a structure like that there. So the question I'm coming to is this. We know that certainly 9000 years ago people were beginning to build quite large structures in some parts of India – for example level 1A at Mehrgarh in the Indus valley. Now, admittedly that's 2000 years later than the proposed inundation date for this structure but it's in the same general ballpark – back at the end of the Ice Age. So my point is that if people were building permanent structures at Mehrgarh in the north-west 9000 years ago, then what is the objection in principle to the possibility that people could have been building permanent structures here in the south-east 11,000 years ago on lands that were flooded?'

'Well, because we don't see any such structures in the archaeological record for south India or any part of India 11,000 years ago!'

'But maybe that's precisely what we're seeing here, Gaur! We haven't seen it before because it's been underwater, but now that it's been found surely we have to allow it to speak for itself? It seems to me that the archaeology needs to be done on the site first before we make any definite statements about the level of culture that was here 11,000 years ago.'

'From what I have studied and from what I understand about Mehrgarh,' Gaur replied, 'if you go back to level 1A it was simply mud walls and they were concentrated in one area – they were living in a group and the village community started. But when we come to Poompuhur – well, if you see the U-shaped structure, it is such a big one. And it is part of a complex with other big structures spread over a wide area. So it means if human beings made this then they must have had very great technology at that time. I don't think it can be compared with the simple mud-brick structures of Mehrgarh . . .'

'In other words, if the U-shaped structure is 11,000 years old and was made by human beings it would be rather disturbing for our view of history.'

'Yes. Obviously.'

Dive 2: impatience and haste

Martin takes over from Stefan on the camera for the second dive but Stefan comes down as well, just in case.

We all descend the buoy-line and are back at the entrance – which is oriented north. I swim south as before, heading for the curving passageway near the far end of the 'U' that I'd noted on the first dive and forgotten to re-examine.

Sundaresh is a metre or two behind me, still looking for his courses of masonry and I'm steaming ahead when I feel him reach out and grab my fin. He points to something that he clearly regards as noteworthy, but whether it's because I am disoriented diving on an unfamiliar structure, or whether it's because of the appalling visibility, or because I'm in too much of a hurry, or because of my migraine, *I just don't see what he's showing me.*

Behind us Martin doesn't either – but he keeps shooting, recording the relevant incident in twenty-six seconds of videotape that I'm not able to review until late that evening. The first twenty-four seconds show me being impatient and hasty. The last two seconds show something that I should under no circumstances have allowed myself to miss through impatience and haste – something that I should have examined thoroughly on the spot and had filmed and photographed from every different angle.

Instant replay

26 February 2001, 15.37.02–15.37.28:

Hancock and Sundaresh swimming north-to-south, along western wall of U-shaped structure, Hancock in lead, depth approximately 22 metres.

Sundaresh pauses to examine area of wall, attracts Hancock's attention, then returns to wall.

Hancock joins Sundaresh, who points to area of interest on wall.

Hancock gives it cursory glance and seems keen to get a move on.

Camera tilts down to base of the wall, just above the surrounding sea-bed, then begins to tilt up for point-of-view shot.

Shot holds for two seconds on a narrow section of the wall about 1 metre high that is clear of growth and reveals in lower right of frame an ordered pattern of small blocks arranged in four distinct courses with the edge of a possible fifth course partially visible under marine growth. The blocks are brick-sized but irregular in cross-section and appear to be set into some kind of matrix.

Camera tilts up to top of wall, rediscovers Hancock who is swimming determinedly away, and follows . . .

Excursion to the mound

At this point – frustratingly still before I have reached the curving passageway that branches inside the structure at the southern end of the 'U' – the other divers signal us back, wanting to stick to the plan that we all agreed in advance for this dive and that I had forgotten as soon as I hit the water. The plan is to spend most of our fairly limited bottom-time at this depth exploring a second major structure that lies close by (about 45 metres away according to Sundaresh when we had discussed the matter in Dwarka the year before).[1] One of a pair of 'mounds' lying to the north of the U-shaped structure, it was identified during the NIO's 1991 and 1993 seasons at Poompuhur, and Sundaresh had spoken of seeing 'perfect cut blocks' scattered on the sea-bed beside it.[2]

Bandodkar, whose word is law amongst the NIO divers, has insisted that this second dive should be limited on decompression grounds to half an hour or less – a prudent but in my view unnecessarily zealous interpretation of the nitrogen tables for the depth we are working at. I suppose it was because I was feeling rebellious about this time-limit on the whole dive that I rushed off so fast at the beginning to attend to my interest in the curved passageway.

Now I am rightly brought to order so that we can all proceed as a group to the second structure. I can see the point of being safety-minded in these conditions. The visibility is extremely bad – almost like being lost in an immense sandstorm but with a different texture. And although the NIO divers have previously rigged a yellow nylon rope as a guideline I still feel disoriented as I follow it. North? South? East? West? Up? Down?

Down is easy. In fact I'm so close to the sea-bed that I'm practically slithering on it and yet it gives me no points of reference because it consists of absolutely and uniformly smooth, flat and unbroken fine-grained sand. The contrast with the stony textures and the bulky solidity and complexity of the U-shaped structure could not be more pronounced.

Then we reach the 'mound'. Like the U-shaped structure it has an isolated position in the middle of the flat plain with no slope or build-up. A lot of silt and sand has been deposited on it and around it, but there's no doubt that its core is a massive stony pile. I can make out what seems to be the edge of a wall a metre thick and similar in general appearance to the enclosure walls around the U-shaped structure. Festooned in scorpion fish it rises to a height of about 3 metres above the sea-bed before disappearing into the larger mass of the mound behind it.

Martin shoots this scene but then signals that he is unwell and must return to the boat. Sticking to Bandodkar's safety rules the entire group leaves the mound and makes the trek back along the rope with him until we come again to the entrance of the U-shaped structure where the buoy-line is anchored. Martin and some of the other divers then ascend. Stef, who has the camera again, follows me.

Blocks in the passageway

I'm still determined to explore that curving passageway, so I swim south as usual along the west wall. Sundaresh and Stef both keep pace.

I can see the narrow entrance to the passage coming up on my left when I notice something on the bottom to my right less than 2 metres west of the base of the wall. It looks like a small splintered tree stump protruding upwards out of the sand. But it proves to be made of badly damaged and eroded stone. Two more similar objects are near by but none of them in itself seems particularly interesting. Feeling pressured for time, I do not examine them further.

Next I'm into the passageway. Been there. Done that. I want to see where it leads to.

So I follow it all the way through this time and find myself in something like a room, very roughly defined, that seems to be free of the otherwise all-pervasive stony aggregate that so confuses the picture elsewhere on the structure.

Platform? Or enclosure? It would be a funny sort of platform that had an open-roofed room carved out in the middle of it – maybe more than one room for all I know.

For my money, therefore, this is yet another good reason to conclude that the U-shaped structure is an enclosure, that it probably has several internal walls that are presently hidden from view, that it has its main entrance to the north and at least one subsidiary entrance in the west wall, and that either through human or natural agency it has at some point been partially filled up with stony rubble.

Ah, the freedom and manoeuvrability of diving. On a whim I adjust my buoyancy by breathing in and ascend out of the 'room' to a point a few metres above the structure hoping to get a plan view – but once again the awful visibility defeats me and I can see almost nothing.

I drop back down and work with Stef to complete a little sequence of me looking around the 'room' then swim out of shot while he finishes filming inside. A moment or two later I see him emerge backwards from the curving passage, still filming, with the camera seeming to focus mainly on the floor and the lower part of the side walls.

On that footage too I will later note something else of interest that I missed in the rush and stress of the day. It's on just eight seconds of tape.

Instant replay
26 February 2001, 15.56.33–15.56.42

Shot tracks unsteadily along floor of passage and passes across net draped over and partially obscuring change of level and possible step up in floor.

Camera ascends about a metre, shot tracks left of net and picks up a clear line of five blocks emerging from under marine growth. They are dark, almost charcoal black, and brick-sized like those seen on the first dive, but here much more regular in cross-section.

Shot wavers, returns to net, then tracks left again passing the same line of blocks which is now seen to continue to the left by at least a further six blocks, with other courses in outline above and below it, before it disappears under the heavy marine growth again.

Ascent
On the way up we do the routine five-minute stop at 5 metres to reduce our nitrogen levels. The water is very still and warm, the visibility worse than ever, and I drift in neutral buoyancy slowing my breathing, just thinking things through.

It feels strange to have been privileged to see a structure hidden from human eyes for 11,000 years.

A structure more than 7000 years older than the Great Pyramid of Egypt.

A structure for which no archaeological context exists.

A ruined net-draped structure.

A ghost in the water . . .

More blocks on tape

This was turning out to be a good day for Glaxo Wellcome. After the second dive I took my fourth injection of Immigran at $50 a shot. Then I had to collapse again, sprawled out like a landed fish on the wooden deck of the trawler while the pain in my head gradually dulled and withdrew – only people who suffer from severe migraines will understand the sense of relief and release that I felt as the drug did its work.

By 5.30 I was back on my feet drinking tea and chatting to Kamlesh Vora. At around the same time Martin went down for a short dive in the last of the daylight accompanied only by Gudigar in the hope of getting relatively clean, undisturbed shots of the structure.

When I later came to review these shots I found that they contained a third brief sequence showing construction blocks, this time of better quality than the previous two. The sequence is timecoded 17.36.15–17.36.29 and Martin seems to be standing on the sea-bed near the enclosure wall:

> The shot starts focused on a small white shell lying on the sand then quite slowly pans across to the base of the wall and holds steady for several seconds on four distinct courses of masonry. Again the size of modern household bricks, perhaps a little larger, the blocks here are extremely regular and almost cylindrical – or cigar-shaped. The exposed sections of each course can be seen to continue horizontally over a width of approximately a dozen blocks until they either vanish out of shot or disappear beneath thicker marine growth.

Mysterious

By profession Kamlesh Vora is a geologist, not a marine archaeologist, but geology plays an increasingly important role in modern marine archaeological research and is one of several important skills necessary to distinguish whether a disputed structure is natural or man-made. Moreover, Kamlesh had been involved in the very first work that the NIO had ever done at Poompuhur way back in 1981 – long years before the 1991 and 1993 campaigns – and had carried out the initial sonar surveys on which much of the later work plan was based.

I kicked off our interview on the boat with a leading question: 'The ocean is a big place and we see that there are some possible structures here. Have you done any kind of surveying from the surface?'

Kamlesh replied: 'In 1981, when we started marine archaeological explorations in Tamil Nadu, we began with Poompuhur. And we scanned the sea-bed using echosounder and magnetometer. What we found interesting was that otherwise the sea-bed was flat, even and smooth as far as the echosounder was concerned. But there were a number of anomalous features scattered in the area – some a bit oblong in structure, some like pinnacles – and the echosounder showed the elevation of these features to be in the range of 2 to 5 metres above the bottom. Such outcrops and elevations are not at all to be expected from local geology and we could not comprehend how they had been formed. If they are to be natural extensions of bedrock, then we should see different topography. For example, off the west coast of India we have found pinnacles or things like this because of a number of reasons, and we have collected samples and then done our investigations.'

'And on the west coast they're a natural extension of the rock?'

'There are basaltic rocks,' Kamlesh clarified, 'which may have extensions. And we have found man-made structures underwater in the north-west like Dwarka which, as you know, have come in the last 5000 years . . .'

'But the story is different here on the east coast?'

'This is totally different because we could not give any logical explanation for them. So even during those times we considered them as anomalous.'

'So, looking at them as a geologist, as you are, you find it surprising that these features are sticking up if they're purely natural?'

'Yes,' Kamlesh replied with a shrug. 'Only thing during that time is we didn't have support of diving team. So we could not collect samples and do analysis of the rocks. Even now when we collect it, we could not get the proper rocks for different kinds of test, so we don't have samples enough to go to some logical theory on that.'

'This U-shaped structure that we've just been diving on,' I asked, 'was it identified in that survey?'

'Yes. And totally up to twenty structures were identified round about.'

'But you've not had a chance to dive on the other ones?'

'No,' said Kamlesh, 'we didn't get the opportunity to come back and work like this. So maybe in future we shall come and concentrate on them. Then also we should seek information and try side-scan sonar surveys and diving to see if there are other structures in other areas along the coast. Because this one place may be in isolation. But if there are three or four other major groups of structures in other locations . . .'

He looked out to sea and stopped speaking without completing his sentence.

'It feels to me like a very exciting area,' I offered after a moment, 'with so many, as you say, anomalous structures And they are anomalous. We don't know what they are. But it seems to me an area that deserves more attention.'

'Mysterious,' Kamlesh replied after a moment more.

The mound at 27 metres

When we were parting company with our NIO friends well after nightfall on the darkened beach, Gaur took me aside to tell me that he had remembered a dive done during 1993 at Poompuhur that might be of interest to me. The dive had been a first exploration, never subsequently followed up, to check out one of the anomalous mounds in 27 metres of water – 4 metres deeper than the U-shaped structure. Gaur had not dived on this deeper structure himself but had been told about it by colleagues who had: 'It was a heap,' he said, 'of things . . . It's quite high. I mean 2 metres high.'

'Is it in the same general area as the U-shaped structure?' I asked.

'No,' Gaur replied. 'It's further out. A 4 metre difference in depth here means you have to go out at least another 500 to 600 metres.'

'All the more obvious, then, that there's a need for a really extensive survey and much more marine archaeology here . . .'

'I agree,' said Gaur, 'even if only to prove that these things are not man-made.'

Secrets of the Reinal map
February/March 2001

Readers will recall that three days before our dives at Poompuhur I had received an e-mail from my researcher Sharif Sakr concerning an intriguing Portuguese map – the Reinal map of the Indian Ocean, dated 1510. But not until I was back in England at the beginning of March did I have the time to consider in detail what Sharif had to say about it or compare his attached scan of the Reinal map and other maps that he mentioned with Glenn Milne's sequence of inundation maps covering the end of the last Ice Age.

Sharif Sakr to Graham Hancock
23 February 2001

Hi Graham,

I've noticed an interesting correlation between the Jorge Reinal map of 1510 (see attached scan from facsimile in Hapgood, fig. 77) and Glenn Milne's inundation maps of India. It is perhaps not immediately obvious, so please let me know what you think. (There is a good facsimile of the Reinal 1510 in vol. 1 of the *Portugaliae Monumenta Cartographica* in the Bodleian Library, Oxford and I've ordered a reproduction. Until it arrives we must rely on the tracing in Hapgood, which lacks detail but is basically accurate.)

I was first attracted to the Reinal by its remarkable accuracy, and its obvious relationship to the Cantino 1502, and also the Ptolemaeus Argentinae 1513. While the map is not as accurate as the Cantino in terms of the ratio of India's long and lat extensions, it is nevertheless an amazing development relative to the older Ptolemaic model, especially considering that Portuguese naval exploration of India only began after 1498. E. Kemp (*Asia in Maps*) suggested that Cantino's depiction of India came not from Portuguese observation but from contacts with the traders of Calicut – perhaps Reinal's map of India was based on the same sources (and perhaps these sources were the Indian Ocean nautical charts mentioned by Polo?).

Despite the map's general accuracy, there are a number of glaring mistakes. Firstly, at the precise latitude of the mouth of the Indus there is a large gulf rather than the delta which exists today. Secondly, moving south along the map, Reinal makes the same mistake

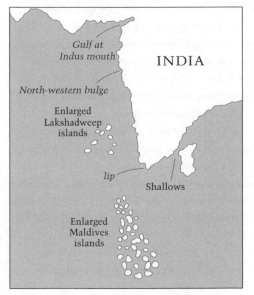

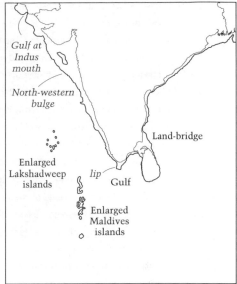

Outline of India's coastlines in Jorge Reinal's map of AD 1510, based on tracing by Charles Hapgood (1966).

Outline of India's western coastline as it was 21,300 years ago.

as the author of the Cantino, and fails to show the important Kathiawar peninsula or the gulfs (Kutch and Cambay) that flank it. Instead, Reinal has given this north-west corner of India a distinct bulge, such that it appears 'fatter' than it should. Thirdly, Reinal has apparently ignored the proper portolan convention of depicting very tiny islands (too small to be drawn to scale) as crosses (or some other diagrammatic symbol) and has instead drawn the Lakshadweep and Maldives as rather large islands – far larger than they really are. Lastly, Reinal has failed to give the southern tip of India its proper south-easterly orientation. Instead, he has given it a south-westerly orientation, and distinct 'lips' which make it look like an open mouth, ready to bite off the top of the Maldives.

While these deviations are all errors relative to a modern map of India, they in fact match up extremely well with Glenn Milne's map of India 21,300 years ago at LGM. This inundation map shows a large indent at the mouth of the Indus, a bulge obscuring completely the Kathiawar peninsula, enlarged Lakshadweep and Maldive islands, and, most surprisingly, a SW-pointing 'mouth' shape at India's southern tip that is virtually identical to that shown by Reinal. (Note that the 'errors' match up even better with a basic bathymetric map of India that shows the very distinct outer shelf, which I use as a kind of benchmark for the basic shape of India's coastline around LGM.) As you travel in time through the sequence the correlation is still good 16,400 years ago but is gone by 13,500 years ago when a large island appears south of the Kathiawar peninsula.

The correlation is not perfect – the inundation maps show a clear land-bridge between India and Sri Lanka, whereas Reinal has not drawn a land-bridge. Being Portuguese and living during the exciting time of the Portuguese discovery of India, Reinal would have been a laughing stock if he'd failed to depict the island of Ceylon. Curiously, however, Reinal has drawn dots in the shape of the land-bridge across the Palk Strait, giving the impression that Ceylon is too close to the mainland. Perhaps Reinal was indicating dangerous shallows. But a glance at my bathymetry data suggests there are no such shallows – most of the Strait is over 6 m in depth. Alternatively, Reinal may have wished to indicate tiny islands, but even this would have been inaccurate, as the real distribution of islands in the Palk Strait today is nothing like the shape of Reinal's dots or of the land-bridge that would have existed at LGM. So I wonder why Reinal drew these dots between India and Sri Lanka – was he perhaps trying to reconcile common knowledge of Ceylon as an island with other sources that depicted a land-bridge?

A final point of interest is that as the years went on, after 1510, Reinal began to correct all the mistakes described above (for example he added the Gulfs of Kutch and Cambay). But as he made these corrections, the basic outline of India actually *worsened* rather than improved. To me, this suggests that the earlier 1510 map was based on the same unknown sources as the Cantino (very accurate in terms of long and lat, but with some strange features), whereas the later maps were based on contemporary Portuguese observational mapmaking and all its inherent weaknesses.

Regards,

Sharif

Although Hapgood had reproduced the Reinal map, he had analysed it only from the perspective of its mathematics and inclusion of anachronistic geographical knowledge (e.g., of Australia, not discovered at that time).[3] He had not considered the possibility of a correlation between the way in which it portrayed India and the actual appearance of the Indian coastline during the Ice Age. On the contrary, he concluded:

> It seemed evident to me that this map showed much more geographical knowledge than was available to the Portuguese in the first decade of the sixteenth century, and a better knowledge of longitudes than could be expected of them. The drawing of the coasts, however, left much to be desired. The map looked much like a map, once magnificently accurate, that had been copied and recopied by navigators ignorant of the methods of accurate mapmaking.[4]

So Sharif's approach to the Reinal map did not duplicate Hapgood – something that I was determined to avoid – but looked at its depiction of India in the light of the new science of inundation mapping that had already provided us with an extremely effective and revealing research tool.

I agreed with Sharif that in the light of that science Reinal had in fact drawn a weirdly accurate map of the south-west, west and north-west coasts of the Indian subcontinent between roughly 21,000 and perhaps 15,000 years ago. It was also potentially the strongest lead that I had seen for a long while on the extraordinary possibility that accurate maps could have been made of the world during the Ice Age and that some copies of these maps could have survived and got into circulation again – always in use and subject to constant modification – during the European Age of Discovery.

Maps of the Mediterranean and the Indian Ocean.

Maps of the Pacific and the Far East.

Maps of the North.

Maps of Africa.

Maps of the Americas and the Atlantic – perhaps including the map, never found, that Columbus is rumoured to have used to guide his journey to the New World in 1492.

Even maps of Atlantis . . .

I decided to investigate further.

Cambay: another ghost rising from the deep?
May 2001

Our final filming trip to India, which would focus on Dwarka, and inland Harappan sites such as Dholavira in Gujerat, was scheduled for November 2001, still many months away.

Then in May, although hardly reported at all by the international media, the following story made headline news in the Indian press:

The Times of India

Saturday 19 May 2001

HARAPPAN-LIKE RUINS DISCOVERED IN GULF OF CAMBAY

In a major marine archaeological discovery, Indian scientists have come up with excellent geometric objects below the sea-bed in the western coast similar to Harappan-like ruins.

'This is the first time such sites have been reported in the Gulf of Cambay,' Science and Technology Minister Murli Manohar Joshi told reporters.

The discovery was made a few weeks ago when multi-disciplinary underwater surveys carried out by the National Institute of Ocean Technology (NIOT) picked up images of 'excellent geometrical objects', which were normally man-made, in a 9-kilometre stretch west of Hazira in Gujerat.

'It is important to note that the underwater marine structures discovered in Gulf of Cambay have similarity with the structures found on land on archaeological sites of Harappan and pre-Harappan times,' Joshi said.

The acoustic [sonar] images showed the area lined with well-laid house basements, like features partially covered by sand waves and sand ripples at 30–40 metre water depth.

At many places channel-like features were also seen indicating the possible existence of possible drainage in the area, he said.

Possible age of the finds can be anywhere between 4000 and 6000 years, Joshi said, adding the site might have got submerged due to a powerful earthquake.

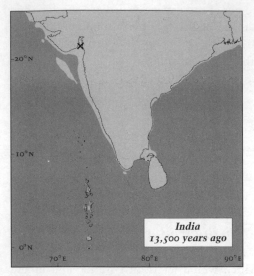

India
13,500 years ago

Cross marks position of Cambay underwater site discovered by NIOT.

This guess seems perfectly reasonable in line with the orthodox chronology of Indian history and prehistory. But it is also perfectly wrong.

What Joshi could not have known without studying inundation maps first is that earthquakes or not (and admittedly this part of India does suffer from severe earthquakes) no site *anywhere* in the Gulf of Cambay could possibly have been above water as recently as 4000 years ago – although 6000 years ago is getting closer. As we have seen, the Gulf of Cambay remained a valley until it was completely flooded by rising sea-levels at some point between 7700 years ago and 6900 years ago.

Then we must consider the scale of the ruins that the researchers from the National Institute of Ocean Technology seem to have identified – this city that is now underwater extends continuously for 9 *kilometres*, meaning that it is many times larger than Harappa or Mohenjodaro or any other city of the Indus-Sarasvati civilization yet discovered.

Think how long it takes to build a city 9 kilometres long. A long time, surely? So even if the Gulf of Cambay was flooded at the latest possible date indicated by Milne – 6900 years ago – we cannot reasonably suppose that the construction of this enormous metropolis could have begun only one or two centuries before that. Surely it would require a millennium, maybe much longer, to build a city so big?

But if we allow a millennium, then that takes us back to somewhere around 8000 years ago – 6000 BC – as the very latest date at which the city beneath the Gulf of Cambay could have been founded.

A city 9 kilometres in extent and more than 3000 years older than Harappa and Mohenjodaro would rewrite not only the history of the Indian subcontinent but of the world.

It was the Holy Grail, all over again.

PART FOUR

Malta

15 / Smoke and Fire in Malta

Lord grant him eternal rest.

SCICLUNA – COMMENDATORE SALVINO ANTHONY, passed peacefully away at
St Luke's Hospital on June 11, aged 73, comforted by the rites of Holy Church.

Sunday Times of Malta, 18 June 2000

There is nothing looking remotely like one of these temples outside the Maltese
Islands.

D. H. Trump

8 November 1999

Some months begin badly, then get worse. November 1999 was like that for
me.

It started when *Horizon*, BBC TV's flagship science series, aired 'Atlantis
Reborn' a one-hour blitzkrieg on my character, my reputation and my work.[1]
But life had to go on, and *Underworld* was not going to research and write itself.

A central part of my research task, as I define it, is to check out personally –
by scuba-diving – any and every sighting of anomalous underwater ruins that
comes to my attention. On 8 November 1999, therefore, just four days after
being blitzed by *Horizon*, a sense of duty compelled me to fly to Malta to
follow up a story that was then circulating on the Internet. Accompanied by
ambiguously blurry colour photographs captured from videotape, the story
concerned the discovery – by a German named Hubert Zeitlmair – of a ruined
megalithic temple 8 metres underwater off Malta's north-east coast.

I had contacted Zeitlmair, and Santha and I had arranged to meet him on our
arrival in Malta later that afternoon. But now, as I passed the flight reviewing
the thin file of documents I had downloaded to my laptop, I had to admit that
the auspices were not encouraging.

A joke or a hoax?

For example, in various unexplained but worrying ways something called the
'Palaeo-Astronaut Society' was involved – thus virtually guaranteeing that the
academic authorities would treat the discovery as a joke or a hoax, irrespective
of any merit it might have. Moreover, it very probably was a joke! By this time
I had done enough diving to know that 99.999 per cent of all mysterious
'man-made' structures sighted underwater prove to be just weird geology or

tricks of the light combined with wishful thinking. Only a tiny fraction check out, and these are usually found by level-headed professional divers with no particular theories to promote.

As he was presented on the official 'Maltadiscovery' website, Hubert Zeitlmair seemed the antithesis of all that. He was described, unpromisingly, as a 'real-estate investor', a 'part-time archaeologist', and a 'fan' of author Zecharia Sitchin (who believes that extra-terrestrial beings had a hand in the construction of megalithic sites around the world). Perhaps this was why Zeitlmair had chosen to announce his discovery of the Maltese underwater temple 'at a meeting of the Palaeo-Astronaut Society' in his home town of Augsburg, Germany on 18 August 1999:

> The final dive that led to the discovery took place on July 13, 1999 at 10:00 AM; and subsequent dives and underwater photography confirmed the nature and megalithic size of the structures.
>
> The temple sits on an underwater plateau about 500 to 900 metres long. The lowest point of the plateau is more than 25 metres below sea level and the highest point of the plateau is about 7 metres below sea-level.
>
> The structure itself shows the same characteristics as the other above-ground temples on Malta. Gigantic stone blocks aligned with astronomical significance, thought to be used as a calendar. The basic diameter of the interior rooms are 6–7 metres and some of the highest walls that are still standing are about 4–6 metres high. There is an avenue that goes up the centre of the structure indicating an orientation to the equinoxes. There are kidney-like formed rooms orientated to an easterly direction, which would coincide with the rising sun and the winter or summer solstices. The main difference is this structure is underwater.
>
> Since the structure, as the others on Malta, had to be first built on solid ground, its present underwater position could result from either the sinking (due to earthquakes?) of coastal parts of the island, or from a marked rise in the sea-level (due to an immense flooding).
>
> Dr Zeitlmair adheres to the second possibility, and wonders whether the cause was the Great Flood described in the Bible and in the lore of many ancient peoples, the so-called Noah's Flood.
>
> He is inclined to this explanation because the west side wall of the structure is more overgrown by sea grass than the east side wall, apparently because there was more sand deposited on that side. Therefore, the stones on the east side are mostly free of sea grass. This could indicate that the destructive water flow came from the west into the Mediterranean Sea, adding confirmation to theories that the water broke through the Strait of Gibraltar, filling the Mediterranean basin. A couple of big stones were lifted up and dropped down in a valley below, apparently by the destructive water flow.[2]

'Great interest amongst foreign archaeologists . . .'

The website also reprinted and translated a number of articles about the discovery that had recently appeared in the press. I'd put these on to my laptop as well and now scrolled through them to see if they had anything to add.

From *Il Mument* (Maltese national newspaper), 31 October 1999:

> Recently, structures that resemble megalithic temples have been discovered on the sea-bed in Maltese waters. These are currently being studied to establish whether they are actually unique megalithic temples.
>
> This discovery has been considered to be of great archaeological importance, and has raised great interest amongst foreign archaeologists . . .
>
> The discovery was made on the 13th of July 1999 at 10 a.m. and was photographed. The diver/cameraman who filmed the structures was Shaun Arrigo, while the photographer who took the photos was his brother Kurt . . .

So two Maltese diver-photographers, the brothers Shaun and Kurt Arrigo, had been involved with Zeitlmair in the discovery – and had in fact taken the blurry photographs that I had seen on the web.

I would need to contact them.

'The age for the megalithic temples must be changed . . .'

What next? I scrolled quickly through another article in the file. It had appeared in the periodical *Maltamag* and contained an interview with Zeitlmair. But in the preamble written by reporter Daniel Mercieca, my eyes were drawn to this paragraph:

> During a meeting with Joseph S. Ellul, a Maltese who has dedicated his life to the study of prehistoric constructions, Dr Zeitlmair was shown a 1933 photo taken by the Royal Navy. This picture seemingly showed a megalithic construction below the surface. Ellul confided to Dr Zeitlmair that he had proposed to the local authorities concerned to start research on site. Unfortunately, his suggestion was never taken up, his numerous letters being left unanswered.

In the interview Zeitlmair commented:

> Following my meetings with Joseph S. Ellul I strengthened my determination and contacted various people about the subject. This led to the formation of a team all set towards one goal – uncovering a temple under sea water. After several futile attempts at locating the site, success came on July 13th, 1999 at 10 a.m.
> *Where exactly is the site of the discovery?*
> It is located some mile and half off the Sliema coast . . . Incidentally, when I first

came to the islands, I was residing at the Diplomat Hotel in Sliema, where I occupied a room with a superb sea-view. Now that the temples have been located, I realize that the answer was lying under my nose for so long!

What accounts for the site being underwater?

Though further investigations have to be made, the Ice Age is most likely the correct answer to this. The last Ice Age ended around 13,000 years ago. Hopefully studies will prove that the 'temples' date back to that period.

Could these findings change Malta's history as we know it?

Most certainly – and not only Malta's! The age for the megalithic temples must be changed to 12,000 or 13,000 years ago. And the same applies to all the artifacts recovered from those periods. Malta may indeed prove that the earth's history as we know it must be changed.

Now I had a new name – Joseph Ellul – to add to the list of contacts who I would need to chase down in Malta, and new doubts about the exact provenance of whatever it was that had been discovered underwater off Sliema. For if the press reports were correct, then: (a) Zeitlmair had *not* shot the original video footage and photographs of the site (these were the work of Maltese divers Shaun and Kurt Arrigo); (b) Zeitlmair had got the idea for the location of the site from a Maltese prehistorian named Joseph Ellul; and (c) Joseph Ellul was in possession of an aerial photograph of the north-east coast of Malta that actually showed the location of the site about a mile and a half off Sliema . . .

'Confused . . .'

The last article in my file was a sarcastic piece by Mark Rose in *Archaeology*, the journal of the Archaeological Institute of America. Entitled 'The Truth, And Some Other Stuff, is Out There' it made heavy weather of Zeitlmair's ancient-astronaut enthusiasms and pointed out that:

> Chronology appears to be somewhat confused in Zeitlmair's interpretation. According to the website, he sees links between the submerged 'temple' and both Noah's Flood and the rise in sea-level following the end of the Ice Age. Furthermore, the presence of deeper sand deposits on the west side of the 'ruins', the side toward Gibraltar, than on the east side is taken as an indication that the flooding of the Mediterranean by Atlantic waters (which really did occur) was involved in the inundation of the 'temple'. The Mediterranean flooding, however, took place some five million years ago.
>
> The Maltese Museum Department's archaeology curator Reuben Grima has visited the site, and was unconvinced that the stones on the seafloor are indeed a temple, according to archaeologist Anthony Bonanno of the University of Malta. Bonanno himself is skeptical of the find, noting that even if there is a submerged structure it does not mean the temples need to be re-dated.[3]

Two more names for my list: Reuben Grima and Anthony Bonanno.

The complete list now included Shaun Arrigo, Kurt Arrigo, Joseph Ellul, Reuben Grima, Anthony Bonanno. And, of course, Hubert Zeitlmair – whom Santha and I had arranged to meet in the coffee lounge of the Diplomat Hotel in Sliema soon after our arrival.

Our plane was coming in over Malta now, gear down, ready to land. The island blazed white with reflected light from its limestone outcrops and cliffs. The sky was clear. The surrounding sea was deep blue and flat calm. Despite warnings that November is an unpredictable month in this part of the Mediterranean, I had every reason to hope that we might be able to dive the next morning and settle the matter of the underwater temple once and for all by thoroughly exploring and photographing it.

But it wasn't going to be quite as easy or as straightforward as that.

Bird's-eye view ... (1)
Malta, 24 June 2001

I'm on board a helicopter – an old Soviet Mi8 with masses of room inside for troops and great visibility out of the open door and rear window. It's been converted for commercial use and I know for a fact that it served for several years as an air taxi in Bulgaria before ending up in Malta. Normally it flies passengers between Malta and Gozo but this afternoon, thanks to Channel 4, we have the exclusive charter of it for an hour.

We take off from Luqa airport, hop straight up into the air 50 metres, circle widely, then head north-east across the township of Paola that separates two of Malta's extraordinary prehistoric monuments – the Hypogeum of Hal Saflieni (fully carved out of the living rock *underground* and thus not visible from the air) and the majestic Tarxien temple complex with its apsidal ('kidney-shaped') rooms, graceful spirals carved in relief, looming 'mother-goddess' figures and gigantic megaliths.

Archaeological consensus dates Tarxien to between 3100 and 2500 BC while the Hypogeum is thought to be a few hundred years older – with parts of it perhaps going back as far as 3600 BC.[4] Such a range of dates ranks these structures amongst the very oldest examples of monumental architecture yet to have been discovered anywhere on earth.

And the problem is that they are clearly not the work of beginners. The megaliths, some weighing 20 tonnes, perfectly balanced and integrated with one another in complex walls and passages, are hewn from the hard coralline and globigerina limestone with which Malta is plentifully endowed and which to this day affords the inhabitants their primary source of building materials. But now it is sawn up into manageable blocks weighing only a few kilos and barely half a metre in length.

We continue north-east across Grand Harbour to hover at 200 metres above the fairytale city of Valletta. It is much younger than the temples, belonging in

every sense to a different epoch of the earth, with most of its labyrinth of narrow alleyways and shadowed courtyards dating from the sixteenth century AD or later. Yet Grand Harbour, now gleaming with gantry cranes unloading great container ships, was once itself the site of a megalithic temple – the remains of which are believed to lie underwater, buried in deep silt and rubble, at the foot of Fort Saint Angelo.[5] According to an eye-witness report by Jean Quintinus, this prehistoric temple extended over 'a large part of the harbour, even far out into the sea' as late as AD 1536. In 1606 Megeiser could still see enough to note that it was constructed of 'rectangular blocks of unbelievable sizes'. And even in the nineteenth century visitors reported 'stones five to six feet long and laid without mortar'.[6]

That nothing is left of the temple today does not surprise me. Since my first research visit to Malta in November 1999 I've learned that objects – and even places – of archaeological importance can and do disappear here in mysterious ways. For example, ancient remains of an estimated 7000 people were found in the Hypogeum of Hal Saflieni, buried in a matrix of red earth, when it was excavated by Sir Themistocles Zammit at the beginning of the twentieth century.[7] Today only six skulls are left, stashed out of public view in two plastic crates in the cavernous vaults of Malta's National Museum of Archaeology. Nobody has the faintest idea what has happened to all the rest of the bones. They've just 'vanished', according to officials at the Museum.[8]

And the six skulls? After much pressure and protest I have been allowed to see them only this morning and they are – I must confess – extremely and unsettlingly odd. They are weirdly elongated – *dolichocephalic* is the technical term but this is dolichocephalism of the most extreme form. And one of the skulls, though that of an adult, is entirely lacking in the *fossa median* – the clearly-visible 'join' that runs along the top of the head where two plates of bone are separated in infancy (thus facilitating the process of birth) but later join together in adulthood. I should be paying attention to the fantastic views and seascapes unfolding beneath the helicopter but I keep on wondering: what would people with skulls like that have looked like during life? How could they have survived birth and grown to adulthood? And did the other skulls from the Hypogeum – the lost skulls, the lost bones – also show the same distinctive peculiarity?

Still at 200 metres, the helicopter is now flying north-west from Valletta to Sliema, following the contours of the coast, taking me over waters that I've dived in many times since November 1999 following the trail of Hubert Zeitlmair's elusive temple . . .

Hubertworld . . . (1)
Malta, 8 November 1999

Zeitlmair met us, as we had arranged, in the coffee lounge of the Diplomat Hotel in Sliema. He proved to be a tall, rather dashing man in his mid-forties with long, well-groomed, salt-and-pepper hair, stylish clothes, a soldierly bearing and an impressive moustache. Within a few minutes it had also become obvious that he was severely sight-impaired, if not entirely blind, and he explained, without rancour, that this was the result of a viral infection of the eyes that he had suffered during a period of military service.

I ventured that his disability must have made diving very difficult – when he was searching for the underwater temple. But he shrugged off my concerns. 'Of course,' he explained, 'I didn't dive myself. I wouldn't have been able to see a thing. I guided the divers to the site and they went down to take the photographs and get the evidence.'

'You mean Shaun and Kurt Arrigo?'

'Yes,' Zeitlmair exclaimed in the manner of a man suppressing a sneeze, 'the Arrigos.'

Until that moment I thought I had come to Malta to dive with Hubert Zeitlmair, the discoverer of the submerged temple off the Sliema coast. Indeed, we had discussed the matter by telephone and he had confirmed that a boat and tanks for up to four dives had been arranged for the following day for that specific purpose. The fact that Zeitlmair himself turned out to be a blind non-diver did not necessarily jeopardize those arrangements, of course. Nevertheless, I thought it was time for some clarification.

'So we'll be diving with Shaun and Kurt Arrigo tomorrow?' I asked. 'They're the ones who know the location?'

'I know the location,' asserted Zeitlmair into his cappuccino. 'It was I who led the Arrigos to it in the first place . . .'

'No offence,' – I had to ask – 'but how did you do that? I mean, since your eyesight is so poor, how did you manage to lead them there?'

At this point Zeitlmair conjured from his briefcase a magnifying glass and a large black-and-white aerial photograph of the coast of Malta between Valletta and Sliema. As he rolled the photograph out on the table between us he said, 'I was able to lead them to the site because of the indications . . . here.' Squinting his eye to the glass and lowering his head he eventually found what looked to me like a pattern of white dots on the photograph in an area of open sea north-east of Sliema. 'This is the site of the temple,' he announced. 'The photograph was taken by the British Royal Navy some time before World War II. The sky and sea were exceptionally clear, and the site became visible to the camera through the water . . .'

Well . . . Maybe. Or maybe it was just light reflecting off dust on the lens.

'Is this the photograph you got from Joseph Ellul?' I asked

'Yes, from Ellul. That's right.'

We then entered into a long, rambling, muddled discussion about who had discovered what. I was on autopilot through most of this, but the gist of it was Zeitlmair's claim to have developed a theory concerning the locations of Maltese megalithic sites which predicted the presence of a structure underwater off Sliema. The theory had to do with the well-known 'pairing' of temples in Malta, one on high ground and the other in the valley below it (as is the case at Skorba and Mgarr, for example, or at Hagar Qim and Mnajdra).[9] To this day I cannot understand which temple exactly Zeitlmair has in mind for the high ground around Sliema, and I am not clear whether his theory takes into account the ancient traditions of a megalithic temple in Grand Harbour. Still, what he's getting at is completely obvious: when sea-level was lower 12,000 or 15,000 years ago, the reefs around Malta, now submerged to depths of 100 metres or so, would all have been above water and the pleasantly sloping valley below Sliema might have seemed an ideal spot in which to build a temple.

As Zeitlmair told it, he was already geared up to fund a diving expedition off Sliema in order to test this theory; indeed, had bought a flat in Sliema to use as a base for the expedition, when his providential meeting with Joseph Ellul occurred. Ellul showed him the Royal Navy photograph which, he was convinced, pinpointed the exact place off Sliema in which the expedition should dive – roughly 2.5 kilometres from land along a bearing 65 degrees north-east off Saint George's Tower.[10]

'Although the location is quite far from shore,' Zeitlmair continued, 'where the water is generally more than 40 metres deep, I reasoned that there must be some sort of reef or shallows there to show up so clearly on the photograph – maybe a little sea-mount, or something like that, a high point standing above the surrounding valley, just the sort of place the temple builders would have appreciated ... Then I hired the Arrigos to get me to the site in their boat and to search the bottom with an echo-sounder. I figured if the echo-sounder suddenly started giving shallow readings in an area of generally deep water, and if we were about 2.5 or 3 kilometres from shore, then we would have found the right place.'

I frowned: 'But why did you need the echosounder? Surely a shallow spot like that would show up on nautical charts? If it's charted you should be able to set a course straight to it. No need to search.'

Zeitlmair shrugged: 'It is not charted ... But still it is there. You will see tomorrow.'

Bird's-eye view ... (2)
Malta, 24 June 2001

The helicopter is at 200 metres, flying north-west from Valletta to Sliema about 1 kilometre from shore. To our right is the open ocean – and somewhere out there the 'sea-mount' that shows up as a glimmer of pale dots on the Navy photograph. Was it ever a real place? Or just a trick of the light?

Despite the bad start that I undoubtedly made with Zeitlmair and the Arrigos in November 1999, my confidence has been growing for more than a year now that there could, after all, be something solid behind all the rumours of an underwater temple off Sliema ...

The case of Commander Scicluna
Malta, 15 June 2000

Joseph Ellul looks as old and as sturdy as a megalith, and his house in the sunlit village of Zurrieq is named after the nearby temple at Hagar Qim – to the study of which he has devoted most of his life. He speaks loudly, has certain eccentric mannerisms, and once launched on the subject of Malta's prehistory increases enormously in size and becomes unstoppable.

Ellul's particular theory – based in some obscure way that I do not understand on the differential weathering-rates of coralline and globigerina limestone – is that the megalithic temples of his native islands were originally built more than 12,000 years ago by a prehistoric civilization, and were much later destroyed by the biblical deluge (which, he reckons, took place 5000 years ago). Ellul sets out this theory in his 1988 book *Malta's Prediluvian Culture at the Stone Age Temples* – a book that has been entirely overlooked by archaeologists because of its cranky Creationist approach and unfortunate emphasis on an impossible mechanism for the deluge. This mechanism, in Ellul's opinion, was a cataclysmic penetration of the Straits of Gibraltar by the Atlantic Ocean 5000 years ago, resulting in instant flooding, from the west, of the previously dry Mediterranean basin. Such a penetration (as Michael Rose points out in the journal of the Archaeological Institute of America cited earlier) did indeed occur – 5 million years before Ellul suggests.

Other aspects of Ellul's theory are less far-fetched and he has some well-reasoned arguments about flood damage at Hagar Qim – but this was not what I had come to talk to him about that day in June 2000 on the second of my three big research visits to Malta. Having failed to make contact with him in November 1999, I was here now exclusively to find out if he could shed any fresh light on the mystery of Zeitlmair's missing underwater temple. It immediately became obvious, however, that Ellul did not regard the temple as in any way being 'Zeitlmair's', or missing, and that he clearly felt aggrieved about how his own role in the discovery had been interpreted.

Muttering in Maltese, he shuffled to a wardrobe positioned in the hall outside his kitchen and took down from it a rolled photographic print. It proved to be another, larger version of the aerial view of the Sliema coast that Zeitlmair had shown me the previous November. At the foot of it Ellul had drawn in a scale by hand and had typed the following legend: 'Undersea Prehistoric ruins situated at Direction Bearing 65 degrees NE of St George's Tower, 2.5 kilometres from land at a depth of 25 feet'.[11]

I was puzzled by one of the figures and asked: 'You got the depth from Zeitlmair, I suppose, after the Arrigos dived on the temple in 1999?'

Ellul favoured me with a sinister smile. 'No,' he replied, 'I got the depth from another Maltese diver, Commander Scicluna, in 1994.'

He shuffled off and returned with a much marked and tagged copy of his book in which he had been incorporating corrections for a new edition. He opened the book and from a small stack of papers folded inside the front cover pulled out a press clipping. The clipping, from the letters page of the *Sunday Times* of Malta, was dated 20 February 1994 and was a response to an article on the subject of sea-level rise that had appeared in the paper on 13 February 1994:

SEA-LEVEL CHANGES
From Comm. S. A. Scicluna

THE ARTICLE 'Sea-level changes of the past and present' by Peter Gatt (the *Sunday Times*, February 13) indicates that Malta's shores are going down at the rate of 2 mm a year . . . This is taking place in many Mediterranean countries, especially in Sicily, which is very close to us. At Marsameni and Motya, the evidence is very clear because both of them are now underwater.

In Malta this evidence is also clear. There are three sites which are now completely under water: the oil wells at Saint George's Bay in Birzebuga (mentioned by P. P. Castagna in *Malta u il-Gzejjer Tagha*), a rock-cut tomb in Sliema (exactly like the ones in Bingemma) – this is now in 25 feet of water; *and a prehistoric temple I located last summer under 25 feet of water, also at Sliema.*

I myself reported this find to President Tabone, to Dr Michael Frendo, Minister of Youth and Arts, and to Dr Tancred Gouder, Director of Museums.

S. A. Scicluna,
Sliema

Commander Scicluna, eh? Another name for my list. Plus of course President Tabone, Dr Michael Frendo and Tancred Gouder. It would be interesting to learn if any of these three, presuming they were still with us, had done anything at all to follow up Scicluna's claim to have found a temple underwater off Sliema.

Because unlike Zeitlmair, whose zany associations with ancient astronauts must not be held against him – but who unfortunately could not dive – it transpired that Scicluna was an archaeological diver of some renown who had led several underwater expeditions and received commendations from the British Navy and from the British Committee of Nautical Archaeology.[12] When such a suitably qualified and experienced man chooses to state in a national newspaper that he has found a prehistoric temple underwater, it is appropriate that he be taken seriously.

But had he been? After parting company with Ellul and returning to the flat that Santha and I had rented that June, I tried directory inquiries for Commander Scicluna's number. They couldn't help me. Then I called Manjri Bindra, a friend of ours in Malta who is very good at finding people, and within an hour she had the number for me.

I dialled and waited. There was a long delay, then a woman's voice answered the phone: 'Hello.'

'Oh. Yes. Hello. Er . . . My name is Graham Hancock. Is this Commander Scicluna's residence?'

Another delay, then: 'Yes.'

'Oh, good. Look, I'm sorry to disturb you, but please may I speak to him?'
Silence.

'I'm an author,' I gabbled, 'I'm researching a book about underwater ruins, and I understand that Commander Scicluna is a great expert in this field. I would like to speak to him about a temple, underwater, that he discovered off Sliema . . .'

'I'm afraid that will be impossible.'

I was nonplussed: 'Why?' I protested. 'I just need to speak with him for a few moments, to confirm something.'

'I regret that my husband passed away four days ago', the lady replied.

Now, all at once, I understood the sadness and fatigue in her voice and stammered my apologies for disturbing her.

'It is all right', she said wearily.

Hubertworld . . . (2)

Malta, 9 November 1999

Santha and I sat in the coffee lounge of the Diplomat Hotel in Sliema drinking cappuccinos with Hubert Zeitlmair. We had been there since 8 a.m.; it was now 9 and there was still no sign of the Arrigo brothers showing up in their truck to take us diving. This was annoying, as we were already partly dressed for the water, had our mesh bags packed at our feet, and could observe that the sea in which we had been expecting even now to be preparing to dive, was calm, windless and generally perfect for our enterprise.

'I don't understand it,' Zeitlmair was saying. 'We had a firm agreement that they would pick us up this morning at eight. Everything was arranged. I spoke with them myself just yesterday.'

We had already tried to phone the Arrigos' dive shop, and their mobiles, but without success. Admittedly it was still early, but it was odd that they were so uncontactable – and so not here. Was Malta going to be a bust? I was beginning to think so. Because, after all, even if an underwater temple did exist at Sliema, why should the Arrigos take me to it? In the event that it was archaeologically important, then it was sooner or later going to become a hot media property; meanwhile, the Arrigos' interests, and the interests of the site, might be best served by keeping its location confidential.

It was obvious even then that the matter of 'proprietorship' was far from settled. Zeitlmair had a strong claim, to be sure, but it was by no means free of encumbrances – and who was to say that he would ever be able to relocate his 'temple' should the Arrigos decide not to cooperate? Even in the best circumstances objects found underwater are easily lost again unless accurate shore-bearings have been taken from the boat – impossible for the blind – or a GPS has been used to record the precise latitude and longitude of the entry point.

'Do you have GPS numbers for the site?' I now asked Zeitlmair.

'No,' he confessed, 'but I told you already it is very simple to find it. We just go out 2.5 or 3 kilometres from Saint George's Tower and use the echo sounder . . .'

'Until we come to a reef that is shallower than the surrounding water?'

'Exactly. Then we will be on the spot.'

Around 11 a.m. we finally managed to get a call through to Shaun Arrigo's mobile phone.

It transpired that the two brothers and their father – who ran the diving business together – were on a boat off Gozo and would not be back in Malta until the evening of the following day. Although they knew of me and my visit, they claimed that no arrangement whatsoever had been made by Zeitlmair for them to guide me to the underwater temple that morning, and that they wanted to meet me first in order to discuss the matter further before deciding whether they wished to guide me at all. Besides, it was the law of the land that I should be certified medically fit by a Maltese doctor before I would be allowed to dive in Maltese waters. Had I yet obtained such a certificate? No? Then that too needed to be arranged. They proposed that I call round to their dive shop in two days time, on 11 November, to see if we could 'work things out'.

Silently fuming at myself for not having dealt directly with the Arrigos from the beginning in a matter as important as this, I turned to Zeitlmair: 'Are you sure you can find the site again?'

'Sure!' he barked.

He did sound sure.

'OK, then, Hubert, here's what I suggest we do . . .'

Bird's-eye view . . . (3)

Malta, 24 June 2001

We've left Sliema behind and now the helicopter is rushing rapidly west along the north coast of Malta. Dropping our altitude to 100 metres, we soar over White Rocks and head for Qawra Point – a finger-shaped promontory dividing Salina Bay from Saint Paul's Bay.

There we circle and hover above the spot in the sea where two days before Chris Agius, a new friend who has come to our aid in Malta within the past month, led us on a dive to a remarkable straight canal cut out of the solid limestone of the sea-bed at a depth of 25 metres. A low bridge, also hewn out of the bedrock, spans the canal at one point, and Chris has identified tool marks on its inner walls . . .[13]

We fly on, crossing Saint Paul's Bay and Mellicha Bay, crossing the Gozo Channel from Cirkewwa with the tiny midway island of Comino to our right. And I remember that here, too, somewhere in the channel between Malta and Comino, a prehistoric stone circle is rumoured to exist. In fact it is rather more than a rumour, since I have talked directly with one of the commercial divers who saw the structure before – as he claims – it was buried by developers beneath concrete pilings . . .

It would not be the first time in Malta that an archaeological discovery has been conveniently hushed up to allow a construction project to go ahead. The same thing happened at the Hypogeum of Hal Saflieni, which was entered and looted by labourers renovating houses above it at the end of the nineteenth century at least three years before archaeologists ever learned of its existence. The initial discovery was very deliberately *not* reported to the authorities for fear that they would sequester the site.[14]

Hubertworld . . . (3)
Malta, 9–10 November 1999

After the failure of communication over our dive plans with the Arrigos for the 9th I felt superstitious. I therefore made the decision to hire a boat and dive support from another dive shop and mount an entirely new search for the underwater temple without the Arrigos' help. Zeitlmair agreed and issued several more cheering statements to the effect that he would lead us straight to the spot, having got the Arrigos there before without any difficulty, etc. . . .

In character with the general pattern of annoyance and frustration that seemed to have draped itself around me that November, I then took our business to a dive-shop staffed by pessimists and safety fanatics who began issuing dire warnings about the weather, and various dangers associated with diving in Malta in the winter months, virtually from the moment that I walked in their door.

It took all of the rest of the 9th to sort out the medical certificates, find and hire the right type of boat, and tie up the arrangements for dive assistance the next day.

But the 10th dawned grey, stormy and windblown, with white-caps breaking out in the open sea in front of Sliema. Santha and I looked gloomily at the waves from our balcony in the Diplomat Hotel and decided that we would chance it. We had dived in worse. And the boat that we had hired was a 50 foot motorized *lutzu* (traditional Maltese fishing vessel) that should, in theory, be able to handle these conditions without too much difficulty. We might take a bit of a pounding getting back on board after each dive, but that was acceptable. While we were submerged we should face no problems.

Our new dive suppliers did not agree. What if there was a current and we were to get swept away from the boat? It was sturdy but not very fast and in high seas it might lose us completely. Sliema was not some enclosed bay, after all. The next landfall was Sicily, 90 kilometres to the north.

More badgering followed along these lines and I was eventually obliged to concede that diving was probably not a very good idea that day . . .

Bird's-eye view . . . (4)
Malta, 24 June 2001

Our hour in the helicopter is rapidly ticking by. We've passed Comino and hover over Gozo's Mgarr Harbour before heading into the heart of the island. There, south of Xaghra – itself the site of a huge semi-subterranean stone circle – is the necromancer's castle, the 'Giant's Tower' of Gigantija, the greatest and the oldest of the megalithic temples of the Maltese archipelago, reckoned to have been built around 3600 BC.

Looking down on it from above, I am struck not only by its enormous size but also by the way in which it faithfully and exactly reproduces what may be

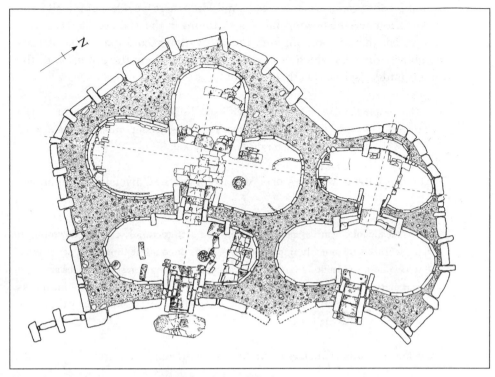

Floorplan of Gigantija temple. Based on Evans (1971).

thought of as the 'canon' of all the Maltese megalithic temples – an outer retaining wall of cyclopean blocks, some up to 5 metres high and many in the range of 15 tonnes or more, set out in a series of expansive, graceful curves to enclose an irregular space that feels more organic than architectural. This inner space contains a series of altars, shrines and large apsidal rooms interconnected by axial passageways, all of which are also lined with huge megaliths of mixed coralline and globigerina limestone.

Unlike other simpler temples, Gigantija features two distinct and not quite parallel axial passages oriented east of south which dominate the whole complex. By means of imposing stone gateways, each of these passages penetrates a concave megalithic façade defining the only two 'entrances' to the structure. The easternmost axis leads to four large apsidal rooms arranged in two pairs of opposed lobes. The westernmost axis leads to five apses – two arranged as an opposing pair and the remaining three in the form of a clover-leaf.

Orthodox scholarly opinion holds that the islands of the Maltese archipelago remained *entirely uninhabited* until 5200 BC – 7200 years ago – when they were settled by Neolithic agriculturalists from nearby Sicily.[15]

Orthodox scholarly opinion dates Gigantija to 3600 BC – 5600 years ago.

The time lapse between settlement 7200 years ago and the construction of

Gigantija 5600 years ago is 1600 years. And while there is evidence of small-scale construction and the hewing out of rock tombs in the Maltese islands during this period, there is nothing from the excavation record that archaeologists are able to show us which in any way seriously charts the *evolution* of the temple-building phase. On the contrary:

> The temple builders did not begin with small-scale structures. Gigantija ... is a tremendous work of architectural design and of engineering, built a thousand years before the date usually given for the Great Pyramid.[16]

To this Colin Renfrew, Professor of Archaeology at the University of Cambridge, adds:

> The façade [of Gigantija], perhaps the earliest architecturally conceived exterior in the world, is memorably imposing. Large slabs of coralline limestone, set alternatively end-on and sideways on, rise to a height of eight metres; these slabs are up to four metres high for the first course, and above this six courses of megalithic blocks still survive. A small temple model of the period suggests that originally the façade may have been as high as 16 metres.[17]

Cyclopean walls 16 metres high? At first sight, admits Renfrew,

> it seems inconceivable that such monuments could be built without the organization and the advanced technology of a truly urban civilization ... Yet according to the radiocarbon chronology, the temples are the earliest free-standing monuments of stone in the world. In the Near East at about this time, 3000 BC and perhaps even earlier, the mud-brick temples of the 'proto-literate period' of Sumerian civilization were evolving: impressive monuments in themselves but something very different from the Maltese structures.[18]

How are we to explain the fact that the oldest free-standing stone monuments in the world, which by virtue of their size and sophistication unambiguously declare themselves to have been built by a people who had *already* accumulated long experience in the science of megalithic construction, appear on the archaeological scene on a group of very small islands – the Maltese archipelago – that had not even been inhabited by human beings until 1600 years previously? Isn't this counter-intuitive? Wouldn't one expect a 'civilization history' to show up in the Maltese archaeological record documenting ever-more sophisticated construction techniques – and indeed wouldn't one also expect an extensive 'civilization territory' capable of supporting a reasonably sized population (rather than tiny barren islands) to surround and nourish the greatest architectural leap forward of antiquity?

Dr Anton Mifsud, President of the Prehistoric Society of Malta, who we will be hearing from a great deal in the coming chapters, offers this succinct summary of the problem: 'Malta is presently too small in size to have sustained the earliest architectural civilization; its civilization territory is missing.'[19]

We circle over Gigantija one more time, then bank sharply to the south-east, cross the Gozo Channel again and hover over a rugged spot called Marfa Point at the extremity of the main island of Malta.

Here, underwater, two days previously, we saw further strange channels cut in the rock, some running in distinctive parallel tracks, leading to the edge of a drop-off at 8 metres. Beneath the drop-off we were shown a terrace of three large right-angled steps cut into the interior of a cave at 25 metres.

Could the 'civilization territory' of Malta be missing because it is now underwater?

Hubertworld . . . (4)
Malta, 11–13 November 1999

I didn't keep the loose appointment that I had made to try to 'work things out' with the Arrigos on 11 November, but I also did not go diving that day; 2 metre waves whipped up by the strong prevailing wind from the north-east still prohibited that.

On the 12th and 13th, however, much to the astonishment of our pessimistic dive hosts, the north-easterly lulled, the angry seas subsided and we were able to take the *lutzu* out and begin searching with the echo-sounder for Zeitlmair's uncharted sea-mount between 2.5 and 3 kilometres from shore.

Within an hour of zigzagging back and forth across water generally 40 to 70 metres deep we suddenly stumbled upon a shallow point where the echo-sounder gave a depth of just 7 metres – more or less exactly as Zeitlmair had promised. It was with the air of a man vindicated, therefore, that he stood by beaming short-sightedly as the *lutzu* was anchored and we prepared to dive.

But we couldn't find his temple – only a series of disparate features that in some way resembled, but did not actually seem to be, the features that Shaun Arrigo had videoed a few months previously in July 1999.

I felt incredibly disappointed, crushed and depressed after those dives, which had seemed so promising initially, and began to believe that we might never find the site if we went on this way – for the same reason that the man on the beach can never count all the grains of sand. By close of business on the 13th, therefore, I had decided to set pride aside, go back to the Arrigos cap in hand, and beg them to take me to their – or Zeitlmair's – or whoever the hell's temple it was.

In my opinion no one owned the temple . . . if it existed at all. I certainly had no desire to own it or lay any kind of claim to it. I just wanted to dive it.

Reuben Grima's short dive in a thunderstorm
Malta, 19 June 2000

The 'Zeitlmair file' in my laptop during my first visit to Malta in November 1999 had contained a report from the journal *Archaeology* that seemed to write off the significance of the underwater temple right from the start. According to that report Reuben Grima, archaeology curator at Malta's National Museum, had dived at the Sliema site and was 'unconvinced that the stones on the sea-floor are indeed a temple'. Quoted alongside Grima was Professor Anthony Bonanno of the University of Malta, who made the point that even if a ruined temple had been found underwater, its submersion did not necessarily mean that all Maltese temples had to be redated.[20]

Bonanno's observation was completely correct. It would be necessary to establish the mechanism of submergence of the site (land subsidence versus sea-level rise) before jumping to any conclusions about the age of any structures on it – and this had not been done yet. On the other hand there would be little point in establishing anything at all about the site if the 'megaliths' and 'kidney-shaped rooms' that had been seen and photographed there were not in fact parts of a temple at all but just natural formations that had been misinterpreted by excited amateurs – as Reuben Grima seemed to have concluded after his dive.

In November 1999 I had been too depressed to do anything much but stubbornly and repeatedly go diving myself in the cold waters off Sliema – trying to get some hands-on experience of the structure so that I could form my own opinions. I hadn't contacted Reuben Grima then, so he was still on my agenda when I returned in June 2000 to resume the dive search.

I had arranged our appointment for 19 June – rather than any other day – with a certain ulterior motive. Santha and I wanted permission to be inside the 'lower temple' at the megalithic site of Mnajdra at dawn on the 20th, the summer solstice and the longest day of the year. Reuben Grima was one of the few people who had the power to grant this very rare privilege – and he did so with good grace and one telephone call to supervisory staff at Mnajdra. 'I understand the effect is spectacular,' he said with a smile, 'but you should be there before 5 a.m. The watchmen will be expecting you . . .'

I told him that I wasn't any kind of archaeologist, just a popular writer, so he should excuse me in advance if I seemed ignorant of archaeological procedures and facts or if I asked naive questions. There was, however, something bothering me about the dating and 'sequencing' of the megalithic temples of Malta within the period 3600 to 2500 BC, and the dating of the first human habitation of Malta to 5200 BC. 'How have you arrived at these dates?' I asked.

As I had been expecting, Grima explained that the primary tool in establishing Malta's prehistoric chronology had been radiocarbon-dating (based on the rate of decay of C-14 stored in all formerly living matter).[21] My views about C-14 are

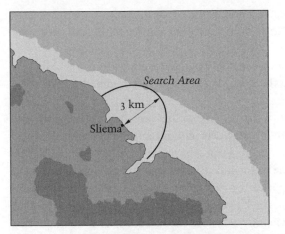

on the record.[22] I think it should be only one amongst several tools and techniques brought to bear on the dating of megalithic or rock-hewn sites. It is a truism, but worth repeating nevertheless, that C-14 *cannot date stone* – only such organic materials as are found around or in association with stone ruins. It is an *assumption* (more or less safe depending on the stratigraphy and general circumstances of the site but still, at the end of the day, an assumption) that organic materials found close to megalith B or trilithon A or dolmen C, etc., do in fact date from the same period as the quarrying and erection of the megaliths concerned.

To this extent the excavation of a megalithic site is a bit like a crime scene. If the scene has been properly protected from contamination and intrusive elements, then the results of any forensic tests are likely to be much more accurate and useful than they will be if the scene has been disturbed. C-14 dating is a forensic test. And looked at as crime scenes, Malta's megalithic temples are pre-eminently 'disturbed' – since they have been used as quarries and goat pens by local farmers for millennia, in some cases arbitrarily reconstructed on a whim,[23] and dug over with great enthusiasm and little skill by amateur archaeologists for at least 200 years before the introduction of carbon-dating in the mid-twentieth century.

But when I put these objections to Grima he brushed them aside: 'Look, of course it's possible that new evidence might yet be unearthed which would require some revision of our chronology for Maltese prehistory. But I think, after all these years and the application of so many eminent minds to the problem, that we've probably got things pretty well right. If we're wrong it will be at the most by a few centuries, not by millennia. So we're not expecting any big surprises.'

'How many carbon-dated samples does the orthodox chronology here actually depend on?' I asked.

'For the temples?'

'Yes, and the Hypogeum too.'

'Well, very few actually.'

'Do you remember how many?'

'Off the top of my head, I don't. But I can easily check. I know it's not a large number.'

'And out of this not very large number of carbon-dated samples from the temple period how many were actually taken from *underneath* undisturbed megaliths?'

'As far as I know none were,' replied Grima.

This seemed a good moment to turn the conversation to the subject of the underwater temple off Sliema.

'I understand you dived on it,' I said. 'What did you make of it?'

Grima raised his hands in a theatrical shrug: 'Not very much. But then to be fair, I didn't see it properly.'

He had gone to the site with Shaun Arrigo, he explained, rather late one afternoon with a thunderstorm brewing. The conditions had looked bad. Moreover, Arrigo claimed not to be sure of the precise location of the 'temple'. Then soon after they dropped into the water and began to look for it, Grima discovered that by mistake he had strapped on a half-empty tank. Bearing in mind the deteriorating surface conditions he had therefore been obliged to abort the dive after only ten minutes. 'The visibility was awful,' he added, 'and we might not even have been in the right place, but what I saw looked like pretty much just ordinary sea-bottom to me.'

'It might well have been. But the question is – was what you saw the same thing that Zeitlmair is claiming is a temple?'

Grima clearly had some difficulty taking Zeitlmair and his ancient astronauts seriously and I could understand why he might be sceptical of any claims emanating from such a source. However, irrespective of his, or my, or anyone else's views on Zeitlmair, I felt that the proposition of a submerged, man-made prehistoric structure off Sliema was an eminently testable hypothesis which could be proved or disproved empirically by diving on it, thoroughly photographing it and collecting samples.

Grima's ten minutes in a thunderstorm didn't even begin to qualify as a test. So no matter how wacky its proponents might seem to be, the hypothesis that a temple could be there had still not been refuted as far as I was concerned. Besides, there had been nothing wacky about Commander Scicluna.

As I was leaving his office at the National Museum in Valletta, I asked Grima if he was aware that six years before Zeitlmair, Scicluna had also reported the existence of a megalithic temple underwater off Sliema, and at pretty much the same depth.

Grima said he knew nothing of the case and asked me to whom it had been been reported.

'To Tancred Gouder, amongst others. I understand he was Director of Museums at that time. Scicluna mentioned the discovery in a letter to the *Sunday Times* of Malta in March 1994. I'm really surprised it wasn't followed up . . .'

Bird's-eye view ... (5)

Malta, 24 June 2001

We've left Marfa Point and are flying over the sea parallel to the wall of sheer cliffs, in some places hundreds of metres high, stacked up along the western coast of Malta. I'm told that these cliffs exist because this side of the island has been slowly but steadily rising over several millions of years as a result of geological upheavals along the submarine Pantalleria Rift – levering itself up out of the sea-bed at an annual rate of a millimetre or two and causing the eastern side of the island, by the law of equal and opposite force, to tilt downwards.[24] That means that the Sliema coast, with its rumours of an underwater temple, has experienced some degree of submergence during the past 17,000 years not only on account of rising sea-levels at the end of the Ice Age but also because of the longer-term process of land-subsidence that is still underway today.

We skip over Paradise Bay and then, in quick succession, Anchor Bay, Golden Bay, with its beach umbrellas and racks of lobster-pink tourists, Ghajn Tuffieha Bay and Gnejna Bay. Then we turn inland over the Bahrija valley and the Wied ir-Rum with the twin medieval towns of Mdina and Rabat to our left and the sea to our right.

Malta's landscape is everywhere rugged and stony, sliced through with plunging valleys, crumbling escarpments and dark defiles – a racked and tortured topography twisted, moulded and scoured out by extreme natural forces over aeons. It is easy to overlook the implications of so much rocky ruggedness and drama being compressed into such a small space, but as Anton Mifsud explains:

> The present surface area of the Maltese islands is not sufficient to account for the extensive valley formations such as the Wied il-Ghasel, Wied il-Ghasri and Wied ix-Xlendi, amongst others. The creation of such deep and precipitous valleys would have required a very extensive land surface to hold the waters which dug them out over the millennia.[25]

And Mifsud is right. The Maltese archipelago was once much bigger – indeed so much bigger that it wasn't an archipelago at all. Around 17,000 years ago, at the Last Glacial Maximum when sea-level was more than 120 metres lower than it is today, the three main islands of Malta, Comino and Gozo, as well as little Filfla in the south, were all joined into one landmass, itself joined by a wide and extensive land-bridge to Sicily 90 kilometres to the north – which was in turn joined to the 'toe' of the Italian mainland. Glenn Milne's inundation maps, as we shall see in chapter 19, leave no doubt about the overall picture while more detailed bathymetric studies reveal the antediluvian central Mediterranean to have been an area of potentially enormous interest to the story of human civilization that has been almost entirely neglected by the responsible scholars.

Hubertworld . . . (5)
Malta, 15 November 1999

Malta is a small place, word of the search that Zeitlmair and I had been conducting off Sliema had got around, and I used the same *lutzu* and crew for diving with the Arrigos that I had used to try to locate 'their' site with a competitor dive-shop just a couple of days previously. None of this heavy-handedness helped to promote good relations, and I am certain that Shaun and Kurt Arrigo, and their father whose name I presently forget, must have regarded me as an entirely unpleasant and untrustworthy customer and a complete idiot into the bargain.

We spent the 14th engaged in angry discussions, recriminations and speeches of self-justification but on the 15th we went diving. Kurt couldn't make it, nor could Arrigo senior, so I dived with Shaun Arrigo, who looks like a pirate. He is young – about thirty and physically fit, with long black hair, a hawk nose, hooded eyes and seven days of stubble. To my surprise, however, he claimed that he was not sure of the exact location of the site and that we would have to search for it. With a sense of *déjà vu* I stood by as the boat zigzagged back and forth over a range of depths, bearings and distances from shore with Shaun Arrigo repeatedly asserting that the site was not as far out as Zeitlmair still believed.

'Well, how far out is it?' I asked.

'Three kilometres,' interjected Zeitlmair.

'One kilometre,' insisted Arrigo.

We used the echosounder to chart the bottom at both distances and at all points in between, but couldn't find the right profile anywhere. Meanwhile, the weather, which had been calm a little earlier, had changed character, assuming an ominous tone as clouds massed overhead. Beneath the keel of the *lutzu* all of us could feel the long rolling upsurge of a heavy swell – more scary in a way than breaking waves because of its aura of suppressed violence and power. The waters that had been blue just half an hour before were now transmuted to dark grey, almost black, and the air temperature had plunged. Even wearing a wetsuit I shivered. The shoreline between Sliema and Saint Juliens seemed far off across the heaving sea. Was I seriously planning to dive in this?

Then the captain called out from the cabin that the echosounder was giving a depth of 20 metres . . . 19 . . . 18.5 . . . 18 metres.

'We'll go in here,' yelled Arrigo, peering wildly over the side and already strapping on his tank and BCD.

I hurried to follow suit while the boat was brought to a standstill. By then, however, we had drifted off the 18 metre contour and the captain announced that we were now in between 25 and 30 metres of water.

'We'll go in here,' Arrigo repeated. 'If it's the right place we'll find that the reef slopes up fairly steadily from 25 metres to 7 or 8 metres. All we should have to do is follow the slope of the reef as it gets shallower and that will bring us to the plateau where the temple is . . .'

'But what if it isn't the right place?' I asked plaintively.

Shaun Arrigo clasped his mask and regulator to his face, jumped overboard and disappeared silently beneath the waves.

Bird's-eye view . . . (6)
Malta, 24 June 2001

The helicopter passes above Dingli now, where the golfball domes of a modern radar station overtop the steep cliffs. Then we come to a sloping area of exposed limestone between Buskett Gardens and the sea. Approximately 2 kilometres square, it is incised with a tremendous network of curving parallel tracks – one of the few surviving tableaux of Malta's famous 'cart-ruts'.[26]

I have walked here several times during previous visits in 1999 and 2000 and know that the ruts are often sheer-sided, sometimes a metre or more deep and up to two hands-breadths wide at the base. Nicknamed locally 'Clapham Junction', the area is preserved as a tourist attraction today. And as we hover 120 metres above it – I can see that it does indeed resemble a junction point where multiple railway lines converge and diverge. Some of the pairs of tracks run straight; some curve; some cross over one another. But there is no particular sense of organization or pattern – which is one among many reasons why no universally accepted explanation of this peculiarly Maltese phenomenon has ever been given.[27] Archaeologists don't even have a clue how *old* the 'ruts' are, although it is certain that those at Clapham Junction were already in place 3000 years ago when datable Punic tombs were cut through a number of them.[28] It is certain, too, that they were not simply worn away in the tough limestone by the passage of cart-wheels over periods of centuries, as many have wrongly theorized; on the contrary, there is no proof whatsoever that cart-wheels *ever* ran in these ruts – which were initially carved or cut out of the bedrock with the use of tools.[29] Some archaeologists associate them with the megalithic temples;[30] others believe that they date from the Bronze Age, between 4000 and 3000 years ago after the culture of the temple-builders had collapsed.[31] The truth is nobody really knows anything at all about what they are, or who made them, or when, or why.

As with so much in Maltese prehistory their origins may belong in an underworld that scholars do not seem anxious to explore. However the existence – to which we can now attest with photographs and film – of 'cart-ruts' on a gigantic scale underwater at Marfa Point raises the possibility that this phenomenon may have much older origins that any scholar has previously suspected.

Hubertworld . . . (6)
Malta, 15 November 1999

I jumped immediately after Shaun Arrigo but he was already far below me and it took me a moment or two of hard finning to catch up with him. Contrary to the indications of the echo-sounder – unless we had already been carried far from our entry point by what was proving to be quite a brisk current – the bottom here was deeper than 25–30 metres. In fact, as we continued to sink, it became clear that it was deeper than 40 metres . . .

Arrigo was a strong swimmer and I found it hard to keep up with him, but we forged ahead against the current until we did finally encounter a reef of bedrock gradually sloping up from 30 metres or so through 28 metres, then 24 metres, before levelling off into what seemed to be a vast submarine plain covered with undulating fronds of seagrass, at about 22 metres. Because of the stormy overhead light, visibility at this depth was poor – like diving at dusk – and even if the plain did lead to an eminence at some point it was obvious that we would only stumble across it by chance.

Besides, we had been down for quite a while now, quite deep – 38 metres at the outset, then a long hard swim for twenty minutes or so at between 30 and 22 metres. I checked my air pressure gauge and found, as I had expected after burning so much energy, that I was already below 100 bar on what was only a moderate-sized 12 litre tank. Another 50 bar – definitely less than twenty minutes at this rate unless we got into shallower water – and I was going to have to ascend, allowing enough air for at least a five-minute rest-stop (and preferably a bit more) at 5 metres. Arrigo seemed to be making a personal statement of some sort by staying ahead of me in the water at all times so I couldn't see his guage. But I could be reasonably sure that his air consumption would be better than mine, since he was twenty years my junior and dived for a living.

We swam on for a while at 22 metres, still against the current, then I caught up with Arrigo with another titanic effort, grabbed one of his fins to get his attention, showed him my guage – now down to 70 bar – and signed that I was going to start doing this dive in shallower water.

He indicated that he preferred to stay deep for a bit longer – making the 'search' signal as he did so.

Hmm . . . Interesting . . .

Very slowly, remaining parallel with Arrigo but now above him, I began to ascend.

I realized that I was exhausted, almost gasping for breath as though the wind had been knocked out of me, but my ego would not allow me to show it or make any sign of distress. So I tried to relax, calm my breathing, reduce my heart-rate. Like other bad, fruitless dives that I had done, I told myself, I was going to get through this one.

I did the rest-stop and had 50 bar left when I reached the surface – all fine and orderly. No panic. The only problem, as I looked around from the peaks and valleys of the billowing waves upon which I now bobbed like a cork with my BCD fully inflated, was that there seemed to be no sign at all of the *lutzu*.

I couldn't see it anywhere. It had gone.

Moments later, blowing like a seal, Arrigo joined me from the depths with 70 bar on his gauge. So at least I would have someone to talk to while I waited to drown or die of exposure.

Bird's-eye view . . . (7)
Malta, 24 June 2001

We're still hovering over Clapham Junction while Colin Clark, the Channel 4 cameraman, and Santha with her Nikons continue to occupy the open door and window, trying to get clean shots of the cart-ruts to compare with what we have seen underwater at Marfa Point.

The complicated question of which parts of the island are rising and which are sinking because of activity along the Pantalleria Rift must, of course, be factored into the equation along with sea-level changes – but theoretically it ought to be possible to calculate a fairly accurate date for the submergence of the Marfa Point 'ruts'. That would then give us a *terminus ante quem* for the cutting of the ruts by human beings – i.e., we could be sure that the ruts had been cut before the date of their submergence and must therefore be at least that old.

Interestingly, Anton Mifsud's tireless research in the archives has unearthed an obscure account published in 1842 of the travels in Malta of a certain Dr J. Davy, who

> observed cart ruts between Marfa and Wied il-Qammieh in northwest Malta, and from their interrupted nature at the edge of the cliffs, inevitably concluded that the Maltese islands had once been significantly larger during the presence of man in Malta.[32]

Now it may well be that the submerged ruts we've dived on off Marfa Point will ultimately prove to pose no problem to orthodox chronology. That is perfectly possible if land subsidence has been the major factor in their inundation. But even so, they should be seen in context of the wider phenomenon of submerged ruts – contiguous to many different stretches of the Maltese coast – which have been reported in the past. Indeed, Anton Mifsud demonstrates that 'before their gradual disappearance over the past few decades' the ruts were 'repeatedly and validly associated' by scholars and travellers with a former extension of Malta's landmass.[33] 'In several maritime sites around the island of Malta,' wrote Sanzio in 1776, 'one could see deep cart ruts in the rock that extended for long distances into the sea.'[34] And in 1804 De Boisgelin believed he had found evidence that:

Some serious disruptions and subsidings have taken place on the island ... An extraordinary subsidence ... must have occurred on the coast not far from the pleasure grounds of Boschetto [Buskett] ... on the southern side of which vestiges of wheels have cut into the rock, and may be traced to the sea ... and the ruts may be perceived underwater at a great distance, and to a great depth; indeed as far as the eye can possibly distinguish anything through the waves ...[35]

Father Emmanuel Magri, the first official excavator of the Hypogeum at Hal Saflieni, recorded the presence up until the end of the nineteenth century of cart-ruts on the tiny uninhabited island of Filfla[36] – which lies some 5 kilometres south of the twinned megalithic temples of Mnajdra and Hagar Qim in the same general area of Malta's south coast. And in 1912, R. N. Bradley commented on cart ruts near Hagar Qim – noting that they ran 'over the precipitous edge of the cliff towards Filfla'.[37] In subsequent years the ruts in both places have been completely obliterated (in the case of Filfla by sustained naval bombardments – the island was for a long while a favoured spot for target-practice). Nevertheless, as Mifsud observes, the combined effect of Magri's and Bradley's testimony is to suggest that cart-ruts once ran all the way from Hagar Qim to Filfla passing across a land-bridge that has therefore been submerged *since* human beings first came to the islands.[38]

In what he would be the first to admit is an untested hypothesis, Mifsud proposes a cataclysmic collapse of the Malta–Filfla land-bridge as a result of rifting processes in relatively recent prehistory – just over 4000 years ago – and he links this hypothetical cataclysm with the seemingly abrupt demise of the temple-building civilization of Malta around 2200 BC.[39]

We have finished our work at Clapham Junction and the helicopter is now running east at 150 metres along Malta's south coast between Ghar Lapsi and the Blue Grotto. To our left, nestled into the slope of the island, is the colossal edifice of Mnajdra and above it on the hilltop stands Hagar Qim. To our right, across the open waters of the Mediterranean, is Filfla.

No diving is presently allowed around Filfla, and the entire area has been designated a closed nature reserve. But I can't help wondering – what lies beneath those waters other than unspent ordnance from the years of bombardment? Could there be the remains of a lost civilization there? Perhaps on the sea-bed between Hagar Qim and Filfla – as on the sea-bed off the Qawra and Marfa Points and off Sliema too – some of the mysterious antecedents of Malta's extraordinary temple-building culture are waiting to be found?

Hubertworld ... (7)
Malta, 15 November 1999

The *lutzu* was there after all, but it had drifted far away. It was obvious, since we could hardly see it, that Santha and the others on board certainly could not see us, especially when the swell carried us down – as it often did – into deep troughs in the waves. I knew that Santha would be beginning to be concerned by now, although she might not be expecting us to surface for some minutes yet if she had been assuming a shallower dive than we had in fact made.

Time passed and the sea was getting higher. Arrigo and I bobbed a few metres apart, beginning to feel cold, not talking because that required energy. Although my BCD was fully inflated, I found that I was constantly inhaling sea-water as waves splashed into my face or rolled me momentarily under. At the same time I found myself reluctant to take air through my regulator from the miserable 50 bar or less that was left in my tank; I might need that for a real emergency.

We tried waving – futile, of course in waves so high. We tried blowing the pathetic little whistles that manufacturers attach to BCDs and that cannot be heard at 5 metres if there's a wind blowing. There was a wind blowing.

Then Arrigo connected up a power-whistle that had been concealed in an emergency kit somewhere on his person to the inflator hose on his BCD and pressed the button. For two seconds the air was filled with an ear-splitting howl that could have been heard on the other side of the island. Then the noise suddenly stopped.

Arrigo cursed: 'Not enough pressure. It's supposed to work down to 50 bar.'

There was no sign of the distant *lutzu* charging course. If they had heard us it had not been long enough to get a bearing.

'But you've got 70 bar,' I pointed out.

Arrigo shook his head. 'Don't think so. Maybe a faulty guage. How much do you have?'

'Less than 50 bar.'

'Shit! Still, give it a try and see what happens.'

I took the whistle from him, connected it to my inflator hose, pressed the button. Nothing.

'Shit.'

We decided that we had better start swimming towards the shore, which by now seemed tremendously far away – had a current been carrying us out to sea all along? After ten minutes of effortful paddling, however, it became obvious that we had made no forward progress at all.

I floated on my back to catch my breath and, on the off-chance, decided to try the power whistle again. This time it worked at full blast and I kept the button pressed for several seconds, joyously observing as I did so that this time the *lutzu* was turning towards us. For a moment the whistle stopped, then started again, and I got three more good blasts out of it before it packed up completely. But the emergency was over. We'd been spotted and, after some manoeuvrings, were recovered into the *lutzu* from the increasingly wild sea.

Back on board, still in my wetsuit and drinking hot tea, I did not realize how close our escape had really been until I saw the massive Valletta-to-Gozo car-ferry bearing down relentlessly on our last position in the water before the recovery.

We had been snatched out of its path with just a few minutes to spare.

Bird's-eye view ... (8)
Malta, 24 June 2001

After the helicopter has made the run over the Ice Age valley long since inundated by the waters of the Mediterranean that once plunged between the two high points of Hagar Qim and Filfla, we circle back to take a closer look at Hagar Qim and at its 'paired' temple Mnajdra.

In total the remains of twenty-three megalithic structures classified by archaeologists as temples have been found in Malta – of which, according to Dr David Trump's authoritative *Archaeological Guide*,

> six stand alone, ten are in pairs, and there is one group of three and one of four. Five more structures of similar type have irregular plans, and there are at least twenty scatters of megalithic blocks . . . which could represent the last vestiges of former temples . . . It is on the whole unlikely that many more remain to be discovered. The number destroyed without trace we shall never know.[40]

All the temples were supposedly built between 3600 and 2500 BC,[41] with the bulk of the work completed before 3200 BC.[42] The best known on the tourist circuit today are Gigantija on Gozo, and Tarxien, Hagar Qim and Mnajdra on Malta. Other important temples, though smaller and less often visited, include Mgarr and Skorba, Tal Qadi and Bugibba. In a peculiarly Maltese compromise, the latter, near our dive site at Qawra Point, has been engulfed and partially ingested by the modern Dolmen Hotel.[43]

The pilot holds the helicopter stationary over Hagar Qim, giving us a bird's-eye view of its impressive perimeter megaliths, which include one 7 metres high that is estimated to weigh more 20 tonnes.[44] As at Gigantija the shape of the temple is defined by graceful curves and re-entrants and it contains a series of paired apsidal rooms, also lined with megaliths. From above, the oval arrangement of the apses make them seem almost like enormous eggs lying in a huge stone nest and I am struck again by the strangeness and uniqueness of this design and by the odd fact, pointed out with some bemusement by David Trump, that 'There is nothing looking remotely like one of these temples outside the Maltese Islands.'[45]

We circle several times, then bank downhill towards the coast where Mnajdra lies – the last stop on our magical mystery tour. Although it is a spacious conglomerate of three temples (the 'Small Temple', the 'Middle Temple' and the 'Lower Temple'), Mnajdra can at first sight seem almost inconsequential, tucked away as it is in rugged terrain against a hillside. The lower temple and the middle temple each have four of the characteristic megalithic apses arranged in two opposed pairs. The small temple is 'trefoil' in plan – with three apses arranged like a three-leafed clover.

I remember how, a year previously – on 20 June 2000 – I'd watched the summer solstice sunrise from within the lower temple at Mnajdra courtesy of Reuben Grima. It was then as the rays of the sun were projected on to a great megalith flanking the south side of the central axis that I understood for the first time how subtle and pure, how understated and yet how purposive, was the architectural genius of its builders. These people, who could achieve the most precise and painstaking alignments in the medium of cumbersome and gigantic stone, had not only been master architects and engineers – and first-class observational astronomers – but also excellent practical mathematicians and geometers. And all of this, presumably, had been harnessed to something else, some greater or transcendant objective that was somehow expressed in the temples.

Our hour is nearly up. The pilot banks away from Mnajdra and we head back towards the airport. In the last few minutes of the flight I find myself returning to the basic conundrum that has exercised my imagination in Malta since 1999, when I first involved myself here. It's the absence of background to the temples, the fact that they're suddenly just there, almost ready-made – without any obvious antecedents. And the fact that ancient megalithic or rock-hewn structures appear to exist underwater at several points around the archipelago – suggesting an older episode of construction that prehistorians have not yet taken account of.

Despite archaeological and C-14 evidence to the contrary, the existence of which I freely acknowledge, I think the time has come to consider the possibility that the origins of Malta's megalithic temples and its mysterious Hypogeum might not be confined exclusively to the fourth millennium BC, as we have hitherto been taught, and that these amazing structures might have far older and far more mysterious roots.

16 / Cave of Bones

> To sleep within the Goddess's womb was to die and to come to life anew.
>
> Marija Gimbutas

There are places in the world made by people gone before us – hallowed places, places of power – in which the art and architecture serve as mantras that dilate the spirit. In some cases it is possible to trace back a sacred history of the site that long predates any surviving structures and symbolism there – suggesting that we may be in the presence of something numinous in the location itself to which human beings of all epochs and faiths can respond.

Without any intention of giving an inclusive list I might mention Chartres Cathedral and the prehistoric painted caves of Lascaux and Chauvet in France, Altamira in Spain, the Dome of the Rock in Jerusalem, the Temple of Seti I and the Osireion at Abydos in Upper Egypt, the Great Pyramid of Giza in Lower Egypt, the Bayon at the heart of Angkor Thom in Cambodia, the Temple of Apollo at Delphi in Greece, the rock shrines of Mount Miwa in Japan, Machu Picchu in Peru, Stonehenge in England . . .

And the Hypogeum of Hal Saflieni in Malta.

Imagine yourself at the entrance to an underground labyrinth with a footprint of half a square kilometre in the horizontal dimension measured out across three irregularly shaped levels stacked on top of one another in the vertical dimension – and the whole plunged in sepulchral darkness. This labyrinth, descending into the bowels of the earth, is the Hypogeum. It is thought by archaeologists to have been created earlier than 3000 BC. Some have speculated that its hive of interconnected chambers may first have begun to take shape naturally millions of years ago as solution cavities in the bedrock which were later expanded and reshaped by man. But the late J. D. Evans, formerly Professor of Prehistoric Archaeology at the University of London and a great authority on Malta, argues that the Hypogeum was entirely man-made from top to bottom and from the very beginning of the enterprise. Evans points out that even the crudest, most cave-like chambers exhibit certain features, 'such as the clever use of natural faults in the soft rock to provide ready-made walls and ceilings' that 'point to a human rather than a natural origin'.[1]

There is controversy about the Hypogeum, as we shall see. But one matter about which there has been no disagreement is that the people who carved it

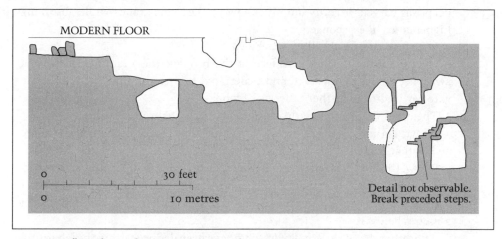

MODERN FLOOR

0 ├─────────────── 30 feet
0 ├─────────────── 10 metres

Detail not observable.
Break preceded steps.

Hypogeum floorplan and cross-section. Based on Evans (1971).

out were the same people who built the great megalithic temples like Gigantija and Hagar Qim above ground on the Maltese islands. Even the general architectural style of the rock-hewn features within the Hypogeum self-evidently belongs to the same 'school' as the free-standing temples. Indeed, fragments of pottery from almost all the recognized phases of the temple-building period – and even from before it in the so-called Zebbug phase thought to date back to 4000 BC have been excavated from within the Hypogeum.[2]

But next to nothing is known about the temple-builders themselves. We do not know what language they spoke. They have left us no script to decipher that might shed light on their rituals, customs, history and beliefs. There are no records elsewhere in the world from so ancient a period that refer to them. So their extraordinary works of art and stone that have endured the passage of the ages are now the only means we have to access all that is most interesting about them – in other words, their religious and philosophical ideas and the level of intellectual development of their culture.

The spaces within the Hypogeum, like the clover-leaf lobes of the megalithic temples, feel womb-like rather than strictly 'architectural'.

Some of the chambers were washed from top to bottom in red ochre, enhancing the organic effect.

Others were gracefully painted with spirals, disks, volutes, honeycomb-patterns, animal figures, hand-prints and ideograms – the majority in red ochre, a few in black manganese dioxide pigment.

Here a cavernous circular hall was hewn out of the bedrock.

There a 'window' was cut at eye-level into the wall of a passage and then an area beyond was hollowed out with infinite care to create an ovoid cist about the height of a man that can only be accessed through the window.

A few paces to the west along the same wall, an elliptical hollow a metre deep was carved. It eerily amplifies low-pitched voice tones while absorbing higher notes like a sponge.

Over here a graceful gallery was hewn.

Over there the rough, blank face of the bedrock was first chiselled into a sweeping curve, then carved and penetrated to create a lintelled megalithic gateway leading to further galleries beyond.

The lintel was painted with a pattern of twelve disks in red ochre.

Above, ceilings were cut here so lofty that they recede from view and there so low that you must stoop to pass beneath them.

Below, the floor was left rough in places, chiselled smooth in others, treacherous curbs and drops were created, and a stairway descending into the lowest depths was left hanging in mid-air after six steps down with a straight fall of 2 metres below it.

Altogether thirty-three major 'rooms' have been defined within the labyrinth. Of these eight are on the upper level, nineteen on the middle level and six on the lowest level. Some of the rooms have as many as four subsidiary chambers branching off them and multiple entrances and exits connecting to the wider network weaving through the entire edifice.[3]

The result, in the end, as we may still experience it, is a surreal underworld of stairways and chambers, galleries, pits, and tunnels interconnected with sinuous passages and shafts – like a game of three-dimensional snakes and ladders.

'No special importance was attached to it . . .'

I have explored the Hypogeum twice.

The first time was in June 2000 when it had been closed to the public for almost a decade (as with my entry to Mnajdra at dawn on the summer solstice in the same year, this private visit was arranged at short notice, courtesy of Reuben Grima of the National Museum).

My second opportunity came when I was in Malta in June 2001 with the Channel 4 film crew. Although the Hypogeum had been reopened by then, we were allowed to work in it out of hours under the benign supervision of Joe Farrugia, the curator.

There is ambiguous evidence that someone, or several people, might have entered some parts of the Hypogeum in the nineteenth century, and possibly even earlier in the seventeenth century,[4] but the official story today is that it was discovered in 1902 after being sealed off for millennia. Two blocks of houses were being built on the land immediately above it in the township of Paola. Bell-shaped water-tanks cut out of the bedrock were a standard feature of Maltese homes of the period and the discovery was made by construction workers cutting one such tank. They broke through into a rock-hewn chamber below the cistern and from there were able to enter 'the main halls of the monument'.[5] Subsequently other parts of the Hypogeum were also exposed as more cisterns were cut:

The builder did not report his discovery to the authorities immediately, but used the underground chambers as handy dumping grounds for stones and debris to save himself the trouble of carting away the useless material. When the houses were ready the owners in a casual way informed some Government officials of the existence of the Hypogeum. The place was visited, but being full of rubbish and swamped with water no special importance was at first attached to it. The Government, however, appointed a Committee to report on the discovery, and in 1903 the place became Public Property.[6]

The doctor and the Jesuit

The first scholar to visit the Hypogeum was the eminent Maltese medical man and polymath, Dr A. A. Caruana, who spent 29 December 1902 there at the request of the British authorities.[7] Caruana was not able to excavate, merely inspect, but he commented particularly on a rather macabre sight. The lowest level of the underground labyrinth proved to contain 'a great quantity of human skulls and bones . . . heaped over each other and at random'.[8]

In 1903 official excavations started under the supervision of Father Emmanuel Magri, a Jesuit priest and one of the members of the management committee of the Valletta Museum. Magri began by sinking a shaft deep into the rock to create the modern entrance to the Hypogeum in its middle level. All the rubbish left behind by the builders was then removed via this shaft. After that followed tonnes of 'dark dank earth' that seemed to have been deposited throughout the structure at some time in antiquity. According to contemporary observers, this deposit was uniformly 'full of fragments of bones, pottery and other small objects'.[9] The pottery and small objects were saved; the bones were placed in a heap for daily disposal by the works foreman and never heard of again.[10] Thus began a story of neglect, muddles and bizarre losses of prime archaeological evidence from the Hypogeum – a story that continues to the present day.

Soon after clearing the central chambers, Magri was called away by the Jesuits to save souls abroad and died suddenly at Sfax in Tunisia in 1907. He had not yet published any report on his work in the Hypogeum and the notebooks that he was known to have kept in which he had recorded the details of his excavations mysteriously disappeared after his death.[11] Perhaps the Jesuits have them.

The consequence at any rate, as David Trump admits, is that though most of the objects and pottery excavated by Magri have been preserved, 'no record of their context or associations survives'.[12] Since full details of provenance are essential if an informed archaeological judgement is to be made, or a chronological sequence proposed, the value of the finds is thus greatly reduced.

The godfather

After Magri came Themistocles (later Sir Temi) Zammit, the renowned 'godfather of Maltese archaeology', who was at this time Curator of the Valletta Museum. His careful and systematic excavations at the Hypogeum removed the remaining deposits uncleared by Magri, including the bone-filled earthy mass in the lowest storey which Caruana had noticed in 1902. The nature of this mass was described at some length by Zammit in the official report of his excavations published in 1910:

> A dark compact deposit was found which showed nowhere signs of having been disturbed. In this old deposit no stratification was observed and in caves which were cleared inch by inch, the deposit was always of the same type and contained objects of the same quality. The deposit of the large caves, about a metre in depth, was made of the red earth one finds in our fields and in this, bones and potsherds were intimately mixed ... disjointed and confusedly massed ... Very few bodies were found lying in a natural position and no special arrangements such as trenches, sepulchres, stone enclosures etc., were met with, anywhere, intended to receive a body.[13]

For example in one cave:

> Not a single [skeleton] was found lying with bones in position ... At least 120 skeletons were buried in a space of 3.17 by 1.2 by 1 m. This is enough to show that a regular interment was out of the question as not more than 12 bodies could be laid in such a limited space.[14]

In a separate publication in 1912, coauthored with T. E. Peet and R. N. Bradley, Zammit confirmed that:

> No complete skeletons came to light, and the bones lay in confusion through the soil as in the rest of the Hypogeum, except that occasionally an arm with fingers, and a complete foot, and several vertebrae would be found lying with the parts *in situ*. From the upright position of an isolated radius it might be judged that the filling up of the cave was of a wholesale nature, rather than that individual burials took place in it ... unrelated bones and also implements were found in the interior of skulls ... Animal bones were found mingled with human.[15]

Altogether, Zammit calculated, the skeletons of somewhere between 6000 and 7000 individuals lay tangled and mashed up together within the Hypogeum.[16] One of his students, W. A. Griffiths, who wrote a report on the excavations in *National Geographic* magazine in 1920, put a higher figure on the record:

Most of the rooms were found to be half filled with earth, human bones and broken pottery. It has been estimated that the ruins contained the bones of 33,000 persons . . . Practically all were found in the greatest disorder . . .[17]

Let's assume Griffiths' figure, not repeated elsewhere in the literature, is a mistake and stick with the lower total of 6000 to 7000 individuals. What were they doing there? And how (other than with howls of outrage and disbelief) are we to receive the official admission, already reported in chapter 15, that almost none of this vast horde of prehistoric bones has been preserved? Professor J. D. Evans was by no means overstating the gravity of the matter when he described the disappearance of the remains as 'an irreparable loss to Maltese archaeology'.[18] And that was in 1971 when the National Museum still had eleven of the Hypogeum skulls in its possession.[19] By 2001, as we've seen, only six were left.

Travel plans
June 2000

I first went to Malta in November 1999 because of the rumours of an underwater temple off Sliema reported in chapter 15. My dives that November were arduous and unproductive. But I kept an open mind and determined that I would return the following summer in better weather. I rarely plan things far in advance, but it was obvious that we should be there in June, and very specifically around 21 June – the summer solstice – in order to see the wondrous light effect, contrived by the ancients, that occurs at sunrise at the megalithic temple of Mnajdra. At least that was a sure thing, and worth making the journey for in its own right, even if the diving turned out, as I feared it would, to be a bust for the second time running.

Since solstice alignments usually work equally well on 20, 21 and 22 June (the sun's rising point in the east and setting point in the west hardly change at all during the entire three days), Santha and I scheduled to be at Mnajdra on the 20th and then to fly on to Tenerife in the Canary islands to observe some more solar magic on the 21st – this time at sunset – that had been reported in a group of mysterious pyramids in the little town of Guimar recently excavated by the explorer Thor Heyerdahl. We would meet Heyerdahl at Guimar for the very first shoot of my Channel 4 TV series on the 21st. Afterwards the film crew would return to England but Santha and I would stay on in Tenerife for a few days to check out claims by local divers to have seen 'strange things' underwater at several points around this volcanic Atlantic island – including 'towers made of huge blocks of stone' and a cross (also 'huge') formed by two straight channels intersecting at right angles and seemingly carved into a lava flow on the sea-bed at 27 metres.

From Tenerife the final leg of our June 2000 journey, now spilling into July, would take us to Alexandria in Egypt. There, as reported in chapter 1, we had arranged to meet Ashraf Bechai for 10 days of diving to see if we could relocate the parallel walls of giant regular blocks that he remembered seeing years before underwater off Sidi Gaber.

A temple, or a tomb . . . or something else?

What was the Hypogeum of Hal Saflieni *for*? Presumably its makers must have had a specific function in mind when they invested so much time, energy and human labour in its creation. But what?

J. D. Evans, the most influential of the group of archaeologists who made their names in Malta during the second half of the twentieth century, is reticent on this subject. Concluding a 15,000-word dissertation, which guides us through every room and corridor of the Hypogeum with all the verve, passion and originality of a refrigerator manual, he writes: 'This completes the description of the monument. A few words must now be said about its nature and purpose. In later years Sir Themistocles Zammit was of the opinion . . .'[20] We then get a summary of Zammit's opinions. In 1910, notes Evans, the great man had believed that 'the Hypogeum was in part used as a sanctuary in which religious ceremonies were conducted, and in part as a burial place in which the bones of the dead were deposited after being deprived of the flesh'.[21] In later years, however, he

> was of the opinion that it was an underground temple, roughly analogous in function to the stone-built ones above ground, though perhaps also used for special initiation rites, and that only at some later time was it used for the burial of the large number of people whose remains were found in it.[22]

And what of Evans' own opinion, set down in his authoritative 1971 survey, *The Prehistoric Antiquities of the Maltese Islands*: 'In point of fact, there is no cogent reason against, and much evidence in favour of, the primary use of the Hypogeum as a place of burial. It is its use as the locus of a cult which, if anything, may be secondary . . .'?[23] He only momentarily allows himself to speculate, but when he does so he gets interesting:

> Even admitting that a certain amount of cult activity must have gone on in the inner halls of the Hypogeum, the number of persons involved must have been very small. The Hypogeum was at no time a place of public worship, as the stone temples seem to have been. Had it been so the smoke of the flares and torches necessary to provide adequate light must have stained and blackened the porous limestone of the walls and ceilings, whereas in fact no traces of this can be seen. The Hypogeum was in all probability never fully illuminated in antiquity; its magnificently carved and painted halls were perhaps only half apprehended in a flickering and uncertain light by a few privileged or dedicated persons.[24]

Dr David Trump, another of the acknowledged experts on Maltese prehistory, speculates that the Hypogeum 'began as a simple rock-cut tomb [and] became elaborated to include a funerary chapel at its heart'.[25]

Colin Renfrew, in *Before Civilization*, describes Hal Saflieni as 'a great charnel house' but also notes: 'The main chamber has an imitation façade which almost certainly mimics the temples above the ground.'[26]

So some sort of a combination between a tomb and a temple, with perhaps just a smidgeon of dimly lit cultic or initiatory behaviour grafted on, seems to be a fair summary of the gamut of orthodox opinion as to the function of the Hypogeum.

The Goddess and the Sleeping Lady

Zammit, Evans, Trump and Renfrew do represent orthodox opinion on this matter. They're the heavy hitters. Centre Court at Wimbledon. In their league only the late Marija Gimbutas, formerly Professor of European Archaeology at UCLA, takes a divergent approach – and even she does not question the basic, seemingly obvious, assumptions that the Hypogeum was used as a burial place and that rituals of some kind must have been performed within it as well. She likewise accepts, without examination, the orthodox chronology for the construction of the labyrinth (3600–2500 BC).[27] For these reasons, though radical, her view is not so divergent from the mainstream position as it can sometimes appear. Rather, she works within the same framework but places less emphasis in her analysis on burial at the Hypogeum than on the cultic activities and initiation rituals that she believes were also performed there.

Gimbutas, who passed away in 2001, is one of the leading proponents of an intriguing hypothesis about who was who and what was what in prehistory. It concerns the distinctive carved and/or painted figures of enormously fat women that have been found in many European Neolithic sites (c.7000–4000 BC) and the almost equally numerous and *virtually identical* examples going far back into the world of Palaeolithic cave art (the Venus of Laussel, c.30,000 BC; the Venus of Lespugue, c.25,000 BC, etc.).[28] According to Gimbutas and others who have entered this fray, these figures are the symbols and representations of an archetypal 'Mother Goddess' figure – simultaneously the Goddess of Fertility, the Goddess of Death and the Goddess of Rebirth – whose worship was ancient and must once have been extremely widespread.[29] Whether we find her painted, carved in relief out of the rock wall of a cave (as in the celebrated example of Laussel), or in the form of a free-standing sculpture, the Goddess is usually represented as an imposing, hugely fat woman with dangling breasts, egg-shaped buttocks and bulging calves and forearms. It is therefore noteworthy that many figures exactly matching this description have been excavated from Malta's megalithic temples, including two in repose – usually referred to as 'the Sleeping Ladies' – that were found in the Hypogeum itself.

'The Hypogeum', notes Gimbutas:

with its rooms painted liberally with red ochre wash, represents the Goddess's regenerative womb . . . An indication of the religious use of these womb-shaped chambers are the figurines of Sleeping Ladies lying stretched out on low couches, associated with two cubicles opening into the Main Hall. The more articulate one, known as 'The Sleeping Lady of the Hypogeum', is a true masterpiece. This generously rounded lady with egg-shaped buttocks lies on her side, asleep, almost visibly dreaming. Why is she sleeping in the tomb? One explanation is that this represents a rite of initiation or incubation. To sleep within the Goddess's womb was to die and to come to life anew. The Sleeping Lady could also be a votive offering from one who successfully passed through the rite of incubation in the Hypogeum . . .[30]

I have stood before the Sleeping Lady of the Hypogeum many times. Her exact provenance within the labyrinth is not as simple a matter as Gimbutas thinks because she was excavated by the ill-fated Father Magri. All we know, and that is hearsay, is that she was found in a 'deep pit of one of the painted rooms'.[31] These days she occupies a glass case mounted on a slender plinth in a cubicle at the rear of the National Archaeological Museum in Valletta. The cubicle is dimly lit and the tiny clay figure, just 12 centimetres long, seems to float in space, sleeping if she is sleeping, dreaming if she is dreaming . . .

But can anyone really claim to know what was in the mind of the prehistoric sculptor who moulded her from clay, arranged the pleats of her figure-hugging midi-skirt over her ample thighs, and positioned her in lifelike repose upon an oval couch with her right hand wedged under her ear for a pillow and her left arm draped forward, supported by her huge breasts?

Now you see it, now you don't
Malta, 6–20 June 2000

During the two weeks we were in Malta before the June 2000 solstice we devoted an intensive week to diving. A Maltese friend, George Debono, supplied the boat – a small, comfortable cabin cruiser that is his pride and joy – and he, his son Chris and his sister Amy spent days with us tracking back and forth on the thankfully calm seas off Sliema. Dive support, tanks and refills were provided by Andrew Borg, a friend of George's and a top-flight diver who worked with us untiringly. We were lucky enough to have with us from Britain Tony Morse, a professional geologist and a PADI dive instructor in his own right. And Hubert Zeitlmair was on board as well, his confidence renewed each morning as we set out that that this would be the day on which we would relocate his missing underwater temple.

But we never did. We dived and dived and dived again yet we could not find it – as though it had dissolved in the sea or, like some magical castle, had the power to appear and vanish, appear and vanish . . .

In the Grail Castle Parsival fails to ask the right question and the Fisher King and his Knights and all the maidens of the procession, and the Holy Grail itself, and the castle too disappear without a trace. Was that what I did off Sliema? Did I fail to ask the right question?

I had certainly become over-focused on Zeitlmair's notion that his temple was on a sea-mount 3 kilometres from shore. That, at any rate, is what I kept us looking for, even though I remembered Shaun Arrigo insisting the previous November that the site he had filmed for Zeitlmair was not 3 kilometres out but just 1. I would have liked to conduct a thorough search at both distances. But the problem was that I could only afford to devote a few days to speculative diving around Malta – a week at the most – and it made better sense to investigate one area well than two areas badly. So I had to gamble. One kilometre or 3?

I liked the level of conviction Zeitlmair radiated that the temple ruins stood on a shallow spot surrounded by deep water and I felt reasonably confident that such a place (with or without a temple on it) did exist off Sliema. Part of it was the possible uncharted reef on the Royal Navy aerial photo that Zeitlmair had shown me at our first meeting in the Diplomat hotel. And more provocatively, although it is difficult to judge distances accurately at sea, the very first of my November 1999 dives seemed to have been in exactly the right place on a reef with exactly the right profile – which, unfortunately, I had not searched properly.

So surely all we needed to do was find that reef a second time, which shouldn't be too difficult, I reasoned, since we'd already found it once – get its GPS bearings and then search it thoroughly from end to end until we came to the temple.

But neither the temple nor the uncharted reef wanted to be found twice – at any rate obviously not by us. We abandoned the diving on the 14th. On the 15th I met Joseph Ellul and saw his original of Zeitlmair's aerial photograph and the press-clipping that he kept of Commander Scicluna's modest 1994 report of having found a temple underwater off Sliema. And this shifted my perspective on the whole problem. Because nowhere in Scicluna's understated letter to the *Sunday Times* of Malta had he said what distance from the shore he had been diving at when, in his own words, he had located 'a prehistoric temple . . . under 25 feet of water . . . at Sliema' (see chapter 15). It was Joseph Ellul's lively mind that had put the two things together – on the one hand, Scicluna's testimony and, on the other, the general location off Sliema of the 'reef' indicated in the aerial photograph – and it was Joseph Ellul who had concluded, not necessarily correctly, that the temple Scicluna had seen must be located on that reef. Zeitlmair had then taken the inquiry to the next logical stage by hiring the Arrigos to dive the site for him by proxy. And lo and behold, when they had done so they had found and filmed something that looked quite a lot like a temple.

But the opportunities for miscommunication between Zeitlmair with his heavily accented German English and the Arrigos would have been legion and the whole business of agreeing on the exact area in which to pursue the search would have been doubly complicated by Zeitlmair's blindness. Now, over two seasons, I had looked where Zeitlmair had said I should look, and dived where he had said I should dive – pretty thoroughly, I should add – and had failed to find his temple.

Was this because it wasn't there? I would have thought so if it hadn't been for Scicluna's letter. Or was it because we'd been looking in the wrong place? Maybe Zeitlmair and I should have listened more carefully to Shaun Arrigo in November 1999 when he'd insisted that the site was just a kilometre from shore.

More Fat Ladies

If the Sleeping Lady is a form of the Goddess then it is probably significant that two such figures were found in the Hypogeum while none have been found elsewhere . . . But other 'Fat Ladies' – sitting down or standing up, sometimes miniature and sometimes carved on a fairly grand scale out of limestone – were found by the excavators at all the major megalithic temples of Malta. The original of one of these sculptures, from Tarxien (a replica remains on site at the temple) has been moved to the Museum and dominates the room next door to the two Sleeping Ladies. This obese figure is reckoned by Colin Renfrew to be 'the earliest colossal statue in the world'.[32] David Trump believes that she must surely, from her 'size and position', be 'the Goddess herself'.[33]

> When complete she stood about 2.75 metres high, but time, weather and above all the local farmers have reduced her to waist height . . . She wears a very full pleated skirt. It would be ungentlemanly to quote her hip measurements, and her calves are in proportion. She is supported, however, on small, elegant but seriously overworked feet.[34]

The section of the Museum overlooked by the Tarxien colossus is lined with long glass panels. Arranged behind these, like Bangkok prostitutes, a harem of Fat Ladies in varying stages of undress lounge and slouch – all of them disconcertingly headless (although no significance should be placed on this since the evidence suggests that the heads have simply been lost with the passage of time).

The group includes figures from the temple of Hagar Qim thought to date to around 3000 BC retrieved from a strange cache, a time capsule, found 'secreted under an inner threshold step'.[35] Of particular note are the so called 'Seated Goddess' and the 'Venus of Malta'. The former, 23.5 centimetres high,[36] has luxuriously corpulent hips, buttocks and thighs; her ankles are crossed in front of her – crossing the legs would be impossible for a person so fat – and her bulging arms are folded. The Venus of Malta, 13 centimetres high and fashioned from clay,[37] has been praised by many observers for its anatomical exactness and 'startlingly realistic style'.[38] Again, the Mother Goddess attributes of huge breasts and thighs are unmissable.

The remaining figures on display are summed up nicely by David Trump:

> Some are standing, naked or wearing only a pleated skirt, others also skirted, seated on some kind of stool, with legs to the front, yet others naked with the legs tucked up to one side. One or both arms are usually across the chest, the other may hang at one side.[39]

Origins in the Palaeolithic?

I have never visited any of the painted caves of Palaeolithic Europe – Lascaux, Chauvet, Laussel, Peche Merle, Lespugue, Altamira, Cosquer, and dozens upon dozens of other sites – although I still hope, in this lifetime, to have the opportunity to do so. The majority are permanently closed to the public with no possibility that they will ever be reopened and in some cases, as at Lascaux, there is even a long waiting list for access to the (apparently rather good) walk-through model that has been built near by. But I recoil at the idea of touring a model and don't think it is necessary to do so, or even to be an 'expert' on the extraordinary artistic achievements recorded inside these caves, to recognize that the Venus figures found there – dating back as far as 30,000 BC – do bear close comparison to the big-breasted, big-hipped Venuses of Malta, the 'Fat Ladies' represented again and again in the megalithic temples, and the Sleeping Ladies of the supposedly Neolithic Hypogeum.

My choice of the word 'supposedly' here is deliberate. The Hypogeum is *supposedly* – not definitely – a Neolithic structure.

However, it has been *assumed* to be Neolithic since its discovery and has been regarded as securely dated – to between 3600 and 2500 BC – since the introduction of calibrated radiocarbon-dating more than a quarter of a century ago.[40] The habit of viewing it in the Neolithic time-frame is therefore deeply ingrained and not a single scholar within the mainstream has considered the alternative possibility that is suggested by the Mother Goddess figures, the cave-like subterranean labyrinth, the use of red ochre and black manganese pigment – and many other curious and notable features. This is the possibility that the Hypogeum, or parts of it, as well as the ideas and symbolism it enshrines, might have been misdated to the Neolithic 5000 years ago – might in fact date back to the Palaeolithic more than 10,000 years ago.

It is thanks solely to the efforts of three determined Maltese scientists, all medical doctors with a deep and abiding 'amateur' interest in prehistory, that this electrifying possibility, brushed under the carpet for a century, is today on the agenda for serious discussion.

Anton Mifsud is senior consultant in Paediatrics at Saint Luke's Hospital, Malta, and President of the Prehistoric Society of Malta. His son, Simon Mifsud, is a senior registrar in Paediatrics at the Gozo General Hospital. Charles Savona Ventura is a consultant in Obstetrics and Gynaecology at Saint Luke's Hospital, Malta. Together and separately they have presented a devastating critique of the comfortable archaeological consensus, reported in the last chapter, that the Maltese islands remained entirely *uninhabited* by human beings until around 5200 BC.

Recently, to their credit, some archaeologists have begun to pay attention and to do so publicly. Writing in 1999, for example, Anthony J. Frendo had this to say:

> The earliest human inhabitants on these islands are currently thought to have come here around the end of the sixth millennium BC during the Neolithic period. This quasi-dogmatic stance was severely put to the test when Anton and Simon Mifsud claimed that this date had to be pushed back to a much earlier period, namely the Palaeolithic.[41]

After reviewing the detailed findings presented in their 1997 book *Dossier Malta* Frendo concludes that the Mifsuds' claim, though revolutionary, is in fact correct and that their work has proved 'beyond any reasonable doubt' that human beings were present in Malta during the Palaeolithic as early as 15,000 to 18,000 years ago and that 'Malta's history is thus extended backward by eight millennia'.[42]

Reopening the question of temple origins

As Frendo is the Head of Department and Senior Lecturer in Archaeology at the University of Malta, this is no lightweight endorsement. If it is supported by other archaeologists – and it becomes broadly accepted that there were indeed humans on Malta after roughly 15,000 to 18,000 years ago – then the result, ultimately, can be nothing less than a complete rewrite of Maltese prehistory.

In chapter 18 we will weigh up the hard empirical evidence that underwrites the Mifsuds' case. Meanwhile, I doubt whether archaeologists have yet properly understood the ramifications of their profession's inevitable (and I suspect imminent) official adoption of the much earlier date of first human habitation that the Mifsuds propose. At any rate, if they have understood, I see no sign of it in the literature other than Frendo's monograph.

For example, isn't it obvious, once the presence of Palaeolithic humans in Malta is widely acknowledged, that this must force a radical revision of the perspective from which the Hypogeum and the megalithic temples like Gigantija, Hagar Qim and Mnajdra have traditionally been viewed? For even if further investigation reconfirms the conventional wisdom that these great structures were indeed built in the Neolithic between 5600 and 4500 years ago, the proof of a Palaeolithic presence in Malta must raise question-marks over the obviously sophisticated and well-developed *architectural heritage* that all the temples incorporate and express from the outset. It would no longer be entirely safe, or logical, to look exclusively outside Malta for the origins of the skills, knowledge and ideas invested in them – e.g. as part of the intellectual baggage carried by the presumed first settlers (the so-called 'Stentinello culture', thought to have arrived from Sicily 7200 years ago).[43] On the contrary, an accepted Palaeolithic presence would raise the possibility that the temple heritage was *not* an import from Sicily but was instead the product of very long *in situ* development in Malta itself – perhaps in parts of Malta that have so far evaded detailed archaeological scrutiny and particularly in areas that have been submerged by the sea.

This is emphatically not to suggest that the wave of Neolithic settlement which archaeologists have detected in Malta around 7200 years ago did not occur – because it certainly did! It is to suggest instead a parallel hypothesis (my own, not the Mifsuds', I hasten to add) that when Neolithic settlers first entered Malta from Sicily 7200 years ago they may have encountered the remnants of a much older, pre-existing culture which possessed and gradually passed on the secrets of how to build and align the temples.

Let's not even dignify such wild speculation with the label 'hypothesis'. Still it seems to go some way towards resolving the paradox noticed by David Trump that 'though building in stone was introduced to Malta by the first settlers . . . the use of huge blocks, so-called megalithic architecture, is not known before the temple period'.[44] Could this be because the stone-working culture of the 'first settlers' was fundamentally different, and inferior, to an architectural tradition that already existed in Malta before their arrival and which was the true author and ancestor of the Maltese megalithic temples?

17 / The Thorn in the Flesh

> We amateur archaeologists do it for the love of it, and the excitement and
> adventure, whereas the so-called professionals are caught up in the ruts of the
> establishment. Above all, they have no right at all to claim any monopoly of
> interpretation.
>
> Anton Mifsud, July 2001[1]

Malta, 16 June 2000

Anton Mifsud is in his early fifties, of medium build, olive-complexioned, heavily tanned, with a lot of experience and humour and a nice combination of strength, tolerance and intelligence in his face. He is exceptionally open-minded and lateral-thinking by nature – telling me once that he didn't automatically dismiss *any* idea, even if it seemed absurd. The point, he said, was to submit problems in history and prehistory to rigorous inquiry, find out the facts about them and then draw the conclusions indicated by those facts.

I first met Anton on 16 June 2000 when he signed my already much annotated copy of his explosive little book *Dossier Malta*. Just two days previously, on the 14th, I'd concluded that I wasn't going to throw any more money into diving off Sliema. We'd looked, it hadn't worked, the temple didn't exist, and Malta didn't love me.

Then on the 15th I met Joseph Ellul and read Commander Scicluna's letter. So by the 16th, when Anton Mifsud came to visit me at the seafront apartment Santha and I had rented in Sliema, I was already more upbeat about the prospects of an underwater discovery than I had been for several months. I'd also recently acquired and carefully read *Dossier Malta* and begun to digest the implications of Mifsud's research, hitherto unknown outside Malta.

Accompanying Anton that day was Charles Savona Ventura, with whom he has co-authored several books. He's a big bear of a man who looks like a Mexican bandit and is a mine of information about Maltese prehistory.

How, I found myself wondering, had these two obviously busy and successful consultants in hospital medical practice managed to keep their day jobs and learn so much about the past as well? Because clearly they were not just interfering 'amateurs' in the world of archaeology . . . You only had to listen to them for two minutes to realize that they knew their stuff.

Malta: echoes of Plato's island
Malta, 16 June 2000

As the conversation unfolded, Mifsud and Ventura got round to telling me about the latest slice of provocative unorthodox prehistory they were working on – *Malta: Echoes of Plato's Island*[2] – which would argue that Malta is a remnant of the lost island of Atlantis.

'You're not going to like our date for the flood, though,' said Mifsud, who had read *Fingerprints of the Gods*, in which I first began to set out my case for a lost

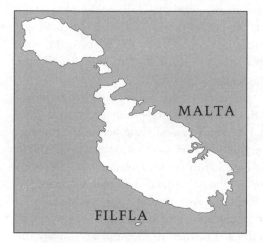

civilization destroyed at the end of the Ice Age more than 12,000 years ago – a lost civilization of the Palaeolithic, in other words.

'Why won't I like it?'

On the one hand, Mifsud explained, he had strengthened and added to his evidence for a human presence in Malta during the Palaeolithic in the three years since the publication of *Dossier*. On the other, however, his new research for *Echoes* (with Charles Savona Ventura and two other co-authors) had led him to a distinctly *non*-Palaeolithic date for the deluge that he believed had destroyed a formerly much larger Malta – the prehistoric Malta that was, in his scenario, the source of the Atlantis myth.

Reduced to its barest essentials, Mifsud's proposal is that a great land-bridge that once joined Malta to Filfla collapsed cataclysmically through faulting of the submarine Pantelleria Rift at around 2200 BC.[3] He links this event, which would have generated massive tidal waves capable of flooding the entire archipelago, to the sudden demise of the temple-building culture that is well attested in Malta's archaeological record at the end of the third millennium BC.[4] And, in an elegant argument, he suggests that it was this lost megalithic culture, and its overnight destruction by earthquakes and floods *c.*2200 BC, that was recorded in ancient Egyptian annals, passed on to the Greeks, and in later times remembered as 'Atlantis'.[5] Mifsud points out that the relative chronologies for ancient Egypt and Atlantis given by Plato – with the latter said to be a thousand years older than the former[6] – coincide with the relative chronologies for ancient Egypt and Malta (the former began to build with megaliths in the Pyramid Age *c.*2600 BC; the latter began to build with megaliths a thousand years earlier at Gigantija, *c.*3600 BC).

'You're quite right,' I told Anton after I'd thought through his reasoning, 'I don't like it at all.'

As he looked at me expectantly, I raised my left hand and began to enumerate the counter-arguments on my fingers.

'Firstly, there's the issue of the relative chronology. To make your argument work – I mean about the megalithic civilization of Malta being a thousand years older than the megalithic civilization of ancient Egypt – you have to buy into the orthodox archaeological datings for both places. But you ought to be the first to know that orthodox archaeological datings may not always be correct. In the case of Egypt we have actual structures, such as the Sphinx and the megalithic temples beside it, which may be much older than the third millennium BC[7] – I'm sure you're familiar with the debate. There's the megalithic stone circle at Nabta, 200 kilometres west of Abu Simbel, which is at least 7000 years old.[8] And then there are the accounts of the ancient Egyptians themselves – the Abydos King List, the Turin Papyrus and so on – which trace the origins of their civilization back 30,000 years into the past. Again, your relative chronology only works if you accept the orthodox position that all such accounts are baseless fictions – which I certainly don't.'

Secondly, I continued, Anton's argument involved not taking Plato seriously on the epoch in which he had set the Atlantis events – supposedly 9000 years before Solon's time, i.e., 9600 BC, i.e., about 11,600 years before the present, i.e., the end of the Palaeolithic. And I could see no good reason *not* to take Plato seriously on that – indeed, he could have hardly set his global deluge (described as affecting both the Mediterranean and the Atlantic) in a more flood-prone and cataclysmic epoch than the end of the Ice Age around 11,600 years ago. To conclude that Plato had not meant 9000 years before Solon's time (9600 BC) but 1600 years before Solon's time (2200 BC) seemed to me arbitrary, to say the least.

Thirdly, the notion inherent in Mifsud's reasoning that Plato must have been speaking of the Mediterranean west of Malta when he referred to the 'true ocean' leading to an opposite continent seemed to me to be highly suspect. I told Anton I was convinced that when Plato said this ocean was the Atlantic and placed Atlantis in it 'opposite the Pillars of Hercules' he knew exactly what and where he was talking about. So to my mind this on its own made Malta in the central Mediterranean an unlikely candidate for Plato's island.

But I hastened to add – and not just out of politeness – that none of this meant Mifsud was necessarily wrong. I could be the one who was wrong. Atlantis had been placed at other sites in the Mediterranean by other scholars – also at relatively late dates.[9] And it had been placed almost everywhere else in the world from Indonesia to the South Pole.[10] I happened to be one of those who believed in taking Plato as literally as possible – if I was going to take him seriously at all – but I recognized the validity of other approaches.

Terminus ante quem
Malta, 16 June 2000

Besides, there was no real contradiction between our positions – for the simple reason that we seemed to be talking about entirely different things. To state, as Anton did, that Malta underwent a flood/earthquake cataclysm about 4200 years ago neither weakened nor strengthened my proposition that it would also have been subject to flood cataclysms – probably several times – during the meltdown of the last Ice Age between 17,000 and 7000 years ago. Likewise, Anton's belief that a land-bridge between Filfla and south-west Malta collapsed through rifting processes 4200 years ago in no way contradicted the well-established fact that Malta's north-eastern coast was connected to Sicily by a 90 kilometre land-bridge before it was swallowed up by the rising seas at the end of the Ice Age.

Both periods are interesting for different reasons. But, I pointed out, Anton's own research indicated that there had been human beings on Malta during the period of the great Ice Age floods at the end of the Palaeolithic. And although he seemed to accept the orthodox radiocarbon 'sequence' and chronology for the temples and the Hypogeum (of 3600 BC down to about 2200 BC), hadn't he himself also written in *Dossier Malta* that:

> The *terminus ante quem* Carbon-14 dates given for these sites . . . are of *before* such and such a year in the Neolithic; whether this period was a year or several centuries cannot be established by the Carbon-14 date alone. The most logical explanation is successive utilization of such sites initially by Palaeolithic and subsequently by Neolithic Maltese.[11]

And a few pages later:

> Since the megalithic temples have been assigned a *terminus ante quem* carbon date of before 3000 BC, nothing precludes that they were a carryover of a tradition which had started in the Palaeolithic. Indeed the bas relief images of bulls and a cow on the large blocks of stone lying just outside the Tarxien temple complex are themselves diagnostic features of Palaeolithic art.[12]

I could sense Mifsud's reluctance as a rigorous scientist to get drawn into idle speculation. But surely he was aware of the general direction in which his arguments were tending? If he was saying that there had been humans on Malta in the Palaeolithic, which he certainly was, and if he was suggesting that these Palaeolithic humans had initiated the development of the megalithic temples, which, again, he certainly was, then weren't the floods at the end of the Palaeolithic of at least as much potential significance to Maltese prehistory as the floods and earthquakes that might also have occurred 4200 years ago?

Skeletons in the Hypogeum

Anton Mifsud's attack on the orthodox chronology and interpretation of prehistoric Malta is made across several different fronts and sometimes produces contradictory data. This doesn't seem to bother him. Once launched on an inquiry, he pursues the quest for data ruthlessly, as an end in itself, not to support particular arguments or positions.

In the case of the Hypogeum, Mifsud's approach, at first, was not directly concerned with chronology. Poring with the eye of a doctor over the early excavation reports of Zammit, Bradley and others, he was puzzled by what they had to say about the state of the human remains found inside the labyrinth.

In summary, as we've seen, all the excavators and all subsequent archaeologists propose slightly different versions of the same theory that this great mass of remains had been *ritually buried* in the earthy matrix that was found filling the Hypogeum's lower levels to a depth of about a metre when it was opened. Yet the instinctive reactions that they set out in the original reports that Mifsud had assembled, referred to in chapter 16, show that they were clearly startled, and in a few cases troubled, by the complete chaos and disorder in which the bones were found, commenting, for example: 'From the upright position of an isolated radius it might be judged that the filling up of the cave was of a wholesale nature.'[13] But how do we explain such a 'wholesale' filling-up of the Hypogeum with the remains of thousands of human bodies, all seemingly just dumped there 'in a haphazard way'[14] with no anatomical disposition? Isn't it a bit of a mystery?

Not according to the archaeologists who say they've seen mass 'catacomb-style' burials before in other parts of the world and on these islands – for example at Burmeghez, a natural cave in Malta,[15] and at the Borchtorff Circle on Gozo, where rock-cut subterranean tombs encircle megaliths.[16] So that makes this sort of 'funerary behaviour' part of a pattern that legitimate experts in the subject can already claim to understand. None of them would deny that the Hypogeum's labyrinthine character is utterly different from the rock-tomb character of the other sites, or that the bones it contained were indeed in such an extreme state of disarrangement that any form of 'regular interment was out of the question'.[17] But the problem can easily be resolved within the prevailing 'burial-place' paradigm by proposing that excarnation – i.e., the removal of the flesh from the bones – was practised before interment and that the Hypogeum must therefore have been 'a burial place in which the bodies were laid or heaped as skeletons'.[18]

Oh really? Heaped up and tossed about so casually? As Mifsud counters, the Hypogeum cannot be legitimately compared to either of the two other significant sites of prehistoric mass burial in Malta:

> At Burmeghez there is a predominance of anatomical relationship between body parts, a left-sided flexed position of the body, an orientation along the main axis of

the cave, and, by way of a lithic assembly, a stony arrangement [large, purposefully laid slabs] protecting the upper body parts . . .[19]

Like at the Borchtorff Circle, all the burials are of an evidently (and uncontested) ritual nature and were disposed in two phases – a pair of rock-cut tombs with a shared central shaft dated to the Zebbug phase[20] (about 4000 BC – some centuries before the supposed beginning of the temple period at Gigantija c.3600 BC), and further subterranean rock-cut tombs of the Tarxien phase arranged in an approximate circle around a subterranean 'megalithic assembly'.[21]

In fact, the only site in Malta that Mifsud regards as comparable to the Hypogeum of Hal Saflieni in general appearance – and that was found on excavation to contain exactly the same sort of chaotic deposit – is the nearby Hypogeum of Santa Lucia (less than a kilometre away) that was excavated in the early 1970s and has since been sealed up, presumably for ever – at least so the authorities must intend – because it has been covered over by a modern cemetery.[22]

Mifsud describes the Santa Lucia Hypogeum as:

> a smaller version of that at Hal Saflieni, with a megalithic entrance and an internal architecture similar to the temples above ground. The deposit inside this hypogeum consisted of human remains admixed with Neolithic pottery and amulets, in a matrix of red earth soil; the context is similar to that at Hal Saflieni. In the words of the Director of Museums at the time, the deposit inside the Santa Lucia Hypogeum was *'as if the mass had been dumped inside the monument from the surface'*. F. S. Mallia could not have been more precise, and the close proximity of the two hypogea enhances even further a similar mechanism operating in both monuments in the creation of the deposit in question.[23]

Which brings us to the heart of the matter. Since Mifsud clearly does not believe that the carpets of disarranged human bones littered and dumped inside the hypogea of Hal Saflieni and Santa Lucia arrived there as a result of burial, then what 'mechanism' does he think was operating?

A flood

Like other good ideas that no one has ever had before but that everybody immediately gets the point of once the secret is out, Mifsud's explanation for the mass of bones inside the Hypogeum is extremely simple:

> The accumulation of human remains at the Hypogeum in Hal Saflieni were not related to primary ritual burial, but were brought down into the Hypogeum labyrinth through the action of floodwater in a matrix of red earth and soil.[24]

The first and most obvious evidence for this novel hypothesis comes in the massively disordered nature of the remains described in the excavation reports.

The presence of these disarticulated, non-anatomically disposed remains in an entirely 'unstratified' deposit 'made of the red earth one finds in our fields' that was 'always of the same type and contained objects of the same quality', cannot in Mifsud's view be explained by *any* form of deliberate burial – with or without prior excarnation. Only one agency, he argues, is capable of creating such a conglomeration in an unstratified earth matrix in which 'the same quality of shards were found on the surface, at the bottom and in the space in between',[25] and in which 'fragments of shards in parts of the Hypogeum fitted other fragments deposited in other caves far away'.[26]

That agency is a massive flood – and such events, from varying causes, have not only been known to occur in the Maltese islands but also have left distinct traces of their passage in the form of animal and human bones, as well as assorted other materials, all muddled up together and evenly spread throughout deposits of silt or earth trapped inside caves or rock fissures. The classic example is Ghar Dalam, an extensive natural limestone cave near Birzebbuga in eastern Malta, which contains six distinct layers of flood deposits swept into its depths at different periods over the last 200,000 years. Exactly as in the Hypogeum, notes Mifsud, the organic remains in Ghar Dalam 'were not distributed in an anatomical manner as they would have been in a ritual burial, but they were dispersed in random fashion inside the stratum of earth they lay in'.[27]

What Mifsud is proposing in the case of the Hypogeum, therefore, is a one-off, one-time deluge that swept over the surrounding fields and habitations, and finally over a great surface-level necropolis that then existed in the area, carrying away all its mouldering dead in one fell swoop and dumping their skeletons and their grave goods, promiscuously mixed with fragments of pottery, the bones of large and small animals (including those of frogs and hedgehogs)[28] and a motley collection of other objects, into the nearest possible sinkhole – in this case the Hypogeum itself.

Moreover, Mifsud believes that this was the same flood – caused by the collapse into the sea of his proposed Filfla land-bridge and the resulting *tsunami* – that brought the temple-building culture of Malta, and all its activities within the Hypogeum, to an abrupt and permanent halt c.2200 BC. Since carbon-dates from Malta are as scarce as ice cubes in hell, as we shall see, it is interesting that the first ever radiocarbon-dating of the Hypogeum's few surviving human remains – carried out in 1999 – does place them at the end of the Tarxien phase, c.2200 BC just as Mifsud argues.[29] This new evidence from the Hypogeum, he concludes, further strengthens 'the feasibility of a sudden cataclysm accounting for the sudden termination of the Tarxien people'.[30] And he points to the well-known fact that the Tarxien temple itself was sealed at the end of the late Tarxien phase under a metre-deep layer of sterile silt.[31] After several centuries of abandonment a new culture then appeared – one that had nothing to do with the temple-builders – and began building – on top of the silt layer.[32]

Old Stone Age

Although nobody in the world of archaeology seems to have noticed yet, the late Tarxien date (of between 2470 BC and 2140 BC)[33] for the Hypogeum's human remains contradicts the long-established convention, entered into dogma by J. D. Evans, that 'the primary use of the Hypogeum [was] as a place of burial'. Since all archaeologists accept that the construction of the labyrinth began significantly earlier than 3000 BC, perhaps as early as 3600 BC (and since it even contained pottery of the Zebbug phase prior to 4000 BC), its 'primary' purpose can hardly have been to receive human remains that were not deposited in it until around 2200 BC (whether or not one accepts that they were deposited there by flood). It must, therefore, have had some other quite distinct function at the time of its origin – a function that scholars may hitherto not have guessed, since no serious attempt has ever been made to investigate alternatives to the burial scenario.

It may have been an underground version of a temple, of course – and its 'temple-like' features have always been recognized – but if so, why is it such a unique and unusual temple? Why does it need all those winding corridors and levels and cists secreted within rock walls, and spooky sound-effects and red-painted chambers, and falls and traps?

Whatever its purpose – probably it will never be known fully, or known at all – certainty that the Hypogeum was *not* primarily designed as a central place of burial for the dead of the temple-building culture as had hitherto been thought left the way open for Mifsud to explore other possibilities about its function and identity. And about its age. Because, as with the megalithic temples, so with the rock-hewn labyrinth, the argument of *terminus ante quem* applies, and nothing rules out the possibility that the Hypogeum may be a 'carryover of a tradition which had started in the Palaeolithic' – to quote again Mifsud's own words, cited earlier. By the same token the flood that he believes swamped it with bones and debris at around 2200 BC tells us nothing whatsoever about the origins and antiquity of the structure itself – only that it was already there to be flooded in 2200 BC (and certainly not how long it had existed before that).

Long but ultimately relevant excursion to two potentially unpronounceable temples

Maltese is a lovely, lilting language to hear. Structurally it belongs to the Semitic family and is thus closely related to Arabic and Hebrew – indeed, Maltese friends tell me that their language and Arabic are often mutually comprehensible without need for interpretation. Modern Maltese also includes great numbers of Indo-European loan words that, for historical reasons, come mostly from Italian and English. Written Maltese uses the Latin alphabet but the pronunciation of the letters is often quite unusual in order to allow full expression of Semitic and uniquely Maltese cadences in speech. Thus Hagar

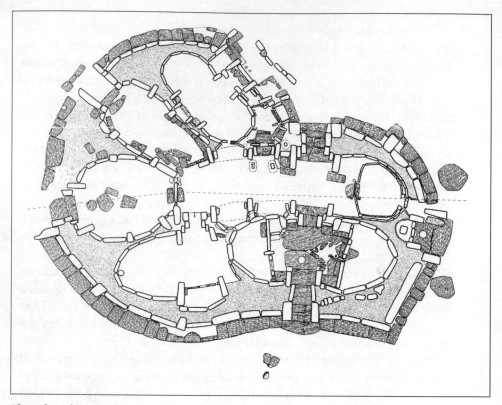

Floorplan of Hagar Qim temple. Based on Evans (1971).

Qim is pronounced something like *Hajar-iim*, Tarxien is *Tarshien*, Mgarr is *Umjaar*, Zebbug is *Zebooj*, and the potentially unpronounceable Mnajdra reaches the ear as a soft and mellifluous *Munaidra*.

About 8 kilometres south-west of the Hypogeum, but separated from one another by less than a kilometre, the temples of Mnajdra and Hagar Qim stand on Malta's south coast, overlooking a spectacular panorama of deep blue sea and Mediterranean sky in which the craggy little island of Filfla – blasted to smithereens by centuries of artillery practice – floats like a mirage. By night, roofless (though thought to have been roofed in antiquity),[34] they gaze up at the wheeling constellations and take an interest in the peregrinations of the moon. By day they use a variety of shadows, peep-holes and cunningly contrived alignments to follow and to record the path of the sun.

Hagar Qim is the higher and northernmost of the twin temples. Occupying a flattened promontory of glaring white limestone, it is thought to have been built between 3500 and 3300 BC.[35] As with other surviving sacred architecture of archaic Malta, it seems to abhor straight lines, seducing the eye with patterns of curves and waves. Its flowing perimeter, flung out in a great irregular ellipse, is defined by a picket of enormous upright megaliths, deeply gnarled and weathered, some laid side-on, some face-on, some broken, some missing, some restored. What seems like its primary entrance, framed by an imposing trilithion, is on the south-eastern side of the structure in a gently concave section of wall made of large, finely fitted blocks. On the north side, to the east of a second trilithion, a narrow, tapering monolith, like a chimney or an obelisk, towers 7 metres tall; in the very top of it, only visible from a helicopter or a crane, is a carved basin, function unknown.

Inside the temple there are the usual clusters of lobed, egg-shaped rooms arranged in pairs – but I will not describe these further here other than to refer the reader to the relevant plans and photographs. With the notable exception of their astronomical and solar alignments, which were deliberately and precisely hard-wired into the architecture and from which certain deductions may legitimately be made, all ideas of function that have been proposed for them, and for the rooms of Malta's other temples, are entirely speculative. For example, we might say that this feature here is an 'altar', that that feature in the wall over there is an 'oracle hole', while this one in the floor at our feet is a 'libation hole'; that here the priests met in convocation; that there public gatherings were held . . . and so on and so forth. But it would all be guesswork, fantasy, invention. Since we don't have the texts of the temple builders, the truth is that we don't know why they built the temples, or why they built them with megaliths (rather than smaller, more manageable stones), or how these structures were used, or even if they were 'temples' at all in anything like the traditional meaning of the word.

Hagar Qim offers several alignments on the summer solstice. One, at dawn, is on the north-east side of the structure, where the sun's rays, passing through the so-called oracle hole, project the image of a disk, roughly the same size as the perceived disk of the moon, on to a stone slab on the gateway of the apse within. As the minutes pass the disk becomes a crescent, then elongates into an ellipse, then elongates still further and finally sinks out of sight as though into the ground. A second alignment occurs at sunset, on the north-west side of the temple, when the sun falls into a V-shaped notch on a distant ridge in line with a foresight on the temple perimeter.

I suspect in some way connected with astronomy is an object, unknown from any other site in Malta, in Hagar Qim's south-western apse. Described as 'a mysterious column altar',[36] it is a smoothly hewn white limestone pillar, almost circular in cross-section, with a circumference of about 1 metre and a height of 1.5 metres. The pillar stands upright within the curve at the south-western end of the apse – which has been identified as an 'inner sanctum'[37] – so it seems to have been accorded a special significance.

Were such an object to be found amidst the ruins of a south Indian temple it would instantly be recognized as an ancient *Sivalingam*, the symbol and the manifestation of the god of knowledge, measurement and astronomy. But India

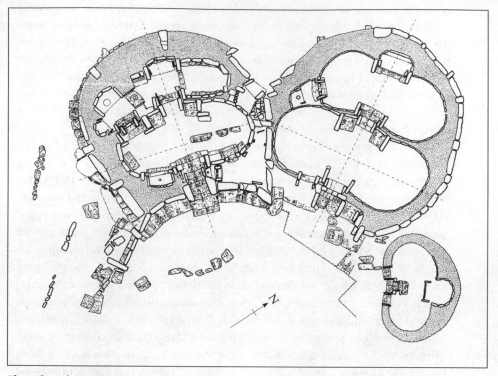

Floorplan of Mnajdra. Based on Evans (1971).

is among the few places in the world where a culture of vast antiquity is still alive today. In Malta the thread connecting the present to the past is broken and the voices and ideas of the temple-builders have not been heard for millennia . . .

Below the promontory on which Hagar Qim stands, the land falls away steeply in a south-westerly direction towards Mnajdra and the sea. It is rough land, heavily overgrown by wild thyme in the summer, with knolls and ridges of bedrock poking everywhere through the thin topsoil. These days, however, the walk down takes less than ten minutes on a concrete footpath installed by the Museums Department – which is in overall charge of the two sites.

Mnajdra is not one temple but a complex of three. Of these the easternmost, with three delicate apses disposed as a clover leaf, is the smallest and is believed to be the oldest – about 3600 BC, the same period as Gigantija. Archaeologists think that the westernmost, 'lower' temple was built next, around 3400 BC. Finally, at around 3200 BC, the middle – or 'upper' – temple was squashed in between its elder predecessors.[38]

All are megalithic and all demonstrate a very high degree of architectural, engineering and mathematical competence on the part of the builders, but the lower temple is particularly imposing, with several courses of cyclopean masonry still intact on top of enormous dressed boulders at ground level. It was described in 1993 as the 'best preserved of all the Maltese temples'.[39] How long it can remain so is open to question, since the Museums Department's custody of the site has not yet run to the provision of full-time night-watchmen. In consequence, in 1996 and again in April 2001, Mnajdra was severely vandalized – at night – by well-organized gangs of men armed with spray paint and sledge-hammers. It beggars belief that this could have been allowed to happen – even once – on an archaeological site of acknowledged global importance that is more than 5000 years old. But for it to have happened twice?

In such ways, either by accident or by design, Malta rends and devours her own past.

This cannibal feast shows no sign of coming to an end, and, although the megalithic temples are strong and massive none of them can last for ever. As noted in chapter 15, the archaeologist David Trump recognizes twenty-three groups of ruins in the Maltese archipelago as 'classifiable temples'. But he also comments on the existence of at least twenty further 'scatters of megalithic blocks . . . which could represent the last vestiges of former temples' and accepts that we can never know how many there may once have been.[40]

So if the raw materials that the archaeologists have to work with in order to understand and date the temple-building culture have been so radically reduced – reduced almost by half from forty-three to just twenty-three sites (not to mention other sites that may have disappeared completely with the passage of time, or await discovery underwater) – then how can we be sure that their interpretation of Maltese prehistory is correct? And if it is difficult enough to explain how twenty-three megalithic temples appear with no architectural antecedents at the dawn of history, on tiny arid islands that had only been inhabited for 1600 years, then how much harder it is to account for forty-three of them.

Perhaps the answer lies in the Palaeolithic.

How to tell your Palaeolithic from your Neolithic

Palaeolithic is one of those supposedly exact 'scientific' terms in anthropology and archaeology that promotes inexact thought. Meaning 'Old Stone Age', it is defined – arbitrarily – as having come to an end 12,000 years ago, and to have been followed by the Neolithic, 'New Stone Age', from 12,000 years ago (10,000 BC) onwards. After about 7000 years of Neolithic culture, the metal 'ages' of copper (roughly third millennium BC), bronze (roughly second millennium BC), and iron (roughly first millennium BC) then followed.

In summary, the term Palaeolithic is generally applied to all human remains and activities prior to 12,000 years ago while the term Neolithic is generally applied to all human remains and activities between roughly 12,000 years ago and 5000 years ago. However, on closer examination it turns out that the definitions are not purely chronological – since it is possible to find certain isolated societies that may be said to be at a 'Palaeolithic' or more often 'Neolithic' stage of development even today.[41]

Thus, as well as referring, somewhat vaguely, to periods of prehistory, 'Palaeolithic' and 'Neolithic' are also terms that say something about the lifestyles of the people to whom they are applied. Typically, archaeologists focus on the types of stone tools used at a newly discovered Stone Age site (its 'lithic assembly'), on its art, on any evidence concerning its inhabitants' means of sustenance, and of course on any materials that can be dated by radiocarbon or other techniques, to get a first sense of how it should be classified.

Since we are dealing with the Stone Age here, study of the lithic assemblages is a definitive exercise. Archaeologists skilled in this field are often able at a glance to assign stone tools not only to the Palaeolithic or the Neolithic, but also to sub-categories of those broad divisions. Moreover, it is generally true to say that flints, scrapers, axe-heads, arrowheads and spear-points from the Neolithic end of the Stone Age spectrum are smaller, more delicate, more refined, better made and more skilfully worked than their counterparts from the Palaeolithic.

Although this fits comfortably with modern notions of progress and natural selection (i.e., the glorious and unbroken ascent of Man, via ever finer technology, from a 'primitive' to a sophisticated creature), other evidence suggests that the arrival of the Neolithic entailed a cultural Fall. Look at the extraordinary art that Palaeolithic humans left behind, much of it painted or engraved on the walls of inaccessible European caves between roughly 30,000 and 12,000 years ago. Nothing so beautiful, so technically accomplished, or so 'sophisticated' was ever attempted again by any known culture until the time of the Renaissance – and Picasso is said to have commented on emerging from Lascaux: 'We have invented nothing.'[42] Yet Palaeolithic art is Palaeolithic art. It did not survive into the Neolithic.

Another indicator is the presence of pottery – with the general rule being none in the Palaeolithic and a gradual introduction during the Neolithic. However, the absence of pottery does not necessarily mean that a site automatically belongs to the Palaeolithic. Many Neolithic cultures passed through a pre-ceramic phase, such as the first inhabitants of Mehrgarh in Pakistan, for example (Level 1A), and the first two habitation layers at Jericho (pre-pottery Neolithic A and pre-pottery Neolithic B).[43]

Archaeologists also look at how the inhabitants of a Stone Age site got their living – because here they identify another important difference between the Palaeolithic and the Neolithic. In the Palaeolithic, though they sometimes lived in fixed communities, our ancestors are thought to have been hunter-gatherers with no agriculture or systematic food production of any kind. In the Neolithic, on the other hand, indeed at the very beginning of the Neolithic, agriculture was 'invented' (apparently independently at several locations) and food-production rapidly became the engine of expanding human culture.

But here any precise system of definitions or chronology begins to break down. As some of the new research reviewed in earlier chapters suggests, there is not a clean 'start-line' in the agricultural story 12,000 years ago at the arbitrary 'beginning' of the Neolithic. Agriculture does not seem to have taken root in some areas until thousands of years later, well inside the 'Neolithic' in chronological terms.

None of these qualifications are supposed to matter very much in Malta, where the Palaeolithic is treated by archaeologists as simply irrelevant to the human story.[44] As we've seen, the orthodox view holds that the Maltese islands were not inhabited by humans until 7200 years ago, a Neolithic date, and that the very first people were indeed Neolithic farmers – immigrants from Sicily – with a typical Neolithic 'tool-kit' and Neolithic pottery, etc. So when Anton Mifsud *proves* (as Anthony Frendo of the University of Malta conceded in 1999)[45] that humans were after all present on Malta in the Palaeolithic, and has even gone so far as to suggest a possible Palaeolithic origin for such complex 'Neolithic' structures as the megalithic temples and the Hypogeum, it should be obvious that he is stepping very far out of line.

'Regarding the antiquity of the Hypogeum,' Mifsud e-mailed me on 15 July 2001 after I had asked him to reconfirm his position, 'my gut feeling is that there is strong evidence to show that it had originated in its function subserving the ancient Maltese in the Palaeolithic . . .'[46]

What is that evidence?

The strange case of the bison-bull (1)

In David Trump's authoritative *Archaeological Guide* to Malta, most recently updated and extended in March 2000, the visitor to the Hypogeum is urged to:

> Pause to look at the wall opposite the stairs down to the lower storey. Dark lines of black paint outline what is apparently intended to be a bull. It is crudely done, and the head and shoulders have not survived. That it is ancient and intentional is shown by the fact that the ochre wash on the wall ceases exactly at the black line.[47]

There is an amazing allegation explaining why the head and shoulders of the Hypogeum bull have 'not survived' – and why most of the rest of its body has now also faded to a ghostly shadow that few visitors will be able to make out at all. The reason, reports Mifsud, is that 'The bison-bull at the Hypogeum was removed at the express directive of the Director of Museums.'[48]

What Trump calls simply a 'bull' Mifsud described as a 'bison-bull' (an extinct species) for very specific anatomical reasons:

> Besides the multitude of drawings in red ochre at the Hypogeum, there are also drawings in black manganese dioxide pigment, and one of these measures 1.15 by 0.95 metres. It represents a bovid, the Pleistocene European bison-bull, 'with a hunch on its back, with short horns and tail' [Megary, T., 1995, *Society in Prehistory*, page 261] and is situated on the left wall at the entrance of the Holy of Holies . . . The red

ochre wash on the same wall is a later feature for it terminates just short of the figure. The red wash itself is a recognized feature of early Upper Palaeolithic cultures; [for example] at Tito Bustillo [northern Spain], red wash covers the entire surface of the walls, and this has been dated to the Magdalenian [15,000 to 10,000 BC] ... Paintings in black were dominant in earlier forms of cave art and considering the simple crude design of this Hypogeum bovid, together with its frozen aspect, the lack of perspective and infill, and the non-differentiation between foreground and background, its dating in the Upper Palaeolithic is therefore estimated to be very early on in the pre-Magdalenian period.

Together with the horse, the bison was a dominating theme in European Palaeo-lithic art. Regular bulls also feature significantly in the same culture, with entire halls of bulls being represented such as at the classical Palaeolithic site of Lascaux, which is dated to the early Magdalenian.[49]

If there is any possibility that Mifsud could be right about the Palaeolithic identity of the Hypogeum 'bison-bull' then the alleged act of scrubbing it off the wall represents not just a desecration of the integrity of an ancient archaeological site but something more sinister. The result was the destruction of scarce physical evidence which potentially contradicts teachings about Malta's prehistory that are at the heart of the orthodox world view – i.e., that Malta was not inhabited by humans during the Palaeolithic, that it remained in this condition until 5200 BC, when it was settled by a Neolithic people from Sicily, and that the Hypogeum is, therefore, a Neolithic structure, wholly a Neolithic structure, and nothing but a Neolithic structure ...

The strange case of the bison-bull (2)

Anton Mifsud's extremely serious charge – effectively of official vandalism against what is now a UNESCO World Heritage site – was first put on public record in *Dossier Malta* in 1997. I was unable to find any official rebuttal of it or even a comment upon it from the appropriate authorities and when I contacted Mifsud in July 2001 to confirm that he still stood by the story, he replied that he did, 100 per cent. But, I asked, what possible motive could F. S. Mallia, the Director of Museums at the time, have had for issuing such an extraordinary order?

The motive, suggested Mifsud, was just plain stubbornness. It was well known that in the 1960s David Trump had believed the then recently discovered bull figure to be of archaeological significance. And it was well known that Mallia, a pupil of J. D. Evans who was at that time being trained to take over Trump's position, had disagreed. Much later, when Mallia was in authority at the Museum, he decided upon a final solution to the vexatious matter of the bull: 'Mallia ordered one of the employees to scrub the representation of the wall, and he thus imagined that he had settled the issue once and for all.'[50]

The strange case of the bison-bull (3)

I felt that it would be wrong to leave an allegation as grave as this unresolved
and on 17 July 2001, and again on 12 November 2001, I sent the following e-mail
to Dr Anthony Pace, Malta's current Director of Museums:

Dear Dr Pace,

Thank you for your department's cooperation during my recent visit with the Diverse
Productions film crew to shoot the Malta segment of a three-hour television series
(*Underworld*) that we are making for Britain's Channel 4 about the origins of civilization. I
am the writer and presenter of this series. I am also writing a book of the same title, to be
published by Penguin. Both book and television series are scheduled to come out at the
same time early in 2002.

In connection with these projects and in the interests of ensuring that what I write is
accurate and fair, I would be grateful if you would e-mail me by return with your official
on-the-record comments on the following – potentially rather grave – issues concerning
the Museum:

In their 1997 book *Dossier Malta*, Anton and Simon Mifsud speak of a 'bison-bull' figure
in the Hal Saflieni Hypogeum (Trump describes it simply as a 'bull' and notes that 'the
head and shoulders have not survived'). The figure is or was painted in black manganese
dioxide pigment on the wall opposite the stairs down to the lower level of the Hypogeum
(Trump, *Archaeological Guide*, 72), but Mifsud and Mifsud state on page 168 of *Dossier
Malta* that 'THE BISON-BULL AT THE HYPOGEUM WAS REMOVED AT THE
EXPRESS DIRECTIVE OF THE DIRECTOR OF MUSEUMS':

(1) Is this extremely serious charge true?

(2) If it is true, in what circumstances and for what reasons did the former Director of
Museums [F. S. Mallia] order the removal of this prehistoric painting?

(3) How much of the painting was in fact removed and how much still remains visible
today?

(4) If this charge, on the record since 1997, is NOT true could you please direct me to
the place where I can find your department's official rebuttal and refutation of it.

Additionally, I have received a more detailed account of the alleged 'removal' of the
Hypogeum bull than that given in *Dossier Malta*. According to this account, F. S. Mallia
apparently engaged in arguments about the significance of the bull with D. H. Trump: 'At
one point Mallia ordered one of the employees to scrub the representation off the wall,
and he thus imagined that had settled the issue once and for all.'

Would you like to comment on behalf of the Museum about this account of the events?

I would also be grateful if you could supply me with Dr Mallia's present contacts so that I
may invite him to comment on this matter directly.

I look forward very much to hearing from you.

Yours sincerely,

Graham Hancock

Despite sending the e-mail twice to Dr Pace, and once to another member of staff at the Museum to pass on to him directly, I have not, at time of writing (15 November 2001) received any reply. I read nothing sinister into this. Dr Pace, having only been Director of Museums since 1999, may have no knowledge of the issue and was certainly in no way involved in the events themselves. However, it is disappointing not to have the benefit of his comments on this important question. Nor have I been able to confirm or refute the story by questioning F. S. Mallia, the former Director of Museums alleged to have ordered the removal of the bull figure. Unfortunately Dr Mallia passed away some years ago.

The strange case of the bison-bull (4)

The next step was to talk to D. H. Trump, now retired in Cambridge, so I prepared a list of questions for him and asked my assistant Sharif to find him and interview him. The recorded interview, which sheds some further light on the mystery, took place on 26 October 2001:

> *Sharif:* In your *Archaeological Guide* – this is the main source I'm going on, the updated edition – you mention a bull in the Hypogeum and you say, 'Pause to look at the wall opposite the stairs down to the lower storey. Dark lines of black paint outline what is apparently intended to be a bull. It is crudely done, and the head and shoulders have not survived. That it is ancient and intentional is shown by the fact that the ochre wash on the wall ceases exactly at the black line.' Do you remember the bull I'm talking about?
>
> *Trump:* I do indeed.
>
> *Sharif:* OK, now in *Dossier Malta*, Mifsud alleges that . . .
>
> *Trump:* That this was scrubbed out.
>
> *Sharif:* Yes, he alleges that it was scrubbed out.
>
> *Trump:* The very simple answer to that is what on earth would Francis Mallia have wanted to scrub it out for? Absolutely no motive for this. It was very slight indeed in the first place. It is known that there has been deterioration of the paint under the Hypogeum – this is what all the recent restoration work has been doing to try to stabilize the situation as it is now.
>
> *Sharif:* So what's the cause of the deterioration? Is it the tourists visiting the site, something in the air?
>
> *Trump:* Presumably, yes.
>
> *Sharif:* Mifsud says that Mallia was a pupil of J. D. Evans.
>
> *Trump:* Yes, he was sent back to study under Evans at the Institute of Archaeology in London, to give him the qualifications to take over the job.
>
> *Sharif:* And the suggestion is that because Mallia was a pupil of J. D. Evans,

he had a position that was somewhat contrary to your own position, such that you two entered into a disagreement about the significance of this bull. And it was following this disagreement between you and Mallia that Mallia ordered an employee of the Museum . . .

Trump: I don't think the Museum knows anything about him.

Sharif: He was a nobody in terms of academia?

Trump: We don't know who he was.

Sharif: Right, but what do you have to say about this general picture of a dispute between yourself and Mallia?

Trump: Well, as with all scholarship, we had slightly different views of this. I was more willing to accept this very faint figure than Mallia was. The bull figure. I wouldn't regard this as a disagreement, we certainly didn't squabble over the issue.

Sharif: So it was a difference in academic viewpoint?

Trump: Well yes. I was prepared to accept – by the way it was our curator there who pointed it out to us; no one had noticed it before; it was as faint as that. I looked at it and thought, 'Well maybe there's something in it.' I wanted to put it into the *Guide* so that people could . . .

Sharif: Look for themselves . . .

Trump: Have a look and make up their own minds. Whereas Mallia was rather more dubious of it. But I wouldn't put it more strongly than that. And to call it a disagreement is quite misleading.

Sharif: OK, so really the disagreement was that you thought it was of archaeological significance . . .

Trump: I wouldn't even put it as strongly as that. I thought it might be, he thought it probably wasn't.

Sharif: So his view was that it was actually impossible to take anything from it – even to be sure that it was an ancient piece of art?

Trump: Yes.

Sharif: And your view was that it might be?

Trump: Yes.

Sharif: But you'd never seen it in a state of better preservation – from the outset it was rubbed off?

Trump: From the outset it was extremely faint. As I say, no one noticed it until our curator, who obviously was up and down passing it every day for years, spotted what he thought might be something, and pointed it out to the authorities at the museum. We went and had a look and said, 'Well, maybe' – but it was never any clearer than that.

Sharif: You've seen his figure yourself – what remains of it?

Trump: It was barely perceptible then, I wouldn't . . . well it's even less perceptible now.

Sharif: So have you seen the changes?

Trump: Oh yes.

Sharif: And those are the changes that the restoration project is trying to stop?

Trump: Yes.

Sharif: These are not deliberate changes – they're changes that all tourist sites have to think about?

Trump: Yes, the question of the air conditioning and the like . . .

Sharif: Is there any part of this bull figure which leads you to think about Mifsud's suggestion that it actually represents an extinct species? Is there enough of it left for you to tell that?

Trump: No.

Sharif: What do you think Mifsud is basing that on? He actually takes it as suggestive evidence of a Palaeolithic presence by saying that this is a Palaeolithic species painted in a Palaeolithic style.

Trump: Frankly, rubbish! The site wasn't there – wasn't excavated until long after the Palaeolithic.

Sharif: Right, how do we know that?

Trump: Well, from the archaeological content.

Sharif: From radiocarbon-dating of that content?

Trump: Well, not directly from the Hypogeum, which was excavated back in 1910 – long before radiocarbon. But there was no archaeological material, no pottery or anything out of the Hypogeum earlier than the Zebbug phase. Which, with radiocarbon, we'd now put at about 4000 BC. The chambers were deliberately excavated, but not before 4000 BC. So there's no question of extinct Pleistocene species.

The strange case of the bison-bull (5)

Mifsud's position, while the complete opposite of Trump's, is not contradicted by the presence in the Hypogeum of materials only of the Zebbug phase and younger. As we saw earlier in this chapter he disputes the view that the Hypogeum was constructed as a place of burial and has presented evidence that the materials and skeletal remains found inside it by archaeologists were *not* deliberately placed there but are a *flood deposit* carried in from surrounding Neolithic burial sites. The dating of those remains to the Neolithic Zebbug phases and younger is therefore exactly what Mifsud's theory predicts and leaves effectively unchallenged the revolutionary possibility that lies at the heart of his analysis – i.e., that the structure itself may long pre-date the Neolithic. 'Regarding the antiquity of the Hypogeum,' he confirms:

> my gut feeling is that there is strong evidence to show that it had originated in its function subserving the ancient Maltese in the Palaeolithic, *and the bovine representation constitutes one of the main arguments for this*.[51]

A re-evaluation

As well as the unresolved (and now probably unresolvable) question of the bison-bull with its possible pre-Magdalenian associations, Mifsud points to Malta's Goddess cult as further support for his view that the islands' prehistoric culture may have developed from very ancient Palaeolithic roots. The so-called 'Sleeping Lady' statues found in the Hypogeum and numerous 'Venus' figurines found throughout Malta's megalithic temples leave little doubt that a form of Mother Goddess was the supreme deity worshipped in these mysterious places. But these artifacts 'have all been attributed arbitrarily to the Neolithic',[52] even though they are distinctly characteristic of European Palaeolithic art forms, dating as far back at 30,000 BP.

In brief, Mifsud also draws attention to the following points:

- Modern research into the Palaeolithic cave art of Europe 'includes the study of wall configuration and their adaptation to the drawings, and to the significance of human voice resonance, a feature which immediately brings to mind the Oracle room of the Hypogeum'.[53]
- The art forms in the Hypogeum call for a re-evaluation. 'The designs in red ochre and black pigment draw strong parallels with Palaeolithic sites abroad. The red ochre designs have hitherto been traditionally assigned to a "tree of life" nature and dated arbitrarily to the Neolithic.'[54]
- At the entrance to one of the Hypogeum's painted rooms, the faint engraved impression of a large human hand, also arbitrarily assigned to the Neolithic, may still be seen. It 'has parallels in similar designs in Palaeolithic sites at Gargas, El Castillo, and particularly with Montespan in the Franco-Cantabrian region.'[55] The impression shows a hand with six fingers[56] [a condition known as polydactyly that is also seen on at least one of the 'Fat Lady' figures on show in the National Museum of Archaeology].[57]
- Also of great interest is another Hypogeum design. It 'is in the form of an ideogram and comprises a black and white chequered pattern; this simple geometric design is considered to represent an early stage of Palaeolithic art'.[58]
- Last but not least, tests have been conducted on the red ochre pigments in the Hypogeum for their constituent mineral components. In 1987

samples were taken of red ochre pigment on rock from the north corner of the Oracle room, together with a rock sample without pigment from the same room. On the 26th of July these were examined at the Smithsonian Institute, Washington DC, at the Conservation Analytical Laboratory. Both samples were submitted to x-ray diffraction studies and the red ochre sample was also viewed through a scanning electron microscope. In keeping with the routine composition of Palaeolithic art pigments, these samples confirmed the presence of the oxides of Silicon, Iron, Aluminium, Calcium, Potassium, Sodium and Magnesium.

An earlier study, carried out by Janusz Lehman in 1979, tested two samples of red ochre pigment from the decorations in the Hypogeum's middle level. As well as all the above ingredients these samples contained traces of manganese dioxide, the main component of black. 'This finding confirms that the red ochre design examined by Lehman had been super-imposed upon an even earlier design in Palaeolithic black pigment.'[59]

None of this is to insist that all or even most of the designs inside the Hypogeum do in fact date back to the Palaeolithic – only that there is a significant *possibility* that *some* of them do.

That the Hypogeum was extensively used, and perhaps even developed and expanded during the Neolithic, and that this happened in more or less exactly the time-frame allocated to it by archaeologists (i.e., 3600–2500 BC) is not, I repeat not, in dispute here. But what is contested is any attempt to claim that the scholarly consensus explains everything about this dark and powerful labyrinth beneath the ground and that the 'minor mysteries'[60] of the Hypo-geum's true origins and antiquity have long been solved – 'cleared out of the system',[61] by leading academics.

The consensus may be correct. But I believe Anton Mifsud has successfully demonstrated that important evidence contrary to the consensus does exist, has been overlooked and, in at least one case – the bison-bull – may actually have been extirpated like an idol brought before the Inquisition.

A pattern?

If a failure to preserve and consider potentially controversial evidence has frustrated a full understanding of the Hypogeum, then the same is also true for the megalithic temples and even the prehistoric cave sites in Malta. Thus, Mifsud points out that archaeologists excavating Ghar Dalam cave in the early twentieth century (see chapter 18 for a fuller treatment of Ghar Dalam) 'discovered several knives, scrapers, borers and burins in previously undisturbed deposits, and although stratigraphically Pleistocene, they have been arbitrarily attributed to the Neolithic'.[62]

Likewise, there is the matter of twenty-six flint implements (flint is not native to the Maltese islands) which were excavated at Hagar Qim, also in the early twentieth century:

> They are illustrated in Zammit's *The Valletta Museum* [1931, plate facing page 21] *but have since gone missing*. The implements comprised blades and bladelets, microliths, scrapers and burins, all datable to the Upper Palaeolithic.[63] [My emphasis.]

Probably there's nothing to it. Still, it does seem bizarre that so much evidence with the potential to support a Palaeolithic human presence on Malta gets lost or damaged.

Finally, together with Charles Savona Ventura, Mifsud draws attention to the little-known Ghar Hasan cave located on a precipitous cliff-face on Malta's south coast not far from the more famous Ghar Dalam.[64] This cave was investigated in 1987 by a high-powered team of Italian archaeologists from the Centro Camuno di Studi Preistorici led by Emanual Anati, Professor of Palaeoethnology at Lecce University and a world authority on cave art. Anati has since issued a series of publications concerning Ghar Hasan, the most recent in 1995:[65]

> For the first time in the long history of the cave, a repertoire of Palaeolithic art forms were partially uncovered from beneath the stalagmitic encrustations which covered them for the past fifteen millennia. The figures numbered altogether approximately 20 designs, and they are painted in red, brown, dark brown and black. They represent various animal figures, an anthropozoomorphic design, several handprints and an array of ideograms . . .
>
> In Panel One, at least two of the animal figures represent the elephant, 'two heavy quadrupeds with a long muzzle'. These animals were extinct in Malta before the end of the Pleistocene.[66]

The so-called 'Pleistocene/Holocene boundary' in geology coincides quite closely with the Palaeolithic/Neolithic boundary in archaeology. So what Anati's expedition seemed to have found with these representations of extinct species in Ghar Hasan was more evidence of a Palaeolithic human presence on Malta.

Soon after news broke about these published conclusions and their stark contradiction of the orthodox view on Malta's prehistory, the Italian team distanced itself from its initial Palaeolithic leanings and claimed instead that the depictions in Ghar Hasan are 'out of context' – which indeed they are if one is only prepared to countenance a Neolithic context for the earliest human presence in Malta.

Another development at about the same time was that the Ghar Hasan cave began to be vandalized, and the paintings defaced or completely removed, a process that continued over a long period. The result, which would have caused an international furore anywhere else but Malta, is that today:

> The only depictions which have survived, unless more are obscured by stalagmitic material on the cavern walls, are the two handprints in red pigment in Gallery D . . . Vandalism not of the popular type has destroyed and obscured the entire repertoire of images on the accessible areas.[67]

The best paintings described, photographed and published by Anati,[68] were in the 'Gallery A' section of Ghar Dalam. Within a few weeks of the arrival in Malta of Anati's publication, a steel gate was erected that restricted access to this section. Officially, the gate had nothing to do with Anati's publication or the vandalism of the paintings, but was 'for the protection of a small colony of bats'.[69]

The ghost of Piltdown Man

Rigorous scientist that he is, Anton Mifsud would be the first to admit that none of the clues, hints, anachronisms, anomalies and whispers of conspiracy that he has amassed from the Hypogeum and the megalithic temples of Malta are proof that these structures had a Palaeolithic origin. Certainly they are suggestive! But they prove nothing and they run entirely contrary to increasingly accurate C-14 evidence that archaeologists have had at their disposal since the 1950s – revolutionized by dendrochronology in the 1960s[70] – which places the temple-building period within a definite time-band in the Neolithic (3600–2500 BC) and finds no evidence of *any* human presence in Malta at any date prior to 5200 BC, let alone as far back as the Palaeolithic.[71] The earliest radiocarbon evidence of a definite human presence in Malta is from Ghar Dalam and gives a Neolithic date of around 5200 BC.[72] The orthodox position is that no samples taken anywhere in the Maltese islands suggest any earlier date.

So it is clearly not enough, if one wishes to propose something so radical and upsetting as a Palaeolithic human presence in Malta, merely to offer apparent similarities in religious iconography, apparent similarities in artistic styles, apparently similar types of pigments used, etc. Such impressions are all very well, and even helpful, but the interpretation of them is bound to be subjective. What is needed in addition to all this is solid empirical evidence – from scientific tests supported by reliable provenance and stratigraphy – that confirms a more ancient presence of man.

Naturally Anton Mifsud would not have embarked on his course of confrontation with the archaeological authorities over the basic terms of Maltese prehistory if he did not possess such evidence. He does. And in the process of acquiring it, as we shall see, he has uncovered some very strange and disturbing archaeological behaviour that took place during the 1950s and 1960s. This was the precise period in which the foundations of Maltese prehistory were being laid down by Professor J. D. Evans. It is not an accident that this was also the period when the islands became defined as 'apalaeolithic' – i.e., not inhabited by humans before the Palaeolithic – a definition that has been taught to later generations of archaeologists as dogma. The scandal has even gone so far as to entangle the Natural History Museum in London in its clutches and to resurrect the restless ghost of Piltdown Man.

29. *Arunachela temple, Tiruvannamalai.*

27. Mahabalipuram seashore at dawn with people gathering to watch the sunrise. Shore temple is in the background to the left. Local traditions speak of extensive underwater ruins.

28. The Meenaksi temple, Madurai, with its sacred Tank (right foreground).

30. *Arunachela, the sacred red mountain of Tiruvannamalai, embodying the presence of Lord Siva. The temple nestles at its foot.*

31. *Siva devotees on the slopes of Arunachela overlooking the great rectangle of the temple.*

32. *The author interviewing fishermen, Mahabalipuram. Stories of underwater ruins are commonplace along this coast.*

33. *The author with NIO team and fishermen at Poompuhur on the way out to the dive boat.*

34. *Local fishermen directly over the U-shaped structure at Poompuhur. Underwater structures provide attractive shelters for fish.*

35. *Side-wall of the U-shaped structure, Poompuhur, looming out of the murk. The structure was submerged about 11,000 years ago.*

36. *The author diving on the U-shaped structure, Poompuhur, at a depth of 23 metres and 5 kilometres from shore. Diving conditions here are difficult, with poor visibility.*

37. *U-shaped structure, Poompuhur.*

38. Curved trench or passage in U-shaped structure, Poompuhur.

39. *The author diving on the U-shaped structure, Poompuhur.*

40. *The author with the NIO's team of marine archaeologists. Kamlesh Vora is at the extreme right. The author is flanked by Sundaresh and A. S. Gaur.*

18 / The Masque of the Green Book

We have no reason to suppose that Palaeolithic man ever set foot on Malta.

J. D. Evans, 1959[1]

The conspiracy of silence over the decades represents the triumph of prejudice over logic.

Anton Mifsud, 1997[2]

The whole thing is in limbo, really . . .
David Trump, October 2001[3]

When did people first live on the Maltese islands? The question seems innocent and simple enough – almost routine – but evidence that has a bearing on it has been tampered with and lost, and the search for the correct answer to it is the fundamental issue of Maltese prehistory. Because of Malta's special place in the wider story of civilization it is a fundamental issue of global prehistory as well. For how can we claim to have understood the origins of civilization if we have failed to unravel properly the processes and motivations, the skills and the ideas, that led up to the creation of humanity's first ever works of monumental religious architecture?[4]

And what architecture it is!

- Not simple rough-and-ready building experiments as one might expect, but beautiful, accomplished, harmonious structures that were the work of master architects, planners and stone masons, from the very beginning.
- Not monuments that were easy to make, but monuments that were extremely difficult to make – and that would be difficult to make in any epoch, with any technology.
- Monuments like Gigantija, described in chapter 15, with its walls of 5 metre tall megaliths.
- Monuments like the Hypogeum, an incredible achievement of troglodytic burrowing and hewing to create a mysterious labyrinth beneath the earth.
- Monuments like Hagar Qim and Mnajdra that feature astronomical and solar alignments requiring years of careful observations and measurements to confirm and install.

So what was going on in Malta that led to all this? Why did the first megalithic temple-builders in the world choose to make things so difficult for themselves? Why didn't they start with *small* megaliths (if that is not too serious a contradiction in terms)? Why didn't they start simple? Why did they plunge straight into the very complicated stuff, like Gigantija and the Hypogeum? And, having plunged, how did they manage to produce such magnificent results? Was it beginner's luck? Or were their achievements as humanity's pioneering architects the product of some sort of heritage?

Beginner's luck is possible, but having studied the earliest temples, and their level of perfection, archaeologists agree that heritage is the right answer. The only problem is what heritage? And where is it to be looked for? Since it is the received wisdom that no human beings lived on Malta before 5200 BC, and since this is a 'fact' that is at present unquestioned anywhere within conventional scholarship, archaeologists from roughly the mid-twentieth century onwards have simply seen no reason to explore the possibility that the heritage of the Maltese temples might be older than 5200 BC. To do so would be the research equivalent of an oxymoron – like breeding dodos, trying to conduct an interview with William Shakespeare or seeking evidence that the earth is flat – and would invite the ridicule of one's peers.

The result, necessarily, is that archaeological inquiry into the origins of monumental civilization in Malta has been confined to the narrow chronological band between 5200 BC – the supposed date the islands were first settled – and 3600 BC, the supposed date that Gigantija was built. Whatever alchemy transformed the rude and unimpressive stone and brickwork of the Maltese of the fifth millennium BC into the awe-inspiring cyclopean temples of the fourth millennium BC is therefore – again necessarily – to be traced *within* this period, not outside it. The only possible external and earlier influence that might reasonably be countenanced by proponents of this model could, as noted in chapter 16, lie in 'intellectual baggage' that the original Neolithic settlers presumed to have first colonized Malta from Sicily in 5200 BC – the Stentinello culture – might have brought with them. But the evidence is against this, since the Stentinello people of Sicily did not develop a megalithic culture and, indeed, there are 'no true megalithic monuments' at all anywhere in Sicily.[5]

So we're back to Malta again, confronted by the massive physical presence of the world's first monumental architecture. And since it assaults common sense to suggest that such huge and accomplished temples could be the work of people who had never built with megaliths before, we're searching for the intermediate structures on which the Maltese stone masons presumably must have learned their craft during the first 1600 years that there were people at all on Malta – i.e. between 5200 and 3600 BC.

5200–3600 BC, the archaeologists' story
Ghar Dalam 5200–4500 BC

The first phase of human settlement that archaeologists recognize, 5200–4500 BC, is known as the Ghar Dalam phase. The name is from the type site, Ghar Dalam cave itself, but the 'Phase', defined by its pottery and tools, is represented at sites throughout Malta and Gozo. This phase has left no evidence of any large-scale construction activities at all. Nor is it easy to make out any of the signs of organized cultic and religious behaviour that normally precede full-blown temple worship. All that has come down to us are a few traces of rudimentary huts and shelters and a stumpy wall, 11 metres long but less than a metre high, made of two rows of small upright slabs with a filling of rubble in between.[6]

Skorba, 4500–4100 BC

Archaeologists identify a second phase immediately following Ghar Dalam, of which the type site is Skorba (not to be confused with the megalithic temple at Skorba – itself of the Gigantija phase and later! – which stands near by). The dates of the Skorba phase are 4500–4100 BC and there is no doubt that the architecture does get bigger and more impressive during these 400 years. Indeed, Trump proposes, and some of his colleagues agree with him, that two oval rooms at Skorba which have been carbon-dated to 4100 BC may actually be the first precursors of the later temple architecture.[7] The pebbled courtyard of the northern room is partially covered by the eastern side of the temple, providing a clear sequence, in full accord with orthodox chronology, in which the Skorba phase, and its C-14 date, precede the Gigantija and later phases of the temple – with all the phases nicely stacked up one on top of the other on the same site.

Trump thinks that the two rooms may have been basements, 'as the southern one had no doorway through its massive walls'. The northern room, on the other hand, was entered by way of the pebbled courtyard mentioned above.[8] Within the rooms:

> The irregularity of the floors and the unlevelled surface of the bedrock argue against domestic use, and the group of figurines . . . from the northern room also suggest that this building had a religious function, a true predecessor, then, of the temples which appeared some centuries later. The main difference in construction was that the upper walls had been built in mud-brick shaped from Maltese blue clay.[9]

I must protest in passing that while irregular unlevelled bedrock floors do argue against domestic use they do not inevitably suggest that that building had a religious function (all the later temples had levelled floors). Conversely, the figurines that Trump mentions do suggest a religious function similar to that of the temples and the Hypogeum since they include 'female figurines, stylized and with greatly exaggerated buttocks'.[10]

On the basis of the oval shape of the rooms and the versions of the familiar goddess images found within them it is hard to disagree that there is a connection here. On the other hand, there is no trace whatsoever of megalithic architecture in these oval rooms from 4100 BC. Their walls may be 'massive', as Trump suggests, but the architectural and engineering challenges faced in building them are not to be compared in any way with the challenges that faced the temple-builders.

So, yes, the people of the Skorba phase did build big structures. And yes, they do seem to have venerated the Goddess within them. But the walls of these structures were made of small, easily handled stones and rubble packed together and surmounted by mud-bricks,[11] so they can hardly be described in architectural and enginering terms as 'intermediate' steps on the way to the megalithic temples.

Zebbug, 4100–3800 BC

The next phase of Maltese prehistory is named Zebbug – as usual after the type site – and is dated from 4100 to 3800 BC. Archaeologists classify this phase as being within the 'Temple period' since it occupied that last five centuries before the construction of the first megalithic temples – during which time, it is assumed, Maltese society must have been gearing itself up in various ways for the colossal effort that lay ahead. The evidence for this gearing-up process is, however, not overwhelming. The Zebbug phase produced no megalithic architecture and no rock-hewn temples, but is identified by stylistic changes in pottery and distinguished by what are thought to be the first rock-hewn tombs in Malta – a group of five rather unimpressive dish-shaped depressions discovered in a field in the parish of Zebbug in 1947.[12] A few are somewhat elliptical[13] and might be said to bear comparison with the elliptical 'theme' of the temples and the Hypogeum – although none of the labyrinthine or subterranean characteristics of the latter are present at Zebbug.

A twin-chambered rock-cut tomb from the Zebbug phase has also been found on Gozo close to the temple of Gigantija. It is a collective tomb and consists of a vertical shaft approximately 1 metre deep opening into two low-roofed, shallow, rock-cut chambers that had been filled up, over many centuries, with the bones of fifty-four adults and eleven children:

> Most of the bones were disarticulated and pushed to the back and sides of the chambers, as if to make way for a more recent burial. Indeed at the entrance to one of the chambers lay the contracted almost complete skeleton of an adult male, presumably the last of the burials.[14]

> A stylized human bust of stone was placed at the entrance to one of the chambers, as if intended to guard the tomb.[15]

Other Zebbug phase tombs, at Xemxija, which feature kidney-shaped and 'clover-leaf' rock-hewn chambers devolving off a central shaft just under a metre deep, have been proposed by J. D. Evans as possible models for the characteristic kidney-shaped apsidal rooms of the megalithic temples – a view that David Trump also believes has 'much to recommend it'.[16]

Mgarr, 3800–3600 BC

After the Zebbug phase – again rather loosely classified as being within the temple period but still before a single example of megalithic architecture had appeared, archaeologists insert the Mgarr phase, 3800–3600 BC.[17] Essentially irrelevant to the quest for intermediate structures on which the temple-builders practised and honed their skills, Mgarr is classified by its pottery – 'a transitional phase named after the site in Malta where a development in style of the Zebbug pottery was first noticed'.[18]

Gigantija

And then suddenly, around 3600 BC, the fireworks start to fly with the Gigantija phase (3600–3000 BC). Here, as we know, the type site is not a pottery heap, a mud-brick wall, or a few rock-cut tombs, but Gigantija herself – the 'tower of the giants' – literally the mother of all temples if the orthodox chronology is correct, built with megaliths that are consistently amongst the biggest ever used in Malta.

Know-how has to start somewhere

How are we to explain such a sudden and dramatic leap forward as the appearance in the Gigantija phase not only of the 'blueprint' for the archetypal Maltese megalithic temple – to which, with adaptations and refinements, all later temples adhere – but also, at the same instant, the complete suite of organizational and technical abilities necessary to build such temples when, we are told, none had ever been built before?

In a recent paper on the architecture of the Maltese temples, Trump admits there is a problem:

> Know-how has to start somewhere. Though building in stone was first introduced to Malta by the first settlers, as was shown at Skorba, the use of huge blocks, so-called megalithic architecture, is not known before the temple period. The skills must have been built up slowly, over time.[19]

I am not an archaeologist, but after reviewing what archaeology has found out about the 1600 years between the supposed date of first settlement and the beginning of temple-building at Gigantija – 5200 BC down to 3600 BC – I personally see no convincing evidence of any build-up of skills 'slowly, over time' that would have been relevant to the construction of the megalithic temples.

I note that Trump and Evans both hint that the temples may somehow have evolved out of the *shape* of Zebbug phase tombs, and there is an undeniable resemblance. But even if we accept that the shape of the tombs of 4100 BC is related to the shape of the temples of 3600 BC – giving us 500 years of 'evolution' to explain the phenomenon of Gigantija – this still leaves unanswered the bigger question of how and where the ancient Maltese learned to reproduce such shapes, above ground, in megaliths weighing many tonnes?

Could the solution be that another wave of settlers arrived in 3600 BC bringing the temple blueprint and the necessary building skills with them? This was once a fashionable idea that has gone out of favour as the archaeology of Malta and of the Mediterranean as a whole has improved. As David Trump has recently affirmed, 'There is nothing looking remotely like one of these temples outside the Maltese islands, so we cannot use "foreign influence", to explain them away.'[20] Likewise, far back as 1959 J. D. Evans wrote:

> It is abundantly clear ... that the Maltese temples and tombs were something indigenous, rooted in the beliefs and customs of the people whose religion they express, and they evolved step by step with these. There seems no question of their having been introduced as a result of influence from other cultures.[21]

So we have come full circle back to Malta again, still searching for the baby temples, the kid temples, the adolescent temples – or if not temples then other kinds of structures requiring the same skills – that ought to precede the mature, prime-of-life temples of the Gigantija and later phases. And they aren't there.

Could this be for the same reason that Malta lacks what Anton Mifsud calls a 'civilization territory' big enough to account for the impressive manifestations of civilization scattered all over the Maltese islands? Could we be missing the evolutionary phases of the great megalithic temples because the land on which those phases are represented is now underwater? And, the corollary of this, is there any evidence that submergences on a sufficient scale to obliterate the entire hinterland of a culture have ever occurred in the Maltese archipelago?

It is here that settling the date of Malta's first inhabitation by humans becomes pivotal to our inquiry. Because if we accept the orthodox academic view that these islands were entirely without a human presence until 5200 BC, then we would have no reason to be interested in floods that might have occurred earlier than that date. But suppose there is reason to doubt the academic verdict? Suppose, for example, it were to transpire that Malta had in fact been peopled during the late Palaeolithic, from as early as 18,000 years ago. Then the possibility would have to be seriously countenanced that these Palaeolithic inhabitants and their descendants could have been responsible for the evolution and development of the architecture of the 'Temple period' – with the more populous Neolithic settlers merely participating in and merging their identity with its last phases.

This is why the misrepresentation and possibly even manipulation of evidence by party or parties unknown to give a falsely late date for the earliest human presence in Malta that Anton Mifsud has exposed is, potentially, of explosive significance.

The leavings of violent floods

The story begins at Ghar Dalam, a spacious natural cave more than 7 metres wide, 5 metres high and 120 metres long that opens into the wall of one of Malta's many precipitous valleys, the Wied Dalam, located in the south-east of

Ghar Dalam

the island. Though it is arid today, the valley was gouged out by a great river and floods that have flowed violently through it at various times in the past. It continues for just over half a kilometre beyond the cave mouth before finally plunging beneath the sea in Saint George's Bay.

The cave is thought to have begun its existence as a solution cavity dissolved in the bedrock by percolating groundwater that was later broken into from above, penetrated and extended by these palaeofloods. In the process the ongoing erosion of the river bed cut down the valley floor still further so that it now lies 6 metres below the level of the cave mouth. There have been several occasions during the past quarter of a million years when the flooding has been on such a scale as to overtop the valley sides and completely inundate the cave, leaving behind layers of muddy earth, clay and pebbles mixed with a fantastic assortment of animal remains that were carried along in the flood waters. Archaeologists say that the last of these cataclysmic flood deposits was laid down during the melting of the Ice Age. John Samut Tagliaferro of the University of Malta dates the event to 18,000 years ago.[22] David Trump goes for a slightly more recent estimate: '[This] level in the cave yielded great numbers of red deer bones and was probably laid down in the cool wet period of the closing stages of the last Ice Age some 10,000 years ago.'[23] After that terminal Ice Age event between 18,000 and 10,000 years ago no further flood deposits were laid down. The river in the valley floor ran dry and the cave remained undisturbed, gathering dust, visited only by grubbing wild animals, for almost 3000 years.

Finally, so the official story goes, human beings began to register their presence there with the earliest traces of their occupation – supposedly the oldest in Malta – radiocarbon-dated to around 5200 BC. The dates are from the so-called 'Cultural Layer' of the cave: a thin deposit containing beads and other ornaments, buttons, tools, weapons, bones and rubbish – the usual detritus of human habitation – and also fragments of the distinctive incised pottery, excavated here and at other sites in Malta, by which the Ghar Dalam phase as a whole is classified.[24]

Because of the earlier flood epochs, however, archaeologists excavating beneath the Cultural Layer found *five other layers of deposits*, providing a complete archive of climate, ecology and fauna in Malta over approximately the past quarter of a million years. In brief, and in descending order (with the youngest layer, of course, being the highest), we have:

6. *The Cultural Layer*: traces of Neolithic man from 5200 BC onwards.
5. *The Calcareous Layer*: a thin, sterile, chalky deposit that usefully 'seals' the older Pleistocene (Ice Age) layers beneath it and serves as a clear separator between them and the post-glacial Cultural Layer above.
4. *The Cervus Layer*: the most recent of the flood deposits, dated to between 18,000 and 10,000 years ago and so named because it contains the bones (in immense quantities) of the Pleistocene European red deer (*Cervus elephas*, an extinct species). Other Ice Age faunal remains in the Cervus Layer include those of wolf, brown bear and fox.
3. *The Pebble Layer*: just that, a stratum consisting almost entirely of stones and pebbles swept into the cave by water action and strewn across its floor.
2. *The Hippopotamus Layer*: in which the remains of extinct species of dwarf hippopotamus and dwarf elephant predominate.
1. *The Clay Layer*: immediately above bedrock. This layer, the oldest, forming the bottom of the Ghar Dalam sequence, is sterile and contains no remains whatsoever.

It is certain anomalous discoveries that were made by archaeologists excavating the Cervus (Deer) Layer during the first half of the twentieth century – and the subsequent fate of these discoveries – that threaten to turn the prehistory of Malta on its head.

Cooked hippo, a human hand bone, and some stone tools

In fact, points out Mifsud in *Dossier Malta*, the first anomalous discovery was made much earlier than that – by the Italian scholar Arturo Issel in the 1860s. He began an excavation at Ghar Dalam at an arbitrary 100 paces from the cave entrance:

> The remarkable finds of his, the first official excavation of Ghar Dalam, included the burnt remains of hippopotamus, whose bones had apparently been cooked and opened up to extract the marrow for consumption.[25]

The burnt remains of a hippo with its bones in the condition described do strongly suggest a human presence. However, little attention has ever been paid to Issel's finds, which have not been preserved and have never been considered part of Malta's archaeological story.[26]

Mifsud's research also revealed that further excavations had been conducted in the 1890s by a certain John H. Cooke, a teacher with a systematic approach towards archaeology. He dug a series of eight trenches at regular intervals throughout the cave from its deep interior to a point just 10 metres before the entrance:

> The main finds were in two trenches. A human hand bone was found in his Trench IV, in the Cervus Layer, whilst a human implement was discovered in Trench VI, also in the deer layer. For the first time human implements and remains lay in the same horizon below the cultural layers of Ghar Dalam, precisely in the Cervus Layer.
>
> Immediately overlooking Cooke's layer 'e', the fifth layer from the surface, and equivalent to the Cervus Layer, at a depth of two feet three inches, a stone implement was discovered by Cooke. According to Dr A. A. Caruana, he was 'of the opinion that it has undoubtedly been fashioned by man'.[27]

The next digs followed in 1912–13, coordinated by Napoleon Tagliaferro and Guiseppe Despott. Their project was taken over a year later by the British Association, but Despott remained involved, conducting further digs with Temi Zammit in 1914, and leading the digs in 1916 and 1917.

The rise and fall of Despott's and Rizzo's big teeth
During the 1917 dig the discovery was made, again in the Cervus Layer, of two human teeth of a very special type known as 'taurodont' (the word means literally 'bull-tooth' and refers to the supposed 'bull-like' appearance of the tooth, with a large heavy body and extremely short or non-existent roots). Anton Mifsud takes up the story:

> The breakthrough came about in the summer of 1917, in one of the two trenches Despott had excavated that year. Trench I was situated 50 feet from the entrance, and the crucial Trench II lay 60 feet further inside the cave. It was in the latter, Trench II, that two taurodont molars were discovered in the stratum of red cave earth. The curator Giuseppe Despott and a Mr Carmelo Rizzo were supervising their men digging in Trench II, when the latter's workers came across a large bull-shaped human molar tooth amongst several deer teeth obtained from the Deer Layer of this trench; a few days later Despott himself discovered a similar molar a few feet away, several inches deeper in the cave earth . . .
>
> Despott's molar was registered as lying one foot deeper in the cave earth of the Cervus Layer, and separated by seven feet from Rizzo's; the pair of molars possibly derived from two individuals, but their relative proximity cannot exclude a single

source. The teeth had an unusually large pulp cavity so that the roots were very small.[28]

By chance, just a few years earlier, the famous British palaeoanthropologist and anatomist Sir Arthur Keith had described unusual teeth of exactly this sort found at sites elsewhere in Europe. He had attributed these teeth to Neanderthal Man and it was he who had coined the term 'taurodont' for them.[29] Now Rizzo and Keith submitted photographs of their molars to Keith for examination and were delighted when he diagnosed them, without hesitation, as taurodont:

> In size and form such teeth have been seen in no race of mankind except *H. Neander-thalis*; in condition of fossilization and in the fauna which keep them company, in the red cave earth in Ghar Dalam, they are in their proper Pleistocene setting.[30]

Keith followed this up by writing a letter to *Nature* on the subject and, in subsequent years, the hypothesis began to be widely accepted, indeed orthodox, that at least Neanderthal humans had been present on Malta during the Palaeolithic and had left their remains at Ghar Dalam. The view was strengthened in the 1920s by the discovery of a large number of definitely late Neolithic human remains and artefacts, including 2250 teeth, in the Burmeghez cave. Keith examined the teeth carefully and could not find a single example of taurodontism among them. This he took as further support for his hypothesis of the more 'primitive' Palaeolithic origin of the Ghar Dalam teeth.[31]

Soon afterwards the hypothesis came under attack from various quarters. Scattered reports had begun to appear in the dental literature of taurodont teeth in modern humans, which, if confirmed, would much reduce the probability that Despott's and Rizzo's molars were very old Neanderthal teeth – rather, for example, than more recent ones that had somehow (perhaps through burial) been introduced into the Cervus Layer. However, further investigation of the evidence demonstrated that while some of the modern teeth were genuinely taurodontic none of them showed anything like the degree of taurodontism evident in Despott's and Rizzo's molars.[32] Under a classification that Keith had already proposed the latter were as 'hypertaurodontic' as any Neanderthal teeth from other parts of Europe, whereas the modern human teeth were mesotaurodontic or more commonly hypotaurodontic (the least severe form of the condition).[33]

Keith's defence held, there was support from other worthy authorities, a third taurodontic tooth was discovered in 1936 by Dr J. Baldacchino (then the Curator of the Museum) in the same Cervus Layer as the two 1917 molars[34] – and through the combination of all these favourable auspices, 'A slot was secured for Neanderthal humans in the Maltese history books, albeit for a few decades.'[35]

Why only for a few decades? 'In the early 1950s,' explains Mifsud, the person in charge of archaeological surveys in Malta, J. D. Evans, defined the Maltese Neolithic ... as the start of Malta's history, at the same time that he discarded the taurodont molars as unreliable evidence on the basis of their isolation. Three years later, in 1962, a Maltese dental surgeon, J. J. Mangion, reported upon the incidence of taurodontism in modern Maltese, and thus seemed to discredit the validity of the Ghar Dalam molars as diagnostic ... evidence for Neanderthal humans.

But, as Mifsud says, the *coup de grâce* was delivered in 1964, when a report from Malta's Museum of Archaeology misrepresented the results of chemical dating tests carried out on the taurodonts by claiming them to be Neolithic. Within a decade Neanderthal man was out of the Malta history books and the taurodonts were totally discredited as evidence for a Palaeolithic presence in the Maltese islands.[36]

Trump and Evans on the record on the taurodonts

Before we look into the grave charge of misrepresentation that Mifsud is lodging here, and find out whether the claim of taurodont teeth amongst the modern Maltese is as significant as Evans made it out to be, let's clarify the 'official' position on the taurodont controversy today.

In addition to J. D. Evans' comprehensive 1971 survey, *The Prehistoric Antiquities of the Maltese Islands*, which still forms the foundation for all orthodox teaching about Malta, an important channel through which the voice of orthodox Maltese archaeology reaches the general public is David Trump's highly regarded *Archaeological Guide* – most recently updated in March 2000.[37] On page 91 of this updated edition Trump gives the visitor helpful information about the Hippotamus Layer and the Cervus Layer of the Ghar Dalam cave and then concludes: 'No trace of human occupation has been found in either of these levels.'[38] Interestingly, however, Trump then directs the reader to: 'see below'.[39]

What he actually says 'below' is framed as an attack on Mifsud's investigations in *Dossier Malta* that reopened the taurodont controversy in 1997 – although Mifsud is not acknowledged by name. Trump begins his statement on page 92: 'Two human teeth gave rise to much controversy, which has recently been reopened. For present purposes I hold to the former version, but see p. 19.'[40] On page 19 we find a passage in which Trump appears to be hedging his bets, ever so slightly, on the dogma that there was no human presence in Malta before 5200 BC:

There is very little to suggest that man reached the islands until something like 7000 years ago, and nothing secure ... though there is always at least a faint possibility that material of the Old Stone Age [i.e. the Palaeolithic] may yet come to light ...[41]

We then turn back to page 92, where Trump continues that the two Ghar Dalam teeth:

> were of taurodont form, with a single large hollow root, found commonly in Neanderthal man. But this form is known, if rarely, in modern man too – one was extracted from the jaw of a living Maltese only a few years ago – and so does not prove the presence of Neanderthalers here. Careful chemical analysis at the British Museum (Natural History) . . . confirmed that these teeth were contemporary with the bones of domestic animals, more recent than the deer bones and much more recent than the fossil fauna. It was similar analysis which suggested that the hippopotamus tooth implicated in the Piltdown forgery probably came from the same site.[42]

It is also worth reminding ourselves of Evans' position on the taurodont matter as he set it out in his *Prehistoric Antiquities* – although this, of course, refers to the earlier episode of the controversy in which Despott had proposed that the teeth 'could be used as evidence of the presence of man in Malta during the Middle Palaeolithic period'.[43] Evans replies that the suggestion does not fit the facts:

> Dr Baldacchino has since pointed out that taurodontism occurs in teeth definitely assignable to the Neolithic period of Malta (for instance, some from the Hypogeum). The two teeth from Ghar Dalam, therefore, could quite easily belong to a later period. A few other human teeth and bones which have been found at depths of up to 6 ft (1.80 m) all appear to be of the modern type. In view of these facts, then, the two taurodontic molars can hardly be accepted as good evidence for the existence of man in the Maltese islands in pre-Neolithic times.[44]

Earlier on the same page, while reporting the 1917 discovery of the Ghar Dalam teeth, Evans describes them as: 'two very large human molars, both exhibiting the characteristic of taurodontism, or fusion of the roots'.[45]

Truth and fiction (1)

It is disturbing that, in the passage cited above, Evans wrongly equates taurodontism, a condition in which the tooth either has extremely small roots or no noticeable roots at all, and in which the pulp cavity of the tooth is correspondingly enlarged, with an entirely different condition known as 'fused' roots.[46] This is a mistake of some significance, Mifsud reminds us:

> For while the condition of fused roots was commonly found in Neolithic and in modern man, taurodontism was not. Hence the reason [i.e. confusion of fused roots with taurodontism] for Evans' assertion further down the same page that taurodontism was described by Baldacchino as being common in Neolithic teeth: 'Dr Baldacchino has since pointed out that taurodontism occurs in teeth definitely

assignable to the Neolithic period for Malta (for instance, some from the Hypogeum).'[47]

But is Evans' confusion genuine? Or is it sleight of hand to persuade us that the possibly very ancient human teeth from Malta – said on the basis of their taurodontism to be Palaeolithic – are not after all diagnostic of the Palaeolithic because the same taurodontic morphology 'occurs in teeth definitely assignable to the Neolithic period for Malta'?

If it is not sleight of hand then it is bad scholarship. For no taurodontic teeth have ever been recovered from the Hypogeum. And although Evans might have been confused, Baldacchino himself knew very well how to distinguish taurodontism from fused roots. After studying the thousands of teeth in the Neolithic deposit from the Burmeghez burial cave he wrote:

> No trace of taurodontism was found in these specimens; the only form of degener-
> ation which was present was that with which we are familiar in modern teeth –
> fusion and maldevelopment of the roots, particularly in those of the third or 'wisdom'
> molars.[48]

Another less ambiguous and more annoying example of prestidigitation that Mifsud draws attention to concerns Evans' misrepresentation of the position of the British archaeologist Gertrude Caton-Thompson on the subject of the Palaeolithic in Malta with specific reference to the two taurodont molars that had been discovered in Ghar Dalam in 1917. Here, writing in 1925, is what Caton-Thompson actually said:

> The discovery of possible Palaeolithic man appeared to me of considerable impor-
> tance to prehistory . . . Apart from the discovery in the red earth of the two taurodont
> teeth, in circumstances incapable of satisfactory interpretation, there are but two
> other records in the island of possible relics of Palaeolithic man.[49]

In this passage there can be no doubt that Caton-Thompson is treating the taurodont teeth – along with the two 'other records' she mentions – as 'possible relics of Palaeolithic man'. Moreover, when she says that they were found in 'circumstances incapable of satisfactory interpretation' she means that they cannot be satisfactorily interpreted within a Neolithic framework.[50]

But this is not what Evans has her saying in his *Prehistoric Antiquities*. It's there, discussing the 1917 taurodonts on page 19, where he argues that 'the teeth from Ghar Dalam could . . . quite easily belong to a later period . . .' He then reinforces this point with a footnote in which the reader is informed: 'Miss Caton-Thompson remarks that that the discovery of the molars was made "in circumstances incapable of satisfactory interpretation".'

Thus, by smoke and mirrors, we are led to believe that the taurodonts were not found in a good Palaeolithic context, whereas Caton-Thompson herself had originally stated almost the opposite. As Mifsud puts it:

> Evans ... misinterpreted Caton-Thompson when he extracted one phrase of hers out of its context and quoted it in another; he thus created the impression the validity of the molars was being questioned by her as archaeological evidence, whereas the contrary is correct ... Evans' inaccuracies were perpetuated through repetition by later authors ... who have accepted Evans on the weight of his authority ... including anatomists, archaeologists, medical historians, and other historians, until errors crystallized into accepted facts.[51]

Truth and fiction (2)

But Mifsud has much bigger game in his sights than scholars misrepresenting one another. What he's really after lies in the *interpretation* that Trump and other archaeologists have subsequently put on the 'careful chemical analysis' undertaken at the Museum of Natural History in London. This is the analysis which supposedly confirms that the Ghar Dalam taurodont teeth were not contemporary with the Cervus (Deer) Layer in which they had been found but, on the contrary, were 'more recent than the deer bones'. Such an interpretation, Mifsud demonstrates, though honestly held, is quite unjustified. Because, although they have only ever been published in a highly abridged form – which does lend itself strongly to the erroneous interpretation innocently put on them by others – the Natural History Museum tests did *not* confirm the Ghar Dhalam teeth to be 'more recent' than the Cervus Layer of deer bones washed into the cave in a cataclysmic flood of the late Palaeolithic between 18,000 and 12,000 years ago. On the contrary, as we shall see, the results of the tests are highly ambiguous. Nevertheless, to the extent that any interpretation can legitimately be put on them at all, these results suggest much more strongly that the human teeth are *contemporary* with the Cervus Layer – and thus in every sense part of the ancient Ice Age deposit.

We will look into this in more detail in a moment. Meanwhile, although this error has very large ramifications for our views about when humans first settled in Malta, I want to make it absolutely clear here, in the plainest possible language, that David Trump is not to blame for it in any way. As he told us when Sharif interviewed him in October 2001 he himself is not a chemist and he had therefore relied on the proper authority for the opinion he had expressed in the most recent edition of his *Archaeological Guide*. That authority had been none other than the Natural History Museum's Kenneth Oakley, celebrated in the 1950s for his uncovering of the Piltdown Man hoax, and the top scientist in his field at the time:

> *Sharif:* First of all, I'd like to get your general opinion on the work of Anton Mifsud and colleagues who, particularly in the book *Dossier Malta*, have

alleged that the orthodox view of Neolithic being the earliest habitation of Malta is firstly wrong and secondly based on gerrymandered evidence. What's your general opinion about that?

Trump: That on a matter such as this I trust Dr Kenneth Oakley and his followers far more than I trust Dr Anton Mifsud.

Sharif: OK, so by referring to Oakley you're specifically referring to the same chemical tests carried out at the Natural History Museum in London that Mifsud reported upon at length in *Dossier Malta*?

Trump: Yes.

Sharif: Have you read *Dossier Malta*?

Trump: Yes.

Sharif: And so you're aware of all the specific allegations and claims?

Trump: Yes.

Sharif: And you don't accept Mifsud's evidence and allegations particularly regarding the chemical tests.

Trump: Frankly, no. I'm not a chemist, I can't give an expert opinion on the details of this. But I certainly trust Dr Kenneth Oakley much further than I trust Dr Anton Mifsud in his arguments.

Sharif: I'd also like to ask you, do you accept any of the claims in *Dossier Malta* about humans being there before the Neolithic? Is any of that likely or plausible?

Trump: The latest evidence suggests that it would have been much easier than we had allowed for Palaeolithic humans to have reached Malta. But, and it's a very big but, we have no evidence whatsoever that they actually did. I'm quite prepared to believe it's possible. If evidence were advanced I would give it all consideration. I would certainly not rule it out of court out of hand. To my mind, no reliable evidence has yet been advanced . . .

Sharif: To come back to the tests done at the Natural History Museum: In your own *Archaeological Guide* – there's an updated edition from March 2000 – you refer to them as a 'careful chemical analysis' and state that this analysis confirms that the Ghar Dalam taurodont teeth were not contemporary with the Cervus Layer.

Trump: These are the Oakley analyses.

Sharif: Yes. Can I ask, what was your source for that view, that the human teeth and the deer samples are not contemporary?

Trump: Yes, well the stratigraphic evidence such as it was – there was a certain amount of disturbance there – that there were three layers of interest in the cave. The lower one with your pigmy hippopotamus, elephant, etc. – no evidence whatsoever of human activity. Layer two, with the deer bones – still no confirmed human activity there. And then the upper level, which was largely mixed, with everything from Neolithic down to modern all jumbled up.

Sharif: Sure, I understand that. My question is specifically what was the academic source . . .

Trump: . . . for the analyses? Now, if I remember rightly, the first test done suggested that the teeth could have been contemporary with the deer bones at least – not with anything earlier. But the – I'm speaking from memory here . . .

Sharif: Sure, I accept that.

Trump: . . . further tests were done which, if not categorically disproving, strongly suggested that the teeth belonged with that uppermost level – could be as early as Neolithic, but not as early as the deer bones.

Sharif: OK, now, as far as I know, there are only two places that give these results. One is a review – a summary – in the 1964 Museum *Scientific Report*, which quotes a letter from Oakley, that's a 1964 source – is it that which you used to actually know what the results were?

Trump: No, it was personal communication from Kenneth Oakley himself.

Sharif: Oh, so did he give you a full list of the chemical results or just a summary?

Trump: No, he just discussed them in general terms.

Remember the 'Missing Link'?

There will always be some archaeologists who behave as though they are omniscient about prehistory. But though it has been said that Piltdown Man could never happen again, the amazing success and longevity of this extraordinary hoax – which began in 1912 and was not exposed until 1953 – is a reminder that when things do go wrong in the study of any area of the past they can go very wrong indeed. In the Piltdown case a false and (with hindsight) obviously absurd idea about the sequence of human evolution was sustained for forty years because it fitted in with the deep-seated prejudices and preconceptions of the British Empire (the Piltdown skull – claimed to be that of the 'missing link' between apes and men – was, naturally, British!). For the entire period until it was unmasked this counterfeited skull enjoyed all the prestige of a full scientific classification (*Eoanthropus dawsoni* – literally 'Dawn Man, found by Dawson') and pride of place in a display case in the Natural History Museum in London. So Piltdown was an embarrassing episode. And although, to their credit, the fraud that had taken in scientists for so long was also exposed by scientists, the net effect was to shake the public's confidence in the infallibility of science and of scientific judgement.

Here are the rudiments of the story, which is little spoken of today:

Fossilized fragments of cranium and jawbone were found [in 1912] by Charles Dawson in a gravel formation at Barkham Manor, on Piltdown Common, near Lewes, England. Together with these were fossil remains of extinct animals, which suggested an early Pleistocene age for the site . . . In 1953 and 1954, as an outcome of later discoveries of fossil man and intensive re-examination, the remains were shown to be skilfully disguised fragments of a quite modern human cranium and an ape (orang-utan) jaw fraudulently introduced into the shallow gravels ... The animal bones were found to be genuine remains of extinct species, but they were not of British provenance ... The eventual exposure of the fraud clarified the sequence of human evolution by removing the greatest anomaly in the fossil record. At the same time, a series of valuable new tests were developed for palaeontological study.[52]

And here are the connections with Ghar Dalam:

1. Amongst the remains of extinct animal species that the hoaxer had introduced into the Piltdown gravel in order to give authentic Pleisto-cene 'context' to the skull was a hippopotamus tooth. It is now thought that this tooth had come from Ghar Dalam.[53]

2. The same 'valuable new tests' which proved that the different fragments of bone assembled in the Piltdown skull *were not* contemporaneous with one another or with the animal remains introduced into the gravel were also run on the Ghar Dalam taurodont teeth in 1952 [the 'careful chemical analysis' referred to earlier by Trump] and suggested very strongly that they *were* contemporary with the deer remains in the cave's Cervus Layer.[54]

Or, to put it another way, the very tests that were accurate enough to prove Piltdown Man young and a fraud indicated that the Ghar Dalam taurodonts must be old and genuine.

Beyond truth and fiction

But if the Ghar Dalam taurodonts are genuine then why aren't we told this in Evans' *Prehistoric Antiquities*, the canonical text of Maltese archaeology that was published almost twenty years after the results of the 1952 tests were known? Or was Evans correct in 1971 when he promulgated the dogma that 'the two taurodontic molars can hardly be accepted as good evidence for the existence of man in the Maltese islands in pre-Neolithic times'?[55]

Anton Mifsud's approach to this investigation was to set aside all precon-ceptions and prejudices – both his own and those of the archaeologists – about whether or not Malta could have been inhabited by humans in pre-Neolithic times. He took the view, consistent with his personal philosophy, that all that should matter, and be weighed up, were empirically verifiable facts. In the case of the Ghar Dalam taurodonts the 'best' facts (i.e., those that most clearly speak for themselves without requiring interpretation) fall into two categories, both of which are well understood by archaeologists.

On the one hand there is the superb *stratigraphy* of the site – the distinct layers of deposits laid down one on top of the other at different times. Archaeologists all over the world routinely derive dates and sequences of dates from stratigraphy such as this. And, indeed stratigraphically, the human remains at Ghar Dalam lie contemporaneously with Pleistocene red deer and other extinct fauna in the deer layer'.[56]

Secure stratigraphy on its own should have been enough to confirm the presence of Palaeolithic man on Malta. From the beginning, however, J. D. Evans would not accept the obvious implications, raising the objection that the teeth must be intrusive. So the question now, as Mifsud explains, is not whether the teeth were really found in the Deer Layer – because they certainly were – but whether they were there as a result of 'an intrusive later burial by Neolithic humans, or else an actual deposit of the remains of Palaeolithic humans together with the remains of the deer layer fauna during the late Pleistocene'.[57] And in order to answer that question stratigraphy on its own, no matter how good, is not enough. What's needed is the record of the *scientific tests* that were done on the Ghar Dalam teeth in 1952 at the Natural History Museum in London.

Mifsud travelled to London and, after some detective work, managed to find the original records in the vaults of the Natural History Museum. To make sense of them we first need to know more about the so-called FUN (*F*lourine, *U*ranium, *N*itrogen) tests that the Museum conducted on the Ghar Dalam teeth in 1952.

Oakley's FUN

Although some of them had a long prior history, the FUN tests had been modified and developed by the British palaeontologist Kenneth Page Oakley of the Natural History Museum, apparently with the specific intention of confirming or denying the antiquity of the Piltdown skull.[58] But it is a little-known fact, now clarified by Anton Mifsud's research, that these tests were first applied in 1952 to human and animal remains from Ghar Dalam (and also, as we shall see, from the Hypogeum in Malta) – i.e., a year *before* the same tests were used with such devastating effect on the Piltdown skull in 1953. Mifsud notes that Oakley 'was in Malta on several occasions, on holiday, and as the guest of (the Maltese palaeontologist and geologist) George Zammit Maempel, with whom he shared common scientific interests'.[59]

In assessing the Piltdown skull, Oakley began by measuring the concentrations of *fluorine*. The surprise, notes Mifsud, was that:

> The skull and the jaw gave readings that set them wide apart in time by several tens of millennia. The other scientific tests, including *Nitrogen [and] Uranium Oxide* ... confirmed the hoax ... Pursuing the matter further Oakley then sought the origins of the associated remains of the Piltdown assembly. The hippopotamus molar gave a low fluorine reading which immediately suggested its source from a Mediterranean limestone cave, such as a Maltese one, typically Ghar Dalam. Tests on Ghar Dalam hippo molars confirmed the suspicion.

Malta thus became involved and this led to the performance of the same repertoire of chemical tests on the other finds at Ghar Dalam ... These chemical tests had by this time established themselves as the most reliable indices for the purposes of relative dating of archaeological specimens elevated from the same horizon ...[60]

So the tests that were conducted on the Ghar Dhalam teeth and other material from the Deer Layer were the best and most appropriate tools available in the 1950s for settling what is indeed 'the basic question' of the taurodont controversy: were the human teeth deposited in the Deer Layer at the same time as the rest of the layer was laid down, i.e., between 18,000 and 10,000 years ago, or were they introduced into it later than 7200 years ago in the form of a burial by the Neolithic people responsible for the Cultural Layer?

Here are the bare minimum of details about the tests necessary to understand the results:

Flourine and uranium

These two tests work because after death and excarnation the bones and teeth of animals and humans deposited together in the same environment – as well as such substances as deer antler – absorb fluorine and uranium from their surroundings. From environment to environment the *supply* of fluorine and uranium changes – the less there is, the less the bones, teeth and antlers can absorb, and vice versa – but *within* any given context the rate of absorption of the available local fluorine and/or uranium will be the same for any bones, teeth and antlers deposited there. Thus 'the estimation of fluorine confirms or refutes contemporaneity of bones and teeth in the same horizon'.[61]

Example: if human teeth and deer bones and/or antlers are excavated from the same stratum ('horizon' in archaeology-speak), and if the teeth prove on testing to contain much *lower* levels of fluorine or much *lower* levels of uranium (or much *lower* levels of both) than the deer remains, then the implication would be that the teeth must be much *younger* than the deer remains and are thus intrusive to the horizon. If, however, the environment is one known for its particularly low levels of natural fluorine, such as limestone cave-systems like Ghar Dalam, then the fluorine test obviously becomes less useful the lower the local level of fluorine is – and of no use at all once that level reaches zero. But with this proviso, and with the passage of time:

> Both elements accumulate in greater amounts. When bones are buried in different levels at the same location, older bones positioned in lower levels show greater amounts of fluorine and uranium than do those positioned above them. The accumulation of both elements is dependent on time and water-action present at the location. In view of the low concentrations involved, fluorine estimation may not be ideal for limestone environments, but once measurable amounts are present conditions are more suitable than if the percolating water is saturated with the mineral.

Levels of uranium oxide in modern bone are practically nil, but in ancient buried bone these may rise as high as 1000 ppm [parts per million] depending on the concentration of uranium oxide in percolating water. Aitken gives the range in fossil bone as lying between 1 and 1000 ppm. Trace amounts of fluorine are present in modern bone, ranging from 0.01 to 0.1 per cent in human bone, from 0.024 to 0.07 per cent in adult dentine of tooth, and between 0.02 to 0.1 per cent in Red Deer bone. Thus the maximum ever in modern specimens of tooth and bone in man and deer is 0.1 per cent.[62]

Nitrogen

This test works the opposite way round from the other two. Unlike fluorine and uranium, which can only begin to accumulate from the surrounding soil and its percolating water after the death and burial of the organism, nitrogen is accumulated in bones and teeth etc. only during life and then begins to dissipate. Thus after death the general rule is that 'nitrogen decreases with increasing bone age':[63]

Bone and teeth contain a certain percentage of nitrogen, averaging 3.4 per cent in teeth and 4 to 5 per cent in bone. Following death and burial organic remains lose their nitrogen with time, once the requirements of its breakdown are available. These include the absence of glacial conditions, an alkaline medium, absence of surrounding clay, and the presence of a specific bacterium, the Clostridium histolyticum . . .[64]

Example: if human teeth and deer bones and/or antlers are excavated from the same horizon and if the teeth prove on testing to contain much *higher* levels of nitrogen than the deer remains, then the implications would be that the teeth must be much *younger* than the deer remains and are thus intrusive to the horizon. The provisos, however, are many: if the environment lacks the bacterium necessary for nitrogen breakdown, or is glaciated, or surrounded by clay, then nitrogen is retained in any buried teeth and bones and the depletion of nitrogen becomes less useful as a test of relative antiquity.

'In effect, therefore,' Mifsud concludes,

a low nitrogen is useful to indicate antiquity, whereas a high nitrogen is not significant unless it is associated with a low fluorine and uranium oxide, which will definitely indicate a recent specimen. Conversely, the presence of a low fluorine and uranium oxide is not significant in the presence of a low nitrogen, for there are factors which impede fluorine and uranium oxide uptake, particularly in limestone caves of which Ghar Dalam is one. On the other hand a high fluorine and uranium oxide is significant in reflecting antiquity.[65]

Politics and ambition

Mifsud's view is that the results of the 1952 FUN tests on the Ghar Dalam teeth got caught up in a number of matters incidental to the proper concerns of archaeology that made it expedient for them either to be ignored, or better still discredited, as evidence of a Palaeolithic human presence on Malta. Of these the two most important were local politics on the one hand and the academic ambitions of the late Professor J. D. Evans on the other.

As to politics, Malta in the early 1950s was pursuing an integrationist policy with Britain. Absurd as it seems now there was embarrassment in official circles that the taurodont teeth might prove the modern Maltese to be directly descended from primitive Neanderthal ancestors (although, as we've seen, taurodontic teeth on their own, even specimens as large as the Ghar Dalam molars, do *not* necessarily prove that the original owners of those teeth were Neanderthals, since the condition still exists to varying degrees in modern, non-Neanderthal humans today).

As to the second matter, Mifsud notes that J. D. Evans had graduated from Cambridge in 1949 and that in the early 1950s he was 'in desperate need of a PhD'.[66] The thesis that the future Professor of Prehistoric Archaeology at the University of London chose to develop, influenced by the Italian archaeologist Barnarbo Brea, was that the very first human inhabitants of the previously unpeopled Malta had been immigrants from the Neolithic Stentinello culture of Sicily – a theory that is still part of the conventional academic wisdom about Malta today. In pursuing this thesis, Mifsud suggests, it was not convenient to the young Evans to have to deal with the evidence of the Ghar Dalam teeth that suggested a prior, Palaeolithic, human presence in Malta.

This, then, either as a conscious or unconscious motive, could explain why Evans was so vehement in his attacks on the antiquity of the taurodonts and so economical with the truth in his published statements about them. He wanted them out of the way – permanently – of his own theory about Malta's first inhabitants.

A tale of two museums

To get to see the records of the 1952 tests Anton Mifsud at first expected that he would need to travel no further than the distance from his own home to the Valletta headquarters of the National Museum – on behalf of which the Natural History Museum in London had carried out the tests in 1952.

This turned out to be rather a naive expectation. But what Mifsud did discover in Valletta was that:

> On 3 March 1952 Dr J. G. Baldacchino registered the sampling of the taurodont molar discovered by Despott in 1917. Other remains from Ghar Dalam cave included another tooth which was picked up by Caton-Thompson in 1924, a taurodont molar elevated by Baldacchino in 1936, and one sample each of a hippo molar and deer

longbone. *There is no record in the Museum of Archaeology Reports [for 1952/3] of these tests being carried out.*[67]

At this point Mifsud flew to London, where he found, to his relief that:

> The Green Book at the Museum of Natural History is still available and contains the original readings of the entire repertoire of tests carried out between 1952 and 1968/9 on the 'Malta Samples'. Two teeth from the Hypogeum were also included.
>
> The five human teeth submitted to the Natural History museum were therefore Caton-Thompson's (Ma.1), Despott's (Ma.2), Baldacchino's (Ma.7) and two molars from the Hypogeum (Ma.5 and 6).

The results in the Green Book were not what Mifsud expected. The fluorine test for Despott's molar (Ma.2) gave the highest results of all the samples tested, including the Pleistocene deer and hippo samples. This result did not jibe with the official position that the 'careful chemical tests' had proven the tooth to be Neolithic. The nitrogen result for Despott's molar was 1.85 per cent. Were it not for the fluorine readings, this result would have been compatible with the official position. But as it stands, the disparity revealed in the Green Book only suggests that either the fluorine tests or the nitrogen tests – or both – were unreliable. The official position should therefore have been that the results for this tooth were internally inconsistent and hence ambiguous.

Baldacchino's molar gave very similar nitrogen results to the two hippo molars from Ghar Dalam (0.44 per cent compared with 0.4 per cent for the hippos), clearly suggesting contemporaneity between the human samples and the Pleistocene animal samples in the Cervus Layer. Again, this result is incompatible with the official claim that these chemical tests proved the human teeth in the Cervus Layer to be Neolithic intrusions.

The tooth Ma.1 was discovered in the mid-1920s by Gertrude Caton-Thompson, the British archaeologist cited earlier whose views were misinterpreted by Evans. According to Caton-Thompson's notes, it was found in an 'unstratified layer' in the company of hippo, horse, deer, thirty potsherds and the end of a flint blade. This tooth yielded fluorine results (0.2 and 0.3) equivalent to those of the Pleistocene deer samples (0.25 and 0.3). It also yielded similar nitrogen results to Baldacchino's molar, with two different tests yielding results of 0.39 per cent and 0.79 per cent. That the same tooth yielded such different results highlights the unreliability of the nitrogen testing in a similar manner to the fluorine-nitrogen inconsistency of Ma.2. Nevertheless, these nitrogen results overlap with the readings for the Pleistocene hippo samples (0.4 per cent), and further raise the possibility of a Pleistocene date. This is especially interesting because the tooth bore no signs of taurodontism, and might therefore be taken as evidence of a Palaeolithic presence of humans on Malta who had normal, non-taurodontic teeth.

It is worth noting here that another non-taurodontic tooth was discovered in Ghar Dalam in the 1920s, this time by George Sinclair, a civil engineer with the British Admiralty. The tooth was buried almost a metre deeper than Despott's 1917 molar,[68] and it is unfortunate that it was not also submitted for chemical testing.

Ma.6, a human tooth from the Hypogeum, gave a nitrogen reading of nil. If we were to base everything on Oakley's nitrogen test, we would have to conclude that the owner of this tooth was alive way back into the Palaeolithic. However, it must be pointed out that Mifsud later managed to get this tooth carbon-dated through the Natural History Museum. As noted in chapter 17, the carbon-date put the tooth into the Late Tarxien phase, around 2200 BC. Along with the inconsistencies noted for Ma.1 and Ma.2, this further highlights just how unreliable Oakley's nitrogen dating technique can be.

The 1964 Report: erasing the Palaeolithic peril

No official record of the chemical testing was published until the Museum of Archaeology's 1964 report – a decade after the nitrogen and fluorine results had been achieved. During this lengthy hiatus only a very small number of people knew that the 1950s tests had ever been carried out at all and even fewer could have been aware of their results.

Such a delay in the publication of important modern dating evidence confirming a Palaeolithic human presence in Malta is plainly odd in itself. But that the gist of the evidence should subsequently have been misrepresented by the omission of crucial data when publication finally came about is far more extraordinary. Moreover, Mifsud believes the timing of publication in 1964 was not accidental. In that year there was already a dating furore in the air following the discovery that C-14 *underestimates* the age of materials that are more than about 3000 years old – and that there is a progressively larger underestimation the older the sample is. By 1964 this 'built-in' error had been accurately calibrated millennium-by-millennium by means of 'dendochronology' (comparison with the annual ring counts of very ancient species of trees). The implications of the new 'calibrated' dates for Malta were that the entire Temple Period suddenly had to be shifted a full millennium back in time. For example, before 1964 Gigantija was thought to be no older than 2500 BC; after 1964 and 'the tree-ring revolution' the date was pushed back to the presently accepted figure of 3600 BC.[69]

It is interesting to note that Evans was very slow to accept the implications of dendochronology for his carefully worked-out sequence for the Temple Period (the beginning of which he had hitherto set at 2500 BC) – and even as late as 1971 he was still refusing to let go entirely of his pre-calibration scheme.[70] But the tree-ring revolution was an irresistible force, like a rising tide – with implications for radiocarbon-dates all around the world – and even Evans in the role of King Canute could not hold back the waves.

The FUN tests in the 1950s were quite a different matter, done behind closed doors, strictly between the Natural History Museum in London and the National Museum in Malta. By 1964 the extremely annoying and inconvenient results of these tests had been withheld from the public for ten years with no one else any the wiser. Accordingly, there was no basis for protest when the National Museum published an abridged and unfortunately highly misleading version of the results in its 1964 Scientific Report. Whether by accident or by design, the net effect was that only information which supported the Evans paradigm was available on the public record.

The relevant passage from the 1964 Scientific Report itself reads as follows:

> Considerable help has been received from foreign experts in the analysis of Maltese material of various sorts.
>
> Dr K. P. Oakley of the British Museum, Natural History, analysed a number of bone samples for their collagen content, expressed as a percentage of nitrogen. The figures obtained were – hippopotamus bone, nil; deer antler 0.13 per cent; normal human tooth 0.7 per cent; taurodont human tooth (these four all from Ghar Dalam) 1.85 per cent . . . This proves conclusively that the taurodont tooth is later than the material from the other prehistoric sites, and so cannot possibly be of Neanderthal man.[71]

This statement contains paradoxes which seem all the more bizarre because they are left unacknowledged.

Firstly, the report only makes a conclusion about one human taurodont – Despott's molar classified as Ma.2 – and yet, as we've seen, Trump has used the results of these 'careful chemical tests' to draw a conclusion about two different Ghar Dalam taurodonts.

Secondly, why were the obvious inconsistencies in the data ignored? Caton-Thompson's molar is reported as having a nitrogen reading of 0.7 per cent, when in fact, according to the Green Book, it yielded the two very different results of 0.39 per cent and 0.79 per cent. More importantly, why did the report provide only the nitrogen reading for Despott's molar, and ignore the contradictory fluorine result? The nitrogen test had already proved itself capable of producing variable and hence unreliable results, so why was it given automatic and exclusive preference over the fluorine results?

Thirdly, if the nitrogen content of 1.85 per cent is supposed to be appropriate to the Neolithic, as we are effectively being told here, then does it not follow that the reported reading of 0.7 per cent from the normal human tooth from Ghar Dalam is indicative of a much older, pre-Neolithic date? And what does it say about Baldacchino's molar, which gave a nitrogen percentage of 0.44 per cent?

The results published in the 1964 Report misrepresent the actual set of results recorded in the Green Book in London. Had the complete set of results been included, or properly summarized, then it would have been clear that the results were to a large extent ambiguous, but also suggestive of a Palaeolithic human presence on Malta.

This misrepresentation of the actual results of the 1950s tests has subsequently had a pivotal effect on public perceptions of Maltese prehistory and on what university archaeology departments do and do not see as valid and worthwhile research on Malta. Unedited, the results from the chemical tests might have inspired a new generation of archaeologists to break away from J. D. Evans' 1950s 'Neolithic' paradigm and pay more attention to the possibility of much older relics around the Maltese islands – even underwater. But in the distorted form in which the results finally reached the public in 1964 (a mere decade after the tests had been carried out) there could be no danger that they would do any such thing.

Let's note in passing that seven years after the tests – and five before the misrepresented test results were first put on the public record in the Museum of Archaeology's 1964 Report – J. D. Evans had begun to talk as though conclusive results were already on the record. Here are three characteristic passages from his *Malta* (1959):

> There are as yet no trustworthy traces of the presence of man in Malta before the Neolithic period . . .

> We have no reliable evidence that any of them [Palaeolithic humans] made their homes in Malta . . .

> We have no reason to suppose that Palaeolithic man ever set foot on Malta.[72]

'The logic of Evans' conclusions', comments Mifsud, 'was founded on false premises and a significant iota of misrepresentation . . . The weight of authority established his hypothesis as semi-dogma; the consequence was bad history.'[73]

The uranium control

We've seen that the fluorine and nitrogen results for Despott's molar (Ma.2) contradicted each other. The former suggested a Palaeolithic date, whereas the latter – the one which was published – suggested a Neolithic date. But a third test was later carried out on Despott's molar that had not been carried out on the other two teeth in 1952. This was the uranium oxide assay – a more sophisticated procedure which was not yet fully established in 1952 and which was only applied to Despott's molar in 1968. This later test took place, Mifsud has discovered, at the specific request of Kenneth Oakley – who also asked that it be carried out on Baldacchino's (1936) taurodont molar at the same time.[74]

The uranium oxide result for Despott's molar supported the flourine result and embarrassingly contradicted the high nitrogen result that had been published in the Museum's Scientific Report in 1964 as proof that the tooth was

Neolithic. The result of the uranium assay was 13 ppm, compared with 0.1 ppm or less in living bone and levels of between 4 and 12 ppm in various Pleistocene hippo and deer samples from Ghar Dalam. Ghar Dalam is an environment with low levels of uranium oxide (and fluorine) in the percolating water, so it is very hard to see how Despott's molar could have accumulated so much of it within just 7000 years. As Mifsud sums up:

> The dating to the Neolithic in the 1964 Report could not be sustained in the face of the 13 ppm reading . . . Despott's molar has survived to tell its tale . . . Its fluorine and uranium content ranks it contemporaneous with the fossil fauna of the Cervus Layer.[75]

As one might expect, knowing all the facts, this uranium oxide result was never published. The problem posed by the chemical results to the Neolithic date of first human settlement favoured by orthodox theory and confirmed in the 1964 Report has been efficiently dealt with by archaeology by simply ignoring the disturbing fluorine and uranium results whilst focusing only on a highly select-ive group of results from unreliable nitrogen assays. In consequence, until Mifsud rooted them out from the pages of the Natural History Museum's Green Book and published them in *Dossier Malta* in 1997, neither the fluorine nor the uranium levels of any of the Ghar Dalam teeth were known outside the narrow circle of the two museums.

'Adjustments'

In *Dossier Malta*, Mifsud claimed that the inconsistency between the uranium and fluorine results on the one hand and the high nitrogen result on the other could be explained by forgery. He took photographs of the Green Book during his visit to the Natural History Museum and spotted that there appeared to be two layers of ink in the box containing the nitrogen result for Despott's molar. The bottom layer gave '.8 per cent' (i.e. 0.8 per cent, but without the zero). The top layer, in a different shade of ink, added a 1 and 5 to this result to give '1.85 per cent'.

Anthony Frendo, Head of the Department of Archaeology at the University of Malta, initially concluded that the nitrogen results published in the 1964 Report effectively demolished any possibility that Palaeolithic humans had lived on Malta.[76] But in what amounts to an extraordinary endorsement from the heart of the establishment, Frendo concedes that Mifsud's research has now shown those nitrogen results to have been 'tampered with' and the fluorine and uranium oxide tests suppressed so as to create a false Neolithic chronology for the human teeth from Ghar Dalam:[77] 'This means that early man must have come to the Maltese islands in pre-Neolithic times.'[78]

How have other archaeologists reacted to Anton Mifsud's accusation of for-gery and its seismic implications for the orthodox paradigm of the Neolithic origins of Maltese civilization? On the latter point there has simply been no reaction. Maltese prehistoric archaeology continues on its Neolithic way, seemingly untroubled. On the former point, like Frendo, John Samut Tagliaferro of Malta's Museum of Archaeology agrees that the final figure now to be seen in the Green Book 'of 1.85 per cent of nitrogen content for the molar Gh.D/2 [Despott's molar, coded by the Natural History Museum as Ma.2] was superimposed on the original result, namely that of 0.8 per cent'.[79]

Unlike Frendo, however, Tagliaferro says he sees nothing sinister in the superimposition. He argues that all the samples from Malta were subjected to more than one nitrogen assay at the Natural History Musuem – and these sometimes produced different results, quite properly leading to 'adjustment', after the second test, of the figures yielded by the first. In the case of Despott's molar the original figure had been written as '.8 per cent' (with no zero preceding the decimal point). The fact that this figure was then overwritten so that it would read '1.85 per cent' could be easily explained as the result of such an 'adjustment' after retesting.[80]

I had already seen ample evidence that the chemical test results had been misrepresented and so was prepared to consider the possibility of forgery. But I had also seen enough to convince me that the chemical tests were capable in themselves of producing inconsistent and ambiguous results. I was therefore not prepared to accept Mifsud's allegation without following it up and offering the Museum a chance to rebut it. I also wanted to see the Green Book for myself and, with the help of Channel 4 and permission from the Museum, get its data – which were obviously controversial even if one disregarded the forgery allegation – on film.

Tackling the Natural History Museum (1): controlled access

Our contacts with the Museum unfolded over a period of several months and were handled primarily by my research assistant Sharif Sakr with occasional back-up when needed from Roy Ackerman, Head of Programmes at Diverse Productions (the company making my TV series for Channel 4). Here is the transcript of Sharif's opening (11 July 2001) telephone conversation with an official (name withheld) who deals with access to records at the Museum's Palaeontology Department:

> *Sharif:* Hi, my name is Sharif Sakr. I just spoke to a colleague of yours in the archives department, and she recommended I speak to you. I'm calling from Diverse Production, a TV company in London, and we're making a documentary that's going to involve some Maltese prehistoric archaeology, and as part of the research and filming, I'd like to know if it's possible for me to get access to this thing called the 'Green Book', which contains records of bone analyses done between 1952 and, I guess, the late 1960s, on some Maltese teeth.

Official: I don't know if it's possible or not.

Sharif: You don't know if it's possible or not?

Official: I don't know if it's possible, because it relates, it has information about human remains, and that would be available really only to academics – people who are doing academic research, such as, we've had people from Malta do research on that in here . . .

Sharif: Really? People have already come to look at the stuff I've talked about?

Official: Yeah, that's right, Dr Anton Mifsud has looked at that book, but he was an academic. But merely for the sake of filming – you know, what's the point? What's the use of that?

Sharif: You know, to get the actual numbers on camera, if possible.

Official: Erm . . . No . . . I think that will not be possible, basically, without permission from a much higher level than me, basically.

Sharif: Well, who would that be?

Official: Well, a letter from your head of department or head of, whatever you are, to Dr Louise Humphrey here, who deals with access to human remains.

Sharif: Do you have any contact details for her?

Official: Yes, Dr Louise Humphrey, at the address of this museum, which is the Natural History Museum, Cromwell Road, London SW7 5BD.

Sharif: OK, thank you. What about, well, cameramen aside, what about the possibility of me, you know an individual without any cameras or anything, coming in to browse through this book?

Official: Good heavens, no! No documentation relating to human remains is available for browsing by non-academics. I mean you're not doing academic research, so you don't get to see the documentation, it's as simple as that.

Sharif: OK, that's quite clear.

Official: That's basically the rule that we're following now, in relation to the anthropology collection.

Sharif: And that rule exists for reasons of preservation or ethics?

Official: For reasons of ethics, I guess, more than anything else.

Sharif: What if I said that the samples I wanted to see weren't solely human – in fact a number of them were hippopotamus bone . . .

Official: Yes, but it's documentation within the anthropology section. So it gets regarded as . . . er . . . and also it might be unpublished, even if it's fifty years old, I'm not sure if the information's been published or not. If it's been published, then why would you want to see the original notebooks where the results are recorded?

Sharif: Well I can answer that, because this man, I've never met him, but Anton Mifsud is claiming that in fact the results have been ignored, i.e., not published, misrepresented when they were mentioned, and in fact he's even claiming that there was some tampering going on, such that the only real place where you're going to find these results in their original form from 1952 is in the Green Book, and that's why it's so important to see that book, rather than secondary evidence, for example in the National Museum in Malta.

Official: Another thing that comes to mind is that, er, if the reputation of the Museum is at stake, then probably the director of science would have to look at this first, you know . . .

Sharif: Well, it's not at stake – your museum is supposed to have the untampered-with evidence . . .

Official: It's just that you may misrepresent whatever we have, and that would mean that we get embroiled in all kinds of funny tests going on, about whether it's one pen or three pens or five pens on a piece of paper, whether it was written in 1950 or 1960 or 1970, which would go on and on for months and weeks and there'd be no end to it . . .

Sharif: I was wondering . . . you talk as if you've had experience of Anton Mifsud . . .

Official: Oh, I do . . .

Sharif: Was he annoying, was he dishonest?

Official: No, he was very pleasant, without a doubt. You know he was . . . But perhaps what I would say is that erm, between, since then the whole climate in relation to human remains has changed . . .

Sharif: So it's controlled access now . . .

Official: It's controlled access. The ethics of human remains.

Tackling the Natural History Museum (2): one of our pages is missing

As the official suggested, Sharif made contact with Dr Louise Humphrey concerning our request to film the relevant page in the Green Book containing the altered nitrogen figure for Despott's molar (code number Ma.2). On 26 October 2001, Dr Humphrey presented us with an astonishing piece of news. We would not be able to film the page containing this test result – or at any rate not at the Natural History Museum – because it was 'missing'.

Ironically, the only place in the world where a true copy of it could now be found was in Anton Mifsud's 1997 *Dossier Malta* where he had reproduced his photographs from the Green Book. Perhaps, Dr Humphrey suggested, we would like to film his photographs instead? While we were at it, she added, could we please ask Dr Mifsud to send a photograph to her as well so that she could use it to replace the missing page in the file?

It was in this e-mail that Dr Humphrey offered the Museum's rebuttal to Mifsud's allegation of forgery. Humphrey had managed to find the original laboratory reports from which the results had been taken and entered into the Green Book. These lab records contained a reading of 1.85 per cent for Ma.2, effectively proving Tagliaferro's suggestion that this later result was genuine and a proper substitute for the original figure of 0.8 per cent.

From: Louise Humphrey
To: Sharif Sakr
Sent: Friday, October 26, 2001 12:43 PM
Subject: Green Book

Dear Mr Sakr

Thank you for your e-mail of 18 October. The page listing results for Ma.1–Ma.7 in the Green Book is missing and I have not been able to find any evidence for where it might be. We know that the page was still present in 1995 since Dr Mifsud states in the acknowledgements of his book that he photographed the page when he visited this Museum on 10 August 1995. Fortunately, Dr Mifsud does have photographs and, according to his acknowledgements, photocopies of the relevant page of the Green Book. It would therefore be possible for you to film these copies for your programme. I would also be grateful if you could ask Dr Mifsud to send me a copy of his photographs, photocopies or both to replace the missing original page in our files.

Some of the results in the Green Book were compiled from other primary sources, for example, correspondence between Museum staff and staff in the laboratories where the analyses were conducted or forms completed during the process of analyses conducted here. I have not found primary records to back up all of the results summarized in the Green Book, and it is possible that some analysis results were entered directly into the Green Book. The departmental archives include two files of correspondence between Dr Oakley and staff at Microanalytical Laboratory in Oxford where nitrogen determinations were carried out, including letters detailing the results for all of the samples from Malta. For example, one letter, dated 17 June 1955, gives the analytical result for Ma.2 (1.85% N). Dr Mifsud's claim (e.g., page 96 of his book) that Dr Oakley deliberately and fraudulently altered this result is evidently erroneous. The departmental archives also include forms completed during the process of uranium analyses, including those for several Maltese samples. The analysis of Ma.2 was carried out on 23 February 1967 and yielded a result of 13 +/− 1 [parts per million].

I should reiterate that each of the analytical techniques used to investigate the samples from Malta between 1952 and 1969 can yield anomalous or ambiguous results . . . Dr Oakley had many years experience working with these techniques and was probably better qualified than anybody to interpret the results and identify anomalies . . . Ma.6 is a very clear example of an anomalous result. The nitrogen reading is nil, indicating that the tooth had been buried for long enough for all the organic materials to be lost. Taken in isolation this result could suggest an early (e.g., Pleistocene) date, yet the radiocarbon date for this tooth is 4130 +/− 45 (see *Archaeometry* 41: 421–431).

[NB Ma.6 is *not* one of the contested Ghar Dalam teeth but one of the Hypogeum teeth also assayed by Oakley. Mifsud's theory does *not* dispute but in fact *predicts* the dating to the Neolithic of the Hypogeum teeth, which he believes to have been swept into the underground structure from surface-level Neolithic graveyards by the agency of a flood – see chapters 16 and 17.]

Fluorine, uranium and nitrogen tests have fallen into disuse because more reliable and accurate dating techniques are now available. If the aim of your programme is to provide accurate scientific information, it would not be appropriate to rely on unpublished information using out-of-date techniques taken from historical archives. Results that are unpublished have not been submitted to peer review and do not carry the same weight scientifically as those that have been scrutinized by independent reviewers. I understand from your e-mail that it may not be possible to remove samples from the Maltese taurodont teeth for radiocarbon dating. Nevertheless, I think it is important to point out that the dating of the human teeth is insecure without this additional evidence.

Kind regards,

Louise Humphrey

Question-marks persist

Humphrey's e-mail helps to answer some questions, but leaves others unanswered and raises yet more.

On the forgery issue, Mifsud's allegation as it stands is clearly weakened by the proof that 1.85 per cent is a genuine test result. But does this necessarily mean that 0.8 per cent, written in the original layer of ink, wasn't also a genuine result? 0.8 per cent would make much more sense given the results of the fluorine and uranium oxide tests. Dr Louise Humphrey made it clear to us in a later e-mail that if there was a lab report containing a nitrogen result of 0.8 per cent for Ma.2, then she probably – but not definitely – would have found it. But it should also be pointed out that Anton Mifsud – who was kept informed of our correspondence with the Natural History Museum – stands by his allegation and expects to publish further evidence to support it in 2003. He intends to prove that whereas Ma.2 was tested for fluorine in 1952, *no sample* was taken from this tooth for later nitrogen testing, such that the 1.85 per cent reading actually corresponds to a different tooth that was substituted by someone – and Mifsud intends to show who – outside the Natural History Museum. It was knowledge of this dishonest switch, thinks Mifsud, that led an honest and concerned Kenneth Oakley to resubmit the original tooth for uranium oxide testing in 1968.

Forgery allegations aside, it is bizarre, and indeed rather disquieting, that the extremely important and controversial page from the Green Book containing results which were misrepresented by scholars should have been present in 1995 and should have gone 'missing' subsequently – without any trace or explanation, as Dr Humphrey admits. For whereas one might expect items of primary evidence to 'disappear' during a mobster trial, it seems inappropriate for the same sort of thing to happen in an archaeological dispute. Moreover, staff at the Museum are obviously well informed about the very serious allegations made in *Dossier Malta* in 1997. It therefore seems contrary to human nature that they would not at that time have opened up the Green Book to have a look at the page Mifsud claimed had been misrepresented and 'corrupted'. If so, does it not follow either that the page was still present in the Green Book in 1997 or – if it was found to be gone then – that no report was made of its disappearance at the time?

But the biggest question to arise from all of this concerns the chemical tests themselves. If Oakley's chemical tests on the Ghar Dalam teeth are really as obsolete and insecure as Dr Humphrey claims, why were they still being used in 2000 to contradict the stratigraphic context of the teeth and demonstrate that they are Neolithic? And if the tests aren't as useless as Dr Humphrey claims, can we really accept her assertion that only the late Kenneth Oakley was sufficiently versed in his own esoteric techniques to be able to interpret their results? If the orthodox position rests on nothing more than a missing page of numbers that have been subject to highly misrepresentative publication and deeply unfathomable interpretation, then does this position deserve to be considered 'scientific'?

An interpretation with feet of clay

Let me reiterate that the real issue in this saga is not the allegation of forgery but the *interpretation* that has consistently been put by archaeologists on the whole suite of results from Kenneth Oakley's chemical tests. Proponents of the 'Neolithic-first' theory of Maltese prehistory have claimed the results prove the human teeth from the Cervus Layer of Ghar Dalam to have been Neolithic, and thus several thousand years younger than the Cervus Layer and probably introduced by intrusive burial. This is the interpretation that has entered the history books and become orthodox. Yet we now know that it is based on disputed, ambiguous and internally contradictory evidence – which may be highly suggestive but which is frankly nowhere near good enough to settle such an important matter. Worse still, when we look closely at the FUN test results, as Mifsud has enabled us to do by publishing the full set of elusive figures from the Green Book, we find that what they are highly suggestive of – according to the standard rules of interpretation – is *not* the Neolithic date for the Ghar Dalam teeth claimed by the National Museum of Malta. Instead, the predominant overall pattern of high fluorine, high uranium and low nitrogen that these teeth manifest is, as reported earlier in this chapter, highly suggestive of a date in the Palaeolithic.[81] It becomes legitimate, therefore, to wonder why the 'Neolithic-first' hypothesis for Malta continues to be promulgated at all.

Sharif came at the problem in a roundabout way in a recorded telephone interview with Louise Humphrey:[82]

> *Sharif:* Do you mind me asking, what's the nature of the Green Book? Are these things filed as attached pages or they on separate pages that are easily removed.

Humphrey: It's a ring-binder.

Sharif: Would a page have to be ripped in order to be removed?

Humphrey: No, it could probably be opened, but it would be a hassle, because you'd have to pull out half the pages in the book – M being in the middle of the alphabet, because they're filed by country.

Sharif: Because, you know, the basis of our whole story here is this idea that something is going wrong with the preservation of records relating to Malta. And that's why, I mean, I understand it would be much better to go and see these results in a published . . . in an academic periodical with comments about them, but they never made it that far except in very abridged and misleading form in a 1964 Scientific Report of the National Museum of Malta.

Humphrey: The reason they might not have been published is because they were considered suspect.

Sharif: Fine, that's what someone would say who's not directly involved. But the real story is why and how these exact same test results were allowed to be used from the beginning to support a Neolithic date. Because in 1968 and earlier, these were valid dating techniques and what they suggested on balance was that the human teeth from Ghar Dalam were not Neolithic but Palaeolithic.

Humphrey: It was the best that they had then . . .

Sharif: Yes. So it's really a question of representation, rather than truth. It is a question of what would happen if evidence was ignored – and it has been ignored in this case. It's almost not the point of the argument to use this as proof that orthodox opinion is wrong, as much as it is to show that certain personalities responsible for forming orthodox opinion about Maltese prehistory did not give proper consideration to evidence that might have contradicted their own position.

Sharif asked Dr Humphrey whether she herself did not feel it would be interesting to follow up the 'anomalous' uranium oxide reading of 13 parts per million for Despott's molar (Ma.2) – a reading, as we've seen, that is indicative of Palaeolithic antiquity for this tooth.

Humphrey: My interpretation – now, don't forget these techniques went out of use before I was born, I didn't even learn them at university because they were obsolete. But my interpretation of them now, as a non-expert, is that they're inconclusive, that they're ambiguous. Because, for example, for Ma.2 there is that seemingly very high uranium result that would suggest an early date. But you've also got a very high nitrogen result [the contested figure of 1.85 per cent] that would suggest a recent date . . . Erm . . . so to me that would be unsatisfactory. I would consider that inconclusive.

Sharif: OK, it's not my position to agree or disagree. But it is your position to stand up, not for what makes good TV and what doesn't, but for what's scientifically valid. So, I just wish the Green Book page was there . . .

Humphrey: So do I . . .

Sharif: To end it all . . .

Humphrey: Well, it wouldn't . . .

Sharif: No it wouldn't – what would end it would be to carbon-date the contested tooth.

Humphrey: Yeah, I think that's the only way you'd get any truth out of this.

Unfortunately, however, the Maltese authorities remain resolutely opposed to any carbon-dating of Despott's molar and have recently been reluctant even to grant access to it.

Limbo

Since David Trump had in good faith regarded the 'careful chemical tests' on Despott's molar carried out in the 1950s at the Natural History Museum as reliable, we thought it would be interesting for him to hear Louise Humphrey's view of the tests as 'inconclusive' and 'ambiguous'.

Sharif: I just want to ask one more question about these chemical test results. Now, the problem I have with it is, I've interviewed Dr Louise Humphrey at the Museum . . .

Trump: The Natural History Museum in South Kensington?

Sharif: Yes, exactly. Now, she's seen all the results in the Green Book, but only in Mifsud's book, which is today the only published record of these results in the whole world because, for some reason, the Museum don't know why, but they've lost the one specific page in the Green Book which has the human chemical test results for the Ghar Dhalam teeth.

Trump: That is a pity. Of course the people who are arguing against it will probably suggest this was all part of the conspiracy.

Sharif: Yes, that's basically what will happen.

Trump: These glorious conspiracy theories!

Sharif: I'm in no position to claim there's dishonesty or whatever. It's not really my interest. My interest is that in 2000, in your book, you described these as careful chemical analyses which effectively proved that the teeth were more recent than the deer bones. Now, Dr Louise Humphrey is saying, in 2001 – just last week – she's saying that these results are completely ambiguous and aren't really worth the paper they're written on. Therefore, she's saying, even if we found the lost page, it's not really relevant to archaeological inquiry on Malta. In other words, she's completely against Oakley's FUN testing because she regards it as obsolete and disreputable, basically. What would your view be on that, considering what you wrote in 2000? The reliability of these chemical test results, as they stand – and you've said you aren't an expert on them and I accept that . . .

Trump: I er . . . don't quite know what to say. Erm . . . the only thing to do would be to . . . get directly myself, before changing anything, the scientific opinion on these. And if that is exactly as you say, to admit that those tests did not prove what it was thought at the time they did. Could we please get some more tests done?

Sharif: Sure, particularly carbon-dating . . .

Trump: Well now of course we've got the AMS that can do it with very small samples, that might be possible.

Sharif: Mmm, particularly, there's this one tooth sample, Despott's molar . . .

Trump: But the deer bones are not . . . they're not Pleistocene, are they?

Sharif: Yes, the layer is Pleistocene. And the layer above it is, well Mifsud claims it is a relatively coherent stratigraphic layer . . .

Trump: I see . . .

Sharif: So that does make a barrier above the Cervus Layer which establishes that the Cervus Layer is Pleistocene. Obviously that doesn't rule out intrusion of later material into it . . .

Trump: No . . . Yes . . . If we could get a direct date on the teeth . . . That would put human occupation back earlier, well before the date we've got. But at the moment then, the whole issue is unproven.

Sharif: Yes, that is exactly my feeling. Particularly, Dr Humphrey would draw attention to this nitrogen reading of 1.85 per cent for tooth sample Ma.2 . . . You probably don't remember any of this . . .

Trump: I don't, and I wouldn't know what it meant.

Sharif: Oh sure, well basically there is a large amount of internal inconsistency in the results that are reported in the Green Book. These are the results done on the Maltese samples between 1952 and 1969. Now, from what I can tell from Dr Louise Humphrey who does seem to know her stuff, I must say, is that at best these results are ambiguous. And if you look at Mifsud, he does make quite a good case that particularly with the uranium oxide reading that, yes, they don't prove anything, but if they are *suggestive* of anything, it's of a Pleistocene date. So what Mifsud is actually alleging is that what Oakley reported in the official Museum publication in 1964 was not representative – and I'm not saying it's dishonest, maybe he took what he felt was representative – but the modern opinion is that those results he gave were not representative of the full set of results, the majority of which actually suggest a Pleistocene date. The fluorine and most of the nitrogen and particularly a uranium oxide reading for one of the tooth samples from Ghar Dalam [Despott's molar] are extremely suggestive of a Pleistocene date. Obviously, we'd much rather have carbon-dates – but unfortunately carbon-dates are not available for any of the Ghar Dalam teeth. So to what extent do you feel that what I'm saying – and I'm your only source for this apparently right now – but how do feel about this whole 'Neolithic-first' thing in the settlement of Malta if this point is made?

Trump: If this point is made, I would accept that we've got to reconsider the argument that the Neolithic settlers were the first on the island. I would do that quite willingly, if secure evidence is put forward. I've nothing against Pleistocene settlement of the island . . .

Sharif: Sure, sure, it's an academic question really, not a question of religion. But implicit in that point you just made is that you think that Oakley's chemical test results are quite pivotal to the 'Neolithic-first' case. They're an important strand of evidence supporting that orthodox model that there were no humans before the Neolithic. Is that right?

Trump: I think so, yes.

Sharif: Is there any other pivotal evidence that supports that?

Trump: Only the complete absence of any other evidence. And one has to admit that negative evidence is never reliable. It may just not have been found. But until either this evidence comes through securely or other evidence comes to light . . .

Sharif: Well I think the future lies with getting the National Museum in Malta to give access to these most controversial tooth samples from Ghar Dalam to allow them to be carbon-dated. I think that's the future – this is just my view – but until that's been done things are rather up in the air.

Trump: The whole thing is in limbo really, yes. Yes.

The miraculous transmutation of Baldacchino's molar

There are other matters that add to this sense of Maltese chronology in limbo. Readers will recall that, as well as two teeth with normal roots excavated in Ghar Dalam in the 1920s by Caton-Thompson and George Sinclair, there are altogether *three* taurodont teeth – Rizzo's and Despott's molars, both discovered in 1917, and Baldacchino's molar, discovered in 1936. Where reference codes have been applied to these teeth they are prefixed 'Gh.D' in the case of the National Museum of Malta, and 'Ma.' in the case of the Natural History Museum. Thus, for example, the Natural History Museum code for Despott's 1917 taurodont molar is Ma.2, for Caton-Thompson's normal tooth Ma.1, and for Baldacchino's 1936 taurodont molar Ma.7. The National Museum of Malta code for Baldacchino's molar is Gh.D/3.

Although Baldacchino's molar was one of the teeth assayed for its nitrogen level at the Natural History Museum in 1952, we've seen that the very low result of 0.44 per cent that it produced was withheld from the 1964 official Report on the tests. Then in 1971, Evans' *Prehistoric Antiquities of the Maltese Islands* somehow failed to mention the existence of Baldacchino's molar at all in its survey of Ghar Dalam, and discussed the taurodont controversy with reference only to the Rizzo's and Despott's molars. Since Evans' text remains the basic work of reference on prehistoric Malta, the net effect of this omission was to consign Baldacchino's molar to a research limbo – where it stayed until Anton Mifsud focused attention on it again in 1997 when he published the suppressed 1952 test results in *Dossier Malta*.

The odd thing is that when Baldacchino discovered the tooth in 1936 he described it as being heavily fossilized. Today the very few people who have been allowed to see it in the vaults of the National Museum of Malta report that it is *not* fossilized – and this mysterious transmutation is confirmed in Anton Mifsud's 1997 photographs 'where it is evidently identical in shade to modern molar teeth, rather than to the 1917 molars'.[83]

Even odder is the fact that a startling discrepancy exists between the very low result of 0.44 per cent obtained from the tooth in the 1952 *nitrogen* assay and the result of the *uranium oxide* assay that was carried out on it in the 1960s at Kenneth Oakley's request (this was at the same time that Oakley also ran the uranium assay on Despott's molar). The nitrogen result makes Baldacchino's molar very old – definitely Palaeolithic. But the uranium assay gave 'a nil reading for uranium oxide',[84] indicating that the tooth was most probably modern.[85] Last but not least, although the tooth now coded Gh.D/3 in the vaults of the National Museum of Malta is a taurodont, Mifsud points out that its degree of taurodontism is relatively minor – mesotaurodont or hypotaurodont, and that it certainly does not attain the very large hypertaurodontic type of the two 1917 molars.[86]

What should we conclude from these paradoxes? The obvious answer, Mifsud suggests, is that Baldacchino's molar was old when it was described as fossilized in 1936, and still old when it was assayed for nitrogen in 1952, but it was *no longer old* when it was assayed for uranium in the 1960s. In other words, a modern taurodont tooth – perhaps one of several that are known to have been extracted in Malta during the early 1960s[87] – was substituted for Baldacchino's Palaeolithic molar some time after its nitrogen test in the 1950s and before its uranium test in the 1960s.

It is impossible to guess who might have actually carried out the switch but it was undoubtedly facilitated by the peculiar lack of documentation that afflicted the tooth after 1952. As we've seen, Evans failed to mention it in 1971. Mifsud points out that it was also:

> omitted in subsequent references to taurodontism in archaic human remains. J. L. Pace (1972) and G. Zammit Maempel (1989) do not mention it in their contri-

butions.[88] It has never been published in a photographic form, so that a substitution was all the more easily possible . . .[89] Baldacchino's molar was kept in a box of its own separate from Despott's and Rizzo's molars. It was replaced by a modern taurodont and labelled as Gh.D/3. The same could not be done to Despott's and Rizzo's for they had been studied, photographed and radiographed by several workers.[90]

In the light of Mifsud's evidence about the switching of Baldacchino's molar, how can we be sure that Despott's molar wasn't also swapped for a modern tooth before being sent off for the nitrogen test in which it gave an anomalously high reading? Perhaps we should regard the switching – and apparent loss – of Baldacchino's molar as plain negligence, along with the ignoring and misrepresenting of crucial data in the Green Book. Still, I have to be suspicious of the fact that negligence in Maltese archaeology has always tended to remove evidence that threatened the 'Neolithic-first' theory of Maltese prehistory.

Anthony Frendo is courageous enough to stick his neck out from the ivory towers of the University of Malta to acknowledge, albeit carefully, that something is amiss:

> The evidence marshalled by Mifsud indicates that the tooth examined in 1968 . . . is not the same as that examined originally in 1952. There is no direct evidence to affirm that an intentional switch did take place, but it is well-nigh conclusive that the tooth in question is not the same.[91]

19 / *Inundation*

> One hears frequently of Malta's 'land-bridges'. Such there certainly were, at least
> north to Sicily – they are needed to explain the fossil fauna of Ghar Dalam for
> example – but not, as far as we know, at a period when there were men to take
> advantage of them. They are of great interest to the geologist and palaeontologist,
> but none to the archaeologist.
>
> Dr David Trump, 2000

Anton Mifsud and his colleagues have exposed the Palaeolithic skeleton
(and teeth!) in the cupboard of Maltese prehistory. But their investigation has
taken years of dedicated effort, patiently cutting through the misrepresen
tations, the omissions of contradictory data and the strange disappearances of
pivotal evidence that have allowed archaeologists to persist for so long with
the fiction that no humans reached these islands until the Neolithic around
5200 BC.

Since 1997, the National Museum of Archaeology has been embroiled in an
unwelcome local media controversy – that has refused to die down – about
the very grave charges set out in *Dossier Malta*. And since 1999 the evident
preference of senior officials that the 'Mifsud problem' should (like the Ghar
Dalam teeth?) just 'go away' has been further frustrated by the visible support
now being given by prominent archaeologists such as Anthony Frendo to the
demand for a complete review of the prehistory of Malta in the light of the
confirmed presence of Palaeolithic man.

But this ferment – which is really a struggle for the soul of Malta's past – has
so far remained very much an internal Maltese problem. Beyond the shores of the
islands, where *Dossier* has never been published or circulated, the international
community remains ignorant of the scandal – and the prehistory of the world's
oldest free-standing megalithic temples continues to be taught without any
reference at all to the Palaeolithic.

The tampering and selective loss of anomalous evidence that Mifsud alleges
is only part of the problem. Much damage, in my view, has already been done
by two generations of archaeologists 'conditioned' in the school of J. D. Evans,
who have tended to filter, or redefine, or file as 'out of context' any hints or
traces of human activities before 5200 BC that they might have come across in
their fieldwork in Malta. And I want to be clear that I am not attributing these
tendencies to any conspiracy. It's just a matter of how the rational mind works:

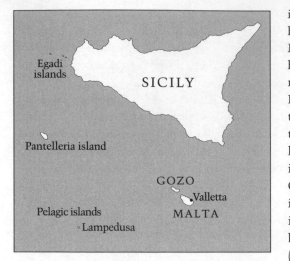

Egadi
islands

SICILY

Pantelleria island

GOZO
Valletta
Pelagic islands   MALTA
Lampedusa

if the foundation of everything you have been taught and believe about Malta is that it was first inhabited by humans in the Neolithic then this makes it much more difficult to see the Palaeolithic, even if it's there. Perhaps the most significant consequence, certainly until very recently, is a profound lack of interest amongst archaeologists in the fact that Malta, Comino and Gozo were joined to form one large island in the late Palaeolithic – an island that was in turn joined to Sicily by a land-bridge 90 kilometres long (and thence to the Italian peninsula).

David Trump says it all when he writes:

> One hears frequently of Malta's 'land-bridges'. Such there certainly were, at least north to Sicily – they are needed to explain the fossil fauna of Ghar Dalam for example – but not, as far as we know, at a period when there were men to take advantage of them. They are of great interest to the geologist and palaeontologist, but none to the archaeologist.[1]

This is from the updated edition (March 2000) of Trump's *Archaeological Guide*, published three years after the revelations in *Dossier*. The loud and clear message that it sends is that there is simply no point in looking underwater along the now-submerged land-bridge to Sicily to increase our knowledge of Maltese prehistory. On the contrary, Trump emphasizes, the land-bridge is of no interest to archaeologists because, 'as far as we know', there were no humans to take advantage of it.

However, it is clear that David Trump, unlike some of his colleagues, is an open-minded man. Interviewed in October 2001, he did not prove to be a stubborn or dogmatic adherent to the orthodox 'Neolithic-first' model of Maltese settlement, was genuinely disturbed to learn about the ambiguities and uncertainties in the full gamut of results from the FUN tests carried out by Kenneth Oakley in the 1950s and 1960s and concluded (see chapter 18) that until modern C-14 tests could be conducted to confirm the age of Despott's molar and the other Ghar Dalam samples, 'the whole thing is in limbo, really'.

Since this is also the view of Louise Humphrey at the Natural History Museum in London (again see chapter 18), it seems to me – though no one has perhaps noticed – that a Rubicon has already been crossed. As Trump admitted in his October 2001 interview with Sharif Sakr, the FUN results were so fundamental to the construction of the 'Neolithic-first' paradigm of orthodox Maltese archaeology that – if they are discredited – there remains *no positive evidence whatsoever for that paradigm*: 'only the complete absence of any other evidence. And one has to admit that negative evidence is never reliable. It may just not have been found.'

Trump's openness to the notion that evidence for a Palaeolithic human presence in Malta might simply not yet have been found prompted a question on his current views concerning the land-bridge issue.

> *Sharif:* OK, we'll move on. In your *Archaeological Guide*, you state some-where . . . oh yes . . . the land-bridge idea. You state that the land-bridge is of no interest to the archaeologist – have you changed your opinion on that? It might be relevant to help you gauge the likelihood of finding evidence of Palaeolithic man on Malta in the future.
>
> *Trump:* Well, we accept that . . . But I'd use the word possibility, not likelihood . . . If there was a land-bridge, that means the sea-level was very much lower – so all the most desirable countryside, coastal plains, etc., are deep underwater and there's no hope of finding evidence of it.
>
> *Sharif:* Well, what about marine archaeology? Would you be in favour in principle of marine exploration to see if there's anything . . .
>
> *Trump:* Not a hope. I mean if you've got shipwrecks or even drowned buildings then fair enough, but if you're looking for a scatter of flints on the bed of the sea, I don't think there's a remotest possibility of ever finding them.
>
> *Sharif:* Because Palaeolithic archaeological evidence is so . . .
>
> *Trump:* Scanty. I mean it's difficult enough, I won't say impossible, but it's difficult enough above water. Below water there's not a remotest hope.

Woven into Trump's view is the perception, shared by the vast majority of orthodox archaeologists, that human activity in the Palaeolithic was limited to a very simple material culture that left only scanty remains such as scatters of flints. The perception is a reasonable one, since this is all that any sites definitely recognized as Palaeolithic on land anywhere in the world have ever shown to the excavator. But this reasonable perception is also a self-fulfilling prophecy. It predicts that nothing surprising or unusual about the Palaeolithic – perhaps even 'drowned buildings' – who knows? – can be expected to be found at the bottom of the sea. And since this is the case, and the remains of Palaeolithic material culture are in general so scanty, there would not be 'a remotest hope' of finding them underwater.

It is easy to see how out of this perception flows the untested conclusion, at least where Maltese prehistory is concerned, that it is not worth looking underwater at all. Yet the possibility cannot be ruled out that the study of archaeological remains submerged at the end of the Ice Age could shed light on the mysterious origins of Malta's megalithic civilization with its apparently unprecedented temples, unlike any others known in the world, its elaborate goddess cult – distinctively Palaeolithic in general style and symbolism – and its few surviving traces of cave paintings executed in the same pigments of red ochre and black manganese oxide that were favoured by Palaeolithic artists.

Refuge Malta

The closing millennia of the Ice Age, between around 17,000 years ago and the arbitrary 'end' of the Palaeolithic 12,000 years ago, were not only a period of rapidly melting ice-caps and rapidly rising sea-levels but also a period in which climate conditions across Europe were wildly unstable and frequently extremely cold and arid (see chapter 3). In the high latitudes, until the kilometres-thick ice-sheets had melted, human life would have been impossible – while even in lower latitudes many of the vast areas of inland Europe that were nominally ice-free were reduced to bleak and inhospitable tundra.

Given such conditions it would have been natural for human beings – at any level of social development – to migrate towards warmer and more congenial climes. And we can tell from the distribution of fossil remains that this was certainly the survival strategy adopted by all 'cold-intolerant' animal species of the period – including game species such as red deer (*Cervus elephas*) that we know were hunted by Palaeolithic humans. Places of refuge where the local climate was for one reason or another less harsh – scientists studying the Ice Age use the technical term 'refugia' for such sanctuaries of life – were inevitably sought further and further south during the worst episodes.

Straddling the thirty-sixth parallel, *Malta is the southernmost point of Europe* – indeed it is further south than the cities of Tunis or Algiers in North Africa. And while Malta today is a small archipelago 90 kilometres from Sicily – which is itself separated from the Italian mainland by the Straits of Messina – we know that this was not the case at the Last Glacial Maximum 18,000 years ago.

We would know this even without the modern science of inundation mapping to show us the changes that transformed the antediluvian Siculo-Maltese landmass between 18,000 and 10,000 years ago. We would know it, as Trump rightly points out, because of the presence of large quantities of fossil fauna in Ghar Dalam such as the Pleistocene European red deer, wolf, brown bear and fox, which were not big swimmers and could only have come on foot to Malta by way of a land-bridge. Indeed, there is no dispute from any authority that during the extremely cold and arid periods that occurred several times between 17,000 and 10,000 years ago:

> man and animals could migrate from the Italian peninsula, by land, to the warmer climates of the Siculo-Maltese district. Herds of red deer left northern latitudes and settled in all parts of present-day Sicily, the present-day Egadi islands of Favignana

and Levanzo, and the Maltese archipelago, the latter site being the warmest of the Siculo-Maltese district during the Pleistocene.[2]

So here is the puzzle. On the tiny islands of Favignana and Levanzo, which, like Malta, were joined to Sicily (and hence to the mainland) during the Ice Age, there is abundant and undisputed evidence, including cave graffiti carbon-dated to 12,000 years ago, for the presence of Palaeolithic man.[3] Sicily, today the largest Mediterranean island, presents even more abundant evidence of an even more ancient human presence. As Anton Mifsud reminds us,

> Humans have indubitably inhabited it for much of the Palaeolithic, and it has a clear sequence of carbon-dated lithic implements, in places reaching back to the Acheulean [between 600,000 and 75,000 bp].[4] The caverns hold the same faunal assemblage as that at Ghar Dalam, namely Pleistocene hippo-elephant-deer fauna. Upper Palaeolithic cultures have been identified in all regions of Sicily, including the south-eastern region of the Hyblean plateau which abuts the Siculo-Maltese land-bridge of the Pleistocene . . .[5]

Only an attitude of stupefied indifference to the implications of the land-bridge for the mobility of Palaeolithic humans can explain why archaeologists did not become concerned much earlier by the apparently 'apalaeolithic' status of the Maltese islands – a status that seems acutely anomalous when viewed in its regional context and that becomes even harder to explain when we remember that Malta was the furthest south, the warmest and the most suitable refugium of the entire Siculo-Maltese landmass. Obviously, with the same cold-intolerant fauna roaming freely across the whole of that landmass – very much including Malta, as we know from Ghar Dalam – there is no good reason why Palaeolithic humans, who are everywhere else believed to have followed and hunted that same fauna, should not have reached Malta as well.

And, as we now know, they did.

The drowning of the land-bridge

G. A. Milne to Graham Hancock
13 July 2001, 19:22
Subject: Maps

Graham,
I ran some new high resolution predictions of sea-level change and made maps for the Tyrrhenian[6] and Mediterranean Seas. There are four .pdf attachments showing the coastline at the times
18.3 kyr BP

16.4 kyr BP

14.6 kyr BP

13.5 kyr BP

(these are all calibrated times). You will see that Malta became isolated between 16.4 and 14.6 kyr bp. The large loss of land area between 14.6 and 13.5 kyr bp is associated with the melting event known as Meltwater Pulse 1-A (about 15–20 metres sea-level rise in about 500 years around 14 kyr bp).[7]

I hope the maps are useful. There may have been some significant tectonic motion in this region that is not accounted for in my predictions.

Cheers,

Glenn

This was the second batch of Maltese inundation maps that Glenn Milne had sent me – the first batch, at lower resolution and wider intervals, covered the same Tyrrhenian region of the central Mediterranean as it had looked 21,300 years ago (21.3 kyr BP), 10,600 years ago, 4800 years ago and the present day.

Scrolling through each of the maps one after the other there were a number of immediate and obvious observations to make:

- Until 16,400 years ago Malta was still joined to Sicily by a land-bridge.
- The land-bridge was severed by rising sea-levels between 16,400 years ago and 14,600 years ago. However, the new straits created were at first extremely narrow and most of the mass of the former isthmus remained above water.
- Between 14,600 years ago and 13,500 years ago there were very dramatic losses of land and all the remaining parts of the antediluvian isthmus were swallowed up by the sea.
- Despite these losses Malta, Comino and Gozo were still joined to form a single larger island 13,500 years ago. But other than an extension a few kilometres in width along parts of the north-east coast, the surface-area of that landmass had been reduced to dimensions only a little larger than those of today.
- By 10,600 years ago the separation of Malta, Comino and Gozo had occurred and the islands were virtually indistinguishable from their modern appearance.

Before the flood: 18,300 years ago

The map opposite presents the region as it would have appeared 18,300 years ago. As well as revealing the much greater extent of Italy when global sea-level was at its lowest, particularly on the Adriatic side of the peninsula,[8] and the enlargement of Corsica, Sardinia and the North African coast, it demonstrates that the situation of the Maltese islands was utterly different from their situation today. Instead of being a tiny archipelago lost in the central Mediterranean,

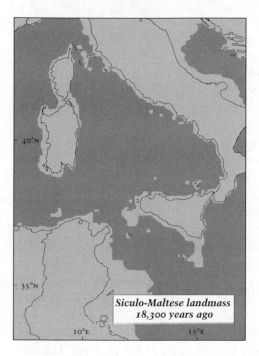

Siculo-Maltese landmass 18,300 years ago

Malta 18,300 years ago formed an integral part of the Italian mainland through the isthmus then connecting it to Sicily. The isthmus was approximately *twenty times* larger than the present Maltese islands, running not only 90 kilometres to the north but also extending more than 70 kilometres further to the south and east.

All along the north-east coast of antediluvian Malta, effectively an extension of the land-bridge, there was an exposed shelf, approximately 8 to 12 kilometres in width.

Along the south-west coast, although it is beyond the limits of resolution of the map to determine very fine details of the inundation sequence, it is certain that Filfla – which today is separated from Malta by a strait 3 kilometres wide – was not isolated. With that reservation, however, it is notable how relatively minor the changes along this coast seem to have been during the past 18,000 years – a product of the steep cliffs and sheer drop-aways to depths greater than the maximum fall in sea-level of around 120 metres.

I emphasize that the changes *seem* to have been minor quite deliberately in view of Glenn Milne's explicit warning that his model cannot take into account 'significant tectonic motion in this region'. The caveat is important because the central Mediterranean is one of the world's tectonic and seismic hotspots and has experienced massive volcanic eruptions and earthquakes routinely throughout the historic and prehistoric periods.[9] Sudden elevations and subsidence of land, which might have had dramatic effects on relative sea-levels at specific locations, are entirely possible in such an area. Indeed, as we've seen, this is precisely what Anton Mifsud suggests did happen in south-western Malta 4200 years ago following a cataclysmic fault collapse along the submarine Pantalleria Rift.[10] In addition to its well-documented Ice Age extensions to the north-east, north, east and south-east, we should therefore keep our minds open to Mifsud's suggestion that antediluvian Malta may have possessed a substantial extension to the south-west during the Palaeolithic that may have remained above water until it subsided catastrophically into the sea at the end of the temple-building period.

A final point of observation comes when we zoom out of the above map. With Malta and its land-bridge extending far to the south of the eastern tip of Sicily and with a similar southerly extension to Sicily's western tip – almost like two horns reaching out to touch the North African coast (itself also greatly enlarged)

– the eastern and western sides of the Mediterranean came very close 18,300 years ago to being divided into two separate seas. This enclosing and funnelling of great waters through narrow spaces could have enormously intensified the effects of the post-glacial floods when they hit the region. Indeed, as readers may recall from chapter 3, it has been suggested that at times the meltdown of the European ice-sheet into the Mediterranean was so severe that the Mediterranean 'bath tub' filled up more rapidly than the excess waters could drain out through the Straits of Gibraltar (which were reduced to a width of only 8 kilometres at the Last Glacial Maximum).[11] It has been proposed that such meltwater surges 'could have temporarily raised the Mediterranean by some 60 metres'.[12]

However, this calculation is based on the bottleneck effect caused by the narrow Straits of Gibraltar alone. Now we know that there would have been a second bottleneck between the Siculo-Maltese landmass and the North African coast which would certainly have made things worse – although how much worse is difficult to calculate. In addition, the consolidation of Corsica and Sardinia into one large island enclosed much of the Tyrrhenian Sea – and this would have further exacerbated the local effects of the meltdown there.

But hardly a trickle out of the vast reservoirs of meltwater hemmed in on the European ice-cap had yet reached the Mediteranean 18,300 years ago. Hardly a trickle over the previous 3000 years. Hardly a trickle – and spread out over so many generations that individuals would not have noticed the tiny ominous changes taking place.

Minor erosion: 16,400 years ago

Over the 1900 years between 18,300 years ago and 16,400 years ago the map opposite shows that there was only further minor erosion of the coastal margins and a narrowing of the Malta–Sicily land-bridge.

Malta becomes an island: 14,600 years ago

The map opposite documents the isolation of Malta some time between 16,400 years ago and 14,600 years ago. The event was not a particularly dramatic one in terms of land-loss, though it would undoubtedly have held great significance for the Palaeolithic Maltese who we now know were present then. For the first time they were cut off from the mainland. Perhaps this Palaeolithic isolation, rather than the Neolithic invasion that occurred more than 7000 years later, was the real genesis of the distinctive character and achievements of Maltese civilization.

The apocalypse: 13,500 years ago

It is in the map on page 422 that we see the effects of 'Meltwater Pulse 1A' – the first of the three global superfloods into which most of the 10,000-year-long meltdown of the Ice Age was concentrated (see chapter 3). As Milne points out,

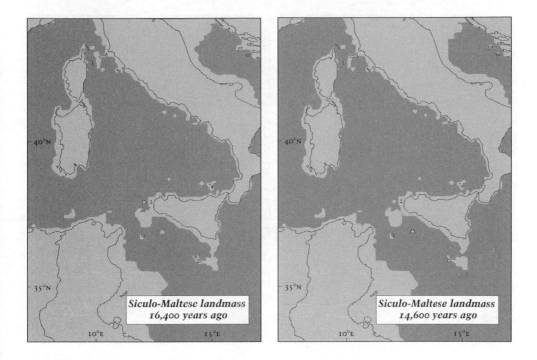

Siculo-Maltese landmass
16,400 years ago

Siculo-Maltese landmass
14,600 years ago

Meltwater Pulse 1A raised global sea-level by 15–20 metres in just 500 years around 14,000 years ago. That sounds bad enough. However, it is not necessarily the case that this very large rise was evenly spread out over the 500-year period resolved by inundation science. In my view the uncertainties regarding post-glacial events make it possible that all or most of it could have been compressed into a single event of much shorter duration anywhere within that 500-year period.

What the map at any rate reveals is that the newly isolated Malta of 14,600 years ago had lost 70 kilometres of its width by 13,500 years ago due to the complete and relatively rapid inundation of its former large extension to the east and south. No marine archaeology has ever been done on these submerged lowlands, which may conceal archaeological evidence of vital importance to the full understanding of Malta's prehistory.

The map also shows that Malta had actually become two islands 13,500 years ago – one, to the west, consisting of the present Malta, Comino and Gozo joined into a single mass, and the other, quite small, lying a little to the east. It is notable that other than this eastern islet nothing was left by this stage of the former grandeur of antediluvian Malta except a reduced extension 2 to 5 kilometres wide along the north-eastern coastal strip.

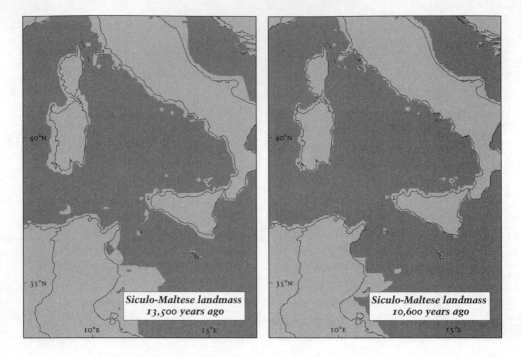

Siculo-Maltese landmass
13,500 years ago

Siculo-Maltese landmass
10,600 years ago

The end of the Palaeolithic: 10,600 years ago

By 10,000 years ago the Maltese archipelago was as it is today. The islet to the
east had gone, and the extension of the north-east coast had also been fully
submerged. It is somewhere on this north-eastern extension, however – the
very last part of antediluvian Malta to go beneath the sea – that the rumoured
underwater 'temple' sighted off Sliema by Commander Scicluna and by the
Arrigo brothers must be located. The implication of the inundation maps,
therefore, is that this structure was submerged between 13,500 years ago and
10,600 years ago – a date that can probably be pinned down more narrowly to
around 11,000 years ago, marking the second of the three episodes of global
superfloods outlined in chapter 3.

I want to re-emphasize here that the inundation maps have no bearing one
way or the other on the matter of Anton Mifsud's proposed extension of Malta
to the south-west – which he suggests was submerged by tectonic subsidence
as late as 4200 years ago. As noted earlier, the inundation maps cannot account
for large unregistered tectonic events in prehistory and such events are probable
in south-west Malta because of its proximity to the Pantalleria Rift.

The evidence that such an event did occur around 2200 BC is strong and may
prove to be the final key necessary to unlock the mysterious origins of Maltese
civilization.

20 / The Morning of the World

Graham Hancock: If we take the dating of a temple like Mnajdra or Hagar Qim, the better-known temples, how many samples of carbon-datable material would this dating be based on?

Anthony Bonanno: Nothing at all.

Mnajdra, 20 June 2000

It is a little after five a.m., and dark, when we park our rented car near Hagar Qim. There are watchmen huddled together, drinking tea. They won't be on duty during the night of 16 April 2001 when Mnajdra gets trashed by a well-organized assault-squad armed with sledgehammers, but right now they're working overtime. Their mission is to keep out any peaceful hippies who might want to commune with the solstice at the temples before they open officially at eight – although apparently it's tomorrow that most of the would-be meditators and pagans are expected to show up.

Hagar Qim is enclosed by a tall wire-mesh fence, which we now walk around on our way down to Mnajdra. Through the fence the row of big, heavily eroded megaliths on the south-west side of the temple can be seen glowing whitely, like the teeth of an ancient giant disinterred from the earth.

I love dawn in Malta in midsummer, with the smell of wild thyme on the soft breeze, and the sea, dark in its depths, quicksilver at the surface, stretching away beneath the fading stars. It always feels like . . . the morning of the world. As though some wonderful experience – I don't know what – is just about to envelop me and change me for ever.

The sky is lightening as we walk, and way off-shore to our south I begin to make out the distant shape of Filfla rising out of the sea. I am troubled by vague feelings of guilt about not having arranged to dive the strait between here and the little island because I'm genuinely intrigued by Anton Mifsud's theory of major land subsidence in this area in 2200 BC. Since my first meeting with him on 16 June I've consulted my copy of the British Admiralty Chart for Malta and found that it shows a submerged ridge, the top of which is nowhere deeper than 49 metres, running from the rocks of the Hamrija shallows directly beneath Mnajdra all the way out to Filfla. On either side of the ridge, roughly east and west of it, the bottom drops off steeply to 80, 90 and then 100 metres.

It would be extremely interesting, though technically demanding, to explore the ridge, especially inside the zone with a radius of 1 kilometre centred on

Filfla itself that is enticingly marked 'Entry Prohibited' on my chart. If Mifsud is right that a greater landmass collapsed here 4200 years ago, then the shattered remains of man-made structures built in former times along the Filfla–Mnajdra ridge could await discovery at the bottom of the strait. And though they might have been submerged as recently as 2200 BC – as Mifsud's research suggests – who is to say when such structures might have been built?

Theoretically, they might have been built thousands of years earlier than Hagar Qim and Mnajdra, might be the true Palaeolithic predecessors of the great temples of the Neolithic era, and might have survived, revered and imitated, down to Mifsud's date of 2200 BC, when the land on which they stood sank beneath the sea . . .

Theoretically, other relics of Malta's missing megalithic heritage could have been submerged much earlier by the rising seas that followed the Last Glacial Maximum – especially so if they had been built in the north and east during the late Palaeolithic when a land-bridge 90 kilometres in extent connected Sliema to Sicily . . .

And theoretically, of course, pigs might fly, lions might lie down with lambs – anything is possible . . .

Still, there is the problem of the unexplained origins of Malta's remarkable temple-building culture and the sequence which requires us to believe that Gigantija, and the oldest parts of Mnajdra, were that culture's first-ever experiments in free-standing monumental architecture. And there is the problem of the model temples, some fashioned from terracotta, some from stone, excavated from within the temples themselves and now usefully on display in the National Museum in Valletta.[1] While some of these beautiful little models faithfully depict temples of exactly the type that have survived to this day, a few others show an entirely different, highly geometric style of megalithic architecture, where the theme is all straight lines and recurrent right-angles.[2] Why don't we find the ruins of the original structures that these other, rectilinear models supposedly represent? Are they just 'architect's designs', dreamed up but never realized, as David Trump asserts?[3] Or could they preserve the images of temples that once existed and were swallowed by the sea?

I'm deep in these thoughts as we approach the entrance to Mnajdra. It's almost full daylight now, the sun washing the whole sky with a soft, indirect glow, and I can see through the wire-mesh fence that as well as two guards by the gate, there are at least three other people, dwarfed by the ponderous megaliths, already inside the temple precincts. One of them is mounting a video camera on a tripod; one has a clunky garland of SLR cameras slung around his neck; the third is clutching a biro and a spiral-bound notebook.

I groan inwardly. The solstice effect at Mnajdra is supposedly subtle and beautiful, one of the powerful epiphanies of archaic surveying and astronomy. I want to see it with no distractions – just silence, the temple and the sun – so that it can speak most clearly for itself. Now at the bare minimum I am going to have to be polite to strangers, make small talk and exchange opinions while we wait for the effect to begin.

I observe out of the corner of my eye that the man with the notebook is walking towards me and obviously intends to introduce himself. Why do human beings have to *talk*, I find myself wondering. Is it really necessary for us to make these noises?

'It's Graham Hancock, isn't it?' he asks. 'Remember me? I'm Chris Micallef.'

Suddenly I recognize him. He's the nephew of the late Paul Micallef, the Maltese archaeo-astronomer who first undertood that Mnajdra is a solar calendar in stone and began to unlock the ingenious precision of its alignments. I met Chris during the generally disappointing flurry of our previous stay in Malta in November 1999 when he gave me his uncle's book,[4] then lost touch with him afterwards. Far from being a source of unwelcome noise, he's the very best person I could possibly hope to meet at Mnajdra. His uncle's book is why I'm here.

The sea keeps its secrets
June 2000–June 2001

After the June 2000 trip when I witnessed the solstice effect at Mnajdra, exactly a year passed until we were able to get back to Malta again. But, despite the risks, the frustrations and the expense of the previous trips, I remained convinced that the rumours and whispers of submerged structures were worth pursuing.

Part of this new up-beat mood, as I've explained, was my discovery of the late Commander Scicluna's involvement in the matter and the report that he had published in the *Sunday Times* of Malta in 1994 of having found a megalithic temple underwater off Sliema at a depth of 25 feet.

But another part of it came from my growing acquaintance with the work of Anton and Simon Mifsud, Charles Savona Ventura, Chris Agius and others. Their research helped me to realize that although orthodox archaeologists had probably weighed, measured and counted everything Neolithic on Malta, they had done no justice at all to the possibility – no, the certainty – of a human presence here during the Palaeolithic. On the contrary, it seemed that J. D. Evans had gone to great lengths to bury that possibility so deeply that it would never vex his 'Stentinello First' hypothesis again. And while he might not have been the villain who actually switched Baldacchino's molar for a modern taurodont or put a misleading interpretation on the results of the FUN tests carried out in the 1950s and 1960s, these actions signal – at the very least – a ruthless resolve and an indifference to truth on the part of a person or persons highly placed in Maltese and museum circles. In such a murky setting, where I already knew that there had been outrageous tampering with evidence and gerrymandering of records, it seemed to me to be absolutely within the bounds of possibility that much worse could have been done.

Just suppose, for example – speculation only – that traces of an earlier, pre-Neolithic civilization had been found on Malta during the 1950s. Suppose the evidence was fragmentary, small, but clear. Would the discovery ever have been made public? Somehow I doubted it. Indeed, the net of confusion and misdirection woven over the years concerning the Ghar Dalam taurodont teeth seemed to me to demonstrate that such a discovery would *never* have been made public at all if it could possibly have been hushed up.

Still, there remained one place where no evidence could yet have been tampered with and where the ruins of a former civilization, if truly ancient, might have been preserved for thousands of years. That, of course, was under the sea. And that was why it seemed worth keeping an open mind about doing more diving in Malta and paying attention to any sightings by local divers of submerged structures.

Just a month after our June 2000 trip had ended I received an e-mail from Anton Mifsud telling me of two such sightings.

The first was from Audrey and Rupert Mifsud – friends of Anton's but no relation – who own a dive shop called Buddies on Ramla Bay in northern Malta. Leading a dive off nearby Marfa Point on the north-west side of the island, whilst some of their clients were taking souvenir snapshots of each other in an area of interesting underwater scenery, Audrey had swum over a series of parallel 'canals', which had immediately caught her attention as being very unusual and distinctive. Returning to the spot on a second dive, she had discovered several more of these canals cut into the limestone sea-bed at a depth of 8 metres. Beyond the canals, but directly below them near the bottom of a drop-off at 25 metres, Rupert had explored an unusual cave and found three large regular steps carved inside it.

The second discovery, also in northern Malta, had been made by Chris Agius Sultana, an experienced spear fisherman and scuba-diver and one of the co-authors, with Anton, of *Echoes of Plato's Island*. Off Qawra Point on the north-east side of Malta Chris too had found an underwater 'canal', this time surmounted by what he said looked like a low bridge, at a depth of 20 metres.

Storm god
Malta, 18–19 June 2001

Santha and I arrived in Malta on 18 June 2001 for our third research visit. But this was a filming trip, too, so our time would not be our own after the evening of the 20th, when the Channel 4 crew were scheduled to join us. Our plan was to put in a couple of days of diving before they arrived – the 19th for an advance look at the new sites found by Chris Agius and the Mifsuds in the north, and the 20th for a more targeted search off Sliema at 1 kilometre rather than 3 kilometres from the shore, particularly if Shaun Arrigo could be persuaded to guide us. Because both time and money were running too short to allow the luxury of speculative search diving with no definite prior sightings to follow, I made the decision that on this trip we would not try to pursue the tempting question of what – if anything – might lie at the bottom of the strait between Mnajdra and Filfla.

There is a god of stormy weather who likes to follow me around. Honestly, I'm coming to believe this. Ask anyone who dives with me regularly. For the entire week before our arrival the seas around Malta had been flat calm, no clouds, not even a breeze – perfect conditions for really successful diving. But soon after our plane touched down on the afternoon of the 18th, a strong wind began to blow in from the north-west. This was the very worst kind of wind that we could possibly face, as Malta's orientation is approximately north-west to south-east. Nor'westers therefore blow down both sides of the island so that Marfa Point in the north-west, and Qawra Point and Sliema along the north-east coast would all be equally badly affected.

We hoped that the wind would fade during the night but it strengthened and on the morning of the 19th we sat with Chris Agius in his Land Rover looking at the big breaking waves lashing in over Qawra Point.

'I don't get it,' Chris protested in obvious disbelief. 'Until yesterday afternoon the weather was perfect.'

'It's just my storm god,' I replied gloomily. 'He often does this to me.'

We debated going in anyway, but since I had come close to death tackling a similar shore-dive in similar conditions in Tenerife the year before, I finally decided against it. Whatever was underwater off Qawra Point wasn't going anywhere and would still be here on the 22nd, our scheduled day of diving with the film crew. Then we'd be using a 50 foot boat to cover both the underwater sites in the north and wouldn't have to worry about getting smashed to bits against the rocks doing a dodgy entry or exit from shore.

Around Malta with the Viking
19–20 June 2001

The wind continued to blow all day on the 19th and all day on the 20th, boiling up the waves into an angry foam. But at least the sun was still shining, the sky was clear and there wasn't any rain. So instead of diving we spent the two days

driving around Malta with Chris Agius, who is in his mid-thirties with glacial blue eyes and looks like a Viking, and who willingly shared his insights and research with us.

It turned out that it had been Chris who first took the idea that Malta might be a remnant of Atlantis to Anton Mifsud – and Anton had initially been sceptical. But as he had investigated the matter further, he had gradually been won round to Chris's point of view – hence, ultimately, their book *Echoes of Plato's Island*.

I'd last read the book thoroughly when Anton had e-mailed the text to me around September 2000, so this was a good opportunity to clarify a few points.

'If I remember correctly, *Echoes* identifies the Atlantis flood as an event here in Malta caused by land collapse in the south-west around 2200 BC?'

'That's right. But of course the temple civilization was much older than that.'

'How much older?' I asked.

We were sitting in the bar of the Lapsi Waterfront Hotel in Balluta Bay on the evening of the 19th and Chris looked left and right over his shoulder before replying: 'Maybe twelve thousand years older. It was a civilization of the last Ice Age.'

'But how do you know that?'

'I've seen things,' Chris hinted mysteriously. Then he laughed: 'But I can't prove this. Not yet anyway. I'm working on it.'

On the 20th we spent a couple of hours stumbling around Malta's biggest concentration of rock-hewn 'cart-ruts' nicknamed Clapham Junction. Up to a metre deep, and in some cases almost a metre wide at the surface – though narrowing towards the base – they are incised into a big outcrop of bedrock sloping gently uphill between Buskett Gardens and the cliffs at Dingli about 5 kilometres west of Hagar Qim and Mnajdra. But unlike the temples that have come down to us from remote antiquity – near many of which impressive groups of ruts have been found – the ruts themselves suggest no obvious function, either ceremonial or utilitarian.

Some, as we saw in chapter 15, disappear directly into the sea. Others stop abruptly at the edge of cliffs 100 metres above the waves. Others run between two once connected but now separate locations such as Filfla and Mnajdra. The majority, however, are found in tightly packed groups criss-crossing one another as at Clapham Junction. And although it is undoubtedly the case that very large and heavy-wheeled or sledded vehicles would leave parallel ruts looking quite like these if they were to be rolled or dragged through a field of thick mud or clay, it is altogether a different matter to imagine how such tracks – and so many of them – could be impressed into solid rock. And what would the motive have been? What would the motive have been if the ruts had been worn down a pair or two at a time by the runners of wooden sleds (at present a popular orthodox theory)? How long would it have taken, this way, to make all the ruts that scar the island?

And what would the motive have been if, by some mighty effort, all the ruts had been made at more or less the same time?

Wailing and screaming from underground . . .

20 June 2001

After the cart-ruts Chris drove us up to a hilltop named Salib ta Gholia with a view over the twin cities of Rabat and Mdina. The hill was crowned by a sixteenth-century church built out of beautiful limestone ashlars, which glowed gold in the afternoon light. On its wall was a notice in Latin stating that the right of sanctuary formerly accorded to fugitives taking refuge there had been withdrawn. The church seemed closed, its windows and doors boarded up.

'Come on,' said Chris, beckoning that we should follow. 'There's something I want to show you.'

He led us along a zigzag footpath that ran down the side of the hill beneath the church until we came to what seemed to be the mouth of a cave. Blocking our way further was an uncompromisingly locked, thick-barred steel gate.

Chris gestured between the bars: 'Take a look in there,' he suggested. 'I think you'll find it very weird.'

I did. It was.

'What is it?' I asked.

'Nobody knows for sure. The official view is that it was made by early Christians – that it could have been some sort of secret church. But a lot of stuff they can't explain here gets attributed to the early Christians.'

I was peering through the bars. What I could make out in the shadowy recesses beyond seemed to be a roughly circular, very high chamber. And in the centre of the chamber, standing on a broad base, were huge tapering pillars soaring up into the gloom above. Room, walkway, pillars and ceiling were all carved out of the solid bedrock of the hillside. Chris was still talking: 'Here in Malta once the archaeologists say something is early Christian then everyone stops thinking about it.'

'Obviously you don't think it's early Christian.'

'Do you know, a few years ago there were exceptionally heavy rains. A lot of water pooled on the flat ground up above at the top of the hill. Then suddenly a very strange hole opened in the ground and the same moment a huge pile of rubble dropped through into this chamber. It took a week to clear it out. But it wasn't a natural collapse. The hole turned out to be a triangular man-made shaft, with each side of the triangle measuring about half a metre, and it ran vertically about 20 metres through the ceiling of the chamber and all the way to the top of the hill. It had been blocked and filled up over time . . .'

'So . . .'

'So I just don't see any reason why early Christians – or any Christians – would have gone to the trouble to make something like that. What it sounds like to me is ancient astronomy.'

I nodded. Such a shaft, like any vertical shaft, would have marked the bi-annual zenith passage of the sun – here with a spectacular glowing triangle at midday on the floor of the chamber. And it would have made a splendid fixed telescope at night for observing stars at the zenith.

But what also interested me was the further hint that the shaft and the chamber offered of advanced rock-cutting and tunnelling abilities amongst the ancient Maltese – abilities of which the Hypogeum may only represent a fraction. Indeed, there have long been rumours that a vast network of tunnels and passageways of unknown origin exists beneath Malta. And at the beginning of World War II, soon before the islands came under heavy attack from the German and Italian Air Forces, a rather odd report from a gung-ho American cyclist named Richard Walter appeared in the obsessively fact-checked *National Geographic* magazine. After describing the Hypogeum ('where prehistoric man worshipped his deities and buried his dead') Walter wrote:

> While we cycled homeward, our friends told us that the island was honeycombed with a network of underground passages, many of them catacombs. Years ago one could walk underground from one end of Malta to another, but all entrances were closed up by the Government because of a tragedy. On a sightseeing trip, comparable to a nature-study trip in our own schools, a number of elementary school children and their teachers descended into the tunnelled maze and did not return. For weeks mothers declared that they had heard wailing and screaming from underground. But numerous excavations and searching parties brought no trace of the lost souls. After three weeks they were finally given up for dead. Sections of the underground network have been used to protect military and naval supplies. Indeed many of the fortifications themselves are merely caps atop a maze of tunnels . . .[5]

Just another urban legend? Or another tantalizing glimpse of Malta's prehistoric underworld?

The pendulum of the sun
Mnajdra, 20 June 2000

Chris Micallef is around thirty years of age, stocky and dark, quite intense, a typical Maltese. He's wearing a smart white shirt, open at the neck. He has a slightly professorial air, as though teaching comes naturally to him or is often expected of him. And he knows a lot about astronomy. At our first meeting in November 1999 he and his father – the late Paul Micallef's brother – showed me a film they had been preparing for more than a decade which meticulously documents the impressive array of alignments that the massive Lower Temple of Mnjadra has on offer at different seasons of the year.

I look at my watch. It's already 5.50 a.m.

'Don't worry,' says Chris. 'We won't see the effect for about another twenty minutes.' He points to the long sloping shoulder of the hill to our east, at the top of which Hagar Qim is located. 'Of course, nothing happens until the sun's disk begins to appear over the ridge.'

'So this isn't exactly a sun*rise* alignment, then?'

'No. It's much more clever and complicated than that. If the local horizon were completely flat, which it pretty much is up at Hagar Qim, the sun's disk would have already been in view for more than half an hour. But because we're at the bottom of a hill and the hill lies east of us, we don't see it down here yet. So all the sunrise alignments for Mnajdra had to be calculated and observed by the ancients against this sloping local horizon – not an easy thing to do.'

But, nevertheless, a thing that was done. What happens is this:

- As the sun crests the horizon on the spring and autumn equinoxes, 21 March and 21 September (when night and day are of equal length) its rays exactly bisect the huge trilithion entrance to Mnajdra's Lower Temple, projecting a spot of light into a small shrine in the deepest recesses of the megalithic complex.
- On the winter solstice (20/21 December, the shortest day) a very distinctive 'slit-image' – looking something like the illuminated silhouette of a poleaxe or a flag flying on a pole – is projected by the sun's rays on to a large stone slab, estimated to weigh 2.5 tonnes,[6] standing to the rear of the west wall of the Lower Temple's northern apse.
- On the summer solstice (20/21 June, the longest day), the same distinctive slit-image appears – but now with the 'flag' oriented in the opposite direction – on a second large stone slab, this time weighing 1.6 tonnes standing to the rear of the west wall of the Lower Temple's southern apse.

'And it works like that,' Chris Micallef continues, 'like a pendulum, sweeping left to right, then right to left, back and forward throughout the year: summer

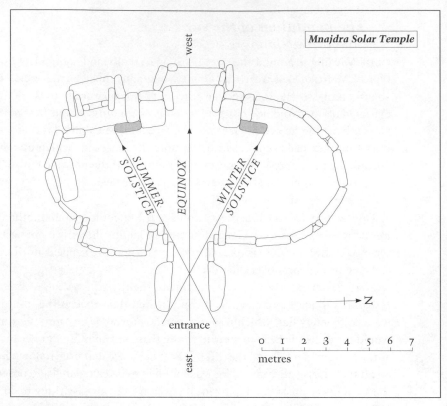

Between the winter and summer solstices, the rays of the rising sun act like a pendulum swinging between the north and south vertical stones (shaded) inside the temple. On the equinoxes, the sun shines straight along Mnajdra's east–west axis. Based on Micallef (1992).

solstice image to autumn equinox to winter solstice image, back to the spring equinox, then to the summer solstice image again and the cycle starts over. There are further subdivisions also signalled by slit images for the cross-quarter days, mid-way between the solstices and the equinoxes, and for the "eighth days" mid-way between the cross-quarter days and the equinoxes on one side and between the cross-quarter days and the solstices on the other.'

Chris tells me about other alignments, notably some very precise lunar alignments that the temple also registers: 'All in all, when we consider the high precision of Mnajdra's alignments, and the many astronomical problems that were solved – way beyond what is required if the only objective was a simple agricultural calendar – we have to conclude that full-time professional observers must have been at work here for many, many years. Then you have to think about the problems of site-selection – and then many more years patiently observing to establish the required back-sights.'[7]

'So these guys were pretty sharp observers?'

'They were,' said Chris. 'And pretty sharp surveyors too.'

'And good engineers,' I added. 'They knew how to move and position the big megaliths.'

'And they had mathematical and measuring abilities . . . Come and see this . . .'

Chris leads me up the slope to the small, south-facing trefoil temple on the northern side of the site, presumed to be the oldest in the Mnajdra complex. On the basis of exclusively Gigantija-phase pottery excavated here, it has been dated to c.3450 BC (as against c.2800 BC for the Lower Temple where mainly Tarxien-phase pottery was found). But Chris doesn't want to talk about dates. He wants to talk about ellipses.

'I've studied the elliptical forms of the temples mathematically,' he says. 'And it seems that some kind of megalithic building or measuring unit was used. It seems so.'

We're inside the trefoil temple now, which is indeed highly elliptical. 'In fact this is the major axis, right?' says Chris. 'And this is the minor axis of the ellipse. There is a property which says that if you take from that point to the centre, and from here to here, and if you square this part plus this part squared then that part comes equal to exactly half the major axis. Eventually it comes. I'm not saying that they invented the Pythagoras theorem, but they had discovered it by chance, so that they could alter the eccentricity of the temples as much as they wanted.'

I tell Chris that a lot of this is going right over the top of my head, but he says the main point is very simple. What it comes down to is that the people who built the Mnajdra complex, and all the other megalithic temples on Malta, worked with a fixed unit of measurement. This unit, of 0.83 metres, is identical to the 'megalithic yard' identified by the Scottish archaeoastronomer Alexander Thom and found throughout megalithic sites that he had surveyed from Callanish in northern Scotland to Carnac in Brittany.[8]

'I calculated the perimeter,' Chris continues with a gesture, 'and it comes out to whole numbers in megalithic yards. It is the same for all the ellipses, though they vary in eccentricity. So they had the facility to arrange the eccentricity of each temple to the precise extent they required and yet keep the measurement of the perimeter in whole numbers of megalithic yards. Somehow these kinds of mathematical concepts must have been in quite wide circulation in the ancient world and were passed on from one society to another society, perhaps by seafarers. A possible harbourage has been suggested, right here below Mnajdra. And though there are no images of ships carved here, such images do appear at Tarxien.[9] So probably this kind of knowledge was passed on by word of mouth, and there was some kind of society that this knowledge was passed on from and to . . .'

'And ships were part of it?'

'Yes.'

'Because this kind of accurate astronomy is also what you want for navigation really. It's the same skill.'

'That's right,' says Chris, but he sounds distracted. He is an engineer by profession, and I can see that he is uncomfortable with speculation and prefers to stick with what he can measure and observe.

Still, there is one other point – much more speculative than ancient navigational skills – that I want to ask him about. When his late uncle was completing his analysis of the archaeoastronomy of the Lower Temple at Mnajdra he had discovered something odd concerning the summer and winter solstice alignments.

It is well known that the sun's rising points at the solstices are not fixed but vary with the slowly increasing and then decreasing angle of the earth's axis in relation to the plane of its orbit around the sun. These changes in what is known technically as the 'obliquity of the ecliptic' (presently in the range of 23 degrees 27 minutes) unfold over a great cycle of more than 40,000 years and if alignments are sufficiently ancient they will incorporate a degree of error, caused by changing obliquity. From the error (assuming they were built accurately in the first place) it is possible to calculate the exact date of their construction.[10]

In the case of Mnajdra, the alignment today is good, but not quite perfect because (to take the example of the summer solstice) the rays that form the slit-image are projected two centimetres away from the edge of the large slab at the rear of the temple. However, Paul Micallef's calculations show that when the obliquity of the ecliptic stood at 24 degrees 9 minutes and 4 seconds the alignment would have been perfect with the slit-image forming exactly in line with the edge of the slab. This 'perfect' alignment has occurred twice in the last 15,000 years – once in 3700 BC ('this is the first consideration of the Mnajdra Temple's age', notes Paul Micallef)[11] and again, earlier, in 10,205 BC ('this is the second consideration of the Mnajdra Temple's age').[12]

But Chris doesn't want to be drawn on the earlier date. He admits it's a 'mathematical possibility' but says he would prefer to stick with the orthodox scheme of things: 'The second age makes the temple 12,205 years old, which is absurd when compared to archaeological history. In my view the archaeological context locking the temples in to the fourth and third millennia BC is reasonably good, so that's the context I work with.'

And he's right. The archaeological context *is* reasonably good – in the sense that no find, or at least none that have been officially logged, conclusively demonstrates that any of the temples are older than the fourth millennium BC. But, that being said, the archaeological context of the megalithic temples of Malta is also, in another sense, appallingly, awfully bad.

Antediluvian temples of the giants?

The essential problem, repeated over and over again, is contamination of the crime scene. Indeed, other than Skorba, which was thoroughly and professionally excavated by David Trump in the 1960s and which is partially built over the top of habitation layers predating the temple's construction,[13] it seems that not a single megalithic temple on Malta has presented itself to archaeologists of the post-radiocarbon era in a sealed and undisturbed condition. Although this includes Tarxien, which was excavated from 1915 onwards (still fifty years before calibrated radiocarbon), the superb stratigraphy and detailed site notes of the commendable Sir Temi Zammit do provide us with a reliable record there.[14]

The same cannot be said for the semi-subterranean Borchtorff Circle excavated at Xaghra on Gozo between 1987 and 1994. It proved to have fallen victim to an earlier excavation in the 1820s by a certain Otto Beyer in the employ of the British Army, who very badly disturbed and redistributed the stratigraphy and kept no records.[15]

Likewise, Mnajdra was first excavated in 1840 by C. Lenormant, who kept no records, followed by a mixed assortment of other diggers, then by Mayr at the beginning of the twentieth century and then by Ashby, who excavated in 1910 'those parts which had not been completely ransacked by the original excavators'.[16]

Hagar Qim has been constantly interfered with by treasurer hunters, amateur archaeologists and self-appointed site-restorers from at least the eighteenth century onwards. Particularly extensive site clearance and restoration took place in 1839 on the orders of the then governor of Malta Sir Henry Bouverie. Only a short and extremely inadequate report accompanied by an inaccurate plan was prepared.[17]

And at Gigantija excavations were begun in 1827, once again by Otto Bayer. True to form, he produced no report and did not preserve pottery and small finds.[18] Oddly enough, however, the first description of the monument following Beyer's excavation (published in Paris later in 1827 by L. Mazzara) bore the title *Temple antediluvien des Géants*.[19]

Carbon-dating Malta: is the chronology secure?

Tas Slig, 25 June 2001

It is certainly the case that not a single carbon-date from Malta supports the presence of *any* humans on these islands prior to 5200 BC, let alone the presence of humans capable of building with megaliths. On the other hand, it must also be observed that the general state of disrupted stratigraphy at the temples has made it difficult for archaeologists to obtain C-14 samples in contexts where they can unequivocally confirm the age of the megalithic ruins – and indeed to obtain C-14 samples at all.

On 25 June 2001 I discussed these problems with the charming and affable Professor Anthony Bonanno of the University of Malta on site at a dig he was supervising at Tas Slig.

GH: If we take the dating of a temple like Mnajdra or Hagar Qim, the better-known temples, how many samples of carbon-datable material would the dating be based on?

Bonanno: Nothing at all.

GH: Nothing at all?

Bonanno: Hagar Qim and Mnajdra were cleared rather than excavated in the nineteenth century, and no proper records were kept, and the excavation methods were far from scientific. So no biological material was kept that could be carbon-dated.

GH: Right. Does that apply to Gigantija too?

Bonanno: That applies to Gigantija as well, yes.

GH: Right. In general, are the megalithic temples founded very close to bedrock, or are they founded on an earth layer on top of the bedrock?

Bonanno: You can tell that in the Maltese context, all stone buildings lie on bedrock. The cover, earth cover, is very shallow . . . and then, of course, you need a really solid base.

GH: But how would they – sorry, this may seem like an ignorant question – but if they put the megaliths on bedrock, how do they make them stand up? Don't they have to bed them into earth or something?

Bonanno: Right. It doesn't mean that the uprights of the temple stand on bedrock. In fact this is another difference between the construction technique of our temples and say the construction techniques of Stonehenge. There, the standing megaliths are inserted into the ground. Here, a platform is normally prepared consisting of megaliths, but horizontal megaliths, and it is on top of those that the lower uprights of the temple are placed.

GH: I see. And the platform itself is on bedrock?

Bonanno: The platform itself is on bedrock.

GH: Interesting, interesting. Have any samples been taken from underneath a megalith?

Bonanno: From underneath a megalith? I don't remember any samples being taken from underneath megaliths.

GH: What's troubling me is with the megalithic temples founded on bedrock and therefore no possibility of strata under the temple itself, how sure can we be about the contemporaneity of the organic samples that can be carbon-dated and the construction of the site? It doesn't worry you about the dating of the megalithic structures themselves?

Bonanno: Not really, because anything underneath the megalith could be as old as 100,000 years. What, as an archaeologist, I would want to find is a stratum, a layer, which would be touching on, therefore sealing, a wall or part of a wall, because that is what would be telling me the date of the wall itself. Anything below could be as old as ever.

GH: Is there any megalithic temple in Malta where you have sealed secure carbon-dates from layers like that?

Bonanno: Skorba, yes, Skorba and the Xaghra [Borchtorff] Circle.

I'm not an archaeologist, but as a journalist it seems to me we are left with an awful lot of temples for which we have no carbon-dates at all and certainly no sealed secure ones. Worse still, the complete repertoire of radiocarbon-dates for the prehistory of the Maltese islands, upon which so many of our notions of the origins and chronology of its megalithic civilization depend, is, overall, extremely limited. I was surprised to discover that there are only twenty-seven official C-14 dates for the entire archipelago and that most of these are of equivocal quality. Moreover, twenty-two of the twenty-seven dates come from only two sites – eight from Skorba and fourteen from the Borchtorff Circle.[20] Of the remaining five, one comes from Mgarr and is relatively secure, being wood charcoal retrieved from the under the floor. Logically, however, the most that it can tell us is the age of the floor itself – which may have been a restoration. It has no bearing on the age of the megalithic uprights since the excavator – J. D. Evans in 1954 – informs us that the sample was found just *above* the level on which the wall foundations were resting.[21]

Prehistoric Malta's final four carbon-dates out of its grand total of twenty-seven are from Tarxien.[22] Of these, one is wood charcoal from the first apse to the right in the South Temple. The remaining three are all described as 'carbonized beans from cinerary urns'. On further investigation it transpires that these samples were found in glass jars in the National Museum of Malta labelled 'Tarxien Cemetery', which were assumed to contain the contents of cinerary urns excavated by Temi Zammit in 1915.[23]

Despite some significant anomalies and inconsistencies,[24] I want to emphasize again that none of these C-14 samples undermine – and all generally support – the orthodox chronology of the rise and fall of Malta's unique temple-building culture. Nor is it my purpose here to challenge that chronology – at any rate, not necessarily with reference to the temples that survive above water. But I do think that in too small a field monopolized by too small a group of archaeologists, too much has been claimed for too long on the basis of too little data. In consequence, the 'out of Sicily' hypothesis that ignores the Palaeolithic has thrived, and its supporters – quite naturally – have focused whatever scarce archaeological resources may be available on the search for further evidence to elaborate and confirm an exclusively Neolithic heritage for Malta.

So I don't mind too much *when* the surviving megalithic temples were built. The counter-hypothesis that I offer for their origins is that they are the end-result of a very long process of development in Malta that began in the Palaeolithic and that has been veiled from us by rising sea-levels, cataclysmic land subsidence, academic mendacity and a self-protecting old boys' club closing ranks.

A god of light and geometry
Mnajdra, 20 June 2001

It's just after 6.05 a.m. and we are all gathered inside the northern apse of the lower temple, waiting for the sun to project an image on to the massive slab – the summer solstice stone – to the left of the central passageway in the southern apse. The image will be formed, Chris Micallef has explained, when half the solar disk is above the sloping natural horizon of the ridge. At that moment the sun's rays, coming out of the north-east, will pass through the trilithion gateway, striking the inside edge of its northern upright and the underside of the lintel, thence diagonally across the entrance passage to strike the inside edge of a megalith at the south-west end of the passage, and finally across the southern apse to strike the summer solstice stone – in our epoch 2 centimetres from its southern edge. So the 'slit' through which the rays pass to form the projected image is not simply a gap in the masonry but a result of the careful juxtaposition of three different megaliths, two upright – but more than 4 metres apart at opposite ends of the entrance passage – and the third horizontal and more than 2 metres off the ground.

The morning light is mellow, warm, no harsh edges yet. There's still some pink of dawn left in the sky. And the moon, almost full, floats high and pale above a great menhir that projects like a finger out of the south-west wall.

'We should see the effect very soon,' announces Chris. 'And I would ask you to remember that it has maybe only a hundredth of the impact that it would have had in antiquity when the temple was fully roofed and dark inside. So you should try to imagine the effect suddenly materializing in a place of darkness.'

A few more minutes pass. I know what I should be looking out for and where I should see it, but I don't see it yet. And in the back of my mind I'm absorbing Chris's point about the roof, wondering how we can expect to see anything special at all under the present conditions. Isn't there already way too much light inside the roofless temple for what is, after all, an effect composed entirely of light? Won't it just wash out against the bright background?

Then I become aware of . . . a presence – a faint, ghostly glimmering, like moonglow, that has appeared on the solstice stone. I don't know how long it lasts, a second or two only I would guess, but while it is there it seems less like a projection – which I know it to be – than something immanent within the stone itself. And it seems to function as a herald for it fades almost as soon as it has appeared and in its place the full effect snaps on – instantaneously. It wasn't there, and then it's there.

As Chris had described, the effect does curiously resemble a poleaxe, or a flag on a pole, and consists of a 'shaft', narrow at the base but widening a little

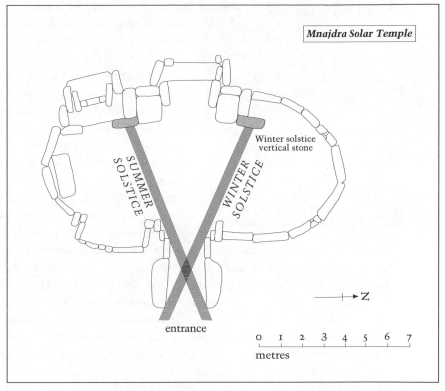

A clear slit image is formed by the sun's rays shining on the vertical stones inside the temple. Based on Micallef (1992).

towards the top, running up the left hand side of the solstice stone, surmounted by a right-facing 'head' or 'flag'. An instant later an almond-shaped spot of light, like an eye, appears a few centimetres to the right of the 'flag' and the effect is complete.

Weirdly – I do not claim it has any significance – this flag-on-a-pole symbol is the ancient Egyptian hieroglyph *neter*, meaning 'god', or 'a god' – and not to be understood at all in the Judaeo-Christian usage of that word but rather as a reference to one of the supernatural powers or principals that guide and balance the universe.

Manifested here, in this strange Stone Age temple, it glows, as though lit by inner fire.

The cave at the foot of the cliff
Marfa Point, 22 June 2001, dive 1

We made three dives, two at the sites off Marfa Point in the north-west of the island that had been found by Rupert and Audrey Mifsud and one at the Qawra Point site in the north-east found by Chris Agius.

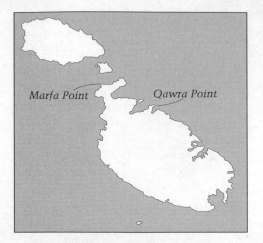

My storm god had taken a break, the day was calm and beautiful, and our dives were safe and unthreatening in seas entirely free of currents and waves. Also, for the first time ever in Malta, we just jumped in the water and went to the suspected man-made sites without any of the fruitless searching with echo-sounders and wasted hours zig-zagging backward and forward that I had come to regard as normal here.

First Rupert led us over a level area covered in fields of waving sea-grass sloping down gently from 7 to about 10 metres. Then we came to the edge of a sheer underwater cliff dropping 15 metres to the sea-bed below. There we launched ourselves into blue space and drifted down the side of the drop-off like slow motion skydivers.

At the bottom, at 25 metres and in the sudden cold of a thermocline, was the opening to a cave. A very strange cave. I have never seen one like it before.

Its entrance was, I suppose, about 5 metres wide at the base and soared, narrowing, to half the height of the drop-off where the two sides came together to make the roof. Inside, I found that the floor of the cave was not horizontal but rose from the sea-bed at an angle of about 45 degrees forming, effectively, a steep ramp. The surface of the ramp, though overgrown, was surprisingly smooth and it was difficult to see how such a feature could have formed naturally in a setting like this. Besides, now that my eyes were becoming accustomed to the gloom, and in the beam of light from Rupert's torch, I could see that some areas appeared to have been deliberately cut and quarried into shape.

Cave diving can be scary if the cave you are in is part of a system with many side branches, like a maze, or if it is so long that you cease to see the light filtering through the entrance behind you. But what makes caves really dangerous, and the reason that they regularly kill divers, is sediment.

Some years ago in Yonaguni, Japan, where I dive regularly, four leisure divers and an instructor were killed by sediment – not killed by it directly, of course, but killed by it because their finning stirred up centuries of silt piled on the cave-floor into a thick suspended mist. In it they became disoriented, confused and, tragically, could not find the exit before they ran out of air.

But this cave in Malta was not like that. It entrance was so wide and the cave itself so relatively shallow that it would be impossible to get lost in it, even in the worst conditions. Nevertheless, it was silty and the visibility was deteriorating steadily despite all our efforts.

About 5 metres inside, up near the top of the ramp, Rupert showed me what we had come to see – three large steps, or terraces, each about half a metre high and extending across the entire width of the cave. They were deeply covered in marine growth and layers of sediment but they seemed to be much too straight-edged and right-angled to have been shaped by any natural process – especially in such a sheltered position.

Two or 3 metres beyond them the cave terminated in a wall penetrated by a gap large enough for me to pass through – which I did without hesitation, since I could see light streaming in from the other side. The gap led to a second cave, in an entirely rough and natural state, with its own separate entrance. I swam back through the gap again and returned to the steps which, by now, were enshrouded in a fog of sediment.

Man-made, or natural? It certainly looked to me as though people had been at work in this cave cutting and shaping the rock to some design or plan of their own – as they had been over the millennia in Malta in so many caves and underground tunnels.

I allowed myself to float up to the roof, an easy move for a diver but something that would have been impossible without scaffolding when the cave was above water. Yet the roof of the cave was nicely cut and squared off, presenting an extremely symmetrical 'frame' of two verticals and a horizontal upright.

Underwater Clapham Junction
Marfa Point, 22 June 2001, dive 2

Our second dive at Marfa Point was on what Rupert called 'the channels'. They were located on the plateau above the drop-off at a depth of about 8 metres and were immediately recognizable as part of the same phenomenon as the unquestionably man-made 'cart-ruts' at Clapham Junction and other above-water locations on Malta.

But there were some differences.

Firstly, these ruts were a good bit wider and deeper than the ruts at Clapham Junction. It was possible to get my whole body down horizontally into most of them and to swim along inside them for distances of 20 metres or so before reaching a break.

Secondly, although most ran in parallel pairs as though left by cart-wheels, exactly as at Clapham Junction, there were indeed several single 'channels' even wider and deeper than the others.

Thirdly, in places where I managed to strip away the thick sea-grass covering the bottom of a rut I came across an extremely odd feature. About twice as wide at the base as the average ruts at Clapham Junction, these proved everywhere I searched to be divided into two 'lanes' by a knife-edged ridge of limestone about a hand's-breadth high that had been left in place running the full length of the rut. In my view there can be no question of this being a natural feature. It is definitely man-made.

Fourthly, the top of the ruts at Clapham Junction lie flush with the bedrock.

Here underwater, although the interior of the ruts had been cut down into the bedrock in the same way, the sides of the ruts also rose about 30 centimetres *above* the level of the surrounding bedrock – like low, narrow parallel walls.

As at Clapham Junction all these features appeared to have been hewn out of the solid bedrock by tools, and not to have been worn down by centuries or millennia of abrasion.

As at Clapham Junction I also found one place where a pair of ruts was interrupted by what almost seemed like a roadway running transverse to them. The ruts stopped completely on one side of the 'roadway', which had been cut through them and thus obliterated them at this point, and then resumed their course on the other. The obvious implications of this state of affairs were that the transverse road had been made after the ruts and that the ruts may therefore have already been ancient when they were submerged.

As at Clapham Junction these ruts didn't seem to be coming from anywhere in particular or going to anywhere in particular. Some of them did lead in general towards the edge of the drop-off but vanished completely into the sea-grass before reaching it.

Canal
Qawra Point, 22 June 2001, dive 3

Diving and filming are both activities that require a lot of time, preparation and messing around with equipment, so it was nearly three in the afternoon before we were finished at Marfa Point and after four when we reached Qawra Point on the other side of the island.

Chris Agius put on his scuba gear while we were anchoring and jumped in to relocate and mark the site before we went down to it. And since this was one of those very rare days in Malta when everything went right he was back up within five minutes waving success.

We descended to a flat, rocky bottom at a depth of about 18 metres and, though it was overgrown as usual with thick sea-grass, I could see the level plain extend to the limits of visibility on all sides of me. It seemed completely natural. But moments later Chris brought us to a clear, clean gap in the sea-grass caused by a channel – perhaps something more like a canal really – that ran straight for tens of metres through the bedrock. A little over 2 metres deep and about the same wide, its floor consisted of pure, white, level sand.

But how thick was the sand? I pushed my gloved hand experimentally down into it until had it had disappeared beyond my wrist. It was deeper than that, possibly a lot deeper. But it would take airlift equipment to find out for sure.

The walls of the canal were cut down vertically into the bedrock on each side and did, very strongly, give the impression of being artificially formed.

Chris and I swam along the bottom of the canal side by side – there was room to do that – for 20 or 30 metres until we came to the place he wanted to show me. Here the canal was spanned by a 'bridge' flat on top and level with the surrounding plain. It too was a rock-hewn feature – a narrow section of the original bedrock that had simply been left in place when the canal was formed, and then hollowed out underneath into an archway through which the contents of the canal could flow.

We swam under the arch several times and Chris pointed out how the vertical side walls bore what looked like tool marks. I agreed with him. And again I found myself looking at something underwater that could not easily be explained as natural. For whereas arches, sometimes on a very grand scale, are found beneath the sea, they almost always prove to be part of collapsed cave systems. That was not the case here, for we were in open underwater country – that would have been open country before its submergence – and because this arch crossed a dead-straight 2 metre deep channel that was completely out of character with everything else that nature had succeeded in doing in the vicinity.

Last but not least, the canal proved to run due north–south – an orientation significant to humans but not to nature.

The missing piece of the jigsaw puzzle

Anton Mifsud cites the underwater channels around Malta as further evidence for his thesis that this was 'Plato's island', since canals or channels also feature prominently in Plato's description of Atlantis.

Mifsud's proposal, as we've seen, is that the world-famous story of the destruction of Atlantis in a 'single dreadful day and a night' that Plato recounted at the beginning of the fourth century BC is an echo, or folk memory, of massive destruction wrought on Malta in 2200 BC by a fault collapse along the submarine Pantalleria Rift. He notes that Malta today has a pronounced 'wedge-like' tilt from south-west (the thick end of the wedge, e.g., the towering coastal cliffs at Dingli and Maghlaq) to the north-east (where the thin end of the wedge disappears under the sea, as at Sliema). This tilt has come about because Malta lies very close to the tectonic collision front between the African and Eurasian continental plates:[25]

> The upwarped shoulders of the Pantalleria Rift bear the Pelagian islands of Lampe-
> dusa and Lampione on the western shoulder and the Maltese islands on the eastern
> one. The still active shoulder upwarping on both sides of the Pantalleria Rift causes
> the tilting. As the island of Lampedusa continues to tilt southerly, the Maltese
> islands tilt in a complementary manner towards the northeast . . .[26]

This is the underlying geological process that created the sheer cliffs of south-west Malta, themselves forming the edge of an exposed fault-line at Maghlaq, near Mnajdra. And as Mifsud points out, the nature of the process makes it highly likely that Malta may once have extended much further to the south-west of Maghlaq than it does today (a continuation of the 'thick end of the wedge' on the upwarping shoulder of the Rift). He proposes the cataclysmic collapse of this hypothetical south-western extension 4200 years ago as the explanation for the mystery of the sudden and apparently overnight extinction of Malta's age-old temple-building culture at the same date:

> Tectonic movements in the Central Mediterranean are still responsible for the continuing separation of the two shoulders of the rift, respectively bearing the Maltese islands on the northeast and the Pelagian group on the southwest shoulder. It is far from inconceivable that [a] landmass joined to the southwest coast of Malta, at the Maghlaq site would have collapsed and submerged at a point in time when its underlying structures gave way to the rifting process. Such a collapse would have occasioned the displacement of massive volumes of seawater on the southwestern coastline, with a rapidly flowing torrential flooding event along a SW to NE direction.[27]

It is this deluge that Mifsud proposes as the source of the grisly avalanche of jumbled and disarticulated skeletons washed out of Neolithic graveyards and into the Hypogeum 4200 years ago, and for the metre-thick deposit of silt that was dumped inside the Tarxien temples at the same time. And while I cannot agree with Anton that the same deluge and instantaneous loss of a large part of south-western Malta was also the source of the Atlantis myth, his notion of a cataclysmic fault collapse in this area is highly plausible and in full accord with the geological evidence. In addition, since Mifsud's hypothetical south-western extension to Malta would have been created by tectonic motion along the Pantalleria Rift and destroyed by the same forces, it would have remained *invisible* to Glenn Milne's inundation-mapping programme which, explicitly, does not account for tectonic motion.

For the purposes of my own quest, the single most intriguing aspect of Mifsud's hypothesis is that it permits Malta to have retained a large extension in the south-west down to 4200 years ago – i.e. more than 6000 years *after* the end of the post-glacial floods that had earlier been responsible for the inundation of huge areas to the north and east (see chapter 19). In the search for the experimental and 'learning' phases of Malta's megalithic temples during the long gap between the end of the post-glacial floods 10,600 years ago and the 'sudden' appearance of the Gigantija phase 5600 years ago, I therefore suggest that we could hardly do better than begin to look here.

Moreover, and again entirely beyond the data and resolution capabilities of Glenn Milne's maps, there are the knock-on tectonic effects throughout Malta and Gozo that would have been caused by a massive collapse of upwarped lands. Many adjustments of the coastline may have occurred that we will never have any knowledge of.

What is certain, however – although the rates are unpredictable – is the continued stealthy *emergence* in the upwardly warping south-west and the continued stealthy *submergence* of the north-east Maltese coast. The particular implication of this process is that sites in the north-east shown on the inundation maps to have been submerged by 10,600 years ago may not in fact have been submerged until much later, when Malta's tilt forced them under. It therefore follows that the inundated north-east, off-shore of Sliema, also remains a prime candidate for the missing archaeological remains of earlier phases of Malta's temple-building culture.

Broken images

I have grossly oversimplified Anton Mifsud's theory of a fault collapse along the Pantalleria Rift and left out much of the detailed empirical evidence that sustains the theory and dates the collapse to 2200 BC. Readers wishing to pursue the matter further are referred to his own book on this subject, *Echoes of Plato's Island*, which presents the case more thoroughly than I am able to attempt here.[28] I do, however, want to draw attention to one particular category of supporting evidence that Mifsud includes in *Echoes*. Unlike his geological and geophysical evidence this material is very hard to measure and assess and might be considered highly speculative. Nevertheless, I believe that it may prove to be of the greatest importance.

In their research, Mifsud and his co-authors came across recurrent references in traditions and classical geographies and maps to a *formerly much larger Malta*. For example: 'Some medieval maps do not speak of Malta, but of a certain *Gaulometin* or *Galonia leta*, and combine Malta and Gozo into one big island.'[29]

We know from Glenn Milne's inundation data that Gozo and Malta were indeed one big island during the Ice Age, down to approximately 13,500 years ago, and that they did not take on their present form as an archipelago of three islands (with little Comino in between) until around 11,000 years ago. Accordingly, if the medieval tradition of Malta and Gozo as one big island is not a complete invention – and why should it be? – then, 'fantastic' though it may seem, it somehow preserves a memory of Malta as it appeared more than 11,000 years ago. It is well known that most medieval mapmakers were only copyists reproducing older maps and, for reasons that we will explore in Part 5, I believe we cannot exclude the possibility that the single large island called Gaulometin or Galonia leta that has somehow survived on certain medieval maps may indeed be a representation of Malta in a much earlier time.[30]

A mental leap is required in order even to consider such a possibility. It is necessary to set aside all preconceptions about the past, and all unexamined notions of how societies evolve. Above all, we have to rid ourselves of the ingrained conviction that (despite some setbacks) the basic story of human civilization has been steadily and reassuringly onwards and upwards from the very beginning.

It may not have been so. There may be tremendous gaps, of which we are blissfully unaware, in the evidence presently available to us concerning the origins and progress of civilization. In particular, there has been no sustained or serious search for very ancient underwater ruins along the millions of square kilometres of continental shelves flooded at the end of the Ice Age.

So it is *possible*, and within the bounds of reason, that a civilization of some sort might have flourished during the closing millennia of the Ice Age and might not yet have been detected by archaeologists. A civilization not necessarily at all like our own but still advanced enough to have mastered complex skills such as seafaring and navigation (that do not call for a large material or industrial base) and to have left behind memories of the world as it looked before the flood and at various stages during the rising of the seas. The sort of civilization, perhaps, that would have built with megaliths and aligned them with navigational precision to the path of the sun. Maybe even a civilization that measured the earth, mapped it and netted it with a latitude and longitude grid.

Until such a lost civilization has been entirely ruled out – and we are far from that – it is rational to keep our minds open to the possibility, however extraordinary it may seem, that certain ancient maps have indeed carried down to us broken images of the antediluvian world.

Thus Mifsud is right to be intrigued that:

> A southern extension of the Maltese islands . . . is recorded in the annals of Claudius Ptolemy, the renowned ancient geographer, mathematician and astronomer . . . He had unlimited access to the ancient documents in the Alexandrine library, and his research included the Mediterranean and Maltese islands. Although his readings outside the Mediterranean were sometimes erroneous, his Mediterranean latitudes in particular were significantly accurate.[31]

Ptolemy (c.AD 90–168) carried out his geographical research at the fabled library of Alexandria in Egypt, the most extensive archive of ancient texts then preserved anywhere in the world. Is it possible that he, too, was drawing on antediluvian sources with his uncharacteristically 'inaccurate' references to a formerly larger Malta?

What is particularly noticeable about Ptolemy's coordinates, Mifsud demonstrates, is that they extend Malta significantly into the sea to the south and west of the present coastline in the vicinity of Filfla – exactly where he believes that massive land-loss occurred through catastrophic faulting 4200 years ago.[32]

> The crucial point [is] that Ptolemy gave co-ordinates for Malta which extended over twenty minutes of latitude (between 34°45' and 34°25'). He was therefore attributing a maximum latitude width for Malta alone of at least 30.82 kilometres. This measurement today is approximately 21.5 kilometres, so that it is evident that in the ancient sources researched by Ptolemy, the Maltese islands still extended southward significantly more than today.[33]

Maps drawn in late medieval and early Renaissance times from Ptolemy's original coordinates contain a variety of anomalies that may also reflect the same ancient sources. For example: 'An early world map of Ptolemy [Ulm 1482]

Phantom Island south-east of Sicily

The Ptolemaic map in Ebner's manuscript of AD 1460 shows a large ghost island south-east of Sicily. Similar islands are also seen on the Klosterneuberg of AD 1450 (which appears to merge the Maltese islands into a single landmass) and the Ulm of AD 1482.

shows a large unidentified island in the central Mediterranean.'[34] Although displaced too far to the east, this large unidentified island bears a strong resemblance to Malta as it would have looked 14,600 years ago, shortly after it first became isolated from Sicily. A similar island is clearly shown on the Ptolemaic world map in Ebner's manuscript of 1460.

Another map, reputedly copied 'from ancient sources' at Klosterneuberg, Austria, in AD 1450, shows a 'significant landmass between Sicily and North Africa'.[35] Again, the possibility that this is a reverberation of ancient information about Malta's former extent, even if distorted through the passage of time, cannot in my view be ruled out.

Malta in this respect is far from unique, but stands as the representative of a wider problem that we will return to in Part 5.

Tantalus

Balluta Bay, 25 June 2001

On the very last evening of our June 2001 filming trip to Malta for Channel 4, Anton Mifsud arranged for us to meet Shaun Arrigo. We hadn't seen him since our disastrous dives in November 1999 and I wanted to clear up the misunderstandings that had occurred between us then. Fortunately, this proved easy to do and, thanks to Anton, we passed the evening in the bar of the Lapsi Waterfront Hotel with a new mood of trust and cooperation in the air.

As we talked it emerged that Shaun had been back to the Sliema 'temple' site several times. Working in a team with Anton Mifsud and other colleagues, he had also filmed a second submerged site in the same general vicinity (which the group had named Janet-Johann site after the discoverers). 'Do you want to see it?' he asked. 'I've brought the tape along with me, if anyone's got a player and a TV.'

Our producer Stefan Wickham offered the facilities of his room and we all crowded upstairs to watch the video.

It became obvious, within moments, that the Janet-Johann site was of great interest. At depths of between 10 and 15 metres off Sliema Arrigo's footage showed a series of very large, almost 'monumental' canals and parallel 'cart-ruts' much wider and deeper than those we had seen at Marfa and Qawra. Some of the canals cut through the bedrock in perfectly straight horizontal lines for more than 100 metres without any break. Then, beyond them, the camera came suddenly into an area of huge scattered megaliths. All were fallen except one which stood partially upright leaning at a drunken angle.

'I found a piece of pottery round there,' Arrigo told us. 'It was lodged in a fissure, and very worn and ancient. I retrieved it and took it to the National Museum, but they just weren't interested – told me I could keep it.'

'And did you tell them about this site as well?' I asked, indicating the images on the TV screen.

'Yes I did. I told them I thought it was a very suspicious, very man-made-looking place. I offered to lend them the tape or guide someone from the Museum there.'

'And?'

'Same story. They weren't interested. In fact they seemed rather annoyed with me. They've been annoyed with me ever since the publicity in 1999 and I still don't understand why.'

There was no time on that trip for us to take a look at the new Sliema site with Arrigo – besides he himself was leaving for Italy the next morning. So we agreed that he would dive it again later in the summer on contract to us and shoot more detailed and more extensive tape of what he'd found there. Then we would decide what to do about it – although frankly, with a book to write, I did not see myself getting back to Malta to pursue the Sliema temple any time soon.

I felt like Tantalus, the thirsty Greek king whose fate it was to stand for ever up to his neck in water that receded whenever he tried to drink it.

PART FIVE

Ancient Maps

21 / *Terra Incognita*

Marinus of Tyre seems to have been the most recent of our students of *geographia* and to have applied himself to the subject with the greatest enthusiasm . . . If we could see that his latest composition lacked nothing, we should even have been happy to complete our description of the known world from these notes of his alone, without researching any further. But as on certain points he himself seems to have composed without reliable comprehension, and as in embarking on his map he has in many places not devoted enough thought either to convenience or to symmetry, we were naturally induced to contribute to his work what seemed necessary to make it more logical and useful.

Claudius Ptolemy (c. AD 90–168)

From the outset portolan charts appear to have been remarkably accurate with little evolutionary development from the earliest-known examples to the later charts made towards the end of the seventeenth century.

John Goss

Maps of the Mediterranean drawn in the fifteenth century AD, according to a table of coordinates devised by the Alexandrian geographer Claudius Ptolemy in the second century AD, show the Maltese archipelago as a single large island, much as it looked in the thirteenth millennium BC . . . Ptolemy, as we will see, based himself on an earlier geographer, Marinus of Tyre – a Phoenician – who in turn had drawn on even older maps and geographical knowledge.

How far back in the human story does the quest for geographical knowledge go? And for how long – either in actual maps and charts, or in tables of coordinates, or in verbal accounts and 'word-pictures' of coastlines and journeys – has such knowledge been preserved and promulgated by navigators?

There has been debate since the 1950s about the significance of certain maps from the late Middle Ages and the Age of Discovery that appear to show Ice Age topography and coastlines – rather than the world as it looked when the maps were drawn. Could these maps have been copied from older source maps that had emanated, ultimately, from a lost civilization of the Ice Age?

I first touched on this mystery in *Fingerprints of the Gods*. But that was in the early 1990s, before I knew about the science of inundation mapping or had been able to explore the hidden world that it revealed. As the new information from Glenn Milne began to come in during the last quarter of 2000, therefore, I set my research assistant Sharif Sakr the task of reopening the investigation – with a brief to stay away from anomalies that I had already discussed in *Fingerprints* and to look only for good, new correlations between the ancient maps and the inundation data now at our disposal. We agreed that this would be a long-term project that should run continuously in the background while Sharif attended to many other day-to-day research matters for me. I warned him that I would sometimes have to take him off the maps for weeks at a time to work on more urgent and immediate issues.

The Reinal map of 1510

I was in India when Sharif e-mailed me in February 2001 with news of his first significant 'hit' – an early sixteenth-century Portuguese map of the Indian Ocean (the Reinal map of 1510), that appears to show the west coast of India as it looked more than 15,000 years ago. Sharif's e-mail discussing the relationship of the Reinal map to other maps of the early sixteenth century, and setting out the initial details of the correlation, is reproduced in chapter 14.

I didn't hear from him again on the subject of Reinal for several months. Then in August 2001 he sent me an update:

Sharif Sakr to Graham Hancock
10 August 2001

Large photos of the Reinal map of 1510 and Cantino map of 1502 have finally arrived from the Bodleian. Not only do they support the correlation I described in my e-mail of 23 Feb but they also suggest that the correlation is even more detailed than I thought – particularly with India at 11,500 BC (not at the LGM, as considered before).

Before detailing the correlation, there are a couple of things I need to explain about the correlation I described on 23 Feb.

Firstly, I suggested that Reinal's map of India omits the Kathiawar peninsula and the Gulfs of Kutch and Cambay that flank this peninsula, such that it correlates with Milne's maps of India before sea-levels had risen to today's levels. The omission of the peninsula is evidently true from the map itself, and I stand by it. But from looking at the maps of Reinal's contemporaries (such as the Cantino 1502 and the Ribiero 1519), I suspect that if we could ask Reinal, 'Why haven't you drawn this important peninsula?' he would reply, 'I have,' and point to a specific peninsula on his map, far away from where the Kathiawar peninsula actually exists. Relative to the surrounding geography, this feature is much too far north and west to be the Kathiawar peninsula, and it's on the wrong side of the Indus river. Nevertheless, this feature was erroneously associated with the Kathiawar peninsula on the Cantino and labelled 'Camba' – i.e. Cambay, which is the name given on modern maps to the long gulf on the south-east side of the Kathiawar peninsula. Reinal may well have made the same mistake. As to where this false 'Camba' peninsula comes from, the answer is quite clear: it comes from the older Ptolemaic model of India, which was highly inaccurate.

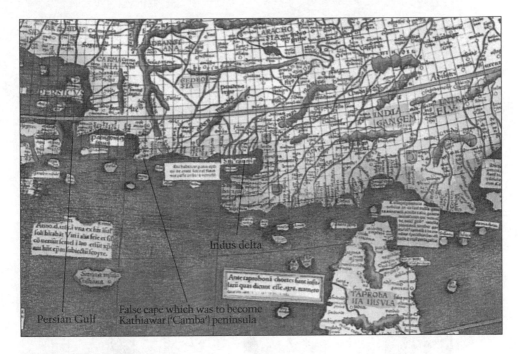

Indus delta

Persian Gulf

False cape which was to become
Kathiawar ('Camba') peninsula

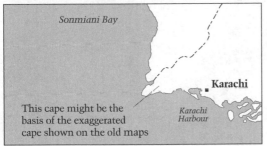

Sonmiani Bay

Karachi

Karachi
Harbour

This cape might be the
basis of the exaggerated
cape shown on the old maps

Waldseemüller's Ptolemaic map of
India.

Modern map of the Pakistani coast.

Though Reinal's map of India is mostly superb for its time, the north-western part is
very *inaccurate* between the Persian Gulf and the Indus river because it closely follows
the old Ptolemaic model – rather than the mysterious and distinctly non-Ptolemaic source
which I speculate was responsible for the rest of India's coastlines on the Cantino and
Reinal maps. That is why Reinal repeats the false north-western peninsula that is shown
on Ptolemaic maps (such as Waldseemüller's 1507 shown above).

The position, shape and orientation of the false Ptolemaic 'Camba' peninsula, as
shown on the Reinal, plus the little island beside it, correlate well with the peninsula on
which the modern city of Karachi is situated, although the scale is vastly exaggerated.
This exaggeration may have originated in the reports of Alexander the Great's
sea-captain, Nearchus, who sailed back from the Indus towards the Persian Gulf and
made specific mention of coastal features and a supposedly 'haunted' island along the
way.

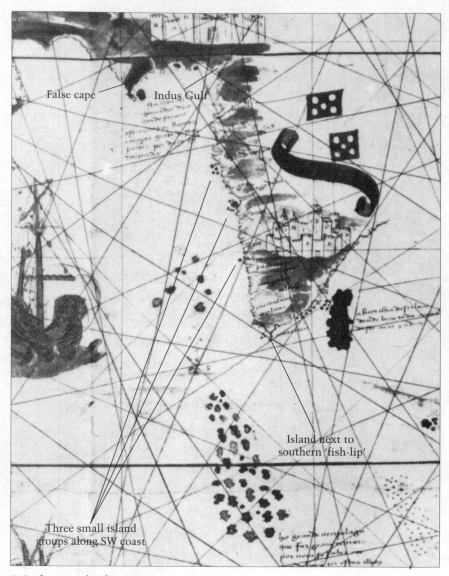

Labels on the map:

False cape

Indus Gulf

Island next to
southern 'fish-lip'

Three small island
groups along SW coast

Reinal's map of India, AD 1510.

The Reinal map departs from the Ptolemaic model specifically at the Indus delta (where Alexander stopped and turned back for home) and then southwards along the entire Indian coastline. As I said before, this coastline is infinitely more accurate than the Ptolemaic model, and strongly suggests that the source from which it is derived was far superior to anything previously available to Western seafarers and mapmakers. This coastline also correlates extremely well with Milne's inundation maps showing India's coastlines before about 12,000 years ago.

Of particular note is Reinal's depiction of four small groups of islands, all close to India's

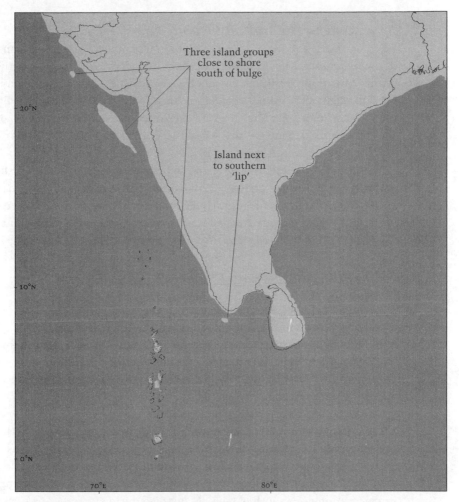

Three island groups
close to shore
south of bulge

Island next
to southern
'lip'

India's coastlines as they were in 11,500 BC.

shoreline and all south of the NW bulge that should have been the Kathiawar peninsula. No such islands exist today, but Milne's maps suggest that there were islands – including one very large one – in roughly the same positions, down to about 10,000 years ago.

Is it possible that what Reinal depicts are the remnants of these islands in the terminal stages of their post-glacial inundation?

Three of the island groups he shows lie along India's west coast, in the right area for such remnants, and one lies immediately next to the southern 'fish-lip' (now less clear than at the LGM) at the very tip of the sub-continent.

Coasting the Indian Ocean

The Cantino and Reinal maps of the Indian Ocean were produced in an epoch of intense competition for trade and a real hunger for geographical knowledge on the part of the European powers that had witnessed – among many other breakthroughs – the rounding of South Africa's Cape of Good Hope by Bartolomeu Dias in 1488,[1] the 'discovery' of the Americas by Columbus in 1492, and the Portuguese encounter with the East that began when Vasco da Gama reached Calicut in south-west India in 1498.[2]

This first European crossing of the Indian Ocean was made from the East African port of Malindi (on the Swahili coast of modern Kenya) where da Gama and his small fleet arrived on 14 April 1498.[3] There they were welcomed by the local chief, who arranged the services of 'a loyal and extremely competent pilot', Ahmed-bin-Majid, described as 'the most famous expert in the navigation of the Indian Ocean in the 15th century'.[4] With this man as their guide they reached India very rapidly, anchoring in front of Calicut on the Malabar coast on 20 May 1498.[5]

There the Portuguese remained for several months, attempting to put arrangements in place to build a trading post, but were foiled at every step by established Arab merchants alarmed at the prospect of European competition ruining their business with the East. Eventually da Gama left empty-handed, 'convinced that only a stronger expedition . . . would have the power to bring negotiations to a successful conclusion'.[6]

On the return voyage there were outbreaks of scurvy, often as few as half a dozen crew were well enough each day to man the ships, and the fleet was alternately becalmed then driven off course by contrary winds. Their zigzag route took them through the Lacadive archipelago – which da Gama named the Santa Maria islands – and then to the small island of Angediva, some 70 kilometres south of Goa.[7] Many died during the crossing to Malindi, which took three times as long as the outward passage, and it was the summer of 1499 before the survivors limped home to Portugal in their two remaining ships.[8]

Almost immediately after da Gama was welcomed back by King Manuel, the Portuguese monarch announced that a new, armed fleet would be sent to India – thirteen ships, with crew and soldiers totalling 1500 men, under the command of Pedro Alvares Cabral. Such a force, it was felt, would be sufficient to set aside the political and commercial obstacles that had confronted da Gama.[9]

The new fleet set sail on 9 March, reaching the Canary islands five days later and the Cape Verde islands on 22 March 1500. There one of the ships was 'eaten by the sea'.[10] The remaining twelve crossed the Atlantic to South America where Cabral made landfall in Brazil on 26 April, claiming it for Portugal. Sending one ship back to Lisbon with news of the discovery of the land that was first known as Vera Cruz, then later Santa Cruz, and finally Brazil,[11] he

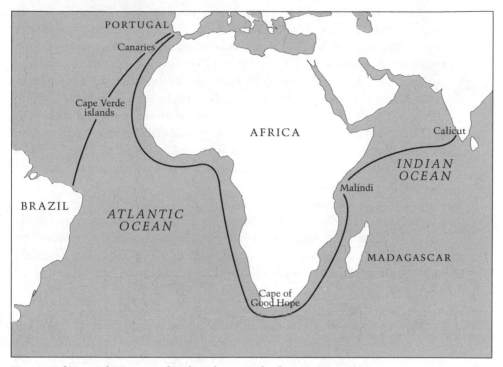

Voyages of Vasco da Gama and Pedro Alvares Cabral, AD 1498–1500.

remained only until 2 May, then turned his fleet south-east and set a course for the Cape of Good Hope.[12]

By this point Cabral's fleet was reduced to eleven ships. Rounding the Cape of Good Hope four more vessels went down with all hands in a violent tempest – among those who perished was Bartolomeu Dias, who thus 'came to be buried in the waters of which he had been the discoverer twelve years before'.[13] A fifth ship, separated from the fleet in the same storm, discovered the island of Madagascar, and then returned to Portugal on its own.[14]

Cabral was therefore down to six ships and less than half his original force when he crossed the Indian Ocean to Calicut. The opposition to a Portuguese trading post still remained strong there and he was now no longer in a position to overcome it. He therefore sailed further south along the Malabar coast looking for a friendlier reception and found it at Cochin where the local rajah permitted him to set up a 'factory'. Cabral then took the fleet to Cananor where they loaded cargoes of spices before returning to Portugal in the early summer of 1501, just over a year after they had left.[15]

Although in both cases under extreme time pressure and in difficult circumstances, the expeditions of da Gama and Cabral undoubtedly did conduct some cursory exploration of several hundred kilometres of the Malabar coast between roughly 15 degrees north latitude (Goa) and roughly 10 degrees north latitude

(Cochin). On the third and fourth expeditions, however, these explorations were not extended:[16] 'It was only with the fifth India fleet in 1503 under Albuquerque that exploration was carried further, as far as Coulon [Quilon], almost on the southern tip of Malabar.'[17]

Cape Comorin – modern Kaniya Kumari, the true southern tip of the Indian peninsula – was first rounded near the end of 1505 by a fleet under Lourenco de Almeida. The fleet had been sent to the Maldives to spy on the sea trade with the Indonesian islands further east but was carried off course to Cape Comorin by winds and currents. From there Almeida sailed his ships to Sri Lanka: 'Thus Lourenco de Almeida and his companions were the first Portuguese to pass into the eastern Indian Ocean.'[18]

In 1506 there was another 'first' – Joao Coelho was the first Portuguese to reach the northern terminus of the Bay of Bengal and 'to drink the waters of the Ganges'.[19] But it was not until 1509 that Diogo Lopes de Sequeira made the first full crossing of the Bay of Bengal to reach Malacca[20] – the Malaysian peninsula known until that time on Ptolemaic maps as Aurea Chersonesus, the Golden Chersonese.[21]

Thus, it can be seen that the focus of Portugal's attention for more than a decade after Vasco da Gama first reached India in 1498 was on the Malabar coast south of Goa and on the eastern Indian Ocean and the Bay of Bengal. The long lines of supply and relative scarcity of men and ships meant that no attention could be paid to the stretch of the Indian coast that runs north-westwards from Goa, at roughly 15 degrees north latitude, past the Gulf of Cambay, the prominent Kathiawar peninsula and the mouths of the Indus, up to the northern terminus of the Arabian Sea at roughly 25 degrees north latitude. As Damiao Peres writes in his authoritative *History of the Portuguese Discoveries:*

> In the first years of Portuguese expansion in the Indian Ocean, the reconnaissance of the Gulf of Arabia [i.e., the Arabian Sea] was limited to a few southern ports of the Malabar coast to the east, and to the coast of Arabia and its neighbouring areas to the west. Included were some island groups lying between the two. The northern part of the Gulf of Arabia [Arabian Sea] and its adjacent waters – the Persian Gulf and the Red Sea – were only visited in the first years of the second decade of the sixteenth century.[22]

The mystery of the Cantino map of 1502

And this brings us to what is mysterious about the Cantino map – so named after Alberto Cantino, the Lisbon-based diplomatic agent of the powerful Duke of Ferrara in Italy.[23] Cantino somehow acquired this beautiful but unsigned world map in Portugal, or had a cartographer there copy it specially from another map, and then smuggled it out of the country, getting it to Italy by or before 19 November 1502[24] (no mean feat, since Portugal was jealous of its discoveries

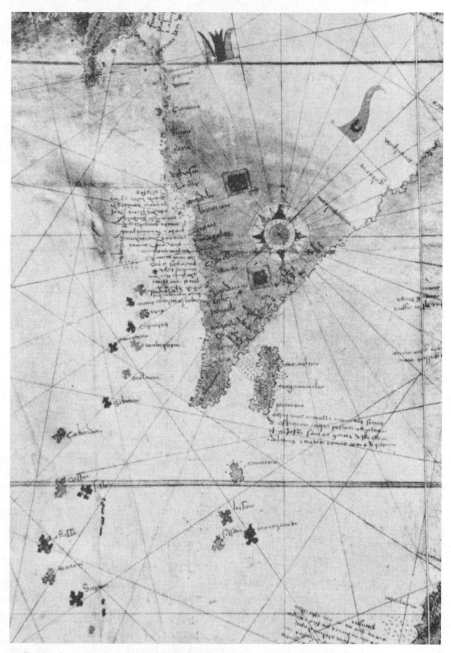

India on the Cantino planisphere of c. AD 1502.

and imposed the death penalty on those caught smuggling maps out of the country).[25]

When was the Cantino map drawn?

Let us start by stating the obvious: it must have been drawn before

19 November 1502, when it reached Italy. Indeed, according to H. Harisse, it typically took craftsmen of the period about ten months to prepare such a map. If this is correct, then it pushes the origins of the Cantino back at least to the beginning of 1502.[26]

Going back still further, there is internal evidence in the map itself which proves that it could not have been drawn much before the summer of 1501. That was when the ships of Cabral's second India fleet returned to Portugal from their voyage – begun a year previously – that had taken them not only to India but also to South America. The evidence survives because the Cantino is a world map that shows – and claims with Portuguese flag-icons – the section of the Brazilian coast discovered by Cabral in 1500.[27] Since similar flag-icons are likewise seen over Cochin and Cananor in southern India – which Cabral reached later in 1500 – it is the inescapable conclusion that the Cantino map expresses knowledge that could only have been acquired on the Cabral voyage.

Indeed this is the conclusion of orthodox historians of cartography,[28] so it is not controversial to restate it here. What is extremely strange, however, is that neither on Cabral's voyage of 1500/01, nor on the earlier 1498/9 voyage of Vasco da Gama, nor on any later Portuguese voyage until after 1510, was the north-western part of India ever visited. Yet the Cantino map shows north-western India very clearly. And although the portrayal is *inaccurate vis-à-vis* India's western coast as it has looked for the past 7000 years – in the single, significant respect that it entirely omits the Kathiawar peninsula – it is still hugely *more accurate* for India as a whole than the grotesque image of the subcontinent provided in the Ptolemaic maps.

Particularly noteworthy is Cantino's representation of the east coast of India. In general (see diagrams) it matches well to what the east coast of India should look like.

I do not deny that the Portuguese were capable of drawing maps as accurate – and indeed more accurate – than this one. But the puzzle to me is how Cantino's Portuguese cartographer could have acquired such accurate knowledge of the outline of eastern India as early as 1501–2, when historical records show that the fleet of Lourenco de Almeida did not even round Cape Comorin and enter the eastern Indian Ocean until 1505? This part of the map also shows Sri Lanka at close to its correct size and very close to its correct location more than three years before Lourenco de Almeida became the first Portuguese to sight Sri Lanka.

Surely, therefore, curiosity should drive us to find an explanation for the existence of this strikingly good chart of supposedly uncharted waters?

'T-O' maps

Good is a relative term. To understand why the Cantino and Reinal maps of the Indian Ocean are 'good', and in fact in some ways close to 'revolutionary', we need to view them in the cartographic context of their place and time – i.e.

The Augsburg T-O map of AD 1472.

Europe and the Mediterranean in the fourteenth to the sixteenth centuries AD.

During this period mariners, merchants, adventurers and armchair travellers had at their disposal four distinctly different types of maps and charts. The simplest of all – far too simple to be of any use to navigators – are the so-called 'T-O' maps. With a long history going back to the seventh century AD, these show an encircling 'O' of water that is often inscribed with the words 'MARE OCEANUM' – representing the 'Ocean Sea' (sometimes 'Ocean River') that was believed in antiquity to surround all the lands of the world[29] (an idea, by the way, that is completely correct, as all the world's oceans do indeed interconnect). Inside the 'O' a 'T' is then inscribed, dividing up the land into the three known continents of Africa, Asia and Europe. The vertical stroke of the T represents the Mediterranean, separating Africa from Europe and adjoining the Ocean Sea at the Atlantic. The cross-bar of the T is the north-flowing river Nile on one side of the Mediterranean, the south-flowing river Don on the other side of the Mediterranean and also, vaguely, the Black Sea, the Bosporus and eastern Mediterranean, beyond which lies the continent of Asia. The Garden of Eden is also often depicted on the 'top' of such maps, which are oriented eastwards. Map historian John Goss points out that frequently 'Four rivers were also described as flowing from the Garden of Eden: Psihon, Gihon, Tigris and Euphrates.'[30]

The T-O maps provide at best a 'shorthand picture of the world'.[31] But the enduring power and pervasiveness of this essentially useless cartographic tradition is illustrated by the oldest surviving printed map of Europe – a T-O map, printed by Gunther Zainer at Augsburg in 1472, that reproduces exactly the original concept as set out by Isidore, Bishop of Seville, in his *Etymologiarum* written in the early seventh century.[32]

Mappamundi

The second category of maps and charts available between the fourteenth and sixteenth centuries is known as the *mappamundi*. It is important to be clear that this is a very distinct and specific type of world map (because, in the texts of those times, other world maps of completely different types were also sometimes referred to as *mappaemundi* or *mappamundi*, when what was meant was just 'world maps' in the rather loose, general sense of 'maps of the world').[33]

So, to be clear, the *mappamundi* to which I refer here were normally handpainted on cloth or vellum (hence the origin of the name *mappamundi* –

Hereford mappamundi, c. AD 1290.

meaning, literally, 'Cloth of the World'). The classic example is the Hereford
mappamundi, attributed to Richard of Haldingham c.AD 1290, but *mappamundi* continued to be made well into the fifteenth century. They retain the
essential design of the T-O maps but greatly increase the amount of detail
concerning mountains, rivers, pilgrim routes, etc., on the three recognized
continents of Africa, Asia and Europe – sometimes taking into accounts myths,
legends and recent traveller's tales. Unfortunately, none of the specifically
geographical details that these maps provide would have been of the slightest
bit of use to travellers or mariners since all the details – all of them! – are wrong,
misguided and misleading.[34] In short, the *mappamundi* promulgate a wholly
incorrect image of the world – an image that is almost all dry land and that

reduces the Ocean Sea covering seven-tenths of our planet to the narrow, ribbon-like rim of the surrounding 'O'. 'The very crudeness of the geography of the Hereford map', comments John Goss, 'reflects a marked deterioration in geographical knowledge from the time of Ptolemy a thousand years earlier.'[35]

Ptolemaic maps

Almost nothing is known about the life of Claudius Ptolemy.[36] His first name is Roman and his second Macedonian.[37] He is thought to have been born in Upper Egypt[38] c.AD 90 and to have died around 168.[39] A scholar at the Library of Alexandria from roughly AD 127 to 145,[40] his two famous surviving works are the *Almagest* (*Ho megas astronomos*), a book of astronomy and cosmology in which he expounds the 'Ptolemaic system' of a fixed spherical earth at the centre of a revolving universe, and the *Geography* (*Geographike hyphegesis*), in which he includes information on how to construct maps of places in Europe, Africa and Asia tabulated according to latitude and longitude.

It is not absolutely clear whether Ptolemy ever in fact drew maps himself, or even had maps drawn to accompany his work.[41] Strictly speaking, they were not necessary because his primary method was to provide the longitude and latitude coordinates of more than 8000 places and topographical features in such a way that: 'the reader can draw for himself regional maps on various suitable scales, and even a general map of the world'.[42]

The *Geography* (its Greek title translates literally as 'Instruction in Map-drawing') 'professes to be concerned solely with the task of scientific mapping'.[43] But what, in fact, have its contributions been to the scientific mapping of the world?

One signal contribution was the memorialization of knowledge of the earth's basic form as a sphere. Just how ancient this knowledge already was in the second century AD remains unclear. Scholars agree that its earliest surviving *documented* appearance is in the work of Pythagoras in the sixth century BC;[44] however, it may have long pre-dated Pythagoras in oral traditions or in documents that have since been destroyed with the passage of time. My personal view, already expressed elsewhere, is that the concept of the spherical earth was well known to the first great historical civilizations such as the ancient Egyptians and the Sumerians 5000 years ago and will ultimately be proved to date back to a much more remote period even than that. But wherever it comes from originally, we owe a debt of gratitude to Ptolemy for its preservation and repromulgation in the second century AD – for, despite the intellectual ravages of the Dark Ages that were to follow, his vision of the earth as a sphere was never quite forgotten. Robert Fuson, Professor of Geography at the University of South Florida, puts it this way:

> The Ocean Sea had now taken the form it was to retain until the 16th century and the aftermath of Magellan's circumnavigation. The earth's sphericity was no longer debated by any practical navigator, cosmographer, or educated person. This fact had

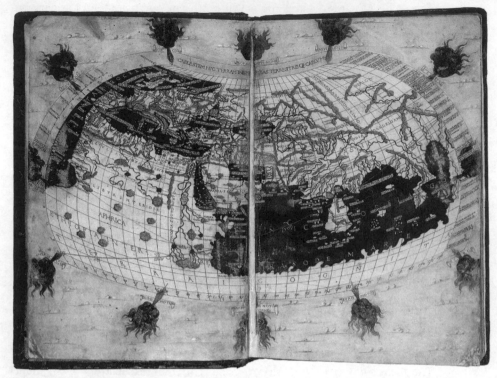

Francesco Berlinghieri's Ptolemaic map, AD 1482.

been established since the days of Classical Greece. The only areas open to serious disagreement were details of coastal configuration, exact location by coordinates, island discovery and location, and the dimensions of the Ocean Sea. By the 1400s [before the discovery of the Americas] no reasonable person questioned the proposition that Asia might be fetched by sailing west from Europe, if the ships and crews could survive the tremendous distance.[45]

Other significant contributions that Ptolemy made to the scientific mapping of the world include the establishment of functional parallels of latitude, and of a prime meridian, passing through the Canary islands, that was to serve as zero degrees longitude for sixteen centuries.[46] Moreover, though maps drawn to Ptolemaic coordinates leave much to be desired, even the worst of them are far superior to the schematic 'T-O' maps and *mappamundi* of the Dark Ages.

A representative selection of Ptolemaic world maps is reproduced herewith. The reader will note that the Mediterranean is at least recognizable, and that in spite of many discrepancies a real attempt appears to have been made to reflect the true shapes and locations of the lands bordering it. Ptolemy and his informants had first-hand, day-to-day knowledge of this central region of what they called the *oikumene* – the habitable world – and clearly, with some peculiar

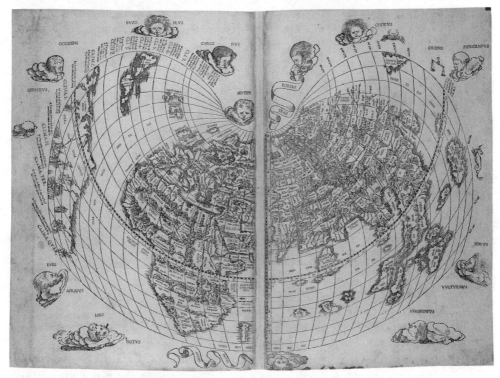

Ptolemaic map from Venice edition of Ptolemy's Geography, AD 1511.

Waldseemüller's Ptolemaic map, AD 1507.

exceptions, they used that knowledge well. But outside the Mediterranean the level of accuracy rapidly falls away.

For example, on the authority of Poseidonius (135–50 BC),[47] Ptolemy *under-estimates* the circumference of the earth at the equator, setting it at 20,400 miles (as against the correct figure of 24,902 miles).[48] At the same time he greatly *overestimates* the east–west extent of Asia and, bizarrely, portrays the South Asian coast above the Indian Ocean apparently *without any representation whatsoever of the great peninsula of the Indian subcontinent*. As though to compensate for this loss, however, Ptolemy places an enormous island, Taprobana (presumed to be Sri Lanka), just off-shore of the stretch of non-peninsular mainland identified as India.

What is going on here? In their major new study of Ptolemy's *Geography*, J. Lennart Berggren and Alexander Jones suggest that the root of the problem is simple. India has this 'flattened out' appearance because Ptolemy, somehow, has managed to turn the subcontinent on its side so that its orientation is roughly west-to-east instead of north-to-south as it should be:

> Asia exhibits greater and greater distortions as one progresses further east, the most obvious faults being the north–south compression of the Indian subcontinent so that its western coast is made to run parallel to the equator, and the exaggerated size of the island of Taprobana (Sri Lanka).[49]

If the subcontinent has indeed been swung eastwards in the way that Berggren and Jones propose, then 'Taprobana' is not only too big to be Sri Lanka but is also positioned in entirely the wrong place. Sri Lanka lies in the Bay of Bengal, off India's south-east coast. Once the reorientation of the peninsula on Ptolemy's map is taken into account, however, then we can see that Taprobana has in fact been portrayed as lying off India's *west* coast – where there are no large islands today.

We will return to the possible implications of this later. Meanwhile, to conclude the description of Ptolemy's world map, let us note that the older examples (e.g. page 466) portray the Indian Ocean as a lake landlocked by the northern edge of a southern continent (Terra Australis on some editions; Terra Incognita on others) that connects southern Africa with the south-eastern extreme of Asia:

> At the eastern edge, where the lands represent central China and Southeast Asia, it is virtually impossible to identify any of the features on Ptolemy's map with real counterparts. At the eastern limit Ptolemy draws the coast of Asia turning south and then west, eventually to join the east coast of Africa, thereby making the Indian Ocean a vast enclosed sea unconnected with the Atlantic.[50]

Ptolemy was not the originator of the *Geography* – as he himself goes to great lengths to point out. Instead, he tells us that his role has been to refine and correct an earlier *Geography* prepared by his predecessor, the Phoenician geographer Marinus of Tyre, who was active around AD 100 or 110 and whose great work was itself called *Correction of the World Map*.[51] In Ptolemy's own words:

Marinus of Tyre seems to have been the most recent of our students of *geographia* [= map-making] and to have applied himself to the subject with the greatest enthusiasm ... If we could see that his latest composition lacked nothing, we should even have been happy to complete our description of the known world from these notes of his alone, without researching any further. But as on certain points he himself seems to have composed without reliable comprehension, and as in embarking on his map he has in many places not devoted enough thought either to convenience or to symmetry, we were naturally induced to contribute to his work what seemed necessary to make it more logical and useful.[52]

As well as the honesty of this statement, what I find particularly striking is the strong suggestion Ptolemy leaves us with that his *Geography* was part of a tradition, and that his predecessor Marinus had been part of that tradition too – but by no means its first student, just the 'most recent' who had 'corrected' an older map.[53] Such a tradition might, theoretically, have extremely ancient roots and it need not necessarily be the case that successive 'refinements' of it over long periods of time must have improved it. An alternative possibility, which it would be unwise to ignore entirely, is that far from being the pinnacle or 'culmination' of ancient geography, as many scholars suggest,[54] Ptolemy's maps may actually have been the end-products of a long process of decline, degradation and accumulated errors introduced by many different hands into a far older and once far superior map-making tradition. Again, this is a theme that we will return to.

Some centuries after Ptolemy's death the Dark Ages descended over the *Geography*, but it was still preserved here and there in a few monasteries in Europe.

In the Arab world Muslim geographers are known to have possessed editions of the *Geography* as early as the eighth century AD, as well as separate editions of the earlier work of Marinus (the latter now all being lost):

In the early ninth century al Ma'amun, Caliph of Baghdad AD 813–833, set up an Academy of Science, which among other things produced a world map [lost] and 'improved tables', – i.e. modernized coordinates.[55]

In Byzantium (Constantinople) in the late thirteenth century it was Maximus Planudes (c.1260–1310) who was responsible for bringing the knowledge enshrined in Ptolemy back to the attention of the world:

He searched for manuscripts of Ptolemy's *Geography*, and his search was rewarded in 1295, but it was not as exciting as he had hoped. As he explains in a letter and some verses, after at last finding what he knew was a neglected work, he was disappointed to discover that it had no maps.[56]

Although there are older manuscript maps (such as the late twelfth- or early thirteenth-century *Codex Urbanus Graecus* 82), the oldest surviving manuscript copy of the *Geography* containing maps based on the descriptions and coordinates given by Ptolemy was made by monks at Vatopedi on Mount Athos in the early fourteenth century.[57] It later formed the basis for the first printed atlas to appear in Europe, published at Bologna, Italy, in 1477.[58]

The Ptolemaic cartographic tradition was at first very successful in adapting to the challenges posed to its world-view by the Age of Discovery. Thus, the original maps based on Ptolemy's own coordinates were added to several times during the sixteenth century to accommodate so-called *tabulae novae* (or *tabulae modernae*) recording the expanding revelation of the Americas and of the East.[59] This could be done without causing serious disturbance to the Ptolemaic concept of the *oikumene* so long as the simple expedient could be maintained of tagging the Americas on to Asia like some vast peninsula. Ultimately, however, these maps, like the dinosaurs, were an evolutionary dead-end doomed to extinction.

It would be wrong to imagine from all this that the few surviving Ptolemaic maps in libraries and archives around the world have nothing to teach us. They may appear distorted and clumsy to the sophisticated modern eye, but it is possible that their very awkwardness and peculiarity could have caused scholars to overlook significant details concealed within them.

Portolan charts

The fourth category of maps circulating in Europe from the fourteenth to the sixteenth centuries, known collectively as portolanos, portolan charts or simply portolans, shows no dependence whatsoever on either Ptolemaic maps or data or the *mappamundi*. The vast majority of the portolans depict only the Mediterranean/Black Sea area and the countries immediately round about, but some are world maps, or world atlases, for which the style and approach of the Mediterranean portolans serves as a basis. These old charts are drawn to the highest cartographical standards and are uncannily accurate – so accurate, though the earliest examples go back to the end of the thirteenth century, that they were not surpassed by new scientific techniques, measurements and observations for almost 500 years.[60]

A. E. Nordenskiold, the great Swedish polar explorer and map historian, had a special interest in portolans. He points out that they were used, almost entirely, by practical mariners and navigators:

> Slight was the attention paid to them in the fifteenth and sixteenth centuries by learned geographers. Thus Munster seems to have totally overlooked them, and in the first edition of the *Theatrum Orbis Terrarum*, Ortelius does not mention a single drawer of portolanos amongst the cartographical authors enumerated in his *Catalogus Auctorum*. At present the investigator into the history of geography

acknowledges them as unsurpassed masterpieces, and reckons them amongst the most important contributions to cartography during the Middle Ages.[61]

Likewise John Goss notes:

The portolan charts were quite unlike contemporary medieval maps. They often incorporated detail of remarkable accuracy, based on close and actual observation, rather than the conventional medieval habit of repeating cartographical and mythical information issued by the Church.[62]

Goss and Nordenskiold also point to other characteristics that make the portolans look and 'feel' different:

- A network of intersecting straight lines (usually called 'rhumb-lines' or 'loxodromic lines')[63] originating from sixteen equidistant points, spread about the circumference of a 'hidden circle' around the map.
- An elaborate 'compass-rose' at one or more of the points of intersection of the lines.
- Place and feature names written perpendicular to the coastline, in sequence along the coast.
- Charts drawn in ink on vellum or parchment with colour conventions, e.g., most important names shown in red, rest in black; lines depicting four main wind directions drawn in black, the eight half winds in green, the sixteen quarter winds in red.
- Coastlines emphasize bays and headlands; hazards such as rocks, reefs and shoals are marked with dots or small crosses.[64]

What all these characteristics have in common is their utility and significance to mariners. The coastal hazards are matters of life and death. The networks of rhumb-lines assist compass navigation. Even the perpendicular place names, inevitably upside-down from some angles, make sense when you realize that they are meant to be viewed in the same direction as that of a vessel following the coast.

It has been suggested that portolans are such an improvement on previous maps because they reflect the earliest introduction of the compass into Europe, thought to have taken place around the end of the thirteenth century[65] (although the use of magnetized needles as a means for sailors to find their bearings is attested earlier than that).[66] But while there is no doubt that such charts in conjunction with compasses do provide very effective navigational guides, it is by no means so certain that compasses and compass-bearings were used to prepare them in the first place. On the contrary, says A. E. Nordenskiold, 'many of them are evidently older than the use of the compass on board ships'.[67]

No map projection is imposed on the portolans, and there is no latitude and

Carta Pisan portolan, c. AD 1290.

longitude grid – although in the expanded 'world portolans' the equator is often shown, together with the Tropics of Cancer and Capricorn and the Arctic and Antarctic Circles. Nevertheless, when relative latitudes and longitudes on these maps are measured, they prove to be extremely accurate. For example, on the Dulcert portolan of AD 1339 the total longitude of the Mediterranean and the Black Seas is correct to within half a degree.[68]

It is wrong to argue, as I myself have done in the past,[69] that mariners and chartmakers of the fourteenth century would have found it *impossible* to achieve such accurate longitudes. Such suspicions arise from the fact that marine chronometers – which made reliable calculations of longitude at sea possible – were not introduced until the second half of the eighteenth century. However, scholars are right to object that there are other, simpler (if vastly more time-consuming) ways to obtain almost equally accurate longitudes. As Gregory McIntosh put it in an e-mail:

> We moderns seem to tend to think that because we now have methods of making very accurate measurements very quickly, those in the past could not make any measurements at all. The Portuguese (and others, of course) made dead reckoning measurements of longitude [i.e. calculations based on empirical estimates of course, speed and time]. It is a method of measurement. Some writers of the Hapgood ilk [Charles Hapgood, *Maps of the Ancient Sea Kings*] would have us believe that dead reckoning is not a valid method of measurement. They would have us believe that the Portuguese did not measure longitude. But of course they did. That's what dead reckoning is – a method of measuring longitude . . . with several such measurements from repeated voyages.[70]

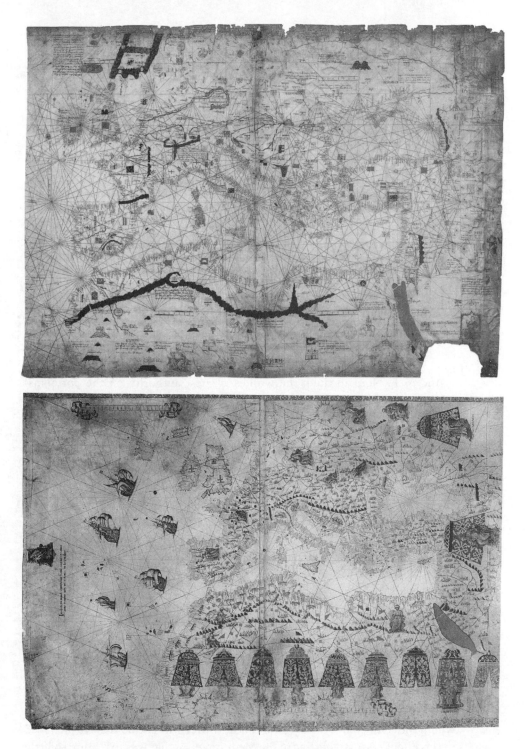

Top: *Dulcert portolan,* AD *1339.* Bottom: *Maggiolo portolan,* AD *1563.*

This seems like an entirely reasonable explanation for the accuracy of the portolan charts – that they are the result of the accumulated observations and measurements of navigators plying the coasts of the Mediterranean over relatively long periods of time. Some have suggested that they may even trace their origins back to the detailed written accounts of sea-journeys, harbour conditions, winds, currents and trade – the *periploi* – that were in favour amongst the ancient Greeks as far back as the fifth century BC.

Still, there is a very large gulf indeed between the crude directions of the *periploi* and the navigational accuracy of the portolan maps. Along any hypothetical evolutionary road from one to another it is reasonable to expect to see intermediate forms – since getting maps right by dead reckoning, as McIntosh points out, is a painstaking, long-term process of trial and error, correction and gradual improvement.

And this is the central problem of the portolans. Quite simply, *there are no intermediate forms*. Indeed, remarks John Goss:

> From the outset portolan charts appear to have been remarkably accurate with little evolutionary development from the earliest known examples to the later charts made towards the end of the seventeenth century.[71]

And A. E. Nordenskiold, the world's greatest authority on the portolans, reminds researchers that:

> Notwithstanding all the progress made during the fifteenth and sixteenth centuries in the art of drawing maps with the aid of newly invented nautical instruments, there was published a chart in Holland in 1595 by one of its most expert mariners which is only a copy, or rather a copy of copies, of portolanos drawn 250 to 300 years earlier. This is an extremely remarkable fact in the history of civilization. But moreover the principal features of the portolanos from the beginning of the fourteenth century are still to be found on Van Keulen's sea-charts of 1681–1722. I suppose that up to the beginning of the nineteenth century the influence of the old portolan charts may yet be traced on the charts of several parts of the Mediterranean and Black Seas.[72]

The good legacy

The 'extremely remarkable fact in the history of civilization' that Nordenskiold draws our attention to here is the ability of maps, apparently produced by dead reckoning in the thirteenth century, to compete on an equal footing with scientific nautical charts from as late as the nineteenth century.

And it is remarkable. Because, yes, we can accept with McIntosh that it was within the competence of navigators of the thirteenth century to have produced the excellent portolan outline of the Mediterranean that was to require so little improvement over the next 500 years – in other words, we can accept that it *could* have been done. We can even accept that it *might* have been done. But it is much harder to agree that this is what actually *was* done, since neither McIntosh nor any other scholar favouring the gradual 'evolutionary' explanation for the very early perfection of the portolan genre has yet been able to provide us with even a single example of charts that illustrate even a single aspect of this proposed 'gradual evolution'.

In my opinion, therefore, Peter Whitfield is right to evaluate the *Carta Pisane*, the oldest surviving portolan in the world, as 'one of the most enigmatic charts in the history of mapmaking'.[73] In his 1996 study, *Charting of the Oceans*, he elaborates on this theme:

> The appearance of this chart (and of the others which survive from the following century) is one of the most mysterious events in the history of mapmaking. A glance at the Pisan Chart immediately reveals two outstanding features: the coastlines of the Mediterranean are drawn with striking accuracy; and the map is covered with a network of lines radiating from two central points, which clearly impose the form of the compass over the whole map. How did this highly accurate map suddenly appear in medieval Italy, and how exactly was it linked to the compass? Was it the original work of a single individual, or was it descended from a line of much older charts which had been developing for centuries? The former is difficult to believe, but the latter cannot explain why there is no shred of evidence for the existence of such maps before 1270.[74]

Whitfield outlines the orthodox scholarly response that the evolution of the portolans must have taken place within the oral lore of mariners and within the textual tradition – going back to the Greek *periploi* – of books of sailing directions:

> One famous example entitled *Lo Compasso da Navigare* was current among Italian mariners and it would be tempting to suppose that the contents of a text such as this had been transformed with the aid of compass bearings into the Pisan Chart. Unfortunately, the places named in *Lo Compasso* differ sharply from those named on the map, even the names in Italy itself. *Moreover, the transition from a list of names and bearings to an accurate map is an enormous one, requiring not only a high degree of geometric and drafting skill, but also an imaginative leap to create a graphic form for which there was no parallel.* Even if the Pisan chart was based on some now-lost portolano, we have no real idea how it was done. Nor can we really answer the most fundamental question of all about the chart – how was it used? We have no independent description of its use, although we do know, from examination of the chart itself, that the compass lines were *plotted* before the map itself was drawn . . .[75]

Concerning *how* it was done, Whitfield notes:

Later practice was to make a running survey, in which coastal features – capes, bays or islands – were sighted from two, three or four positions as the ship sailed by. Starting from the ship's course, the distances run and the angles of sight were used to build up a profile of the coast. This method was in use by the later sixteenth century, but we can only conjecture whether it was known at the time the Pisan Chart was drawn. If it was not, it is extremely difficult to account for the accuracy of some of the coastlines, which would scarcely be improved on this scale until the eighteenth century.[76]

But even if we admit that running-survey and compass techniques *were* somehow being used on ships to produce sea-charts as early as the thirteenth century (which most historians of science would rule out) we still come against the unexplained enigma of the miraculous and fully formed *de novo* appearance of the *Carta Pisane*. As we've seen, not a single chart pre-dates it that demonstrates in any way the gradual build-up of coastal profiles across the whole extent of the Mediterranean that *must* have occurred before a likeness as perfect as this could have been resolved.

It is possible, of course, through the vicissitudes of history, that all the evidence for the prior evolution of portolans before the *Carta Pisane* has simply been lost. If that were the case, however – in other words if the *Carta Pisane* is a snapshot of a certain moment in the development of an evolving genre of maps, and if we accept that all earlier 'snap-shots' have been lost, wouldn't we nevertheless expect that such an 'evolving genre' would have *continued* to evolve after the date of the earliest surviving example?

Whether we set the date of the *Pisane* between 1270 and 1290 (as Whitfield suggests)[77] or a little later – between 1295 and 1300 – as other scholars have argued, we've seen that that there was *no significant evolution afterwards*.[78]

Now kept in the Bibliothèque Nationale in Paris, the enigmatic *Pisane* is an unsigned chart and scholars have no idea who the cartographer might have been.[79]

Next comes what Whitfield rightly describes as the 'startling and precocious' work of the earliest chartmakers known to us by name in the first half of the fourteenth century. These include Vesconte and Pizzagano in Venice, and Dulcert and Valseca in Majorca. None of them seems to have copied the *Carta Pisane* directly, but neither do they add significant cartographical detail in the central Mediterranean/Black Sea area covered by the *Pisane*. On the contrary, what we see in their more lavish maps are only the effects of very minor tinkering and stylistic improvements. The basic template inherited from the thirteenth century remains unaltered and stays that way for the rest of the life of the genre.

So the hypothesis of a gradual evolution of portolan charts out of books of sailing directions does not withstand close scrutiny. Convinced of this, A. E. Nordenskiold sought a more satisfactory explanation and came, after many years of study, to a radical conclusion – that the original model for all the portolan charts, a hypothetical common ancestor that he refers to as the 'normal portolano' is most likely to have been derived from the long lost sea-charts of the Phoenician geographer Marinus of Tyre.[80]

In other words, the *Carta Pisane* and the other early portolan charts that started the genre were not a 'development' of anything. They were a legacy.

The Sea-fish of Tyre

Nordenskiold points out that the same legends and place names, presented in the same way, appear on all portolan charts. He makes a special illustration of this with reference to the Catalan Atlas of the fourteenth century, Giroldis' portolan of the fifteenth century, and one by Volontius of the end of the sixteenth century, but argues that it is true for all portolans:

> When to this is added
> (1) that the Mediterranean and the Black Sea have exactly the same shape on all these maps; (2) that a distance scale with the same unit of length ... occurs on all these maps, independently of the land of their origin; (3) that the distances across the Mediterranean and the Black Sea measured with this scale agree perfectly on different maps; (4) that the conventional shape given to a number of smaller islands and capes included in the maps remained almost unaltered on portolanos from the 14th century to the end of the 16th; then it may be held as completely proved that all these portolanos are only slightly altered and emended 'codices' of the same original which I designate by the name normal portolano.[81]

In his quest 'to determine when and where the normal portolano was composed',[82] Nordenskiold uncovered a previously overlooked passage in a work written in AD 955 by the important Arab geographer Masudi who states that he had: 'seen the maps of Marinus, and that these by far surpassed those of Ptolemy'.[83]

The portolans are the only maps drawn in ancient times or in the Middle Ages that are better than the maps of Ptolemy.[84] We cannot say for sure how ancient their origins are. But they must have a background somewhere. It is Nordenskiold's hypothesis that 'the first origin of the portolanos is to be derived from the Tyrian charts described by Ptolemy under the name of Marinus'[85] and that the world map of Marinus could have been 'a real portolano, provided with a text'.[86] Moreover,

> If Ptolemy himself had not always spoken of Marinus as a definite personality, it could have been conjectured that the name Marinus of Tyre, or the Tyrian sea-fish, had only been a collective name for a certain category of nautical maps ... The numerous editions mentioned by Ptolemy mean that the Tyrian charts were made for a practical purpose, and the improvements, introduced according to Ptolemy in every new edition, constituted the germ of the future masterpiece ...[87]

This is an interesting speculation, for indeed there is no mention of Marinus outside of Ptolemy which independently confirms the Phoenician geographer's existence. Nor is it too much to ask of the facts to suggest that the famous seafaring city of Tyre to which Marinus supposedly belonged might have originated a special category of charts that came to be known, colloquially, by a name something like the 'Tyrian sea-fish'. Perhaps, despite the personalization, it was an atlas of 'Tyrian sea-fish' regional charts and a 'Tyrian sea-fish' world map that Ptolemy 'corrected' and 'improved' in the second century AD, and not the work of any individual geographer?

And I've already noted that we only have Ptolemy's word for it that he actually did improve on Marinus. Maybe he thought he was doing that – while all the time his 'improvements' were only making the Phoenician charts worse. That would explain why Arab mariners of the tenth century still treasured the original Marinus maps that they had somehow managed to preserve and declared them to be so much better than the Ptolemaic ones.

Arabia without maps

Just three Arab portolans, all classic 'normal portolanos' of the Mediterranean and Black Sea area, have ever been found. The earliest dates from 1300, very close to the date of the *Carta Pisane*, and the other two from 1413 and 1461 respectively.[88] This suggests at least two things to me: first, like the Europeans, the Arabs made no attempt to develop the inherited normal portolano (other than putting modern names and legends on their copies of it); secondly, although the Marinus 'normal portolano' had been preserved by the Arabs, as Masudi testifies, and although there was clearly some demand for it, the survival of only three Arab copies suggests that its use never became anything like as widespread in Arab seafaring as it did in the seafaring of the Europeans.

In his discussion of Arab cartography, A. E. Nordenskiold has this to say:

> Various admirable descriptions of distant lands and of extensive voyages written by Arabian scholars and far surpassing the geographical productions of the same period among the Christians, are still extant. But similar perfection was never attained by the Arabian maps, which, if they were original drawings and not, as the planisphere of Idrisi, mere copies or reproductions from Ptolemy, are not only far inferior to the maps of the Alexandrian geographer, but not even comparable to the Esquimau-sketches brought home by English and Danish polar travellers from the icy deserts of the polar regions.[89]

This may seem an over-harsh judgement, since there is no doubt that the Arabs were brave and adventurous explorers. For example, the same Idrisi mentioned in the passage above also indicates that in the tenth century Arab sailors crossed or attempted to cross the Atlantic.[90] But it is true that Idrisi, geographer to King Roger II of Sicily at the end of the twelfth century, did base his beautiful maps

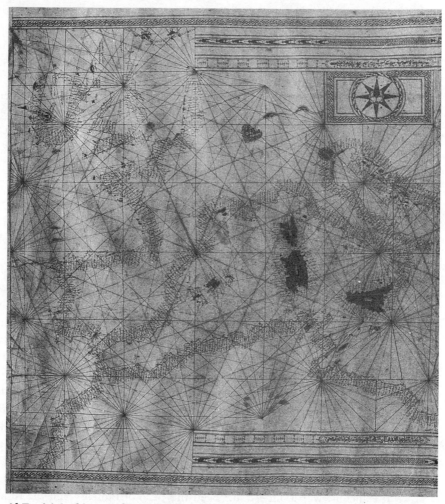

Al-Tunisi Arabic portolan, western section, AD *1413.*

on Ptolemy.[91] And it is true, with the exception of the three rare Arab portolans that have survived (one of them being notably early), that the quality of the rest of Arab cartography in this period was not high.

Regardless of whether the Arabs themselves were good or bad at making maps, however, as Nordenskiold points out:

> It is shown by the passage referred to in Masudi, that the maps of Marinus of Tyre were still extant in the middle of the 10th century, that is to say, shortly before the time when the first portolan maps were drawn. Since that time they have completely disappeared. It might be legitimately concluded from this that the portolanos may have arisen as a modernization of the *Tyrian sea-fish* undertaken during the Crusades, and that they stood in the same relation to the maps of Marinus as the *tabulae-*

modernae in the printed editions of the *Geography* of Ptolemy stood to the Alex-
andrine geographer's own work.[92]

On marvellous things

By documenting the presence among the Arabs at so late a date of good ancient
maps that were (a) attributed to Marinus of Tyre and (b) recognized as belonging
to a distinct tradition superior to Ptolemy, Nordenskiold provides at least the
beginnings of a plausible answer to the riddle of the 'lost evolution' of the
portolans prior to the *Carta Pisane*. Here is the scenario in brief: 'sea-fish' maps,
unadulterated by Ptolemy, that had been carried to perfection by the second
century AD were preserved by Arab culture until the thirteenth century AD.
Then at least part of the legacy – a chart of the Mediterranean and Black Sea
region, Nordenskiold's 'normal portolano' – fell into European hands, providing
the model, with the necessary modernization of place names, etc., for the *Carta
Pisane* and the entire portolan genre.

In my opinion this is a more rational and more parsimonious way to account
for the highly developed state of the normal portolano than to ask us, as most
historians do, to accept that such striking and precocious cartography somehow
'evolved' out of books of sailing directions. And Nordenskiold's hypothesis,
though it leaves unanswered all questions about the roots and antiquity of the
Marinus tradition before the second century AD, is also on sound logical ground
by reminding us of the role of the Phoenicians in all this.

Known to have circumnavigated Africa by 595 BC,[93] 2000 years before the
Portuguese, the Phoenicians maintained fleets throughout the Red Sea, the
Indian Ocean and the Mediterranean (at powerful naval and mercantile cities
like Tyre, Sidon and Carthage), planted major colonies on the Atlantic coasts
of Europe and North Africa, and crossed the Atlantic at least as far as the Azores
and the Canary Islands.[94] They were, without contest, the greatest mariners of
the ancient world. Indeed, between the time of Ptolemy and the time of the
Portuguese one looks in vain for any other seafaring culture of the Mediter-
ranean/Black Sea region that would have had both the capacity *and the incli-
nation* to devise a map like the normal portolano.

Moreover, if the normal portolano is indeed derived from the lost atlas of
Marinus of Tyre, then it follows that other high-quality maps of regions much
further afield than the Mediterranean and the Black Sea, and indeed a world
map, might also have been preserved by the Arabs – for we know from Ptolemy's
testimony that other Marinus maps, including a world map, did once exist. It
will therefore do no harm to keep an open mind to the possibility that the
portolan world maps that began to appear during the century after the *Carta
Pisane*,[95] might also have been influenced by earlier 'Tyrian sea-fish' maps of
Phoenician origin. Christopher Columbus, whose passionate belief in lands
across the Atlantic led to his 'discovery' of the New World, seems to hint at a
Phoenician connection when he describes one of the inspirations for his journey:

Aristotle in his book *On Marvellous Things* reports a story that some Carthaginian merchants sailed over the Ocean Sea to a very fertile island ... this island some Portuguese showed me on their charts under the name Antilia.[96]

Antilia first appears on a portolan chart of 1424. It is a mysterious presence there, a riddle, to which we will return.

What Guzarate showed da Gama

The suggestion has been made that 'world' portolans – indeed, any that show regions outside the normal portolano area – could have been based on the lost world map of Marinus. And if the normal portolano reached Europe after being preserved among the Arabs for many centuries, it could be the case that the Arabs preserved the world map too. We've seen that some Arab portolans of the Mediterranean/Black Sea area do exist – although they are very few in number. So it makes sense to look for traces among the Arabs of a portolan world map as well.

Nordenskiold believed he had identified such a trace. Combing through geographical works from the Age of Discovery, he found a passage in J. De Barros' *Asia* (first Portuguese edition printed 1552) which states that the Arabs in the Indian Ocean possessed sailing charts with degree-lines, 'perhaps comparable in their finish to the portolanos':[97]

> When Vasco da Gama during his first voyage, in April 1498, arrived at Malindi on the east coast of Africa, he there procured a pilot named *Guzarate* to sail his ship to India. Da Gama was much pleased with him, especially since the pilot showed him a map made in the Arabian (Moorish) manner of the whole Indian coast, without compass lines but divided by meridians and parallels into small squares. The pilot also showed him some nautical instruments intended for determining latitude, different to those which da Gama had brought with him.[98]

There are a number of points of great interest in this report:

- The name that De Barros gives for the pilot is quite different from the name of 'Ahmed-bin-Majid' provided by other sources. In fact, Guzarate doesn't sound much like a name at all. What it does sound like is a nickname or familiar term – 'Gujerati' – that may still be heard on Kenya's Swahili coast today in reference to natives of the Indian state of Gujerat. Is it possible that da Gama's 'Arab pilot' was in fact an Indian pilot – a Gujerati?
- The map is said to show the 'whole Indian coast'.
- The map is said to be 'without compass lines' – which takes it far from the standard European presentation of a portolan.
- The map is said to possess meridians and parallels – again far from the

normal portolano, which has no meridians and parallels. However, these meridians and parallels are also said to divide Guzarate's map into 'small squares'. It is of note in this respect, though they do not result from intersecting meridians and parallels, that the *Carta Pisane* has four areas divided up into small squares and two other areas divided into slightly larger squares. Such divisions occur on no other portolan chart known in the west.[99]

- The pilot is said to have used unfamiliar nautical instruments, presumably in conjuction with the map.

We've already seen that neither on da Gama's 1498/9 voyage, nor on Cabral's of 1500/01 – and indeed not until after 1510 – did the Portuguese have the opportunity to chart the north-west coast of India between Goa and the Indus delta. The evidence of this is in the record of the voyages and also, obliquely, in the Cantino map of 1502, which draws on the latest knowledge that the Portuguese had acquired along the way. Ironically, the very *absence* of an accurate portrayal of the Kathiawar peninsula in the Cantino map, an absence that still persisted in 1510 when the Reinal map of the Indian Ocean was drawn, provides further convincing evidence that the Portuguese did not chart north-west India until after 1510 – because if they had they would have done a much better job of it (at least as good a job as they did on the coasts of Brazil also discovered on the 1500/01 voyage). They would certainly not have overlooked such a prominent feature as the Kathiawar Peninsula of Gujerat with its two great gulfs of Kutch and Cambay (the latter offering particularly rich trade potential). If we accept in addition that a Gujerati pilot of some repute seems to have been known to the Portuguese, it becomes all the more incredible to imagine that the most precise navigators and mapmakers of the fourteenth and fifteenth centuries could have charted the coast of their pilot's home region and failed to make an accurate representation of it.

In short, everything suggests that the Portuguese were *not* there, and did *not* chart those coasts until after 1510, and that the representation of north-west India which appears in the Cantino and Reinal maps must therefore have been borrowed by them from a pre-existing local map.

What better candidate for such a map than the very one that Guzarate showed da Gama and that da Gama so admired on his first crossing to Calicut in 1498?

Quick detour to Oceania

One of the several intriguing possibilities suggested by the Guzerate story is that a tradition of accurate map-making with its roots lost in prehistory – perhaps the same tradition that also nourished Marinus of Tyre in the Mediterranean and that eventually expressed itself in the medieval portolans – survived amongst both Arab and Indian navigators in the Indian Ocean right up until the time of the European voyages of discovery.

The quality of the maps derived from the Indian Ocean tradition was recognized in the fifteenth and sixteenth centuries by the great Portuguese mariners like da Gama (and others as we shall see in later chapters). But there is evidence that these maps and the navigational system that lay behind them had also influenced other cultures in much earlier epochs. I note in passing that in his detailed study of the astonishing achievements of Micronesian and Polynesian navigators in their discovery of the Pacific between approximately 2000 BC and 1000 AD Dr David Lewis draws attention to 'some remarkable similarities between what has been recorded of ancient Indian Ocean systems of non-instrumental navigation, unquestionably the older, and their Pacific counterparts'.[100]

Lewis points out that 'the magnetic compass . . . was preceded in the Indian Ocean by a star compass . . . a compass-card marked in star points'.[101] Strangely, the archaic Indian Ocean star compass proves to be very similar to star-compasses of the far Pacific:

> No fewer than eighteen of the thirty-two star points appear to be identical in the Indian Ocean and the Pacific systems . . .[102] [There is] every reason to believe that what we term 'Polynesian-Micronesian' navigation was merely part of a system once practised through all the Asian seas, and which very probably did not even originate in Oceania at all.[103]

Ice Age India?

We will encounter other traces of the same lost system when we reach China and Japan in later chapters. But our concern for the moment remains with its impact on European maps of India produced in the early days of the age of discovery. We've seen that the Cantino and the Reinal maps (1502 and 1510 respectively) were drawn before the Portuguese exploration of India's coastlines was complete and that a likely explanation for this that is that both were copied from a pre-existing local source map – perhaps the very map that Guzarate showed da Gama.

Having a shared common source, or deriving from different but closely similar sources, provides a simple explanation for why the Cantino and Reinal maps are so much alike in almost all respects and also, crucially, why both contain similar mistakes. As I was already aware from Sharif Sakr's first report (see chapter 14) these mistakes include the absence of the Kathiawar peninsula with its characteristic Gulfs of Kutch and Cambay; a distinct bulge in the north-west corner of India; enlargement of many small island groups, and a south-westerly orientation (with what Sharif describes as 'distinct lips') of the southern tip of India. In his e-mail of 23 February 2001 he then makes the crucial observation that:

> While these deviations are all errors relative to a modern map of India, they in fact match up extremely well with Glenn Milne's map of India 21,300 years ago at LGM.

This inundation map shows a large indent at the mouth of the Indus, a bulge obscuring completely the Kathiawar peninsula, enlarged Lakshadweep and Maldives islands, and, most surprisingly, a SW-pointing 'mouth' shape at India's southern tip that is virtually identical to that shown by Reinal.[104]

It seems to me that these correlations, and the others that Sharif reported on 10 August 2001, are obvious, striking and speak for themselves. The only questions that need to be asked about them are: (1) do they result from the workings of coincidence? Or (2) are they there because the source maps for Cantino and Reinal were originally drawn at the end of the Ice Age – perhaps not as far back as the LGM but certainly before the final inundation of the Gulfs of Kutch and Cambay which created the Kathiawar peninsula around 7700 years ago?[105]

We already know, and nobody would dispute, that the maps of Claudius Ptolemy have now survived in human culture for almost 2000 years and that they incorporate far older streams of ideas, some certainly going back as far as the sixth century BC and some probably much further.

In the light of Masudi's testimony confirming the late survival amongst the Arabs of the maps of Marinus, it by no means seems far-fetched to suggest, with Nordenskiold, that the Marinus 'branch' of cartography was never 'lost' at all but simply transformed itself into the portolan tradition. Otherwise we have the paradox of 'the most perfect cartographic work of the Middle Ages'[106] appearing suddenly, from nowhere, with no prior evolution. And since we already accept that Ptolemy incorporated ideas much earlier than his own in the making of his maps, why shouldn't we accept that Marinus did so too?

The Reinal and Cantino maps are portolans that extend far beyond the normal portolano area of the Mediterranean and the Black Sea. And while they do incorporate a few Ptolemaic ideas about the shape of the world, both maps are more distinguished by their stark differences from – and superiority to – Ptolemy. How much of this is due to Marinus? And how old might the oldest information be that could have been included in the Marinus maps? Could some of it have been as old as the last Ice Age when India did actually look the way it is portrayed by Cantino and Reinal?

If there is any possibility that the latter scenario is correct, then it would become interesting to work out what precise period during the 10,000-year post-glacial meltdown between 17,000 and 7000 years ago is portrayed by the Indian coastlines on the Cantino and Reinal maps.

Final report on Reinal

Sharif Sakr to Graham Hancock
15 August 2001

It seems that every time I go back to comparing the Reinal map of 1510 and the Milne map for 11,500 BC, I find that the correlation is even better than I previously thought. My latest revision highlights the great affinity between the latitudinal positions of the 'erroneous' features on Reinal's non-Ptolemaic Indian coastline and the correlating features on Milne's inundation map.

Milne's map, in harmony with bathymetric maps of India's outer shelf, clearly shows a large gulf at the latitude of today's Indus river delta. I call this feature the 'Indus Gulf', simply because before the postglacial period the Indus river may have emptied here. In my first e-mail I correlated the 'Indus Gulf' with the only gulf shown on Hapgood's tracing of Reinal's map in roughly the right place. This correlation is not perfect: the portolan gulf is the wrong shape and it lies too far north (because Reinal's Tropic of Cancer is too far north, continuing a Ptolemaic error). Moreover, this northern gulf on the Reinal might be better matched with Sonmiani Bay (and the mouth of the Porali river), which lies to the north of the Indus and which was well known to Arab geographers of the time because of the important seaport of Daibul. This northern part of the map is so inaccurate that it is difficult to be sure of anything.

But the Bodleian photograph reveals another large gulf on the Reinal, not shown properly in Hapgood's tracing, which exactly matches the Indus Gulf on Milne's map in terms of shape, size and latitude. This gulf lies south of Reinal's erroneous Tropic of Cancer, and at exactly the right latitude relative to, for example, the eastern tip of Oman on the opposite side of the Indian Ocean. It lies well outside the area covered by the old Ptolemaic model and is therefore very likely to have been present on the mysterious non-Ptolemaic source that Reinal used.

When we correlate the gulf shown in Milne's map with this gulf on the Reinal map, the latitudinal positions of Reinal's other 'errors', relative to each other and to this northern landmark, make far more sense. Overleaf is my final matching of 'errors' on the Reinal map to features on the Milne map – just follow the numbers:

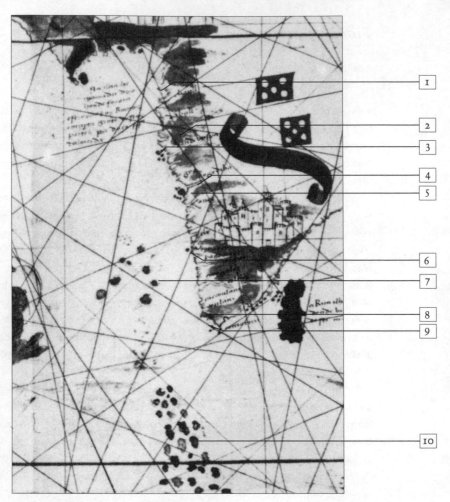

India's coastlines in Reinal map of AD 1510.

1. Today this is the mouth of the Indus river, which is a delta. But on both Reinal's and Milne's maps, it is marked by a wide gulf.
2. A large bulge that in both Reinal's and Milne's maps replaces the Kathiawar peninsula that exists today.
3. An island (or island-group) which is depicted on both maps but which does not exist today.
4. A gulf which on both maps is much smaller than the Gulf of Cambay that exists today.
5. A large island (or island-group) which is depicted on both maps but which does not exist today.
6. An island at the same latitude as the northernmost Lakshadweep island (approximately 12 degrees north) is shown on both Reinal's and Milne's maps. No island exists there today.

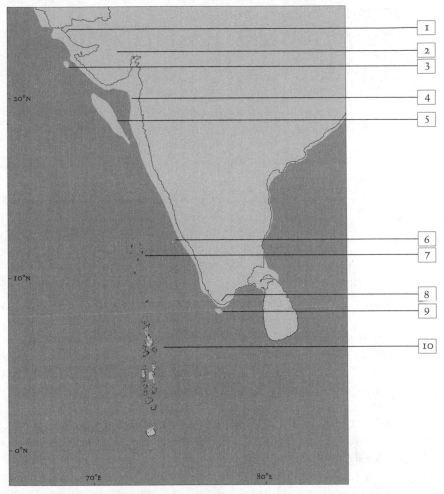

India's coastlines in 11,500 BC.

7. The Lakshadweep islands, which exist today but which are enlarged in both Reinal's and Milne's maps.

8. The tip of the sub-continent. Both maps show the tip of the sub-continent somewhat like a bay, wide but not deep, facing south-west towards the northern Maldives – very different from the south-east-facing tip that exists today.

9. A tiny island which is depicted on both Reinal's and Milne's maps next to the southern tip of the sub-continent. No island exists there today.

10. The Maldive islands, which exist today but which are enlarged in both Reinal's and Milne's maps.

How likely is it that such extensive and detailed correlations could have come about by chance?

22 / *The Secret Memories of Maps*

Polo's explanation of the size accorded Ceylon on the chart was that the chart's geography originated at an earlier time before much of the island had been submerged.

Thomas Suarez

There is a saying that in ancient times the noble isle of Sumatra was joined to the main, until mountainous seas eroded its base and cut it off.

Camoes, *The Lusiads*, 1572

Imagine setting off on a journey along the hippy trail to Afghanistan and the East in 1971 and not getting home again until 1995.

Though more of a merchant adventurer than a hippy, that's what Marco Polo did in the dangerous days of *Kubilai Khan*. He left Venice in AD 1271, travelled to the East via the Black Sea, Persia, Afghanistan and the Pamirs, spent seventeen years in China and seven on the road and at sea, and returned to Venice in 1295. Later he composed a book, *Il milione* ('The Million'), known in English as the *Travels of Marco Polo*, which was to become a geographical classic.[1]

Polo's account of his outbound journey – almost entirely overland – and of his long residence in China, contains little of relevance to the mysteries we are exploring in *Underworld*. His return journey, however, begun around 1292, is of much greater interest to us here. It includes the first-ever notice by a European of the existence of Japan – which Polo called Cipango (or 'Zipangu') from the Chinese Jih-Pen[2] – and it describes the epic sea voyage that he undertook on his way home, beginning at the eastern Chinese port of Ch'uan-chou (modern Quanzhou, opposite the island of Taiwan), sailing south around Vietnam and Cambodia, across the Gulf of Thailand, around the Malay peninsula, through the narrow Strait of Malacca that separates the peninsula from Sumatra, thence across the Bay of Bengal to Sri Lanka, around Cape Comorin, north along the west coast of India to the Gulf of Cambay, and finally across the Arabian Sea to Hormuz at the entrance to the Persian Gulf.[3] Thus it was that Marco Polo made familiar to Europeans the names and descriptions of many places that would not be heard of again until the Portuguese exploration of India and the Indies in the sixteenth century, more than 200 years later.

Though Polo himself states frankly that he has never visited Japan – and thus that what he has to say about it is second-hand and perhaps inaccurate – the

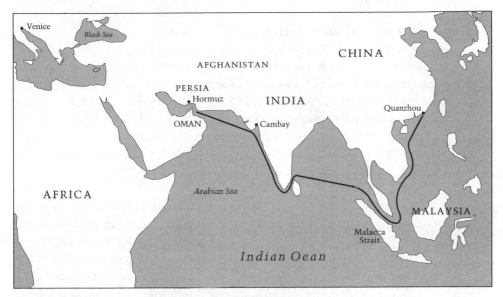

Marco Polo's return voyage from Quanzhou to Hormuz.

notion of the mysterious island kingdom of Cipango that he planted in European consciousness at the end of the thirteenth century was later one of several power-ful influences that spurred Christopher Columbus forward in his crossings of the Atlantic at the end of the fifteenth century. This was so because Columbus – underestimating the circumference of the earth and knowing nothing of the exist-ence of the Americas or of the Pacific Ocean – believed that he could reach Cipango, and thence the Chinese mainland beyond, by sailing directly westwards across the Atlantic from Europe. Columbus is also likely to have calculated that Cipango would be reached after only a relatively *short* journey towards the west – for he had read Marco Polo, who describes Cipango, erroneously, as lying 'far out to sea' fully 1500 miles to the east of the Chinese mainland[4] (the true distance is nowhere much more than 500 miles). Polo goes on to inform us that:

> Cipango . . . is of considerable size; its inhabitants have fair complexions, are well made, and are civilized in their manners. Their religion is the worship of idols. They are independent of every foreign power, and governed only by their own kings. They have gold in the greatest abundance, its sources being inexhaustible . . . The entire roof [of the sovereign's palace] is covered with a plating of gold, in the same manner as we cover houses . . . with lead.[5]

'Gold in the greatest abundance,' echoes Columbus in a marginal note beside this passage in his own copy of Marco Polo's *Travels* – now preserved at the Biblioteca Colombina in Seville.[6]

We will return to Columbus, and his obsessions.

A 'map' of antediluvian Sri Lanka?

After traversing the Bay of Bengal, commenting *en route* on 'the island of Andaman' (described as 'a very big island' inhabited by 'a cruel race' of cannibals with heads, teeth and eyes like those of dogs)[7], Marco Polo's homeward voyage brought him to 'the island of Zeilan – Ceylon – modern Sri Lanka.[8] In his account of Sri Lanka, which further illustrates his already established tendency to exaggerate distances (in this case approximately tenfold) the Venetian travel-ler nevertheless makes certain observations about the ancient geological history of the region that come remarkably close to the truth:

> The island of Zeilan presents itself. This, for its actual size, is better circumstanced than any other island in the world. It is in circuit 2400 miles, but in ancient times it was still larger, its circumference then measuring full 3600 miles, *as the Mappa-Mundi says*. But the northern gales, which blow with prodigious violence, have in a manner corroded the mountains, so that they have in some parts fallen into the sea, and the island, for that cause, no longer retains its original size.[9] (Emphasis added.)

This is the translation of William Marsden (1754–1836) from the Italian of Giambattista Ramusio's printed edition, dated 1553.[10] The more recent (1958) translation of Ronald Latham provides clarification of some elements of the same passage:

> The traveller reaches Ceylon, which is undoubtedly the finest island of its size in all the world. Let me explain how. It has a circumference of some 2400 miles. And I assure you that it used to be bigger than this. For it was once as much as 3600 miles, *as appears in mariners' charts of this sea*. But the north wind blew so strongly in these parts that it has submerged a great part of this island under the sea. That is why it is no longer as big as it used to be.[11] (Emphasis added.)

In yet another translation we read again that Ceylon in Polo's day has a circum-ference of: '2400 miles ... in old times it was greater still, for it then had a circuit of about 3600 miles, *as you find in the charts of the mariners of those seas*'.[12] (Emphasis added.)

Despite slightly differing nuances, and what looks like a tenfold exaggeration for distances, all the translations converge on two very clear and really quite startling messages:

1. Ceylon was believed by Marco Polo to have been one-third larger in the past than it had become by his day – with extensive lands to the north of the present island said to have been 'submerged under the sea'. In the process its circumference was reduced in size from 3600 units of measurement to 2400 units of measurement, i.e. by one-third.

2. Maps were in use amongst mariners in the Indian Ocean when Marco Polo was there – either *mappamundi* or mariners' charts depending on the translation – which continued to show the one-third larger, antediluvian Ceylon.

A reduction by one-third

On the first of the two points above – the one-third reduction in the size of Sri Lanka by flooding – we cannot deny, having studied the inundation history of south India and Sri Lanka in earlier chapters,[13] that the tradition which Marco Polo here preserves and passes down to us is essentially correct when set within the time-frame of the end of the last Ice Age.

Since approximately 7700–6900 years ago, when the last remnants of its land-bridge to south India were inundated, Glenn Milne's maps suggest that there have been no significant changes in Ceylon's size. Prior to 7700 years ago the picture is very different, and as we go back through 8900 years ago, 10,600 years ago, 12,400 years ago, and 13,500 years ago, we note a progressive enlargement of Sri Lanka, *exclusively in the north around the land-bridge to south India*, resulting from the lowered sea-level of those epochs.[14] At its greatest extent the enlargement is of the order of one-third.

Polo's quaint theory about how these former lands were lost through the action of the north wind is wrong. But he is completely right when he tells us that Sri Lanka was much larger 'in old times', right when he tells us that its land-loss took place in the north, right again when he tells us that the lost land was submerged beneath the sea, and right yet again in his information that approximately one-third of antediluvian Sri Lanka was lost in this way.

The question of how a Venetian traveller of the thirteenth century could have equipped himself with such esoteric facts of palaeogeography brings us to point two.

Where did Polo get his information from?

Polo himself tells us only that he had learned of the former extent of Ceylon from an ancient 'Mappa-Mundi' or 'mariners' chart' that he had seen, and he seems to accept without demur the obvious implication that this chart must have originated *before* the epoch of inundation. As historian of cartography Thomas Suarez confirms, 'Polo's explanation of the size accorded Ceylon on the chart was that the chart's geography originated at an earlier time before much of the island had been submerged.'[15] This is quite an extraordinary and interesting explanation, in my view. However, Suarez does not pursue it. He also ignores Polo's clear suggestion that the chart showing a formerly much larger Sri Lanka was actually *in use* by 'mariners of those seas', rejects Polo's explanation for the out-of-date geography of the chart (namely that it had come down from antediluvian times), and rather dogmatically asserts his own theory that the 'Mappa-Mundi' or 'mariners' chart' Polo is referring to must be a Ptolemaic world map.[16]

Suarez admits that Ptolemaic world maps were only in extremely limited circulation in Europe in Polo's time and are most unlikely to have been known to him from any European source. But he is right also to point to the possibility that such maps could have been preserved amongst the Arabs trading in the Indian Ocean, and that Polo could thus have seen a Ptolemaic map – without knowing it to be 'Ptolemaic' or recognizing it as such – during his stay in Ceylon.[17] Moreover, it is true that all Ptolemaic world maps show the very large island of Taprobane in *approximately* the place where Sri Lanka/Ceylon might be expected to be found. Thus, Suarez concludes that the chart referred to by Polo 'followed the Ptolemaic model with its characteristic reversal of the relative proportions of Ceylon and India'.[18]

Return of the Tyrian sea-fish

Suarez's logic is easy enough to follow: (1) Polo has been shown a Ptolemaic world map, probably preserved by Arab seafarers in the Indian Ocean,[19] featuring the giant island of Taprobana, which he takes to be Ceylon; (2) confronted by the much smaller Ceylon of his own day he concludes that the map he has seen preserves an image of Ceylon made before large parts of it were submerged; (3) he is incorrect in this conclusion and his notion of a formerly enlarged Sri Lanka results only from his misunderstanding of a well-known error on all Ptolemaic maps.

Yet this is surely only one possible explanation for Polo's 'knowledge' of obscure palaeogeographic facts – and one moreover that requires us to accept the supposedly firm identification that Suarez makes between Sri Lanka/Ceylon and Taprobana (an identification that is generally but by no means universally favoured by modern scholars and ancient cartographers).[20]

Another explanation for Polo's apparent anachronistic knowledge might be that there is nothing to it at all and that he made the whole idea up, scoring a few correlations with post-glacial reality purely by coincidence.

Still another and by no means impossible explanation might be that Polo's account was in some way informed by the flood traditions of Sri Lanka and south India, reported in previous chapters, that speak of the lost Tamil homeland of Kumari Kandam.

But as Polo does tell us quite explicitly that the source of his ancient geographical knowledge about Ceylon was 'mariners' charts' ('charts of the mariners of those seas' or 'Mappa-Mundi') we should surely also consider another possibility. This is the suggestion first raised by A. E. Nordenskiold and discussed in chapter 21 that a genre of maps older than the Ptolemaic maps and attributed to Marinus of Tyre was in circulation amongst the Arabs at least as early as AD 955 (the date of a direct reference by the geographer Masudi, who, as the reader will recall, had 'seen the maps of Marinus' which 'by far surpassed those of Ptolemy').[21] Nordenskiold argues that these 'Tyrian sea-fish' maps formed the prototype for the mysteriously accurate portolans of the Mediterranean region that seem to appear suddenly in the cartographic record in the late thirteenth century. But, as we've seen, the portolan genre was never confined to the Mediterranean region alone. The greatest number of surviving examples of portolans do depict the Mediterranean, it is true. But from very early on portolan *world* maps also appear. Though sometimes contaminated by Ptolemaic 'inserts' or 'patches' in sections of the globe for which, presumably, the cartographer had no portolan original at hand to copy from, these in their own way are as startlingly precocious as the Mediterranean portolans. To give just one example here, Piedro Vesconte's world map of *c.*1321 shows Africa to be circumnavigable – in complete contradiction of the Ptolemaic tradition – more than one and a half centuries before the Portuguese finally circumnavigated it.

Isn't it possible, therefore, that the chart Polo saw in the Indian Ocean which convinced him that Ceylon had formerly been one-third larger than it was in his day, that its lost lands had lain to the north, and that they had been submerged by the sea, could have been one of these 'Tyrian sea-fish' maps?

Still the best after all those years . . .

Polo was not the only European traveller in the Indian Ocean to have seen very interesting maps in the hands of 'mariners of those seas'. The reader will recall that Vasco da Gama was also shown what seems to have been a highly sophisticated map by the navigator Guzarate, who guided him so rapidly from Malindi in East Africa to Calicut on the west coast of India in 1498.[22]

It is important to stress, contrary to Suarez, that such maps, which were clearly *used* by local navigators – and to all accounts used effectively – could not possibly have been Ptolemaic maps (whatever else they might have been). This is so because of the extreme and indeed almost grotesque *inaccuracy* of all Ptolemaic maps of India/Sri Lanka – arising not only from the peculiar presence of Taprobana (which may require a more complex interpretation than it has hitherto received) but also from the fact that India's west coast is made to run parallel to the equator instead of roughly north–south as it does in reality.[23] Mariners like Guzarate, or those who took Marco Polo to Ceylon, were men whose lives depended on knowing the waters they sailed. Even if they had possessed a Ptolemaic map as a curiosity, we can be quite sure that they would never have taken the risk of actually *using* it for navigation.

This forces Suarez into the paradox – as he wraps up his argument for the Ptolemaic provenance of the map Polo claimed to have seen – of having to take Polo's direct reference to nautical charts ('the charts of the mariners of those seas') as evidence that such charts did not actually exist:

> The fact that the map seen by Polo retained such an incorrect dimension for Ceylon supports the view that native pilots guided their vessels by navigational texts, and did not refer to the charts themselves.[24]

It seems to me that something quite other than this is likely to be the case, since Polo makes no mention at all of navigational *texts* as the source for his notion of a formerly larger Ceylon, but does make very explicit mention of *charts*. We are now also clear that the charts he was referring to could not have been of Ptolemaic origin – simply by virtue of the fact that that they were routinely and successfully used by experienced local mariners in the Indian Ocean. Last but not least we have seen that the issue of the very exaggerated dimensions given to Ceylon (by a chronicler admittedly prone to the exaggeration of dimensions) may be less important than the entirely correct notion Polo preserves that 'in old times' one-third of Ceylon had been swallowed up by the sea.

Isn't it possible that what confronts us here is another trace, like the brief report of Masudi, of a parallel tradition of cartography (parallel, that is, to the Ptolemaic tradition) that survived from antiquity into the Middle Ages and that was associated by some with the works of Marinus of Tyre?[25] From the little that we already know and may reasonably speculate about these 'Tyrian sea-fish' charts, they seem to have been acknowledged and recognized for their overall accuracy and excellence *despite* having been overtaken in certain locations such as the north of Ceylon – as Polo testifies – by geological changes linked to flooding.

It is the circulation of precisely such sophisticated yet curiously out-of-date charts amongst Indian Ocean navigators like Guzarate, as we saw in chapter 21, that could provide the best explanation for the strange anachronistic perfection of the Cantino and Reinel maps drawn by Portuguese cartographers in the early sixteenth century. The reader will remember that these maps not only represent areas of the Indian coast that the Portuguese had not yet visited but also show a number of detailed and inexplicable correlations, particularly around Gujerat and Cape Comorin, with India's Ice Age coastline.

Knowledge of Ice Age topography in Ptolemy too?

When Sharif Sakr first drew Marco Polo's comments on Sri Lanka to my attention he pointed out that 'Polo's primary assertion is that Sri Lanka had changed in size since ancient times, and that the old topography is preserved in nautical charts.'[26] In the same report Sharif also notes:

> Other historical characters apparently believed that Ptolemy's maps depict an ancient topography, for example with respect to a former land-bridge between Malaya and Sumatra, across the present Strait of Malacca.
>
> The Dutch adventurer Linschoten (1596) stated that some believed that Sumatra was the *Chersoneso Aurea* [Golden Chersonese] of old, and that 'in times past it was firme land unto Malacca [Malaya]'.
>
> Camoes in his famous epic poem *The Lusiads* (1572), dealing with the birth of Portugal as a nation, writes: 'There is a saying that in ancient times the noble isle of

Sumatra was joined to the main, until mountainous seas eroded its base and cut it off.'[27]

Abraham Ortelius explained in a legend on his 1567 map of Asia: 'It is true that Samotra is not now a peninsula, but it is very likely that it was torn from the continent by the force of the Ocean after Ptolemy's time. Moreover, if you imagine Samotra being joined to Malacca with an isthmus, it will agree very well with the shape of the Golden Chersonese as described by Ptolemy.'

I think it is absolutely fascinating that this basic belief, that old maps could depict ancient and hence different topography, is so apparent in the writings of adventurers who visited the Indian Ocean and must surely have been in contact with 'the mariners of those seas'. That Ortelius takes the contemporary separation of Sumatra from Malaya as evidence that the land changed since the time of Ptolemy merely indicates his eagerness to try to understand whatever source information he had, and also his ignorance of the real geological processes that led to the separation of Sumatra from Malaya – at least 6000 years before Ptolemy.[28]

Readers who have come this far will already know enough inundation science to realize that there was indeed a time, at the end of the Ice Age, when the Strait of Malacca did not exist (as all the traditions quoted above correctly assert), when there was 'firme land' between Sumatra and the Malaysian peninsula, when 'the noble isle of Sumatra' was 'joined to the main' – and so on and so forth. For this area was all part of a continuous, near-continent-sized peninsula that geologists call Sundaland, a once fertile exposed shelf of well-watered low-lying plains – extending as far south as Surabaya, as far west as the Philippines and as far north as Taiwan – that was inundated in a series of catastrophic floods between 15,000 and 7000 years ago.[29]

How likely, therefore, is it to be an accident that the Ptolemaic world maps – said by Ptolemy himself to have been based on those of Marinus – do appear to present a fair image of Ice Age Sundaland in the form of the great peninsula that is labelled on those maps sometimes as the Golden Chersonese and sometimes as the peninsula of Mangi? Isn't it at least equally probable, as Ortelius was already more than half way to suggesting 500 years ago, that this 'mythical' peninsula is a genuine echo of Ice Age topography?

Likewise, it may be significant that the Cantino world map of 1502, which we have suggested could have come down to us directly through the Marinus-to-portolan 'line' (rather than indirectly via Ptolemy's abridged and 'corrected' version of Marinus), also shows a vast peninsula reminiscent of the exposed Sunda Shelf.

Given the highly anomalous traditions cited by Linschoten and Camoes concerning the flooding of the Strait of Malacca – traditions that are anomalous purely and simply because of their remarkable convergence with palaeogeographic facts – it seems almost perverse not to consider the possibility that certain maps, too, might have preserved reflections of the Ice Age world.

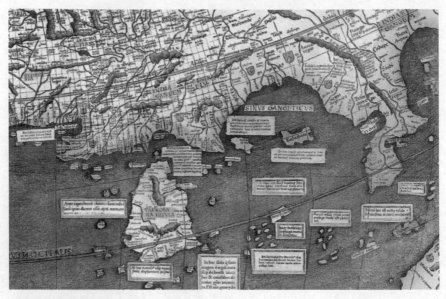

Waldseemüller's 'Golden Chersonese', AD 1507.

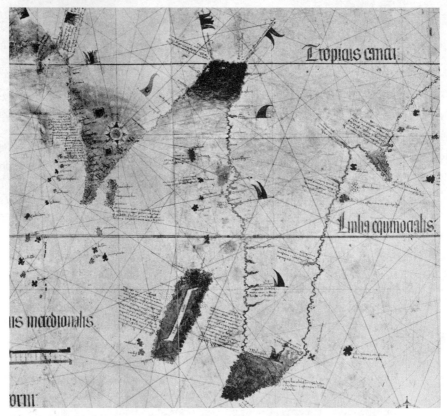

The Golden Chersonese as shown in the Cantino planisphere, c. AD 1502.

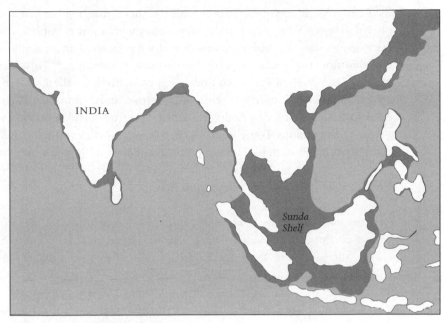

Exposed Sunda Shelf at the LGM.

But traditions, with all their folksy redolences, are relatively safe matters for scholars to speculate about. Maps and nautical charts on the other hand – especially accurate, sophisticated maps of the kind used by Guzarate to chart Vasco da Gama's course from Malindi to Calicut in 1498 – are quite another matter. If maps have indeed come down to us containing recognizable representations of Ice Age topography – as arguably may be the case with the depictions of India and of the long-submerged Sundaland peninsula by Cantino and Reinal and with the depiction of the 'Golden Chersonese' by Ptolemy – then prehistory cannot be as it has hitherto been presented to us.

If they are what they seem, such maps mean a lost civilization. Nothing more. Nothing less.

'A piece of a map . . .'

In 1937 the eminent Portuguese map historian Armando Cortesao, an indefatigable searcher after lost cartographical treasures, discovered – in Paris – 'the long-sought codex containing the *Suma Oriental* of Tome Pires and the *Book* of Francisco Rodrigues'.[30]

During the years 1512–15 when he wrote his *Suma* (now recognized as 'the most important and complete account of the East produced in the first half of the sixteenth century')[31] Tome Pires had been the first official Portuguese ambassador to China.[32] For some inexplicable reason, however, his great work lay 'forgotten and practically unnoticed', until Cortesao brought it to light again in the twentieth century.[33] This was all the more puzzling because the *Suma* proved to be bound together in the same codex with another volume which, far from being forgotten, had been sensationally republished (in an abridged, illustrated edition) in the 1849 *Atlas* of the Viscount de Santarem.[34] This second volume was the *Book* of Francisco Rodrigues, containing detailed written sailing directions and 'precious maps' (with compass roses and rhumb lines) drawn in the early sixteenth century by Rodrigues himself – a true portolan in other words.[35]

Unlike the famous Tome Pires – with whom it was nevertheless his fate to end up bound between two covers – and despite the publicity given to his maps in Santarem's *Atlas*, Francisco Rodrigues is virtually unknown. Indeed, says Cortesao, so little is known about him that:

> It is impossible even to attempt a biographical sketch. Besides the information we can gather from Rodrigues' *Book* itself, he is mentioned in two letters of Alfonso de Albuquerque to King Manuel of Portugal written from Cochin, 1 April and 20 August 1512.[36]

The suspicion that European travellers in the Indian Ocean in the sixteenth century may from time to time have stumbled across charts and maps containing the remnants of a lost geography (perhaps even the maps of Marinus of Tyre, said to have been superior to those of Ptolemy) is intriguingly enhanced by the first of Alfonso de Albuquerque's two letters. It introduces a 'piece of a map' that Albuquerque has acquired in his travels in the Indian Ocean and that he is sending to King Manuel. This fragment, he explains, is not the original but was 'traced' by Francisco Rodrigues from:

> a large map of a Javanese pilot, containing the Cape of Good Hope, Portugal and the land of Brazil, the Red Sea and the Sea of Persia, the Clove Islands [effectively a world map, therefore], the navigation of the Chinese and the Gores [an unidentified people, thought by some to be the Japanese, or the inhabitants of Taiwan and the Ryukyu archipelago][37] with their rhumbs and direct routes followed by the ships, and the hinterland, and how the kingdoms border on each other. It seems to me, Sir, that *this was the best thing I have ever seen, and Your Highness will be very pleased to see it*; it had the names in Javanese writing, but I had with me a Javanese who could read and write.[38]

This report of the tracing by a Portuguese cartographer (Rodrigues) of a map owned and used in the Indian Ocean by a Javanese pilot – and for no less a person than the Portuguese king himself – casts a very unusual sidelight on cartographic history. The events unfold in the early sixteenth century when Portugal was at the height of its maritime power and believed to be surpassed by none in its mapmaking sciences and achievements. Yet here we have a Portuguese emissary proudly sending back to his monarch a mere tracing of a mere fragment of a map owned by a Javanese pilot as though it were a classified military document of the highest order!

Remember that this is 1512 – a full decade after the superb Cantino map was created in Portugal. Some map scholars believe that the Cantino may have greatly resembled the *padrao* – the top secret 'master map', incorporating all the latest known discoveries, as well as relevant information from ancient charts, to which the kings of Portugal had special access. At the very least we can be absolutely confident that in 1502 the Portugese monarch would have had a map *at least as good* as the Cantino – and probably much better. Likewise, we can be certain, with continuous feedback from ever-widening Portuguese expeditions, that the *padrao* of 1512 would have been far superior to the *padrao* of 1502.

So it is against Manuel's privileged access to such a superb Portuguese world map as the *padrao* that we must weigh the enthusiasm with which his emissary Albuquerque sends him a tracing of a fragment of a Javanese pilot's map acquired in the Indian Ocean, describing it as 'the best thing I have ever seen' and assuring the King that 'Your Highness will be very pleased with it.'

Good enough to have faith in

Nothing – absolutely nothing at all – makes any sense of Albuquerque's letter unless the Portuguese themselves had reason to believe that maps were available in the Indian Ocean, in the hands of pilots of various nationalities, that might be better than their own. And, as we've seen before with such rumours of sophisticated ancient maps, there is also the recognition that they will sometimes have been outdated by geological changes. Thus, in the *Suma Oriental* Tome Pires informs us that:

> The Gujaratees were better seamen and did more navigating than the other people of these parts, and so they have larger ships and more men to man them. They have great pilots and do a great deal of navigation.[39]

Yet mysteriously he also tells us that it has only been since about 100 years before his own time that these Gujeratis (the countrymen of da Gama's highly skilled pilot Guzarate), had found the route through the Strait of Malacca between Sumatra and the Malaysian peninsula.[40]

This is strange because (a) the Gujeratis described by Tome Pires obviously knew a thing or two about navigation; and (b) the Strait of Malacca was being used by ships long before the fifteenth century – in the thirteenth century, after all, Marco Polo had sailed through it.

How are we to explain this anomaly? 'Could it be,' suggests Sharif Sakr,

> that the Gujeratis possessed maps (Ptolemaic or otherwise) which failed to show the Strait of Malacca, such that they had either lost knowledge of it, or such that Pires had speculated, having seen such maps, that the Strait was only recently discovered?[41]

In other words, could the Gujeratis have been working with maps showing Ice Age topography?

We've already seen that the 'mistakes' on otherwise technically excellent maps of India such as the Cantino of 1502 and the Reinal of 1510 can be explained this way – as the results of Portuguese borrowings from Ice Age maps somehow in the hands of Gujerati navigators. So maybe the anomalous and unexpected Gujerati ignorance of the Strait of Malacca reported by Tome Pires is part of the same syndrome? Maybe the Gujerati navigators used maps that showed the Strait as 'firme land' from Sumatra Malacca – as it last looked about 8000 years ago – and simply didn't bother to find out that things had changed. Maybe the old maps were generally quite good enough, despite such faults, to justify faithful reliance? That would make a strange kind of sense of the way in which the Gujeratis are reported to have adhered for so long to a much more roundabout route than the one through the Strait that was used by their competitors.

But is there any other evidence, except in maps of the Indian Ocean, which really suggests the survival of Ice Age topography?

The legendary Hy-Brasil – a glacial reality
Report by Sharif Sakr, 10 March 2001

Irish folklore tells of a small but significant island called Hy-Brasil, lying in the Atlantic Ocean not too far off the western coast of Ireland. The tale is at least as old as AD 1110, which is the date of the first written record of it (*The Voyage of Maeldiun*). The tale almost certainly existed prior to this, for an unknown length of time, as an oral record. Gaelic legends appear to hold that the land was lost to the ocean, but makes a brief reappearance once every seven years, such that it can be seen from the Irish mainland if one is standing in the right place.

Happily for us the legend of Hy-Brasil made its way on to the portolan charts of the fourteenth and fifteenth centuries. These graphic representations give a far more detailed and precise insight than verbal or written traditions ever could into what was believed about the size and location of the island.

The first recorded depiction of Hy-Brasil in a map is in the Dulcert portolan of 1325 or 1330. It appears again on Dulcert's 1339 portolan – opposite. Although faint, it should be obvious that the map is generally very accurate. It even shows the tiny lump of rock known as Rockall, which was occupied by Greenpeace recently as part of a demonstration against oil-drilling in the area. Note, however, that the tiny land of Rockall is somewhat enlarged on the Dulcert map.

There are very similar depictions of the legendary island of Hy-Brasil on many other portolan charts, which probably represent copies (or copies of copies) of some original (perhaps the Dulcert, but probably some older portolan chart).

Below is part of the Catalan Atlas of 1375. Its representation of the British Isles is typical of all portolans, including the Dulcert, and, in addition to the legendary Hy-Brasil, its characteristic errors include a dry Donegal Bay on the north-west corner of Ireland.

The next map, overleaf, comes from the Ptolemaeus Argentinae collection of 1513, which represents a successful hybridization of the Ptolemaic and portolan traditions.

Can it be coincidence that there is a relatively shallow submerged bank – it is marked on modern sea-charts as the Porcupine Bank – in exactly the same place as the legendary island shown on all these ancient maps?

Glenn Milne is currently unable to produce reliable inundation maps of this region with the required zoom and detail, partly because the region is so close to the ancient British ice-sheet – the exact behaviour of which has not yet been fully incorporated into the model. However, for our purposes bathymetrical maps will serve just as well. The one overleaf is state-of-the-art, with a resolution of 2 minutes. Depth can be gauged through the shading as well as by the contour line which I have placed at a depth of 55 metres beneath today's sea-level.

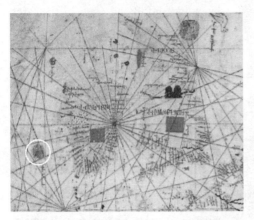

Hy-Brasil (circled) as shown on the Dulcert portolan, AD *1339.*

The light-shaded Porcupine Bank can easily be seen directly west of Ireland, in exactly the same place, and roughly the same size, as the legendary Hy-Brasil on the portolan charts. The entire bank lies between 40 and 200 metres beneath the surface, and most of it (probably more than 600 square kilometres) would have been exposed at the Last Glacial Maximum, 21,000 years ago.

Hy-Brasil as shown on the Catalan Atlas of AD *1375.*

The correlation between Porcupine Bank and Hy-Brasil on the portolans is, in my view, too close to be coincidental. Even Robert Fuson, Professor Emeritus of Geography at the University of South Florida, is convinced that Hy-Brasil is based on real observation. But rather than consider an Ice Age origin for the legend, he suggests it is based on some unknown but recent tectonic event. But I do not think it is necessary to speculate about recent tectonic cataclysms, or even to go all the way back to the LGM, in order to find a good correlation between past geography and the portolans. The black contour line is set at 55 metres below current sea-level and reveals that there would have been a significant island, with an area of perhaps 100 square kilometres, in the location of the legendary Hy-Brasil even in the later stages of the glacial meltdown – around 12,000 years ago.

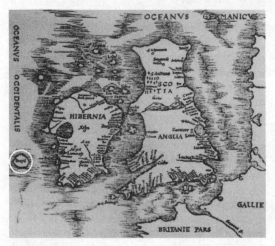

Hy-Brasil as shown on the Ptolemaeus Argentinae of AD 1513.

Other features of the portolans correlate better with Ireland as it looked at this later period than with the geography at the Last Glacial Maximum. The island of Rockall was enlarged, such that it had roughly the same size as shown in the Dulcert and Argentinae maps. (Note that there would probably also have been two much smaller islands in the vicinity of Rockall, which are not shown on the portolans.) Also as shown on the maps, the Bay of Donegal, at the north-west shoulder of Northern Ireland, would have been dry land and there would have been a large island immediately off this coast. The many islands that today lie off the west coast of Ireland and between Northern Ireland and Scotland would have been incorporated into the Irish and Scottish mainlands respectively, but would have been replaced by other small islands further to the west which are now submerged but which are in keeping with the islands shown on the old maps. The same goes for the Isle of Man, which would have been replaced by a similar-sized island slightly further to the south. The Outer Hebrides would have been a single massive landmass, as represented on

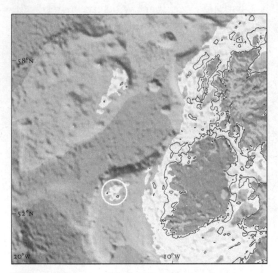

Bathymetric map of Ireland, with grey contour line at depth of 55 metres.

the Dulcert portolan (although this map has the island slightly too far south and east).

41. *The towering ruins of Gigantija, Malta, thought to be the oldest free-standing temple in the* world.

42. *Ghar Dalam cave, Malta – site of an extraordinary archaeological controversy.*

43. *The Hypogeum, Malta.*

44. *Surviving part of monumental 'Goddess' figure from Tarxien temple, Malta.*

45. *'Sleeping Lady', Malta.*

46. *Five of the six skulls that have survived from the remains of more than 7000 people found in the Hypogeum, Malta.*

47. *The Mnajdra temple complex, Malta, from the air.*

48. *Mnajdra: summer solstice light effect.*

49. *Mnajdra with the island of Filfla in the background.*

50. *Withered megaliths of Hagar Qim temple, Malta.*

51. *Entrance to Hagar Qim, Malta.*

52. *The author, left, with Chris Agius, Malta.*

53. *The author, right, with Anton Mifsud, seated in ancient cart-ruts, Malta.*

54. *The author diving on submerged cart-ruts, Malta. The submerged ruts are larger and deeper than their counterparts on land.*

55. Cart-ruts at 'Clapham Junction', Malta.

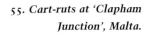

56. Submerged cart-ruts, Malta.

57. Submerged channel and archway, Malta.

58. *Underwater wall with battlement, Taiwan.*

59. *Exposed masonry blocks in the Taiwan underwater wall.*

60. *Giant megaliths of the Bimini Road.*

The Antilia mystery

Hy-Brasil is by no means the only mysterious island looking for a home in ancient maps of the Atlantic Ocean. Even stranger, as we shall see in the next chapter, are two other islands – the fabulous Antilia and Satanaze – which beckon like the Holy Grail. They first appear on an anonymous portolan chart of 1424, and subsequently on many other maps of the fifteenth and sixteenth centuries. Yet the islands themselves have never been found.

Is this because they never existed? Or might there be a better explanation?

23 / *Looking for the Lost on the Road to Nowhere*

There lies in the Ocean an island which is called *The Lost*. In Charm and all kinds of fertility it far surpasses every other land, but it is unknown to men. Now and again it may be found by chance; but if one seeks it, it cannot be found, and therefore it is called *The Lost*.

Honorius of Autun, *De Imagine Mundi*, about AD 1130[1]

Give me some ships and I will find a new world for you.

Christopher Columbus, about AD 1480[2]

For some reason that has never been explained properly there was, for a very long while before the time of Christopher Columbus, a firm and entirely correct belief amongst mariners in ancient Europe and around the Mediterranean that vast lands and extraordinary islands awaited discovery and colonization somewhere to the west across the wastes of the Atlantic Ocean. The belief was expressed in legends and traditions, some of which have been preserved down to modern times, and also in graphic form on maps and nautical charts.

The mystery of Hy-Brasil, introduced in the last chapter, is part of this very thorny unsolved problem of anachronistic geographical knowledge and, at the same time, a microcosm of the whole issue:

- Ancient references to Hy-Brasil exist both in legendary and traditional oral and written sources and in maps dating back as far as the fourteenth century – for example the Dulcert portolan.
- Belief in the existence of Hy-Brasil – i.e. physically, in the Atlantic Ocean somewhere – was strong enough to have inspired expeditions to find it. Records have survived of two such expeditions, the first led by a certain John Lloyd, that were sent out from the port of Bristol in the west of England in AD 1480.[3]
- The location given to the 'legendary' island of Hy-Brasil by medieval mapmakers correlates strongly and closely with the location of the submerged Porcupine Bank – which was unknown in medieval times but parts of which, as we've seen, would have been exposed as an island at the end of the Ice Age.

The trouble with Hy-Brasil

If it were simply a matter of an old legend of a lost Atlantic island somewhere to the west of Ireland, and modern bathymetry showing a shallowly submerged bank in roughly the same vicinity, the most probable explanation would be coincidence. The appearance of Hy-Brasil on maps, however, cannot be accounted for so easily. Scholars universally conclude that these representations are no more than imaginative graphic expressions of pre-existing written and oral traditions. The consensus view is that medieval cartographers referred to many sources in constructing their maps, including legends. Since Hy-Brasil is obviously a 'legendary' island, it follows that the shape and location given to it on the maps must have come from legendary sources. But if the cartographer who placed Hy-Brasil on the Dulcert portolan were working only from legends he would have been free to draw it anywhere to the west of Ireland – giving him wide scope. What, therefore, must be the odds against his having imagined an island that is not only approximately the right size to match the antediluvian Porcupine Bank but that is also placed in exactly the spot on the map where the Porcupine Bank would have been exposed at the end of the Ice Age?

It could all be the result of some sort of extraordinary coincidence, I admit. Or it could be that the cartographer worked from a source map – like the hypothetical source maps behind the Cantino and Reinal world portolans – that somehow depicted genuine Ice Age topography and coastlines?

As we've seen in previous chapters, it is not unreasonable to suppose that maps belonging to the tradition of Marinus of Tyre could have been preserved in pockets in the Indian Ocean and elsewhere alongside the better known maps of Claudius Ptolemy. Nor is it impossible, as Arab eye-witnesses as late as the tenth century attest, that the original maps of the 'Tyrian sea-fish' might have been *superior* to those of Ptolemy (despite Ptolemy's own propaganda to the contrary). It is not wild speculation on my part, but the argument of the distinguished historian of cartography A. E. Nordenskiold, that the preserved maps of Marinus may have formed the original corpus out of which emerged the astonishingly sophisticated portolan tradition in the thirteenth and fourteenth centuries. And it is a fact that the earliest representations of Hy-Brasil – Dulcert, Benincasa and many others – all appear on portolan charts.

There will be ways for scholars to underplay the significance of this, I'm sure. But what Hy-Brasil looks like to me is evidence not only for the survival of an ancient non-Ptolemaic mapmaking tradition but also for the preservation within the tradition of accurate records of Ice Age topography and coastlines. That in turn more or less automatically makes the tradition itself extremely ancient; logically it must be at least as old as the Ice Age features it represents. Moreover, despite its great antiquity, it is a mark of the respect accorded to the general accuracy and reliability of this tradition by mariners down the ages that expeditions to find Hy-Brasil – and other 'ghosts' of Ice Age topography – were still being launched as late as the fifteenth century. Though there seems to have been an inkling that cataclysmic changes and floods had intervened, as we saw in the last chapter, I think it is unlikely that the seafarers who set out from Bristol in 1480 to search for Hy-Brasil could have imagined that the island given that name on their portolan charts had been swallowed up by the sea more than 11,000 years previously.

I anticipate the objection that it is *inconceivable* for a mapmaking tradition to have survived for 11,000 years. But why should it be inconceivable? Don't we already have in Ptolemy a mapmaking tradition that has survived – verifiably – for 2000 years? And doesn't Ptolemy himself state that his *Geography* is a correction of the earlier work of Marinus of Tyre, who in turn was supposedly only the 'most recent student' of this ancient discipline? Nothing compels us to imagine, therefore, that the 'Marinus' tradition began with Marinus a few decades before Ptolemy. On the contrary, Ptolemy's references suggest that Marinus of Tyre (if this was not actually a generic term that was used to refer to a certain category of nautical maps) was simply the latest custodian and redactor of a body of geographical knowledge preserved from a far more remote antiquity.

Perhaps it was their custodianship of this knowledge that made the Phoenicians such inquisitive explorers of the margins of the Atlantic (which later navigators feared and called 'the Sea of Darkness')[4] as though they were searching, always searching, for something that lay just beyond the next horizon . . .

Hints of a lost Atlantic geography
According to the Greek historian Diodorus Siculus, writing in the first century BC,

> There lies out in the deep off Libya [Africa] an island of considerable size, and situated as it is in the ocean it is distant from Libya a voyage of a number of days to the west. Its land is fruitful, much of it being mountainous and not a little being a level plain of surpassing beauty. Through it flow navigable rivers . . .[5]

Diodorus goes on to tell us how Phoenician mariners, blown off course in a storm, had discovered this Atlantic island with navigable rivers quite by chance. Soon its value was recognized and its fate became the subject of dispute between Tyre and Carthage, two of the great Phoenician cities in the Mediterranean:

> The Tyrians . . . purposed to dispatch a colony to it, but the Carthaginians prevented their doing so, partly out of concern lest many inhabitants of Carthage should remove there because of the excellence of the island, and partly in order to have ready in it a place in which to seek refuge against an incalculable turn of fortune, in case some total disaster should overtake Carthage. For it was their thought that since they were masters of the sea, they would thus be able to move, households and all, to an island which was unknown to their conquerors.[6]

Since there are no navigable rivers anywhere to the west of Africa before the seafarer reaches Cuba, Haiti and the American continent,[7] does this report by Diodorus rank as one of the earliest European notices of the New World?

Likewise, what did Lucius Annaeaus Seneca have in mind in his *Medea* (*c*.AD 50) when he wrote:

> In later years there will come a time when *Oceanus* [the Atlantic] shall loosen the bonds by which we have been confined, when an immense land shall be revealed and *Tiphys* [the pilot of Jason's legendary ship *Argo*] shall disclose new worlds.[8]

Seneca's strange observation reads like a weirdly accurate prophecy of the inevitable discovery of the Americas. But is it *too* accurate to be guesswork? Had he seen a map that showed an immense land literally waiting to be revealed on the far shores of the Atlantic?

The opposite continent

The suspicion that certain ancient authorities possessed good knowledge of the real shape of the Atlantic and its islands, and of the lands on both sides of it, must also arise from any objective reading of Plato's world-famous account of Atlantis.

As we have seen in earlier chapters, this story is set around 11,600 years ago – a date that coincides with a peak episode of global flooding at the end of the Ice Age. The story tells us that 'the island of Atlantis was swallowed up by the sea and vanished', that this took place in 'a single dreadful day and night' and that the event was accompanied by earthquakes and floods that were experienced as far away as the eastern Mediterranean.[9] But of more immediate interest to us here is what Plato has to say about the geographical situation in the Atlantic immediately *before* the flood that destroyed Atlantis:

> In those days the Atlantic was navigable. There was an island opposite the strait [the Strait of Gibraltar] which you [the Greeks] call the Pillars of Heracles, an island larger than Libya and Asia combined; from it travellers could in those days reach the other islands, and from them *the whole opposite continent which surrounds what can truly be called the ocean*. For the sea within the strait we are talking about [i.e. the Mediterranean] is like a lake with a narrow entrance; the outer ocean is the real ocean and the land which entirely surrounds it is properly termed continent . . . On this island of Atlantis had arisen a powerful and remarkable dynasty of kings who ruled the whole island; and many other islands as well, and parts of the continent . . .[10]

Whether or not one believes that an island called Atlantis ever existed in the Atlantic Ocean, Plato's clear references to an 'opposite continent' on the far side of it are geographical knowledge out of place in time. It is hard to read in these references anything other than an allusion to the Americas, and yet historians assure us that the Americas were unknown in Plato's time and remained 'undiscovered' (except for a few inconsequential Viking voyages) until Columbus in 1492.

The mysterious book of Columbus

A curious anteroom to the Columbus story exists. It is prefigured in the Irish legend of the voyage of Saint Brendan – the earliest surviving version of which appears in Adamnan's *Life of St Columba*, written before AD 704.[11] Brendan is said to have sailed across the Atlantic from Ireland in the sixth century AD with a group of monks on an eventually successful expedition to find 'an immense region in the west . . . the Land of Promise'.[12]

Once again we are reminded that the ancient seafaring nations of Europe and the Mediterranean were imbued through and through with the same geographical idea that enlightened Plato – the idea that a rich and almost limitless opposite continent awaited those daring enough to attempt the Atlantic crossing. And once again the obvious questions arise. Where could the idea of the opposite continent have come from? Why should it have arisen in the first place? How do we account for its persistence down the ages in so many different cultures from the Phoenicians to the Irish?

In 1513, in handwritten notes on an enigmatic map that he had prepared showing the newly discovered Americas, the Turkish Admiral Piri Reis offered an intriguing answer to all these questions – at any rate for the particular case of Christopher Columbus, the most recent and most renowned of the ancient Atlantic dreamers. Piri's note, one of many on the same map, is written over the interior of Brazil:

> Apparently a Genoese infidel, by the name of Columbus was the one who discovered these parts. This is how it happened: *a book* came into the hands of this Columbus from which he found out that the Western Sea [i.e. the Atlantic] has an end, in other words that there is a coast and islands on its western side with many kinds of ores and gems. Having read this book through, he recounted all these things to the Genoese elders and said, 'Come, give me two ships, and I shall go and find these places.' They said, 'Foolish man, is there an end to the Western Sea? It is filled with the mists of darkness.'[13]

It seems to me that there are two points of enormous interest about this reported 'book' of Columbus. First, we are told that it showed the opposite continent, with its coast and islands, on the western side of the Atlantic. Taken at face value, therefore, what we have here is a clear reference to the existence of a pre-Columbian map of the Americas – a notion that runs completely contrary to the accepted history of science. Secondly, we are led to understand that it was on account of what he had learned in this remarkable book – no other cause is mentioned – that Columbus began to tout his proposed expedition to potential sponsors.

One might question the *bona fides* of a Turkish admiral claiming to have any

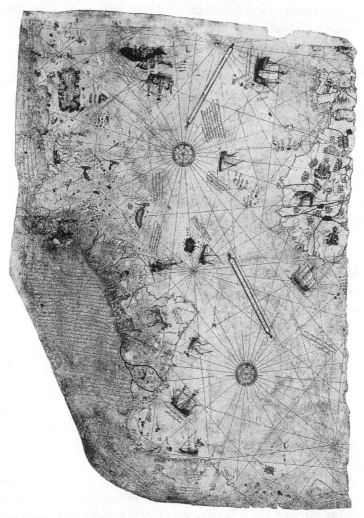

Piri Reis map, 1513.

inside knowledge at all of the voyages of Christopher Columbus; however, in this case such questions appear to be misplaced. Recent scholarship by map historian Gregory McIntosh has confirmed that one of the twenty or so source maps to which Piri Reis tells us that he referred to compile his own map was almost certainly – as Piri claims – a chart of the Caribbean that had been drawn by Columbus himself.[14] The implication is that some fairly direct link must have existed between the two men and Piri informs us of such a link. He says that he acquired his inside information about Columbus from a Spaniard captured by Turkish corsairs after a naval battle in the Mediterranean. This 'Spanish prisoner', as Piri calls him, had sailed with Columbus on three of his four voyages to the New World.[15]

Piri's reference to the mysterious 'book' of Columbus can therefore be traced back to a reliable source. But I have yet to find a single orthodox map scholar, Gregory McIntosh included, prepared to look further into the potentially controversial and important revelation that the book contained a pre-Columbian map of the Americas. On the contrary, the revelation is dismissed as manifestly incorrect. In consequence those few scholars who have devoted any thought at all to the 'book' have ignored the one definite lead that Piri gives us about it – namely that it showed how the Atlantic Ocean came to an end in an opposite continent with its own coast and islands – and instead have speculated that it might have been a copy of Cardinal D'Ailly's *Imago Mundi*, or of Marco Polo's *Travels*: 'books which influenced Columbus's plan of sailing west to reach Asia'.[16] To this Gregory McIntosh adds: 'In the *Bahriye* Piri Reis refers to the book that influenced Columbus in terms that indicate it may have been Ptolemy's *Geographia*.'[17]

Needless to say, the orthodox paradigm of the discovery of the New World is safe if the mysterious 'book' that supposedly motivated Columbus can be reduced to a known, non-threatening quantity like the *Geography* or the *Travels*. And it is possible, since all the texts named above recognize the earth to be a sphere, that any one of them, and probably all of them, might have played a part in shaping Columbus's well-known conviction that Asia could be fetched by sailing west from Europe.

None of this, however, permits the conclusion that the 'book of Columbus' to which Piri Reis refers was *in fact* one of these texts. Indeed, though the point is passed over in silence by McIntosh, it seems extremely unlikely that it could have been. The named texts were already well known in Europe when Columbus was seeking support for his expedition and were not viewed by anybody as proof positive that either a New World, or Asia, lay on the other side of the Atlantic. If all he had to impress sponsors was information that they already had at their disposal from those texts, then he would not have convinced anyone. In other words, if there was a 'book of Columbus' which played the important part that Piri gives to it, then it must have been a much rarer and less familiar text than any of these and it must logically have contained new and more persuasive information about the far coasts of the Atlantic.

Why not take Piri at face value?

Piri Reis is not only remembered for his 1513 map but for another slightly later work, a manual of sailing directions known as the *Bahriye*, which also contains references to the book of Columbus.[18] Reported above is McIntosh's impression from comments made in the *Bahriye* that the 'book' Piri is speaking of might have been Ptolemy's *Geography*. Yet the Turkish scholar Svat Soucek points out that this is not the obvious deduction from the text of the *Bahriye* where it touches on 'the great story of the discovery of America':

The country's name is Antilia, and it was discovered by a Genoese *muneccim* (astronomer-cum-astrologer) named Columbus . . . The story goes all the way back to Alexander, who had roamed the whole earth and written a book about it. The book remained in Egypt until the Muslim conquest, when the Franks fled the country, taking the book with them. Little attention was paid to it until Columbus read it and realized the existence of Antilia to the west of the Atlantic. He convinced the king of Spain of the possibility of its discovery and colonization, which he then successfully carried out.[19]

I find it difficult to agree with McIntosh that Piri might have had Ptolemy's *Geography* in mind as the book that inspired Columbus – for the *Geography* consists of dry and uninspiring coordinates mapping out the *Oikumene* (the inhabited world as known to the ancient Greeks) and has nothing to say one way or the other about the western terminus of the Atlantic nor of any place such as Antilia. Moreover, McIntosh's conclusion requires us to ignore Piri's own very clear and unambiguous attribution of the original authorship of the 'book' to Alexander the Great and to accept instead that when Piri wrote 'Alexander' he really meant 'Claudius Ptolemy'.[20] The argument for this truly outrageous act of second-guessing, and denigration of the intelligence and education of Piri Reis, goes something as follows: (1) Alexander the Great, a Macedonian, invaded Egypt and established the city of Alexandria; he was very famous; (2) after Alexander's death his general Ptolemy Soter, also a Macedonian, also very famous, declared himself pharaoh and founded the Ptolemaic dynasty; (3) almost 400 years later the astronomer Claudius Ptolemy (no relation to Ptolemy Soter, but famous too) compiled his *Geography* at the library of Alexandria; (4) Piri Reis mixed up all the facts about these famous people and places in his own mind and churned out the hilariously incorrect conclusion that the book that had convinced Columbus of the existence of the New World had originally been written by Alexander the Great.[21]

Rather than going through such convolutions, which ultimately just pour scorn on him, I fail to understand what is so terribly wrong with taking Piri at face value. Why not simply credit him with enough learning and intelligence to have known the difference between Alexander and Ptolemy? Why not explore the possibility that Columbus really could – exactly as Piri says – have been motivated to attempt his Atlantic crossing after having seen a very old book, a survival from the time of Alexander the Great, in which the western shores of the Atlantic were shown?

The questions are purely rhetorical and there is one answer for all of them. Scholars cannot take Piri Reis at face value on the subject of the book of Columbus because this would mean accepting the possible existence not just of a pre-Columbian map of the Americas (itself a historical heresy of the highest order), but of a *pre-Ptolemaic* map of the Americas dating back at least to the time of Alexander the Great – i.e. to the fourth century BC.

The maps of Marinus of Tyre were pre-Ptolemaic and have not come down to us. Thus we do not know and can only speculate about their true antiquity, their origins, their contents, and about what they showed and did not show before the 'improvements' and 'corrections' that Ptolemy implemented. But if A. E. Nordenskiold is right to suggest a genetic link between the lost corpus of Marinus and the remarkably advanced portolan charts that began to appear from the late thirteenth century onwards then, in a sense, anything is possible.

We have seen that these portolans contain strange echoes of the Ice Age world – suggesting that some of the source maps on which they were based may have been drawn thousands of years ago, before the post-glacial sea-level rise. If that is the case, then why shouldn't the as yet unidentified prehistoric culture or cultures that made these maps have 'discovered' and charted the Americas as well?

The survival of such maps, or copies of copies of copies of them, among mariners in the Mediterranean and along the Atlantic seaboard of Europe since time immemorial would explain the ancient yearning to discover an 'immense land' in the west. It would explain the ancient certainty that such a land was there. And it would explain why, down the generations, hard-headed seafarers and adventurers were again and again prepared to mount hazardous expeditions to try to find the great continent and islands that the maps told them lay out in the Atlantic.

So what about the most famous Atlantic island of all? What about Atlantis?

The Atlantis-Antilia mystery

Plato's story of Atlantis, though it contains no diagrams, nevertheless summons up an accurate mental picture of the Atlantic Ocean – bounded to the east by Europe and Africa and bounded to the west by the vast enclosing arc of the 'opposite continent'.

In the midst of the Atlantic Plato then presents us with another geographical image, this time supported by quite specific *chronological* data. The image is of the great island of Atlantis, no longer extant, that was swallowed up by the sea 9000 years before the time of the Greek lawmaker Solon. This suggests a date of around 9600 BC for the submergence of Atlantis – a date that falls in the midst of the cataclysmic meltdown of the last Ice Age.

We've seen that the topographical ghosts of other inundated Ice Age islands, like Hy-Brasil and the unnamed island off the southern tip of India portrayed on the Cantino and Reinal maps, mysteriously begin to appear on portolan charts and world maps from the fourteenth century onwards. If Atlantis was also an island submerged by rising sea-levels at the end of the Ice Age, and not just a figment of Plato's imagination as many suppose, then is it possible that its spectre too could haunt the portolans?

A number of researchers believe that they have found the ghost of Atlantis manifesting as a large, roughly rectangular, 'mythical' island named Antilia that began to appear on portolan charts in the first half of the fifteenth century. The earliest surviving example was drawn in Venice in 1424 and is attributed

Pizzagano chart, 1424.

to the cartographer Zuane Pizzagano.[22] It is not known what source maps he may have been working from. Together with a second large 'mythical' island – named Satanaze – that Pizzagano portrayed lying to the north, Antilia went on to enjoy a long and ubiquitous life in global cartography and was not finally exorcized from most charts and atlases until the eighteenth century.[23] As was the case with Hy-Brasil (which in fact survived on one nautical chart until the middle of the nineteenth century)[24] there was also at one time a firm belief amongst mariners in the physical existence of Antilia – firm enough at any rate to have inspired several voyages of discovery.[25]

Map sleuth George Firman points out that the positions of Antilia and Satanaze on the 1424 and later charts lie extremely close to, if not exactly on top of, the huge subterranean mountain range, connected to the world's tectonic system, that geologists today know as the Mid-Atlantic Ridge.[26] Proposing what is essentially an amplified version of the 'forebulge effect' described in chapter 3,[27] Firman suggests that downward pressures on the continental landmasses of Northern Europe and North America during the Ice Age could, through isostatic compensation, have forced the mid-Atlantic Ridge upwards, perhaps far enough upwards to have elevated its highest peaks and plateaux above water for as long as 40,000 years before the ice-sheets went into meltdown.[28] Conversely, with the removal of the downward pressure exerted on the continents by the ice-masses as the meltwaters poured back into the world ocean, the temporary uplift of the Mid-Atlantic Ridge would have ceased and subsidence pressures would have begun to build. As sea-levels rose, and as the isostatic rebound of

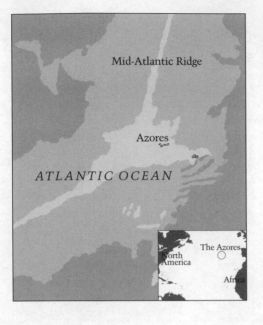

the continents continued, it is then theoretically possible that the entire ridge, as Firman puts it, could have been plunged into the depths of the Atlantic Ocean 'at the same approximate time'.[29]

Firman believes that such an event did occur 'between the years 9500–8000 BC',[30] that the 1424 chart contains antediluvian information, and that in its portrayal of Antilia and Satanaze it provides us with:

> the original location of the last two main islands of Atlantis. Both islands conform to the bottom topography of the Mid-Atlantic Ridge and of the adjoining Azores Plateau ... The largest island, to the south, is the main island on which the capital cities of the empire were located.[31]

It is true that Plato speaks of more than one island in the Atlantean empire.[32] And I have come across some peculiar reports (mainly from Soviet oceanographic sources between the 1950s and 1980s) of deeply submerged underwater ruins – including such features as stone columns, buildings and stairways – on the Mid-Atlantic Ridge near the Azores.[33] Since none of these reports ever seems to have been followed up, the possibility remains that Firman could yet be proved right and that the existence of sunken cities will one day be confirmed in the Mid-Atlantic. But the search area and the search depth are far too great for individual divers to be of any use. It will take a well-funded oceanographic institute with submersibles and a lot of time at its disposal to settle this matter.

Recent investigations indicate that there is 'weird geology' down there which may perhaps provide a simple explanation for the Soviet sightings of alleged ruins. On 13 July 2001, for example, ABC News in the United States released the following science story picked up from the 12 July issue of *Nature*:

> More than 2000 years ago, the Greek philosopher Plato wrote about a splendid city named Atlantis, with fertile soil and glorious temples, that 'in a single day and night of misfortune . . . disappeared into the depths of the sea'.
>
> Now researchers probing the ocean bottom have found 18-story-high towers of stone deep in the ocean near a section of volcanic fault ridges that extend for 6200 miles along the Atlantic floor [the Mid-Atlantic Ridge].
>
> The majestic height of the two dozen stone structures and their location on a seafloor mountain named Atlantis Massif inspired the scientists to name the area 'Lost City' in honor of the fabled flooded city referred to by Plato.

The underwater stone spirals are unusual for their composition and location . . . 'It was clear these were unlike anything we'd ever seen before,' said Deborah Kelly, an oceanographer at the University of Washington . . . The Lost City is also strikingly bright – brighter than the usual conditions in which things can generally be seen using artificial light a half-mile below sea-level. Although other rock formations around volcanic ridges have appeared black, the newly discovered formations are gleaming white because they are made up of materials similar to those of pale concrete, such as carbonate minerals and silica.[34]

Could there be more to this story than meets the eye? Could this be a real lost city that is being mistaken for weird geology? Highly unlikely, I should say – but, honestly, who knows what's really down there, seen and unseen?

Meanwhile, geological opinion, with good reason, remains solidly set against any involvement of the Mid-Atlantic Ridge in the Atlantis mystery. Galanopoulos and Bacon sum up the consensus nicely:

> There never was an Atlantic landbridge since the arrival of man in the world; there is no sunken landmass in the Atlantic; the Atlantic Ocean must have existed in its present form for at least a million years. In fact it is a geophysical impossibility for an Atlantis of Plato's dimensions to have existed in the Atlantic.[35]

This statement is certainly correct – and doubly so if we are to envisage 'Atlantis' actually sinking into the ocean through some abrupt isostatic event (as opposed to being inundated by rising sea-levels). Yet while it is indeed impossible for a landmass 'larger than Libya and Asia combined' to have existed in the Atlantic,

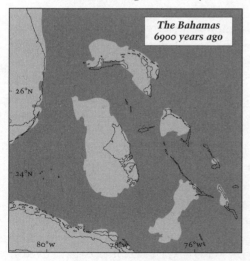

**The Bahamas
6900 years ago**

it is also only fair to point out that the ghost islands of Antilia and Satanaze depicted on the 1424 chart are in the range of just 500 kilometres long by 200 kilometres wide and thus come nowhere near Plato's extraordinary dimensions for Atlantis. Moreover, inundation science reveals that three islands the size of Antilia and Satanaze – islands that are today completely submerged or that have survived only in the form of tiny remnants still above sea-level – did in fact exist in the Atlantic down to as late as 6000 years ago (although much nearer America than the Mid-Atlantic Ridge).[36]

One of these lost islands was formed by a large section of the Great Bahama Bank, which stood more than 120 metres above sea-level at the Last Glacial Maximum. Today all that is left of this imposing antediluvian landmass is the rugged island of Andros to the southeast and tiny Bimini to the north-west, facing the Gulf Stream and the Florida peninsula.

Off the north-west coast of Bimini, running parallel to the Gulf Stream, is what appears to be a huge submerged man-made structure – an impressive megalithic engineering work made of enormous blocks laid side-by-side to form an underwater 'road' more than 800 metres long. At its southern end the structure curves shoreward, giving it the shape, quite visible from the air, of a reversed letter 'J'. Towards its northern end it divides into two parallel tracks separated by open sand. Closer to shore two additional smaller sections of 'road', each about 300 metres long, run parallel to each other at an angle to the main axis of the 'J'.

Some people say the whole complex is a vestige of Atlantis. Others say it's just three outcrops of natural blocky beachrock. But neither side has yet seriously considered the problem in the light of inundation science and what it has to tell us about sea-level changes and land-loss in this corner of the Atlantic Ocean at the end of the Ice Age.

The rise and fall of the Bimini Road

The 'Bimini Road' varies between 5 and 7 metres in depth. Situated in an area of generally calm blue water that reaches a temperature of 30 degrees centigrade in the summer months it therefore represents just about as unthreatening a dive as it is possible to experience in scuba gear. A kilometre to your south is Paradise Point on north Bimini island. A kilometre to your east is a beach of picturesque white sand. To your west, were you to follow it over a distance of 3 kilometres, you would find that the sea-bottom slopes down in gradual increments to a depth of about 100 metres before the abyssal drop-off into the Gulf Stream is reached.

This deepwater channel between Bimini and Florida was always there and filled with the ocean, even at the Last Glacial Maximum. But the submerged site of the Bimini Road and much of the sea-bed between it and the channel were above water then – and may have remained so until about 6000 years ago. Whether natural or man-made, therefore, the site would have enjoyed a spectacular and significant antediluvian location at the top of a long gentle slope overlooking the Gulf Stream.

The Road was discovered in 1968 by a team of volunteers, all of whom were connected with varying degrees of closeness to an organization called the Association for Research and Enlightenment (ARE). A harmless, good-willed but dottily eccentric American cult with Christian and spiritualist values and an ageing membership, the ARE has its headquarters in the coastal resort of Virginia Beach, overlooking the Atlantic Ocean, and promotes the teachings of the healer and psychic Edgar Cayce (1877–1945). Cayce claimed to have lived a past life as an Atlantean more than 12,000 years previously and before his death he prophesied that the ruins of Atlantis would begin to emerge from the sea in 1968 or 1969. He was quite specific about where this would take place –

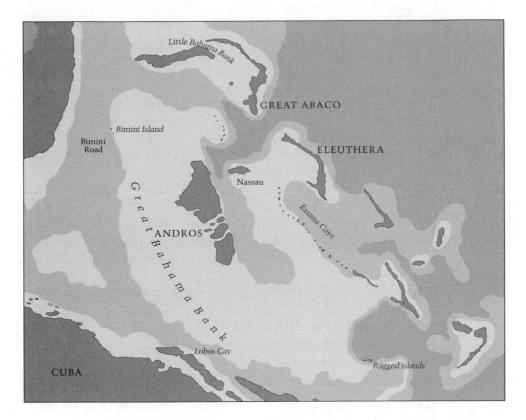

somewhere near Bimini. The apparent fulfilment of the prophecy with the 1968 discovery of the great rows of underwater megaliths off Bimini's Paradise Point therefore made for sensational headlines.[37]

Initially, high hopes were raised that the irrefutable proof of Plato's lost civilization had at last been found. Then there came a devastating scientific backlash which seemed to demonstrate coolly and professionally that the Bimini Road was not a man-made megalithic structure after all, but an entirely natural feature that could be explained simply by geology without any need to invoke psychic archaeology or the master masons of an imaginary lost civilization.

Mahlon Ball, Rosenstiel School of Marine and Atmospheric Science, University of Florida, and John A. Gifford, University of Miami, writing in National Geographic Research Reports, *vol. 12, 1980, pages 21–38*

The rise of sea level from 15,000 BP to the present produced a succession of beaches that formed on the outer platform off the west coast of North Bimini as the shoreline transgressed eastward over the Great Bahama Bank. Along these transient beaches deposits of beachrock formed and subsequently were submerged as the water over them deepened . . . [After several thousand years] the shoreline migrated to a position approximately one kilometre north of the present Paradise Point. Here over a period of perhaps 700 years, three successive beaches were the site of the formation of three parallel, linear deposits of beachrock . . . [38] The following observations were made during our initial field investigation:

1. The three features are unconnected at the southwest end; scattered blocks are present there, but do not form a well-defined linear feature connecting the seaward, middle, and shoreward features.
2. No evidence exists anywhere over the three features of two courses of blocks, or even a single block set squarely atop another.
3. Not enough blocks lie in the vicinity of the three features to have formed a new-destroyed second course of blocks.
4. Bedrock closely underlies the entire area of the three features, eliminating the possibility of excavations or channels between them.
5. No evidence was found of blocks being cut into or founded on the underlying bedrock surface.
6. No evidence was found of regular or symmetrical supports beneath any of the blocks.
7. We saw no evidence on any of the blocks of regular or repeated patterns of grooves or depressions that might be interpreted as tool marks.
8. [None of the features] is well founded or continuous enough to have served as some kind of thoroughfare.

In fact the only attributes of the three linear features that suggest a human origin are the regular shapes of the blocks. These are also attributes of natural beachrock deposits.[39]

W. Harrison, Environmental Research Associates Inc., writing in Nature, *vol. 230, 2 April 1971, pages 287–9*

The blocks are believed to have originated as follows. A shell-hash gravel was deposited in shallow water as relative sea-level fell during the most recent emergence of the Bahama Banks, and later brought into the fresh water environment. The materials were cemented and joints formed in the material as is usually the case with limestones. After two sets of practically right angle joints had developed, submergence of the area brought the jointed coquina limestone first into the breaking zone of waves and then the offshore zones. Wave action probably caused much of the initial separation into blocks, but when the formation was further offshore the destructive activity of marine life would have become dominant.

The overall result is a field of blocks that at first sight appear to have been fitted together, and this has led to statements such as 'some human agency must have been involved'. The blocky remains of the limestone outcrop are, however, no more enigmatic than other subaerial or subaqueous outcrops of jointed limestone found in various stages of fracture and decay in the north-western Bahamas.[40]

Marshall McKusik, University of Iowa, and Eugene Shinn, US
Geological Survey, writing in Nature, vol. 287, 4 September 1980,
pages 11–12

Amateur enthusiasts have claimed that the Bimini blocks were quarried by ancient Atlanteans and laid out in an ancient 'Cylopean, megalithic roadway'. . . However, the limestone structures observed off Bimini in 15 feet [5 metres] of sea have all the features of natural beachrock. The limestone is in a narrow band and extends for a considerable distance along a former shoreline . . . The tabular fractures are natural and the original slope to the sea is present. A sample of 17 oriented cores obtained by Shinn and Tomkins has been examined with X-radiographs. Two areas of the formation were studied, and both show slope and uniform particle size, bedding planes and constant dip direction from one block to the next. If the stones had been quarried and relaid there is no reason to suppose bedding planes would carry stratigraphically from block to block. The sedimentary laminations clearly show that these were not randomly laid stones but a natural, relatively undisturbed formation.

Although under 15 feet of water, the beachrock is of recent geological origin. One C-14 date on shell has already been published as 2200 plus or minus 150 years BP. Jerry J. Stipp (Radiocarbon dating lab, University of Miami) has run seven bulk samples from cores as a class project and gives slightly older dates for the Bimini submerged beachrock [varying from 2745–3510 BP].[41]

The road to nowhere

The last point cited above, the carbon-dating of organic materials in the stone of the Bimini Road, is potentially the most devastating of all the evidence presented by science against the claimed 'Atlantean' origin for the site. Plato put the submergence of Atlantis sometime close to 11,600 years ago and the ARE prophet Edgar Cayce proposed 12,500 years ago. Either way, the C-14 dating of the Road to between 2200 years ago and 3500 years ago seems, at a stroke, to rule out any Atlantean or indeed any very ancient connection.

Despite the apparently overwhelming and self-evident case for a natural and recent origin of the site, there were fightbacks and rebuttals by some of the original discoverers of the Road, including the oceanographer Dimitri Rebikoff and Dr J. Manson Valentine of the Miami Museum of Science. A Ph.D. from Yale University (in zoology, palaeontology and geology), and latterly Research Fellow in entomology at the Bishop Museum in Hawaii, Dr Valentine was a polymath who emerged as the unlikely spokesman for the pro-Atlantean group. Writing in the Explorers Journal in December 1976, he acknowledged the hostile response of other academics (mainly marine geologists) but argued that the sceptics had so far fallen 'far short of explaining':

1. why the stones of the Bimini complex are of flint-hard micrite (unlike soft beachrock, it rings when struck with a sledge and will not cleave under the same treatment);
2. why the three short courses of closely fitted stone are so straight-sided, mutually parallel and terminate in corner stones;
3. why the long avenue lies at a slight angle to the others and is composed of a double series of small blocks interrupted by two expansions containing very large, flat stones propped up at their corners by vertical members (like the dolmens of western Europe);
4. why the southern end of this great, wide track swings into a beautifully curved corner; and, finally,
5. how to account for all the rectangular shapes, right angles and rectilinear configurations associated with this complicated site as seen from the air.[42]

Likewise, in 1978 Dr David Zink, another pro-Atlantean with academic credentials, presented evidence questioning the uniformity at the microscopic level of adjacent beachrock blocks at Bimini (suggesting deposition in an entirely natural way) that had been alleged in the scientific reports:

> The cementing of the sections – composed of marine life forms and crystalline forms of calcium carbonate – was not alike. One sample was dominated by aragonite crystals, another by sparry calcite. This implied that adjacent stones were formed in different chemical environments.[43]

Together with Terry Mahlman, David Zink also presented a paper at a conference on underwater archaeology held at the University of Pennsylvania in January 1982. The paper raised a number of serious reservations about anomalies in the sequence of very young carbon-dates that had been published in *Nature* and elsewhere. The authors pointed out that these dates, between 2200 and 3500 years ago in the case of *Nature*, and between 3200 and 6000 years ago in the case of another study, do not square with solid information now in the hands of marine geologists concerning Atlantic sea-levels since the end of the Ice Age:

> The radiocarbon dates of the site, when matched with known Atlantic sea-levels at the same dates, put the megalithic blocks either above or below the tidal zone at the time of their formation. Because of the need for a tidal environment in which to form beachrock, and because sea-levels in the Atlantic for the past thirteen thousand years are the most solid elements of the Bimini problem, we are left with the likelihood that the dates are unreliable.
>
> [For example] two of the megalithic blocks dated by an early investigator, the first from the seaward side of the site and the second from a position 100 metres toward

the beach, yielded dates which conflict with the theory of an *in situ* origin for them. The seaward block was dated by radiocarbon to *c*.4000 BC. In its present position it would have been about 23 feet *above* the tidal zone. Clearly it would have been impossible for it to have formed as beachrock by the known process. The second block, located 100 metres closer to the present beach and at the same depth, was dated by radiocarbon to *c*.1200 BC. In its present position at that date this block would have been about eight feet *below* the tidal zone.

The literature on dating methods suggests that even ground-water contamination on land can render radiocarbon dates too young. How much greater an error might be introduced by the continuous addition of calcium with an ever-increasing proportion of the C-14 isotope as occurs in micritization of beachrock? For all these reasons the dates presently assigned to these blocks would appear to be unreliable.[44]

Despite these and other reasoned attempts to keep interest alive in the Bimini Road as a possibly man-made and possibly very ancient site, the *Nature* and *National Geographic* reports had hit the scientific credibility of the subject like cruise missiles. Likewise, souring after their initial flirtation with Atlantis and the Cayce prophecy, the tabloid media soon lost interest and moved on.

In such a way the Road to Atlantis became the road to nowhere.

24 / *The Metamorphoses of Antilia*

It's just a fact of life in this case that no one and no organization is going to fund a prehistoric underwater archaeological survey of the Bahamas.

John Gifford, University of Miami, July 2001

Friends, come, come with us on this voyage! Here you're creeping about in poverty; come and sail with us! For with God's help we're going to discover a land that they say has houses roofed with gold.

Martin Alonso Pinzon, Captain of the *Pinta*, recruiting crews for Columbus, 1492

Before I spent two weeks diving at Bimini in August 1999 this was my honest opinion: David Zink and Manson Valentine were wrong and the marine geologists from Florida were right; the Bimini Road was a natural formation. But after the diving I wasn't quite so sure.

I still felt the force of the scientific arguments, but now I'd also experienced the force of the great structure underwater and my reaction to it was not the same as the reaction of the geologists. Where they'd seen a 'natural' formation of tabular beachrock with uniform particle sizes, constant dip direction and no tool marks, artefacts or other signs of human intervention, I'd seen something that looked like a majestic work of art or sculpture – perhaps a colossal mosaic – something, at any rate, that felt coherent, organized, purposive, planned, idiosyncratic and *designed*. It is true that beachrock does fracture into jointed blocks, and that examples of this process can be seen in Bimini today and around many other Bahamian islands (in fact it forms so quickly that bottle tops and other modern items are frequently found cemented in the matrix). However nothing I have ever seen that is definitely and unassailably beachrock, either on Bimini or anywhere else, really looks like the Bimini Road.

We dived with Trigg Adams, a salty old sea dog and former Eastern Airlines pilot who'd been one of the original discoverers of the Road back in the days of Manson Valentine. We used his yacht the *Tryggr*, which he brought over the Gulf Stream from Miami under motor power, for the duration of our trip. And we also took advantage of Trigg's flying skills to go tearing around the skies in a chartered plane for a couple of hours so that we could see the Road and other mysteries of Bimini from the air.

Despite haze and cloud that morning we had no difficulty in spotting the 800 metre long, 20 metre wide main axis of the reverse-J with its characteristic shoreward curve to the south-east. It was also easy to make out the point at which the axis bifurcated into two narrower parallel piers, each 5 metres wide, separated by a 10 metre wide strip of sand running all the way to the northern terminus of the structure. Through the crystal-clear water we could even see individual blocks – some of them gigantic, some much smaller, all seemingly arranged and oriented in a highly organized manner. The two shorter segments shoreward of the 'J' ran absolutely parallel to one another and again showed interesting combinations of small and large blocks – including seven particularly enormous megaliths lying side by side near the southern end of the inner segment.

Trigg took the plane higher and circled several times over the enormous underwater mosaic. It reminded me, I realized, much less of a road or any kind of thoroughfare than it did of the great earth diagrams – the long straight lines and the animal, insect, bird and fish figures – of the Nazca plateau in southern Peru. Whether by accident or by design these works of geometry and stone sprawled out on an ancient Atlantic beach and, long since submerged beneath the sea, had something of the same sense of scale and grandeur when viewed from the air. I was therefore intrigued to discover, as we continued the flight over Bimini's two main islands and lagoons, that in several densely wooded and uninhabited areas there were stony mounds with exposed surfaces the size of tennis courts on which nothing grew. The surface of one mound, only visible from the air, took the shape of a huge sea-horse. The surface of another was shaped like a giant fish complete with realistic fins and tail and, again, could only ever have been seen from the air. A third mound was geometrical, offering a rectangular surface to the sky.

In all the discussions and academic papers I have read in which the Bimini Road is described as a natural beachrock formation I have never once seen any comment, one way or another, on these peculiar and distinctive mounds. Are they also to be dismissed as natural formations of no interest to the archaeologist? And if not – if they are man-made – then shouldn't they be taken into account in any attempt to judge the provenance of the nearby 'Road'?

Diving the Bimini Road

Shallow dives sometimes don't feel like real dives. There's not that sense of challenge, that frisson of danger, that you get when you're down deep. Just 5 or 10 metres below the surface you would have to be very stupid and very persistent to risk the bends or a lung-expansion injury. So Bimini was a gentle and kindly place to be underwater. Even the occasional nurse shark sulking in the shelter of one of the great blocks just looked like he might be dangerous but wasn't really. And at these depths a full tank of air went a very long way.

The typical Bimini block is of dark, extremely hard stone, measures about 2 metres in length by a metre in width by half a metre high, weighs about a tonne, is pillow-shaped, slightly convex, and rounded off at the corners and

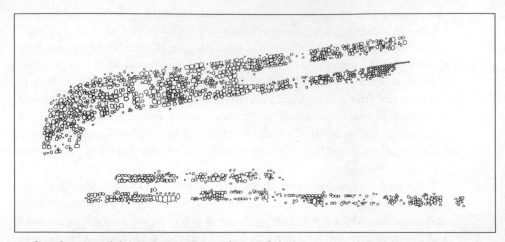

Outline drawing of the Bimini Road. Based on Zink (1978).

edges. Many others are much smaller but there are dozens of true monsters of 5 tonnes or more, with a few selected individual blocks verging towards 15 tonnes.

Contrary to the National Geographic Society research report I found that certain blocks in the 5–15 tonne range – some exceptional examples of which measured as much as 5 metres across – were propped up on small vertical supports, apparently of a completely different stone type, resembling stubby pillars. The effect of these supports – sometimes as many as five at a time – was to lift the big blocks completely clear of the bedrock foundation so that you could see underneath them from one side to the other.

I supposed that these were the 'dolmens' that Manson Valentine had spoken of in one of his reports – certainly there was nothing else on the Bimini Road that fitted this description. But despite a superficial resemblance – big blocks propped on top of smaller blocks – these structures obviously weren't dolmens. I wondered if the little vertical 'pillars' were just bits of loose rock that had been lying around on the sea-bed and that had been washed under the big blocks by tides or storm swells. But if so, why were they only under the biggest and heaviest blocks – the ones that would have been hardest for storms to shift around – and not under the smaller, lighter ones?

I spent days drifting up and down the Road, trying to get my bearings on it and to figure out what it is. Around noon with the sun most directly overhead and the underwater visibility at its best, the long straight avenues of blocks seemed to stretch away for ever in either direction. Mostly they lay directly on top of the extensive plateau of exposed limestone bedrock but sometimes they would disappear completely under sand-drifts, only to reappear on the other side, keeping the same heading.

Within the overall theme of parallelism other recurrent patterns were also evident – blocks arranged in circles, groups of three blocks of different shapes combined to form a triangle, seemingly deliberately fashioned cornerstones 'finishing off' a square or rectangular arrangement of dozens of blocks – and so on and so forth. There were also groupings of similar-sized blocks such as the seven very large megaliths near the southern end of the inshore pier laid side by side next to much smaller blocks pursuing the same axis. In this case the seven large blocks crossed the full width of the axis. The smaller blocks next to them continued along the same axis and to the same width but were arranged in two parallel rows separated by a cleared area.

Natural and young, or man-made and old?

So what is the Bimini Road? Is it a natural formation and not very old? Or, in spite of all the scientific objections, could it be a man-made megalithic structure – even a remnant of Atlantis – covered by rising sea-levels many thousands of years ago?

To begin with the natural-versus-artificial debate, I do not think that the scientists have either proved that it *is* a natural formation or proved that it is definitely *not* a man-made formation – which would amount to the same thing.

For example, the research report from the National Geographic Society quoted in chapter 23 claims that there is no evidence anywhere on the site of courses of blocks having been piled on top of one another and that not enough scattered blocks lie in the vicinity to have formed a now-destroyed second course. This is taken as evidence in favour of the natural origin of the Bimini Road; however, I see no good or logical reason why humans should not have chosen from the outset to construct a structure one course high. Moreover, no consideration is given to another option – which is that the immense structure did have more than one course in the past but that the blocks are no longer there because the vast majority of them have been removed. Although there may be no connection, elementary research amongst elderly islanders has uncovered several eye-witness reports of barges from Florida that used to *quarry stone* underwater off Paradise Point during the 1920s and take it back to Miami for use in construction projects. As the islanders tell it, the barges repeatedly visited the area to carry off stones over a period of several years.[1]

Another example of the scientific criticism of the proposed artificiality of the Bimini Road that I find disappointing is the National Geographic Society's claim that there are no regular or symmetrical supports beneath any of the blocks. This is flatly contradicted by my own experiences diving on the Road.

We've even seen that the evidence for microscopic uniformity within the stones, which plays such a key part in the scientific argument for a natural origin of the site, has not gone uncontested. Zink and others have had quite different and equally bona fide results from their own drill cores, which indicate blocks adjacent to one another in the formation that were not formed side by side but in different chemical environments. The implication of this is that, while there is no doubt that the material used in the Bimini Road *is* beachrock (none of the pro-artificiality researchers have ever argued that it is anything else), it remains *possible* that beachrock deposits were cut, shaped, manipulated and arranged by human hands.

In their 1982 paper for the Society for Historical Archaeology's Conference on Underwater Archaeology at the University of Pennsylvania, Terry Mahlman and David Zink sum up the central thrust of the pro-artificiality defence:

> The most controversial aspect of this site is the history of the megalithic blocks. More directly put, are they beachrock blocks cut and shaped by man or were they formed naturally in situ? Their composition, most agree, is micritized shell hash, or beachrock, which through the continued process of solution and recrystallization of its cement by sea water rich in calcium carbonate has become extremely hard in comparison with modern beachrock. The authors of this paper theorize that, after their original formation in a beach environment, these blocks were removed, shaped and placed above water by human agency. Later as the sea-level continued to rise after the last glacial period, the blocks were again covered and micritization commenced. Newly formed beachrock is easily worked in comparison with the blocks of the site. Their extreme hardness caused the destruction of the diamond bit of our 80mm core barrel after only 12 cores had been taken.
>
> Micritization, once again the on-going replacement of the calcium carbonate cement binding the shell hash, also contributes to the problem of dating these blocks. This is because the new cement contains an increasingly higher proportion of Carbon 14, thus making the sample appear younger than it actually is.[2]

This brings us to the question of the age of the structure. Have orthodox scientists at least proved their case, as McKusik and Shinn claim, that some of the stones used in the Bimini Road might be less than 3000 years old?

Again, I don't think so. The situation of the megaliths is ideally conducive to the production of falsely youthful radiocarbon dates – and these young dates are further contradicted by the depth of submergence of the sites. As McKusik and Shinn themselves admit:

> Testing of submerged features in Florida and one test on North Bimini island shows that the sea level has risen at a rate of about one inch every 40 years for the past 5000 years. This rate of submergence over 2200 to 3500 years [the range of radiocarbon dates for the stones published by McKusik and Shinn] would account for 5.58 to 7.22 feet of the 15 feet of sea observed over the beachrock.[3]

Ignoring the fact that the depth of the Bimini Road is generally greater than 15 feet, McKusik and Shinn account for 'the remaining 7 to 9 feet of sea' by 'the undermining of sand, allowing the beachrock to gradually settle'.[4] This explanation, however, cannot work in the case of the seaward block cited earlier and carbon-dated to *c*.4000 BC – i.e., around 6000 years ago. At that date the block would have been well above the tidal zone and thus unable to form as beachrock at all. Mahlman and Zink's suggestion that there could have been contamination leading to falsely youthful carbon-dates from the tests on the Bimini Road therefore seems a reasonable one.

The mystery of Cabo San Antonio: a possible underwater city off Cuba

In my opinion a mistake shared by the polarized and mutually suspicious communities that have studied the Bimini Road – both those who favour an artificial origin for the site and those who believe it to be entirely natural – has been to confine the arguments solely to dry debate about drill cores, micritiz-

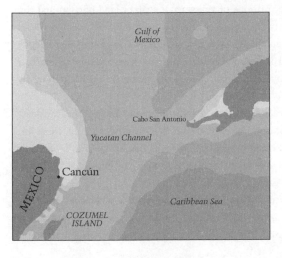

ation, shell hash, bedding planes, C-14, and suchlike. Meanwhile other – mainly contextual – issues have been underplayed.

One entirely new issue, which could well prove to be contextual to the Bimini problem if it checks out, was put before scientists on 14 May 2001, when Reuters News Agency published an astonishing report of the apparent discovery of a complete city submerged in more than 700 metres of water off the west coast of nearby Cuba.[5] The team that had made the discovery were not psychic Atlanteans but a consor-

tium of scientists and salvage experts who had secured an exclusive concession from the government of Cuba to conduct searches for shipwrecks in Cuban waters. Such a search has never before been permitted and, though expensive to mount, is likely to prove very lucrative – since experts believe that billions of dollars' worth of sunken Spanish treasure ships lie in the deeps off Cuba.[6]

What one would not expect to find in water anywhere near as deep as 700 metres would be a sunken city – unless it had been submerged by some colossal tectonic event rather than by rising sea-levels. Mind you, the two are not necessarily contradictory and a colossal tectonic event occurring *amidst* an epoch of global sea-level rise seems to be exactly what is suggested in the Atlantis myth.

Some soundbites from the Reuters report:

'It's a new frontier', enthused Soviet-born Canadian ocean engineer Pauline Zelitsky, from British Columbia-based Advanced Digital Communications, poring over video images of hitherto unseen seafloor taken by underwater robots.

'We are the first people ever to see the bottom of Cuban waters over 50 meters . . . It's so exciting. We are discovering the influence of currents on global climate, volcanoes, the history of formation of Caribbean islands, numerous historic wrecks and even possibly a sunken city built in the pre-classic period and populated by an advanced civilization similar to the early Teotihuacan culture of Yucatan,' she said.

The report then tells us that, ADC, Zelitsky's company, is 'the heavyweight among four foreign exploration firms here' and that, merely while testing its equipment in Havana Bay, it successfully located the wreck of the USS *Maine* which blew up and sank mysteriously in 1898:

> ADC has also been exploring a string of underwater volcanoes about 5000 feet deep off Cuba's western tip, where millions of years ago a strip of land once joined the island to Mexico's Yucatan Peninsula.
>
> Most intriguingly, researchers using sonar equipment have discovered, at a depth of about 2200 feet, a huge land plateau with clear images of what appears to be urban development partly covered by sand. From above, the shapes resemble pyramids, roads and buildings.
>
> 'It is stunning. What we see in our high resolution sonar images are limitless, rolling, white sand plains and, in the middle of this beautiful white sand, there are clear manmade large-size architectural designs. It looks like when you fly over an urban development in a plane and you see highways, tunnels and buildings,' Zelitsky said.
>
> 'We don't know what it is and we don't have the videotaped evidence of this yet, but we do not believe that nature is capable of producing planned symmetrical architecture, unless it is a miracle,' she added in an interview in her office at Tarara, along the coast east of Havana.[7]

As the first edition of *Underworld* goes to press, the status of Cuba's underwater city remains unresolved. Is it a city? Or is it just a wonderful sonar hallucination? Presumably time will tell.

To clarify matters as much as possible in the meantime I asked Sharif to do a couple of telephone interviews. The first was with Paul Weinzweg, co-founder of Advanced Digital Communications (and husband of Pauline Zelitsky), who confirmed:

> The sonar images we have are very extensive, the structures extend over several kilometres. They're very large. Some as long as 400 metres. Some are up to 40 metres high. They're of different shapes. But there's a good deal of architectural symmetry. We've shown them to scientists in the US, Canada and Cuba . . . and they tell us that it's not geology, or that it's a great mystery . . . And we have very extensive bathymetry of that area as well, and it is very interesting that the shelf terraces down in even gradations. And it's obvious that if it is a major settlement of say a pre-classical or Atlantean

nature, then the whole thing just sank, altogether, during some disastrous geological event. There are a couple of fault lines there, and an ancient volcano . . . It's off the coast of Cabo San Antonio, off the Western tip of Cuba . . .[8]

One of the scientists named by Weinzweg as supporting a possibly non-geological nature for the structures in the sonar images is Dr Al Hine, a marine geologist at the University of South Florida. He described what he'd seen on the images that Pauline Zelitsky had shown him as:

> Just bizarre. I couldn't provide an explanation for it but on the other hand there might be a reasonable explanation. They want to turn it into an archaeological site. I suppose that's possible but there are just as many alternative interpretations that could be valid as well. It's something worthy of further study, I suppose . . . But it was something that didn't really jump out at you. It was kinda vague and it might be something real or it might not be. That's the way it is with looking at acoustic geophysics on the sea-floor.[9]

Additional relevant comment came from Grenville Draper of Florida International University, an expert in the neotectonics of Cuba and its region, who thought it highly improbable that tectonic subsidence sufficient to have plunged several square kilometres of land to a depth of 700 metres below the sea could have occurred any time during the known human occupation of Cuba:

> Nothing of this magnitude has been reported, even from the Mediterranean. The only other possibility is that the 'objects' were carried into position by an underwater landslide, something possible, even probable, in the Cabo San Antonio region.[10]

Inundation history

The odds in favour of the Cuban underwater city actually turning out to be anything of the sort don't look particularly good to me. But it would be nice to be surprised and we shall have to wait and see.

Meanwhile there are other more immediate contextual issues surrounding the Bimini Road that have never been examined. For example, no serious attempt has been made to explore the possibility that some sort of cultural relationship might exist between the 'Seahorse' and 'Shark' mounds above water on Bimini and the geometrical mosaic of the now-submerged Bimini Road. Likewise, there has been a failure by both sides to consider the topography of Bimini and its changing relationship to the sea since the end of the Ice Age. For until as recently as 6000 years ago, as I was to discover when I received Glenn Milne's inundation maps for the region in the summer of 2001, Bimini remained part of a large antediluvian island lying across the Gulf Stream from Florida. Very close to the north-western tip of this palaeo-island, overlooking the Gulf Stream then as they do today, were what is now Paradise Point and the present site of the Bimini Road.

My question is this. Doesn't the existence of a large and perhaps inhabited island in the immediate vicinity of the Bimini Road until around 6000 years ago suggest the possibility that vital information concerning the Road's origins could now also be underwater? How can anyone arrive at certainties about this enigma when, as remains the case today, no extensive underwater archaeology has ever been done on the Great Bahama Bank?

In July 2001 after my second series of dives at Bimini, this time with the Channel 4 film crew, I flew to Florida to put these doubts to Dr John Gifford of the University of Miami, co-author of the National Geographic Society research report quoted earlier, and one of the leading scientific proponents since the early 1970s of an entirely natural origin for the Bimini Road.

> GH: John, when did your involvement with Bimini begin? When did it all start, and why?
>
> Gifford: I came to the University of Miami as a graduate student in September of 1969, and at that time there were articles in the local newspapers describing a discovery that had just been made off the coast of North Bimini, which was described as Atlantis, and the Dean of the school at the time, F. G. Walton Smith, decided that this would be a great project for someone who was interested in both archaeology and geology, as I was, so he essentially told me to go over there and study it and find out whether it was archaeological or geological.
>
> GH: Right. And was that the main focus of your research on Bimini – that specific question? Or was it wider?
>
> Gifford: It was, because at the time, again, in the fall of '69, it was a major, major news story, and people were calling this place and saying, you know, 'What can you tell us about Atlantis?' and so we wanted to be on top of things.

I was interested to note, when I questioned Gifford on the age of the Bimini Road, that he did not rely on the disputed carbon-dates from the cores, but instead on the dates of seashells found under the blocks.

> GH: Setting aside for a moment the argument about whether the Bimini Road is in any way artificial or not, how old do you think it is?
>
> Gifford: That particular deposit is somewhat less than about 6000 or 7000 years old.
>
> GH: And that's based on what? How do you arrive at that?
>
> Gifford: One of the things we accomplished in my fieldwork there back in the early '70s, was to excavate underneath the blocks at a number of locations, and we recovered very well-preserved marine shells, mollusc

shells. We radiocarbon-dated those and the dates on the shells all fell between 6000 and 7000 years old.

I next pointed out to Gifford that our inundation maps showed a large island behind the Bimini Road down to about the same period – a solid mass of land quite different from the tiny strips of rock and sand that are all that remain of it today. 'I don't know what kind of landmass it was,' I said. 'Has your work ever touched on that?'

Gifford: No, no.

GH: But it strikes me that it might have been quite a habitable place at the time, when North America was covered in a vast ice-sheet . . .

Gifford: Sure.

GH: And therefore possibly a place where people lived?

Gifford: Well, that's something that has occurred to a number of people, including myself, and so the first step, of course, would be to go to the Bahamas and look for very early archaeological sites not only underwater but on land.

GH: On land too. Yes.

Gifford: But out of all the archaeological surveys that have been done today on all the islands in the Bahamas, the oldest site that has ever been found on land is only about 3000 years old. There is simply nothing older than that.

GH: How much marine archaeology has been done in the Bahamas?

Gifford: Well, prehistoric marine archaeology, very, very little. Certainly there's been a lot of treasure-hunting for shipwrecks and so forth, but only within the last decade or so have some people begun to do things like explore the Blue Holes in the Bahamas. Those are obvious places where one might look for prehistoric remains. And I've heard reports of human bones being found at great depth in some of these Blue Holes, but I think in most cases the bones have been introduced much, much more recently and they've simply fallen down in the slopes. So my point is though, you see, if you've got an exposed Bahama Bank – thousands of square kilometres – and you've got people wandering around, at least some of those people are going to leave some traces on the high points, which are then going to become the islands, which would then be places where land archaeologists would have found some traces.

GH: Now that's a fair point. But it's not a conclusive one. If we treat the Great Bahama Bank as an Ice Age island, the archaeology that has been done on it – even if you thoroughly archaeologized every bit of land that's above water, you'd still be only touching about 10 per cent or 15 per cent of the former island. So that means say, 90 per cent of the former island has never been looked at at all.

Gifford: That's true.

GH: Don't you think that's a bit unparsimonious, to jump to conclusions without doing the archaeology first?

Gifford: Well, it, it's . . . it's just a fact of life in this case that no one and no organization is going to fund a prehistoric underwater archaeological survey of the Bahamas . . .

A late flood

This is the way with self-fulfilling prophecies. The scientific consensus that there is nothing particularly worth looking for underwater around the Bahamas inevitably affects research priorities and the result is that no serious underwater research gets done. Naturally, in consequence, nothing is found. This in turn reinforces the view that there is nothing worth looking for – and so on *ad infinitum*.

But coming at the problem from the point of view of inundation science introduces the possibility of a different perspective – one that tends to excite curiosity about the past. Rather than simply being underwater, inaccessible and unlikely to attract research funds, the inundation maps show that the Bimini area once contained not just one but in fact three principal islands, as well as several smaller islands, that are likely to have enjoyed a favoured climate during the Ice Age. The inundation map for 12,400 years ago (see opposite) shows, to the north, a crescent-shaped island around present-day Grand Bahama, Great Abaco and Little Abaco. Clockwise to the south-east from there we come to a second lost island. This island fills in what is now Tarpum Bay under Eleuthera, then connects via the thin but very probably unbroken line of the Exuma Cays to an even larger exposed area stretching almost as far south as Cuba – itself significantly larger than it is today. Third, to the north-west in the direction of the Florida peninsula covering present-day Andros island and occupying most of the Great Bahama Bank, is the largest antediluvian island of all, with Bimini and the Bimini Road right at its tip.

The inundation map for 6900 years ago (page 534) shows some coastal erosion of the three main islands but otherwise the picture remains basically unchanged – indicating that the islands survived beyond the last of the three great episodes of global postglacial flooding around 7000 years ago. However, in the next inundation map in the sequence, for 4800 years ago (page 535), all the islands have gone. The most likely culprit for their inundation is the so-called Flandrian transgression, the final spasm of the Ice Age meltdown, which took place between 6000 and 5000 years ago.

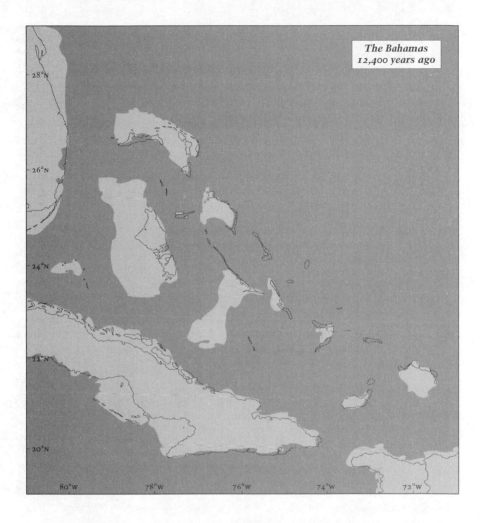

The Bahamas
12,400 years ago

Speculation 1: the chart shared by Columbus and Pinzon

To the strictly limited extent that inundation science can accurately reconstruct former coastlines I find it interesting – nothing more than that – that the formidable antediluvian island of which Bimini formed a part until about 6000 years ago does bear a loose resemblance in size, shape and general orientation, to the 'mythical' island of Antilia on the 1424 Pizzagano chart. Like Antilia on that chart, antediluvian Bimini even has a smaller island lying to its west – occupying the position of the present-day Cay Sal Bank.

Is it possible that the mysterious 'book' said to have inspired Columbus to cross the Atlantic by showing him that it had an end could have contained a chart of the Atlantic Ocean of the kind proposed by Nordenskiold – a chart dating back to the mapmaking tradition of Marinus of Tyre? Other charts linked

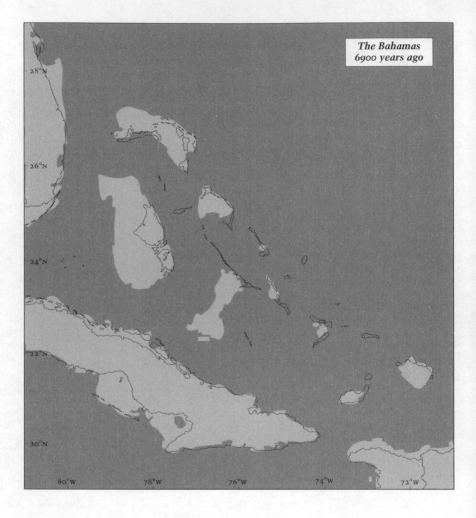

to this tradition, such as the Cantino and Reinal maps of the Indian Ocean and the numerous portolans featuring Hy-Brasil, contain memories or 'ghosts' of Ice Age topography and coastlines. So perhaps the coast and islands that the 'book' of Columbus was said to have portrayed on the western side of the Atlantic were also shown as they looked before being inundated by rising sea-levels? If Bimini on the original 'Tyrian sea-fish' source map had been depicted as it looked at almost any time between 12,000 years ago and 6000 years ago, then it could theoretically have provided the model for the 'mythical' island of Antilia that began to appear on portolan charts during the seventy years prior to Columbus's voyages of discovery.

For whatever reason, we do know that Columbus had a special interest in Antilia. Cited earlier, he is on record with a comment that suggests he recognized a specific Phoenician (in this case Carthaginian) heritage behind the appearance of Antilia on fifteenth-century nautical charts:

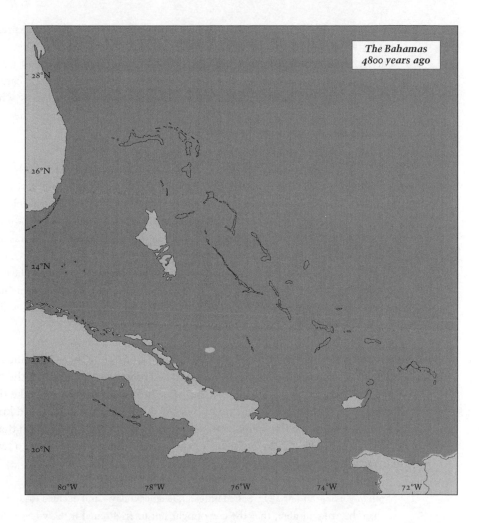

The Bahamas
4800 years ago

Aristotle in his book *On Marvellous Things* reports a story that some Carthaginian merchants sailed over the Ocean Sea to a very fertile island ... this island some Portuguese showed me on their charts under the name Antilia.[11]

Indeed, prior to winning the Spanish sponsorship that financed his expedition to the New World, it is reported that 'Christopher Columbus was making himself a nuisance at the Portuguese court with persistent requests for an expedition to enable him to verify the marking of Antilia' on certain maps.[12]

We've already explored some of the issues raised by the alleged 'book' of Columbus and the hints that it may have contained an ancient nautical chart of the Atlantic that also showed certain parts of the New World. The suspicion that a map had indeed fallen into Columbus's hands is further strengthened by certain passages from his own *Journal* of the first voyage – an abridged version of which, edited by his friend the friar Bartolome de las Casas and often expressed in the third person, has come down to us.[13]

The Atlantic crossing began from the port of Gomera in the Canary islands on 6 September 1492. Three weeks later Columbus and his three little caravels were deep into the terrifying, unknown reaches of the Ocean Sea – where, supposedly, no man had ever gone before to make a map. It is strange, therefore, to read the following entry:

> *Tuesday 25 September 1492.* The Admiral [Columbus, upon whom the King and Queen of Spain had bestowed the title 'Admiral of the Ocean Sea'] spoke with Martin Alonso Pinzon [Columbus's second-in-command], captain of the caravel *Pinta*, regarding a chart which the Admiral had sent to him three days before in which, it appears, he had certain islands marked down in that sea. Martin Alonso was of the opinion that they were in the neighbourhood of those islands, and the Admiral replied that he thought so also but, as they had not found them, it must be due to the currents which had carried them to the NE . . . The Admiral called upon him to return the chart and, when it had been sent back on a rope, the Admiral with his pilot and sailors began to mark their position on it.[14]

In my opinion this entry leaves very little room for doubt that Columbus and Pinzon did indeed possess a chart – or charts – showing some areas of the New World and suggesting a route across the Atlantic Ocean that would take them directly to it. This might also explain why Columbus consistently and knowingly underestimated the distance travelled each day in the information that he gave to his crew. He did so every day of the outbound voyage. Here are a few of the relevant entries from the *Journal*:

> *Sunday 9 September 1492.* Sailed nineteen leagues today – and decided to count less than the true number, that the crew might not be frightened if the voyage should prove long.[15]
>
> *Monday 10 September.* In that day and night sailed sixty leagues . . . Reckoned only forty-eight leagues, that the men might not be terrified if the voyage should be long.[16]
>
> *Wednesday 26 September.* Sailed day and night thirty-one leagues and reckoned to the crew twenty-four.[17]
>
> *Wednesday 10 October.* Day and night made fifty-nine leagues progress to the West-south-west; reckoned to the crew forty-four.[18]

Is it possible that Columbus adopted this practice of under-reporting the actual distances travelled because he had, from the outset, a very *good* idea from his chart about how long the voyage was likely to be and knew that the men would never have set out at all, and would want to turn back, if he had been more honest with them?

Speculation 2: the world according to Columbus

For all the reasons outlined in previous chapters, let's speculate that Columbus did – somehow – come into possession of an old nautical chart showing the New World, and that he was sufficiently convinced of its veracity to risk crossing the Atlantic on the strength of it. Moreover, we've seen that Columbus promoted his expedition to potential sponsors on the explicit grounds that he had a chart which showed the coast and islands at the end of the Western Sea. Unless Columbus was completely mad, it follows that this chart must have possessed some quality (perhaps to do with the 'book' in which it was incorporated) that left him in no doubt that it was accurate. Certainly, it must have distinguished itself in a significant and obvious way from any other maps or charts (for example the Behaim globe – see overleaf) that would have been available to Columbus and already known by his sponsors in 1492. Let's also speculate that this vitally important and convincing chart did *not* show the entire Atlantic coast of the Americas but was a fragment featuring only the mainland and islands between the Florida peninsula and Venezuela on the western side of the Atlantic (probably combined, with a typical portolan portrayal of the coasts of southern Europe and north Africa on the eastern side of the Atlantic).

What mainland and what islands would Columbus have been most likely to have believed were shown awaiting discovery by anyone daring enough to cross the Ocean Sea? Everything suggests that, far from a 'New World', what the Admiral actually expected to find at the end of his first crossing of the Atlantic was the remote and fabulous eastern extremity of the *Old World* – quite specifically Japan and China as they had been described in Marco Polo's *Travels* and other sources.

This was not a zany, way-out idea on Columbus's part but was the consensus view of geographers, mariners and merchants of his day. All accepted that the earth was a sphere and that it should be possible to sail round it in both directions. None knew of the existence of the Americas. All accepted, at least theoretically, that this meant Japan and China in the extreme east might be fetched more quickly, safely and easily by sailing west across the Atlantic from Europe, than by means of the arduous overland route that Marco Polo had taken to the Court of the Great Khan in the thirteenth century . . .

Such ideas were in wide circulation and had been expressed in clear visual form on maps and globes prepared before Columbus ever crossed the Atlantic. The classic example is the Behaim globe, completed at the beginning of 1492 – which Columbus is known to have seen in the months before his first voyage.[19] Redrawn here in plan form (overleaf) this globe by the geographer Martin Behaim (Martin of Bohemia) shows the British Isles, Spain, North Africa and the Canary islands separated from Cipango (i.e., Japan), China, 'Greater India' and the Indonesian archipelago by an Ocean Sea *about one-third wider* than the Atlantic.[20] In between there is no sign of the New World – of course, because

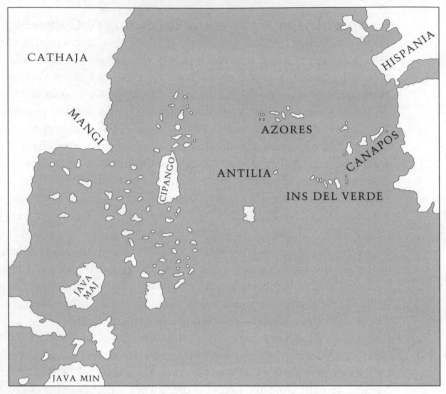

Coast outlines from Martin Behaim's 1492 globe. Based on Fiske (1902).

Columbus would not discover it until later in 1492 – but Behaim has installed for good measure some 'mythical' islands, including the island of Saint Brendan and also Antilia. It is noteworthy that he represents the island which he labels Antilia as rather small and insignificant – nothing like the large and roughly rectangular landmass shown under the name of Antilia on the 1424 chart. But, weirdly, a landmass with something of the traditional rectangular, north–south shape of Antilia does appear much further west on the Behaim globe – lying in the Ocean Sea off the Chinese mainland. Behaim has labelled it Cipango (Japan) and surrounded it with numerous smaller islands.

Other maps of the period that depict Cipango in the same Antilia-like manner include the Yale-Martellus world map of 1489 and the Contarini-Rosselli world map of 1506.[21]

What all have in common is an Ocean Sea far wider that any sailor of the fifteenth century would have dared to cross, Columbus included. All the more reason to suppose that the chart upon which he relied to make the crossing did indeed show the width of the Atlantic accurately – still a formidable enough distance to travel, but possible . . . possible.

Speculation 3: to Asia with a map of the Americas?

I want to reinforce the point here that Columbus, in possession of our hypothetically accurate but outdated 'Tyrian sea-fish' chart of certain parts of the eastern seaboard and islands of the Americas, could have used it successfully to guide his little fleet of caravels to the New World while yet remaining absolutely convinced that the coast and islands he had reached were parts of eastern

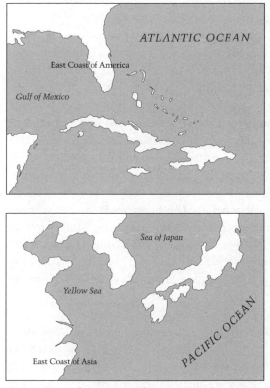

Asia. Conditioned by Polo and Ptolemy, Columbus's conception of the eastern extremity of Asia is likely to have been close or identical to that shown on the Behaim globe. Sailing west, he would, in other words, have been expecting first to find Antilia, then the island of Cipango (set amongst numerous other islands as Polo had indicated). And after Cipango he would have expected to arrive at the great curving peninsula of the Chinese province of Mangi, described by Marco Polo, with its fabulous capital of Zaitun.

Meanwhile – please remember this is speculation – what Columbus and Pinzon actually had to guide them was an antediluvian chart *not* of the coast and islands of Japan and China but of the Americas and the Caribbean between Florida and Venezuela. The chart showed pre-deluge Bimini (connected to Andros and the exposed Great Bahama Bank) as a large island

Comparison of east coast of America with east coast of Asia.

with a shape and orientation roughly similar to that of Antilia on the 1424 Venetian portolan, *but showing an even closer resemblance to Cipango on the Behaim globe.*

We know from the inundation maps that antediluvian Bimini was also surrounded by other islands – as Columbus expected Cipango to be. Imposing his preconceptions on the chart, it is therefore quite possible that he mistook for Cipango what was in fact the cartographic ghost of antediluvian Bimini 6000 or more years ago, and that he mistook the Central American mainland that lay beyond for the Mangi peninsula.

It is usually argued that Columbus's dreamlike and almost hallucinatory misunderstandings of the geography of the region he discovered arose out of his deep belief that he was sailing to Asia on the one hand and his actual experiences in the New World on the other. But I suggest – again, speculation only – that the real source of the dissonance between expectation and experience was that Columbus's chart showed antediluvian features that had been long submerged by 1492 and that therefore could not be found no matter how frantically he searched for them. Despite these 'maladjustments', however, the islands and mainland of the part of the New World he had arrived in matched up well enough with his expectations of the islands and mainlands of Asia (see page 539) for him to convince himself that he was indeed in Asia.

Entries from the *Journal* of the first voyage make this extremely clear. First landfall was made at San Salvador on 12 October 1492,[22] a point very close to the group of large antediluvian islands that existed around Bimini down to 6000 years ago. If the chart that Columbus used to get his fleet across the Atlantic had shown these ghost islands – the largest of which he believed to be Cipango – then he would have been disappointed and disoriented when he failed to find any large islands at all in the area. He might well have concluded that the chart on which he had placed so much reliance was after all inaccurate, or he might have concluded that he had failed to follow his course properly.

The *Journal* suggests that Columbus believed his fleet could have been carried too far to the north-east by currents on the transatlantic crossing.[23] It is therefore of interest that on leaving San Salvador he chose to sail a compensatory route south and west, through the characteristic tiny cays and sandbars that dot the seascape today, trying to pick up intelligence *en route* about the whereabouts of the large island of Cipango:

> *Sunday 21 October 1492.* I shall presently set sail for another very large island which I believe to be Cipango according to the indications I receive from the Indians on board. They call the island Colba [Cuba]. [From there] I am determined to proceed on to the mainland, and visit the city of Guisay [Qinsai] and deliver the letters of Your Highnesses [Ferdinand and Isabella of Spain] to the Great Khan, demand an answer and return with it.[24]
>
> *Tuesday 23 October.* It is now my intention to depart for the island of Cuba, which I believe to be Cipango from the indications these people give of its size and wealth, and will not delay any further here . . .[25]
>
> *Wednesday 24 October.* This must be the island of Cipango, of which we have heard so many wonderful things. According to the globes and maps of the world I have seen, it must be somewhere in this neighbourhood.[26]

Columbus did not complete the exploration of Cuba on his first voyage, and on his second voyage he changed his mind about its identification with Cipango and decided that it was part of the mainland of south-eastern China instead. This was because islanders told him that 'Cuba had no end to the westward', and referred him for further particulars to 'the people of Mangon, a province towards the west'.[27] As Charles Duff explains:

> The name Mangon inflamed the imagination of Columbus, who immediately identified it with the Mangi of Marco Polo, the southern province of China, 'the most magnificent and the richest province that was known in the eastern world,' according to Polo.[28]
>
> [Columbus] was now – as it happened – within two or three days sail of the western end of Cuba, the discovery of which would have disillusioned him concerning its connexion with the mainland of Asia. As it was, he turned back firmly convinced that Cuba was the eastern extremity of the Asiatic continent. And in that belief every person on board expressed his concurrence by a solemn signed deposition. Columbus never afterward abandoned his conviction – he remained unshaken to the end of his life. The dream or fantasy was to him a reality . . .[29]

Despite the constant stream of new discoveries and rapidly improving maps that followed the voyages of Columbus, the dream remained a reality for many others as well. Thus, an inscription placed next to the coast of Asia on the Contarini-Rosselli world map of 1506 informs us that 'Columbus sailed westward to the province of Ciamba, the region of China opposite Cipango.'[30]

Last but not least, improbable though it may seem, it is known that Columbus finally decided that the island of Hispaniola was the Cipango of his dreams.[31]

As noted earlier, Gregory McIntosh has presented a compelling case that a copy of an original map drawn by Christopher Columbus in which Cuba is represented as part of the Central American mainland is incorporated into the world-famous Piri Reis map of 1513. It is therefore intriguing to note – on exactly that section of Piri's map that was derived from Columbus – that a large 'ghost' island with approximately the same shape, dimensions and north–south orientation as antediluvian Bimini is prominently depicted. What seems to seal the identification with antediluvian Bimini – as the reader may confirm at a glance from the zoomed window overleaf – is that this ghost island is clearly marked with a row of huge stone slabs laid out in a manner that strongly resembles the layout and appearance of the slabs in the now-submerged Bimini Road. McIntosh does not comment on this peculiar megalithic image on the 1513 map; however, he does believe that he can explain the presence there of the non-existent island itself without any recourse to ghosts from before the flood.

It's all terribly simple, he argues. This large north–south oriented island cannot be found today because it is just the result of a dishonest – or at any rate self-deluding – representation by Columbus of the island of Hispaniola to make it look more like the island of Cipango that he had convinced himself it was.[32]

Now a glance at any modern atlas will show that Hispaniola (today divided between Haiti and the Dominican Republic) has an east–west rather than

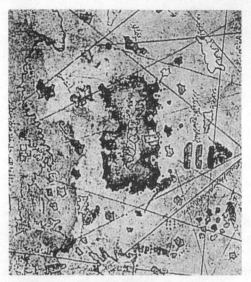

Piri Reis's ghost island in the Caribbean.

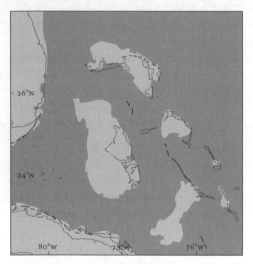

Exposed Great Bahama Bank, 6900 years ago.

north–south orientation and that an island of roughly the right size, in roughly the right place and with roughly the right east–west orientation to be Hispaniola does actually appear on the Piri Reis map. However, McIntosh ignores that option and reminds us (correctly) that Columbus shared the general conception shown on the Behaim globe, etc., of Cipango as an island with a north–south orientation.[33] So wedded was the great explorer to that idea, alleges McIntosh, that in maps made on his second voyage (one of which Piri Reis copied) he simply flipped Hispaniola 90 degrees so that it now lay north–south – with the end-result that: 'The shape and orientation of Hispaniola on the Piri Reis map is strikingly similar to that of Cipango shown on maps of the fifteenth and sixteenth centuries.'[34]

'It is difficult to accept the fact,' adds McIntosh:

> that Columbus, perhaps the greatest navigator of his time, would contort the orientations of Cuba and Hispaniola as much as 90 degrees out of place . . . [Yet] for Columbus in 1495–96, when the map was made that Piri Reis was later to use, to have turned Hispaniola 90 degrees to its correct position was to admit it was not Cipango and that his 'Enterprise of the Indies' was a failure.[35]

This is actually a moot point and McIntosh passes over it too quickly. Columbus may well have shared the general preconception, apparently based on nothing more than vague reports of Marco Polo, that Cipango was a north–south-oriented island, but he was a practical man as well as a dreamer, and a good navigator by all accounts. I do not think he would have persisted in the notion that Cipango was oriented north–south if the island that he believed to be Cipango turned out, in practice, to be oriented east–west. Either he would have decided that he had not yet found Cipango, or that the old travellers' tales about Cipango being oriented north–south had been wrong and that his own scientifically measured east–west alignment for the island should be substituted on future maps.

Hispaniola was discovered by Columbus on his first voyage, when he named it not Cipango but 'La Isla Espanola' – 'The Spanish Island'. McIntosh's identification of Hispaniola with the strange north–south island on Piri's map is therefore strengthened by the fact that Piri labels it, literally, 'The Island Named The Spanish Island'. Additional support comes from a place-name on the island – Paksin vidad. 'This name,' says McIntosh, 'is undoubtedly Navidad, the name of the first settlement founded in the New World on the north coast of Hispaniola.'[36]

The peculiarly ambiguous identification of Hispaniola with Cipango that McIntosh believes Columbus was anxious to make filtered through to others and survived long after it was known that Hispaniola was definitely *not* Cipango. Thus, in a legend on the Ruysch map of 1507 we read that 'what the Spanish have named Hispaniola is also Cipango'.[37] Likewise, Ononteus Finnaeus, in his world map of 1534, labelled Hispaniola as Cipango.[38]

The picture is then further complicated by other sources in which we find Hispaniola being identified not with Japan but with Antilia – for example in letters from the explorer Amerigo Vespucci published in 1506.[39] And this, in turn, may well relate to what was apparently a widely held opinion after Columbus's first voyage, particularly amongst the Portuguese, that 'the islands he had discovered were the islands of the legendary Antilia and not the coast of Asia'.[40] Indeed, this is the reason why the Caribbean islands on modern maps are still called the Antilles today.[41]

Amidst such cartographic confusion over place-names and attributions I think there is room to respect the quality of the theory that McIntosh has put forward while remembering that it is only a theory and that there are other possible explanations of the island thought to represent Hispaniola/Cipango on the Piri Reis map. It is possible, for example, that the place-names on the island which so strengthen its identification with Hispaniola ('The Island Called The Spanish Island' and 'Paksin vidad') were not present on the Columbian original but were put there speculatively by Piri Reis himself.

In view of correlations with Ice Age topography identified on other maps of the period, and of the special importance given to the anachronistic map showing the end of the Western Sea that the Admiral was said to have possessed *before* discovering the Americas, I remain open to the possibility that all along what Columbus thought of as representations of Cipango and its surrounding islands on his mysterious chart could have been ghosts of the antediluvian islands of the Great Bahama Bank.

Taken to the limit, this line of reasoning might even suggest that the model for early cartographic representations of Cipango (conceived of as an island that could be reached by sailing west from Europe) was not provided by vague travellers' reports sent back across the breadth of Asia as has hitherto been supposed, but was in fact derived from the representation on 'Tyrian sea-fish' maps (one of which had fallen into the hands of Columbus) of the large antediluvian island of Bimini.

But if the ghost of antediluvian Bimini did provide the model for early representations of Cipango, then, logically, it could not also have doubled up as the model for the legendary island of Antilia (which often appears on the same maps as Cipango).

Is there a model for Antilia?

That thing between Columbus and Pinzon again . . .

We've seen evidence, both from the inscriptions of Piri Reis and from the *Journal* of the first voyage, that Columbus possessed a chart of the Atlantic – and that it was considered so important as a guide to the crossing that it was passed back and forth between Columbus in his flagship the *Santa Maria* and his second-in-command Martin Alonso Pinzon, captain of the *Pinta*. The presence and exact character of this chart seem enigmatic when we remember the intensive demand for it from the two captains (who shared it but apparently did not possess individual copies), its obvious practical utility to them throughout the voyage, and the fact that it got them to the New World. Such a useful chart of the Atlantic cannot be explained against the background of the cartographic knowledge of the time. On the contrary the part that it played in the success of the Columbian voyages must be weighed up against the background of the abysmal ignorance of even the greatest mapmakers in Europe of the true circumstances of an Atlantic crossing and of the real appearance of the coast and islands on the western side. To have followed the speculative vision of Behaim in his famous globe, or of others like him,[42] would have been disastrous, even though their work represents the cream of fifteenth-century mapmaking and was known to Columbus. Indeed, as one commentator has observed, if his chart had been based on the Behaim scenario, 'Columbus could not even have known of the whereabouts of the New World, much less discover it.'[43]

Yet not only does he seem to have known where he was going but, on some accounts, when he was going to get there:

> Now and then Pinzon and Columbus consult and deliberate – mutually discuss their route. The map or chart passes not infrequently from the one captain to the other; the observations and calculations as to their position are daily recorded, their conduct and course for the night duly agreed upon.
>
> On the eve of their due arrival Columbus issues the order to stay the course of the armada, to shorten sail, because he knew that he was close to the New World and was afraid of going ashore during the obscurity of the night . . .
>
> How does he know the place and the hour?
>
> 'His Genius' says the Columbus legend in explanation. But the Map? The critics will ask, what did it contain? Whose was it? What did that map contain that was so frequently passed from Columbus to Pinzon during the voyage?[44]

I've presented my case that what the map may have contained was an accurate but ancient, and indeed antediluvian, representation of the coast and islands of Central America, notably the north–south-oriented Great Bahama Bank island, which Columbus – no less ignorant than any of his contemporaries about the existence of the Americas – took to be an accurate map of part of the coast of China and the islands of Japan.

An interesting sidelight on this story concerns Pinzon himself. In 1515, nine years after Columbus had died, the Pinzon family brought a lawsuit against the Admiral's estate on account of promises of benefit-sharing that he was said not to have kept. During this lawsuit it emerged that Pinzon too claimed to have had prior information of the route to the New World:

> Arias Perez Pinzon, the son of Martin Alonso testified that his father had definite indications concerning the Lands to the West, which indications he had found in documents in the library of Pope Innocent VIII. The witness said that he saw given to his father a document which contained the necessary information for the discovery. His father took it and carried it away with him, and upon his return to Castile from Rome he decided to set out to discover the said lands, and often talked with the witness about the voyage. Meanwhile the Admiral arrived . . . with a plan to discover the same lands. The father of the witness, hearing of it, went to see this Christopher Columbus and told him that his plan was a good one, that he was sure of it, and that if the Admiral had delayed a little longer he would have found Martin Pinzon already started with two caravels to make the discovery himself. The Admiral, knowing that, put himself on intimate terms with the father of the witness and brought about an agreement whereby the said Martin Pinzon was engaged to accompany him.[45]

It is not obvious from the proceedings exactly what Pinzon found in the Papal Library in Rome or how it set forth 'the necessary information for the discovery', but Gregory McIntosh argues that it must have been 'an old document (a manuscript book or portolan chart?) that told of a mythical expedition that sailed west to Cipango . . .'[46]

Cipango again. And here are the words that Pinzon is reported to have used to recruit crews for Columbus's ships:

> Friends, come, come with us on this voyage! Here you're creeping about in poverty; come and sail with us! For with God's help we're going to discover a land that they say has houses roofed with gold.[47]

Houses roofed with gold are diagnostic of the fabulous island of Cipango described in Marco Polo's *Travels*.[48] It is therefore clear that whatever posthumous disagreements may have occurred over their relative roles in the discovery, Pinzon and Columbus had been absolutely of one mind from the very beginning that Cipango was to be their first destination and that the old charts or documents that they possessed showed the way there. They were not to know that their 'Cipango' was the outline of a ghost island amidst a ghost archipelago drowned 6000 years previously or that the mainland it lay off was not the end of the old world but the beginning of a new one.

The previous sentence, of course, is pure speculation on my part – just a hypothesis launched to provoke inquiry into neglected possibilities. And it still leaves the problem of Antilia unresolved.

Professor Fuson's lateral thinking about Antilia and Satanaze

The identity, location, size and orientation of the 'mythical' island of Antilia underwent continuous bewildering changes on all kinds of maps and charts over a period of hundreds of years. There is, however, a definite beginning to this energetic metamorphosis and that is marked by the 1424 Venetian portolan on which Antilia first appears – presumably in its purest, least-changed form. On that chart, a smaller island is also shown lying to the west of Antilia. And it is important to remember that a second large 'mythical' island, Satanaze, is shown lying to the north-east of Antilia, again with a much smaller island (named Saya) near by, this time to the north.

The identification of the two larger islands by Professor Robert H. Fuson of the University of South Florida – in his 1995 book *Legendary Islands of the Ocean Sea*[49] – is, in my opinion, a masterpiece of historical detective work. And it illustrates, better than any other example I know, how the ghosts of islands can migrate not only through time but also through space, and sometimes through both dimensions simultaneously.

What Fuson has demonstrated, conclusively I think, is that Antilia and Satanaze, marooned in mid-Atlantic on the 1424 Venetian chart, are in fact the earliest true maps to appear in the West of the Pacific islands of Taiwan and Japan. His argument in brief is that the mapmaker Pizzagano had somehow come into possession of Chinese nautical charts of Taiwan and Japan and – being as ignorant as Columbus and others of the existence of the Americas – had placed these islands in the Mid-Atlantic with the assumption that the mainland of China lay somewhere beyond.

Why Antilia is Taiwan

Fuson begins provocatively:

> A number of large, Asiatic islands were charted by the Chinese during the active maritime period of the first two decades of the 15th century. One of these islands, Antilia, is known today as Taiwan.[50]

As was said in many of the legends about Antilia, Fuson points out that Taiwan has gold-bearing sands.[51] Moreover,

Taiwan also has something else that Antilia must have, and that is a small island to the west. On the 1424 Pizzagano chart it was called *Ymana*. Today it is the *Peng-Hu* group, or *Pescadores* (Islands of the Fishermen). There are 64 islands totalling 50 square miles.[52]

Some quotations from Fuson gave a taste of the quality of his proofs and the strengths of his arguments:

- Antilia on the 1424 nautical chart is about the right size and its shape articulates well with modern Taiwan.
- Every one of the eight or nine river mouths of Antilia matches one of the principal river mouths of Taiwan.
- The five largest rivers are correctly placed on the 1424 map of Antilia. Of Taiwan's ten major rivers, seven are indicated on the map of Antilia and in approximately the correct locations.
- Every significant coastal feature is plotted: embayments, capes and peninsulas. Antilia and Taiwan also share a unique north-eastern coastline. There the tip of the island terminates in a sharp, narrow cape. To the north-west the coastline is smooth and rounded.[53]

Why Satanaze is Japan

Fuson's case for Japan is equally well made and again I will give the gist of it briefly and in his own words:

- North of Antilia on the 1424 chart are two islands: Satanaze and Saya. Without question these are the Japanese islands. Saya . . . the Japanese word for 'bean pod' . . . is Hokkaido, while the three main islands (Honshu, Shikoku and Kyushu) are represented by the single island of Satanaze. The channel between Kyushu and Shikoku/Honshu is well defined.
- The origin of the name Satanaze is easy to understand . . . The southern tip of Kyushu is Cape Sata (Sata-Misaki). Approximately 300 kilometres to the south, in the northern Ryukyu islands, is the city of Naze.
- The most important bays in Japan are depicted on the Satanaze/Saya chart . . . and two of them merit special notice. The entrance to the Inland Sea at Bungo Strait is the largest oceanic indentation (as it should be) and Tokyo Bay is guarded by the volcanic island O Shima, one of the most prominent harbour landmarks on earth. From a mariner's perspective it is quite appropriate to exaggerate a feature such as O Shima.
- Saya [Hokkaido], which was not even mapped by the fifteenth-century Japanese, was depicted in its bean-pod shape for more than 300 years. Its 1424 rendering by the Venetians reveals all the important features along the south coast and is every bit as detailed as Portuguese examples in the seventeenth century.[54]

After first appearing on the 1424 chart, notes Fuson, the Antilia group of islands found their way onto at least seventeen other charts and one globe (the Behaim globe):

> Nomenclature was chaotic and occasionally one or another island was omitted. Antilia was mapped as an island in the Ocean Sea until at least 1508 (the Ruysch map), but Japan had captured its form in 1492 on the Behaim globe ... The old Antilia/Taiwan shape continued to appear in what had become the Pacific Ocean and in 1546 (Munster map, Basel) carried the label 'Zipangu'. A major problem had arisen as the shapes and locations of Antilia/Taiwan [and] Satanaze/Cipango ... became entangled ... When the West Indies became the Antilles in the middle of the sixteenth century, Antilia-the-island was no longer needed. The original island was relegated to mythological status and Japan was free to use its body. By 1570 the magnificent atlas *Theatrum orbis terrarum* (by Abraham Ortelius) placed Japan in its proper location and labelled it 'Iapan' (Japan).[55]

Ghosts of a drowned world

It is Professor Fuson's view that Chinese charts of Taiwan and Japan were the source of the 1424 portrayal of Antilia and Satanaze. He makes a very persuasive case that such charts are likely to have originated from the seven spectacular voyages of discovery made by the famous Ming admiral Cheng Ho between 1405 and 1433.[56]

Cheng Ho was a giant of a man, 'seven feet tall with a waist of 60 inches',[57] and is worth a giant of a story in his own right – though unfortunately this is not the place to tell it. Much suggests, however, that Robert Fuson is correct to deduce that the charts of Taiwan and Japan that somehow found their way into the hands of Zuane Pizzagano in Venice in 1424 must have originated from the voyages of Cheng Ho.

Yet there is a problem. As we will see, Antilia and Satanaze on the 1424 chart don't show Taiwan and Japan as they looked in the time of Cheng Ho, but rather as they looked approximately 12,500 years ago during the meltdown of the Ice Age.

Is it possible that Cheng Ho, too, like Columbus, was guided in his voyages by ancient maps and charts, come down from another time and populated by the ghosts of a drowned world?

PART SIX

Japan, Taiwan, China

25 / *The Land Beloved of the Gods*

As a tradition which began in the High Heavenly Plain,
I humbly speak before the sovereign Deities
Who dwell massively imbedded like sacred massed rocks
In the myriad great thoroughfares . . .

Ancient Japanese ritual prayer[1]

The highest peak of Mount Fuji . . . is a wondrous deity . . . and a guardian of the land of Japan.

The Man' yoshu[2]

The identification of the 'legendary' Atlantic islands of Antilia and Satanaze with Taiwan and Japan is the hypothesis of Professor Robert Fuson, and is delivered as the punchline to his utterly convincing book *Legendary Islands of the Ocean Sea*.[3] He further suggests that the source-map from which the outlines of Antilia and Satanaze were derived must have come from China and would most probably have been drawn up during the voyages of the great Chinese admiral Cheng Ho.

What Fuson does not notice – there is no reason why he should – is that Antilia and Satanaze on the 1424 Venetian chart do not portray Taiwan and Japan as they looked in the early fifteenth century, the epoch of Cheng Ho's voyages, but as they looked around 12,500 years ago during the meltdown of the Ice Age. One would have to go back to around that date, for example, to find the three main Japanese islands – Honshu, Shikoku and Kyushu – joined together into one larger island, as is the case with Satanaze. I will substantiate this statement and pursue this mystery to its conclusion in due course.

Meanwhile, by a strange, roundabout route I had found my way back to Japan, encountering it where I had least expected it – in the middle of the Atlantic Ocean. Long before I learned that it had been shown in its Ice Age configuration on a 1424 chart, however, I was already acutely aware of another Japanese mystery centred on the end of the Ice Age when rapidly rising sea-levels inundated a series of massive rock-hewn structures around the coasts of the Ryukyu archipelago of southern Japan.

I've outlined some of the background to this in chapter 1 – how I first heard of Japan's underwater ruins in 1996 and how the generosity of an extraordinary Japanese entrepreneur enabled me to explore all the main sites between 1997

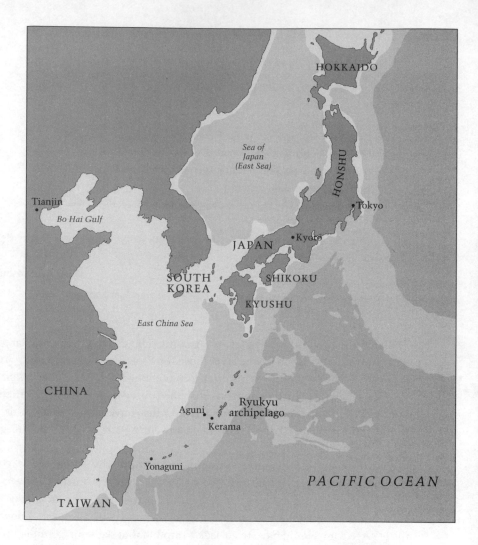

and 2001. I made close to 130 dives at Yonaguni with logistical assistance from
Seamen's Club, Ishigaki, and with the best and most knowledgeable local teams
led by men like Kihachiro Aratake and Yohachiro Yoshimaru. Then there were
around eighteen dives at Kerama (on five different visits there) again in the
company of great local specialists like Kuzanori Kawai, Mitsutoshi Taniguchi,
Isamu Tsukahara and Kiyoshi Nagaki. I successfully dived twice at Aguni, in a
most forbidding, inaccessible and difficult spot, and around a dozen times at
Chatan off the west coast of Okinawa – once again, in both places, with
exceptional local support.

In chapter 1, where I briefly describe the four main Japanese underwater sites,
I also suggest that the solution to the mystery of these places – and to the
interminable wrangle about whether they are natural or man-made – cannot be
arrived at purely by a consensus of geologists. This is not only because there is,

in fact, no consensus of geologists on the character of these structures (on the contrary, opinions are polarized) but also because geological opinion alone is not adequate to settle the matter. One need not be a specialist in anything to see that Japan has cultivated a unique sensitivity to the beauty that is immanent in natural forms and to realize that such a refined intimacy with rock and mountain, forest and valley is likely to have extremely ancient roots. Sculpting in rock and the placing of sculpted rocks in artistically manipulated landscapes remains a distinguishing Japanese passion – and an intensely spiritual one – to this day. It therefore makes sense, in pursuit of reasoned conclusions about the underwater rock structures of Japan, to take into account not only geological considerations but also what is known about the character, the level of development and the artistic and religious culture of the ancient Japanese at the end of the Ice Age when those rock structures (whether natural or man-made) were not yet submerged.

Preconceptions about the Jomon

At first glance I could see nothing encouraging about prehistoric Japan. The consensus view for the past half century has been that during the period from 17,000 years ago (roughly the end of the Last Glacial Maximum and the beginning of the global meltdown) down to about 2000 years ago the islands were populated exclusively by a culture of hunter-gatherers, the Jomon, who were in most respects extremely primitive.

The 'Stone Age' image of the Jomon put me off the idea of researching them. How could so backward a people, who supposedly never discovered agriculture, have anything to tell me, one way or another, about my central interest – the possibility of a great lost civilization of antiquity? Small tribal bands wandering from place to place, grubbing around in the mud for nuts and berries, spearing the odd fish or mammal, did not fit with my idea of what I was looking for.

Nevertheless, I knew that I could not afford to discount the Jomon entirely – if only because their culture seems to have emerged very suddenly in Japan around 16,500 years ago, at which remote date it is attested to by fragments of the oldest known pottery in the world. The pottery itself at such an early date is highly anomalous. And whatever the end of the Ice Age really meant – then and for thousands of years afterwards – the Jomon witnessed it, went through it, were part of it, and triumphantly survived it down almost to historical times. I still felt a definite reluctance but I realized that sooner or later I was going to have to learn more about this prehistoric people whose story was veiled by the mists of the past.

The prehistoric city and the man-made mountain

In 1998, on the suggestion of Japanese friends, I visited the Jomon site of Sannai-Muryama in Aomori Prefecture, and was surprised to discover how large and how well-organized the ancient settlement had been at its peak

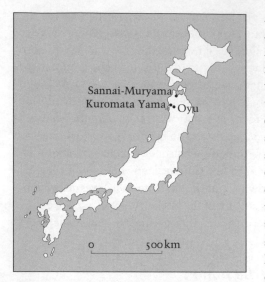

4500 years ago – the same epoch exactly as ancient Egypt's 'pyramid age'. Sannai-Muryama, with its spacious public buildings, wide streets and planned sanitation, was not at all what I had expected of primitive hunter-gatherers. These were the obvious signs of permanent settlement, stability, order, organization and economic success. And they were accompanied by equally clear indications of a society with evolved spiritual ideas. In particular, the use of grave goods by the ancient inhabitants, and of symbolic burial patterns, are suggestive of complex beliefs in the afterlife of the soul. A ceremonial pathway that dominates the site proved, on excavation, to be lined on each side by tombs with the feet of the dead pointing towards the path and their heads away from it.

On the same trip I learned that certain pyramid-shaped mounds, hills and mountains are regarded as sacred beings in Japanese mythology and saw evidence which suggests not only that this belief is rooted deep in Jomon times but also that it sometimes led the Jomon into 'artistic manipulation' of the landscape on an even larger scale than the disputed structures now underwater at Yonaguni, Chatan and Kerama.

In Akita Prefecture, for example, two hours' drive from Aomori, I climbed the cedar-covered slopes of an 80 metre high mound which juts emphatically out of the surrounding plains. Its name is Kuromata Yama (Mount Kuromata) and according to local legend it is 'a pyramid built by an ancient people'.[4] Geologists remained sceptical until a multi-disciplinary team of scientists from the Japan-Pacific Rim Studies Association led by Professor Takashi Kato of Tohoku Gakuin University produced detailed radar maps of Kuromata Yama in the 1990s. The maps show that the interior of the mound

> consists of seven terraces with stones laid out on each terrace. This is a clear indication that it was shaped by man and is certainly very different from a natural mountain formed by volcanic eruptions or natural weathering.[5]

The experts concluded that a natural hill had indeed once stood on the site but that this had been deliberately quarried, sculpted and reinforced with stone blocks to create a pyramidial core with seven terraces that was finally covered with ramped earth and then overgrown by vegetation. Thus, 'Although the mountain is not a pyramid in the Egyptian sense, it was nevertheless made into the shape of one for religious purposes.'[6] Since work of such ambition and scale has never previously been associated with the Jomon, it was at first assumed that the construction work was unlikely to be very old – perhaps no older than the eleventh century AD. The Motomiya Shrine of Japan's indigenous Shinto religion that stands on its summit seems to be linked to that epoch, since it is named after a physician who served Sadato Abe (AD 1019–1062), a local ruler of north-east Honshu.[7] On the other hand, since Shinto shrines are completely rebuilt according to a pre-existing pattern every twenty years on sites that in most cases have been sacred ever since records began, this perhaps proves less than it should. At any rate, the excavations by archaeologists on Professor Kato's team settled the matter with the discovery of fragments of Jomon pottery in the mound and other archaeological evidence which confirmed beyond serious dispute that Kuromata Yama had indeed been landscaped into its pyramidal form 'in the Jomon era'.[8]

Equally important was another discovery published a year earlier by team-member Masachika Tsuji of Doshisha University in Kyoto. He showed that four Shinto shrines positioned around the base of Kuromata Yama lie in direct lines pointing north, south, east and west from the summit and incorporate solstitial alignments datable through the accepted formula for changes in the obliquity of the ecliptic to 4000 years ago: 'The shrines were built relatively recently on what are known to be sacred sites dating from ancient times, suggesting the shrines may have maintained that link since the Jomon Period.'[9]

Surviving ancient texts enable us to trace the recorded story of Shinto back at the most about 2500 years and realistically probably less than 2000 years; however, at that stage it seems already to have been fully formed. All authorities therefore agree, though lost in prehistory, that Shinto's origins must be much older than 2000 years. As far as I know, however, the discoveries at Kuromata Yama are the first to demonstrate such a clear relationship between the religious architecture of the prehistoric Jomon and the Shinto religion as it survives and expresses itself to this day – a religion, it is worth reiterating, that is unique to Japan and that is of unknown age and origin.

Perhaps the clearest sign of a family relationship to emerge from the excavations is that the Motomiya Shrine shares the summit of Kuromata Yama with the remains of a previously unknown stone circle constructed by the Jomon. Kuromata Yama is also clearly visible from two further Jomon stone circles that have been excavated at Oyu, 2.2 kilometres to the south-west. Both of these are more oval than circular, one about 35 metres in diameter, the other 20 metres in diameter. Both are about 4000 years old, a little younger than Britain's Stonehenge. By European 'megalithic' standards they are not large and the stones actually used in their construction are puny when compared with Stonehenge or Carnac. Still, they are 'stone circles' in every meaningful sense of the term.

The mystery of the pots

Nor were these the only surprises that the Jomon had in store for me. As we've already noted, what is truly outstanding and unexplained about these 'primitive hunter-gatherers' is that they were the first people in the world to invent pottery – one of the great leaps forward in human culture which, in their case, took place not just hundreds but thousands of years before anybody else. As recently as 1998 most scholars believed that the oldest Jomon pottery was made about 12,500 years ago – itself a staggeringly early date – but so rapid is the pace of new discovery in this field that that the origins of Jomon civilization have had to be continuously revised backwards.

In May 2000, on my second visit to the Aomori area, I held in the palm of my hand four fragments of a broken Jomon pot 16,500 years old. Excavated at a site known as Odayamamaoto No. 1 Iseki, the potsherds had been dated using state-of-the-art AMS technology.

It is still a little-known fact that the Jomon of Japan are the world's oldest pottery-making culture. But even less well known is the extent to which this prehistoric people maintained a distinct identity as a single, homogeneous group. According to Dr Yasuhiro Okada, the Aomori Prefecture's Chief Archaeologist at Sannai-Muryama, 'they were one culture, from beginning to end'.

Imagine that – one culture, probably one language, probably one religion, staying intact for more than 14,000 years. That's the time-span between the oldest Jomon pottery – 16,500 years old – and the youngest examples – which are about 2000 years old.

Genius or influence?

What happened to the Jomon? If their culture could survive for 14,000 years, how come they aren't still with us today?

The archaeological record points to the influx into Japan – probably from Korea and probably between 2700 and 2300 years ago – of a larger, more populous and more economically competitive group of people. Named the 'Yayoi' by modern scholars (we do not know what they called themselves), these were sophisticated, highly organized rice-growers and it is generally supposed that their way of life simply overwhelmed that of the indigenous hunter-gatherers. Although the Yayoi were a martial culture and the Jomon were not, there is no evidence of military conflict or of genocide. The Jomon were not 'wiped out'. If anything, the latest archaeology prompts us to envisage something more like an effortless merging and mixing of peoples into the new synthesis that would cross from prehistory into history – from forgotten time into remembered time – in the already complete form of classical Japanese civilization. In a sense, therefore, Jomon culture is still with us and may never have come to an end.

Does it have a beginning? The archaeological record is constantly subject to revision by new evidence. But to the extent that the Jomon are defined by and identified with their pottery-making skills, then the earliest definite evidence for their existence that has so far been discovered consists of that little group of pottery fragments from 16,500 years ago.

Did something happen in Japan at that time that could explain why the Jomon invented pottery millennia before anybody else? Shimoyamu Satoru of the Ibusuki Archaeological Museum on Kyushu island suggests: 'maybe there was just a Jomon genius who figured it out – you know, clay, open fire, pot. He saw the potential.' On the other hand, Professor Sahara Makoto, Director-General of the National Museum of Japanese History, believes that 'there must have been some influence'. Sitting cross-legged on the floor of his office, he drew up a map of Japan, China and Siberia. 'Here in Japan', he explained, 'we have high levels of development – new roads, new houses, even new cities are constantly being built. This means that the soil must be broken and turned over – and every time this happens there is the possibility of archaeological discovery. But in China such activity is much less and in Siberia less still. So it is possible in Siberia, for example, that archaeologists might one day find the traces of an even earlier pot-making culture that influenced the Jomon.'

Technology transfer

What neither scholar appears to take into account is the peculiar coincidence in dates between the earliest Jomon pottery, about 16,500 years ago, and the end of the Last Glacial Maximum, about 17,000 years ago – which was followed by thousands of years of ice-sheet meltdown and by global sea-level rises. Is this just a coincidence or could there be some weird causative link between the post-glacial floods and the pottery?

Sahara Makoto has already expressed his views on the subject of influence. He thinks the Jomon were influenced by an earlier, probably Siberian, pot-making culture. But to be fair, that is just his guess. It is undoubtedly correct that pottery was being made at a very ancient date in Siberia[10] (though not as ancient as the oldest Jomon); however, the idea of pottery, the essential mental work to make the great leap forward, does not *require* contact with a hypothetical mainland tribe – and what counts against this hypothesis is the palaeo-geological evidence. As the archaeologist Douglas Kenrick points out, 'When the earliest recorded pottery was made, the sea had engulfed any landbridges that might have remained and had created a natural barrier between Japan and the mainland.'[11]

In other words, if the Jomon *were* 'influenced' 16,500 years ago – to become potters and whatever else – then that influence is more likely to have entered Japan by sea than by land. It could, theoretically, have been passed on by a single survivor, or a handful of survivors, of a shipwreck. And since those were times of global floods the possibility cannot be ruled out that such a ship could have come to Japan from very far away – could, theoretically, have been blown in from almost anywhere. But whether the mariners marooned in Japan were

Siberian tribesmen or highly sophisticated survivors of a hypothetical lost civilization it is unlikely that they would have been able to pass on more than a handful of useful 'civilized' skills to the primitive local inhabitants.

It goes without saying that the skill of pottery would always be ranked near the top of the list in any such emergency technology transfer.

Time and space

Whatever the source of the original inspiration, there is no doubt that Jomon pottery is very distinctive. Its most characteristic decoration is the cord-mark (indeed, Jomon means 'cord-mark' in Japanese and is another name given by archaeologists; as with the 'Yayoi' we do not know what the 'Jomon' called themselves). This decorative technique requires the potter to press lengths of knotted twine down into the clay before firing and sometimes to roll the cords to produce additional effects. The range of possible combinations is large and these 'cord-marks' in their turn are only a tiny part of the full Jomon repertoire of extravagant and unusual designs.

This repertoire, it is worth remembering, exists in four dimensions – in time as well as in space. I say this because on the one hand Jomon pottery is scattered geographically throughout Japan, from the far south, including the Ryukyu archipelago, to the far north, including Hokkaido, and on the other is spread out in time, connecting the world of relatively recent and comprehensible history (2000 years ago) with the world of remote prehistory, 16,500 years ago, when the Ice Age went into meltdown.

Genie in the bottle

Archaeologists in Japan are more accommodating than their Western counterparts. Whereas most of the latter would rather be mummified than have me in their museums, the Japanese are much less snobbish and judgemental. In Japan I have again and again been given the incredible privilege of handling very ancient artefacts – national treasures that in some cases are more than 12,000 years old. At the Sato Haramachi Archaeological Centre near the city of Miyazaki this privilege extended to holding in my hands the oldest piece of painted pottery ever found in the world – part of a fine Jomon pot, painted red on the inside, securely dated to 11,500 years ago.

To touch it was like boarding an express elevator on the way down to the depths of time. I could almost see the ancient artist at work on the same object that now rested in my hands. In a peculiar way, I realized, he – or she – was still alive in this potsherd, like a genie in a bottle. For a moment the 11,500 years that separated us – more than twice the age of the Great Pyramid of Egypt – seemed a small matter.

'It requires imagination,' says Douglas Kenrick, 'to comprehend the length and vitality of the Jomon pottery age. Age leaves its mark on vessels buried for so long, but a feeling of awe at the age of a vessel should not blind us to its beauty.'

In my travels in Japan I have seen a great deal of beautiful Jomon pottery of all epochs. Made without recourse to the potter's wheel, and always in open fires, it takes on a fantastic variety of forms – from the spectacular 'flame pottery' of 5000 years ago, with its grotesque and elaborate rim-work, to austere and simple rounded bowls, more than 12,000 years old that are decorated only with cross-hatch or shell-scrape patterns. The cord-mark motif keeps cropping up again and again. And other patterns repeat, such as distorted human faces sculpted into the shoulders of vases. Pottery masks have been found that replicate their gargoyle expressions and one particular style of mask, with its nose bent at right angles to the side of its face, seems weirdly futuristic; it could almost be a contemporary work in a gallery of surrealist art; instead, it is 4500 years old, as old as the Great Pyramid, and part of an ancient Jomon tradition of representing the human form.

Dogu

Although I have not personally seen examples more than 8000 years old, archaeologists I have talked to in Japan assure me that simple pottery representations of the human figure have been found in strata dating back more than 12,000 years. These earliest figures, and all the later examples, are known in Japan by the generic term *dogu*.

The best-known *dogu* date from around 3000 years ago and are better described as 'anthropoid' than human – since it is by no means certain that the figures they represent are human beings. They have hands and feet, legs and arms and a head, like human beings, but their features are weirdly distorted – almost as though they are concealed behind some kind of face-mask or helmet. The eyes of these figures are most disconcerting, being depicted as large ovals each with a single horizontal slit.

Other *dogu* are very different, some seeming to freeze a tortured human face in the act of screaming, some imposing the features of an animal – a cat for example – on to an otherwise human form, some creating the appearance of mythological beings with the body unnaturally elongated or the face lozenge-shaped. There are multiple examples of exaggerated female figures, notably the 5000-year-old 'Venus of the Jomon' found recently at Tanabatake Iseki in Nagano Prefecture. With her gigantic thighs and hips, this 'mother goddess' is similar in proportion and general appearance (and possibly in function as well) to stone Venus figures found in the megalithic temples and underground labyrinths of the far-off Mediterranean island of Malta (see chapters 16–20).

It is difficult to guess what the Jomon were trying to achieve through the production of so many different kinds of *dogu* over an unbroken period of at least 10,000 years. It is very likely, but not certain, that these were religious icons of some kind and were meant to stand in alcoves or niches. But it is also obvious, looking at them – and indeed at the whole range of Jomon pottery – that they are the work of a prosperous culture with sufficient surplus to support a full-time, professional artisan class dedicated exclusively to the production of beautiful and sometimes awe-inspiring objects.

The rice bombshell

The next surprise I had about the Jomon concerned their way of life. Since visiting Sannai-Muriyama in 1998 I'd been aware that these 'hunter-gatherers', somewhat anomalously, sometimes chose to live in large, permanent settlements. I had assumed, wrongly, that Sannai-Muryama, built about 4500 years ago, was the earliest of these.

Then in April 2000 I visited Uenohara, a much older Jomon site on the island of Kyushu. Kuzanori Aozaki, one of the prefecture's archaeologists, explained that Uenohara had been a continuously inhabited settlement over a 2000-year period from roughly 9500 to 7500 years ago. 'They had their lives pretty well worked out,' he explained. 'At any one time they had more than 100 people living here. They were comfortable . . . I would even say prosperous. All their basic needs were met. They had ample food, good shelter, comfortable, elegant clothing.'

'And this was a permanent settlement, like a village or a small town?'

'Yes.'

'But doesn't that contradict the idea of the Jomon as simple hunter-gatherers?'

'Yes it does, because the idea is wrong. The more you get to know the Jomon the more you know that they were many things as well as simple hunter-gatherers.'

Aozaki went on to tell me how in his opinion the Uenohara community had managed to support itself through a kind of organized 'agriculture' and 'harvesting' of the forest – not quite farming, but certainly a planned husbandry of nature aimed at sustained, long-term survival.

This was not to be the last time during a seven-week journey through Japan in April and May 2000 that I would hear hints of agriculture. At Ofuna C Iseki on Hokkaido the chief archaeologist, Chiharu Abe, told me he was convinced that the Jomon had 'farmed' chestnut trees: 'They imported seedlings from Honshu and then cultivated them here. To all extents and purposes they were doing agriculture.'

Another intriguing recent discovery is that as far back as 8000 years ago the Jomon were cultivating a non-indigenous plant, the bottle-gourd, which palaeo-biological studies indicate must have been imported from Africa. There is also some evidence of the cultivation of beans at a very early date. Indeed, according to Profesor Tatsuo Kobyashi, the Jomon made effective use of nearly every species of available plants and animals – 'a conscious and rational use of nature's bounty with a low-level use of less desired species to avoid depletion of preferred ones'.

Since it was for a long while more or less automatically assumed that the Yayoi brought rice cultivation to Japan it is also highly significant that archaeologists have now found undisputed evidence of paddyfield rice cultivation by the Jomon at Itazuke on the island of Kyushu. This evidence has been firmly dated to around 3200 years ago and thus is older than the Yayoi period by several hundreds of years. Matsuo Tsukada of the Quaternary Ecology Laboratory of the University of Washington summed up the findings this way:

> The oldest evidence of rice pollen [in Japan] . . . comes from the well-known Itazuke site, Fukuoka, which dates to about 3200 BP. Since the plant is not a Japanese native, its presence provides definite evidence that rice cultivation began in Late or Latest Jomon in Kyushu. Phylolith studies also support the fact that rice cultivation began at this time. It has been clear for some time that the notion that its cultivation appeared in Japan at the beginning of the Yayoi is outdated. Yet this idea persists in the writings of many specialists in East Asian archaeology![12]

But it was Sahara Makoto, the Director-General of the National Museum of Japanese History, who dropped the biggest bombshell on my preconceptions about the Jomon. When I met him on 17 May 2000 he told me quite casually of new evidence that had just come his way, unconfirmed as yet but startling if true, which suggested that the Jomon could have been cultivating rice as early as 12,000 years ago.

Revolution

So first rice was thought to be a Yayoi introduction to Japan. Then it was discovered that the Jomon grew rice hundreds of years before the arrival of the Yayoi. Now suddenly here was the dizzying possibility that the Jomon could have been growing rice deep in the Old Stone Age, thousands of years before anybody else . . .

'If that's true it's a revolution, isn't it?' I stuttered.

'Yes, in a sense,' replied Makoto, 'but then you see with the Jomon you always have to be ready for a revolution.'

There was other evidence, Makoto now told me, tiny particles of rice that had somehow got into the potters' clay before firing. Known to Jomon scholars for a decade, this evidence concerned several different pieces of pottery and several different sites, all of them in the range from 5000 to 3000 years old. Some archaeologists had gone to great lengths to underplay the significance of these finds, even arguing that the rice fragments had been brought over from China on the wind, or on the feet of grasshoppers – any logical contortion would be worthwhile, it would seem, rather than question the central paradigm of the Jomon as 'simple hunter-gatherers'.

Yet the more I looked into these matters the more obvious it became that increasing numbers of Japanese archaeologists are abandoning the 'hunter-gatherer' paradigm and are moving towards a new view of the Jomon as a sophisticated and very ancient culture – perhaps even as a 'civilization'.

Everything is up for grabs

Because we keep on learning new things about the Jomon at a very rapid rate it is inevitable that our impression of them will constantly have to be revised. We have seen how cherished views about their primitive hunter-gatherer economy are being challenged by new evidence of rice-growing. When a find of pottery like the 16,500-year-old fragments at Odayamamaoto No. 1 Iseki is made, it can push back previously accepted dating schemes by thousands of years. Indeed, almost everything is subject to revision. The excavation of sophisticated, well-planned urban settlements like Sannai-Muryama and Ueno-hara (the latter going back almost 10,000 years) has forced revision of the old idea that Jomon society was nomadic. Likewise at Sakuramachi Iseki, near Oyabe City in western Honshu, archaeologists have recently excavated examples of 4000-year-old Jomon carpentry using complex joints, dovetails and corners of a type not previously thought to have been introduced into Japan before AD 700.

Another example of historians radically misdating and misattributing inventions, ideas and icons concerns the classic curved jewel of the Japanese nobility – the comma-shaped (or foetus-shaped?) *magatama*, often carved from jade. References to *magatama* in Japan's national epic, the *Nihon Shoki*, which was compiled at the end of the seventh century AD, and the frequent finds of *magatama* in archaeological sites of that period have led most Japanese to an unquestioned assumption that the *magatama* is an invention of the so-called 'Yayoi' and 'Kofun' periods, roughly from 300 BC to AD 800. Yet on my travels through Japan archaeologists showed me dozens of beautiful *magatama* from Jomon times, some of them more than 8000 years old.

This speaks of more than just the antiquity of Jomon craftsmanship. The real point is the way in which a very ancient Jomon religious symbol survived the arrival of the Yayoi in the first millennium BC and continued to be regarded as a revered object at the time when the earliest texts of Japan's unique Shinto religion were written down.

Rock temples of the sea?

In how many other ways did the prehistoric culture of the Jomon impose itself on the culture of the invaders? How much more of the Jomon story remains to be told?

There is one obvious line of inquiry to pursue. Archaeologists admit that areas of the Japanese islands that previously stood above water and that at one time were almost certainly inhabited by the Jomon were inundated at the end of the Ice Age. The flooding was less massive and rapid than elsewhere. But since the Jomon were and remained for more than 14,000 years predominantly a coastal people, it is entirely possible – probable even – that this remorseless rise in sea-levels could have concealed important parts of their story. Had they carved structures out of rock along the ancient sea-shores, for example, then these would have been the first to be covered by the waves.

So, alongside the theory that they are freak natural phenomena, and alongside the theory that they are the work of a lost civilization, I think there is also room to ask whether the underwater ruins of Japan might not be the work of a known civilization – the Jomon – in a hitherto unknown and perhaps extraordinary phase of their culture.

Big stones; sacred mountains

There is a curious reverence for big stones – *iwakura* – which persists in Japan to this day. Auspiciously shaped and positioned stones are thought of as junctures between heaven and earth, places where a god can descend from sky to ground. In 1998, following a kind invitation from the Governor of Gifu Prefecture, I was able to spend several days exploring *iwakura* in the beautiful mountainous district of Ena, located near the centre of the Japanese landmass on the island of Honshu.

I was guided by a delightful group of local enthusiasts from the township of Yamaoka, who had formed themselves into a society in order to study all of the megaliths in their region. There were a great many *iwakura* that they wanted me to see. But here, as so often in Japan, the problem that immediately presented itself was whether the megaliths in question were in any sense man-made (or even 'man-arranged') as opposed to being just striking and unusual natural formations. Many of the towering rock piles, weathered boulders and huge, strangely shaped arrangements of stone that I was shown were, I am certain, entirely natural. However, in Japan – where the wonderful and awe-inspiring in nature have been noted and worshipped for millennia – such a provenance does not contradict ancient beliefs that the stones are sacred shrines descended from the Age of the Gods. Indeed, traditions state that it was here, amongst the rocks and trees of Mount Ena, that the placenta of Amaterasu O-Mikami, the sun-goddess and ancestral mother of the Japanese Imperial Family, was enshrined.[13]

Much less frequent than the natural *iwakura* of the Ena district are several that are undoubtedly the work of humans. These include a bizarre chain of grey granite tetrahedrons up to a metre high that run in a straight line through forests and valleys between the bases of two neighbouring mountains, finally culminating at a conspicuously large rock which, authorities confirm, was 'worshipped as a deity until recent times'.[14]

Another man-made *iwakura*, revered by local villagers as 'the sacred rock deity, the object of worship' and recently classified by archaeologists at the Ena Municipality as an Important Cultural Property, was excavated by the late Ryuzo Torii, Professor of Archaeology at the Imperial University of Tokyo, who dated it to the Jomon period.[15] It consists of a parallel pair of upright granite megaliths 1.6 metres high that stand isolated in a forest on the slopes of Mount

Nabeyama in the southern part of the Ena Basin. The megaliths, which are massive in cross-section and roughly squared off, have a gap of a few centimetres between them that is aligned with spectacular effect on the summer solstice sunrise.[16] More curiously, a straight line joining the tops of the two megaliths and extended northwards culminates at the sacred mountain of Kasagi, where numerous Jomon artefacts have been excavated by archaeologists.[17] An archaic ceremony of unknown origin that was conducted there until recent times involved the procession of a huge model serpent with scales made of leaves of magnolia *hypoleuca* followed by villagers praying to the mountain itself for rain.[18]

There are many sacred mountains in Japan. They are known as *reizan* (which means simply 'the sacred mountain') and also as *shintaizan*, which means 'the mountain as object of worship'.[19] The evidence from the radar-mapping and excavations at Kuromata Yama raises the possibility that at least some of these mountains might have been 'landscaped' by the Jomon in similar ways. Whether entirely natural, or touched by man, however, much suggests that they were sacred first to the Jomon and then inherited by later cultures.

Take the case of Hakuzan ('white mountain') in western Honshu. A focus of active pilgrimage today, the roots of its sanctity seem to be extremely ancient. This at any rate is a legitimate interpretation of recent archaeological evidence from the Jomon site of Chichamori Iseki near the modern city of Kanazawa. As well as many beautiful pieces of Jomon pottery, *dogu* figures and *magatama*, excavations at Chichamori Iseki have revealed the remains of two spacious 'wood-henges', built by the Jomon, which are thought to be about 3600 years old. The uprights consist of the split trunks of twelve huge chestnut trees arranged in a circle. Each circle has a ceremonial entrance aligned exactly on Hakuzan.

And just as the Jomon seem the most likely source of the sacred mountain idea, so it seems increasingly obvious that the origin of the *iwakura* idea must be theirs too.

After visiting Kuromata Yama and the Oyu stone circles in 1998 I returned to the Aomori region in May 2000 on hearing the news that seven small stone circles had been uncovered by archaeologists at the great Jomon settlement of Sannai-Muryama. They had been measured and catalogued and then immediately buried again by the excavators. A few kilometres away another, much larger Jomon stone circle (more exactly, it should be described as an oval, since it is somewhat elliptical in shape) had also recently been excavated and had been left exposed. Its name is Komakino Iseki. Climbing on to the top of a plinth to get an overview of it, I could see that the outer circle, or oval, built up out of distinctive rounded river stones, had a diameter of about 150 metres and that it in turn surrounded a series of inner rings arranged concentrically, with groups of smaller ovals – touching at the edges like the links of a chain – sometimes scattered across the width of a ring.

Komakino Iseki, which will have an important part to play later in this story, is thought to be about 4500 years old.

From Aomori I travelled further north to the island of Hokkaido. There within half an hour's drive of the modern port of Otaru I visited three more stone circles. Two of them, Nishizaki-Yama and Jichin-Yama, crown hill-tops, the former with a mass of small, interconnected stone circles, the latter with a ring of mid-sized megaliths. The third, Oshoro, is the largest intact circle in Japan and includes on its south side twenty stones that are in the half-tonne range. Like Komakino Iseki, Oshoro is arranged in concentric circles. Excavations suggest that it is about 4000 years old.

A helping hand

We had a strange experience at Oshoro – which I visited, as always, with Santha. Two friends were also with us there – the historian Akira Suzuki and Shun Daichi, the Japanese translator of my books. Had they not witnessed what happened I would hesitate to report it.

From the perspective of a photographer the problem with Oshoro is that it is

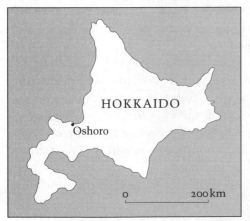

too big. As at Komakino Iseki it was therefore necessary to climb on top of something in order to get an overview. And, as at Komakino Iseki, the local government had conveniently provided a stone plinth, bearing a carved inscription, on top of which it was possible to climb. At Oshoro the plinth was a thick granite column about a metre and a half high mounted on top of a metre-high stone and concrete base.

Hanging on to the branches of a nearby tree, I scrambled on to the base, with my feet on either side of the column, then hauled myself up the column and perched unsteadily on top of it. I spent five minutes sitting up there with my video camera shooting panoramics of the stone circle, wanting to remember the scale and flow of the great outer ring, looking at the soft shadows the megaliths cast, at the manner in which the sunlight descended on them, trying to put my finger on the special way they felt they belonged among the tall cedars that grew around and amongst them. Wind blew, still a cold wind in May at this latitude in Japan, rustling through the trees, whistling down from the still snow-covered mountains of Hokkaido. And it was easy to visualize the spirit of the wind, the spirit of the trees, the spirit of the stone, the spirit of the sun – as I knew the ancient Japanese had – not just as poetic metaphors for natural powers but as real transdimensional entities capable of operating in both the spiritual and material realms.

The characters of these *kami* – let us call them by their Japanese name – are not always consistent or predictable. They are more than spirits. But, although the word *kami* is often translated into English as 'god', a *kami* is less than God in the Judaeo-Christian sense. *Kami* are supernaturally powerful, but not omnipotent. They can be killed. Sometimes they do good for mankind, sometimes they might harm us. They are everywhere, in everything. And it always pays to treat them with respect.

I lowered myself down from the granite column, placed my feet on the ground, and turned to put the video away in our rented car which was parked right behind us just outside the northern edge of the circle. Meanwhile, Santha, Suzuki and Shun had also been standing just outside the circle immediately behind the column. Now that I'd finished filming from it, Santha handed her Nikons over to Shun, stepped forward, climbed on to the platform, then wrapped her arms around the granite pillar and tried to shin up.

I'm quite heavily built and I think that when I climbed I must have loosened the cement that held the pillar to its base. Now as Santha began her climb the pillar rocked dangerously then broke off completely. For a moment she and the pillar seemed to hang in mid-air, locked in a deadly embrace; then the solid granite mass, weighing perhaps 100 kilos, bore her down to the ground and smashed her into it with an awful thud.

It all happened so quickly, that Shun, Suzuki and I were dumbfounded, stunned, confused. For a moment none of us could move, then we rushed to lift the pillar that lay diagonally across Santha's body pinning her from her pelvis, across her ribs and over her left shoulder, just missing her neck and her face. It took all our strength and a determined effort to move the big, sharp-cornered stone and as we hefted its weight I had a horrified premonition of the terrible internal damage that it must have done in such a fall.

Santha was gasping with shock, her eyes rolling upwards, exposing the whites. A couple of times she cried out, 'I'm dying, I'm dying.'

While an ambulance was called I gingerly felt her ribs, her collarbone, her hip, finding nothing broken, trying to reassure her. Gradually she quietened then informed me in an almost normal voice. 'Somebody caught me as I was falling. A hand came from behind me, over the top of my shoulder, and supported the stone. Another hand pressed into my back as I went down. It stopped me hitting the ground too hard.'

I presumed it must have been Shun or Suzuki since it certainly hadn't been

me – I hadn't even seen the whole thing, let alone been fast enough to lend a helping hand. But I paid no further attention to the matter then and didn't remember it until Santha brought it up again later that day after being discharged from the excellent private hospital in Otaru where her injuries were thoroughly scanned, x-rayed, examined and found to be minor. Bruised ribs and a twisted neck were about the worst of it – although strangely as I write these words eighteen months after the accident, Santha's ribs are still bruised, still tender and painful, though they long ago should have healed.

Amazingly there was no further damage and everyone, particularly the ambulance paramedics who had seen the size of the object that she had fallen under, regarded her escape as a miracle. Santha put it down more simply to the fast actions of Shun or Suzuki reaching out from behind her to take the weight of the pillar and cushion her fall.

But this was where the mystery began. Because as we talked the whole incident through with Shun and Suzuki the next day it emerged that neither of them had reached out a hand to catch Santha. Shun had been standing too far back and holding her cameras; Suzuki had been looking the other way when she fell. But Santha remained adamant that she had seen a man's hand coming in over her left shoulder to support the pillar and had felt a hand cushioning her back as well . . .

As we inquired into the matter further a curious story began to unfold. It seemed that we had arrived at Oshoro one day later than planned and that our original schedule had included a visit to a private house near the stone circle, where a small museum of objects from Oshoro was kept. This house belonged to the family of a farmer, now deceased, who had spent almost half a century as the self-appointed guardian and caretaker of the stone circle, which he was known to have loved and venerated. The objects in the museum had been his own collection.

When we did arrive a day late, the family was not there to receive us, so we went ahead with our visit to Oshoro without meeting them. Santha's accident occurred and she had a powerful personal experience of some sort of miraculous intervention. What we learned later was that the family had been away attending the memorial service for the farmer whose death, it transpired, had occurred eight years previously *on that very day*.

At Santha's request Suzuki telephoned the farmer's daughter from our hotel. She had already heard of Santha's accident at the stone circle and wanted us to know that she was angry with the spirit of her father for having failed to prevent it. Suzuki then told her of Santha's experience of being rescued and saved from serious harm by the strong and gentle hands of a man no one had seen, and translated Santha's honest question – did she think the rescuer could have been the spirit of her father?

Of course she thought that. We all did. For no matter how modern, rational and scientific Japan has become, it is still a land in which powerful and ineffable

spiritual forces are perceived to move in secret behind all things, to pervade all things, and to underlie the very fabric of reality.

Isn't it obvious that such ideas are extremely old?

The god in the mountain

Far away from Oshoro in Nara Prefecture on the island of Honshu, there is a sacred mountain called Miwa-Yama. In a pattern with which I was now becoming familiar, this entire pyramid-shaped mountain is considered by Japan's indigenous Shinto religion to be a shrine, possessed by the spirit of a god who 'stayed his soul' within it in ancient times.[20] His correct name is Omononushi-no-Kami (although he is also popularly known as Daikokusama) and according

to the ancient texts he is 'the guardian deity of human life' who taught mankind how to cure disease, manufacture medicines and grow crops.[21] His symbol, very strikingly, is a serpent – and to this day serpents are still venerated at Mount Miwa, where pilgrims bring them boiled eggs and cups of *sake*.[22]

In May 2000 the Shinto priests of Miwa processed me through the elaborate purification and blessing ceremony that is necessary for any pilgrim wishing to climb the mountain. Among other procedures this involved a ritual washing of my hands and mouth from a pure-water spring – over which reared the serpent icon of the god.

The climb itself, on a beautiful sunlit morning, took about two hours. From the beginning the way was steep and the path frequently led beside a tumbling stream.

Near the base of the mountain at the side of the path was a shrine consisting of a group of megaliths, each weighing a tonne or more and some showing signs of having been quarried or cut. On the right-hand side of this shrine, under a towering cedar tree, the devout had placed a dozen small statues of serpents.

My guide was a young Shinto priest. Seeing my interest in the rock shrine, he pointed out several other examples to me on the way up. In each case these shrines consisted of a single boulder or a group of boulders adorned with loops of thick rope. Some of the boulders seemed to have been arranged artificially; others appeared to be in entirely natural dispositions.

At the summit of the mountain we came upon a huge collection of *iwakura* forming a spacious, filled-in circle. It was hard to believe that these massive boulders had all just congregated here on this high point by chance. On the contrary, from what I had already seen of the Jomon obsession with stone circles

and with the landscaping of mountains, Miwa-Yama's summit shrine looked like something that would have been well within their repertoire. Indeed, in many ways it was typical of their open-air 'rock temples'. It felt strange, therefore, to see modern pilgrims assembled here wearing white smocks over jeans, and to realize as they chanted the name of Omononushi-no-Kami, the god whose spirit had possessed the mountain, that in many ways Japan is still a Jomon country.

'As to mountain worship,' writes Professor Hideo Kishimoto of Tokyo University,

> Its significance may change as the ages pass away, and its interpretation may vary according to the individuals. But people's feeling of admiration and reverence to the mountain will not be affected by time so long as it soars sublimely into the sky with infinite mystery breeding solemn atmosphere. In Mount Miwa, a Shinto faith, based on such feeling, shows living force.[23]

Cult of stone

Surrounding Mount Miwa is the district of Asuka – a treasure house of tombs and ruins. Here there are hundreds of the keyhole-shaped mounds, known as *kofun* – the name Kofun is also applied to the culture that built them, the immediate successors to the Yayoi. The mounds are thought to have served as the tombs of the earliest members of Japan's imperial family – roughly from the fourth to the eighth centuries AD – and of the nobility of that period. Even in our own enlightened twenty-first century, the emperor does not permit intact *kofun* to be excavated, and so archaeological understanding of these mysterious structures remains sketchy. All that can be said for certain is that their dating to the first millennium AD seems to be securely based on a wide range of evidence from a few *kofun* that had been opened for one reason or another during past centuries.

Under the pyramidial central earth mound it is now clear that all *kofun* conceal an inner megalithic burial chamber and a megalithic passageway, usually oriented south. One of the most spectacular of these 'barrow' structures, Ishibutai, thought to date to the seventh century AD, can be visited today because erosion long ago exposed and isolated its megalithic core. The two giant stones that form its ceiling weigh close to 100 tonnes each while the lesser stones of the side walls and the passageway are still enormous megaliths by any standards, weighing between 10 and 20 tonnes.

Near by are dozens of other megalithic ruins, all of which are thought to date to the same period around 1400 years ago. One, Kameishi Iwa, is a large rock carved into the form of a turtle. Another, Sakafune-ishi, is a granite slab into which has been cut, with remarkable precision – but as yet undetermined purpose – a circuit-like pattern of geometrical grooves and channels. A third consists of the upper and lower parts of a rock-hewn tomb (known locally as

Onino Sechin – literally 'demon's toilet'! – and Onino Manaita, literally 'demon's chopping board'). The two parts were separated in an ancient earthquake and now a modern road runs between them.

But by far the most strikingly enigmatic of the Asuka megaliths is the Masuda-no Iwafune, the 'boat stone' (so-called because it is thought to resemble a capsized boat), which juts out from a densely wooded hillside. Consisting of one mass of granite weighing in the region of 1000 tonnes, it is 10 metres long, 8 metres wide and almost 4 metres high. One puzzling characteristic is that in places it is rough and unfinished, seeming entirely like a work of nature, and in other places beautifully cut into right-angled planes.

Although there are theories, and a date in the seventh century AD is preferred by most scholars, no archaeologist is in a position to state with certainty how old the Masuda-no Iwafune really is or what its original function might have been. There are some indications of astronomical orientation but these are too vague to be of any use,[24] and, as the Asuka Historical Museum admits, the 'actual purpose' of the great megalith 'remains a mystery'.[25]

All that can be said for sure is that its presence testifies to the long-term persistence and vigour of a cult of stone in Japan – stone on a gigantic scale, either natural or artificially cut (or both at the same time), serving as an interface between earth and heaven. It is not difficult to imagine how such a cult that was in one place and time responsible for the Masuda-no Iwafune and in another for the Jomon stone circles could, in yet another, have made monuments like those later inundated by the sea at Yonaguni and Kerama.

Kerama: entrance to the underworld

Diving well is all about relaxation. It's like good sex. If your body and mind are relaxed you can go on for ever ... But how can you be relaxed when you're almost 30 metres under a deep blue sea in a place where a powerful current can suddenly whip up, like a gale-force wind, and have you fighting for your life in seconds? How can you be relaxed if you pause to think even for a moment about the vastness of the ocean and the improbable smallness of yourself, or about the fragility of your body, or about your life-or-death reliance on your equipment with its fallible valves and tubes?

I made my first dives at Kerama in April 1999 and found it a dark and scary place. In April 2000, a year later, I came back for more.

We worked from a rented cabin cruiser owned by Isamu Tsukahara, a local diver who has made a speciality of exploring the underwater stone circles. Also with us was another professional diver, Mitsutoshi Taniguchi, who discovered 'Centre Circle' more than twenty-five years ago and who has written a book on the subject. We were joined by Kiyoshi Nagaki from Chatan, a brilliant diver who had saved Santha's life the year before at Pohnpei in Micronesia when she accidentally began to descend into deep water with her air turned off. Another member of the team at Kerama was our old friend Shun Daichi, the translator

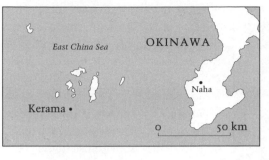

of *Fingerprints of the Gods*. In addition, Tsukahara had two of his staff divers with us underwater at all times – so we were a large group.

But it was one of those perfect days that every diver dreams of. Although the current was still running strong when we first arrived at the site, it had dropped to nothing an hour later when we rolled into the water. We then sank down, in absolute stillness, through a cool, blue column of ocean lit by sparks and splinters of sunlight.

I had experienced before a certain dizziness at Kerama and I experienced it again now as I descended in a wide spiral over the megaliths of Centre Circle. Dropping into the circle itself, I reached bottom at the base of the central monolith, where my gauges registered a depth of 27 metres (as against a depth at the top of the monolith of 23 metres).

While the other divers pursued their own interests I sat at the base of the monolith looking up at the ring of huge cut stones towering above me. Then I swam several times around the circle and followed side channels out of it, some of which lead to a second monument – the one that local divers call 'Small Centre Circle' – while others lead nowhere. The whole place felt like a maze, or a labyrinth in which it was extremely easy to become disoriented. Glad of the open sea above me I relaxed and allowed myself to ascend until I floated weightless about 3 metres above the top of the circle, looking down through blue water at the bizarre and out-of-place structure.

From this perspective, in this light, it seemed like the entrance to a fairy-tale kingdom, a spiral stairway into the underworld . . . I was filled with a sense of awe and wonder, and of numinous dread. I have experienced the same feeling at other religious monuments – the great Gothic cathedrals of Europe, the Pyramids of Egypt, Stonehenge, the Hypogeum and the megalithic temples of ancient Malta . . .

'If the Jomon built this,' I found myself wondering, 'then what else might they have built?' But did the Jomon build it? Was it even man-made at all?

Iseki Point

When we consider the Kerama and Yonaguni underwater monuments in the light of what is known about the veneration of sacred mountains and stone circles in prehistoric Japan, they make perfect sense and do not in any way appear outlandish or improbable. With the notable exception of Kuromata Yama, they are on a larger scale than any Jomon structure previously known on land; however, they are of the general type of shrine that we know the Jomon could and did make. The circles speak for themselves in this matter – since no one can deny that stone circles played an important role in Jomon culture – and

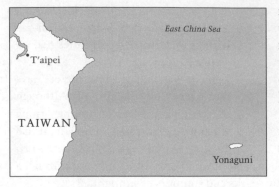

it takes little imagination to see the terraces of the Yonaguni main monument as the result of an extension of the *iwakura* and sacred mountain principles that were so well established in Jomon times.

Unlike Kerama, where the dive site is far out in the open sea, Yonaguni's main monument lies close to the present southern shore under a glowering mudstone cliff. The locals call it 'Iseki Point' ('Monument Point') and make much of its terracing; however, this is not the only aspect of the site that impresses me. Less obvious, but more persuasive, is the way that the whole layout seems to be organized cardinally and ceremonially.

Tucked in behind the north-west corner of the monument and oriented east–west, two huge, clean-cut megaliths, thought to weigh about 100 tonnes each, lie stacked side-by-side like slices of toast. It is obvious that they bear a striking resemblance to the parallel megaliths of Mount Nabeyama in Gifu Prefecture (see diagram page 620). I suggest that they are unlikely to have fallen into such a position by chance, that they are intended as a focal point, and that the gap between them, as at Gifu, may prove to have a solar alignment (in this case equinoctial rather than solstitial). They are approached through a narrow tunnel of big, symmetrical boulders piled on top of one another in two courses.

To the south and west are what appear to be the ruins of a walled complex with a curved ramp.

A clearly defined path or causeway runs from west to east along the monument's south face.

At the extreme western end of the causeway the diver comes to a classic *iwakura* shrine, part natural rock, part man-made.[26] If this shrine were to be moved to the slopes of Mount Miwa it would blend in seamlessly with what is already there.

Riding the Black Current

Japan is not a small country, but it was bigger 17,000 years ago at the end of the Ice Age, just before the global floods began. Once the meltdown was properly underway, however, the land-bridges to the mainland were rapidly inundated and the islands began a long process of shrinkage that still continues to this day.

As recently as 9000, possibly 8000, years ago the island of Shikoku was still part of a continuous landmass with neighbouring Honshu. Then the remorseless sea-level rise cut it off and ever since has continued to whittle away at its boundaries. Looking at the modern map, it is instructive to remember that the Jomon were there to witness the sea rush in to fill the lowlands between

**The Black Current between Japan and the
Americas. Based on Meggars et al. (1965).**

Takamatsu and Tamano – as indeed they witnessed all the strange phenomena and earth changes that marked the end of the Ice Age.

Perhaps it was this experience of rapid and invincible floods that led them to become navigators or perhaps they inherited their knowledge of the sea from the same unidentified 'influence' that brought them pottery and stone circles. Either way, it has for a long while been generally accepted amongst scholars that the Jomon traded extensively throughout the Japanese islands, and with the mainland, and must therefore have been using sea-going boats from a very early date.

More controversially, there is a growing body of evidence which suggests that the Jomon may not have confined themselves to exploring their own region. According to the findings of an international team of researchers led by C. Loring Brace of the University of Michigan's Museum of Anthropology, migrants entering North America across the Bering land-bridge at the end of the Ice Age were 'people closely resembling the prehistoric Jomon of Japan'.[27] Published in the 31 July 2001 edition of the *Proceedings of the National Academy of Sciences*, the findings provide:

> strong evidence supporting earlier work suggesting that ancient Americans . . . were descended from the Jomon, who walked from Japan to the Asian mainland and eventually to the Western hemisphere on land-bridges as the Earth began to warm up about 15,000 years ago at the end of the last Ice Age.[28]

But perhaps they didn't always walk. There is at any rate evidence from a later period, approximately 5000 years ago, that they may have undertaken transoceanic voyages, reaching as far as the shores of South America. The most famous, though still disputed and controversial, case is the discovery at Valdivia in Ecuador of what has been claimed to be Jomon pottery in deposits more than 5000 years old. But Jomon pottery has also turned up in almost equally ancient layers across the South Pacific – at Fiji, for example, and at Vanuatu. 'It's reasonable to conclude', says Professor Yoshihiko Shinoto of the Bishop Museum in Hawaii, 'that the Jomon travelled very widely in the Pacific area. Of course they could only have done so by sea.'

One route of migration that would have been open to them runs past Cape Ashizuri, the southernmost point of Shikoku, and then flows northwards from there up the eastern side of the Japanese archipelago, swings out across the Pacific with the Kuryl and Aleutian islands, comes close to land again along the northern California coast and runs south from there past the Pacific coast of Central America until it reaches Ecuador. This route, a kind of 'highway in the sea', is known in Japan as the 'Black Current' (Kuroshio) and is most visible where it passes Shikoku at Cape Ashizuri – running like a river at a steady 40 nautical miles per day. Given sufficient time and the survival of its crew, it is easy to see how a boat could ride the black current from Japan to South America.

Covering unimaginable distances

Words cannot express the degree of similarity between early Valdivia and contemporary Jomon pottery . . . Not only techniques of incision but motifs and combinations of motifs are the same. In most categories of decorative technique examples can be found so similar in appearance that they might almost have come from the same vessel.[29]

With these observations, Smithsonian Institution anthropologists Betty Meggers, Clifford Evans and Emilio Estrada sparked off a storm of controversy that is still blowing today. 'Early Valdivia' means at least 5000 years ago and, according to the orthodox model of history, Japanese hunter-gatherers, even strange ones who made pots, are not supposed to have had the capacity to sail across the Pacific 5000 years ago. Yet what Meggers, Evans and Estrada found in Valdivia – thousands of pieces of Jomon pottery in securely dated strata – seems to prove the orthodox model wrong. Once their results were properly codified it became clear that 'twenty-four of the twenty-eight major characteristics of the Valdivian pots were found in Jomon pottery. Their decorative elements and the construction of their spouts were among the most striking of the similarities.'

Meggers, Evans and Estrada published their revolutionary thesis in *Smithsonian Contributions to Anthropology* in 1965. Their ideas, which they themselves stick by, have been neither universally accepted by scholars nor conclusively rejected.

In Japan I found that Sahara Makoto was not a supporter of the 'Valdivia' connection, preferring to put down to coincidence all the numerous similarities between Jomon and Valdivia pottery. Conversely, Yasuhiro Okada, Chief Archaeologist at Sannai-Muryama, feels it is 'very likely' that the pottery of Valdivia was influenced by Jomon migrants 5000 years ago. 'More and more', he told me, 'I am coming to realize that we cannot understand the Jomon if we view them only in the context of Japan. They were Pacific voyagers. They used the sea.' Professor Mozai Torao of Tokyo University agrees:

It may be assumed that, before the dawn of history, ancient peoples were well-travelled, going far and wide on the earth by way of navigation or drift, that they in fact covered distances quite unimaginable for modern people.

Stone boat

I visited Cape Ashizuri on the invitation of a Japanese politician, Senator Sadao Hirano of Kochi Prefecture on the island of Shikoku. He had heard of my interest in the possible Jomon origins of Japan's veneration of big stones and wanted to draw my attention to the existence of large groups of *iwakura* scattered like

sentinels along the hilly margins of the Cape, all of them overlooking the Black Current. Showing us around were volunteer members of a local amateur historical association, the Ashizuri Jomon Kyoseki, who are making a long-term study of the megaliths and who are convinced that they were the work of the Jomon.

On several occasions over those two days as I was guided from rock shrine to rock shrine in the tree-covered hills I had the strange sense that I was diving again. This was because many of the Ashizuri megaliths are lost in the depths of forests where even at midday the bright light of the sun hardly penetrates. Standing on the floor of such a forest feels like standing on the bed of a deep green sea.

In one enchanted glade I came upon the carved figure of a turtle's head jutting out of a boulder. Elsewhere, a group of twenty megaliths, like smaller versions of the sarsens of Stonehenge, lay scattered around, overgrown by weeds and grass. In a clearing I found a stone circle made up of six large slabs. Near by, at the bottom of a narrow defile, a phallus-shaped menhir stood erect, surmounted by a second smaller boulder seemingly representing the glans. I walked on, climbed a forested hillside and arrived eventually at a grey stone block, 10 metres long, that had been carved into the shape of a boat with a high prow.

As I stood silently amongst the trees and the rock, looking up at the distant sun, I felt the prow of the stone boat beneath my fingers and was reminded again of the very many ways in which the Jomon are still alive today – alive through their pottery, alive through their sacred mountains, alive through rock shrines in deserted forests and in the depths of the sea, alive as great and powerful ancestral *kami*, alive as ideas embedded within the mysteries of the Shinto religion. And as I thought through everything I had learned about the Jomon I realized how far I had moved from the original preconceptions I had held about them. For here were a people who had explored their world by land and sea – reaching the Americas at least twice between 15,000 and 5000 years ago. Here were a people who had used pottery millennia before anyone else and gone on to refine it into a beautiful art form. Here were a people who engineered their landscape to create sacred mountains, circles of stone, temples of rock. Here were a people who lived in harmony with their environment, who made use of an intelligent mixture of strategies to ensure comfortable survival and security for the future, and who successfully avoided the pitfalls of militarism, materialism, conspicuous consumption and overpopulation that so many other cultures of the ancient world lost their way in. Here, above all, was a people whose civilization remained intact and flourished – decently, humanely, even generously, as far as we can know these things from the archaeological record, for more than 14,000 years.

If they could only speak to us, despite the lapse of time, what secrets would the Jomon have to tell of the true story and mystery of ancient Japan?

26 / Remembrance

> When Sosano went up to Heaven, by reason of the fierceness of his divine nature
> there was a commotion in the sea, and the hills and mountains groaned aloud.
>
> *Nihongi*

A haunting refrain, played softly, winds its way through the myth-memories of ancient Japan. It is the story of a journey to the realm of gloom that lies beyond death – to the Land of Yomi, the Underworld of the oldest Shinto texts. It is also the story of a sojourn on an enchanted island. And it is the story of a voyage underwater to the Kingdom of the Sea King.

The plots and characters differ. However, the story always involves a love affair; the female partner always remains in the mystical kingdom; and the male partner always returns to the sublunary world. Such shared details do not feel accidental. But where do they come from? More specifically, is some sort of association implied between the enchanted island, the submerged 'towers' and 'palaces' of the underwater kingdom, and the Underworld of Yomi in which the soul must tarry after death?

In my search for the Jomon, described in chapter 25, I travelled very widely around Japan visiting a series of important Jomon sites all the way from Kyushu in the south to Hokkaido in the north and listening to the wisdom of the leading field archaeologists. These experiences equipped me with some sense of the Jomon way of life, of their relationships with each other and with nature, of their unique ceramic art, of their spiritual system centred upon the veneration of stone and mountains, and of their belief – attested in burial practices at Sannai-Muryama and elsewhere – that the soul survives death.

But still I seemed to be only scratching the surface: the Jomon did not make use of a written language, and a thousand years of Yayoi and Kofun culture separated the end of the Jomon period from Japan's earliest surviving written collections of scriptures, myths and traditions. It seemed impossible, therefore, for the Jomon to 'speak for themselves' – and I often felt as though I was dealing with a civilization that was completely mute.

Or was I missing something?

The unrecognized legacy of 14,000 years

Japan, of course, has texts, scriptures, myths and traditions in abundance, but scholars have consistently treated them as irrelevant to the problem of Jomon – and the Jomon as irrelevant to the texts.[1] And while there is no archaeological evidence that a complete 'cultural replacement' took place in Japan at the transition from Jomon to Yayoi (quite the contrary, it was a long process of assimilation and syncretization), most scholars and members of the public nevertheless continue to behave as though a complete cultural replacement *did* in fact occur – and thus feel justified in ignoring or underplaying the possibility that 14,000 years of continuous Jomon culture must surely have left some mark, and perhaps a very deep one, on almost everything and anything that is truly Japanese.

As I began to suspect the extent to which Japan might still be a Jomon country I therefore also began to look from a new perspective at the handful of Shinto texts, centred around the famous *Kojiki* and *Nihon Shoki*, which together provide almost the only surviving repositories of authentic Japanese myths, legends and traditions. If so much else goes back to the Jomon, including some of the root concepts of Shintoism itself, as we saw in chapter 25, then it is absurd that the stories and ideas preserved in the ancient texts should continue to be treated as though they are exclusively the innovations of the Yayoi or later cultures – and thus immune from Jomon influence. This habitual posture of scholars has the effect of compressing Japan's entire classical 'myth bank' – and the search for its origins – into that period of just over 1000 years that separates the earliest archaeological evidence of the Yayoi in Japan, at around 400 BC, from the first written codification of the myths in the *Kojiki*, at around AD 712. Within such boundaries, scholars happily discuss influences that have come from as far afield as China, the South Pacific and India. But the possibility that some of the classical myths might have Jomon origins has never been seriously considered.[2]

Are we to suppose then that this extremely old and gifted culture accumulated *no* mythology of its own during the vast span that it held complete possession of the land of Japan? That doesn't seem reasonable. Yet how else are we to explain the alleged silence of the Jomon in the historical and mythical testimony of that land?

One possibility might be that the gods, myths and spiritual ideas of the latecoming Yayoi were so powerful that they not only displaced Jomon mythology but also annihilated it so completely that not a word of it would ever be remembered again.

Alternatively – in this as in Japan's age-old veneration of divine mountains and sacred rocks – the myth-memories preserved in the ancient texts might conceal a profound Jomon legacy.

The surviving records and their limitations

First and foremost, if we discount rumours of two texts said to have been compiled in more ancient times but unfortunately lost,[3] it is essential to register that nothing of a mythical nature seems to have been written down – nothing *at all* – until the early years of the eighth century AD.[4]

Before that, as was the case in India, the old stories, religious teachings and histories were preserved and constantly repromulgated within what seems to have been entirely an oral tradition. Although a professional corporation of 'reciters' (*Katari-be*) did exist in Japan,[5] giving cause for hope that much might have been saved, it is not clear how reliable or systematic the oral tradition was, to what extent it was subject to change and corruption, or at what pace such processes may have occurred. However, by the year 682, the fortieth emperor Temmu-Tenno, who reigned from 673 to 686, was sufficiently concerned to order the collection from all reliable and accepted sources of 'true traditions and genealogies'.[6] Before Temmu-Tenno died in 686 the compilation had been committed to memory by a professional reciter who, it was said, could 'repeat with his mouth whatever his eyes saw, and remember in his heart whatever struck his ears'.[7] The project was then shelved for twenty-five years. Then at last the Empress Gemmei ordered that such of the ancient lore as the reciter still remembered should be written down.[8]

The end result, completed around the year 712, is the *Kojiki* (Records of Ancient Matters), the fundamental scripture of the Shinto religion.[9] Although it expounds at length on the 'Age of the Gods' before history began, and on legendary emperors whom archaeologists do not recognize, it is also a historical document that tells the story of historical Emperors and of the Japanese people down to 628.[10]

Second in importance to the *Kojiki* is the *Nihon Shoki* (also known as the *Nihongi*), which issued forth from the court in 720.[11] Conceived of as a history and royal chronicle, it presents the annals of Japan from the earliest times down to 697.[12] In practice its subject matter is often identical or very close to the subject matter of the *Kojiki*, however:

> the older material is amplified and reclassified, and the whole recital is perceptibly tinctured with Chinese philosophy. Some few legends are omitted and others added, while variants are given of the main episodes.[13]

Other texts that carry down small or large fragments of the myths that were circulating orally in Japan in the eighth century are the *Manyoshu* (the first great anthology of Japanese poetry, which includes mythological tales) and the *Fudoki* (Records of Wind and Earth). Though only five *Fudoki* have come down to us intact, these texts were once part of a huge archive of books compiled by regional authorities to record local traditions after a government edict in 713.[14]

Early in the ninth century the *Kogo-shui*, or 'Collection of Omitted Sayings', was compiled by Imibe-no Hironari. As well as giving eleven myth stories not included in either the *Kojiki* or *Nihongi*, it continues the history of Japan down to 807.[15]

Finally, though their contribution is not so large, the *Shojiroku* (ninth century) and the *Engi-sheki* (tenth century) are the other principal sources of authentic Japanese myths.[16]

We need to understand the limitations of these sources:

- They are not and cannot be comprehensive. What they do is 'flash-freeze' a particular selection of Japanese mythology – no doubt driven and shaped by the subjective concerns of the individual compilers – at a particular moment in history.
- There is no way of knowing how representative they are of the whole body of Japanese myths just prior to the era of codification. Most authorities agree that a great deal must be missing.
- Likewise when even rice – so long assumed to have been exclusively a Yayoi innovation in Japan – turns out to have been grown by the Jomon in pre-Yayoi times (see chapter 25), then one has to wonder on what basis it is possible for scholars to conclude anything worthwhile at all about the epoch or epochs in which the original myths originated. The problem is worsened by the ways in which – like a badly damaged archaeological site – the strata of the traditional stories were ploughed and jumbled by corrupted retellings while they were still within the oral tradition, with further confusion and even political agendas introduced at the stage of compilation.[17]

Myths and memories

Mythology has been described by Robert Graves as 'the study of whatever religious or heroic legends are so foreign to a student's experience that he cannot believe them to be true. Hence the English adjective "mythical", meaning "incredible".'[18]

This strikes me as quite an accurate general description of what most scholars who study myth think they are doing and also of their fundamental attitude towards their subject matter – i.e. that myths are 'incredible' fictions composed in the ancient world either 'to answer the sort of awkward questions that children ask' or 'to justify an existing social system and account for traditional rites and customs'.[19] In consequence, most published analyses of myth all the way back to Sir James Frazer tend to focus on its social, economic and psychological functions. There have been a very few notable exceptions,[20] but as a rule those foolish enough to suggest that myths might in any way provide us with factual historical data have been ridiculed, abused and in some cases effectively excommunicated by their peers.[21]

As a non-scientist with no peers to excommunicate me, and as an author who earns his own keep, I'm free to pursue any line of inquiry that I'm enlightened by and to find my own position on any matter. I have therefore often taken myths seriously – with good reason I believe.

In particular I have tried to show that the universal myth of the deluge simply cannot be accounted for intelligently by the usual fatuous dismissals of professional mythologists, and that its manifestations again and again show remarkable correlations with what is known of the global meltdown at the end of the Ice Age. I can't 'prove' my view that the flood myths are garbled memories of those events any more than the experts can 'prove' theirs that the flood myths are a universal archetype of the foetus floating in the womb – or whatever.[22] Theirs is just a theory. Mine is just a theory. But time will tell which is right.

Meanwhile, contrary to the orthodox line on these matters, I continue to look upon the myths of the world as an archive of treasures, among which the most precious of all may be a kind of 'history of prehistory'. It is not so in the case of all myths, nor is it even necessarily so in the case of all flood myths. But my own experiences and research over many years – the research of a curious layman, not of a 'scientific expert' – have convinced me that the worldwide testimony of cataclysm, flood, and geological and climatic change preserved within the human heritage of myth is a precious thing indeed and may be the only memory and record of any kind that our species has managed to preserve of the great and terrible events that overtook our ancestors at the end of the Ice Age.

The many faces of cataclysm

On a global scale these events were undoubtedly dominated by flooding – horrific floods from the land to the sea as the great ice-sheets melted and the boundaries of glacial lakes gave way, and equally dreadful reverse floods, from the sea to the land, as the oceans inexorably swelled. But we saw in chapter 3 that flooding was only part of the story. During the same 10,000-year epoch in which the ice melted and global sea-level rose by 120 metres – roughly from 17,000 down to 7000 years ago – our planet also experienced dramatically increased volcanism, dramatically increased frequency and magnitude of earthquakes, and a dramatically unstable climate that seesawed rapidly and unpredictably between extremes.

Japan has no flood myth.

Unlike so much of the rest of the blighted northern hemisphere Japan was never covered by an ice-cap – and even on the most northerly island of Hokkaido at the last glacial maximum only the mountain ranges were glaciated.[23] This means that no part of Japan and none of the ancient inhabitants of Japan ever found themselves in the way of the sort of terrifying meltwater floods, 50 or 100 metres high, that rolled out periodically from the collapsing European and North American ice-sheets between 17,000 and 7000 years ago – and scoured the lands across which they flowed. Moreover, although Japan's surface-area was significantly reduced by rising sea-levels – with the most notable effect being the birth of the three islands of Honshu, Shikoku and Kyushu out of a single, much larger antediluvian island – a glance at the inundation maps reproduced in chapter 28 reveals that Japan was, in general, much less severely affected by post-glacial flooding than most other parts of the world. This was so, in the main, because its antediluvian coastlines were naturally precipitous, with few low-lying plains of the kind that were rapidly inundated (even by relatively minor sea-level rises) elsewhere in the region – for example around south-east Asia, where the Sunda Shelf was subjected to repeated catastrophic flooding, and in the basin between the Korean peninsula and the present coast of China that is now filled by the Yellow Sea.

Indeed, all in all it seems that we must regard Japan as having been a blessed land – as its mythology claims – throughout the rigours of the end of the Ice Age. For not only was it sheltered by its own topography from the worst effects of the post-glacial floods, but also it was screened from the most violent extremes of continental climate, thus enabling it to develop the lush and plentiful natural environment in which the Jomon could continue to pursue, across fourteen millennia, their near-idyllic lifestyle as affluent hunter-gatherers, fishermen, horticulturalists and, latterly, farmers.

I therefore do not find it surprising that Japan has no indigenous flood myth. On the contrary, it is exactly what I would expect of Japanese mythology if it is rooted and grounded in the myth-memories of the Jomon (no matter how disguised these may be beneath later influences). For what Japan actually lost during the post-glacial floods of the Jomon era were its 'beachfront properties' – including, I will endeavour to prove, several great coastal temples and sacred sites that now lie as much as 30 metres underwater. But it never lost its heart and soul to the rising seas nor was it ever smashed down to total destruction in the way that other areas of the world were.

Against such a background a strong flood myth would be anomalous.

Yet Japan, though 'blessed' in so many ways, did not entirely escape the upheavals of the meltdown epoch. We know, for example, that even here – though much reduced in frequency – the wild post-glacial climate flips did have their effects. Likewise, as was the case elsewhere around the world, we know that the 10,000 years after the end of the Last Glacial Maximum were accompanied in Japan by greatly increased volcanism.

I have seen evidence of the latter first-hand in early Jomon settlements such as Uenohara on Kyushu, where the ancient habitation layers are interspersed with thick carpets of volcanic ash. Moreover, I think most archaeologists specializing in the Jomon period would agree that in general their tasks of sequencing and stratigraphy are greatly facilitated by the presence of such volcanic layers in a great many Jomon sites around Japan.

So if the *Kojiki*, the *Nihongi* and the other ancient Japanese texts do preserve important Jomon memories alongside the many later ingredients which we know they also include, then it would be reasonable to expect that some of those memories might touch on the experience of volcanic and seismic cataclysms.

All interpretation of myth is speculative – mine and everybody else's. But listen to the story of the 'ravages' or 'havoc' of Sosano-wo-no-Mikoto, the great Kami-deity called Brave-Swift-Impetuous-Male.

The havoc of Sosano

The story is set in the Age of the Gods – more than 10,000 years ago as the chronicles inform us.[24] Whether by coincidence or because it is a memory of those times, this places it in the midst of the epoch of post-glacial tumult.

We are to picture the storm god, Sosano. At this stage he is a young man 'of a fierce temper and a wicked disposition' with 'an eight-hands-length beard'.[25] He has been appointed to rule over the 'Plain of Ocean'[26] by his father Izanagi but, adult though he is, he remains disconsolate at the death of his mother Izanami many years before. Sosano will not accept her loss but howls and rages, howls and rages, seeking to join her in the Land of Yomi:[27] 'The fashion of his weeping was such as by his weeping to wither the green mountains into withered mountains and by his weeping to dry up all the rivers and seas.'[28]

To restore cosmic harmony Izanagi intervenes, ordering Sosano to remove himself from the earth. Sosano replies that he will go down to the Land of Yomi to join his mother but that first he wants to ascend to the High Plain of Heaven to bid farewell to his sister Amaterasu, the Sun Goddess, the 'Great-Sky-Shiner':[29]

> When Sosano went up to Heaven, by reason of the fierceness of his divine nature there was a commotion in the sea, and the hills and mountains groaned aloud. Amaterasu, knowing the violence and wickedness of this Deity, was startled and changed countenance, when she heard the manner of his coming.[30]

These are the words of the *Nihongi*. In the same vein the *Kojiki* tells us that during Sosano's ascent 'all the mountains and rivers shook, and every land and country quaked'.[31] Likewise, both versions note alarming effects on the 'countenance' of the sun – Amaterasu's hair stands out like a corona, or 'in knots', she winds ropes of curved magatama jewels around herself, she stamps her feet and sinks up to her thighs in the hard earth, which she kicks away 'like rotten snow', and she utters a mighty cry of defiance.[32]

Sosano is offended:

> From the beginning my heart has not been black. But as in obedience to the stern behest of our parents, I am about to proceed for ever to the Land of Yomi, how could I bear to depart without having seen face to face thee my elder sister? It is for this reason that I have traversed on foot the clouds and mists and have come hither from

afar. I am surprised that my elder sister should, on the contrary, put on so stern a countenance.[33]

Amaterasu is mollified and a temporary calm descends upon the world. The two deities cooperate in the magical reproduction of further deities. But behind the scenes all is not well and Sosano's troublesome nature is beginning to manifest again. The end result is a cataclysm so great that the sun disappears entirely from view. Here's how the *Nihongi* tells the story:

> Sosano's behaviour was exceedingly rude . . . When he saw that Amaterasu was about to celebrate the feast of first fruits, he secretly voided excrement in the palace. Moreover, when he saw that Amaterasu was in her sacred weaving hall, engaged in weaving the garments of the Kami, he flayed a piebald colt of Heaven, and breaking a hole in the roof tiles of the hall, flung it in. Then Amaterasu started with alarm, and wounded herself with the shuttle. Indignant at this, she straightaway entered the Rock-cave of Heaven, and having fastened the Rock-door, dwelt there in seclusion. Therefore constant darkness prevailed on all sides, and the alternation of night and day was unknown.[34]

Most attempts by professional mythologists to explain this strange story are founded on the alleged 'primitive' fear that ancient peoples supposedly felt around the time of the winter solstice, during the shortest days of midwinter, that the sun would never return to its full power.[35] Somehow Amaterasu's disappearance into the rock-cave is to be taken as a symbol of this seasonal anxiety (which apparently our ancestors were too stupid to overcome), while her eventual re-emergence is naturally thought to symbolize the renewal of growth as the sun moves towards the spring equinox.

This is neat and tidy but in my view nonsense. People born in seasonal climes don't need myths to tell them that winter will end! They know that already from their own life experience, from the experience of their siblings, from the experience of their parents. It's obvious that fear is not the appropriate reaction to such a routine and predictable phenomenon. But fear *is* appropriate where terrible, infrequent and unpredictable disasters are concerned – disasters that shake the earth, roil the sea and blot out the sun from the sky in the violent and terrifying manner that the myths recount. It's this sort of reasonable fear, connected to the geological and climatic violence that Sosano represents, that I think is reflected in the story of his 'ravages' and of the darkening of the sun.

As usual the language of the *Kojiki* is slightly different from that of the *Nihongi* and adds texture to the same tale. After Amaterasu has retired within her cave and made fast its rock-door, we read:

> The whole Plain of High Heaven was obscured and all the Central Land of Reed-Plains darkened. Owing to this, eternal night prevailed. Hereupon the voices of the evil[36]

Kami were like unto the flies in the fifth moon as they swarmed, and a myriad portents of woe all arose.[37]

This sounds more like the end of the world to me than the winter solstice! Or if not in fact the end of the world, then something that obviously felt very much like it to those living at the time. Surely what the texts are asking us to envisage here is not less than a sustained period of cataclysm during which the whole land of Japan was plunged into 'constant darkness'. And if so, then is it coincidence, or because the texts contain a true report, that cataclysms of this magnitude did occur in Japan during the meltdown of the Ice Age when earthquakes and volcanic activity were at their peak? Even the relatively puny volcanic eruptions of the modern era have been known to darken skies across whole regions and provoke intimations of the end of the world.[38] How much more likely it is that the multiple large-scale eruptions that Japan experienced in the Jomon era could from time to time have combined their effects to produce a total blackout of the skies and real fears of the onset of 'eternal night'.

Even the longest volcanic winter does end, however. So as we would expect with such a scenario Amaterasu eventually does emerge from her rock-cave. She is tempted forth by some wonderful commotion and trickery of her fellow Kami, which need not detain us here, and once again: 'The radiance of the Sun Goddess filled the universe.'[39]

But the story is not yet over. What is to be done with the rebellious Sosano, who caused all this trouble in the first place? The assembled Kami would have their revenge on him. He is fined heavily. His toenails and fingernails are pulled out. His beard is cut:

> After this the Kami upbraided Sosano, saying 'Thy conduct has been in the highest degree improper. Thou must, therefore, not dwell in Heaven. Nor must thou dwell in the Central Reed-Plain Land. Thou must go speedily to the Bottom Nether Land [the Land of Yomi].' So together they drove him away downwards.
> Now this was a time of continuous rains . . .[40]

Remember that all interpretation of myth is speculative, but in summary, if we storyboard the ravages of Sosano, I suggest we get something like the following:

1. *A period of extremely arid climate during which the 'green mountains' become 'withered mountains' and the rivers and seas dry up.* Comment: a good, shorthand description of conditions at the Last Glacial Maximum, when global sea-level was at its lowest and north-east Asia, along with many other parts of the world, experienced thousands of years of extreme aridity.[41]

2. *A commotion in the sea; mountains and rivers shake and groan.* Com-

ment: the meltdown has begun in earnest; as the earth's crust readjusts under the changing stresses Japan experiences earthquakes of phenomenal intensity and its network of colossal volcanoes grows restless.

3. *A change in the countenance of the sun: the episode of the piebald colt.* Comment: atmospheric effects from increased volcanism.

4. *The disappearance of the sun into the 'Rock-cave of Heaven'.* Comment: skies darkened and sun obscured by massive volcanic eruptions and prolonged local volcanic fallout combined with global circulation of ash in the upper atmosphere.

5. *Return of the sun followed by a period of continuous rains.* Comment: the sky clears, the sun is seen again; as the meltdown of the far-off ice-sheets continues and more water is made available for atmospheric circulation, global precipitation increases and Japan experiences heavy rains after a long period of drought.[42]

So, yes, I am speculating. And, yes, I do realize that there might be dozens of other far more worthy explanations. Yet Japan did pass through such conditions at the end of the Ice Age.

And the Jomon were there to experience them.

The Land of Yomi

Sosano's long-running story does not quite end even with his expulsion from heaven. Contrary to the command of the assembled Kami, he has deeds to do on earth before he joins his mother Izanami in the Land of Yomi. Most of these are good deeds and feature the killing of an eight-headed serpent-monster that threatens a damsel in distress and the recovery from its tail of an Excalibur-like sword.[43] After having married the damsel and produced more children, 'Sosano-wo-no-Mikoto at length proceeded to the Land of Yomi.'[44]

This brings me back to the point at which I started this chapter – the mysterious journey to the Underworld, to the enchanted island, to the Kingdom of the Sea King, that recurs in the Japanese myths.

Sosano's case touches only tangentially on the issue. It is the story of his mother, the great procreator goddess Izanami (She-Who-Invites) and of his father Izanagi (He-Who-Invites) that will lead us along the correct path. Izanami and Izanagi are the archetypal divine couple, progenitors of gods and men, whom we first encounter in the ancient texts standing on the 'Floating Bridge of Heaven', gazing down into the swirling, oily, cloud-covered mass of the primeval universe in formation:

> Izanagi-no-Mikoto and Izanami-no-Mikoto stood on the Floating Bridge of Heaven, and held counsel together saying: 'Is there not a country beneath?'
>
> Thereupon they thrust down the jewel-spear of Heaven, and groping about therewith found the ocean. The brine which dripped from the point of the spear coagulated

and became the island which received the name Ono-goro-Jima ['Spontaneously-congealed-island'; identified with a small island near Ahaji].

The two Deities thereupon descended and dwelt in this island. Accordingly they wished to become husband and wife together, and to produce countries.

So they made Ono-goro-Jima the pillar of the centre of the land.[45]

It is impossible to pass such symbolism as 'the pillar of the centre of the land' without noting its obvious family resemblance to the notion of the *omphalos* or 'navel-of-the-earth', found as far afield as ancient Peru, Easter Island, India, ancient Egypt and Greece. I have discussed this problem in earlier chapters and in another work,[46] and will not repeat myself; still, the sense of an intrusion into the *Nihongi*'s text at this point of what many as well as myself believe to have been an international geodetic *technical* terminology is overwhelming.

As well as the Sun Goddess Amaterasu, and her troublesome brother Sosano, Izanagi and Izanami become the parents of many other children, several of whom are islands (perhaps even the islands of post-glacial Japan that were formed by rising sea-levels?), while others are Kami of every variety.

In a curious episode, the first-born of the divine couple is described as a leech-child (later identified with the god Yebisu),[47] whom 'they straightaway placed in a reed boat and set adrift'.[48] And just as Sosano's killing of the serpent to rescue a damsel in distress recalls the Greek myth of Perseus and Andromeda, so too this story of a child set adrift in a reed vessel has bizarre similarities to the stories of well-known civilizing heroes who were 'saved from water' in the same way – such as Moses in the Old Testament and Sargon the Great of Mesopotamia, who claimed in the third millennium BC:

> My mother was a priestess. I did not know my father. The priestess, my mother, conceived me and gave birth to me in hiding. She placed me in a basket made of reeds and closed the lid with pitch. She put the basket in the river ... The river carried me away.[49]

Returning to the myths of Japan, the last of Izanami's children is the fire-god Kagu-tsuchi (Fire-Shining-Swift-Male).[50] As he enters the world her uterus is burnt and soon afterwards she sickens, dies and her spirit travels to the Land of Yomi.[51]

Now, another scene from universal myth unfolds – here powerfully reminiscent of the Underworld quests of Orpheus for Eurydice and of Demeter for Persephone.[52] The ancient Japanese recension of this mysteriously global story is given in the *Kojiki* and the *Nihongi*, where we read that Izanagi, mourning for his dead wife, followed after her to the Land of Yomi in an attempt to bring her back to the world of the living:

> Izanagi-no-Mikoto went after Izanami-no-Mikoto and entered the Land of Yomi . . .
> So when from the palace she raised the door and came out to meet him, Izanagi
> spoke saying; 'My lovely younger sister! The lands that I and thou made are not yet
> finished making; so come back!'[53]

Izanami is honoured by Izanagi's attention, and minded to return. But there is
one problem. She has already eaten food prepared in the Land of Yomi and this
binds her to the place, just as the consumption of a single pomegranate seed
binds Persephone to hell in the Greek myth.[54]

Is it an accident that ancient Indian myth also contains the same idea? In the
Katha Upanishad a human, Nachiketas, succeeds in visiting the underworld
realm of Yama, the Hindu god of Death (and, yes, scholars have noted and
commented upon the weird resonance between the names and functions of
Yama and Yomi).[55] It is precisely to avoid detention in the realm of Yama that
Nachiketas is warned:

> Three nights within Yama's mansion stay
> But taste not, though a guest, his food.[56]

So there's a common idea here – in Japan, in Greece, in India – about not eating
food in the Underworld if you want to leave. Such similarities can result from
common invention of the same motif – in other words, coincidence. They can
result from the influence of one of the ancient cultures upon the other two, i.e.
cultural diffusion. Or they can result from an influence that has somehow
percolated down to all three, and perhaps to other cultures, stemming from an
as yet unidentified common source.

The parallel idea of not looking or not looking back after a successful quest in the
Underworld is strong in the myth of Orpheus and Eurydice. In their case Eurydice,
killed by a snakebite, is permitted to return to life after Orpheus has journeyed to
the land of the dead to find her. But there is a condition: neither he nor she should
look back as they depart the Underworld: 'The couple climbed up toward the opening into the land of the living, and Orpheus, seeing the sun again, turned back to
share his delight with Eurydice. In that moment, she disappeared.'[57]

The Japanese recension passed down to us from unknown antiquity in the
Kojiki and the *Nihon Shoki* is hauntingly different and yet hauntingly the same.
The reader will recall that Izanagi has reached the Land of Yomi and has just
addressed Isanami: 'My lovely younger sister! The lands that I and thou made
are not yet finished making; so come back!' And she has informed him that she
has eaten food cooked in the Underworld and thus cannot depart: 'My lord and
husband, why is thy coming so late? I have already eaten of the cooking furnace
of Yomi.'[58] Nevertheless, she says that she will return within and discuss the
matter with the resident Kami. Perhaps an exception can be made and she can
be freed. But she issues one warning: 'Look not at me!'[59]

She goes back into the palace to negotiate her freedom and remains there a long time without giving any sign. Izanagi, waiting outside, becomes impatient. He improvises a torch and follows her within. There, unfortunately, the first thing he sees is Izanami covered in putrefaction and seething with maggots:

> Izanagi-no-Mikoto was greatly shocked and said, 'Nay! I have come unawares to a hideous and polluted land.' So he speedily ran away back again. Then Izanami-no-Mikoto was angry, and said: 'Why didst thou not observe that which I charged thee [i.e not to look at her]? Now am I put to shame.'[60]

Like a vengeful harpy, and accompanied by 'the eight Ugly Females of Yomi', she sets off in pursuit, determined to punish Izanagi for dishonouring her. Just ahead of them he reaches 'the Even Pass of Yomi', the exit to the upper world, and blocks it behind him with 'a thousand-men-pull rock'.[61] This rock, we read, 'is called the great Kami, Land-of-Night-Gate-Block'.[62] On one side stands Izanami, permanently relegated to the Realm of Yomi. On the other stands Izanagi, He-Who-Invites, who still has tasks to complete and powerful Kami to create in the upper world.

Amongst the great Kami brought into being as he performs the necessary ablutions and purifications after his journey are his children Amaterasu and Sosano – whom we have met already and need say no more about here . . .

The enchanted island

I've suggested there is a theme running through Japanese myth of a love affair, a journey to a mysterious parallel realm, and a return to the world.

The first example, the story of Izanami and Izanagi, is set in the distant epoch that the *Kojiki* and the *Nihongi* call the Age of the Gods. But the second example that I will cite, superficially very different, is set in the Age of the Earthly Sovereigns. Here we read of a fisherman later revered as a deity named Urashima:

> He was handsome of feature . . . He went out alone in a boat to fish with hook and line. During three days and nights he caught nothing, but at length he caught a turtle of five colours. Wondering, he put it in the boat . . . While he slept the turtle suddenly became transformed into a woman, in form beautiful beyond description . . . He said to her, 'This place is far from the homes of people, of whom there are few on the sea. How did you so suddenly come here?' Smiling, she replied, 'I deemed you a man of parts alone on the sea, lacking anyone with whom to converse, so I came here by wind and cloud.'[63]

She is, of course, a Kami, as he quickly understands, from a magical land that 'lasts as long as sky and earth and ends with sun and moon'.[64] And she tempts him:

'You can come to that region by a turn of your oar. Obey me and shut your eyes.' So presently they came to a broad island in the wide sea, which was covered with jewels. [On it was a great mansion.] Its high gate and towers shone with a brilliance which his eyes had never beheld and his ears had never heard tell.[65]

They enter the mansion and are received and greeted in a loving fashion by her parents: 'Seated they conversed of the difference between mankind and the Land-of-Spirits, and the joy of man and Kami meeting.'[66] Eventually the fisherman Urashima and the beautiful sea Kami are married. Thereafter: 'For three years, far from his aged parents, he lived his life in the Spirit capital, when he began to yearn for his home and for them.' Observing the change in him, his wife asks: 'Do you desire to return home?'

He replies: 'To come to this far Spirit Land, I parted from my near of kin. My yearning I cannot help . . . I wish to return to my native place to see my parents for a while.' Then we read:

> Hand in hand, they walked conversing . . . till they came to where their ways diverged and where her parents and relatives, sorrowing to part with him, made their farewells. The princess informed him that she was indeed the turtle which he had taken in his boat, and she took a jewel-casket and gave it to him saying, 'If you do not forget me and desire to seek me, keep this casket carefully, but do not open it.' Thus he parted from her and entered his boat, shutting his eyes as she bade him.[67]

In a trice Urashima finds himself back in his home village again but a terrible surprise awaits him. During the three years that he has spent enchanted on the Spirit island 300 mortal years have passed and everything has changed beyond recognition. Stumbling around dazed and disconsolate, discovering from a passer-by that his own disappearance three centuries previously is itself now the subject of a village legend, he forgets the warning about the jewel box and opens it to remind himself of his Kami wife: 'But before he could look into it, something in the form of a blue orchid soared up to the blue sky with the wind and clouds. Then he knew that, having broken his oath, he could not go back and see her again.'[68]

It is already apparent from the narrative that lines are blurred between the enchanted island and the Spirit Land of Yomi. But the blurring goes even further in another variant of the myth where the Kami princess is revealed as no lesser figure than 'the daughter of the Dragon King of the Sea' and in which Urashima is taken not to an island but to an underwater kingdom.[69]

How do we explain such ambiguity? Perhaps it means nothing. But taken at face value what it seems to suggest is that the Mansions of the Sea King did not always lie beneath the waves.

The Kingdom of the Sea King

The same implication is there to be grasped in an earlier cycle of the myth also found in the *Kojiki* and the *Nihongi* and set in an era very near the end of the Age of the Gods – indeed just two generations before the birth of Jimmu-Tenno, part man, part Kami, the legendary first emperor of Japan.

As the story unfolds we are introduced to two brothers. The elder is Ho-no-susori no Mikoto (whose name is usually translated into English as 'Fire-Glow' or 'Fire-Shine') and the younger is Ho-ho-demi no Mikoto ('Fire-Fade' or 'Fire-Subside'). The *Nihongi* tells us, somewhat opaquely, that the former had 'a sea-gift', while the latter had by nature 'a mountain gift'.[70] But the *Kojiki* makes matters clearer:

> His Augustness Fire-Glow was a prince who got his luck on the sea, and caught things broad of fin and things narrow of fin. His Augustness Fire-Fade was a prince who got his luck on the mountains, and caught things rough of hair and things soft of hair.[71]

In other words Fire-Glow, like Urashima, was a fisherman and Fire-Fade was a hunter – occupations that are very far indeed from the 'fighting farmer' stereotype of Japan's later Yayoi and Kofun cultures but that do reflect and idealize the hunter-gatherer lifestyle, always strongly dependent upon fishing, of the earlier Jomon period.[72]

As the *Kojiki* tells it, Fire-Fade the hunter persuaded Fire-Glow the fisherman that they should 'mutually exchange and use each other's luck'.[73] In practice this meant that Fire-Fade was to take Fire-Glow's fish-hook and try his luck in the sea; Fire-Glow was to take Fire-Fade's bow and arrows and try his luck as a hunter in the mountains. Although Fire-Glow was not in favour of the scheme, 'at last with difficulty the mutual exchange was obtained':[74]

> Then His Augustness Fire-Fade, undertaking the sea-luck, angled for fish, but never got a single fish; and moreover he lost his fish-hook in the sea. Thereupon His Augustness Fire-Glow asked him for the fish-hook, saying, 'A mountain luck is a luck of its own, and a sea-luck is a luck of its own. Let each of us now restore to the other his luck. To which the younger brother His Augustness Fire-Fade replied, saying, 'As for thy fish-hook, I did not get a single fish by angling with it; and at last I lost it in the sea.'[75]

Fire-Glow had looked after and returned Fire-Fade's bow and arrows[76] and was insistent that his fish-hook should likewise be returned – although 'there was no means of finding it'.[77] Hoping to settle the matter, Fire-Fade made a new hook, which he offered to his elder brother. But Fire-Glow refused to accept it and again demanded the old hook.[78]

So the younger brother, breaking his ten-grasp sabre that was augustly girded on him, made of the fragments five hundred fish-hooks as compensation; but he would not take them. Again he made a thousand fish-hooks as compensation; but he would not receive them, saying: 'I still want the real original fish-hook.'[79]

The *Nihongi* takes up the story:

Fire-Fade's grief was exceedingly profound and he went and made moan by the shore of the sea. There he met Shihi-tsutsu no Oji ['Salt-sea elder']. The old man inquired of him, saying, 'Why dost thou grieve here?' He answered and told him the matter from first to last. The old man said, 'Grieve no more. I will arrange this matter for thee. So he made a basket without interstices, and placing Fire-Fade in it, sank it into the sea . . .[80]

I introduced the mystery of Fire-Fade's prehistoric diving adventure in chapter 1 because he soon comes to an underwater palace and because its description in the *Nihongi* reminds me so much of the towering underwater ruins I have seen off the island of Okinawa at Chatan and, 50 kilometres further to the west, at Kerama. Here is the passage that first caught my attention:

Forthwith he found himself at a pleasant strand, where he abandoned the basket, and, proceeding on his way, suddenly arrived at the Palace of the Sea God. This palace was provided with battlements and turrets, and had stately towers.[81]

Fire-Fade then loitered outside the gate until he was spotted by a beautiful princess, the daughter of the Sea God, who arranged with her father that this 'rare stranger' should be brought within. In the ensuing encounter the Sea God questioned Fire-Fade as to his purpose and the story of the lost fish-hook came out:

The Sea God accordingly assembled the fishes, both great and small, and required of them an answer. They all said, 'We know not. Only the Red-woman has had a sore mouth for some time past and has not come.' She was therefore peremptorily summoned to appear, and on her mouth being examined the lost hook was actually found.[82]

Mission accomplished? Perhaps. But now that he had experienced the delights of the Sea God's palace Fire-Fade did not want to leave. Instead he married the Sea God's daughter, Toyo-tama-hime, 'and dwelt in the sea-palace':[83]

For three years he enjoyed peace and pleasure, but still had a longing for his own country, and therefore sighed deeply from time to time. Toyo-tama-hime heard this and told her father, saying Fire-Fade often sighs as if in grief. It may be that it is the sorrow of longing for his country.[84]

Fire-Fade admitted that this was so and the Sea God granted him permission to return to the world above the waves, handing over to him Fire-Glow's fish-hook to take back and also gifting him with two magical jewels – 'the jewel of the flowing tide and the jewel of the ebbing tide' – with which he would be able to control the waters.[85] The plan was that he should use these jewels to punish and subdue his elder brother (presumably for being so unreasonable about the fish-hook in the first place):

> If thou dost dip the tide-flowing jewel, the tide will suddenly flow, and therewithal thou shalt drown thy elder brother. But in case thy elder brother should repent and beg forgiveness, if, on the contrary, thou dip the tide-ebbing jewel, the tide will spontaneously ebb, and therewithal thou shalt save him. If thou harass him in this way, thy elder brother will of his own accord render submission.[86]

Before Fire-Fade set off on his journey he was approached by his young sea wife Toyo-tama-hime, who informed him that she was pregnant and that she would follow him soon – for she wished to bear his child above water, in his homeland:

> Thy handmaiden is already pregnant, and the time of her delivery is not far off. On a day when the winds and waves are raging, I will surely come forth to the sea shore, and I pray thee that thou will make for me a parturition house, and await me there.[87]

After his return Fire-Fade, armed with the remarkable jewels that could raise and lower sea level at will, quickly subdued his elder brother, just as the Sea God had promised. Then the time came for Toyo-tama-hime to fulfil her promise and ascend from the underwater kingdom to give birth to their child on land. So she 'bravely confronted the wind and waves, and came to the sea shore' – where Fire-Fade awaited her.[88]

From the *Kojiki*:

> Unable to restrain the urgency of her womb she entered the parturition-hall. Then, when she was about to be delivered, she spoke to her husband, saying, 'Whenever a foreigner is about to be delivered, she takes the shape of her native land to be delivered. So I now will take my native shape to be delivered. Pray look not upon me!'[89]

The *Nihongi*, too, repeats the same warning: 'When thy handmaiden is in travail, I pray thee do not look upon her.'[90] But of course, just as Orpheus had to look back at the gates of hell and just as Izanagi had to look at Izanami in the Land of Yomi:

> Fire-Fade could not restrain himself, but went secretly and peeped in. Now Toyo-tama-hime was just in childbirth and had changed into a dragon. She was greatly

ashamed, and said, 'Hadst thou not disgraced me I would have made the sea and land communicate with each other, and forever prevented them from being sundered. But now that thou hast disgraced me, wherewithal shall friendly feelings be knit together?' So she wrapped the infant in rushes, and abandoned it on the sea shore. Then she barred the sea-path and passed away.[91]

The sequel to this story is that the infant abandoned on the sea-shore grows up to wed his maternal aunt, sent from the underwater kingdom to care for him, and among their offspring is Jimmu-Tenno, the first Emperor of Japan,[92] founder of the imperial line that survives to this day. In a sense, therefore, are we not to understand that the historical civilization of Japan, bound up with the line of the Emperor, is to be traced back through Jimmu-Tenno – by way both of his grandmother and his mother – not only to the lineage of Amaterasu and the great gods of the High-Plain of the Sky, but also to the lineage of the Sea God and to a kingdom of palaces and mansions that now lies beneath the sea?

R'yugu

The ambiguity in the story of Urashima's enchanted island – which has it sometimes above and sometimes below the waves – also occurs in the story of Fire-Glow and Fire-Fade. For whereas the *Nihongi* has Fire-Fade descend to the sea-bed in a waterproof basket, the *Kojiki* has him make the journey *above water* in 'a stout little boat without interstices'. He is told: 'Go on for some time. There will be a pleasant road; and if thou goest in the boat along that road, there will appear a palace built like fishes' scales.'[93] Likewise, later in the story when Fire-Fade is taking his leave he refers to the Sea God's kingdom quite explicitly as an 'island', and the translator Basil Hall Chamberlain feels obliged to explain: 'The Sea-God's dwelling is called an island because it is beyond the sea.'[94]

Otherwise, the versions are virtually identical but in these curious differences I wonder if we are seeing, once again, a before-and-after effect summarized in two different layers of myth – in the earlier of which the Kingdom of the Sea God is remembered as an island, in the later as an underwater sanctuary of walls and palaces and mansions? In crude and simplistic terms, could it be a memory that great structures with 'turrets and tall towers of exceeding beauty' once stood above water but are now beneath the waves?

That seemed like wild and unjustified speculation to me until I discovered exactly where Japanese legends say that the Kingdom of the Sea God is to be found . . .

It seems that its name is R'yugu, and that it lies hidden from the sight of man somewhere amongst the Lu-Chu islands.[95]

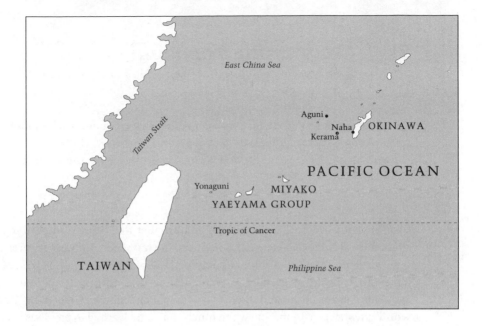

'X' marks the spot

Today the Lu-Chu islands (the old Chinese name) are part of Japan and are better known as the Ryukyu archipelago (from the Japanese pronunciation). The archipelago consists of three separate island groups – the northernmost around Okinawa, including the Keramas and Aguni; then Miyako in the centre; finally the Yaeyama group with Yonaguni in the extreme south-west.

I suggest it is not a matter to be ignored that (a) Japan has a tradition of spectacular underwater structures that may only be reached by diving; (b) there are some indications of a memory that these structures were once above water; (c) the tradition is clearly associated with a hunter-gathering and fishing culture that idealizes much of what we know about the Jomon lifestyle in Japan after the end of the Ice Age and down to about 2000 years ago; (d) the tradition places the underwater structures amongst the Ryukyu islands; (e) divers in recent years have indeed observed a series of spectacular underwater structures in the Ryukyu islands – extending all the way from Yonaguni to Okinawa.

It was time to go diving again.

27 / *Confronting Yonaguni*

> The question was, or is still, is it and, if yes, to what extent is it made by man or overworked by man? This is the question.
>
> Dr Wolf Wichmann, geologist, Yonaguni, March 2001

I was in Tokyo in 1996 when the photojournalist Ken Shindo showed me the first images I had ever seen of an awe-inspiring terraced structure, apparently a man-made monument of some kind, lying at depths of up to 30 metres off the Japanese island of Yonaguni at the remote south-west end of the Ryukyu archipelago. This was the moment, if there ever was just one moment, when the 'Underworld' quest began for me and when much that I had learned in previous years in many different countries began to swing sharply into focus and make sense. I felt an immediate compulsion to explore the beautiful and mysterious structure that beckoned so alluringly from the photographs. And I realized that it would rewrite prehistory if it could indeed be proved to be man-made.

I described in chapter 1 how Santha and I learned to dive, and the remarkable synchronicities and good fortune that brought us to Yonaguni in March 1997 to begin a systematic programme of underwater photography and research there that was to continue until mid-2001. I also described some of the other rock-hewn underwater structures that we dived at with our Japanese colleagues at other locations in the Ryukyus – notably at Kerama, Aguni and Chatan at the northern end of the archipelago.

The most complex and intractable problem shared by all of these otherwise very different structures is also the simplest and most obvious question that anyone might wish to ask about them: were they shaped and carved by human hands or could they have ended up looking the way they do as a result of natural weathering and the erosive weapons of the sea? Though they have an important role to play, geologists are not the only people qualified to decide the answer to such a question. Likewise, though they too are indispensable, archaeologists cannot be the final arbiters. On the contrary, if ever a multi-disciplinary approach was called for then it is here! For, as I've tried to show in the previous chapters, Japan confronts us with a prehistoric cultural and mythological context into which the rock-hewn structures fit snugly like the missing pieces of a jigsaw puzzle. This context includes a clear tradition of unknown antiquity – still manifest in the present day – in which huge rocks are carved and rearranged amidst sacred natural landscapes. Since this is precisely the puzzling and ambiguous aspect – part natural and part seemingly man-made – of the underwater structures scattered around the Ryukyu archipelago, it is foolish and irresponsible to ignore the possibility of a connection.

Yet it is equally foolish and irresponsible to ignore what geology and archaeology have to say on the matter. So it is time, I think, to provide a thorough reckoning.

The three geologists

Three qualified geologists – Masaaki Kimura, Robert Schoch and Wolf Wichmann – have dived at Yonaguni, acquired first-hand experience of the underwater structures, and commented publicly on what they saw. So far as I know, they are, at the time of writing, the *only* geologists ever to have dived there. Therefore, when we speak of 'geological opinion' concerning the Yonaguni anomalies, it is important to be clear that we are referring to the work and ideas of just three men who, moreover, do not agree with one another – so there is no consensus. Other geologists who have expressed views without diving at Yonaguni hardly qualify to participate in the debate.

Since there are grave issues at stake concerning our understanding of prehistory and the story of human civilization I propose to devote the necessary space in this chapter to an accurate summary of the views of the three main geological protagonists.

Dr Kimura

The doyen of the group, and in my view the hero of the Yonaguni saga for his determination, persistence and refreshingly open-minded intellectual approach, is Dr Masaaki Kimura, Professor of Marine Geology at the University of the Ryukyus in Okinawa. He and his students have completed hundreds of dives around the main 'terrace' monument at Yonaguni as part of a long-term project in which they have thoroughly measured and mapped it, produced a three-dimensional model, taken samples of ancient algae encrusted on its walls for carbon-dating, and sampled the stone of the structure itself. Professor Kimura's unequivocal conclusion, based on the scientific evidence, is that the monument is man-made and that it was hewn out of the bedrock when it still stood above sea-level – perhaps as much as 10,000 years ago. The principal arguments that he puts forward in favour of human intervention are on the record and include the following:

1. 'Traces of marks that show that human beings worked the stone. There are holes made by wedge-like tools called *kusabi* in many locations.'
2. 'Around the outside of the loop road [a stone-paved pathway connecting principal areas of the main monument] there is a row of neatly stacked rocks as a stone wall, each rock about twice the size of a person, in a straight line.'

3. 'There are traces carved along the roadway that humans conducted some form of repairs.'
4. 'The structure is continuous from under the water to land, and evidence of the use of fire is present.'
5. 'Stone tools are among the artefacts found underwater and on land.'
6. 'Stone tablets with carving that appears to be letters or symbols, such as what we know as the plus mark "+" and a "V" shape, were retrieved from under water.'
7. 'From the waters near by, stone tools have been retrieved. Two are for known purposes that we can recognize, the majority are not.'
8. 'At the bottom of the sea, a relief carving of an animal figure was discovered on a huge stone.'[1]
9. On the higher surfaces of the structure there are several areas which slope quite steeply down towards the south. Kimura points out that deep symmetrical trenches appear on the northern elevations of these areas which could not have been formed by any known natural process.
10. A series of steps rises at regular intervals up the south face of the monument from the pathway at its base, 27 metres underwater, towards its summit less than 6 metres below the waves. A similar stairway is found on the monument's northern face.
11. Blocks that must necessarily have been removed (whether by natural or by human agency) in order to form the monument's impressive terraces are not found lying in the places where they would have fallen if only gravity and natural forces were operating; instead they seem to have been artificially cleared away to one side and in some cases are absent from the site entirely.
12. The effects of this unnatural and selective clean-up operation are particularly evident on the rock-cut 'pathway' (Kimura calls it the 'loop road') that winds around the western and southern faces of the base of the monument. It passes directly beneath the main terraces yet is completely clear of the mass of rubble that would have had to be removed (whether by natural or by human agency) in order for the terraces to form at all.[2]

Dr Schoch

The second geologist to dive at Yonaguni, Professor Robert Schoch of Boston University, has vacillated tenaciously in his opinions – but I take this as a sign of an open-minded scholar ever willing to revise his views in the light of new evidence. Thus, when we first dived there together in September 1997, he was sure that the structure was man-made.[3] Within a few days, however, he had changed his mind completely:

I believe that the structure can be explained as the result of natural processes . . . The geology of the fine mudstones and sandstones of the Yonaguni area, combined with wave and current actions and the lower sea-levels of the area during earlier millennia, were responsible for the formation of the Yonaguni monument about 9000 to 10,000 years ago.[4]

A few days later, Schoch softened his position again:

> After meeting with Professor Kimura, I cannot totally discount the possibility that the Yonaguni monument was at least partially worked and modified by the hands of humans. Professor Kimura pointed out several key features that I did not see on my first brief trip . . . If I should have the opportunity to revisit the Yonaguni monument, these are key areas that I would wish to explore.[5]

Schoch did have an opportunity to revisit the structure in the summer of 1998, carrying out several more dives there. Then in 1999, in an interview given to the BBC science programme *Horizon* for a documentary attacking my work – and in the same year in his own book *Voices of the Rocks* – he expressed what sound like two very different, even contradictory opinions about the structure. Here is the relevant section from the BBC *Horizon* transcript:

> *Narrator:* Yonaguni looked as if it could be a spectacular discovery and Hancock needed corroboration. He invited the Boston University geologist Robert Schoch to inspect the site. Professor Schoch has taken a keen interest in unorthodox views of the past and he welcomed the chance to examine the underwater discovery. Schoch dived with Hancock several times at Yonaguni.
>
> *Prof. Robert Schoch* (Boston University): I went there in this case actually hoping that it was a totally man-made structure that was now submerged underwater, that dated maybe back to 6000 BC or more. When I got there and I got to dive on the structure I have to admit I was very, very disappointed because I was basically convinced after a few dives that this was primarily, possibly totally, a natural structure . . . Isolated portions of it look like they're man-made, but when you look at it in context you look at the shore features, etc., and you see how, in this case, fine sandstones split along horizontal bedding plains that gives you these regular features. I'm convinced it's a natural structure.[6]

Well, that seems straightforward. But then here is what Schoch says in *Voices of the Rocks*:

> Possibly the choice between natural and human-made isn't simply either/or. Yonaguni Island contains a number of old tombs whose exact age is uncertain, but that

are clearly very old. Curiously the architecture of the tombs is much like that of the monument. It is possible that humans were imitating the monument in designing the tombs, and it is equally possible that the monument was itself somehow modified by human hands. That is, the ancient inhabitants of the island may have partially reshaped or enhanced a natural structure to give it the form they wished, either as a structure on its own or as the foundation of a timber, mud or stone building that has since been destroyed. It is also possible that the monument served as a quarry from which blocks were cut, following the natural bedding, joint and fracture planes of the rock, then removed to construct buildings that are now long gone. Since it is located along the coast the Yonaguni monument may even have served as some kind of natural boat dock for an early seafaring people. As Dr Kimura showed me, ancient stone tools beautifully crafted from igneous rock have been found on Yonaguni. Significantly, Yonaguni has no naturally exposed igneous rocks, so the tools, or at least the raw materials from which they were made, must have been imported from neighbouring islands where such rock is found. The tools could have been used to modify or reshape the natural stone structures now found underwater off the coast of Yonaguni. The concept of a human-enhanced natural structure fits well with East Asian aesthetics, such as the *feng shui* of China and the Zen-inspired rock gardens of Japan. A complex interaction between natural and human-made forms that influenced human art and architecture 8000 years ago is highly possible.[7]

As further evidence for a very ancient human role in the construction of the Yonaguni monument, Schoch then sets out an argument of mine, advanced in my 1998 book *Heaven's Mirror*, that the structure is not only man-made but could also have served a specific astronomical function – since calculations show that around 10,000 years ago, when it was above water, it would have stood on the ancient Tropic of Cancer.[8] Writes Schoch:

> The ancients, I suspect, knew where the tropic was, and they knew that ... its position moved slowly. Since Yonaguni is close to the most northerly position the tropic reaches in its lengthy cycle, the island may have been the site of an astronomically aligned shrine.[9]

In summary, therefore, Schoch has not come down definitively either on one side of the fence or on the other but seems to be wavering in the direction of a compromise in which the structure is both natural and man-made at the same time. I cannot avoid adding that *all* rock-hewn structures, whether the weird terraced granite outcrop at Qenko near Sacsayhuaman in Peru,[10] or the wonders of Petra in Jordan, or the temples of Mahabalipuram in south India are, by definition partly natural – the base rock out of which they are hewn – and partly man-made. They can't help but be anything else.

Dr Wichmann

The third geologist, German science writer Dr Wolf Wichmann, has definite opinions and expresses them with certainty. In 1999 he informed *Der Spiegel* magazine – who had taken him to Yonaguni – that he regards the underwater monument as entirely natural. He made just three dives on the main terraces and then declared: 'I didn't find anything that was man-made.'[11] Japan's marine scientists 'haven't got a clue' what the terraced underwater structure at Yonaguni is, reports *Der Spiegel*:

'It is unlikely to be anything natural,' said the oceanographer Teruaki Ishii from Tokyo. Masaaki Kimura, a marine researcher at the Ryukyus University (Okinawa) talks about 'a masterpiece'. He thinks the structure is a sacred edifice built by a hitherto unknown culture possessing advanced technical abilities.

The debate going on in the Orient has awakened the curiosity of the West. People with second sight find themselves magically attracted by Iseki Point ('ruins'). At the beginning of 1998 the geologist Robert Schoch, who believes the Sphinx was built by the people of Atlantis [*sic* – completely untrue; Schoch does not believe any such thing], swam down to the site and declared it to be 'most interesting'. The guru of ancient antiquity and best-selling author Graham Hancock was also investigating the site. After an excursion in a submersible he records that at the base of the monument can be seen a 'clearly defined path'. [Actually I have *never* been in a submersible at Yonaguni and I do not consider my four years of hands-on diving there as any kind of excursion; there is, however, a clearly defined path at the base of the monument.]

The rock expert Wolf Wichmann could not corroborate these conclusions. In the company of a team from SPIEGEL TV he returned to explore the coastal area, under threat from tsunamis. In a total of three diving operations he gathered rock samples and measured the steps and 'walls'. He was unconvinced by his findings: 'I didn't find anything that was man-made.'

During the inspection it was revealed that the 'gigantic temple' is nothing but naturally produced bedded rock. The sandstone is traversed by vertical cracks and horizontal crevices. Perpendicularity and steps have gradually developed in the fracture zones. The plateaux at the top are referred to by Wichmann as typical 'eroded plains'. Such flat areas occur when bedded rock is located right in the path of the wash of the waves.

Suggestive pictures rich in detail and contrast may indeed reveal something else, but in general the mass of rock looks like a structure rising out of a sandy bed, with no sign of architectural design. The plateaux have gradient sections, and there is no perpendicular wall. Some of the steps just end nowhere; others are in a spiral, like steep hen-roosts.

The stony blocks show no signs of mechanical working. 'Had the "ashlars" been hewn by tools, they would have been studded with flutes and cuts and scratches,' said Wichmann. Three circular recesses on the topmost plateau, referred to by

Kimura as column foundations, are nothing but 'potholes'. These occur when water washes through narrow spaces.

Facts like these fail to stem the current epidemic of mystery-fever. The Yonaguni monument has for some time played a key role in the world picture of archaeological dreamers.[12]

The one archaeologist

One archaeologist has dived at Yonaguni and studied its underwater structures first-hand. Others in his profession who have commented have done so from their desks after browsing through photographs or looking at videotape of the structures. As is the case with the armchair geologists, their opinions can only be of limited value until they have dived there themselves. By contrast the opinion of the only experienced marine archaeologist in the world who has ever dived at Yonaguni must count for a great deal more.

That archaeologist – whose official report is reproduced in part below – is Sundaresh from the National Institute of Oceanography in Goa, India. The reader will recall that we dived with him and other NIO archaeologists at Dwarka in March 2000 and again at Poompuhur in February 2001. Between these expeditions in India, Sundaresh participated with us in an expedition to Yonaguni in September 2000 that had been supported once again (as had Robert Schoch's visit in September 1997) by Seamen's Club.

Also participating in the September 2000 expedition was Kimiya Homma, a businessman from Hokkaido, whose firm owns two very useful high-tech ROVs (remotely operated vehicles) for unmanned exploration in water too deep to be readily reached by divers. So that an effective search for further structures around Yonaguni could be mounted in the short time available, Homma had brought one of the ROVs with him and also an expert team of support staff and technical divers.

Because it is a unique document of reference, being – so far – the first and only evaluation of a wide range of Yonaguni's underwater structures by a marine archaeologist, I reproduce below several sections from Sundaresh's expedition report. Some of the specific submerged sites that we visited with Sundaresh during the expedition are not yet familiar to the reader from the brief account given in chapters 1 and 25 but will be described shortly:

THE STUDY OF SUBMERGED STRUCTURES OFF
YONAGUNI ISLAND OF JAPAN:
THE PRELIMINARY RESULTS FROM RECENT EXPEDITION
1–12 September 2000
BY SUNDARESH NATIONAL INSTITUTE OF OCEANOGRAPHY
DONA PAULA, GOA 403 004
DECEMBER 2000

1.0 Introduction

Yonaguni is the most south-western island of Japan and closest to Taiwan (about 69 nautical miles). This island is almond shaped with 10 km length (from east to west) and 4 km width (north to south). An international expedition was organized by the Seamen's Club, Ishigaki, Japan to further explore the underwater structures in the area. This report describes the archaeological significance of the structures found during the expedition.

2.0 Background information of the area

Underwater massive structures were found initially by Mr Aratake, a local resident of Yonaguni island during 1986–87. He named this point as Iseki ('Monument') Point. He was looking for hammerhead sharks schooling around the island when a massive man-made underwater structure was noticed at a depth of 30 m. This was his first discovery. Later more monuments were found by Aratake and other divers in nearby Tatigami and 'Palace' areas.

4.0 Methodology

4.1 Offshore explorations

Two boats were chartered for explorations off Yonaguni waters from 2 September 2000 to 8 September 2000. The Remotely Operated Vehicle (ROV) was deployed simultaneously with side-scan sonar and echosounder. The ROV was operated with generator power supply. The system was operated in waters between 40 to 80 metres depth around Yonaguni. The survey revealed a rock-cut channel about 1 m wide and more than 20 m long at 2 sea mounts. The ROV observations were confirmed by diving.

5.0 Results

5.1 Terraced structure and canal

A large terraced structure of about 250 metres long and 25 metres height was studied south of the Arakawabana headland. Known locally as Iseki Point, the terraced structure is bound to the northern side of an elongated, approximately east–west trending structure, designated by Professor Masaaki Kimura, University of the Ryukyus, as an approach road. But our observation of the proposed road-like structure suggests that it is more likely to be a canal. The overall width of the terraced structure is around 100 m. From each of the terraces, a staircase leads downwards to the canal (road?).

The length of the canal appears to be more than 250 m, while the canal has a width of 25 m. The purpose or utility of this canal structure is intriguing. Our observation all along the canal indicates that the western end of the structure begins underwater opening away from the terraced structure into the open sea. The width, height and terraced northern side of the canal force us to suggest that the canal structure might have served as a channel for small boats communicating with the Arakawabana headland. The southern natural outcrop wall probably had provided a buffer wall for strong open sea waves. This interpretation appears quite reasonable because the height of the southern wall of natural outcrop and the northern terraced wall are nearly same. The terraces and attached staircases might have been used for handling, loading and unloading boats sailing through the channel. Thus it appears in all probability that the terraced structure and canal might have served as a jetty before submergence to present depth.

5.2 Monolith human head

A large monolith that looks like a human head with two eyes and a mouth was studied at Tatigami Iwa Point. A human-cut large platform in the same monolith extends outwards at the base of the head. An approach way leads to this platform from the shore-side.

The surrounding basal platform is quite large (about 2500 m²), and could easily have accommodated more than two thousand persons sitting. The human head and associated platform with an approach road are suggestive of an area of worship or community gatherings.

5.3 Underwater cave area

Diving operations revealed caves at 8 to 10 m water depth at 'Palace' area. The entry to these caves was possible only through the large 1 metre radius holes on the cave roof. Inside the cave a boulder about 1 metre diameter engraved with carvings was observed. About 100 m towards the eastern side of the caves more rock engravings were noticed on the bedrock. These rock engravings are believed to be man-made.

Once upon a time these caves were probably on the land and were later submerged. The rock engravings inside the cave and on the bedrock were probably carved out by means of a tool of some sort. However, it is very difficult to say that these are rock art of this or that period, or a script.

5.4 Megaliths Point

Diving operations revealed two big rectangular blocks measuring 6 metres in height, about 2.5 metres in width (both) and 4.9 metres thickness which have been located towards the western side of Iseki Point . . . These rectangular blocks are designated by Japanese workers as megaliths. These blocks have been located in between two natural rock outcrops. The approach way to these megaliths is through a tunnel measuring about 3 m long, 1 m high and 1 m width.

The shape, size and positioning of these megaliths suggest that they are man-made. It is believed that the people of Japan's extremely ancient Jomon culture used to worship stones, rocks (Hancock, personal communication, 2000). In light of this practice, it may be worthwhile to suggest that these megaliths might have been used as objects of worship. However, a thorough investigation in this regard is necessary before assigning a definite purpose to these megaliths.

6.0 Conclusion

The terraced structures with a canal are undoubtedly man-made, built by cutting an existing huge monolithic outcrop. The rectangular terraced structure and canal probably might have served as a jetty for handling, loading/unloading small boats before its submergence to present depth.

The monolith rock-cut human head and associated platform might have served as an area of worshipping or community gatherings.

The score so far

By my count so far I have one marine archaeologist, Sundaresh, who is convinced that the Yonaguni structures are 'undoubtedly man-made', and who represents 100 per cent of all the archaeologists who have ever dived there up to the time of writing. I also have one marine geologist, Masaaki Kimura, who believes the same thing, a second geologist, Robert Schoch, who is undecided, and a third, Wolf Wichmann, who is convinced that they are natural.

I decided when I got the opportunity that I should try to dive at Yonaguni with Wichmann and see if I could change his mind. To this end, a few months after the *Der Spiegel* article appeared, I made the following statement on my website:

> I would like to offer a challenge to Wolf Wichmann . . . Let us agree a mutually convenient time to do, say, twenty dives together at Yonaguni over a period of about a week. I will show you the structures as I have come to know them, and give you every reason . . . why I think that the monuments must have been worked on by human beings. You will do your best to persuade me otherwise. At the end of the week let's see if either side has had a change of mind.[13]

'Japanese scientists cannot dive . . .'

In March 2001, on a mini-expedition funded by Channel 4 Television, Wichmann took up my challenge. A small, wiry, dark-haired, unpretentious man, I liked him the moment I met him, and continued to do so throughout the week that we spent diving in Japan and arguing, in a mood of amiable disagreement, about what we were seeing underwater.

Predictably we did not reach a consensus: Wolf left Yonaguni still holding most of the opinions with which he had arrived, and so did I. But I think that we each gave the other some worthy points to ponder. I know that I benefited from what amounted to a very useful field seminar on the natural history of submerged rock and began to understand clearly for the first time exactly how and why a geologist might conclude that the Yonaguni underwater structures are entirely natural – or at any rate (to sum up Wolf's position more accurately) that they all *could* have been formed by known natural forces with no necessity for human intervention.

Before going on to Yonaguni, Wolf and I paid a visit to Professor Masaaki Kimura at his office in the University of the Ryukyus. I started the ball rolling with a general question for Professor Kimura concerning the age of the structure:

> *GH:* People can argue for the next five centuries about whether what we see underwater at Yonaguni is man-made or artificial. But one thing which we can hopefully get clear is how old it is . . . when it was submerged? So the first question I want to ask you is what is your view of the age of this structure? The last time that it was above water?
>
> *Kimura:* This construction has been submerged since 6000 years ago, because the coralline algae attaching to the wall of this structure shows 6000 years.
>
> *GH:* And those coralline algae, because they're organic, you've been able to carbon-date them?
>
> *Kimura:* Yes, carbon 14.
>
> *GH:* Right. So that tells us the age of that biological item . . . it's 6000 years old and it's attached to a stone structure which, therefore, must be older than that.
>
> *Kimura:* It must be older, and so in general 6000 years ago the sea-level at that time [was lower] . . . So if this was made by men, this must be when this area was land . . . it's about 9000 or 10,000 years ago.
>
> *GH:* 9000 or 10,000 years ago? So – again to clarify, because I need to get this straight – you're saying that 9000 or 10,000 years ago, the whole area was above water and the date of submergence would be about 6000 years ago?
>
> *Kimura:* Before 6000 years ago.
>
> *GH:* This is the problem with carbon 14, isn't it? It dates the organism, not the structure. So then you can only say that the structure is older than that, but how much older is not sure. How much work have you done on sea-level change as a dating guide? And how big a factor is the possibility of sudden, maybe recent land subsidence as a result of earthquake?
>
> *Kimura:* Yes, I'm looking for such evidence, that is, geological evidence, but there is no evidence of movement. If this area had subsided by movement it would be due to earthquakes and faulting, but there is no active fault near by, the fringing coast is continuous, and between the beach and Iseki Point, there is no discontinuity or fault.
>
> *Wolf:* I see.
>
> *GH:* That makes things fairly clear, then. It leaves us with the sea-level issue on its own to base a date on, without complicating factors, which is great. At least we can be clear on one thing.
>
> *Wolf:* I think that questions for sea-level rise are very fairly proved by scientific evidence here in the area. I mean, they're experts in their field.
>
> *GH:* So you'd have no problem with the 9000 year date?

Wolf: No, no . . . not at all. No, the question was, or is still, is it and, if yes, to what extent is it made by man or overworked by man? This is the question.

GH: Well hopefully we'll get a chance to investigate that when we go to Yonaguni.

Kimura: We need to research much more.

Wolf: Yes.

GH (speaking to Prof. Kimura): I mean you're practically the only person who's done – you and your team here – have done continuous research for some years. But almost nobody else is working on it, I think, at the moment?

Kimura: Japanese scientists cannot dive.

'A very fine, a very nice thing . . .'

Throughout our discussion Professor Kimura strongly maintained his commitment to the man-made character of Yonaguni's underwater monuments – not simply on the basis of his technical findings, cited earlier, which I need not repeat here, but also, and I found persuasively, because: 'This kind of topography – if this has been made by nature it is very difficult to explain the shape.'

Wolf's riposte was immediate: 'So what I would say to that formation is that I've seen many natural formations, especially coastlines, being worked out by waves and wind, especially with the help of weapons, erosive weapons – sand and so on . . . Seeing with the eye of a geologist or a morphologist it is, OK, a very fine, a very nice thing, but possibly made by nature.'

I asked Wolf whether in fact he had ever seen anything like the Yonaguni 'formation' anywhere else in the world.

'Not in that exact combination,' he replied. 'This is what is surprising me; it's a very strong, compressed combination of the different shapes and the different figures you can find naturally in the world somewhere.'

'But you don't usually find them in combination like this?'

'No, I haven't seen that. So that is a marvel. It is a very beautiful formation.'

'Or the work of human beings?' I prompted.

'Or of that. So that's what we're here for.'

The ramp

On our first dive at Yonaguni I took Wolf to a very curious structure that I had discovered in late June 1999. It stands in 18 metres of water 100 metres to the west of the terraces of the main monument. When it was above sea-level 8000 or 10,000 years ago I suggest that it was originally a natural and untouched rocky knoll rising about 6 metres above ground level. A curving sloped ramp 3 metres wide was then cut into the side of the knoll and a retaining wall to the full height of the original mound was left in place enclosing and protecting the outside edge of the ramp.

I led Wolf to the base of the ramp, and as we swam up it I pointed out how the outer curve of the inner wall – which rises 2 metres above the floor of the ramp and is formed by the body of the mound – is precisely matched by the inner curve of the outer wall, which also rises to a height of 2 metres above the ramp floor, so that both walls run perfectly parallel. Moreover, when we swam up and over the rim of the outer wall we could see that its own outer curve again exactly matches the curves within and that it drops sheer to the sea-bed – as it should if it is indeed a purposeful wall and not simply a natural structure.

I showed Wolf that the ramp floor itself, though battered and damaged in places, must originally have had a smooth, flat surface. I also showed him what I believe may have been the function of the ramp. As one continues to follow it round it leads to a platform offering an impressive side-on view of the two huge parallel megaliths, tucked into an alcove in the north-west corner of the main monument, that constitute a spectacular landmark in the Yonaguni 'underworld'. Later we discussed what we'd seen:

> *GH:* OK, Wolf, the first dive we did I brought you to a structure (*attempts to draw ramp structure on notepad*) – I'm sorry, I'm hopeless at drawing . . .
> *Wolf:* Me too . . . (*peers at drawing*) OK, so I recognize it.
> *GH:* Hey, you're a geologist, you should be able to draw. (*Continues drawing.*) And here is a rather nice wall going round on both sides, and in the middle is a bedrock channel or ramp. And it rises from here around to this corner and, in fact, if we follow it all the way round it leads us to a view of the megaliths. Now this wall is not a bank. It is a wall. It's actually about half a metre wide. And it's high . . . more than 2 metres high . . .
> *Wolf:* Round about.
> *GH:* . . . Above this . . . above this ramp, whatever you want to call it. So I simply cannot understand the combination of clean bedrock here (*indicates the ramp floor*), admittedly very eroded and damaged, but clean bedrock here, and these heavily overgrown walls, which are definitely wall-like in appearance and rather high in the sense that they have an outer and an inner edge, and the curve of the outer edge matches the curve of the inner edge; and the same on the other wall.

To my surprise Wolf immediately admitted that this rather innocuous-looking and only recently discovered structure, which he had not been shown on his previous visit, was a 'real challenge'. He was later to describe it as 'the most impressive thing' he had seen at Yonaguni:

> The most impressive thing for me was the wall, the wall which is totally covered by living organisms nowadays, which should be removed to have a look at the structure of that wall, which can also be explained as having been done possibly by nature, but to get it sure we have to do deep research on that.[14]

Nevertheless Wolf would not have been Wolf if he had not at least attempted to come up with a calm, level-headed and unsensational geological explanation for the problem. He therefore drew my attention now to a place on land on Yonaguni called Sananudai that we had taken a look at the day before where he had shown me wall-like formations – admittedly only half a metre high – that had been formed entirely naturally:

> *Wolf:* OK, this is a real challenge to solve. But if you remember, the day before we have been on a platform on land – I forgot the name of the point –
>
> *GH:* Sananudai?
>
> *Wolf:* Right, correct. And by chance we went further down near the sea, and I showed you these encrustation patterns and maybe you remember that I . . .
>
> *GH:* I remember distinctly; you told me that a hard patina formed on the outside of the rock and that the water softened out the inside, leaving a wall-like shape in place.
>
> *Wolf:* Correct. And on the other side the relatively soft sandstone had already begun to be removed. So . . . and I told you that this could be a possible way that a wall can be made by nature . . . OK, it's a theory.
>
> *GH:* It's a theory. I mean, what I saw at Sananudai was actually no curved walls running in parallel with each other, but rather straight, and they were about half a metre high.
>
> *Wolf:* They were at the beginning stage. Right. And if you had a look closer down, you would have seen that there was a little curving, not as clear as this, I have to admit. But I mean, that was really the beginning stage so we don't know.
>
> *GH:* So would you want to explain those walls [on either side of the ramp] that way, as a hard patina which was preserved, and the soft part was cut out?
>
> *Wolf:* At first, and then subsequently overgrown by organisms as we saw. But to get clear what that really is, so I underline repeatedly, it is a challenge, and this is the first and only explanation I have for this. But to really get clear of this fact, we should have to remove the encrustation on one spot, or just from top to the bottom . . . This is the only way to find out of what material this wall consists – there's no other way; or to drill a hole through . . . We are obliged to find out what these walls are made of. Are they made of single patterns like stones or something?
>
> *GH:* Well, see, I don't . . . I very much doubt if the walls will turn out to be made of blocks. I think they'll turn out to be cut. I think we're looking at megalithic culture which cut rock. I think they cut down into the living rock, and they created the walls by cutting, and then later on the encrustation came and grew on top of the walls. That's my theory.

Wolf: I mean, if this was the case, then it would still be very useful to have a look on the core of these. It would tell us exactly what sort of material it was – was it soft sandstone, was it hard mudstone, or what else? And we would be possibly able to find any marks on them, which then would give us the clear proof . . .

GH: So what we have here is a bit of a puzzle which needs some serious research done on it.

Wolf: Correct. That's what I would say.

The tunnel and the megaliths

On our second dive we visited the twin megaliths, weighing approximately 100 tonnes each, stacked side by side like two huge slices of toast in a west-facing alcove in the north-west corner of the main monument. As noted earlier, a prime side-on view of these hulking rectangular blocks unfolds from the top of the curved sloping ramp explored on the first dive. And we've seen that the ramp appears to have been cut down (either by natural or human forces) between two parallel walls out of a pre-existing rocky knoll.

The knoll in turn co-joins other massive, heavily overgrown structures presumed to be outcrops of natural bedrock which form an almost continuous barricade, 3 metres high and 5 metres thick, thrown out in a loose semi-circle in front of the megaliths – all at roughly 15–18 metres water depth. The barricade is penetrated at only one point, and there only by a narrow tunnel a little over a metre wide and about a metre and a half high through which a scuba diver swimming horizontally may pass comfortably.

The tunnel itself looks 'built' – as opposed to rock-hewn like so much else at Yonaguni – in the sense that each of its sides consists of two courses of huge blocks separated by straight, clearly demarcated, matching joints. There is insufficient room to stand up within the tunnel, indeed barely enough even to crouch, so when it was above water 8000 or 10,000 years ago any human entering it would have been obliged to crawl through to the other side. What is striking, then, as soon as you emerge, is the way in which you now find yourself directly opposite and beneath the twin megaliths which, from this angle, rear edge-on above you, are like the paired sarsens at Stonehenge or the pair of upright granite megaliths worshipped since antiquity in Japan's Ena region as 'the sacred rock deity, the object of worship' (see chapter 25).

The swim ahead to the base of the megaliths is a matter of 20 metres and you observe immediately at this point that they do not stand on the sea-bed but are elevated about 2 metres above it, with their bases resting on a platform of boulders, and framed in a cleft. The side of the cleft to your right is formed by the rear corner of the main terraced monument; the side to your left is formed by a lower ridge of rock which also shows signs, though to a lesser degree, of terracing. Both megaliths slope backwards at the same angle against the cleft and both are the same height (just over 6 metres). The megalith to the right is distinctly thicker than its otherwise near 'twin' to the left. Both megaliths taper at top and bottom so that the gap between them, about the width of a fist at the midpoint, is not constant. Although roughened, eroded and pitted with innumerable sea-urchin holes, the megaliths can still be recognized as essentially symmetrical blocks, all the faces of which appear originally to have been smoothed off to match – although, again, whether the process that brought about this effect was entirely natural, or at some point involved the input of human skill and labour, remains thus far a matter of a very few contradictory professional opinions and no facts.

I allowed myself to float up, towards the surface, along the slope of the megaliths, resting my hand in the gap between them as a guide. The light was good and I could see right into the gap; looking back at me from the far recesses a plump red fish eyed me with horror and hoped that I would go away.

As I neared the top of the megaliths, submerged under just 5 metres of water, I began to feel the ferocious wash of waves pounding against the surrounding rocks. I clung on and for a few moments allowed my body to be tugged back and forth by the swell. Enshrouded in a cloud of foam I could see the north-west corner of the main monument still rising above me the final few metres towards the surface.

After the dive Wolf and I again discussed what we had seen and quite soon, after some fruitless trading of opinion, our argument began to focus around a single – potentially decisive – issue. Had these very striking parallel megaliths been quarried, shaped and lowered into position beside the north-west corner of the main monument by human beings? Or had they arrived there through wholly natural processes?

I had drawn another rough sketch map to which I now pointed:

> *GH:* There's the two blocks, and we see above them here, not very high above them, the mass of the structure which leads round to Iseki Point. Explain to me how those blocks got there.
>
> *Wolf:* OK. You have seen lots of blocks fallen down –
>
> *GH:* All over the place.
>
> *Wolf:* On the shoreline we saw from the ship –
>
> *GH:* Many fallen blocks, yes.
>
> *Wolf:* – lots of blocks have fallen down from higher parts –
>
> *GH:* Agreed.
>
> *Wolf:* – from beddings which have been broken, which were harder than the underlying layers; because what happens is that you get an undercurving and undercutting of softer material under harder banks. So in my belief, these two blocks have been once one block of two sandstone banks, with either softer material in between or nothing in between, just only the bedding limits.
>
> *GH:* Well, I want to know how they got where they are now.

Wolf: OK. My opinion is that these blocks have fallen down from a very, very high level, relative to their present situation.

GH: But no high point overlooks them. You would have to go back –

Wolf: Nowadays.

GH: Well, yes, fair enough, nowadays. Nowadays you would have to go back in a northward direction some 50 or 60 metres, maybe more, horizontally, before you reached the cliff.

Wolf: Right, that's clear for nowadays. I'm talking about a time-range of at least 10,000 years . . . maybe more.

GH: That we agree on.

Wolf: So then there could have been places of a higher position from which these stones could have fallen down.

GH: So you are hypothesizing a pre-existing higher place from which these fell?

Wolf: What I'm hypothesizing is that they have fallen down, so . . . and this must have happened from a, let's say, sufficiently higher place. So what this may be then –

GH: Do you agree with me that this place (*indicates top of north-west corner of main monument 3–4 metres above top of megaliths*) is not sufficiently high? The place we see immediately above it now?

Wolf: I don't have it in mind clearly, so I just can imagine from –

GH: But do you remember when we came to the top of these columns, of these blocks, we were coming close to the surface. You could feel the swell hitting you quite hard and the foam above your head very strong. In fact, it's like looking into clouds almost. And you can see the mass of the rock above you, probably not more than another 4 metres above, and you're going to hit the surface there.

Wolf: Yes, I would think this would not be high enough.

GH: No?

Wolf: No.

GH: So we need a hypothetical high place to do it?

Wolf: Yes.

GH: And I, of course, need a hypothetical civilization –

Wolf: Yes.

GH: – capable of moving it here.

Wolf: Yes, of course, yes, yes . . . no doubt about it

GH: So we have two hypotheticals there.

Wolf: I'm not going to discuss any presence or absence of any civilization, because that's not my field . . .

But the problem I feel – and shall continue to feel – is that the very odd combination of major stone structures lying underwater at Yonaguni, and the very odd combinations of characteristics found within every one of those structures, simply cannot be said to have been properly evaluated until the possible 'presence or absence' of a civilization – specifically the Jomon – has been very thoroughly taken into account.

The path and the terraces

Our third and fourth dives were spent examining the 'pathway' or 'loop road' which runs along the base of the main monument directly beneath the terraces in its south face at a depth of 27 metres; and the terraces themselves, which begin 14 metres vertically above the pathway.

The terraces

At this level a spacious patio about 12 metres wide and 35 metres in length opens out and in its north-eastern corner, at depths decreasing from 13 metres to 7 metres, the structures known to local divers as 'the terraces' are found. There are two main 'steps', both about 2 metres high with sharp edges and clean near-right-angle corners. Above them there are then three further smaller steps giving access to the top of the monument which continues to rise northwards until it comes close to the surface.

Here, very clearly, I could see the basis for the argument advanced by Wolf in *Der Spiegel* that the whole mass of the structure – with all its striking and emphatic terraces and steps, its perpendicular and horizontal planes – could be explained by the effects of high-energy wave action on a large outcrop of naturally bedded sedimentary rock. When it first began to form, aeons ago, the sandstone (or more correctly in this case 'mudstone') of the body of the monument was deposited in layers of varying thickness and consistency, traversed 'by vertical cracks and horizontal crevices'. As sea-level rose and turbulent waves began to strike progressively higher levels of the structure, these cracks and crevices were gradually exploited and opened up – with the softer layers separating into flat slabs of assorted shapes and sizes which could then be washed out by the sea. In such a fashion, explains Wolf, 'perpendicularity and steps' gradually developed in the fracture zones creating, entirely without human help, the most striking effects of the structure as we see it today.

According to this reasoning, therefore, I was to envisage the 12 × 35 metre flat-floored patio as having been cut out of the side of the original outcrop by wave action which removed the sedimentary mudstone layers in slabs – with the terraced sections being formed out of the surviving harder members of rock after the softer layers had been washed away.

I helped Wolf measure the two highest steps, then drifted off to the edge of the patio and looked down the sheer 14 metre wall that drops to Professor Kimura's 'loop road' – the flat, rock-floored 'pathway' that runs along the bottom of the channel immediately to the south of the monument. Although 25 metres wide at the depth of the terraces, the channel narrows to a width of less than 4 metres at the depth of the path. Its north wall is the sheer south face of the

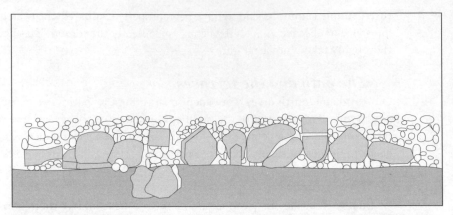

Front view of the 'stone wall' surrounding Iseki Point (looking south from the patio). Based on Kimura.

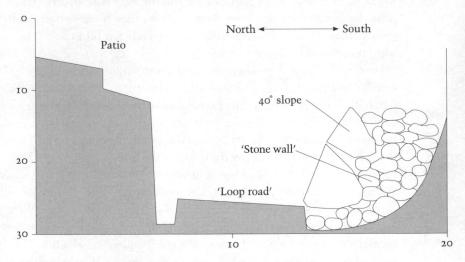

Cross-section showing, from left (north) to right (south), the sheer edge of the patio, the 'loop road' and the 'stone wall'. Based on Kimura.

monument; its south wall is at first not sheer but slopes for some distance further to the south at an angle of about 40 degrees before rising more steeply towards the surface. The 40 degree section is heavily but rather neatly stacked with blocky rubble that consists of an infill of smaller stones supporting a façade of a dozen much larger blocks arranged, as Professor Kimura points out, in a straight line 'as a stone wall'. Kimura is in no doubt that this wall is the work of human beings.

But because it is 27 metres down, and our dive computers didn't like the decompression implications of doing it as the fourth dive of an already hard day, we decided to leave it till the following morning.

The pathway

We dropped in near the twin megaliths, then followed the clearly demarcated rock-hewn pathway that seems to start (or finish?) here, veering to the left of the 'entrance tunnel' that we had passed through the day before, winding gradually to the south into deeper water around the western side of the main monument, then finally turning eastwards into the channel in front of the terraces at a depth of 27 metres.

As we entered the channel I pointed out to Wolf a pattern of three symmetrical indentations, each 2 metres in length and only about 20 centimetres high, cut at regular intervals into the junction of the northern side of the path and the base of the main monument. I also indicated two other details that I find particularly impressive in this area: (a) the way that the floor of the path appears to have been deliberately flattened and smoothed to give almost a paved effect; and (b) the way the path is completely free of any rubble until a point about 30 metres to the east of the terraces (where several large boulders and other stony debris have fallen or rolled).

When Wolf and I later discussed the path and the terraces he remained adamant that all the anomalies in these areas could have been produced by the effects of local erosive forces, mainly waves, on the 'layer-cake' strata of the Yonaguni mudstones. In short, while he could not absolutely rule out human intervention, he did not feel that it was *necessary* in order to explain anything that we had so far seen underwater.

At this point I drew his attention to a project done by Professor Kimura and his team from the University of the Ryukyus in cooperation with the Japanese national TV channel TBS. The result had been a high-quality six-hour documentary, aired over New Year 2001, that made many useful and original contributions to the debate on the Yonaguni controversy.[15] I wanted to acquaint Wolf in particular with the comments and demonstrations of Koutaro Shinza, a traditional Okinawan stone mason who had shown himself to be an expert in exploiting the natural faults, cracks and layers in sedimentary rocks to facilitate quarrying. According to Shinza, whom TBS brought to Yonaguni,

> When I saw the undersea ruins I knew instantly it was a stone quarry. I showed photographs to other stonecutters also and they all said the same. I conclude that it was done by human hands. It's absolutely impossible for something like this to be produced by nature alone . . .[16]

Since Shinza's technique of quarrying along the lines of weakness of existing joints and fractures is functionally identical to the 'method' used by the sea in Wolf's scenario to break up and separate the Yonaguni mudstones into the terraces and steps we see today, I asked him whether he could be absolutely certain that he could tell the difference. He admitted that he could not be

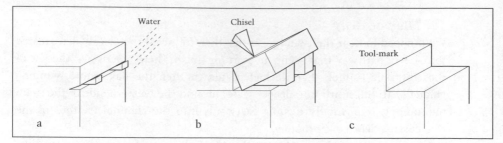

a. Wooden wedges are driven into a natural channel in the stone bed. The wedges are then soaked with water, causing them to expand.

b. As the wedges expand, the stone block splits from its bed. A chisel is used to help split the block.

c. The block is removed leaving flat, smooth surfaces on the bed. A tell-tale tool-mark is left by the chisel on the edge of the upper bed.

certain – although the fact that he had as yet seen no definite tool marks on any of his dives was another reason to assume that humans had not been involved.

> *GH:* Kimura makes a lot of the tool marks issue. He says he has definitely found marks. But I wouldn't be very hopeful after 10,000 years of submersion underwater to find tool marks. It's a long time. This, of course, is hard stone.
>
> *Wolf:* Very hard stone, yes. And it is heavily overgrown with organisms in many places. So we might find some marks, indeed, if we were looking a bit and if we knew where to look exactly and how to identify them clearly. But this I mean is necessary.

Had the sea randomly removed the rock layers to leave the terraces, or had it been ancient stone masons working to a plan? Neither scenario, we realized, could be unequivocally falsified – or proved – by the empirical evidence presently to hand. But there was another way to come at the problem which could at least test the logic of both propositions.

Part of Professor Kimura's evidence for human intervention in the construction of the main Yonaguni monument is the stark *absence* of fallen stony rubble in the pathway beneath the terraces – which he suggests should be cluttered by debris, perhaps even completely buried under it, if the terraces had been cut naturally by waves breaking up the pre-existing bedding planes. Where we do see debris on the path itself it is in the form of a cluster of large boulders (not slabs) 30 metres to the east of the terraces. And the only other area that might be described as debris lies neatly stacked at an angle of 40 degrees against the sloping south face of the channel, touching but never trespassing the southern edge of the path. This is the embankment with a façade of a dozen megalithic blocks arranged in a row that Kimura has identified as man-made. I confess, however, that on all my many visits to Yonaguni – including these March 2001 dives with Wolf – I have regarded this embankment as nothing more than rubble fallen from the south side of the channel and thus paid no special attention to it. It has only been since March 2001, looking back at the photographs and video images, that I have begun to realize how odd it is that not a bit of the supposed 'fallen rubble' transgresses the path itself, how very ordered it seems to be in general, and how very probable it is that Kimura is right.

But on the trip with Wolf I focused only on the issue of the apparent 'clean-up' operation that had been done on the path. I began by reminding him of our earlier discussion about the twin megaliths, each 6 metres tall and weighing 100 tonnes, which he claimed had fallen from above into their present position on the north-west corner of the monument from some hypothetical former high point.

> *Wolf:* I see what you're going for.
>
> *GH:* Well, what I'm going for is the problem of the path as we come in front of Iseki Point, as we come in front of the main monument. There's a sheer wall above the path 14 metres high and then the terracing begins. Now, if ever there was a place on this structure where large slabs of stone should have fallen it is here on the path, directly under where the terraces were created. And so what's bothering me is, if you can accept that the two parallel megaliths fell from a high place and lodged in position in the north-west corner of the monument and stayed there permanently, why don't we find the path in front of the monument littered with the equally big or bigger slabs of rock that must have been dislodged during the formation of the terraces?

I sketched the north and south walls of the channel, with the path at the base, and the embankment of 'orderly rubble' gathered up against the south wall.

> *GH:* Piled up here against the south wall is a huge amount of large stones which continue, in fact, up to this level (*indicates sketch*). And I can very well accept that those stones fell off the top of the south side and found themselves in this position. As a matter of fact Professor Kimura doesn't say that. Professor Kimura says that these stones were placed here by human beings.
>
> *Wolf:* Yes, yes, I know . . . I know.
>
> *GH:* And he may or may not be right on that matter, but I'm prepared to accept that the reasonable possibility, with the forces of gravity as I understand them, is that stones which had been up here along this also rather flat area on top of the south side, may have been washed off in water and tumbled down and piled up here (*indicates embankment*). And

that's what I see. I see stones that fell from up here on the south side. What I can't understand, once we come to the huge main terrace with its steps on the north side of the channel, is why under this nice vertical cliff I don't find any stones at all lying on this 3 metre wide path. And I don't accept that they all rolled from the [north] side into this embankment [on the south side] conveniently leaving the path immediately beside it free. To me that's against logic and nature.

Wolf: We're just guessing. So imagine that this flat area around the terraces was not removed all in one go. What I mean is little small tiny pebbles, cobbles, whatever, over a long time have fallen down and they have somehow been transported and rode supported by gravity here into this part (*indicates embankment area on south side of channel*), being sheltered from further transport, first of all, by these large boulders.

GH: Again I find it difficult to grasp you here. If I stand beside these steps (*indicates the two big steps in the main terrace*), they tower above my head. This means a layer of rock at least 2½ metres thick, all the way around here (*indicates patio area*) has been removed completely to leave behind just the steps.

Wolf: Yes.

GH: I mean this patio is, what, 30 or 35 metres in length?

Wolf: Round about.

GH: And we have a layer of rock 2½ metres thick; that's a hell of a lot of rock.

Wolf: We're not talking about two or three years.

GH: We're talking of a long period of time. So you're explaining this by saying that small pieces were broken off little by little and taken away by the tides?

Wolf: Yes, right . . . in general.

GH: Yeah. I find the more elegant explanation is it was tidied up by human beings –

Wolf: Fine.

GH: – after they finished their job.

Wolf: But where should they put it, then? Somewhere here around?

GH: Wherever they wished.

Wolf: Come on.

GH: If human beings do take material away from sites, they take it right away . . . get it away . . . this is known human activity . . . very normal . . . they don't leave the rubble lying around on the site, this is normal.

Wolf: This is clearly what Kimura says.

GH: It's Kimura's argument, and I find it persuasive.

The Palace

Our fifth dive was at a site several kilometres to the west of Iseki Point that local divers call the 'Palace' and that the Indian archaeologist Sundaresh refers to in his December 2000 report as an 'underwater cave area'. Sundaresh does not comment on the structural characteristics of the Palace itself, which is indeed surrounded by natural caves, but notes that inside it:

> a boulder about 1 m diameter engraved with carvings was observed. About 100 m towards the eastern side of the caves more rock engravings were noticed on the bedrock . . . The rock engravings inside the cave and on the bedrock were probably carved out by means of a tool of some sort.[17]

The entry to the 'Palace' can be made through a number of holes broken in its roof at about 9 metres water depth or through what I suggest may have been its original entrance at a depth of 14 metres. Here the diver has to squeeze through gaps in a jumble of fallen boulders to enter a small, gloomy, gravel-floored chamber oriented roughly north–south with space for four or five adults standing upright. Its south wall is blocked. In its north wall there is a 'doorway', about a metre high, through which visitors would have had to pass crouched, or crawling, when the Palace was above sea-level. The doorway has a rough, damaged appearance with no obviously man-made characteristics, but beyond it is a spacious and beautiful chamber that glows with an otherworldly blue light when the sun projects down through the column of water and illuminates it through the holes in its roof.

Like the cramped antechamber this atmospheric main room is oriented north–south. It measures approximately 10 metres in length and 5 metres in width. Its height from floor to ceiling is also about 5 metres. While there has been a substantial collapse of its eastern side, its western side is undamaged and presents as a smooth vertical wall of very large megaliths supporting further megaliths that form the roof.

Roughly at its mid-point the chamber begins to narrow towards the north until the east and west walls come together in a corridor less than 2 metres wide that culminates in another 'doorway' – this time a very tall and narrow one. Across the top of its uprights, whether by accident or by design, one of the roof megaliths lies like a lintel.

After having passed through this second and more impressive doorway at the northern end of the main chamber, the diver comes into a third and final room of the Palace. It is completely unlike the other two, which were 'built' (either by nature or by man) out of large blocks piled on top of one another. This third chamber, on the other hand, was hewn or hollowed – it is premature to decide by what – out of a mass of ancient coralline limestone that is exposed in this part of Yonaguni. There are no 'blocks' in it at all. It extends only 3 metres in

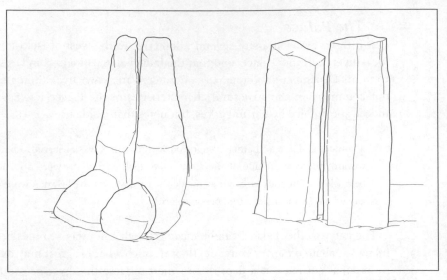

Comparison of submerged megaliths at Yonugani (left) and on land at Mt Nabeyama, Gifu Prefecture, Japan (right, see page 563).

length and a little over a metre in width and culminates at its north end in yet another 'doorway' – this time I insist distinctly 'squared-off' – which leads into a closed alcove that in turn funnels vertically upwards and opens out through a hole in the roof.

All three of the 'doorways' in the Palace, the first at the south side of the main room, the second at the north side, and the third leading into the alcove beyond, are positioned in a straight line creating what is, in effect, an aligned passage/chamber system. And since the rear (northernmost) chamber and alcove door are hewn out of a different kind of rock than other materials in the structure, we must assume that some agency brought these two elements (the rock-hewn element and the megalithic element) together – and in alignment – at some point. But was it nature that did this? Or could it have been the Jomon in a hitherto unrecognized phase of their prehistory when they moved gigantic rocks and boulders with apparent ease and set in train the cult of stone in Japan that still permeates the nation's spiritual life today?

Wolf would have nothing of it. In his no-nonsense view the Palace is, of course, a wholly natural phenomenon and the alignment of the three doorways is entirely coincidental. Very probably he is right. Yet I retain a sense of deep curiosity about this structure and intend, if I can, to do more work in it at some time in the future. On one previous dive near by I came across parts of what looked like a second megalithic passage/chamber system that I would also like to revisit.

Whether they are natural or man-made it is likely, by virtue of their depth of submergence, that both systems are thousands of years older than Japan's

mysterious Kofun era, which is thought to have begun arour
both systems powerfully and eerily remind me of the architect
megalithic passageways and burial chambers of the Kofun ag
structures such as Ishibutai near Asuka, where the megaliths
titanic dimensions and weights (see chapter 25). I remind the re
ologists have as yet uncovered no evolutionary background to the auvan
megalithic skills that suddenly manifest in Japan in the Kofun era, and raise the
possibility for consideration that the knowledge of how to build with megaliths
on such a scale may long previously have evolved in areas around Japan's coasts
that are now underwater.

I realize that this begs more questions than it answers. Still, go figure where
the Kofun tradition came from. Some scholars say Korea, but the evidence isn't
good and others scholars disagree. Nobody pays much attention to Japan's
own earlier epoch of stone architecture – witnessed by the stone circles and
'mountain-landscaping' of the Jomon age – because up till today a prejudice
persists that the Jomon were simple hunter-gatherers and nothing more.

I do not deny that they were simple hunter-gatherers but the deeper I enter
into the labyrinth of Japanese prehistory the more certain I feel that they were
also something much more . . .

The Face and the Stone Stage

On our sixth and final dive at Yonaguni in March 2001 I took Wolf to a place
called Tatigami Iwa 8 kilometres east of the Palace and about 2½ kilometres
east of the main cluster of monuments around Iseki Point.

Tatigami Iwa means 'Standing Kami Stone' and refers to a rock pinnacle
40 metres high, weirdly gnarled and eroded, left behind thousands of years ago
when the rest of a former cliff of which it was once part was washed away.
Understandably revered as a deity in local tradition it now stands lashed by the
Pacific Ocean 100 metres from shore like a ghost sentry for this haunted island.
But it is what is underneath it, in the underwater landscape near by, that really
interests me and that led me to choose it as the site for our sixth dive. For here,
at a depth of around 18 metres, a huge carving of a human face is to be seen –
with two eyes, a nose and a mouth hacked, either by natural forces or by human
agency, into the corner of an outcrop of dark rock that juts up prominently from
a distinctive 'blocky' plain.

I showed Wolf how the 'face formation' manifests a combination of peculiarities. For it is not just a face – or something that looks like one (which nature
provides numerous accidental examples of) – but a grim and scary face, which
seems designed to overawe, carved with care and attention to the lines and flow
of the base rock. Moreover, far from appearing haphazardly with no context, as
one would expect with an accidentally formed natural 'face', it seems framed
within a deliberate ceremonial setting. Thus, a horizontal platform just under
2 metres high and 5 metres wide – called by local divers the 'Stone Stage' –

opens out from the side of the face at the level of the mouth and runs along to the back of the head where a narrow passageway penetrates the whole structure from west to east.

The 'Face', therefore, has to be viewed together with its 'Stone Stage' as a single rock-hewn edifice and I note, as does Sundaresh in his report cited earlier, that the flat area out of which the Stage and Face rise is easily large enough to have accommodated thousands of people before sea-levels rose to cover it. Also noteworthy, however, is the fact that Face/Stage edifice is not alone in this big area but is part of a neighbourhood of anomalous rock-hewn and often rectilinear structures clustered around the base of Tatigami Iwa.

Natural? Or man-made? Or a bit of both? My vote is weird and wonderful nature, enhanced by man, thousands of years ago. But what did Wolf think?

> *Wolf:* First of all we have to mention that this is a totally different sort of sandstone from what we find at Iseki Point. It's very thick – a series of very thick and massive banks which consist, contrary to the Iseki Point material, of quite soft sandstone which is very, very sensitive to erosion and erodes generally in more rounded forms than the Iseki Point sandstone or mudstone. Secondly, erosion of rock, all around the world, often produces forms that look accidentally like human faces . . . So I cannot say very much to the Face. To become clear of that fact, again, you would have to remove all the organisms around because that would give you a free view on the rock and the way it was carved.
>
> *GH:* Did you notice, looking into the eyes, the eye sockets of the Face, that both of them had a central prominence?
>
> *Wolf:* No. No, sorry . . . I haven't looked.
>
> *GH:* You didn't see.
>
> *Wolf:* I saw the Face and I thought, 'Yeah, hmm, what to do with this?'
>
> *GH:* Yes.
>
> *Wolf:* But you see, I'm used . . . I'm not used to go straight to the things but to –
>
> *GH:* Yeah, to stand back, yeah, I noticed that.
>
> *Wolf:* – take a distance and look, hmm, how can this be formed? But it was my first view on that. I don't have an answer on that at the moment.
>
> *GH:* Something else about it too, for me, is the sense that I keep finding these problems – if we look back over our drawings over the last couple of days – well here from our first dive we have, within a short area, parallel curved walls, a ramp, a tunnel, two megaliths. We come round in front of the monument, a clear pathway, and as far as I'm concerned still with the mystery of the missing material – if indeed, as we also agreed earlier, all of this mass of material that we see in the embankment came from the south side – because as you said, it doesn't look like it belonged on the north side –

Wolf: On this view, yes.

GH: – it's the proximity of all these peculiar things, each of which requires a rather detailed geological explanation and, in some cases, requires hypotheticals such as a cliff which once hung over that area and dropped these two megaliths down there. I find – and this is how I felt always almost from the third or fourth visit that I made to Yonaguni – is that this, this fantastic combination of peculiarities in a very compact area – because as you saw today the peculiarities continue as we go further along the coast to the Face and the Stone Stage –

Wolf: That's right, I was deeply impressed when I saw that.

GH: – the thing that's striking is that all of these peculiarities occur along the south and east coasts of Yonaguni, and none of them are found along the north coast – at least, if they've been found, divers aren't talking about them, and divers usually do talk about places like this. So, you know, we find them along the south side but not along the north side. We find them compacted into a relatively tight area, and each one requires a rather different, and to my mind, rather complicated geological explanation, you know, disposing of a mass of rock that is 2½ metres thick and 35 metres in length [and 15 metres wide] is simply banishing it. And attributing that to wave action, to me that's just going a little bit too far –

Wolf: I see what you're getting at.

GH: – on the strength and the variability of geological forces in a small area, and it catches in my throat. I find that I can't, I just can't buy it.

Wolf: OK. I would ask you to have a look into new or even older geological and geographical literature. You'll find all these things precisely described in newly published literature and –

GH: Nowhere in the world – never mind the literature, books are books – but nowhere in the world, not a single place in the world will I find all these things together . . . because one thing's for sure, look at the publicity that this structure has attracted.

Wolf: Because you raised it.

GH: Actually, not me . . . it was –

Wolf: Together with others.

GH: – many other people . . . many other people have raised it. Worldwide it has attracted an enormous amount of publicity. I think it's a fair bet that if something comparable had been found, anywhere else on this planet of ours with its 70 per cent cover by water, if something similar had been found, we would have heard about it by now. And it's the uniqueness of this structure and the series of structures along the south and east coasts of Yonaguni that really leads me towards the involvement of man. Now I believe that the people who were involved in this were a megalithic culture; they understood rock, and they worked just as currents and erosive forces do, that is, they worked with the natural strike of the rock; where there is a fault, it's a good place, let's take advantage of it. Any great sculptor still looks for the natural forms in rock and, indeed, this is an art form in Japan up to this day. So, you know, these are all the factors that lead me to the conclusion that I'm looking at rock that has been overworked by people.

Wolf: And I would say, on the contrary, that it is a natural miracle . . . And just to finish that, my definite point of view is that all that we have seen in the last days could have been made by nature alone without the help of man. That does not mean that people did not have any influence on it. I didn't say that . . . I would never say that. But I say it can have been shaped by nature alone.

Other miracles

There are several other intriguing sites around Yonaguni that I was not able to show Wolf in the time available to us in March 2001 – though I do not think any of them would have changed his mind.

One of these, which takes a form that some recognize as a huge rock-hewn sea-turtle, stands at a depth of 12 metres on the shoulder of the main monument at Iseki Point approximately 150 metres east of the terraces.

A second, badly damaged when Yonaguni was struck by an unusually severe series of typhoons in August and September 2000[18] is found half a kilometre due east of the terraces in about 15 metres of water. Consisting of a one-tonne boulder mounted on a 10-centimetre-high flat platform at the apex of an enormous rocky slab almost 3 metres high, it has all the characteristics of a classic *iwakura* shrine, part natural rock, part man-made. As I noted in Chapter 25, if this shrine were to be moved to the slopes of Mount Miwa it would blend in seamlessly with what is already there.

Two other anomalous sites are located within half a kilometre of Iseki Point, and I would also very much have liked Wolf to see them. One is the extraordinary 'Stadium', a vast amphitheatre surrounding a stone plain at a depth of 30 metres. The other is a second area of very large steps – on a similar scale and of a similar appearance to those of the main terrace at Iseki Point, but much further out to sea, in deeper water, and at the bottom of a protected channel.

Nor does the list of signs and wonders end here, but I think the point has been sufficiently made. Some people with good minds – among them Japanese scientists with Ph.D.'s – are adamant that what they see underwater at Yonaguni are rock-hewn structures that have been worked upon by humans and purpose-fully arranged. Others with equally good minds and equally good Ph.D.'s are equally adamant that they see no rock-hewn structures underwater at Yonaguni at all – only rocks.

Rocks? Or structures? Just interesting geology? Or discoveries that could fix the true origins of Japanese civilization as far back in the Age of the Gods as the *Nihongi* and the *Kojiki* themselves claim? These are grave questions and they cannot be answered at Yonaguni on the basis of available evidence. Wolf is right about that. It is just possible that the remarkable structures and objects that I showed him there underwater are all freaks of nature, which by some amazing additional improbability all happen to be gathered together in one place.

I don't think that is what they are. And I repeat that the balance of first-hand scientific opinion is, at the time of writing, two-to-one against Wichmann in this matter (Kimura and Sundaresh provide two clear votes for the structures having been overworked by man; Wichmann provides one clear vote in favour of the structures being entirely natural; Professor Schoch votes both ways). In the future other discoveries, and other diving scientists, could alter this balance of opinion dramatically in either direction. But we shall have to wait and see. Meanwhile, after a thorough exposure on-site to Wolf Wichmann's relentless empiricism I concede that I am not yet in a position to *prove* that humans were involved in the creation of the Yonaguni structure – any more than Wolf can prove, as he admits, that they were not.

But I believe Wolf came to his conclusions about Yonaguni sincerely, not too hastily, and on the basis of his own vast experience as a marine geologist of how different kinds of rock behave underwater. Although I disagree with him, I therefore resolved as we left the island in March 2001 that I would not base any argument or any claim in *Underworld* on the copious evidence which suggests that the submerged structures of Yonaguni are indeed ancient rock-hewn human sites . . . In this chapter I have simply tried to marshal and present that evidence, and Wolf's purposeful and eloquent counter-views, as clearly and as objectively as possible, as a matter of public record.

But suppose for a moment – an exercise in speculation only – that I and others *are* right about Yonaguni. If so, then what Japan has lost to the rising seas is no small or insignificant matter but a defining episode in world prehistory going back more than 10,000 years. For if the Jomon did make the great structures that were submerged off the south and east coasts of Yonaguni at the end of the Ice Age, then we are confronted by a previously unexpected and as yet completely unexplained dimension of that increasingly remarkable ancient culture. In terms of organization, effort, engineering and ambition, the sheer scale of the enterprise is beyond anything that the Jomon of 10,000 or 12,000 years ago (or any other human culture of that epoch) are thought to have been capable of. Yet it makes a strange kind of sense in the context of the other incongruous characteristics of these strange 'hunter-gatherers' – their permanent settlements, their stone circles, their cultivation of rice, and their navigational and maritime achievements in two different waves of settlement of the Americas (one as early as 15,000 years ago, one more like 5000 years ago).

Wolf and I had just one more day of diving to do after Yonaguni, just one more day for me to find him a major structure in Japanese waters that he could not come up with a natural explanation for . . . For that adventure, and test, I had chosen the great stone circles at Kerama.

28 / Maps of Japan and Taiwan 13,000 Years Ago?

> In part based on Marco Polo's inaccurate figure for Zipangu's distance from the Chinese coast, the Florentine physician and astronomer Paolo Toscanelli, who – like many another medieval scholar – assumed that the world was a sphere, placed Zipangu some 5000 nautical miles west of Europe on his map of the world . . . As early as 1470, Toscanelli proposed to the Portuguese king that one could reach Cathay, Zipangu, and the Spice Islands (the Moluccas) – perhaps even more quickly – by sailing directly westward.
>
> Ulrich Pauly, German East-Asiatic Society, Tokyo[1]

It was the submerged structures of Japan that first awakened me to the possibility that an underworld in history, unrecognized by archaeologists, could lie concealed and forgotten beneath the sea. Then, when I learned to dive and started to look elsewhere, I began to realize how vast this vanished underworld really might be – for its traces seem to have been scattered around the continental margins not only of the Pacific but also of the Atlantic and the Indian Oceans and the Mediterranean Sea.

In five years of diving, following up rumours of anomalous underwater structures wherever they have been sighted, and using the logic suggested by convergences between flood myths and inundation maps to seek out probable sites, I know that I have only scratched the surface of the mystery. I'm just a private individual without any of the institutional infrastructure behind me that is really needed for productive marine archaeology. Even so, there has not been space in this book for me to recount the results and experiences of all my own dives and explorations – let alone all the dives and explorations that *should* be done in the future if we really want to know what's out there.

I've said nothing for example about the underwater enigmas of Tenerife where I dived, and was nearly swallowed by the sea, in June 2000. I learned a lot there . . . about the Kami Great-Ocean-Possessor.

I've not spoken of the work Santha and I did in the South Pacific around the Tahitian islands of Raiatea and Huahine, or of the strange things we saw underwater off the Tongan island of Haapai.

And I've said nothing more about Alexandria, which I introduced in chapter 1. Yet Santha and I spent several weeks diving along the Alexandrian coast with Ashraf Bechai looking for, and eventually relocating, some of the giant blocks of Sidi Gaber that he had first sighted years before. Indeed, we found a carpet of

gargantuan stone blocks in an advanced state of erosion, completely unconnected to any of the known marine archaeological sites in the vicinity, covering a
huge area of the sea-bed at 10–12 metres water depth (see photos 1–3).

But while all this was happening, and as I began to focus more and more
closely on specific regions and specific issues of the 'underworld' problem, it
always remained my intention to seek a final reckoning on the submerged
structures around Japan that had started me on the quest. I took my time – years
in fact – to do the travels and the dives in the Indian and Atlantic Oceans and
in the Mediterranean Sea that I've described in this book. But through it all I was
privileged to be able to revisit Japan frequently, to continue to dive repeatedly at
all the most important sites in the Ryukyu archipelago, and to acquaint myself
thoroughly with their characteristics and peculiarities.

Satanaze and Antilia

So the point we are at now in the story is exactly where I had always intended
that we should arrive. Strangely, however, because quests have lives of their
own, we have arrived here by a route quite different from the one I imagined we
would take. This happened because I did not anticipate the appearance, very
late in the investigation, of a significant intersection between the mystery of
the ancient maps and the mystery of the underwater ruins of Japan. On the
contrary, having pored over many early maps of Japan by both Japanese and
Western cartographers, and having found none that show it in anything like its
Ice Age configuration, I long ago gave up the search.

It was only when I was finalizing the maps argument in Part 5, and in fact
pursuing Bimini, that I read Professor Robert Fuson's breakthrough study,
Legendary Islands of the Ocean Sea, and realized that I'd been looking in the
wrong place all along. If there was indeed a lost cartographical science of the
Ice Age then its best crumbs had been preserved in the portolan tradition in
Europe by pre-Columbian mariners and copyists who themselves knew nothing
about the existence of the Americas or the Pacific Ocean. If an Ice Age map of
Japan – and one of nearby Taiwan – was going to turn up anywhere, therefore,
it made perfect sense that it should do so in a pre-Columbian European portolan
purporting to depict islands in the Atlantic Ocean.

I need to reiterate that Professor Fuson goes nowhere near this far; nor would
he wish to. His breakthrough, which I described in chapter 24, is the discovery
of the compelling series of correlations on the 1424 Venetian chart that link
Satanaze with Japan and Antilia with Taiwan. Fuson plausibly suggests that
the source map – or maps – that the Venetian copyist worked from could have
originated in the voyages of the Chinese admiral Cheng Ho in the early fifteenth
century and could quite easily have found their way to the West from one or
other of Cheng Ho's fleets via Arab intermediaries prior to 1424.

Because the correlations he presents are in general so persuasive, one glaring
mistake on the 1424 chart is not fatal to Fuson's argument. The mistake, as he

admits, is that Japan's 'three main islands (Honshu, Shikoku, and Kyushu) are represented by the single island of Satanaze. The channel between Kyushu and Shikoku/Honshu (modern Bungo-suido and Suo-nada) is well-defined.'[2] But 12,500 years ago, this mistake would not have been a mistake at all because at that time Honshu, Kyushu and Shikoku were indeed consolidated by lowered sea-levels into a single landmass.

While I'm prepared to accept that the Venetian cartographer's source maps probably did come from the voyages of Cheng Ho therefore, it is not inevitable that these maps were necessarily ones that had been newly charted by Cheng Ho's navigators. They could equally well have been amongst the many older maps that Cheng Ho is known to have brought along on the voyages. We will see later that China by Cheng Ho's time, already possessed a cartographical tradition that was hoary with antiquity.[3] It is by no means impossible that the same wellspring of mysteriously anachronistic geographical knowledge from which Marinus of Tyre may have sipped, and that so nourished the portolan tradition in Europe in the late Middle Ages, had also been known all along to the ancient Chinese.

I suggest that the 1424 chart may contain evidence of that knowledge.

The missing waterways

Although sea-level is still rising today, the rate of change is very slow and has made no significant difference to Japan's coastlines during the past 1000 years. It is safe, therefore, to treat the modern map of Japan as an accurate portrayal of the archipelago as it would have looked in the early fifteenth century,

Now compare the map of Japan with the portrayal of Satanaze/Saya on the 1424 Venetian chart (opposite).

At first glance, despite an obvious general similarity of layout, I think one would not immediately leap to endorse Fuson's conclusion that Satanaze represents Kyushu, Shikoku and Honshu (since it is only one island, not three) or that little Saya represents Hokkaido. However, the theory is undoubtedly correct and I have already presented the principal evidence that underpins it in chapter 24. All that remains to be added is the process of 'cartographic devolution' (the gradual introduction of errors and deletions in a series of copies) by which Fuson believes that the Venetian mapmaker managed to turn Japan into Satanaze. This is best expressed in his own diagrammatic way (see page 630).

To focus the discussion here, I will accept Fuson's well-supported argument that most of Hokkaido was simply ignored and reduced to the rump of Saya on the original source map from which the 1424 chart was copied.[4] I will accept too his other suggestion that at some stage in the chain of copying and transmission by which the source map reached Europe a large section of the north of Honshu was missed out thus shortening the distance from the tip of Honshu to the tip of Kyushu.

But it is the other proposed 'deletions' of the copyists that interest me. *All of*

Modern map of Japan.

The island of Satanaze, from the 1424 Pizzagano chart.

*these – every one of them with remarkable consistency – prove to be 'deletions'
of bays and inter-island waterways that have only come into existence around
Japan since the end of the Ice Age.* In other words there was a time, not too long
ago (and certainly well within the enormous span of Japan's mysterious Jomon
culture), when most of the bays and inter-island waterways on the modern map
were dry land and did look pretty much the way the 1424 chart of Satanaze
shows them.

I will focus here on the portrayal of modern Japan's most prominent group of
inter-island waterways around the Inland Sea separating Honshu, Kyushu and
Shikoku. Fuson himself takes special notice of 'the channel between Kyushu
and Shikoku/Honshu' on the 1424 chart, and its 'well-defined' presence there
is undoubtedly helpful to his case. To make use of it, however, he has to
overlook the fact that the equally prominent channel that has separated Shikoku
from Honshu for at least the last 9000 years is not only not 'well-defined' but
actually is not shown at all. Likewise, he must put up with a very poor portrayal
of the segment of Satanaze that he allocates to Kyushu – poor, that is, if it is a
portrayal of Kyushu in 1424. However, either by chance, or because a fragment
of a cartographical tradition from the end of the Ice Age was resurrected in that
1424 chart – or for some other reason – its portrayal of what is now Kyushu does
match very well with Kyushu's actual appearance at the end of the Ice Age.

Let's look more closely at this odd 'coincidence' with reference to the 1424
chart, the modern map of Japan, and inundation maps of the archipelago provided by Glenn Milne and his team at Durham University. The latter model
Japan's coastlines at the following dates: 21,300 years ago (onset of the Last

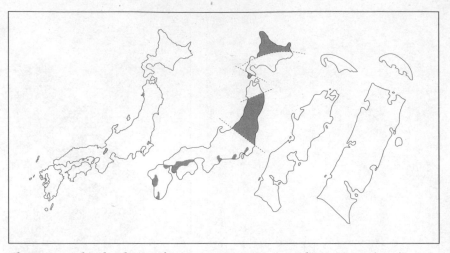

The cartographic devolution of Japan into Satanaze, according to Fuson (1995).
© *R. H. Fuson.*

Glacial Maximum), 16,900 years ago (end of LGM and start of meltdown), and thereafter at roughly millennium intervals for 14,600 years ago, 13,500 years ago, 12,400 years ago, 10,600 years ago, 8900 years ago, 7700 years ago and 6900 years ago (end of meltdown).

Mapping specific Ice Age details

We'll begin with the modern map of Japan, on which we note that Kyushu is only just an island, separated by a very narrow strait from the southern tip of Honshu. Nonetheless, it is an island. The strait widens into the Suo Gulf of the Inland Sea. There it splits into two branches – one trending southwards into the Bungo Strait between Kyushu and Shikoku, the other trending north-eastwards via the Iyo Gulf to the series of further straits that separate Shikoku from Honshu.

Now look at the depiction of the same waterways on the 1424 chart of Satanaze/Japan (opposite). It is obvious immediately that the system is much simpler.

Most notable difference: instead of the narrow strait that today lies between Kyushu and Honshu we observe that the two islands are joined by a land-bridge almost 100 kilometres wide.

Most notable similarity: there is a roughly square inlet on the south-east side of Satanaze which corresponds well with the location and direction of the present Bungo Strait.

But today, as we've seen, the Bungo Strait splits into the Suo Gulf to its north-west and the Iyo Gulf to its north-east. On the 1424 chart, by contrast, the Suo Gulf is completely missing. And although the Iyo Gulf is present, note

61. *Jomon pottery mask, Japan.*

62. *Pottery* dogu *figurine of the ancient Jomon from Sannai-Muriyama, Japan.*

63. *The author inside the megalithic passage grave of Ishibutai, Japan.*

64. *The author standing on top of Masada-no Iwafune, a massive megalithic structure in the Asuka region, Japan.*

65. *Pilgrims at the megalithic rock shrine on the summit of Mount Miwa, Japan.*

66. Jomon stone circle of Oshoro, Hokkaido, Japan.

67. Jomon stone circle, Cape Ashizuri, Shikoku, Japan.

68. Jomon megaliths deep within a forest, Cape Ashizuri, Japan.

69. *The author at Masada-no Iwafune megalithic structure. Note the method of cutting the rock in square sections visible on this side of the structure. Compare with 70, below.*

70. *Stones cut with the same technique as Masada-no Iwafune at a depth of 20 metres, Yonaguni, Japan.*

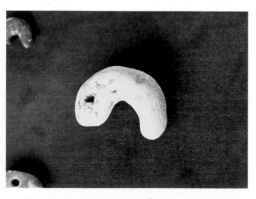

71. *Jomon magatama curved stone ornament, Japan. Compare with 72, right.*

72. *Maltese curved stone ornament in National Museum of Malta. The only one of its kind ever found in Malta, it is identical to the magatama of the Jomon of Japan.*

73. *Small stone circles, Komakino Iseki, northern Japan. Compare with 74, below.*

74. *Small stone circles at a depth of 30 metres, Kerama, southern Japan.*

75. Iseki Point, the main underwater monument of Yonaguni.

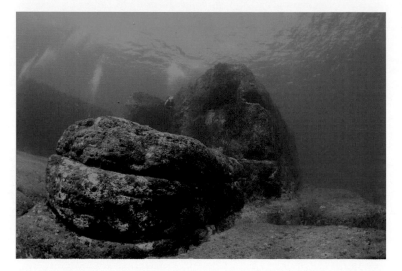

76. The author diving beside the 'Face', Yonaguni.

77. Underwater megalith, Yonaguni.

78. View down on Centre Circle, Kerama, Japan, at a depth of 30 metres.

79. Centre Circle and Small Centre Circle (foreground), Kerama.

80. Centre Circle, Kerama.

81. *The author diving beside the parallel megaliths, Yonaguni, Japan.*

Satanaze from the 1424 Pizzagano chart.

that it is represented only as a very narrow, north-east-trending, fjord-like channel. Opposite its terminus, on the south-west side of Satanaze/Japan, there is a further, much smaller inlet. The neck of land between the two – about 100 kilometres wide – lies along the line of the missing Suo Gulf.

When we compare the 1424 chart with the inundation map sequence (pages 632–4), no obvious correlation emerges down to as late as 14,600 years ago, when Kyushu, Honshu and Shikoku were still so firmly bonded together by lowered sea-levels that even the Bungo Strait did not exist.

Within just another thousand years, however, around 13,500 years ago, the inundation maps show that a squarish inlet topped by a narrow, north-east-trending, fjord-like channel very similar to the portrayal of the Bungo Strait on the 1424 chart had opened up.

The correlation remains much the same on the inundation map for 12,400 years ago, although it is possible to detect a slight opening to the north-west, not shown in the 1424 chart, in what was to become the Suo Gulf.

By 10,600 years ago, however, the correlation is much less precise, with both the Suo and Iyo Gulfs opening up into fat cloverleafs north-west and north-east of the Bungo Strait.

Finally, by 8900 years ago, the submergence of the shorelines around the Inland Sea approaches today's levels, Shikoku, Kyushu and Honshu begin to emerge as separate islands, and the geography of the 1424 chart becomes and remains an anachronism.

Bearing in mind the limitations of inundation science – these maps are models, based on the latest data but do not claim 100 per cent accuracy – there is no doubt that the best correlation between the 1424 chart and the actual appearance of this part of Japan comes not in 1424 but in a specific and clearly demarcated 1100-year time-window between 13,500 years ago and 12,400 years ago.

Coincidence? Or the leavings and memories of ancient world maps preserved

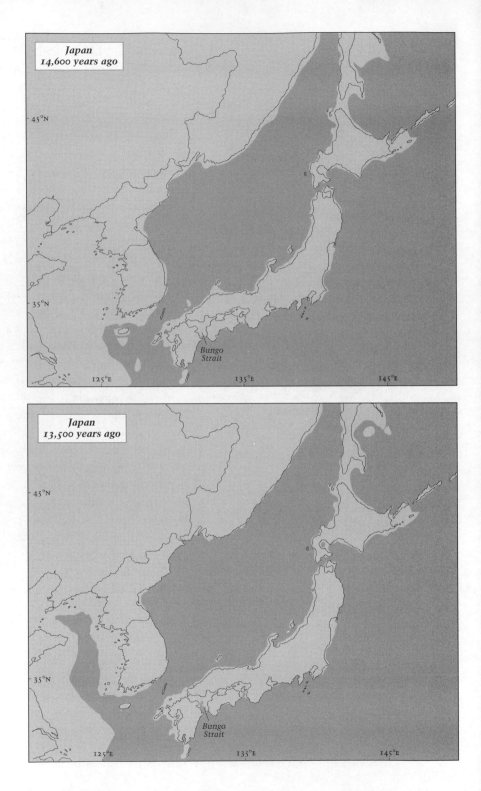

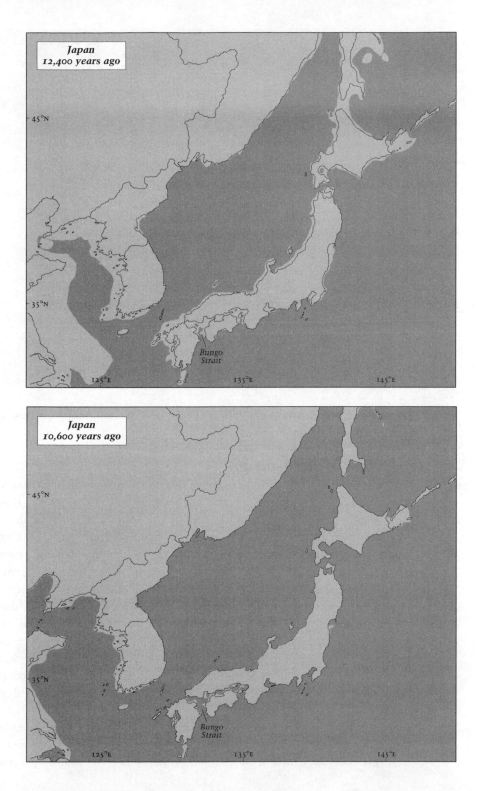

Japan
12,400 years ago

45°N

35°N

125°E   135°E   145°E

Bungo
Strait

Japan
10,600 years ago

45°N

35°N

125°E   135°E   145°E

Bungo
Strait

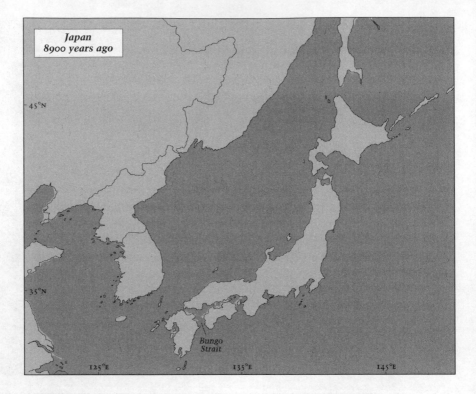

Japan
8900 years ago

45°N

35°N

Bungo
Strait

125°E   135°E   145°E

amongst mariners since the end of the Ice Age in fragments, and copies of
fragments, and fragments of copies?

What about Taiwan?

As I was considering the implications of this interesting problem it occurred to
me – since Satanaze appears together with Antilia on the 1424 chart – that the
two islands probably appeared together on the source map too. In that case
the treatment of Antilia/Taiwan on the 1424 chart could serve as a useful
control to speculation about Satanaze/Japan. If, for example, it should turn out
that Taiwan's 1424 portrayal was best matched by the modern appearance
of the island and bore no resemblance to the inundation maps then it would
make it more likely that any Satanaze/Japan correlations were just coincidences.
On the other hand, if Antilia and ancient Taiwan matched up well to one
another, and especially if they were to do so in the same time-window as
Satanaze/Japan, then I thought this would make it much more likely that the
similarities had been derived from a common source map that had contained
accurate depictions of Japan and Taiwan as they had looked at the end of the
Ice Age.

At the beginning of the meltdown around 16,400 years ago lowered sea-levels
meant that Taiwan was not an island but was, instead, fully integrated with
the east coast of China. The inundation maps show its distinctive narrow

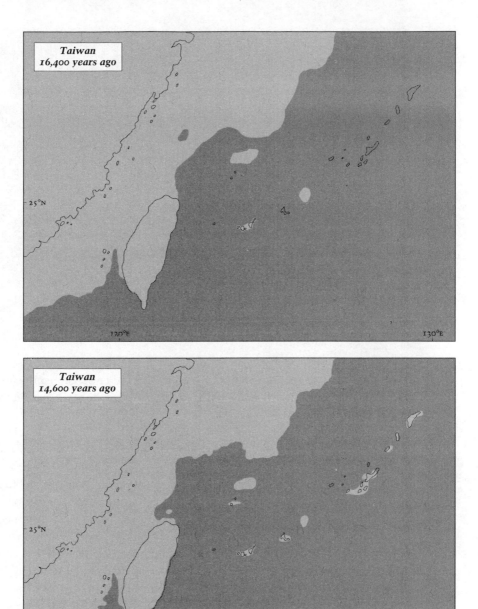

Taiwan
16,400 years ago

Taiwan
14,600 years ago

south-eastern tip, which has changed its appearance very little over time, protruding as a peninsula from a vast antediluvian landmass extending eastwards for hundreds of kilometres from the present Chinese coast. These long-lost coastal plains, fertile with the silt of the ancient Yangtse and Yellow rivers,

635 / *Maps of Japan and Taiwan 13,000 Years Ago?*

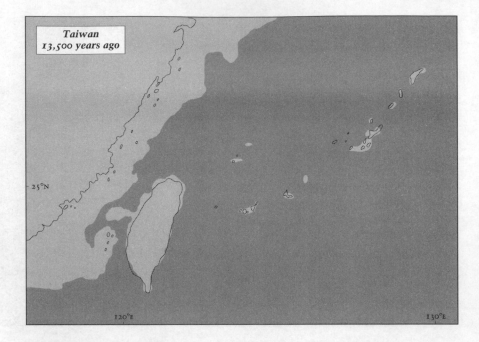

Taiwan
13,500 years ago

25°N

120°E   130°E

were wide and extensive enough to incorporate the entire Korean peninsula much further north, completely filling the basin of the Yellow Sea, and the Bo Hai and Korea Bays (page 635).

The situation of Taiwan had not dramatically changed two millennia later, as represented in the inundation map for 14,600 years ago. We can see there (page 635) that it has made some progress towards its eventual destiny as an island but that it is still very much fixed to the mainland and as such offers no correlation with the 1424 chart of Antilia/Taiwan. In fact the inundation maps show that Taiwan did not become an island, and thus did not even become eligible for comparison with Antilia, until 13,500 years ago (above).

It is notable, therefore, when we compare its appearance at that time to the outline of Antilia, that we immediately find a tantalizing resemblance, but certainly not an exact one. For, although the inundation map shows Taiwan as an island of roughly the right shape, it also shows a distinctive peninsula protruding from the mid-latitudes of its west coast that is not to be seen anywhere on Antilia. Instead, the 1424 chart gives us a second smaller island, named Ymana,[5] roughly where the peninsula on the inundation map ends.

The next inundation map in the sequence, which shows Taiwan as it looked 12,400 years ago, is where things get interesting. Very strikingly the peninsula has vanished and what remains is an island of the right size and in the right location to match Ymana (page 638).

Again, is it a coincidence?

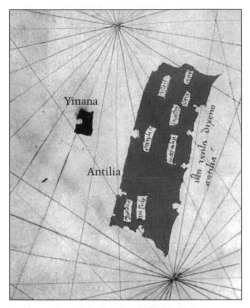

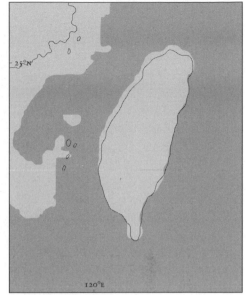

Antilia from the 1424 Pizzagano chart.   *Taiwan 12,400 years ago.*

Here the logic that led me to look for Antilia/Taiwan correlations at the end of the Ice Age (as a control on the apparent correlations I had noticed between Satanaze and Japan in the same period) works in reverse to reduce the likelihood of coincidence still further. Of course it still could *be* a coincidence. The fact is, however, that the representations of Antilia and Satanaze on the 1424 chart not only appear to have captured characteristics of Taiwan and Japan as both looked during the meltdown of the Ice Age but, far more impressively, as both looked during exactly the same 'window' between 13,500 and 12,400 years ago.

Objections

There are two important objections to this line of reasoning, which must be registered and responded to immediately.

First, despite its steep coastlines, Japan did undergo significant changes to its appearance at the end of the Ice Age when one antediluvian island – Satanaze on the 1424 chart – was filleted into segments by the rising seas to form modern Kyushu, Shikoku and Honshu. Taiwan's even steeper coastlines, by comparison, have changed much less since it first became an island around 13,500 years ago. Thus to the extent that Robert Fuson is right at all to identify Antilia as a map of Taiwan, then it could, theoretically, be a map of Taiwan in almost any epoch after 13,500 years ago. As such isn't it too vague and general an indicator to be useful for any particular purpose or to draw any specific conclusions from?

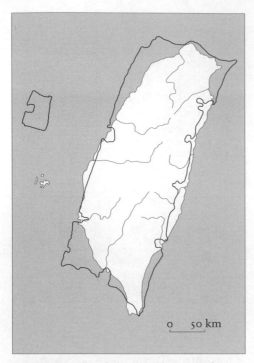

Superimposition of 1424 Antilia on modern Taiwan.

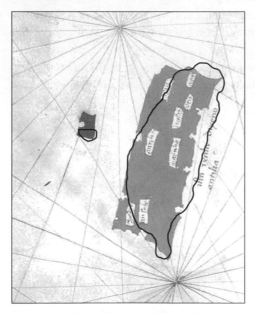

Superimposition of 1424 Antilia on Taiwan as it looked 12,400 years ago.

My response to this objection is that Antilia has more information to reveal than at first meets the eye.

A good starting point is Fuson's own superimposition of Antilia on to a modern map of Taiwan. As the reader will observe, other than the overall dimensions and the roughly rectangular shape being about right, the correlations between the coastlines of the two islands are not in fact particularly good (and would make no case in themselves were it not for the many other convincing comparisons between Antilia and Taiwan that Fuson is able to present).[6]

When compared with the modern map Antilia does best in the south-east – where both it and Taiwan come to a distinct sharply pointed, south-east-facing cape. But in the south-west, north-west and north-east the island depicted on the 1424 chart extends many kilometres beyond the coastal margins of Taiwan as it looks today.

Is it another coincidence that two out of these three supposed 'mistakes' in Antilia's portrayal of the main island of Taiwan would make perfect sense if the source maps showed Taiwan as it looked around the end of the Ice Age? There is no match at any date for the triangle of land that Antilia adds on to Taiwan in the south-west. But the extra lands that Antilia also claims in the north-west and the north-east of Taiwan do correlate closely with extra lands – then still above water in precisely these areas – that show up on the inundation map for 12,400 years ago. Since 12,400 years ago is also the date that provides the best fit for the island of Ymana on the 1424 chart, coincidence seems to me an explanation that is increasingly difficult to defend . . .

However, it is precisely here that a second objection must be registered and responded to. One of Robert Fuson's proofs that Antilia is Taiwan, cited in chapter 24, is that: 'Taiwan also has something else that Antilia must have. And that is a small island to the west. On the 1424 Pizzagano chart it was called *Ymana*. Today it is the Peng-Hu group, or Pescadores (Islands of the Fishermen).'[7] It is the Pescadores, exaggerated into a single larger landmass by cartographers' errors, that Fuson speculates served as the model for Ymana.

My response is that the location of the Pescadores in relation to the main island of Taiwan – marked today by little more than dots on the map – is not identical to the location of Ymana, but considerably further south. By contrast, as we've seen, the inundation map for 12,400 years ago provides a single antediluvian island of the right size and in the right location to be Ymana. The same map shows us that the Pescadores were, at that time, still part of the Chinese mainland and lay at the tip of a peninsula some 200 kilometres south of my antediluvian candidate for Ymana. They did not finally become islands – initially just one island – until around 10,600 years ago and thereafter were gradually broken up into the many much smaller remnants that still survive today.

It remains entirely possible that Fuson is right and that it was the Pescadores that served as the model for Ymana – though I note in passing that they would have more closely resembled Ymana when they were consolidated by lowered sea-level into one island around 10,600 years ago than at any much later date.

I was therefore overtaken by an irresistible feeling of curiosity when news came to me from my Japanese friends that extensive underwater ruins had been discovered in the Pescadores. Lying off the south shore of a tiny island called Hu-Ching – it means 'Tiger Well' – the ruins were said to consist of two gigantic walls crossing each other at right angles extending from a minimum depth of just 4 metres to a maximum depth of more than 36 metres. It was too tempting a prospect to pass up and Seamen's Club were willing to fund one more trip. Santha and I packed our dive gear and flew out to Taiwan at the end of August 2001.

But I'm getting ahead of my story. Before we fast-forward to Taiwan we need to rewind to the end of chapter 27 and the mini-expedition to Japan's Ryukyu archipelago that I made in March 2001 with the German geologist Wolf Wichmann. The reader will recall that Wolf and I left Yonaguni, the westernmost of the Ryukyus, after failing to reach agreement on the provenance of the underwater structures there. Our next destination was Naha, capital city of the much larger island of Okinawa, where we would be a one-hour journey by boat from what are perhaps the most extraordinary and enigmatic underwater structures in all of Japan – the great stone circles of Kerama.

29 / Confronting Kerama

> I agree that this is very amazing and very strange, even to me, how these structural
> buildings could be formed. Patterns like these, I haven't seen formed by nature.
>
> Dr Wolf Wichmann, geologist, Kerama, Japan, March 2001

Although I usually refer, in shorthand, just to 'Kerama', the correct term is 'the Keramas' – for this is in fact a group of small islands, including Aka, Zamami, Kuba and Tokashiki, lying in the Pacific Ocean about 40 kilometres due west of Naha, the capital of Okinawa.

The islands are poignantly beautiful, with verdant hills, rugged, rocky coasts and sand-fringed beaches, and they are separated from one another by expanses of crystal-clear water ranging in intensity from the palest turquoise to the deepest midnight blue. The whole area is a marine nature preserve renowned for the great numbers and varieties of whales and dolphins that congregate there.

And at the end of the Ice Age? The story that Glenn Milne's inundation maps tell is that down to about 14,600 years ago Kerama remained attached to the southern end of Okinawa by a thick, curving tongue of land. Okinawa was itself at that time a much larger and wider island than it is today with many kilometres of low-lying, gently sloping plains extending both east and west of its present coastline. Indeed, it is on these now inundated plains off its south-western coast that Okinawa's own underwater monuments – the 'step-pyramids' and 'terraces' off-shore of Chatan, described in chapter 1 – are located. And at that time there was continuous land between Chatan and Kerama . . .

Looking further through the inundation sequence we find that by 13,500 years ago the Kerama-to-Okinawa land-bridge had been severed and 20 kilometres of water lay between the two. But it is also clear that Kerama at that time had not yet broken up into smaller units. Further detail is difficult to resolve, but the maps indicate that this single, larger Kerama may have survived, with minimal diminution, until as late as 10,000, perhaps even 9000, years ago – though since parts of it were steeply sloping, and parts flat, not all of it would have been submerged at the same moment even then. It would have been around this time, 9000–10,000 years ago, that Kerama's stone circles would have been inundated.

The circles lie under almost 30 metres of water, 10 kilometres south-east of Aka island, at the intersection of latitude 26 degrees 07 minutes north and longitude 127 degrees 17 minutes east. A few jagged rocks just break the surface near by, with waves constantly crashing over them, but otherwise the site is completely exposed in open water.

The constraints
Kerama, March 2001

The March 2001 dives with Wolf Wichmann were funded and filmed by Channel 4 on a rushed, money-saving schedule – two working days for Yonaguni, and one for Kerama. In practice this meant that if the weather turned sour – which it frequently does in the Ryukyus – we would not be able to dive at Kerama at all. And even if the weather god was with us, the sea god might not be: the currents at Kerama are often so severe that you have to fight the water continuously if you want to stay in one place.

When humans fight water, water wins. I've seen divers lose their masks and have their regulators pulled from their mouths by the Kerama currents. I've seen desperate, breathless struggles to stay on top of the site, or to help others stay there, and not get swept away into the wild blue yonder. I've seen fit young adults crawl back on to the boat exhausted, literally trembling with fatigue. So what I've learned, after several unpleasant experiences of that sort, is that it's just not worth diving there when the current flows. It's better to anchor the boat tight with a couple of lines fore and aft, put a buoy in the water, watch how it bobs, and wait for a lull.

If there's a lull.

Briefing
Kerama, March 2001

We set out from Okinawa soon after 9 a.m. on what turned out to be a reasonably fine morning with waves of less than a metre. Once again we were working with the great local diver Isamu Tsukahara and his very professional team, and using his fast, spacious cabin cruiser as our dive boat. Mitsutoshi Taniguchi, the original discoverer of the circles, had come up to join us from his home on Miyako island further to the south. And Kiyoshi Nagaki had also volunteered to dive with us that day.

We began to sight the Keramas after about an hour of steady running to the west, and as we drew closer Wolf explained to me their basic structure evident from areas of bare rock along the coasts and from scars left by earthfalls that had uncovered the underlying strata in the hills. Rather like Malta in the far-off Mediterranean, it seemed that these islands had been formed out of huge

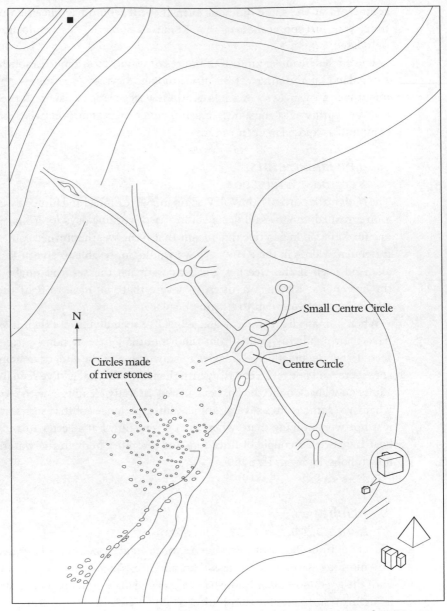

Plan drawing of Kerama's stone circles and associated structures. Based on Kimura.

deposits of coralline limestone (i.e., corals turned to rock) that had been laid down under ancient seas as much as 50 million or 100 million years ago and then subsequently exposed and inundated again, exposed and inundated again, with more coral growth taking place in the epochs of inundation but later itself being fossilized and exposed. In some places sedimentary layers of softer limestones, comparable to Malta's globigerina layers, lay on top of a coralline core. In others coralline outcrops formed the surface layer itself, glaring white in the morning sunlight.

By 10.30 a.m. we were manoeuvring into position over the dive site. Isamu Tsukahara – who always takes the hardest work on himself – went down to set the anchors and the buoy. This must have required an almost superhuman effort on his part, since the current was flowing strongly enough to create visible turbulence on the surface, but he calmly and capably succeeded and was soon back on board none the worse for wear. Then we all sat around and waited, listening to the creaking anchor ropes as the current tried to rip the boat free and send it spinning back to Okinawa. The buoy, rather distressingly, had been sucked completely underwater by the force of the flow, and no diving was going to get done before it popped back up again.

In the meantime I borrowed a DVC player and monitor from the film crew so that I could show Wolf some footage that Santha and I had shot during previous dives at Centre Circle, the largest of the group of structures that lay scattered around on the ocean floor beneath our boat. Every instinct in my body, for years now, had convinced me that these structures *must* be man-made, or at any rate could not have been made entirely by nature – they were simply too bizarre, unique and 'designed'. But secretly I had some doubts. I've learned a fair bit about rocks and reefs underwater around the world since I took up diving, but I'm not a marine geologist and there's a huge amount I don't know. Could it be possible that the strange pillars, the clear pentagonal pathway around the central monolith, and the shaped rock-surround of Centre Circle had all come about as a result of some natural process of which I was ignorant?

I froze the frame at an oblique view from the north-western side of the circle, shot in mid-water about 10 metres above the tops of the megaliths, and pointed out the central monolith to Wolf.

> GH: So this is the top of the central stone or whatever it is, which is then surrounded by a ring of . . .
> Wolf: It's a canyon. It's a sort of canyon.
> GH: It's a sort of canyon, and it runs down into quite a clean-edged pathway round the bottom here at about 27 metres . . . it's a curious mixture of pebbles and sand in the bottom. But it's very clean; there's nothing growing in the bottom at all.
> Wolf (pointing out several of the monoliths): All these single structures are totally overgrown by organisms. So just to have an impression of how they could have been shaped or could have originated, you have to scratch lots off them . . . Do you have any impression about the core material of these?
> GH: It seems to be a mixture of largish, I wouldn't say pebbles, I'd say more like cobbles, you know . . .
> Wolf: Rounded?

GH: Rounded . . . in a, in a sort of concretized mixture of something – I don't know what it is – a rocky, stony mixture.

Wolf: A matrix.

GH: A matrix, yeah. And you can see –

Wolf: So the question to answer is – is the core material consisting of the same matrix and pebbles mixture? Or are the glued pebbles inside this matrix just an outer cover?

GH: On top of something else.

Wolf: Yes. And the only way to find this out is to make some core drillings or something like that. Another way to come closer to the solution of that riddle, that mystery, could be to scratch off the sand on the bottom to see how these structures are linked to the ground rock . . . But what you definitely have to see is the core, the base, of these single structures and how they are fixed to the ground –

GH: So shall we plan to go to the bottom first, and do some of that? You may find that there's some samples that you can get. Have a close look at everything down there and, you know, see if you feel that these kind of curves – the way the outer and the inner curves of the big monoliths match each other – can be natural or can be man-made.

Wolf: So, as far as I can see now, I have really no explanation for this type of pattern.

GH: Here for example (*pointing to screen*). You can clearly see we're looking down on two parallel curved walls . . .

Wolf: Right, right. That's very amazing. So, is the distance between those two walls broad enough to let people walk through?

GH: Yes, yes it is. You can almost, in places, put two divers side by side, but not quite. Well, we'll see when we get down there.

I played the tape forward a few frames, then stopped it at a change of scene – the second circle of great monoliths. Because it is of narrower diameter (not because of the size of its monoliths, which are much the same) local divers call it 'Small Centre Circle'. It lies immediately north-east of and adjoins Centre Circle itself, creating in effect two interlinked rings, the first 8 metres across and the second 5 mctrcs across, contained within what appears to be a huge keyhole-shaped enclosure hewn somehow out of the bedrock that now forms the floor of the ocean.

Wolf: So, how many circles are there in all?

GH: Well, there's these two side by side; one large one and one slightly smaller one. Then there's a third one I guess about 50 or 60 metres further to the north-west, but we don't have shots of that.

Wolf: Yes. And are there other figures? Different from this circle?

GH: In the same area, about 40 metres away to the south, there are quite a number of other circles made of much smaller individual stones, most of them no more than a metre in length. We should be able to look at some of those circles too, on the same dive.

Wolf: But they are built up the same way, of the same material?

GH: Well, they look like some of the cobbles that are compacted into the bigger monoliths.

Wolf: Aha . . . aha.

GH: They look like that kind of –

Wolf: Single cobbles?

GH: Single cobbles.

Wolf: And then positioned . . . ?

GH: But positioned in a ring.

Wolf: That's strange.

GH: It is.

Wolf: Really strange.

I rewound the tape for a few moments then pressed 'play' again. There was one characteristic of Centre Circle which, though obvious enough, I'd so far forgotten to point out to Wolf.

GH: The other thing I feel about it is it's on a human scale. It's monumental, and yet the scale of the thing is human.

Wolf: I'm very astonished about that . . . about the structure, formation. You know, I haven't seen anything like this before.

GH: In years of diving? And nor have I . . . never anywhere in the world.

Wolf: Not only in diving, but also on dry land. There's some . . . some formations at least comparable a bit to this – so-called 'rock castles' or even a certain form of calcite weathering. But they look different. They look totally different and they don't have these canyons with the straight walls going right down.

GH: With straight walls and running all the way round a central stone.

Wolf: Normal calcite weathering is different. It has different wall angles.

GH: You see it's . . . every time I see that, that inner curve matching the curve of this and making this rather nice path, I feel . . .

Wolf: It's very parallely shaped . . .

GH: Yeah, and it feels like a design thing.

Wolf: Strange, yes . . . strange.

GH: And no real research has ever been done here. Not even by Professor Kimura.

Diving on Centre Circle

Finally, in the early afternoon around 1 p.m., the buoy which had been dragged under the surface by the force of the current suddenly popped up again, the pressure on the fore and aft anchor ropes went slack, and it was time to go diving. We were already partially geared up, so it took only a few minutes for us to strap on tanks, fins and masks and jump into the water.

Tsukahara had positioned the boat well and Centre Circle became clearly visible beneath us almost as soon as we were under the surface. There was a small current still running, not strong enough to trouble us, and we allowed ourselves to drift slowly down the main anchor line towards the monolithic structures below.

The word 'monolith' means literally 'single stone' and is used to refer to 'a large block of stone or anything that resembles one in appearance'.[1] But what troubled me most about the monoliths of Centre Circle – a matter related to my secret fear of geological processes known to Wolf but unknown to me – was precisely the question of whether they were 'single stones' or not. I had never done what Wolf intended to do now, which was to scrape off some of the thick marine growth covering the monoliths to see what the core material was made of. But I had handled them many times and had vaguely arrived at the idea that they must consist through and through of the same sort of concretized or aggregated 'matrix' of rounded mid-sized stones – resembling river stones – that seemed to form their exteriors. The problem was I had absolutely no idea whether this was going to be good, or bad, for the proposition – my 'theory' if you like – that Centre Circle is a man-made structure.

In our conversation on deck Wolf had seemed genuinely mystified by the video footage I'd shown him. But perhaps once he was close up he would take one look at the monoliths, chisel out a few samples, and prove beyond argument that they had in fact been formed by entirely natural processes. Perhaps he would even slap himself on the brow as we got back in the boat and announce the obscure but correct geological name for this kind of 'natural formation'. Or perhaps he wouldn't. Either way, I'd know for sure in about an hour. That was when it really dawned on me, I think, that not only Kerama was on trial here but also my whole notion that a phase of higher civilization and monumental construction in Japanese prehistory might be attested by ruins underwater.

Fifteen metres above the top of Centre Circle, as we paused in neutral buoyancy to get a perspective on the edifice, I was glad I'd spent the last couple of hours going through our earlier video footage of the site – because it had forced me to think through issues that I had previously overlooked. It wasn't just the crucial question of what the monoliths were made of that had to be addressed, but also Wolf's observation that they were contained on the floor of something like a 'canyon'.

Looking around from this bird's eye view – for a diver does have some of the freedom of manoeuvre in the water that a bird has in the sky – I began to get a proper sense, for the first time, of the topography that surrounds the two great co-joined circles (Centre Circle and Small Centre Circle) and of how their keyhole-shaped perimeter is formed, and even of the relationships between the fully detached and 'semi-detached' monoliths that make up the circles.

All these structures occupy the summit of a very large, gently sloping outcrop of rock extending away in all directions, gradually disappearing into deeper waters. At the end of the Ice Age, when the outcrop last stood above sea-level, its highest point would have been the place now marked by the top of the central monolith of Centre Circle. From there you could have stood and surveyed the entire area around.

But then – it seemed inescapable – some powerful force must have intervened, perhaps organized human beings, perhaps weird nature, and carved out the flat-floored, sheer-walled, semi-subterranean, keyhole-shaped enclosure now containing the great rock uprights that form the two circles. Marine growth had gnarled and knobbled the contours of the uprights and it would not be clear until the growth was scraped off how smooth and clean-cut – or otherwise – they originally might have been.

I knew that Wolf would be looking for a natural explanation and supposed that much depended on the constitution of the rock. This, hopefully, we would soon be able to establish since he had brought along with him a fearsome little hammer and mesh bags for the collection of samples. Once we had a better idea of the core material, however, the question we had to address ourselves to – the only question in town really – was what *sort* of force could have produced an amazing 'design' like this. Despite lingering doubts, I felt a sudden surge of confidence that nature could not have done it – not unaided anyway. On the contrary, the pattern was a complex and a purposive one, rather difficult to execute in any kind of rock, and the more I studied it the more obvious it seemed that it was deliberate and planned.

With reference to photo 78 the reader will note that directly beneath the diver, on the north side of Centre Circle, is the smallest and lowest of the three completely free-standing uprights. What I noticed for the first time that afternoon was that this 'broken monolith', as I had thought of it before, forms the beginning of a definite anti-clockwise spiral extending through the top of the next monolith (much higher) and of the next (higher still), then winding around the west and south sides of the central upright, where it takes on the curve of the surrounding enclosure wall – itself not continuous but segregated into units separated by deep channels.

The dividing line between Centre Circle and Small Centre Circle is formed by the same low upright where the spiral begins. I swam over it now and looked down on it from the north side with Small Centre Circle just below me. I believe that it shows every sign, as does the entire structure, of having been carved and shaped by man. Though it is a small detail, I have always been impressed by the way it is curved on one side to match the outer curve of the big central upright to its south and on the other to match the curve of the only slightly smaller upright behind it to its north. It is also difficult to imagine how the narrow, clean-edged 'second pathway' that parallels the wider inner pathway around the central upright could have been cut so precisely by any natural force.

Just before we dropped down into the structure I noticed out of the corner of my eye, thanks to the exceptional visibility that afternoon, something I had not seen since our first dives here back in 1999. This was the existence, not far beyond the south-western perimeter of Centre Circle – on the slopes below the summit of the ancient mound – of other circles and ovals and spirals made up of individual stones, large cobbles, boulders, mostly a metre or less in length, all of them rounded and smoothed off at the edges, coiled and intertwined with each other like necklaces or the links of a chain strewn upon the ground. As I had told Wolf earlier, they looked a lot, though not exactly, like the 'river stones' that were also weirdly stuck, or aggregated (or formed part of the bedrock itself?) all over the uprights of Centre Circle.

I made a mental note that we should go and take a proper look at these nearby 'river-stone' or 'river-boulder' circles and try to figure out how they had been formed. Perhaps Wolf would have a sensible geological explanation. But seeing them again for the first time in two years, and having travelled widely overland in Japan since then, they instantly reminded me of the Jomon stone circles such as Komakino Iseki and Oyu that I had visited in the north of Honshu in May 2000. So far as I could remember, those circles too had been fashioned out of river stones and river boulders like these, and disposed upon the ground in exactly the same way.

It was potentially an important connection.

By now Wolf and I had reached the base of Centre Circle and were standing up on the inner pathway examining the monoliths. It was true, as Wolf had observed from the video footage, that they were completely overgrown with a fantastic menagerie of marine organisms. But at the same time, protruding out through them like a harvest of ripe fruit, was this peculiar matrix of individual river cobbles. One that I noticed in particular, about the diameter of a large dinner plate and probably weighing several kilos, jutted sideways from the top of the second monolith in the spiral as though reaching out towards the third. How was this to be explained?

Wolf took samples from some of the more prominent river stones plastered to the exteriors of the uprights, then beckoned me to join him at the foot of the second monolith under the overhanging cobble. This was going to be his attempt to find out what the core material of the monolith was made of – and he showed me how the marine growth thinned out then stopped altogether at the junction with the basal path. Immediately above the path it proved relatively easy – much easier than we had expected – to scrape away a large patch of organisms and begin to expose the core.

Wolf scraped and scraped. Scraped and scraped. And gradually what emerged was not, as I had feared, more of the same stony matix or aggregate that clung to the surface, but rather a hard, bright, white core formed unmistakably of the ancient coralline limestone of the Keramas and fully attached at its base to the bedrock. So far as we were able to make out, the monolith appeared to have been smoothly and perfectly cut down from top to bottom with a beautiful curve incorporated into it to match the curve of the pathways that were defined on either side of it and the curve of the central upright. I could even see, where Wolf had scraped away the growth particularly successfully, the original organisms that had fossilized millions of years ago to form the white coral rock out of which the entire perimeter of the circle and all its uprights had later been cut. Coral rock where it is available is an ideal construction material – and from the little stone blocks used to build private houses in the Maldives today, to the massive 'Trilithion' of ancient Tonga, to the megalithic temples of Malta, you can see the use of white coralline limestones in which the structure of the ancient fossilized organisms can clearly be made out.

I was grateful to Wolf for having done this little and obvious thing – obvious, anyway to a professional geologist – i.e., for having established what the core material of Centre Circle's monoliths actually is. Because this kind of coralline limestone, as well as being visually and aesthetically striking, is also extremely hard. For a natural force to have cut such a material in such a complex way with sheer walls 4 metres deep, and with parallel curves and pathways – the whole hewn out as a semi-subterranean enclosure in the summit of an ancient mound – was, it seemed to me, something that Wolf was going to find very difficult to explain.

Half an hour later we were back on the boat. The principal underwater camera that the Channel 4 team had been using to shoot the dive had malfunctioned, and the director needed us to do it all again. But it was now after 2.30, the current had returned with a vengeance during the last fifteen minutes of the first dive, and it didn't look like we'd be able to get back in the water at all. We decided to sit at anchor until five. Diving much after that, with nightfall coming, would not be safe this far out in the open ocean and we'd have to return to Okinawa with what we'd got. But if the current slackened before, then we would attempt a second dive.

Where has all the debris gone?

We did get our second dive when miraculously, just after 4.30, the buoy popped up from out of the current again. The light below was surprisingly good and we spent a useful forty-five minutes underwater. Certain scenes were shot with me in which Wolf was not needed – during these he went happily off exploring on his own. In other scenes we repeated for the camera what we'd done for real the first time around. Again Wolf scraped off growth from the base of a Centre Circle monolith and exposed the sheer white coralline limestone beneath.

Again I found myself fascinated by this bright underlying stone, cut from almost exactly the same sort of material as the most ancient and enduring megalithic structures of far-off Malta.

Indeed, by visualizing a Maltese temple like Hagar Qim or Gigantija in all its glory, its white coralline limestone megaliths reflecting the dazzling Mediterranean sunshine, I could begin to imagine how the two great rock-hewn circles of Kerama might have looked, in all their glory, when all this area as far east as Okinawa was above water at the end of the Ice Age.

As you approached them from lower down the gentle slopes of the surrounding rocky massif – all of it formed out of the same 100-million-year-old fossilized coral reef – you would at first have not been aware that any structures were present there at all. Only from the rim of the enclosure looking down would you have suddenly found yourself confronted by a majestic and mysterious spiral of glowing monoliths, the tallest more than twice the height of a tall man.

Unlike the uprights of the great Maltese temples, however, which were quarried elsewhere and then transported and erected on the temple sites, these Centre Circle monoliths had been quarried *in situ* out of the bedrock of the ancient mound – to which they were still attached at their bases.[2] That automatically classifies the whole edifice as a rock-hewn structure and, as at Yonaguni, one of the mysteries it confronts us with, if we are to imagine that the 'hewing' was done by natural forces, is – what happened to all the missing rock? The reader may easily verify from photos 78–80 that quite a large amount of this very hard rock would have had to be hewn out to free-up the monoliths and excavate the 4-metre-deep semi-subterranean enclosure in which they are confined. As the photographs also show, none of this excavated rock is present as any form of rubble or debris within the two circles. This is a very troubling anomaly if the circles were made by natural forces, but is exactly what one would expect if they are the work of human beings.

Wolf on Kerama

Much to my surprise – because I had become so used to his hard-nosed scepticism at Yonaguni – Wolf stayed as open-minded on the problem of Centre Circle after our two dives as he had been when I'd shown him the tapes before we got in the water. Moreover, he was able to carry out on-board chemical tests on the samples that he had taken both from the core and from the aggregate of river stones plastered to the outside of the monoliths.

The tests proved on camera – though it was already completely obvious to the naked eye – that these were two entirely different types of rock. The core, as we knew, was very ancient coralline limestone. The rounded cobbles caught up in the aggregate were sandstone and had, as Wolf judged:

> been shaped by waters, by running waters; this is beyond every doubt. These sandstones all show a rounded-out shape, and this leads us to two possible origination

processes for these stones: one would be riverine waters, and the other one would be coastline beach, pebbles or something like that, which have been rolled forward and backward to get this rounded shape.

Wolf added that during the second dive, while I had been working with the cameraman, he had explored outside the perimeter of Centre Circle.

> What was special for me was to discover that these rather large cobbles, pebble stones, made of sandstone, which are glued to the uprights and inner formations, also appear in places outside the circle. So I dived a little sidewards – I don't know the direction – and then I found a field of the same pebbles, not really pebbles, it was really big, big stones, but scattered in a very chaotic way over the surface of the coralline bedrock.

Wolf's suggestion about these 'pebbles' of assorted sizes – which we also referred to in our conversations variously as boulders, cobbles and river stones – was the obvious one but, he warned, a pure guess. At some stage, probably millions of years after the fossilization and exposure of the ancient coral bedrock,

> a river has carried his load here . . . So maybe it sometimes had water and sometimes it dried out, changed the bed, and left the stones in here . . . So it seems that parts of this old coral reef were covered by these boulders somehow transported by river, a very broad river, because the field seems to me to be very broad.

If it was a guess, it sounded like a good one.

But on the larger mystery of the monoliths and uprights of the rock-hewn circles Wolf admitted that he was completely dumbfounded – although he rightly cautioned that he could only speak from his own experience as a marine geologist. Perhaps other geologists had seen natural structures that were the same as or very similar to Centre Circle somewhere in the world and would be able explain the enigmatic curved parallel walls and well-shaped uprights. He could not, however.

> *Wolf:* I have no explanation for these . . . for these . . .
> *GH:* For the circles?
> *Wolf:* For the circles, and for the structures inside them. For sure is that they must have been formatted after the pebbles were laid down on the coralline ground – because some of these pebbles are hanging over the canyons, so and they could not have come earlier . . . But I don't see any force which could have shaped these –
> *GH:* – any force of nature, which could shape the circles and uprights?
> *Wolf:* Yes, of course.

GH: So that leaves us . . .

Wolf: At the moment . . .

GH: That leaves us with one option then? Man-made.

Wolf: I don't know. I would not be . . .

GH: You wouldn't rush so fast?

Wolf: I would not go so far. I mean you have to do really a lot of research to establish that. But what is really strange is these parallel walls running round. It is very strange because if, for example, the erosive force were water, the two edges of a river bed or something like that are not exactly parallel to each other like these. So this is what I can say. And even solution, chemical solution does not leave hints like this, of this accuracy.

GH: Paralleling of walls?

Wolf: Paralleling accuracy.

GH: So what can be said for sure about this structure? Can we be sure about anything?

Wolf: What is clear is that we have an ancient fossilized coral reef and we have these pebbles scattered on top of it which came later. And then a second force started. This was the erosive force which then carved these structures out of the ground – if man or if nature.

GH: Now, you geologists will say 'carved by nature' and we poets will say 'carved by man'.

Wolf: I don't say anything definite. Much more research must be done. But I agree that this is very amazing and very strange, even to me, how these structural buildings could be formed. I haven't seen such structures done by nature. I won't dare say anything else about human activities because I do not know anything about that.

From a geologist as instinctively cautious and phlegmatic as Wolf Wichmann this was as close as I was ever likely to get to a confirmation that the rock-hewn stone circles of Kerama really could be man-made. Still, I couldn't resist pushing for more.

GH: I'll tell you why I think it's man-made.

Wolf: Yes, please.

GH: It's not just the sense of organization of the structure itself. It's the fact that we have an ancient culture on these islands which made stone circles. They are known to have made stone circles and some of those circles still survive – not like Centre Circle, smaller, with the largest blocks about half a tonne, and usually much less. But the idea of a stone circle and, indeed, of interlinked stone circles, was something they did. So you know, when we look at Centre Circle and Small Centre Circle – and we know that we're on a set of islands where we have an ancient culture called the Jomon, who are known to have made stone circles – then to me it's less extraordinary, in a way, to attribute it to them – to the Jomon – than it is to any unknown force of nature. I don't deny that nature often provides a sense of organization, but it's the unique character of this in a land where we have a very ancient culture, actually which existed from 16,000 years ago until 2,000 years ago, the Jomon, who made stone circles . . . you know, I start wondering.

Wolf: OK, I can follow your point. But still it has to be proven that this is really done by the Jomon.

GH: Yes, yes, I agree.

Wolf: And this is very hard to find. You have to scratch and you have to clean it to find marks or to find any evidence, maybe in a series of other monuments being proven to have been constructed by this society.

GH: Yes. Well, we have many stone circles that have been constructed by that society, but this . . . amongst their stone circles, this would rank as the largest and the most unusual. But I repeat, we're on a set of islands here which had an ancient culture, the Jomon, that is recognized by historians. The earliest surviving work of that ancient culture goes back to the Ice Age, around 16,000 years ago. The Jomon were known to make stone circles. We have a stone circle at a depth that is likely to have been exposed at some point during the Ice Age. What's the next logical step?

Wolf: No, no, I mean I agree to that, to that chain, to that chain – it's clear. But the last point . . . this is the point that you must prove. A theory remains a theory unless you have proof.

What I had at that moment was a theory about possible Jomon origins for the underwater monoliths and circles of Kerama, hinting at an early and as yet undiscovered phase of monumental construction in Jomon prehistory. That theory had just passed a very important hurdle, since an on-site investigation by a sceptical marine geologist had been unable to produce any viable natural explanation for the structures.

But it was still a theory.

Komakino Iseki underwater?

Having completed our work at Kerama, we parted company with Wolf the next morning. He flew back to Germany, and Santha and I carried on with the film crew to the north of Japan. There, eventually, we found ourselves at the wonderful Jomon stone circle of Komakino Iseki (see chapter 25) near the big site of Sannai-Muryama in Aomori Prefecture. Though it was by now late March, the weather was still freezing in the north, old snow was still lying on the ground and the whole scene presented a huge contrast to the tropical warmth and blue waters of Kerama.

While the crew were setting up, I paced amongst the stones, shivering with

Komakino Iseki

0   500km

cold. The distinctive, rounded river stones of Komakino Iseki. Boulders, pebbles, cobbles, arranged in a series of concentric circles, the largest with a diameter of 150 metres. And between the rings, groups of smaller circles, touching at the edges like the links of a chain . . .

I'd already made the connection underwater a few days earlier at Kerama. It had struck me as important then and I'd meant to look into it further with Wolf but had been prevented from doing so by shortage of time. It was the phenomenon that he had noticed independently when he'd gone off exploring on his own while I was working with the cameraman and which he'd later described as a boulder field – 'big stones disposed in a very chaotic way over the surface of the coralline bedrock'. But if I was right, the disposal of these big rounded river stones was not nearly so chaotic as Wolf had thought. I was pretty sure that I had seen his 'boulder field' too, and even videoed it briefly in 1999, and glimpsed it again on the first of the two dives we had just completed.

And where he had seen chaos I had seen order. Because when I had filmed them in 1999 some of the big rounded river stones scattered across the coralline plain had definitely been arranged in circles, one stone laid lengthwise next to the other. As at Komakino Iseki, I remembered, these 'circles' were really more oval than truly circular in shape (though I shall continue to refer to them as circles for convenience). And as at Komakino Iseki, the stones had been medium-sized – typically around a metre in length or less.

So Kerama still wasn't finished with me. On this latest trip, as on every previous trip, I had failed to do my job there properly. I'd been lured in by the glamour of the rock-hewn circles with their 4-metre-high monoliths. But I could see now how the proof of the Jomon connection I sought might all along have been lying in that humble 'boulder field' just beyond.

I was going to have to go back.

30 / The Shark at the Gate

> The origin of maps and geographical treatises goes far back into former ages.
>
> Phei Hsiu, Chinese geographer, AD 224–71

The earliest surviving reference to Taiwan in Chinese annals is in the *Sui-Shu* – the history of the Sui Dynasty, AD 581–618.[1] There it is classified amongst the Lu-Chu islands – the old Chinese name for what is now (with the exception of Taiwan) Japan's Ryukyu archipelago.[2] Starting at Yonaguni in the south-west – within sight of the mountains of Taiwan on a clear day – the Lu-Chu/Ryukyus extend through the Keramas and Okinawa almost as far to the north-east as Kyushu, and have been under discontinuous Japanese hegemony since the fourteenth century. However, they did not officially become part of Japan until 1879.[3] It is therefore intriguing that very ancient Japanese legends 'definitely place *R'yugu*, the Sea King's sanctuary, in the Lu-Chu Islands'.[4]

The Japanese notion of the Sea-King's sanctuary, which the *Nihongi* calls 'the Palace of the God of the Sea'[5] and the *Kojiki* calls 'the Palace of the Kami Great-Ocean-Possessor',[6] is a rather complicated one. As we saw in chapter 26, its primary mythical setting is underwater, amidst huge stone structures looming up from the sea-bed, in a place that can only be reached by diving. But it also has elusive connections to an enchanted 'Spirit island', accessed by a magical journey across the sea, where human life dilates towards immortality – or so we may gather from the story of the man who spent three years residing there and then returned to his home only to discover that 300 human years had passed. Last but not least, there seems to be an enigmatic link with the dark and terrifying Underworld of the Land of Yomi where the soul of Izanami fled after her death.[7]

When we turn to the traditions and mythology of ancient China we find the same ingredients – immortality, enchanted islands, the Underworld – often used in the same way. Thus, the oldest dynastic history, the *Shih Chi* (completed about 90 BC), tells us of voyages – sent to the same general area of the 'Great Eastern Ocean' that is occupied by the Ryukyus – in search of magical islands where the inhabitants were immortals thanks to their possession of 'the drug which will prevent death'.[8] And in another text, the *Ling Wai Tai Ta*, we read how 'In the Great Eastern Ocean there is a bank of sand and rocks some myriads of *li* in length, and nearby is the *Wei-Lei*, the place where the water pours down into the Nine Underworlds.'[9]

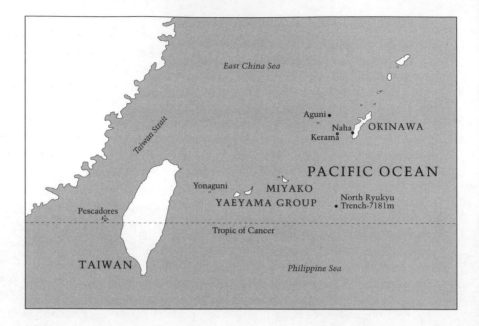

Since a *li* is equivalent to 0.309 of a mile,[10] then thousands of *li* (say 3000 of them?) must equal at least a thousand miles. One wonders *where* in the Eastern Ocean – i.e., the Pacific – such an enormous bank of rock and sand could have been located.

But perhaps it would be better to ask: *when*?

Abodes of the immortals

The Wei-Lei – which might be translated politely as 'the ultimate drain'[11] – has another, more beautiful, name in Japan. There it is called the Kuroshio, the Black Current or the Black Tide,[12] and we saw in chapter 25 how serious scholars at the Smithsonian Institution led by Betty Meggers believe that it may have carried Jomon seafarers all the way across the Pacific to settle in the Americas more than 5000 years ago.[13] There are even indications of earlier Jomon migrations to the Americas going back as far as 15,000 years ago.[14]

I first set eyes on the Black Current – and you really can *see* it; it's an entity; it's alive; it is, I guess, a Kami – from the heights of Cape Ashizuri on the Japanese island of Shikoku, where groups of great megaliths stand gazing down on the rippling waters as though sharing a secret.[15] We know already that from there the Kuroshio runs north past the rest of the Japanese archipelago and thence across the Pacific. South of Shikoku and Kyushu, however, it also flows past Taiwan and the Ryukyu islands immediately to the east of the Chinese coast – the region of the Pacific most directly accessible to Chinese mariners.[16] Could it have been somewhere hereabouts that the ancient Chinese believed 'the Wei-Lei drains into the world from which men do not return'?[17] The distinguished Sinologist Joseph Needham thought it should be further east – perhaps even as far east as the Americas,[18] a quasi-diffusionist view that was ahead of its time in 1971 when he expressed it. But there is no consensus.

Needham also attributed definite historicity to Chinese accounts of searches for magical islands.[19] The *Shih Chi* reports the exploits of a mariner named Hsu Fu, in the late third century BC. Rather like Columbus petitioning the sovereigns of Europe 1700 years later to fund his voyages of discovery *westwards* across the Atlantic in search of Antilia,[20] Hsu Fu petitioned the emperor of China in 219 BC with claims to have special knowledge of a wonderful domain of 'magic mountain islands' to the *east* of China in the Pacific:

> In the midst of the Eastern Sea there are three magic mountain islands, *Pheng Lai*, *Fang-Chang* and *Ying-Chou*, inhabited by immortals. We beg to be authorized to put to sea ... to go and look for the abodes of the immortals hidden in the Eastern Ocean.[21]

The target of this voyage, which did receive the emperor's blessing, is stated to be far off 'in the midst of the Eastern Sea', but again there is no consensus as to its location. Hsu Fu went to look for it with a well-stocked fleet, said to have been carrying large numbers of young men and women and 'ample supplies of the seeds of the five grains'[22] – which suggests settlement plans. The *Shih Chi* records that he 'never came back to China'.[23] But confusingly, the same chronicle also reports other voyages – equally fruitless in terms of any definite discovery – which sought the same islands much closer to the Chinese coast:

> From the time of the Kings of Chhi [c.378 BC] ... people were sent out into the ocean to search for the islands of *Pheng Lai, Fang-Chang* and *Ying-Chou*. These three holy mountain isles were reported to be in the midst of Po-Hai [the Gulf of Bo Hai], not so distant from human habitations ... Many immortals live there, and the drug which will prevent death (*pu ssu chih yao*) is found there, but the difficulty [is] that ... before you have reached them ... these three holy mountain isles sink down below the water – or else a wind suddenly drives the ship away from them. So no one can really reach them ...[24]

Convergence

All taken together it seems fair to say that the Chinese myths contain very much the same sort of strange brew as do their Japanese counterparts – of an entrance to the Underworld, of enchanted islands and of kingdoms beneath the sea. But where Japanese traditions specify the location of the Kingdom of the Sea-King as being somewhere in the Lu-Chu islands, Chinese references to Pheng Lai, Fang-Chang and Ying-Chou, 'the islands of the Sea Mage',[25] are contradictory as to location – varying from Hsu Fu's unspecified destination in the midst of the Pacific Ocean to somewhere extremely close to home like the

Bo Hai Gulf (which lies between the city of Tianjin and Korea Bay at the northern end of the Yellow Sea).

Perhaps the contradiction is less than it seems, however, for Hsu Fu is venerated as a Kami in Japan. There he is the Kami Jofuku whose tomb-shrine exists to this day at Shingu in Wakayama Prefecture of southern Honshu,[26] which, like Cape Ashizuri in nearby Shikoku, overlooks the course of the Black Current. If there is truth to this strange tradition of Hsu Fu's settlement at Shingu then it suggests that the islands of the Sea Mage 'hidden in the Eastern Ocean' to which he had directed his expedition must all along have been somewhere in the vicinity of southern Japan.

Although this cannot be confirmed, I suspect that the convergence of myths from both China and Japan does hint at something very real – perhaps even a shared memory of a lost land with 'palaces and towers', once believed to be enchanted and inhabited by 'immortals', that now lies beneath the sea.

Lu-Chu and Bo Hai

And where are we to look for this lost land, should we wish to rediscover it? Across the two traditions the only clear pointers given to its whereabouts are that it is to be found somewhere in the Lu-Chu islands – effectively anywhere along the arc from Taiwan to Kyushu – or near the northern terminus of the Yellow Sea in the Bo Hai Gulf. These locations are not proximate but are at opposite ends of the same region. Both are highly plausible as potential locations for 'palaces beneath the sea'.

At the end of the Ice Age we know that the Ryukyu islands were larger than they are today. There was therefore ample room along their antediluvian shores for any number of 'palaces' to be built – and later submerged as sea-levels rose. Moreover, as I've endeavoured to show in the preceding pages, a number of extraordinary underwater structures that seem increasingly likely to have been made by humans at the end of the Ice Age *have already been found* around the Ryukyus.

Likewise, if we look at the Bo Hai Gulf on Glenn Milne's inundation maps, we discover that it too has an interesting story to tell. Down to 14,600 years ago it was dry and far from the sea (opposite, above). By 13,500 years ago, however, we observe that the Yellow Sea has penetrated deeply inland towards the modern coast of China and has carved out the Korean peninsula for the first time – but the Bo Hai Gulf is still dry (opposite, below).

Then we come to the map for 12,400 years ago (page 660). In between Shikoku and Honshu we see, still well preserved, the correlation with the Bungo Strait much as it is portrayed on the 1424 chart. And at the northern end of the Yellow Sea we see that the Bo Hai Gulf has at last succumbed to partial inundation. Within it, rather strikingly, an island has materialized. Though it is beyond the resolution limits of Milne's computer model, it is perfectly possible that the single island shown could then or at some stage afterwards have been divided

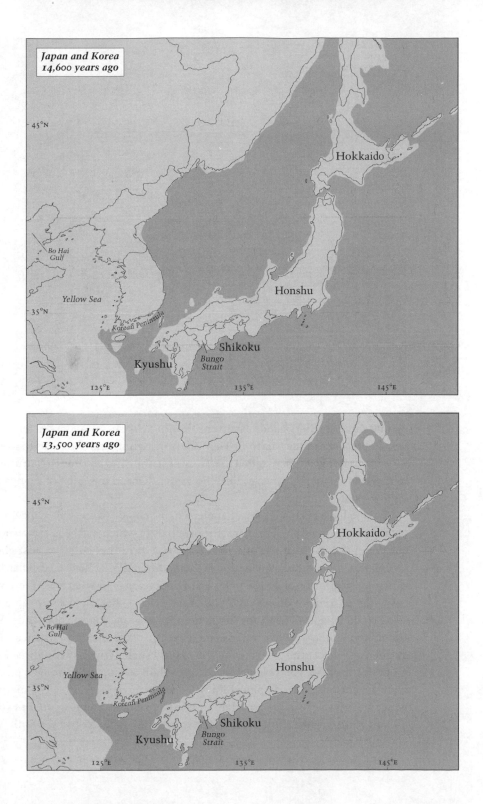

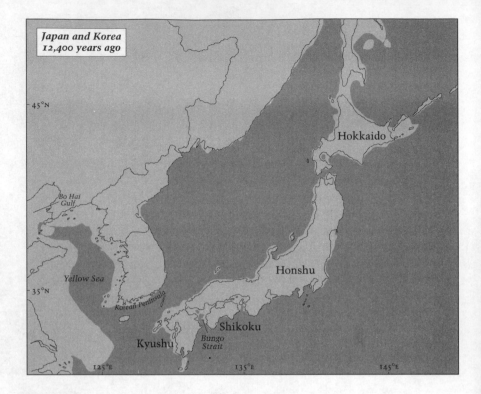

Japan and Korea
12,400 years ago

45°N

Hokkaido

Bo Hai
Gulf

Honshu

Yellow Sea

35°N

Korean Peninsula

Shikoku

Kyushu   Bungo
          Strait

125°E   135°E   145°E

up into three smaller islands exactly as the *Shih Chi* seems to remember. Either way its presence in the palaeo-Bo Hai Gulf is intriguing and obliges us to wonder exactly what it was that inspired the Chinese in the third and fourth centuries BC to make so many real voyages into the Eastern Ocean in search of islands that did not exist.

Could it have been a legacy of ancient maps copied from copies of copies of even more ancient maps, the originals of which had been drawn before the rising sea-levels at the end of the Ice Age gave the world its post-glacial face? If not, what other explanation is there? After all, the Bo Hai Gulf has been in its present form for at least the last 9000 years – so what possible reason could the chronicler of the *Shih Chi* around 90 BC have had to imagine that there could ever have been any dry land in it, let alone a group of islands? Why should the Chinese have gone to such lengths to seek out those islands – over a period of two centuries – when it was plainly such a fruitless enterprise? And must we resort to 'coincidence' again to explain what appears to be a piece of truly anachronistic geographical knowledge in the possession of the mariners who pursued the search – i.e., that an island or islands, which could never be found because it had sunk beneath the waves, did once exist in the Gulf of Bo Hai?

Think about it. What are the odds against the Chinese seafarers of 2300 years

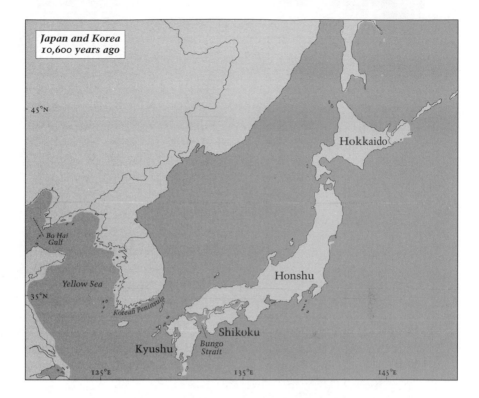

Japan and Korea 10,600 years ago

ago getting their palaeo-geography so right purely by chance? Or we can come at the question from another direction? How likely is it that China's historical quest for the three 'holy mountain isles' in the Gulf of Bo Hai was inspired by meaningless myths – as orthodox historians must conclude – when we now know that an island or islands did exist in the Gulf of Bo Hai 12,400 years ago and did subsequently 'sink down below the water' as the *Shih Chi* maintains?

The palaeo-island was gone within 1800 years, as we see on the next inundation map in the sequence (for 10,600 years ago, above). The maps also show that the entire span of the island's existence did not exceed 3000 years, since it had not yet taken shape 13,500 years ago

And though the *Shih Chi* does not give us a written description of the palaeo-geography of the Yellow Sea for 13,500 years ago (as it seems to for 12,400 years ago), it is a remarkable anomaly of history that something looking very much like a graphic representation of the Yellow Sea and its coastline 13,500 years ago has survived. Now kept in the 'Forest of Steles' in Xian, it is a good Chinese map (Needham describes it as 'magnificent'),[27] carved in stone in AD 1137, called the *Hua I Thu* ('Map of China and the Barbarian Countries').[28] In a refrain familiar from our investigation of the sometimes strangely anachronistic portolans of the West it is known to have been based on older sources.[29] Nobody can be absolutely sure how much older. But if ever there was a land in

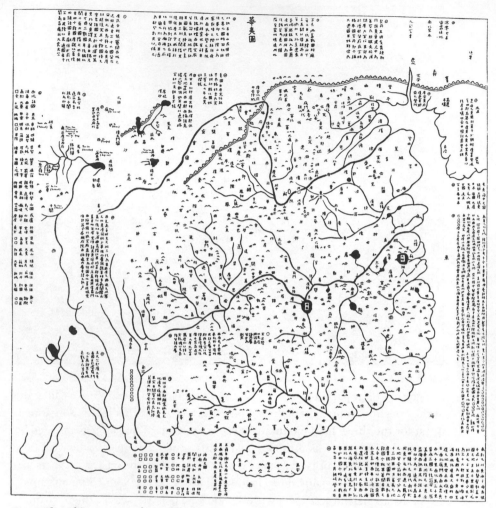

Hua I Thu Chinese map of AD 1137.

which we might expect to encounter an ancient map-making tradition, then that land is surely China.

At around the time that the maps attributed to Marinus of Tyre were being circulated in the Mediterranean, a great Chinese geographer, Chang Heng (AD 78–139) was producing maps of unbelievably high quality in China. Like Marinus, he is often credited by historians with having introduced a grid system for maps – it being said of him that he had 'cast a network of coordinates about heaven and earth, and reckoned on the basis of it'. The title of one of his lost books was 'Discourse on Net Calculations' and there was also a 'Bird's-Eye Map'.[30]

It is clear, however, that Chang Heng, who is considered one of the 'fathers' of scientific cartography in China, must himself have been the 'son' of a much earlier and older tradition – for one does not reach his level of sophistication without a vast store of prior knowledge and experience to build upon. That such a store or archive did exist, and that it did contain extremely ancient material, is confirmed in the dynastic chronicles which also give prominence to the works of another great Chinese geographer, Phei Hsiu (AD 224–271):

> Phei Hsiu made a critical study of ancient texts, rejected what was dubious [outdated by climate change?], and classified, whenever he could, the ancient names which had disappeared [because inundated?]; finally composing a geographical map in 18 sheets. He presented it to the emperor, who kept it in secret archives.[31]

The chronicles also cite the full text of Phei Hsiu's preface to his Atlas, in which he laments the loss of geographical knowledge from earlier times (emphases added):

> *The origin of maps and geographical treatises goes far back into former ages.* Under the three dynasties [Hsia, Shang and Chou,[32] c. 2000–1000 BC][33] there were special officials for this (*Kuo Shih*). Then when the Han people sacked Hsien-yang, Hsiao Ho collected all the maps and documents of the Chhin. *Now it is no longer possible to find the old maps in secret archives,* and even those which Hsiao Ho found are missing; we have only maps, both general and local, from the later Han time. None of these employs a graduated scale (*fen lu*) and none of them is arranged in a rectangular grid.[34]

The implication is not only archives of maps going back thousands of years but also that the rectangular grid was known very early in Chinese history, then fell into disuse under the Han in the first millennium BC, and was then later reintroduced by Chang Heng, the contemporary of Marinus of Tyre, when he cast his 'network of coordinates about heaven and earth'.

So we have a confirmed cartographic science in China from around 2000 years ago (Chang Heng, Phei Hsiu), and references to an ancestral tradition more than 2000 years older than that – which presumably was itself not new in 2000 BC when 'special officials' already existed dedicated to the archiving and probably copying of ancient maps.

It is against this long background, therefore, which disappears into prehistory and has no known beginning in China, that we should evaluate the *Hua I Thu* – the Chinese map of AD 1137 said to have been based on older sources – which I have claimed shows the Yellow Sea and the Korean peninsula not as they looked in AD 1137 but as they looked 13,500 years ago. Although other interesting issues are raised by the *Hua I Thu*, I will confine my remarks here to the north-eastern segment of the map, around the Yellow Sea. In the diagrams overleaf the reader may compare the Yellow Sea and the Korean peninsula as they appear on a modern map with the same areas on the inundation map for

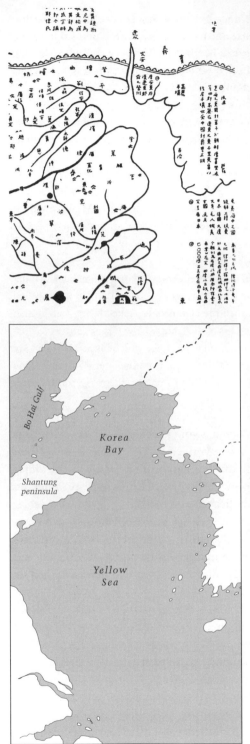

(Left) *The Yellow Sea as shown on the Hua I
Thu map of* AD 1137.

(Below left) *Modern map of the Yellow Sea.*

(Below right) *The Yellow Sea as it looked
13,500 years ago.*

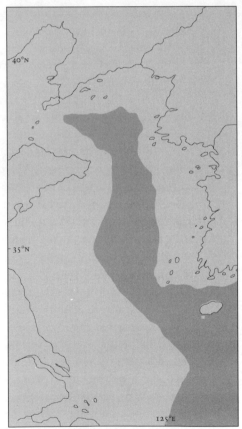

664 / Underworld

13,500 years ago and on the *Hua I Thu*. It will be observed that an excellent level of correspondence does in fact exist between the latter two and that the *Hua I Thu*'s portrayal, though a bad one of the Yellow Sea as it looked in 1137 – and as it still looks today – is rather a good one if it represents the Yellow Sea 13,500 years ago. Particularly noticeable is the absence on both the *Hua I Thu* and the inundation map of the Shantung peninsula, a prominent feature of the northern end of the Yellow Sea, which the rising waters began to carve out some time after 13,000 years ago and which took on its modern form about 10,000 years ago.

Nor can it be claimed that the Chinese of 1137 were simply ignorant of the Shantung peninsula. On the contrary we can prove that they knew it very well – because another map, also carved on stone in 1137 and also preserved in the Forest of Steles at Xian, shows it very clearly and with great accuracy much as it looks today. Called the *Yu Chi Thu* ('Map of the Tracks of Yu'), it too is a copy of an earlier original but, as Joseph Needham observes, 'has a more modern look' than the *Hua I Thu* and seems to belong 'to a different tradition'.[35]

Could this be because the *Hua I Thu*'s portrayal of the Yellow Sea was derived directly from a very ancient source map – perhaps stored with many others in the Imperial archives – while the *Yu Chi Thu* incorporates the results of Chinese maritime expeditions that we know had explored the region thoroughly at least as early as the third century BC?

Cutting Korea down to size

Deferring for the moment our parallel interest in the lost islands and sunken kingdoms of the Ryukyu archipelago, what do we have so far concerning Korea and the northern end of the Yellow Sea?

In summary we have geographical traditions, recorded in the *Shih Chi*, which place the lost 'islands of the Sea Mage' in the Bo Hai Gulf. There was an island in the right place 12,400 years ago. We also have a Chinese map copied from earlier sources on to stone in AD 1137 that anomalously fails to show the Shantung peninsula and that greatly narrows the Yellow Sea between China and Korea. However, there was no Shantung peninsula, and the Yellow Sea was narrowed in exactly this way 13,500 years ago.

Reduced to bare essentials, therefore, what the *Shih Chi* and the *Hua I Thu* both proclaim is that Korea was larger in the past than now. This is completely true. Yet as the inundation maps show, the Korean coastline has remained unchanged for the last 9000 years – having done all its shrinking in the 5000 years prior to that. It follows that if these are memories of a formerly much larger Korea then they must be at least 9000 years old.

Japan too preserves such memories – if they are memories. In the *Fudoki* we read of an exploit of Sosano-wo-no-Mikoto, the great Kami called Brave-Swift-Impetuous-Male, whom we encountered in chapter 26. Seeing that parts of the Korean peninsula are much larger than they should be, he removes them, draws them away ('slowly, slowly, like a river boat') and sews them on to Japan.[36] I have nothing to say about the latter part of the myth, but I do think the bit about the subtraction of land from Korea is interesting: 'Perceiving that it had a portion in excess, he took up a spade, wide and flat like the breast of a maiden, and thrust it into the land, parting it asunder as one cuts the gills of a huge fish, and severing it.'[37]

Sosano repeats this process with several other parts of Korea until he is satisfied[38] and, presumably, the peninsula has taken on its present shape.

The myth is called 'the drawing of the lands'. What it conjures up in my mind are not images of a spade shaped like a maiden's breast, attractive though the concept may be, but inundation maps of this area between roughly 14,600 and 10,600 years ago. These do show lands being 'sliced away' piecemeal as the basin of the Yellow Sea filled up to allow the Korean peninsula to emerge. Therefore, although it may have come about by chance and have no significance whatsoever, what confronts us in this text is another time-capsule of accurate geographical information on the region as it looked during the meltdown of the Ice Age.

The Fall

There are, I think, too many such time-capsules of ancient geography scattered across too many sources from too many lands – myths and folklore, maps and traditions – for every example to be explained away as coincidence. I am convinced that something must lie behind all this and that the odds are rising in favour of a significant forgotten episode in the story of civilization localized in time at the end of the Ice Age. The hypothesis I have followed, which receives virtually unlimited support from world deluge myths, is that the discontinuity – some might call it the Fall – was a direct product of episodes of post-glacial flooding and linked cataclysms. So it follows that the evidence for what we have lost – which might explain how and by whom the world came to be mapped more than 12,000 years ago – should be found on the bottom of the sea.

The entire 'arc' from Taiwan in the south, north-eastwards through all the islands of the Ryukyu archipelago, brushing the tip of Kyushu, leaping across the Korea Strait and thence into the Yellow Sea, Korea Bay and Bo Hai Bay, encloses an area with enormous potential for underwater discoveries.

For me it is an underworld – an ancient domain of forgotten ancestors. Like the others we have entered in this book – in the Mediterranean, in the Atlantic, in the Indian Ocean – I believe it will have to be explored thoroughly one day if we really want to know the truth about our prehistory.

But by March 2001 I was also beginning to feel that I personally had done all I could to initiate the necessary exploration – and after four years of diving in the Ryukyus I had every reason to expect that the trip with Wolf Wichmann would be the last I would need to do. As I reported in chapter 29, however, Komakino Iseki changed my mind. The resemblance of its circles of river stones

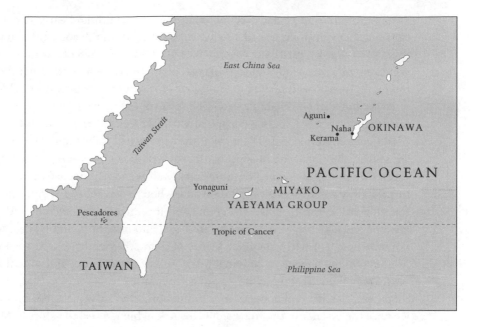

to the circles of river stones I'd glimpsed on the sea-bed at Kerama had to be followed up.

Then came the news that an underwater site had been discovered in Taiwan's Peng-Hu archipelago – the Pescadores Islands – and my Japanese colleagues and I began to plan a short expedition there for late August 2001. Since Taiwan is so close to Okinawa, it obviously also made sense to redive Kerama on the same trip.

Getting from base-3 to point 'D'
August 2001, Taiwan

We're nearing the end of a long story and this is not the place to introduce new characters, plots or locations. But I will mention some of the qualities of Taiwan which place it firmly amongst the usual suspects.

- It was isolated from mainland China by rising sea-levels at the end of the Ice Age when huge areas to its west, south and north were massively inundated.
- It has rich and extensive indigenous flood myths.
- It has megaliths more than 5000 years old positioned on a highly significant geodetic location.
- It has underwater ruins.

I will not delay the reader with lengthy quotations from the very many Taiwanese flood myths that were collected from amongst the indigenous population, primarily by Japanese scholars, in the nineteenth and early twentieth centuries.[39] Typically they tell a story of a warning from the gods, the sound of thunder in the sky, terrifying earthquakes, the pouring down of a wall of water which engulfs mankind, and the survival of a remnant who had either fled to mountain tops or who floated to safety on some sort of improvised vessel.[40]

To provide just one example (from the Ami tribe of central Taiwan), we hear how the four gods of the sea conspired with two gods of the land, Kabitt and Aka, to destroy mankind. The gods of the sea warned Kabitt and Aka: 'In five days when the round moon appears, the sea will make a booming sound: then escape to a mountain where there are stars.' Kabitt and Aka heeded the warning immediately and fled to the mountain and 'when they reached the summit, the sea suddenly began to make the sound and rose higher and higher'.[41] All the lowland settlements were inundated but two children, Sura and Nakao, were not drowned: 'For when the flood overtook them, they embarked in a wooden mortar, which chanced to be lying in the yard of their house, and in that frail vessel they floated safely to the Ragasan mountain.'[42]

So here, handed down since time immemorial by Taiwanese headhunters, we have the essence of the story of Noah's Ark, which is also the story of Manu and the story of Zisudra and (with astonishingly minor variations) the story of all the deluge escapees and survivors in all the world.[43] At some point a real investigation should be mounted into why it is that furious tribes of archaeologists, ethnologists and anthropologists continue to describe the similarities amongst these myths of earth-destroying floods as coincidental, rooted in exaggeration, etc., and thus irrelevant as historical testimony. This is contrary to reason when we know that over a period of roughly 10,000 years between 17,000 and 7000 years ago more than 25 million square kilometres of the earth's surface were inundated. The flood epoch was a reality and in my opinion, since our ancestors went through it, it is not surprising that they told stories and bequeathed to us their shared memories of it. As well as continuing to unveil it through sciences like inundation mapping and palaeo-climatology, therefore, I suggest that if we want to learn what the world was *really* like during the meltdown we should LISTEN TO THE MYTHS.

If you do that you cannot fail to notice, across the 600 or so ancient flood myths known to scholarship, that the events they describe, again and again, were truly terrifying ones. Terrifying. And while we must accept, because the archaeologists say so, that humanity 16,000 or 12,000 years ago was made up entirely of 'primitive' hunter-gatherers, the myths themselves often tell a very different story – when they speak, for example, of the antediluvian cities of the Sumerians or of the Atlanteans before the Fall. If the myths are important memories repackaged as narratives that could be passed down from generation to generation, then what are we to make of memories such as these?

Along with growing numbers of people today I have the uneasy sense that science has not fully understood the peoples of the flood epoch – and that some

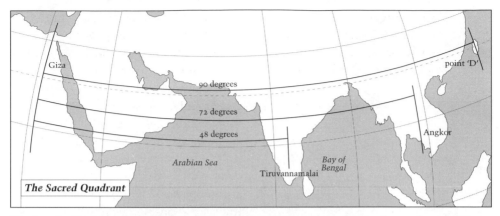

The longitudinal distances, in degrees, from Giza to Tiruvannamalai, Angkor and point 'D'
appear to be based on geometric and astronomical constants.

global cultural development of great significance may have been underway at
that time which was lost or severely dislocated in the inundations. Above all
else it is hints and clues, first to the existence of this lost episode of cultural
development and secondly to its character, that I have sought in the geographical
anomalies of ancient maps – which are not anomalies if they chart the effects
of changing sea-levels at the end of the Ice Age – and in my global search for
underwater monuments that were submerged at the same time. I propose that
the consistent patterns of map anomalies that we have documented – from
Hy-Brasil to India to Japan – bear mute witness to an ancient science of cartography and navigation that explored the world and charted it accurately over a
period of several thousand years during the post-glacial meltdown.

Nor are the maps the only evidence of that conjectured lost geography.
Another point that I have touched on from time to time is relevant here. This
is the apparently planned construction all around the world of sacred, often
megalithic, sites on *specific relative longitudes*. I have commented in this book
on the intriguing longitudinal relationship that exists between the Pyramids of
Giza in Egypt, the great temple of Arunachela at Tiruvannamalai in south India
and the temples of Angkor in Cambodia (Arunachela is 48 degrees of longitude
east of Giza; Angkor is 24 degrees of longitude east of Arunachela; 48 ÷ 2 = 24;
48 + 24 = 72; 5 × 72 = the 360 degrees of a circle). As I have indicated, these
numbers, and others in the same sequence, turn up *repeatedly* in ancient myths
from all parts of the world.[44] The sequence bears a relationship, which may or
may not be causal, to the astronomical phenomenon known as the precession
of the equinoxes (which proceeds, in round numbers, at the rate of 1 degree
every 72 years).[45] But all the numbers that compose the sequence in the myths
also have something else in common – literally their lowest common denominator. They are all divisible by 3.

The number 90 = 30 × 3. In terms of the circle (of which it is exactly one

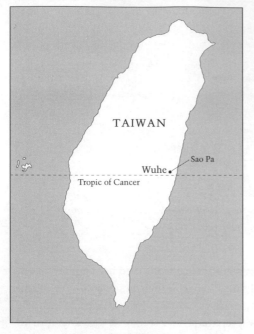

quarter), of geometry (the right angle), and of navigation, there is no doubt that 90 degrees is a significant figure. If point 'A' is 90 degrees of longitude away from point 'D' then the two longitudes ('A' and 'D') are, literally, a quarter of the earth apart from one another. And if there is a sacred site on point 'A' – the Pyramids of Giza – and a sacred site on point 'D' as well, then you would have to be really bad at mental arithmetic not to notice the peculiar long-itudal relationship, based on the lowest common denominator of 3, that seems to link both of them to Tiruvannamalai and to Angkor within the same quadrant. Whether it is by accident, or it is the result of some ancient geodetic survey that founded marker shrines on key longitudes that were later elaborated into monu-ments, the following 3-based relationships do exist: Tiruvannamalai with its Siva cult is 16 × 3 degrees (i.e., 48 degrees) east of Giza, Angkor is 24 × 3 degrees (i.e., 72 degrees) east of Giza; point 'D' is 30 × 3 degrees (i.e., 90 degrees) east of Giza. In addition, point 'D' is 6 × 3 (i.e., 18 degrees) east of Angkor and 14 × 3 (i.e., 42 degrees) east of Tiruvannamalai.

So what and where is this mysterious point 'D' so intricately linked by base-3 geodesy to Angkor, Giza and Tiruvannamalai? It is a spectacular megalithic site in central Taiwan, up in the mountains where the flood survivors went – up on the Wuhe plateau of the central highlands. And not only is it 90 degrees east of Giza. An additional bonus, as I was to discover when I checked its bearings on my GPS, is that it lies almost exactly on the Tropic of Cancer, where at midday on the summer solstice a gnomon – or vertical upright – will cast no shadow.

I didn't even know point 'D' existed when we started our trip to Taiwan in August 2001 – but I had asked our local contacts there to introduce us to any interesting megaliths on the island. They took us to the Wuhe plateau, where by far the most spectacular and truly monumental of Taiwan's many megalithic sites is to be found at Sao Pa, ringed by distant peaks and overlooking a river valley of stunning, simple beauty.

Although folklore has it that two other megaliths originally stood at Sao Pa, only two have come down to us today. Carved in one piece out of black slate, both are classic stele or menhirs, tall and narrow, the larger 7.4 metres in height and the smaller just over 5 metres high. Both show a clean-cut horizontal groove at 'neck' level which is indeed somewhat suggestive of a neck and gives the menhirs a statue-like form.

- In round numbers of degrees and minutes the present latitude of the Tropic of Cancer is 23 degrees 27 minutes north. The location of the Sao Pa menhirs is 23 degrees 28 minutes north. The difference between the two is therefore one minute – i.e., 1/60 of a single degree.
- In round numbers the longitude of the Great Pyramid of Giza is 31 degrees 07 minutes east (i.e., east of the arbitrary and recent Greenwich Meridian); the longitude of the Sao Pa menhirs is 121 degrees 21 minutes east of Greenwich – the difference between the two is therefore 90 degrees, within 14/60 (i.e., less than a quarter) of a single degree.

In summary, if we impose on a map of the earth a 'world grid' with Giza (not Greenwich) as its prime meridian, then hidden relationships become immediately apparent between sites that previously seemed to be on random, unrelated longitudes. On such a grid, as we've just seen, Tiruvannamalai stands on longitude 48 degrees east, Angkor stands on longitude of 72 degrees east and Sao Pa stands out like a sore thumb on longitude 90 degrees east – all numbers that are significant in ancient myths, significant in astronomy (through the study of precession), and closely interrelated through the base-3 system.

So the 'outrageous hypothesis' which is being proposed here is that the world was mapped repeatedly over a long period at the end of the Ice Age – to standards of accuracy that would not again be achieved until the end of the eighteenth century. It is proposed that the same people who made the maps also established their grid materially, on the ground, by consecrating a *physical network* of sites around the world on longitudes that were significant to them. And it is proposed that this happened a very long time ago, before history began, but that later cultures put new monuments on top of the ancient sites which they continued to venerate as sacred, perhaps also inheriting some of the knowledge and religious ideas of the original navigators and builders.

And the original navigators and builders themselves? What direct traces of their civilization are to be found?

This brings us back to the underwater quest – for the traces, anywhere and everywhere around the world, of submerged structures that *do not make sense within the current paradigm of prehistory*. We've followed those traces from the Indian Ocean and the Persian Gulf, through the Mediterranean, into the Atlantic and now finally to the underworld of the East China and Yellow Seas that is bounded in the north by the Korean peninsula and Kyushu, in the east and south-east by the arc of the Ryukyu archipelago, and in the south by Taiwan.

Having explored other anomalous submerged sites in the same region – Aguni, Kerama, Chatan, Yonaguni – I was intrigued, but not surprised, when I first heard that a strange underwater structure had been found off Taiwan's Pescadores islands.

Diving at Tiger Well

I will not repeat the inundation history of the Pescadores given in chapter 28 or of the former island that lay to their north near the spot marked by Ymana on the 1424 Antilia chart. Irrespective of the Ymana issue, however, it is obvious that the Pescadores in their own right – located on the tip of a strategic peninsula of mainland China 13,500 years ago, then later one island, then later still the 64 tiny remnants that are seen today – are a plausible location in which to search for underwater ruins from the flood epoch.

They are plausible for another reason too. Ancient myths of the Pescadores speak of a great castle with huge 'red' walls that lies submerged somewhere amongst the islands. It was precisely these myths that led a government official to ask the brilliant Taiwanese diver Steve Shieh to look for underwater ruins if he happened to be working in the area. Over a period of several years, Steve complied, searching the waters around most of the islands. Eventually he was rewarded with an extraordinary discovery off the island of Hu-ching ('Tiger Well'). This happened more than twenty years ago and has received no attention or publicity in the West. Luckily for me, however, TBS, a large Japanese TV station, ran a report on Steve and his discovery as recently as January 2001. The report was seen by several Japanese friends, who drew it to my attention.

We did two days of diving with Steve Shieh off Hu-ching island at the end of August 2001.

The structure that he showed us consists of two immense walls, hundreds of metres in length, one running due north–south and the other running due east–west, crossing the north–south wall at right angles. At the east end of the east–west wall is a large circular enclosure, part of which has completely collapsed. The east–west wall is in relatively shallow water – 4 to 6 metres depth. The north–south wall starts at 4 metres depth but can be followed down to 36 metres depth. All the walls are a consistent height of 3 metres from the base to the top of the wall; however, some sections are broken.

In a volcanic, earthquake-prone area such as Taiwan one must be conscious of the possibility that such walls could be natural features – specifically basaltic dykes (quite common around the Pescadores). Such dykes form when a wall-like mass of igneous rock intrudes into cracks in older sedimentary rock.[46]

Despite extremely strong currents flowing unpredictably from eight different directions (why are there always currents around underwater monuments?), I was able to examine the walls quite thoroughly. My initial impression is that they are not basaltic dykes. This is mainly so because, after scraping off marine-growth from several sections of the walls, Steve showed me courses of individual blocks laid tightly together side-by-side. The joints between the blocks in some cases admit the point of a knife and it was possible for me to work the knife blade in as far as the hilt and move it entirely around individual blocks. In addition the nice north–south and east–west orientation of the walls,

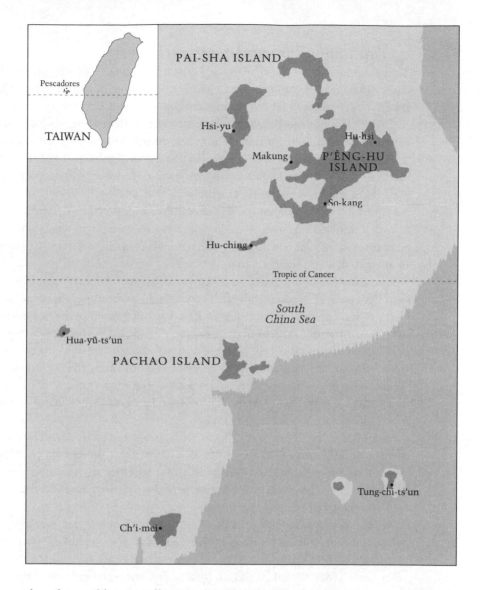

though possible naturally, is a strong indicator that humans were involved. Finally, there is that ancient local legend about a 'castle' that vanished beneath the sea . . .

But here, as everywhere else, more research – much more research – is needed to settle the matter. Any such research should also investigate the submerged bank further north that may be all that remains of the antediluvian island of Ymana marked on the 1424 chart.

One amongst many underworlds

They say that the Kingdom of the Sea God has a gateway that is guarded by a shark. So having never once seen a shark at the stone circles of Kerama I did take it as a good omen when one appeared on my last dive there. It was a sleek reef white-tip, not too fearsome, since it was less than 2 metres long, and it patrolled Centre Circle for several minutes, quite untroubled by our presence.

That was at the beginning of September 2001, after Santha and I had completed the dives in Taiwan and flown up to Okinawa to rendezvous again with Isamu Tsukahara and his team. Shun Daichi, who had been with us in Taiwan, also accompanied us to Kerama, and Kyoshi Nagaki joined us there as well.

We had allowed four days minimum, and assumed we might need more, since September is the typhoon season, but in the end Kerama gave me what I needed in just two dives on the first day.

It gave me Komakino Iseki, 30 metres underwater – not just one but a series of ovals of huge diameter made up of hefty rounded river stones sprawled around Centre Circle on the ancient rocky outcrop at the bottom of the sea. I believe that the similarities evident in photos 73 and 74 speak for themselves, needing very little commentary. Type and size of stone, the method of laying the stones to form the great ovals, the shape of the ovals themselves, the construction of banks consisting of two or three courses of stones piled on top of one another, and the use of patterns of 'chained' interlinked ovals are all identical at Komakino Iseki and underwater at Kerama.

In my view it is necessary for the site to be protected by the Japanese government now, and for excavations to be conducted there as soon as possible by competent marine archaeologists to ascertain whether any pottery fragments and other typical artefacts of the Jomon period are present amongst the stones. I suspect they will be.

But the real mystery that archaeology needs to solve is the relationship between the large-diameter ovals of river stones that are typical of other Jomon spiritual sites like Komakino Iseki – though at this depth they must be *at least* 5000 years older than Komakino Iseki – and the very different and much more ambitious project represented by the weird semi-subterranean complex of Centre Circle and Small Centre Circle.

Logic suggests that the Jomon must have made both the river-stone ovals and the rock-hewn circles, and that if we look we are likely to find other such man-made sites underwater in the region which will further testify to this lost architectural episode in their prehistory.

What else was lost then, in that epoch when we dropped the silver thread of memory that connected us to our own past?

An underworld, I suspect, is truly about to be revealed.

One of many.

Postscript 1 / The Underworld in the Gulf of Cambay

In chapter 14 I reported the claimed discovery by India's National Institute of Ocean Technology (NIOT) of an extensive urban complex underwater in India's Gulf of Cambay. The discovery was announced on 19 May 2001 by Science and Technology Minister Murli Manohar Joshi with the suggestion that the structures dated to the Harappan period of the Indus-Sarasvati civilization. I pointed out that inundation science firmly indicates the Gulf of Cambay to have been submerged in pre-Harappan times and specifically in quite a narrow time-window between 7700 years ago and 6900 years ago. From this it follows, if the structures that have been identified are indeed man-made, that they must date from a rather early phase of the *pre*-Harappan period. Moreover, a date of submergence between 7700 and 6900 years ago only tells us that the city was built at some time before then – not how long before. Since the geometrical structures identified by the NIOT's side-scan sonar readings extend over more than 9 kilometres of the sea-bed, and since a city on that scale could not have grown up overnight, logic suggests that it is likely to be significantly *more* than 7700 years old. I wrote in chapter 14: 'A city 9 kilometres in extent and more than 3000 years older than Harappa and Mohenjodaro would rewrite not only the history of the Indian subcontinent but of the world.'

Archaeologists' reactions to the NIOT's claims were understandably muted and in some cases hostile; in consequence there was virtually no international media coverage of the discovery which, very rapidly, seemed to have been forgotten. Nevertheless, I made contact with the NIOT and provisionally arranged to dive with them in the Gulf of Cambay in November–December 2001 as part of the final filming trip for the television series of *Underworld*.

When we arrived in India in November 2001, I talked with archaeologists at the NIO (National Institute of Oceanography – a completely separate operation from the NIOT) and they told me they did not accept that anything of significance had been discovered. Most likely, they said, the geometrical 'structures' seen on the side-scan sonar readings were merely artefacts of the imaging process. Likewise, R. S. Bisht, a director of the Archaeological Survey of India and a leading expert on the Indus-Sarasvati civilization, told me frankly that he did not believe the NIOT's findings and that they must be 'hallucinations'.

I was therefore not filled with optimism when I arrived in the port of Bhavnagar on the Gulf of Cambay to keep my previously made appointment with the NIOT – and my sense of unease increased when I was informed that Indian

Naval Intelligence had refused to issue me a permit to dive on the alleged underwater site. Still, we set up the cameras for a day of filming on board the NIOT research vessel, the M.V. *Sagar Paschmi*, and waited to see what senior NIOT scientists had to say about the matter.

The story that they told us on camera and the evidence that they presented to us banished any reasonable doubts. Despite the ridicule and cold shoulders to which they had been subjected by the archaeological establishment, they had indeed made a discovery of staggering significance. According to Dr S. Kathiroli, the NIOT's Project Director, and geological consultant S. Badrinarayan, they had been surprised by the hostile reactions of archaeologists to the initial announcement of their findings in May 2001. As scientists, however, they had decided to pursue the mystery further through empirical research and see where it led. Thus, between May and late November 2001 they had conducted further side-scan sonar surveys and backed these up with sub-bottom profiling around the geometrical structures.

The results confirmed their initial impression that extensive man-made ruins did indeed lie on the sea-bed in the Gulf of Cambay at depths of between 25 and 40 metres and at distances of up to 40 kilometres from the modern shoreline. The sub-bottom profiles revealed extensive, well-built foundations to the geometrical structures and in some cases walls rising as much as 3 metres above the sea-bed and extending down several metres below. Moreover, as well as the original 'city-complex' covering a rectangular area roughly 9 kilometres long and 2 kilometres wide, a second city of similar size had been found a little further to the south at similar depths. Both cities lie along the courses of ancient rivers that had flowed here when the area was above water, and in one case the remains of an ancient dam more than 600 metres long have been identified.

Thus far the NIOT had been unable to dive on the sites, due to the great tidal amplitude in the Gulf and extremely hazardous currents. Moreover, when they sent down their remotely operated vehicle (ROV), its cameras were unable to record clear images because of zero visibility due to dissolved solids in the waters of the Gulf. How therefore to get to the sort of 'ground truth' that might impress sceptical archaeologists? The only solution, the NIOT decided, was to lift samples directly off the sea-bed from the heart of the areas identified in their side-scans and sub-bottom profiles.

The results, which they showed us and we were able to film, are spectacular. In just one day of sampling using grabs and trawls *more than* 2000 man-made artefacts were recovered – including jewellery, stone tools, pottery and figurines. The assemblage, which is typically 'pre-Harappan' (and which includes carbon-datable human remains such as teeth), confirms that the underwater structures identified by the side-scans and sub-bottom profiles were indeed large-scale human settlements before their inundation. The extremely ancient character of the artefacts also seems to rule out any possibility that the underwater sites could date from a period later than the pre-Harappan. On the record

S. Badrinarayan told me that in his opinion these submerged city-complexes must, at the very youngest, date to between 7000 and 8000 years ago and that the most likely agent of their inundation was sea-level rise at the end of the Ice Age rather than any kind of fault collapse due to seismic activity – for which the region is nevertheless renowned. Perhaps some catastrophic combination of earthquake activity and sea-level rise could account for the massive and apparently very sudden scale of the submersion which, on the face of things, appears to have obliterated an entire civilization in this region.

More work must be done to establish the dates exactly and to come to terms with the true nature of the enigma of the Gulf of Cambay. Yet we know already, by the very extent of the ruins, that they represent something that orthodox historians and archaeologists have never accepted – the possibility that a lost civilization lies concealed in Indian prehistory and that the Indian flood myth of Manu and the Seven Sages rests firmly on ground truth.

STOP PRESS

On 16 January 2002 India's Minister of Science and Technology released the first results of carbon-dating of the artefacts from the flooded cities of the Gulf of Cambay. The results date the artefacts to 9500 years ago – 5000 years older than any city so far recognized by archaeologists.

Postscript 2 / The Underworld in the Bay of Bengal

3 April 2002, Mahabalipuram, Tamil Nadu

Sinking down through the murky waters of the Bay of Bengal a mile offshore of the south-east Indian town of Mahabalipuram, I found myself amongst huge submerged walls, plazas and pinnacles emerging out of the gloom – structures that seemed more like the work of gods or titans than of men and to belong more in the world of myth than that of history. With a kick of my fins I turned slowly around, and in every direction that I looked I saw extensive and impressive ruins stretching away. Involuntarily my heart began to pound and my breathing speeded up. Because for more than five years I had been diving the world's oceans searching for evidence just such as this – the hard evidence that I had long believed must lie behind mankind's collective inheritance of more than 600 ancient 'flood myths'.

Readers who have come this far will recall that I first visited Mahabalipuram in 1956 when I was just six years old (my father was working as a surgeon at the Christian Medical College in the nearby town of Vellore). My next visit was on a journey of personal reminiscence in 1992. It was then (see chapter 5) that I purchased an anthology of travellers' journals and reports edited by a certain Captain M. W. Carr in 1869 under the title *Descriptive and Historical Papers Relating to the Seven Pagodas of the Coromandel Coast*. The 'Seven Pagodas' is the old mariners' name for Mahabalipuram, and on my third visit to the town in February 2000 (see chapter 11) I took Captain Carr's anthology with me. In one paper J. Goldingham, Esq., writing in 1798, spoke of the part of Mahabalipuram that I remembered best from my childhood – the 'Shore Temple', carved out of solid granite, lashed by waves:

> The surf here breaks far out over, as the Brahmins inform you, the ruins of a city which was incredibly large and magnificent . . . A Brahmin, about 50 years of age, a native of the place, whom I have had an opportunity of conversing with since my arrival in Madras, informed me his grandfather had frequently mentioned having seen the gilt tops of five pagodas in the surf, no longer visible.

An earlier traveller's report, from 1784, describes the main feature of Mahabalipuram as a 'rock, or rather hill of stone', out of which many of the monuments are carved. This outcropping, he says:

is one of the principal marks for mariners as they approach the coast and to them the place is known by the name of 'Seven Pagodas', possibly because the summits of the rock have presented them with that idea as they passed: but it must be confessed that no aspect which the hill assumes seems at all to authorise this notion; and there are circumstances that would lead one to suspect that this name has arisen from some such number of Pagodas that formerly stood here and in time have been buried in the waves . . .

The same author, William Chambers, then goes on to relate the more detailed oral tradition of Mahabalipuram – given to him by Brahmins of the town during visits that he made there in 1772 and 1776 – that prompted his suspicion of submerged structures.

According to this tradition a Raja named Malecheren ruled at Mahabalipuram at some time in the remote past. He encountered a being from the heavenly realms who became his friend and agreed 'to carry him in disguise to see the court of the god Indra' – a favour that had never before been granted to any mortal:

> The Raja returned from thence with new ideas of splendour and magnificence, which he immediately adopted in regulating his court and his retinue, and in beautifying his seat of government. By this means Mahabalipuram became soon celebrated beyond all the cities of the earth; and an account of its magnificence having been brought to the gods assembled at the court of Indra, their jealousy was so much excited at it that they sent orders to the God of the Sea to let loose his billows and overflow a place which impiously pretended to vie in splendour with their celestial mansions. This command he obeyed, and the city was at once overflowed by that furious element, nor has it ever since been able to rear its head.

Go where the fish are

It was this myth that now kept drawing me back to Mahabalipuram. Almost exactly a year later, in February 2001, I was there again – this time to interview fishermen on camera for my Channel 4 TV series, *Flooded Kingdoms of the Ice Age*. My wife, Santha, whose mother-tongue is Tamil – the language of Mahabalipuram – was seated beside me on the beach on a pile of drying nets with a large, gossipy, excited and jocular crowd gathering round us. Everybody in the village who might have an opinion or information to contribute was there, including all the fishermen – some of whom had been drinking palm toddy most of the afternoon and were in a boisterous and argumentative mood. What they were arguing about was their answers to the questions that I was asking and precisely who had seen what, where underwater – so I was happy to listen to their animated conversations and disagreements.

An elder with wrinkled nut-brown eyes and grey hair bleached white by long exposure to the sun and sea spoke at length about a structure with columns which he had seen one day from his boat when the water had been exceptionally clear. 'There was a big fish,' he told me. 'A red fish. I watched it swimming towards some rocks. Then I realized that they were not rocks but a temple. The fish disappeared into the temple, then it appeared again, and I saw that it was swimming in and out of a row of columns.'

'Are you certain it was a temple?' I asked.

'Of course it was a temple,' my informant replied. He pointed to the pyramidal granite pagoda of the Shore Temple. 'It looked like that.'

Several of the younger men had the usual stories to tell about heroic scary dives – lasting minutes, hearts thudding, their breath bursting in their lungs – to free fishing gear snagged on dark and treacherous underwater buildings. In one case, it seemed, a huge net had become so thoroughly entrapped on such a structure that the trawler that was towing it had been stopped in its tracks. In the case of another underwater ruin divers had seen a doorway leading into an internal room but had been afraid to enter it.

One strange report was that certain of the ruins close to Mahabalipuram emit 'clanging' or 'booming' or musical sounds if the sea conditions are right: 'It is like the sound of a great sheet of metal being struck.'

'And what about further away?' I asked. 'If I were to take a boat south, following the coast, what would I find? Are the underwater structures mainly just here around Mahabalipuram or are they spread out?'

'As far south as Rameswaram you may find ruins underwater,' said one of the elders. 'I have fished there. I have seen them.'

Others had not travelled so far, but all agreed that within their experience there were submerged structures everywhere along the coast: 'If you just go where the fish are then you will find them.'

Expedition

The next challenge for me was now somehow to set up a full-scale diving expedition to Mahabalipuram. The responsible authority in India is the marine archaeology division of the National Institute of Oceanography (NIO). However, I knew that the NIO, on its own, did not have sufficient funds or incentive to put an expedition in place.

Fortuitously at about this time – the spring of 2001 – I was approached by John Blashford Snell of the Dorset-based Scientific Exploration Society (SES). Over lunch at my home in Devon, Blashford Snell asked me if any of the mysteries raised by the research for my new book *Underworld* might make a worthwhile SES expedition. Naturally I suggested a quest for Mahabalipuram's supposedly 'mythical' flooded city. I also recommended a second site, Poompuhur, which lies about 200 kilometres south of Mahabalipuram. In chapter 14 I describe my dives there with the NIO in February 2001 to investigate a mysterious 'U-shaped structure' that their own marine archaeologists had found at a depth of 23 metres (about 70 feet). I felt that it was worthy of much closer investigation and that a well-resourced diving expedition to both Poompuhur and Mahabalipuram stood a chance of making some exciting discoveries.

On Blashford Snell's request I introduced the SES to the NIO and over the following months nudged and cajoled the sometimes faltering communications of the two organizations until the plans for the expedition had been fully approved. Former Royal Marines officer Monty Halls, himself a diving instructor, was selected by the SES to lead the expedition, and twelve volunteer divers, mostly from Great Britain, put up all the necessary funding.

Thus it was that in April 2002 I found myself once more in Mahabalipuram – this time supported by all the divers and expertise needed to test my hypothesis that the local myths of a flooded city might actually lead us to a city underwater. Naturally I feared the failure of such a public quest, whilst hoping that it would succeed. Yet even in my wildest dreams I could not have imagined how immediately self-evident it would be, from the moment we got in the water, that following the clues in the myths and local traditions had indeed led to the discovery of a major archaeological site . . .

How old?

But how old are the spectacular underwater ruins lying off the coast of Mahabalipuram likely to turn out to be?

As I sank down through the murky waters on my first dive there and began to notice the huge walls snaking across the seabed, I was struck – despite what the fishermen had told me – by the stark differences in architectural style between these looming submerged structures and the temples of the historical period, such as the Shore Temple, that I knew on land. The material was the same – local red granite – but in general the block sizes of the underwater structures were much bigger. On later dives I was to discover that they included clusters of truly enormous megaliths, weighing up to four tonnes each, that seemed to be the remains of colossal buildings torn apart by some powerful cataclysm – probably the very cataclysm that had flooded this place. Each gigantic cluster proved to lie at the centre and highest point of a large rectangle of ruins and these rectangular areas were spaced out at quarter-mile intervals north and south, parallel to the shore.

I made a couple of dives with Santha, who was shooting underwater stills, and a couple of dives with Trevor Jenkins, the expedition's videographer. On one dive Trevor and I followed a superb, curved wall that ran unbroken for more than 16 metres. On another I was able to expose the core masonry on a different length of wall, revealing the fine jointing between blocks.

On the very last dive, at a site we'd tagged as 'location 4', Trevor discovered what he thought might be a stone carving of a lion's head in profile lying on the seabed. I wasn't so sure: it was too damaged and shrouded in marine growth for me to be certain what it was or even if it was man-made at all. While Trevor filmed it, I tried to clear a section of it with my knife, but all I could establish was that it was solid granite, about the radius and thickness of a car wheel, and roughly crescent-shaped. I couldn't make out any individual features – but that doesn't mean they aren't there. It will have to be found again on some future dive, removed from the sea and studied in the lab before any firm conclusions can be drawn about what it is.

Hasty conclusions

That at any rate would be normal archaeological procedure. So I was confused after our return from Mahabalipuram to learn that the NIO had issued a press statement proposing a possible date and function for the submerged ruins and citing this very same 'lion figure' as evidence:

> Based on what appears to be a Lion figure, of location 4, ruins are inferred to be parts of temple complex. The possible date of the ruins may be 1500–1200 years before present. Pallava dynasty, ruling the area during the period, has constructed many such rock-cut and structural temples in Mahabalipuram and Kanchipuram.

To come to such far-reaching deductions on the basis of ambiguous video footage of an alleged 'lion figure' strikes me as hasty, to say the least. None of the NIO archaeologists saw this 'figure' at first hand as Trevor and I did, and I for one am not convinced that it is any kind of figure at all. Moreover, the use of lion symbolism is widespread in India and indeed in global sculpture of almost all periods. Surely, therefore, a heavily overgrown and damaged object like this ought to be studied thoroughly before it and the submerged ruins all around it can safely be assigned to a specific dynasty such as the Pallavas or to a specific chronology such as 1500 to 1200 years ago?

Moreover, the Pallava temples on land at Mahabalipuram are reliably dated to 1500 to 1200 years ago – precisely the same period suggested by the NIO for the submerged ruins that we now know lie a mile and more offshore. If both groups of structures do date from the same period, however, then the NIO must be able to explain why one group is now submerged beneath the sea to depths of between 5 and 7 metres (15 and 21 feet) whilst the other is still above water.

An obvious explanation, and the one preferred by the NIO, is that there must have been massive coastal erosion in this area, or that perhaps a stretch of the coast just collapsed in some wild, unpredictable tectonic event – perhaps not even tremendously long ago – that submerged the big Pallava constructions on one side of the fault-line while leaving those on the other side intact and still on dry land.

Another possibility, however – one not even considered by the NIO in its statement – is that the submerged group could be significantly *older* than the group on land. As we've seen throughout *Underworld*, sea-level rose more than 100 metres at the end of the last Ice Age (between 17,000 years ago and about

5000 years ago) and if it is sea-level rise alone that has submerged the megalithic structures offshore of Mahabalipuram, then how old are they likely to be?

Bombshell

On our return from India on 6 April 2002 I e-mailed Dr Glenn Milne, the international expert on sea-level rise whose inundation maps have been used throughout *Underworld*, and told him what we'd found at Mahabalipuram. Perhaps he and other geologists at Durham University could help clarify how big a part sea-level rise had played in submergence of the ruins and suggest a date when they had last stood above water.

Glenn's reply, repeated a few days later on BBC Television's national 6 o'clock news, came as a bit of a bombshell:

> I had a chat with some of my colleagues here in the Dept. of Geological Sciences and it is probably reasonable to assume that there has been very little vertical tectonic motion in this region during the past five thousand years or so. Therefore, the dominant process driving sea-level change will have been due to the melting of the Late Pleistocene ice-sheets. Looking at predictions from a computer model of this process suggests that the area where the structures exist would have been submerged around six thousand years ago. Of course, there is some uncertainty in the model predictions and so there is a flexibility of roughly plus or minus one thousand years in this date.

The U-shaped structure of Poompuhur

Prevailing archaeological opinion recognizes no culture in India 6000 years ago capable of building anything much – let alone a series of vast megalithic structures on the scale and extent that confronts us at Mahabalipuram. Nor does the mystery stop here. As we've seen, the fishermen at Mahabalipuram speak of other ruins, even further out from shore in much deeper water, which remain to be identified and explored. On my travels in the region (see chapter 11) I've also heard reports of mysterious underwater structures off Poompuhur, Rameswaram (overlooking the Palk Strait between India and Sri Lanka) and Kaniya Kumari (Cape Comorin) in the far south of the subcontinent. The reader will recall the pervasive Tamil flood myth linking all these areas that speaks of a lost land called Kumari Kandam that was swallowed up by the sea in three terrible deluges, the first of which took place 11,500 years ago.

In the early 1990s the NIO conducted a marine archaeological survey off Poompuhur for the state government of Tamil Nadu and discovered a very large and apparently man-made structure more than 5 kilometres from shore at a depth of 23 metres (70 feet). The story of this discovery is told in chapter 1 of *Underworld* and followed up in chapter 14. Its potential importance arises exclusively from the depth of submergence of the structure – which suggests, as the reader will recall, that it could have been underwater for 11,000 years. If it is man-made, therefore, then the obvious implication is that it must have been built *more than 11,000 years ago* when it still stood on dry land. And just as archaeologists know of no culture in south India more than 6000 years ago that would have been capable of building the now-submerged structures at Mahabalipuram, so also they know of none at Poompuhur (or anywhere else in India, or in the world) that would have been capable of any project on the scale of the U-shaped structure more than 11,000 years ago.

The results of the SES/NIO expedition to Poompuhur in March 2002 were inconclusive. Despite extensive dives over a period of ten days the team could not reach a unanimous verdict on the U-shaped structure. On two shows of hands a clear majority of the group, including one of the two NIO marine archaeologists, concluded that it is man-made. But there were significant exceptions to this view and I therefore do not yet claim to have proved the case for the structure's artificiality that I set out in chapter 14. On my own dives with the team in March 2002 I did, however, notice several features of the structure that had not previously caught my eye. Of these I believe the most interesting are (1) sections of what appears to be a second, lower wall, about 2 metres from the main wall and running parallel to it, that may once have completely surrounded the structure; (2) the impression reported by several divers as well as myself that the main wall of the structure is octagonal or hexagonal in form ('like the old threepenny bits', commented Trevor Jenkins) rather than explicitly 'U-shaped'; (3) well-formed courses of blocks clearly visible beneath marine growth at several points on the structure; (4) evidence from visual examination of a sample that the material of which these blocks are made is laterite – a common construction stone in south India since times immemorial; (5) the presence of large symmetrical slabs (approx 1.5 metres × 1.5 metres × 0.5 metres) scattered on the seabed near a smaller mound 45 metres north-west of the U-shaped structure; (6) evidence from side-scan sonar readings of other anomalous structures nearby, including one identified in March 2002 as a straight wall approximately 100 metres in length, lying in 25 metres of water and almost 7 kilometres from land.

I believe that the U-shaped structure has passed the crucial first test of close scrutiny over a lengthy period by a team of divers and marine archaeologists. It has not yet been decisively proved to be man-made but it has certainly not been proved to be natural either. It remains an anomaly and an enigma. And as is the case with the mysterious underwater ruins of Mahabalipuram, it cries out for further research . . .

Appendix 1 / Report on the Completion of the Joint SES/NIO Expedition to South-east India

Graham Hancock, 6 April 2002

Originally posted on 'The Mysteries' message board at www.grahamhancock.com

Hi folks,

Santha and I flew back in from Tamil Nadu this morning.

As regulars on this MB know, we have been diving at Poompuhur and at Mahabalipuram in south-east India. We have had the privilege of working there with ten first-rate divers from Britain led by Monty Halls of the Scientific Exploration Society and with a great team from India's National Institute of Oceanography led by Kamlesh Vora.

At Poompuhur, despite intensive diving on the mysterious U-shaped structure submerged about 5 kilometres offshore at a depth of 23 metres (see chapters 1 and 14 of *Underworld*), we could not reach a unanimous verdict. On two shows of hands a clear majority of the group, including one of the two NIO marine archaeologists, concluded that it is a man-made structure. But there were significant exceptions to this view and I therefore do not claim to have proved my case there during this expedition.

Over the coming week or so I will set out on this site, supported by photography, the principal pieces of evidence that convince me and others that the structure is man-made. A great deal more work is going to have to be done on it and neighbouring structures, however, before the matter can be regarded as having been satisfactorily settled – one way or the other.

The reason for this continuing uncertainty, despite the best efforts of a large group of determined and objective researchers, lies in the very bad diving conditions and poor visibility at Poompuhur, which hamper and restrict the work underwater at all times.

At Mahabalipuram, the other objective of the expedition, the situation is much clearer. A press conference will be held on 10 April 2002 to announce the extraordinary underwater discoveries that our team made there last week up to 2 kilometres from shore at depths of 5 to 7 metres. Relevant pages in *Underworld* where I describe my research in Mahabalipuram that led directly to these discoveries are 119–122 and 258–261.

Of course, the real discoverers of this amazing and very extensive submerged site are the local fishermen of Mahabalipuram. My role was simply to take what they had to say seriously and to take the town's powerful and distinctive flood myths seriously. Since no diving had ever been done to investigate these neglected myths and sightings, I decided that a proper expedition had to be mounted. To this end, about a year ago, I brought together my friends at the Scientific Exploration Society (SES) in Britain and the National Institute of Oceanography (NIO) in India and we embarked on the long process that has finally culminated in the discovery of a major and hitherto completely unknown submerged archaeological site.

I'll try to find out next week the date that Glenn Milne's model suggests for the submergence of the Mahabalipuram structures. Meanwhile, I want to state very clearly and for the record that I am making no claims as to the age of the structures, or what they are, or who built them, or why and when they were inundated. All this will have to be established through further research – which the NIO estimates will take many years and will involve the participation of experts from many different disciplines. I do, however, feel fully vindicated in the view that I have long held and expressed in my books and television series that flood myths deserve to be taken seriously and can lead to the discovery of significant underwater ruins.

The information that we have gathered at Mahabalipuram up to now will be released at the SES press conference on 10 April.

Appendix 2 / SES Press Release, 5 April 2002, Announcing the Discovery of Underwater Ruins at Mahabalipuram and Inviting Media to a Press Reception, 10 April 2002

The Scientific Exploration Society is proud to announce a major discovery of submerged ruins off the south-east coast of India and invite you to a Press Reception at 10.30 a.m. on Wednesday 10 April 2002, at the Nehru Centre, 8, South Audley Street, London W1K 1HF.

Following a theory first proposed by bestselling author and television presenter Graham Hancock, a joint expedition of 25 divers from the Scientific Exploration Society (SES) and India's National Institute of Oceanography (NIO) led by Monty Halls and accompanied by Graham Hancock have indeed discovered an extensive area with a series of structures that clearly show man-made attributes, at a depth of 5–7 metres offshore of Mahabalipuram in Tamil Nadu.

The scale of the submerged ruins, covering several square miles and at distances of up to a mile from shore, ranks this as a major marine-archaeological discovery as spectacular as the ruined cities submerged off Alexandria in Egypt.

This could prove the ancient myths of a huge city, so beautiful that the gods became jealous and sent a flood that swallowed it up entirely in a single day!

Come and listen to Graham Hancock, Monty Halls and view unique pictures/ video. Further info www.india-atlantis.org.

Contacts: Melissa Dice; Tel: 01747 854898; email: base@ses-explore.org; Sarah Jane Lewis (Press) Tel: 01963 240468.

Appendix 3 / Preliminary Underwater Archaeological Explorations of Mahabalipuram. Statement by National Institute of Oceanography, 9 April 2002

A team of underwater archaeologists from National Institute of Oceanography NIO have successfully 'unearthed' evidence of submerged structures off Mahabalipuram and established first-ever proof of the popular belief that the Shore temple of Mahabalipuram is the remnant of series of a total seven of such temples built that have been submerged in succession. The discovery was made during a joint underwater exploration with Scientific Exploration Society, UK.

The team of archaeologists from NIO, trained in diving, carried out underwater exploration on April 1–4, 2002 and have successfully recorded evidence of presence of ruins underwater off Mahabalipuram. The salient features of the findings are as follows:

- Underwater investigations were carried out at 5 locations in the 5–8 m water depths, 500 to 700 m off Shore temple.
- Investigations at each location have shown presence of the construction of stone masonry, remains of walls, a big square rock-cut remains, scattered square and rectangular stone blocks, big platform leading the steps to it amidst of the geological formations of the rocks that occur locally.
- Most of the structures are badly damaged and scattered in a vast area, having biological growth of Barnacles, Mussels and other organisms.
- The construction pattern and area, about 100 m × 50 m, appears to be same at each location. The actual area covered by ruins may extend well beyond the explored locations.
- Based on what appears to be a Lion figure, of location 4, ruins are inferred to be parts of temple complex.
- The possible date of the ruins may be 1500–1200 years BP. Pallava dynasty, ruling the area during the period, has constructed many such rock-cut and structural temples in Mahabalipuram and Kanchipuram.

To place reasonable arguments on submergence of ruins, a full-scale investigations are underway to record the role of sea-level fluctuations, coastal erosion and neo-tectonic activities in effecting shoreline changes in the area in the recent past.

The site has great potential to explore total lay-out plan of the structures and causes of submergence.

Appendix 4 / Comments by Graham Hancock on the NIO Statement of 9 April 2002 Regarding Preliminary Underwater Archaeological Explorations off Mahabalipuram

I have only two comments to make on the NIO press release, but both of them are grave.

(1) Despite a friendship with the NIO stretching back over two years, I note that the NIO statement makes no mention of my instrumental role in bringing about these exciting discoveries off Mahabalipuram. I regret this oversight, since there can be no doubt that I have earned the right to recognition in this discovery and that my input both in formulating the hypothesis of submerged ruins at Mahabalipuram, in putting that hypothesis forcefully before the public, and in the conception and implementation of an expedition to test that hypothesis has been absolutely decisive.

It is in black and white on pages 119–22 and pages 258–61 of my book *Underworld* (published by Penguin 7 February 2002), and in my Channel 4 Television Series *Flooded Kingdoms of the Ice Age* (broadcast 11, 18 and 25 February 2002) that I have long regarded Mahabalipuram, because of its flood myths and fishermen's sightings, as a very likely place in which discoveries of underwater structures could be made, and that I proposed that a diving expedition should be undertaken there.

It is also absolutely a matter of record that it was I who subsequently took the initiative to bring together the Scientific Exploration Society (SES) and the NIO during 2001 so that the expedition could take place and that I expended considerable efforts putting the two groups in touch and nudging along their co-operation.

I think you will find if you remove Graham Hancock from the equation that another twenty or many more years might have elapsed before the marine archaeology division of the NIO would have dived at Mahabalipuram.

If you remove Graham Hancock from the equation, the SES and the NIO would not have been brought together and the SES would not even have been aware that there was a mystery to investigate at Mahabalipuram.

In other words if you remove Graham Hancock from the equation it is a plain fact, and nothing more nor less than the truth, that neither the NIO nor the SES would have been diving at Mahabalipuram.

The discoveries that we have made might have been made later, or never at all. Such questions are entirely hypothetical, however. The fact is that the discovery has been made now and that my research, initiatives and efforts were instrumental in bringing it about. In any kind of moral or decent universe, in which credit is given where credit is due, I believe that I deserve some recognition for this. I ask nothing more than that.

(2) My second comment on the statement concerns the unwisdom and unfortunate disregard of basic scientific procedure on the part of the NIO in speculating about a possible date of 1500 BP to 1200 BP for the submerged ruins. This speculation seems largely to be based on what is claimed to be a sculpture of a lion at location 4 – thought to be typical of the sculptural art of the Pallava dynasty. Unfortunately, however, neither of the two NIO marine archaeologists who were diving with us actually saw the alleged 'figure'. The only people who did were myself and my dive-buddy Trevor Jenkins. It was Trevor who first spotted it. We then examined it together and Trevor shot video footage of it. All other comments on this lion figure are second-hand, based on viewings of Trevor's video footage only.

My own very much first-hand comment is that if the figure is indeed that of a lion, this by no means confirms a connection with the Pallavas – since lion sculptures are typical of whole swathes of Indian art and symbolism and cannot be regarded as a Pallava monopoly. More importantly, the so-called lion figure is by no means necessarily a lion figure at all. As noted above, I am one of only two divers who have seen it and handled it, and I suspect strongly that it is not a lion's head and perhaps not even part of a statue. I had not voiced that suspicion before now because I thought the scientific community believed that weighty conclusions one way or another about possible archaeological discoveries should only be reached after much further research. But now I see that, without doing any research at all, and without any marine archaeologists ever having examined the alleged 'figure', the NIO rushes in to suggest a possible date in its statement.

In my view the NIO should have refrained from such unwise, premature speculation and simply left the issue of the dating of the site open for the vast amount of further research that does indeed need to be done before anything can be confirmed. As one who has often been accused of prematurely assigning older dates to archaeological sites on the basis of too flimsy evidence, I find it ironic that the NIO should assign a possible date of 1500–1200 BP to this site without any evidence at all. The NIO is not even at this stage aware of the sea-level curve for this part of the south-east Indian coast – surely a crucial factor in any attempt to date the site.

Sincerely,
Graham Hancock
9 April 2002

Appendix 5 / Who Discovered the Underwater Ruins at Mahabalipuram? And Who is Claiming What?

Graham Hancock, 13 April 2002

Originally posted on 'The Mysteries' message board at www.grahamhancock.com

(1) In another thread Martin Stower draws attention to the following commentary:

> Mohapatra, G. P., and M. H. Prasad (1999), 'Shoreline changes and their impact on the archaeological structures at Mahabalipuram'. *Gondwana Geological Magazine*, vol. 4, pp. 225–33. Reading this paper, person finds that they had proposed in print back in 1999 that underwater archaeological ruins lay offshore of the coast of Mahabalipuram. In this case, Hancock is wrong in stating 'But here in Mahabalipuram we have proved the myths right and the academics wrong.' In fact, he has proved the academics, in this case G. P. Mohapatra and M. H. Prasad, were correct in hypothesizing that the remains of ancient ruins lay offshore of coast of Mahabalipuram.

(2) I was unaware of Mohapatra and Prasad's work; had I known of it, I would certainly have referred to it in *Underworld*. Apropos of this, during the recent SES/NIO expedition to Mahabalipuram, Kamlesh Vora informed the team that the NIO too had previously thought of diving there to check out the local flood tradition. This was back in the 1980s under the leadership of S. R. Rao (unfortunately now retired) – a man with a great interest in India's flood myths. (See my interview with Rao in chapter 1 of *Underworld*, where he makes specific reference to the myths of lost lands off the south of India and to the relevance of these for marine archaeological research.) Apparently, however, the water was too 'muddy' when the NIO marine archaeologists arrived at Mahabalipuram and they decided not to dive. The project was never taken up again.

(3) There is no need for speculation about what exactly I'm claiming with regard to the Mahabalipuram underwater discoveries. My views are already on the record on this Message Board. Here are the two definitive passages:

6 April 2002

Of course, the real discoverers of this amazing and very extensive submerged site are the local fishermen of Mahabalipuram. My role was simply to take what they had to say seriously and to take the town's powerful and distinctive flood myths seriously. Since no diving had ever been done to investigate these neglected myths and sightings, I decided that a proper expedition had to be mounted. To this end, about a year ago, I brought together my friends at the Scientific Exploration Society (SES) in Britain and the National Institute of Oceanography (NIO) in India and we embarked on the long process that has finally culminated in the discovery of a major and hitherto completely unknown submerged archaeological site.

9 April 2002

Despite a friendship with the NIO stretching back over two years, I note that the NIO statement makes no mention of my instrumental role in bringing about these exciting discoveries off Mahabalipuram. I regret this oversight, since there can be no doubt that I have earned the right to recognition in this discovery and that my input both in formulating the hypothesis of submerged ruins at Mahabalipuram, in putting that hypothesis forcefully before the public, and in the conception and implementation of an expedition to test that hypothesis, has been absolutely decisive.

(4) It should be clear from the above that I do not claim to be 'the' discoverer of these underwater ruins – the existence of which has been known since time immemorial to the local fishermen. Nor do I even claim to be 'the' theorist who first proposed the hypothesis that there might be ruins underwater offshore Mahabalipuram. As I report in *Underworld*, that 'hypothesis' has been around in scholarly circles since at least the eighteenth century. I and several others have subsequently made input to the elaboration of this hypothesis and the NIO actually set out to test it in the 1980s, but in the end did not go diving. Thereafter, the question of whether or not there were ruins underwater off Mahabalipuram lapsed into obscurity until Mohapatra's and Prasad's work on the one hand, and my own on the other. My path to understanding why the question was worth asking is described in *Underworld*. What I claim is to have been the first person to have followed that path all the way through to its logical conclusion and to have been instrumental in the actual discovery of actual ruins – ruins that had been previously suspected but never proven to exist – underwater off Mahabalipuram.

From my post of 9 April 2002:

It is in black and white on pages 199–22 and pages 258–61 of my book *Underworld* (published by Penguin 7 February 2002), and in my Channel 4 Television Series *Flooded Kingdoms of the Ice Age* (broadcast 11, 18 and 25 February 2002) that I have long regarded Mahabalipuram, because of its flood myths and fishermen's sightings, as a very likely place in which discoveries of underwater structures could be made, and that I proposed that a diving expedition should be undertaken there.

It is also absolutely a matter of record that it was I who subsequently took the initiative to bring together the Scientific Exploration Society (SES) and the NIO during 2001 so that the expedition could take place and that I expended considerable efforts putting the two groups in touch and nudging along their co-operation.

I think you will find if you remove Graham Hancock from the equation that another twenty or many more years might have elapsed before the marine archaeology division of the NIO would have dived at Mahabalipuram.

If you remove Graham Hancock from the equation, the SES and the NIO would not have been brought together and the SES would not even have been aware that there was a mystery to investigate at Mahabalipuram.

In other words if you remove Graham Hancock from the equation it is a plain fact, and nothing more nor less than the truth, that neither the NIO or the SES would have been diving at Mahabalipuram.

The discoveries that we have made might have been made later, or never at all. Such questions are entirely hypothetical, however. The fact is that the discovery has been made now and that my research, initiatives and efforts were instrumental in bringing it about. In any kind of moral or decent universe, in which credit is given where credit is due, I believe that I deserve some recognition for this. I ask nothing more than that.

(5) Credit is also due and should be given to all who have played a part in this discovery – including Santha, Monty Halls, all the individual members of the SES and NIO diving teams and the steadfast Tamil fishermen of Mahabalipuram, who took us on board their little boats and straight out, with unerring accuracy, to each of the submerged sites. I have no idea whether the NIO is aware of Mohapatra's and Prasad's work, or whether the latter are aware of the work of the NIO. But when I come to update *Underworld* later this year I will certainly make reference to it.

Graham Hancock

Appendix 6 / UK Press Coverage of Mahabalipuram Discovery, April 2002

Daily Telegraph, *11 April 2002*

DIVERS FIND REMAINS OF SIX 'LOST TEMPLES'

By David Derbyshire, Science Correspondent

A MYSTERIOUS settlement that sank beneath the waves at least 1,200 years ago has been discovered by divers off the south-east coast of India.

Granite blocks and walls that lie 20 ft below the surface may be the remains of six 'lost temples' that form part of local mythology.

The ruins came to light after the controversial amateur archaeologist and best-selling author Graham Hancock interviewed fishermen for a recent television series.

After hearing accounts of the myth of a submerged city, he and two dozen divers searched the sea bed last week.

India's National Institute of Oceanography, which was involved in the discovery, believes the ruins at Mahabalipuram in Tamil Nadu could be 1,200 to 1,500 years old.

But Mr Hancock, who argues that civilisation predates the ancient Egyptians and Sumerians by thousands of years, believes the city could go back to 3000 BC.

The ruins were discovered half a mile off the coast by a team from the NIO and the UK-based Scientific Exploration Society. They include remains of walls and scattered carved blocks and stones and may cover several square miles.

According to local legend Mahabalipuram was once home to a great city. The gods became so jealous of its beauty that they sent a flood to swamp the city. Six temples were submerged, leaving just one on the shore.

Guardian, *11 April 2002*

DIVERS 'DISCOVER' ANCIENT TEMPLE

James Meek, science correspondent
Thursday April 11, 2002

Indian and British scientists have brought back pictures from the seabed of what they say could be a vast temple complex off the coast of Tamil Nadu – the ruins of a long-lost city, drowned beneath the waves.

The granite ruins, if they are not natural formations, could be what remains of six legendary temples built 1,500 to 1,200 years ago, submerged as a result of natural subsidence.

However, Graham Hancock, the best-selling author of controversial books about lost civilisations, said the ruins could be much older. If they were submerged by globally rising sea levels, their age would be around 5,000 years.

The pictures are the result of a three day diving expedition by India's National Institute of Oceanography and the Dorset-based Scientific Exploration Society. Mr Hancock, who dived with the team, said yesterday that SES had carried out the expedition at his suggestion.

'Our divers were presented with a series of structures that clearly showed man-made attributes,' said Monty Halls of the SES, who led the expedition.

'This is plainly a discovery of international significance that demands further exploration and detailed investigation.'

The site lies at depths of five to seven metres, 500 to 700 metres off Mahabalipuram, the site of a temple on dry land that dates to the first millennium AD.

Mr Hancock, who is not an archaeologist and has infuriated many experts with his theories, said that he had inferred the existence of six temples underwater by collating the stories of local fishermen with a legend that referred to Mahabalipuram as the Seven Pagodas.

Mr Hancock admitted yesterday that the submerged ruins might not be old enough to relate to the kind of post-ice age flooding that destroyed the supposed civilisations of his books.

But he said their discovery vindicated his approach of seeking the substance in local myths. 'I have argued for years that the world's flood myths deserve to be taken seriously – a view that most western academics reject. But here in Mahabalipuram we have proved the myths right.'

Mr Hancock said the site ran for about two kilometres, and contained 'a large conglomeration of large, clean-cut blocks in discrete areas. They seemed like several large ceremonial buildings surrounded by a lot of smaller ones.'

The Times, *11 April 2002*

DIVERS DISCOVER 'LOST CITY' OFF INDIA

By Mark Henderson

SUBMERGED ruins found off India's coast could be those of a legendary city said to have been swallowed by the sea, according to explorers who located the remains.

They are the second set of possible man-made ruins found off the subcontinent this year. Another 'lost city' was found off Gujarat in January, but that claim has been disputed by archaeologists.

The latest underwater stone structures were discovered last week by an Anglo-Indian team, diving a mile off the coast of Mahabalipuram in Tamil Nadu, southeast India. The geometrical patterns that look like a network of walls, roads and ramparts suggest they could have been part of the lost city of Melecherem, which, according to myth, was inundated by jealous gods.

Graham Hancock, an author who believes thousands of ancient civilizations were submerged at the end of the last Ice Age and who took part in last week's expedition by the Dorset-based Scientific Exploration Society and the Indian National Institute of Oceanography, said the ruins had convinced him that the myth was founded in reality.

Daily Mail 'Weekend' Magazine, 27 April 2002

Fantastic tales of lost cities are usually dismissed as romantic myths, but Graham Hancock claims that those very stories led him to a submerged site dating back at least 6000 years to the ice age – far older than any other city on earth. Has the amateur archaeologist really rewritten history? Andrew Wilson investigates.

A few seconds after diving beneath the ocean's surface, Graham Hancock peered through the underwater gloom and saw the distinct outline of an ancient wall rising up from the sands. Swimming closer to the mysterious structure, he took out his diving knife and, in order to test whether this was a man-made building rather than a natural formation, ran his blade through the masonry joints. Stretching out across the ocean floor was an extensive network of walls which ran for at least a mile out into the Bay of Bengal – sunken ruins which stand as evidence of a lost civilization engulfed by the waves.

For years, the hidden underwater city at Mahabalipuram in Tamil Nadu, southern India, had been confined to the realms of mythology. Fishermen spoke of the gilt-edged tops of temples lying beneath the sea, and whispered of the elaborate pyramidal pagodas submerged for thousands of years, but science had dismissed their claims as folklore. However, Hancock's discovery earlier this month of a lost civilization at Mahabalipuram, 30 miles south of Chennai (the former Madras), which Hancock believes could date back at least 6000 years, could force us to rewrite the history books.

Hancock's theory, that civilization began not with the Sumerians in Mesopotamia about 5000 years ago, but in a number of cities submerged by cataclysmic floods between 17,000 and 7000 years ago – has been widely rubbished by academics. Yet research arising out of this new discovery suggests that the maverick writer's views could be rooted in fact. On returning from the dive, Hancock contacted a world-renowned expert in ice age sea levels, who, with the help of a very sophisticated computer, confirmed that the site dated from approximately 6000 years ago.

'If this figure proves correct – and, in truth, a lot more work needs to be done – then it changes everything,' says Hancock. 'We can no longer think of the so-called "Fertile Crescent" of Sumeria as the cradle of civilization. The idea that cities first started to be built around 3500 BC also goes out of the window.

What seems more likely from the large body of evidence I have compiled is that there were a number of cities built before this time which were submerged by rising sea levels at the end of the last ice age. Mahabalipuram, I suspect, is one of them.'

Hancock's detective work begins with a detailed study of an area's flood myths, tales he believes grew up because of a very real phenomenon – the 400-feet rise in global sea levels after the melting of the ice caps. After researching a particular flood myth, Hancock then studies maps to show how the region would have looked at the end of the ice age. If the sea level data matches details passed down through an oral tradition, he believes there's a good chance a hidden city could be lying just below the waves.

Hancock first outlined his theories in his 1995 book *Fingerprints of the Gods*. The title may have established him as a literary Indiana Jones (the book sold a staggering 4.5 million copies), but it incurred the wrath of scholars and academics, who attacked him for what they saw as his selective presentation of evidence, lack of integrity and vulgar sensationalism. 'Scientists asked me to try to substantiate my theories – to find actual sites to support my beliefs – and that's what I've been doing over the past few years, touring the world in search of the lost underwater cities. South India is a black hole in terms of archaeological research, as there doesn't appear to be any trace of human activity between 12,000 and 3000 years ago. But what if the centres of an ancient civilization had once been located along its old coastline, land which was subsequently submerged by flood water?'

His connection with Mahabalipuram stretches back to his childhood when, as a five-year-old boy, he learnt to swim in its sparkling blue water. Born in Edinburgh in 1950, he arrived in India in July 1954 with his parents – his father had been appointed as a surgeon at the Christian Medical College and Hospital in Vellore. 'Imprinted on my memory for years afterwards – until I returned there in fact and was able to overlay old memories with new ones – were images of the eerie rock-hewn temples of Mahabalipuram, overlooking the Bay of Bengal and dating back 1200–1500 years.' In 1992 he travelled to India on a sentimental journey – his family had returned to Britain in 1958 and he wanted to revisit some of the places of his childhood.

During that visit he bought a musty old book, an anthology of traveller's journals, from a shop in Madras, a volume which would later form the first clue in his underwater detective story. Although he didn't read the book until two years ago, its contents forced him to reassess everything he knew about Mahabalipuram. He learnt for the first time of the 'Seven Pagodas' story – the six temples submerged beneath the sea, with the seventh still standing on the shore.

'A Brahmin about 50 years of age, a native of the place . . . informed me his grandfather had frequently mentioned having seen the gilt tops of pagodas in the surf, no longer visible,' wrote one traveller in 1798. According to myth, the ancient ruler of the kingdom constructed a city of such magnificence at Mahabalipuram that the gods grew jealous and orchestrated a tremendous flood to swallow it in a single day. The God of the Sea was ordered to 'let loose his billows and overflow a place which impiously pretended to view in splendour with their celestial mansions,' wrote the traveller. 'This command he obeyed, and the city was at once overflowed by that furious element, nor has it ever since been able to rear its head.'

In 2000 while researching his new book *Underworld* (Michael Joseph, £20), Hancock visited Mahabalipuram once more, where he interviewed a number of local fishermen. Many described having seen underwater 'temples', 'palaces' and 'walls' – even 'roads' – while diving to free trapped nets or anchors. Others talked of hidden doorways and rooms beneath the ocean which emitted strange musical sounds. 'If you just go where the fish are,' one said, 'then you will find them.' Yet the underwater investigator had to wait a further two years before travelling out to Mahabalipuram in an expedition organized in conjunction with the Dorset-based Scientific Exploration Society and India's National Institute of Oceanography. On April 3, half a mile from the shore, Hancock plunged into the blue waters of the Bay of Bengal – and what he saw lying beneath him almost took his breath away.

'What was staggering was that the ruins lay directly beneath the boat,' he says. 'I swam down to a depth of about 20 ft and reached out to scrape the sand away from the stone. It was clear from the masonry joints that the structure was unmistakably man-made, rather than a natural formation. I could see straight and curving walls, all made from clearly defined blocks of stone, and I followed one which was still completely intact for 50 ft. The site contains a conglomeration of large, distinct blocks which seem like several big ceremonial buildings surrounded by a number of smaller ones. My initial reaction was, not surprisingly, one of excitement. This was a man-made site which was new to archaeology, a place where no one had ever dived before. It felt like diving into a lost world.'

Accompanying Hancock on the dive was Monty Halls, a former major in the Royal Marines who led the expedition for the Scientific Exploration Society. During the 17 years he has been diving, Halls, 35, a freelance expedition leader, says he has never seen anything like the majestic underwater structures of Mahabalipuram. 'These enormous granite blocks looked like huge sugar cubes, about 20 ft tall, and there was a cluster of small stones around them,' he said. 'Although it's hard to say with any certainty, what we are seeing could have been a granite shrine surrounded by the remains of four temples.'

Central to the significance of the discovery is the age of the structures. Although mainstream archaeology believes them to date from 1200 years ago – the same time the rock-hewn sculptures and temples on the shore were carved from granite – Hancock believes the underwater ruins to be in the region of 6000 years old. If the flooded city did indeed date from only 1200 years ago, to the time of the Pallava dynasty, one would expect to find evidence of inscription on the stone. Yet during the 49 separate dives done over the course of three days by the team, not one inscription was found. In addition, the two structures differ widely in their architectural styles. The shore sculptures are ornate and highly decorative, while the underwater city is made up of simple, austere, rectangular blocks.

The greatest single piece of evidence so far to date the lost ruins of Mahabalipuram as 6000 years old comes from geophysicist Dr Glenn Milne at Durham University's world-renowned Department of Geological Sciences. Milne has built up a large database of figures and a sophisticated computer programme that can print out images of any shoreline at any period in history. When Hancock relayed data from Mahabalipuram, Milne was able to tell him that the site was at least 6000 years old. 'Assuming there was no tectonic movement at the site, and it looks like there wasn't, then it appears that the area was flooded by a rise in sea levels about 6000 years ago,' says Milne. 'The computer programme is accurate to within 1000 years either side of the allotted date.'

When Hancock heard this, he felt vindicated. 'It proved that the methods I was using – the combination of deciphering ancient myths and new technology – actually worked,' he says. 'Of course, I am still keeping an open mind, but it does suggest I'm on the right track after all. It's mainstream archaeology and science that are blinkered.'

However, this is not the first time Hancock's theories have been bolstered by the application of hard science. In January, it was revealed that the carbon dating of artefacts discovered at two submerged sites in the Gulf of Cambay, off the north-western state of Gujarat, show that these underwater cities are likely to date from 9500 years ago – 5000 years older than any city recognized by mainstream archaeologists. The cities – which are 15 miles apart and lie 12 ft beneath the waves – were discovered in May of last year, during routine pollution testing by India's National Institute of Ocean Technology.

'Since then, of course, archaeology has done everything it possibly can to dispute the evidence,' says Hancock. 'Experts have claimed that the samples could have been contaminated by sea water, and that the wood tested could have sat on the seabed for thousands of years before the cities were built. Scientists will do anything they possibly can to rubbish my name. I'm a threat to them because I'm an amateur – however, I'm an amateur who is able to pinpoint, with remarkable accuracy, a series of lost underwater cities which could force us to rethink everything we have ever learned about the origins of civilization.'

In the next couple of months, Hancock predicts an announcement from Cuba which will reveal the discovery of an ancient man-made city 2200 ft under the ocean. He is also confident that more lost civilizations will be found off the coasts of Malta, Japan, China, Florida, the Bahamas and Central America. 'After all, when the ice caps flooded ten million square miles of land were submerged,' he says. 'Discoveries such as Mahabalipuram are just the beginning – during the next 20 to 30 years I'm sure we will have uncovered dozens of underwater cities. It's not so much the quest for one Atlantis, but the search for many, many underworlds.'

Appendix 7 / Press Report on Paulina Zelitsky's Exploration in Cuba

EXPLORERS TO RETURN TO OCEAN FLOOR

By Anita Show, The Associated Press

Sunday May 19, 2002, 5.10 p.m.

HAVANA (AP) – Floating aboard the Spanish trawler she chartered to explore the Cuban coast for shipwrecks, Paulina Zelitsky pores over yellowed tomes filled with sketches and tales of lost cities – just like the one she believes she has found deep off the coast of western Cuba.

Zelitsky's eyes grow wide as she runs her small hand over water-stained drawings of Olmec temples in a dog-eared 1928 study of Mexican archaeology. The Russian Canadian explorer compares the shapes with green-tinted sonar images captured in March while studying the megalithic structures she discovered two years ago off Cuba's Guanahabibes Peninsula.

Amid piles of sonar-enhanced maps is a well-worn copy of *Comentarios Reales de las Incas*, or *Royal Commentaries of the Incas*, a classic of Spanish Renaissance narrative by the son of an Inca princess and a Spanish conquistador. Zelitsky is particularly fascinated by Garcilaso Inca de la Vega's account of ancient ruins at the bottom of Lake Titicaca, Peru.

'You would not think that a reasonable woman of my age would fall for an idea like this,' chuckled Zelitsky, a 57-year-old offshore engineer who runs the exploration firm Advanced Digital Communications of British Columbia, Canada.

Zelitsky passionately believes the megalithic structures her crew discovered 2310 feet below the ocean's surface could prove that a civilization lived thousands of years ago on an island or stretch of land joining the archipelago of Cuba with Mexico's Yucatan Peninsula, about 120 miles away.

The unusual shapes first appeared on the firm's sophisticated side-scan sonar equipment in the summer of 2000, during shipwreck surveys off Cuba's western coast, where hundreds of vessels are believed to have sunk over the centuries.

The company is among five foreign firms working with Fidel Castro's government to explore the island's coast for shipwrecks of historical and commercial interest. But the mysterious shapes have become the focus of this crew's exploratory efforts.

Puzzled by the shapes with clean lines, the team has repeatedly returned to the site – most recently in March – for more sonar readings, more videotapes of the megaliths with an unmanned submarine. The crew left in mid-May for a month.

Evidence for Zelitsky's hypothesis is far from conclusive, and has been met with skepticism from scientists from other countries who nevertheless decline to comment publicly on the project until scientific findings have been made available. Submerged urban ruins have never been found at so great a depth.

Elsewhere in the Caribbean, the ruins of Jamaica's Port Royal are located at depths ranging from a few inches to 40 feet below the ocean surface. The once raucous seaside community was controlled by English buccaneers before it slid under the waves in earthquakes beginning in 1692.

Located at just 20 feet at the mysterious megalithic structures discovered in the 1960s and 1970s in the sound between the Bahamas islands of North and South Bimini. Scientific expeditions there have produced inconclusive results about the shapes' origins.

Back in Cuba, a leading scientist recently admitted there is no easy explanation for the megalithic shapes found by Zelitsky's crew. The shapes on the sonar maps look like walls, rectangles, pyramids – rather like a town viewed from the window of an airplane flying overhead.

'We are left with the very questions that prompted this expedition,' geologist Manuel A. Iturralde Vincent, research director of Cuba's National Museum of Natural History wrote March 13. At the time he was visiting the area aboard the 270-foot long *Ulises*, the Spanish trawler Zelitsky outfitted with sophisticated computer and satellite equipment for her surveys.

In his written comments, later delivered at a scholarly conference here, Iturralde concluded it was possible the structures were once at sea level, as Zelitsky theorizes.

Because of the large faults and an underwater volcano nearby, Zelitsky supposes the structures sank because of a dramatic volcanic or seismological event thousands of years ago.

Providing some support for that argument, Iturralde confirmed indications of 'significantly strong seismic activity'.

Zelitsky shies from using the term 'Atlantis', but comparisons are inevitable to the legendary sunken civilization that Plato described in his *Dialogues* around 360 BC.

There have been untold, unsuccessful attempts over the ages to find that lost kingdom. One common theory is that Atlantis was located on the Aegean island of Thera, which was destroyed by a volcanic eruption nearly 3600 years ago.

Zelitsky does, however, mention known archaeological monuments when discussing her find.

Numerous photographs are scattered throughout a video show of the megaliths, showing well-known ancient sites: the 1st century fortress of Masada high above the Dead Sea, Britain's circular monument of Stonehenge, the

Roman fortress of Babylon in Cairo, the walls of Chan Chan, Peru, whose inhabitants were conquered by the Incas.

Perhaps, Zelitsky mused, the megaliths off Cuba are remains of a trading post, or a city built by colonizers from Mesoamerica. Those civilizations were far more advanced than the hunters and gatherers the Spaniards found upon arriving here five centuries ago.

Zelitsky admitted much more investigation is needed to solve the mystery.

But that doesn't keep her from believing, or from smiling slyly as she opens her agenda for 2002 to the first page.

Written there are the words Italian astronomer Galileo Galilei uttered under his breath at the height of the Inquisition, right after abjuring his belief that the Earth revolved around the sun.

'E pur si muove,' it reads – 'Nevertheless, it does move.'

Appendix 8 / Press Report from Times of India, 6 July 2002

Submerged structures found off the coast of Mahabalipuram in the Bay of Bengal could well solve the mystery of seven pagodas dating back to the Pallava Period (7th century AD)

By Akshaya Mukul, *Times of India*, *Times* News Network, New Delhi, 6 July 2002

The Archaeological Survey of India's Underworld Archaeology Wing (UAW) has discovered three walls and a number of carved architectural members of ancient temples running north to south and east to west. Also found are seven big submerged rocks 500 metres offshore.

According to UAW in-charge Alok Tripathi, who undertook the diving 500 metres east and north of the Shore temple in November 2001 and March this year, 'the walls are made of thick slabs of granite. Two long stone slabs, each with two verticle slits to receive two other stone slabs, were kept upright. Several such blocks arranged in a row formed a wall.'

The technique of construction, he says, is so effective that these structures are still in place despite violent seas and high-energy surf.

'The remnants are well carved and look like mouldings and pillars of temple. They are similar to the carvings in the existing temples of Mahabalipuram,' he says. Tripathi is hopeful of discovering more structures near the Shore temple. The ASI is planning to undertake diving towards the south of the temple.

'We are planning to dive during the Tamil month of Tai which falls between December and January. We will trace the extension of submerged structures and clean them to reconfirm our conclusion about their nature and purpose,' he says.

Part of the local legend, the story of submerged offshore temples, was first recorded by William Chambers, a British traveller, in the *Asiatic Research Journal* in 1788. He quoted older people having seen the 'tops of several pagodas far out in sea', covered with copper. By the time Chambers visited the place 'the effect was no longer the same as the copper had been incrusted with mould and verdigris'.

What lends credence to the UAW's excavation is a search carried out by divers of UK-based Scientific Exploration Society and Indian National Institute of Oceanography in April. They claimed to have found ruins spread over several square kilometres off the coast. During the expedition, divers came across structures believed to be man-made.

Online Appendices and Photographs

A number of appendices prepared for this book, which could not be included in the printed edition for reasons of space, are available online at my website: *http://www.grahamhancock.com*. Go to the section marked *Underworld*, where a full listing of the appendices appears. In addition, updates to the research, new underwater discoveries subsequent to publication and elements of debate raised by the book will be featured on the website. Many more of Santha Faiia's photographs of the submerged structures explored in *Underworld* will also be made available there.

Graham Hancock
January 2002

Notes

PART ONE: *Initiation*

1 / Relics

1. *Journal of Marine Archaeology*, vols. 5–6, 1995–6, 14–15, Society of Marine Archaeology, National Institute of Archaeology, Goa, India, 1997
2. Ibid., vol. 2, July 1991, 14
3. Ibid., vol. 2, 15
4. Ibid., vol. 2, 15–16
5. Ibid., vol. 2, 16
6. Ibid., vols. 5–6, 14
7. E.g. see *British Museum Encyclopaedia of Underwater and Maritime Archaeology*, 203, British Museum Press, 1997
8. *Journal of Marine Archaeology*, vol. 2, 8–9
9. Dr Glenn Milne, Department of Geology, Durham University, personal communication
10. This structure was badly damaged during particularly heavy typhoons July–September 2000
11. Privately commissioned research report from Akira Suzuki, 1999
12. Models of sea-level rise are generally unable to take account of local tectonic events in calculating sea-levels at specific locations
13. *Encyclopaedia Britannica, Micropaedia*, vol. 9, 356, Chicago, 1991
14. Jean-Yves Empereur, *Alexandria Rediscovered*, 82 and 86–7, British Museum Press, London, 1998
15. See *Fingerprints of the Gods*, *Keeper of Genesis* (US: *Message of the Sphinx*) and *Heaven's Mirror*
16. Ian Shaw and Paul Nicholson (eds.), *British Museum Dictionary of Ancient Egypt*, 23–4, British Museum Press, London, 1995
17. Empereur, op. cit., 37
18. Ibid., 74–5
19. Ibid., 75
20. Ibid., 75
21. For details of the earthquakes see ibid., 86–7
22. Ibid., 80
23. E. M. Forster, *Alexandria – A History and a Guide*, reprinted 1968 by Peter Smith, Gloucester, Mass., USA
24. Roy Macleod, *The Library of Alexandria – Centre of Learning in the Ancient World*, I. B. Tauris, London, 2000; Mostafa El-Abbadi, *Life and Fate of the Ancient Library of Alexandria*, UNESCO, Paris, 1992; Dorothy L. Sly, *Philo's Alexandria*, Routledge, London, 1996; Luciano Canfora, *The Vanished Library: A Wonder of the Ancient World*, Hutchinson Radius, London, 1989
25. These figures are approximate. For the Pyramid's statistics see *Fingerprints of the Gods*, chapters 33–8
26. Empereur, op. cit., 71
27. Cited in E. M. Forster, op. cit., 141
28. Ibid., 141
29. Ibid., 141 and 142
30. Empereur, op. cit., 37
31. Plato, *Timaeus and Critias*, 36, Penguin Books, London, 1977
32. Ibid., 36
33. See J. G. Frazer, *Folklore in the Old Testament*, vol. 1, 104–361, Macmillan, London, 1918
34. Alan Dundes (ed.), *The Flood Myth*, 1, University of California Press, 1988

35. J. G. Frazer, op. cit., 105, 343–4
36. Dorothy B. Vitaliano, 'The Deluge', in *Legends of the Earth: Their Geologic Origins*, 142–78, Indiana University Press
37. Dundes, op. cit., 1
38. Roger Lewin, *Human Evolution*, 74–7, Blackwell, Oxford, 1984
39. *Encyclopaedia Britannica, Micropaedia*, vol. 12, 127
40. Or at least so claim its enthusiasts
41. Robert Kunzig, *Mapping the Deep*, 2, Sort of Books, London, 2000
42. In Goa and Lakshadweep
43. See full discussion in parts 2 and 3
44. G. A. Milne, J. L. Davis, J. X. Mitrovica, H.-G. Scherneck et al., 'Space-geodetic constraints on glacial isostatic adjustment in Fennoscandia', Science, 291, 2001, 2, 381–5. G. A. Milne, J. X. Mitrovica, J. L. Davis, 'Near-Field Hydro-Isostasy: The Implementation of a Revised Sea-Level Equation', *Geophysical Journal International*, 139, pub. 1, 1999, 464–83. G. A. Milne, J. X. Mitrovica, 'Postglacial Sea-Level Change on a Rotating Earth', *Geophysical Journal International*, 133, 1998, 1–19. G. A. Milne, J. X. Mitrovica, A. M. Forte, 'The sensitivity of GIA predictions to a low viscosity layer at the base of the upper mantle', *Earth and Planetary Science Letters*, 154, 1998, 265–78. G. A. Milne, J. X. Mitrovica, D. P. Schrag, 'Estimating past continental ice volume from sea-level data', *Quaternary Science Review*, in press, 2001
45. Sharif's rough guess was very close. The coordinates, per GPS to within 50 metres, are latitude 11 degrees 11.200 north and longitude 79 degrees 54.192 east

2 / The Riddle of the Antediluvian Cities

1. Samuel Noah Kramer, *History Begins at Sumer*, 148ff, University of Pennsylvania Press, 1991
2. Ibid., 148
3. The account is in the Book of Genesis, chapters 6–9
4. Kramer, op. cit., 148
5. Ibid., 148
6. Ibid., 149
7. Ibid., 149
8. Ibid., 149; William Hallo, 'Antediluvian Cities', *Journal of Cuneiform Studies*, vol. 23, 1970, 61
9. I have discussed these texts at length in earlier publications
10. Cited in Kramer, op. cit., 149–51
11. Ibid., 151
12. Ibid., 151
13. Ibid., 151
14. Ibid., 152
15. Ibid., 152
16. Ibid., 152
17. Ibid., 152–3
18. Ibid., 153
19. Ibid., 148
20. See discussion in Gerald P. Verbrugghe and John M. Wickersham (eds.), *Berossos and Manetho*, 15ff, University of Michigan Press, 1999
21. Samuel Noel Kramer, *The Sumerians*, 39–40, University of Chicago Press, 1963
22. Ibid., 39
23. Ibid., 39–40
24. Ibid., 40
25. Ibid., 42
26. Time-Life, *The Age of the God Kings*, 10–11, Time-Life Books, 1989
27. See *www.grahamhancock.com*, Forum, 'The Quantas Mystery'
28. Leonard Woolley, *Ur of the Chaldees*, 21, Pelican Books, 1940
29. Ibid., 21
30. Ibid., 21, 24

31. Ibid., 24
32. Oppenheimer re Flandrian transgression
33. Kurt Lambeck, 'Shoreline Reconstructions for the Persian Gulf Since the Last Glacial Maximum', *Earth and Planetary Science Letters*, 142, 1996, 43–57
34. Ibid., 47
35. Oppenheimer, *Eden in the East: The Drowned Continent of Southeast Asia*, 57, Weidenfeld and Nicholson, London, 1998
36. Ibid., 46. See also Julius Zarins, 'The Early Settlement of Southern Mesopotamia', 57, *Journal of the American Oriental Society*, 112, 1, 1992
37. Oppenheimer, op. cit., 46
38. Georges Roux, *Ancient Iraq*, 4, Penguin Books, London, 1992, citing C. E. Larsen, 'The Mesopotamian Delta Region: A Reconsideration of Lees and Falcon', *Journal of the American Oriental Society*, 95, 1975, 43–57. P. Kassler, 'The structural and geomorphic evolution of the Persian Gulf', in B. H. Preuser, *The Persian Gulf*, Berlin, Heidelberg, New York, 1973, 11–32. W. Nutzel, 'The formation of the Arabian Gulf from 14,000 BC', *Sumer*, 31, 1975, 101–11
39. Kramer, *The Sumerians*, 2 and 31
40. Ibid., 30 and 31
41. Ibid., 31
42. Ibid., 31
43. Roux, op. cit., 60
44. Ibid., 60
45. Ibid., 48, 60. Roux identifies the first stages of construction at Eridu with Ubaid I pottery, a style that he dates to 7000 years before the present
46. Ibid., 108
47. Ibid., 112
48. Zarins, 'The Early Settlement of Southern Mesopotamia'
49. Ibid.
50. Ibid., 57
51. Ibid., 60
52. Roux, op. cit., 111
53. Kramer, *The Sumerians*, 26
54. Roux, op. cit., 109, 112
55. Hallo, op. cit., 61
56. Verbrugghe and Wickersham, op. cit.
57. Ibid., 49
58. Ibid., 49
59. Ibid., 49, footnote 19
60. Ibid., 49–50
61. Ibid., 50
62. Roux, op. cit. See maps, Southern Mesopotamia
63. Hallo, op. cit., 61
64. Edmond Sollberger, *The Babylonian Legend of the Flood*, British Museum Publications, London, 1984, 17
65. The *Epic of Gilgamesh*, Penguin, London, 1972; Stephanie Dalley, *Myths from Mesopotamia*, Oxford University Press, 1990
66. E.g. see Verbrugghe and Wickersham, op. cit., 3–91
67. Hallo, op. cit., 63
68. Ibid., 63
69. Roux, op. cit., 33, 48
70. Ibid., 37–38
71. Ibid., 44–5
72. Ibid., 49
73. Ibid., 51
74. Ibid., 53
75. L. Luca Cavalli-Sforza et al., *The History and Geography of Human Genes*, 215, Princeton University Press, 1994
76. Roux, op. cit., 54
77. *Encyclopaedia Britannica, Micropaedia*, vol. 12, 98
78. Roux, op. cit., 69
79. Cavalli-Sforza et al., op. cit., 215
80. Roux, op. cit., 48, 69
81. Roux, op. cit., 82–3: 'In all respects the Uruk culture appears as the development of conditions that existed during the Ubaid period
82. Roux, op. cit., 48, 76–7. There is an

intervening sub-phase of the Uruk
period known as the Jemdat Nasr
period after the type-site between
Baghdad and Babylon. Roux, op. cit.,
76: 'Between the cultural elements of
that period [Jemdat Nasr] and those
of the Uruk period there is no fundamental difference.'

83. Ibid., 48
84. Ibid., 66
85. Ibid., 66
86. Ibid., 80
87. Ibid., 80
88. Ibid., 80–81
89. Ibid., 80
90. Ibid., 80
91. Cavalli-Sforza et al., op. cit., 215
92. Roux, op. cit., 82
93. Benno Lansberger, 'Three Essays on
    the Sumerians II: The Beginnings of
    Civilization in Mesopotamia', in
    Benno Lansberger, *Three Essays on
    the Sumerians*, Udena Publications,
    Los Angeles, 174; Verbrugghe and
    Wickersham, op. cit., 17 and 44;
    Stephanie Dalley, op. cit., 182–3,
    328; Jeremy Black and Anthony
    Green (eds.), *Gods, Demons and Sym-
    bols of Mesopotamia*, 41, 82–2,
    163–4, British Museum Press,
    London, 1992
94. Verbrugghe and Wickersham, op. cit.,
    43
95. Ibid., 44
96. Benno Landsberger, op. cit., Essay 2
97. Ibid., Essay 3
98. Ibid., Essay 2
99. Ibid., Essay 2
100. Ibid., Essay 2
101. Lambeck, op. cit., 43–53
102. Ibid., 43
103. Roux, op. cit., 48, 60
104. Lambeck, op. cit., 55
105. Ibid., 55
106. Ibid., 56
107. See for example discussion in Wil-
    liam Ryan and Walter Pitman,
    Noah's Flood, 178–9, Simon and
    Schuster, New York, 1998
108. Roux, op. cit., 4
109. Ibid., 4
110. Lambeck, op. cit., 54
111. Ibid., 54
112. Ibid., 54
113. Ofer Bar-Yoseph, 'The Impact of Late
    Pleistocene-Early Holocene Climatic
    Changes on Humans in Southwest
    Asia', in Lawrence Guy Straus et al.,
    Humans at the End of the Ice Age,
    68, Plenum Press, New York and
    London, 1996
114. Ibid., 68
115. Lambeck, op. cit., 54
116. See discussion in Verbrugghe and
    Wickersham, op. cit.
117. See Kramer, *History Begins at Sumer*

3 / Meltdown

1. Elise Van Campo puts the beginning
   of his 'LGM interval' at 22,000
   carbon-14 years ago (approximately
   equivalent to 25,500 to 21,500 years
   ago), based on his Arabian sea-core
   data: *Quaternary Research*, 26, 1987,
   376. Jonathan Adams gives
   17,000–15,000 carbon-14 years ago as
   the period of most extreme glacial
   conditions in several areas of Eurasia.
   This corresponds to a period of 20,300
   to 18,000 calendar years ago. J.
   Adams, *Eurasia During the Last
   150,000 Years*, on-line literature
   review at http://www.esd.ornl.gov/
   projects/qen/nercEURASIA.html
2. Cesare Emiliani, *Planet Earth*, 543,
   Cambridge University Press, 1995
3. Glenn Milne, Dept of Geology, University of Durham
4. China: 9.6 million sq. kms; Europe
   10.3 million sq. kms; Canada 9.9 million sq. kms
5. Lawrence Guy Straus et al., *Humans*

at the End of the Ice Age, 175, Plenum Press, New York and London, 1996

6. Ibid., 175

7. Ibid., 175

8. Ibid., 177 and 188-9, emphasis added.

9. Richard Rudgley, *Lost Civilizations of the Stone Age*, 100, Century, London, 1998

10. Ibid., 100

11. Ibid., 100

12. N. C. Fleming, 'Archaeological evidence for vertical movement of the continental shelf during the Palaeolithic, Neolithic and Bronze Age periods': 'Of the 500 known submarine sites worldwide containing *in situ* remains of buildings, structures, harbour works, quarries, or lithic artefacts, approximately 100 are older than 3 ka BP, that is, in European archaeological terminology, Bronze Age or older.' To assess the preponderance of interest in shipwrecks over the search for ancient structures see the *British Museum Encyclopaedia of Underwater and Maritime Archaeology*, British Museum Press, 1997

13. Thomas J. Crowley and Gerald R. North, *Palaeoclimatology*, 48, Oxford University Press, 1991

14. Ibid., 48

15. R. C. L. Wilson, S. A. Drury and J. L. Chapman, *The Great Ice Age*, 14, The Open University, London, 2000

16. Ibid., 15

17. Ibid., 15

18. Ibid., 14

19. Ibid., 16

20. Oppenheimer, *Eden in the East: The Drowned Continent of Southeast Asia*, 43, Weidenfeld and Nicholson, London, 1998

21. Ibid., 41

22. Wilson et al., op. cit., 17

23. Ibid., 14-17

24. Ibid., 16

25. Plato, *Timaeus and Critias*, 38, Penguin Books, London, 1977

26. Vitacheslav Koudriavtsev, *Atlantis: Ice Age Civilization*, Institute of Metahistory, Moscow, 1997. Koudriavtsev's work may be accessed on the Internet at www.imh.ru

27. Cesare Emiliani, *The Scientific Companion*, 251 and 257, Wiley Popular Science, 1995

28. Cesare Emiliani held a Ph.D. from the University of Chicago, where he pioneered the isotopic analysis of deep-sea sediments as a way to study the Earth's past climates. He then moved to the University of Miami, where he continued his isotopic studies and led several expeditions at sea. He was the recipient of the Vega medal from Sweden and the Agassiz medal from the National Academy of Sciences of the United States

29. Emiliani, *Earth and Planetary Science Letters*, 41, 1978, 159, Elsevier Scientific Publishing Company, Amsterdam

30. Robert Schoch, *Voices of the Rocks*, 147-8, Harmony Books, New York, 1999

31. Ibid., 148

32. Paul LaViolette, *Earth Under Fire*, 183, Starburst Publications, New York, 1997

33. E.g. scenario: population is tempted to migrate to coasts, or to low-lying valleys near coasts, by improved conditions; several thousand years of stability and prosperity; all eggs placed in the one basket of the coastal cities; flooding suddenly resumes and engulfs the cities; there are only a few survivors, etc.

34. Emiliani, *Planet Earth*, 543

35. Ibid., 540

36. Emiliani, *The Scientific Companion*, 251, 257
37. Taped interview with John Shaw, conduced by John Grigsby, research assistant to GH, 1999
38. *Nature*, vol. 389, 2 October 1997, 473
39. Ibid., 473
40. Ibid., 474
41. Ibid., 474
42. Ibid., 474
43. Ibid., 474–5
44. Arch C. Johnston, 'A Wave in the Earth', *Science*, vol. 274, 1 November 1996, 735
45. Oppenheimer, op. cit., 40
46. Johnston, op. cit.
47. Ibid.
48. Ibid.
49. Ronald Arvidsson, 'Fennoscandian Earthquakes: Whole Crustal Rupturing Related to Post-Glacial Rebound', *Science*, vol. 274, 1 November 1996
50. Ibid.
51. Johnston, op. cit., 735 (emphasis added)
52. Re measures of seismic magnitude: the comparative figures for the Parvie quake were given in ML units in the source document. ML units are on the *Local Magnitude* scale, which is the basis for establishing an earthquake's level on the famous Richter scale. ML measures the amplitude of a wave as it appears on a seismograph that has been set up in a particular location, so it is essentially a measure of the extent to which a certain bit of ground moves vertically in an earthquake. The Richter scale is logarithmic, so the magnitude goes up exponentially (by a factor of 10) with each step up the scale. An earthquake measuring 6.0ML has 10 times greater magnitude than an earthquake measuring 5.0ML, and 100 times greater magnitude than one measuring 4.0ML. NB, the units of the Richter Scale are M rather than ML, because the magnitude we're dealing with is no longer local but should be the same everywhere.
53. Johnston, op. cit.
54. *Encyclopaedia Britannica, Micropaedia*, vol. 10, 55
55. *Guardian*, London, 18 January 1995
56. *Encyclopaedia Britannica, Micropaedia*, vol. 10, 55
57. Johnston, op. cit.
58. Arvidsson, op. cit.
59. Wilson et al., op. cit., 19 and 28
60. Straus et al., op. cit., 129–30
61. Schild in ibid., 129–30
62. Schoch., op. cit., 147–8
63. Plato, op. cit., 38
64. Ibid., 35–6
65. Emiliani, *Earth and Planetary Science Letters*, 159
66. Isaac and Janet Asimov, *Frontiers II*, 110–11, New York, 1993, cited in Charles Ginenthal, 'The Extinction of the Mammoth', 266, *The Veilkovskian*, vol. 3, nos. 2 and 3, New York, 1997
67. *Encyclopaedia Britannica, Micropaedia*, vol. 4, 235; LaViolette, op. cit., 203–4: 'Drumlin fields are found widely distributed in both North America and Europe. In North America conspicuous fields are present in regions that once lay at the edge of the ice-sheet, such as those found in central-western New York (about 10,000 drumlins), east central Wisconsin (about 5000 drumlins), south central New England (about 3000 drumlins), south-western Nova Scotia (2300 drumlins). Other fields are believed to be present in intervening districts but to have escaped detection . . .'
68. John Grigsby, interview with John Shaw

69. Ibid.
70. Shaw, 'Drumlins, subglacial meltwater floods, and ocean responses', *Geology*, vol. 17, September 1989, 853–6
71. Ginenthal, op. cit., 267; John Shaw and Donald Kvill, 'A Glacio-Fluvial Origin for Drumlins in the Livingston Lake Area, Saskatchewan', *Canadian Journal of Earth Science*, vol. 21, 1984, 1442
72. Shaw, op. cit., 855
73. John Shaw, *A Meltwater Model for the Laurentide Subglacial Landscapes, Geomorphology Sans Frontiers*, 181, John Wiley and Sons, 1996
74. John Grigsby, interview with John Shaw
75. John Shaw, 'A Qualitative View of Sub-Ice-Sheet Landscape Evolution', *Progress in Physical Geography*, 18.2, 1994, 166
76. Ibid., 164
77. Shaw, 'Drumlins, subglacial meltwater floods, and ocean responses', 854
78. Shaw and Kvill, op. cit., 1455
79. John Shaw, 'Sedimentary Evidence Favouring the Formation of Rogen Landscapes by Outburst Floods', http://www.sentex.nettcc/rogen/main.html, 4
80. Shaw and Kvill, op. cit., 1455
81. Paul Blanchon and John Shaw, 'Reef drowning during the last deglaciation: Evidence for catastrophic sea-level rise and ice-sheet collapse', *Geology*, vol. 23, no. 1, January 1995, 6. See also Wilson et al., op. cit., 113–21
82. Blanchon and Shaw, op. cit., 4
83. Shaw, 'Sedimentary Evidence Favouring the Formation of Rogen Landscapes by Outburst Floods', 4
84. Fletcher and Sherman, 'Submerged Shorelines . . .', *Journal of Coastal Research*, special issue no. 17, 147
85. Ibid., 147 (emphasis added)
86. Scott Fields, 'Metafloods at the end of the Ice Age', cited in Charles Ginenthal, op. cit., 267
87. Reported in Fletcher and Sherman, op. cit., 148
88. Wilson et al., op. cit., 113–15
89. Ginenthal, op. cit., 265
90. Blanchon and Shaw, op. cit., 6
91. Wilson et al., op. cit., 117
92. Ibid., 117
93. Ibid., 117
94. Ibid., 117
95. Crowley and North, op. cit., 61–2
96. Blanchon and Shaw, op. cit., 7
97. Fletcher and Sherman, op. cit., 147
98. Crowley and North, op. cit., 64
99. Fletcher and Sherman, op. cit., 147
100. Ibid., 147–8
101. Ibid., 148. Approximately the same dating for the catastrophic draining of Lakes Agassis and Ojibway through the Hudson Strait is given in D. C. Barber et al., 'Forcing of the cold event of 8200 years ago . . .', *Nature*, vol. 400, July 1999, 344ff
102. David Keys, 'Lethal Floods Ravaged Stone Age Britain', *Independent*, London, 15 October 2000
103. Oppenheimer, op. cit., 35
104. There was, for example, a 25 metre rise in sea-level, followed by a similar fall, during a period estimated at less than 2000 years centred on 8000 years ago. *American Association of Petroleum Geologists Bulletin*, 1995, 1568
105. Oppenheimer, op. cit., 40
106. LaViolette, op. cit., 225
107. Ibid., 206
108. Ibid., 199–200; 202–3

PART TWO: *India (1)*

4 / Forgotten Cities, Ancient Texts and an Indian Atlantis

1. Jonathan Mark Kennoyer, *Ancient Cities of the Indus Valley Civilization*, 70, American Institute of Pakistan Studies, Oxford, 1998

2. Introduced in 1972, the written script for the Somali language is based on the Latin alphabet, with modifications

3. The general view is that the short inscriptions of the Harappan script are trade linked and were probably in most cases used to label merchandise

4. John M. Cooper (ed.), *Plato: Complete Works*, 551, Hackett Publishing Company, Indianapolis, 1997

5. Ibid., 551–2

6. Georg Feuerstein, Subash Kak, David Frawley, *In Search of the Cradle of Civilization*, 29, Quest Books, Wheaton, Ill., 1995

7. Although the possibility cannot be completely ruled out. There is a curious reference in *the Atharvava Veda*, 19.72.1, to placing the *Vedas* back in a chest – which suggests the existence of a written version. The matter is discussed in David Frawley, *Gods, Sages and Kings*, 249, Passage Press, Salt Lake City, 1991

8. Gregory L. Possehl, *Indus Age: The Beginnings*, 6, University of Pennsylvania Press, 1999

9. Ibid., 6

10. Ibid., 7, 8

11. Ibid., 5–6, and see for example Kennoyer, op. cit., 24

12. Possehl, op. cit., 41

13. Personal communication, Professor B. B. Lal, formerly a student of Wheeler's. See also discussion in Feuerstein et al., op. cit., 77ff

14. Possehl, op. cit., 42

15. Ibid., 6

16. Ibid., 6

17. Ibid., 6

18. Ibid., 6

19. Emphasis added. Cited in *Vedavyas, Astronomical Dating of the Mahabaratha War*, 64, University of Vedic Sciences, India, 1995

20. Cited in ibid., 64

21. M. Muller, *The Six Systems of Indian Philosophy*, 34–5, cited in Feuerstein et al., op. cit., 42

22. Kennoyer, op. cit., 29

23. Ibid., 104 ff

24. V. Gordon Childe, *The Aryans*, 1926, 211–12, cited in Possehl, op. cit., 41–2

25. Colin Renfrew, *Archaeology and Language: The Puzzle of Indo-European Origins*, 182, Pimlico, London, 1998

26. Ibid., 188

27. Ibid., 190

28. Ibid., 205

29. Ibid., 205

30. S. R. Rao, *Dawn and Devolution of the Indus Civilization*, Aditya Prakashan, New Delhi, 1991

31. See S. P. Gupta, *The Indus-Sarasvati Civilization*, 91, 97, Pratiba Prakashan, Delhi, 1996

32. Ibid., 146

33. *Vishnu Purana*, vol. 2, 785, Nag Publishers, Delhi, 1989

34. Ibid., 853

35. E.g. see Ananda K. Coomaraswamy, *Myths of the Hindus and Buddhists*, 393, Dover Publications, New York, 1967

36. *Bhagvata Purana*, vol. 2, part 5, 12.3.30, 2139, Motilal Banarsidas, Delhi, 1978

5 / Pilgrimage to India

1. A range of figures for the extent of the civilization are given by different authorities. See Jonathan Mark Kennoyer, *Ancient Cities of the Indus Valley Civilization*, 17, American Institute of Pakistan Studies, Oxford, 1998; S. R.

Rao, *Dawn and Devolution of the Indus Civilization*, 10, Aditya Prakashan, New Delhi, 1991; S. P. Gupta, *The Indus-Sarasvati Civilization*, 1–4, Pratiba Prakashan, Delhi, 1996, gives 2.5 million square kilometres

2. See discussion in Gupta, op. cit., 114
3. Sir Mortimer Wheeler, *The Indus Civilization*, 3rd edn, 1968, 55
4. Rao, op. cit., 49
5. Kennoyer, op. cit., 50. He gives a population of 41,250 for the Lower Town of Mohenjodaro – 76.6 hectares – but notes that the total populated area was much larger – 250 hectares or so. Pro rata this yields a total population of about 150,000.
6. Ibid., 57
7. Ibid., 52
8. Georg Feuerstein, Subash Kak, David Frawley, *In Search of the Cradle of Civilization*, 73, Quest Books, Wheaton, Ill., 1995
9. Ibid., 73
10. Rao, op. cit., 17
11. Ibid., 17
12. Kennoyer, op. cit.,15. The reference comes from Kennoyer's study of blade technologies in the Indus Valley cities. He found plentiful evidence that certain large gastropod seashells (*Turbinella pyrum* and *Chicoreus ramosus*) had been cleanly and efficiently sliced up to make jewellery. These shells are exceptionally hard and he came to the conclusion (page 96) that they must have been cut 'with a specialized bronze saw . . . By studying the depth of each saw stroke on fragments of shell from the ancient workshops, we can reconstruct the basic shape of the saw. It had a very thin serrated edge that was long and curved, similar to the saws still used in shell bangle making in modern Bengal. Even more astounding is the fact that the Indus bronze saw was able to cut the shell as efficiently as the modern steel saws, which suggests that the Indus bronze workers were able to produce a bronze that was as hard as steel.'

13. Gupta, op. cit., i
14. Ibid., i
15. Gupta, op. cit., 141
16. Gregory Possehl, interviewed by Sharif Sakr, 24 October 2000
17. Captain M. W. Carr (ed.), *Descriptive and Historical Papers Relating to the Seven Pagodas of the Coromandel Coast* (first published 1869), reprinted Asian Educational Services, New Delhi, 1984
18. Ibid., 34–5
19. Ibid., 2
20. Ibid., 1
21. E.g. *Vishnu Purana*, vol. 1, 188ff, Nag Publishers, Delhi, 1989
22. Carr, op. cit., 12–13
23. Ibid., 3
24. Ibid., 13
25. Ibid., 13
26. Ibid., 14
27. Ibid., 14
28. Ibid., 14–15
29. E.g. Plato, *Timaeus and Critias*, 145, Penguin Books, London, 1977
30. Rao, op. cit., 141
31. Ibid., 141
32. Ibid., 141
33. Ibid., 126 and 142
34. Ibid., 141
35. Pannikar and Srinivason, cited ibid., 143
36. Ibid., 120
37. Michael A. Hoffman, *Egypt Before the Pharaohs*, 16, Michael O'Mara Books, London, 1991
38. Cyril Aldred, *Egypt to the End of the Old Kingdom*, 35, Thames and Hudson, London, 1988
39. Ibid., 35, 33

40. Kennoyer, op. cit., 114
41. Arthur Posnansky, *Tiahuanacu: The Cradle of American Man* (4 vols.), plate LXXIX.a, J. J. Augustin, New York, 1945
42. Aldred, op. cit., 35
43. Mackay, 1934, 422, cited in Possehl, op. cit., 289
44. E.g. Thor Heyerdahl, *The Ra Expeditions*, BCA, London, 1972
45. Gupta, op. cit., i, 114
46. Possehl, op. cit., 290
47. Ibid., 290
48. Ibid., 290
49. Note in Lothal site museum
50. See *Fingerprints of the Gods*, chapters 24 and 25
51. John Howley, Jada Bahrata Dasa, *Holy Places and Temples in India*, 438, Spiritual Guides, 1996
52. *Bhagvata Purana*, 10, 1571
53. Ibid., 10, 1570

6 / The Place of the Ship's Descent

1. Ralph T. Griffith (trans.), *Hymns of the Atharvaveda*, vol. 1, xxvii, Munisharam Manoharlal Publishers, Delhi, 1985 (first published 1895–6)
2. Gregory L. Possehl, *Indus Age: The Beginnings*, 5, University of Pennsylvania Press, 1999; *Satpatha Brahmana*, part 5, 362, Motilal Banarsidass, Delhi, 1994
3. Possehl, op. cit., 5; Griffith, op. cit., vol. 1, xi
4. Possehl, op. cit., 5; Griffith, op. cit., vol. 1, xi
5. Possehl, op. cit., 5
6. Griffith, op. cit., vol. 1, xii. The *Atharva Veda* is so named, says Griffith (p. xi), 'not from the nature of its contents, but from a personage of indefinitely remote antiquity named Atharvan . . .'
7. M. Sundarraj, *Rig Vedic Studies*, xxi, International Society for the Investigation of Ancient Civilization, Chennai, 1997
8. See *Encyclopaedia Britannica, Micropaedia*, vol. 2, 461
9. See ibid., vol. 1, 517
10. Ibid., 12, 189
11. Ibid., 189
12. J. G. Frazer, *Folklore in the Old Testament*, vol. 1, 85, Macmillan, London, 1918
13. *Encyclopaedia Britannica, Micropaedia*, vol. 7, 693; vol. 9, 920
14. Ibid., vol. 7, 693
15. Ibid., vol. 9, 920
16. Ibid., 804
17. John E. Mitchiner, *Traditions of the Seven Rishis*, xvii–xix, Motilal Banarsidass, Delhi, 1982
18. Ibid., xvii
19. Ibid., xvii
20. E.g. see Ralph T. Griffith (trans.), *Hymns of the Rgveda*, vol. 1, 66, footnote 99, Munisharam Manoharlal Publishers, Delhi, 1987 (first published 1889). There are more than eighty separate references to Manu in the *Rig Veda*
21. Ibid., vol. 1, 66
22. Ibid., vol. 1, 99
23. Ibid., vol. 2, 218
24. Ibid., vol. 2, 513
25. Ibid., vol. 1, 155
26. See discussions in E. A. Wallis Budge, *Osiris and the Egyptian Resurrection*, Dover Publications Inc., New York, 1973 (first printed 1911). See also discussion in Jane B. Sellers, *The Death of Gods in Ancient Egypt*, 21, Penguin, London, 1992
27. Griffith, *Rgveda*, vol. 1, 285
28. Ibid., 286
29. David Frawley, *Gods, Sages and Kings*, 285–6, Passage Press, Salt Lake City, 1991. See *Bhagvata Purana*, vol. 8, 24.10–11, Motilal Banarsidas, Delhi, 1978

30. 'Wash thee away' is Weber's rendering. J. Eggeling has 'cut thee off'. Max Muller has 'cut thee asunder'. See *Satpatha Brahmana*, part 1, 217, note 3.

31. Ibid., 216–18

32. Ibid., 218–19

33. Frazer, op. cit., 185

34. Ibid., 186

35. Ibid., 186–7

36. Ibid., 187

37. Ibid., 191

38. Ibid., 191

39. Ibid., 192

40. Ibid., 192

41. Ibid., 192

42. *Satpatha Brahmana*, part 1, 218, note 1

43. *Atharva Veda*, 19.39.8, translated by Frawley, op. cit., 299

44. Griffith, *Atharvaveda*, vol. 2, 243

45. Ibid., vol. 2, 243, note 8

46. Cited in *Satpatha Brahmana*, part 1, 218, footnote 1

47. Griffith, *Rgveda*, vol. 1, 319

48. Ibid., vol. 1, 319, note 13

49. *Bhagvata Purana*, cited in Frazer, op. cit., 192

50. Ibid., 192

51. *Matsya Purana*, part 1, 7, Nag Publishers, Delhi, 1997

52. See extensive discussion in *Fingerprints of the Gods*, chapter 31

53. *Matsya Purana*, part 1, 7, note.

54. See chapter 2 and Samuel Noah Kramer, *History Begins at Sumer*, 152–3, University of Pennsylvania Press, 1991

55. Mitchiner, op. cit., 206, 208–9

56. 'Formerly in the Svayambhuva age these were the Seven great Rsis. When the Age of Caksusa had passed . . . the Seven Rsis were again born as the seven Mind-born. (The question is asked: How were the Seven Rsis formerly born as the seven Mind-born? And the answer is given: After a long period, the Rsis were born a second time – so we have heard.)' Cited in Mitchiner, op. cit., 33

57. Griffith, *Rgveda*

58. Ludwig, cited in ibid., vol. 2, 624, note 7

59. Mitchiner, op. cit.

60. Ibid., xvi

61. Ibid., 196

62. Lokamanya Bal Ganghadar Tilak, *The Arctic Home in the Vedas*, 425, Tilak Bros, Poona, 1956. I have rendered 'Kalpa' as 'age'.

63. Ibid., 426

64. Ibid., 426

65. See discussion in *Heaven's Mirror*, 156ff

66. Griffith, *Rgveda*, vol. 1, 237

67. Cited in Sundarraj, op. cit., 333

68. E.g. 'he who has eyes can see this, not he who is blind', cited in Mitchiner, op. cit., 10.

69. Tilak, op. cit., 427

70. Lists of different Manvantaras and groups of Manu and the Seven Sages in *Matsya Purana*, for example

71. Tilak, op. cit., 426

72. Mitchiner op. cit., 49–50

73. Tilak, op. cit., 420

74. *Matsya Purana*, part 1, 635. I have rendered 'Kalpa' simply as 'age'.

75. *The Puranas*. See Mitchiner, op. cit., 3

76. Cited in ibid.., 293

77. Ibid., 128–30

78. Ibid., 4

79. Ibid., 5

80. Griffith, *Rgveda*, vol. 2, 538

81. Ibid., vol. 2, 538. See also Mitchiner, op. cit., 10

82. Discussed in ibid., 262–7

83. *Satpatha Brahmana*, part 1, 282

84. Mitchiner, op. cit., 262–3

85. See extensive discussions in Robert Bauval and Adrian Gilbert, *The Orion Mystery*, Heinemann, London, 1994.

See also *Keeper of Genesis/Message of the Sphinx*, and *Heaven's Mirror*

86. Ibid.
87. See discussions in *The Orion Mystery*
88. R. O. Faulkner (trans.), *The Ancient Egyptian Pyramid Texts*, 138, Utterance 419, Aris and Phillips, Wiltshire (first published by Oxford University Press, 1969)
89. Ibid., 155, Utterance 466, 5
90. Mitchiner, op. cit., 253
91. Ibid., 190–91
92. Ibid., 189–90
93. Ibid., 218ff
94. See discussion in *Heaven's Mirror*, chapter 6
95. See discussion in *Heaven's Mirror*, chapter 4
96. Mitchiner, op. cit., 223
97. Ibid., 224
98. Ibid., 225
99. *Mahabaratha*, cited in ibid., 223
100. Ibid., 223
101. See discussion in *Keeper of Genesis/ Message of the Sphinx*, 201–2
102. E. A. E. Reymond, *Mythical Origin of the Egyptian Temple*, 90, 109, 127, Manchester University Press, 1969
103. Ibid., 77
104. Frawley, op. cit., 41
105. Ibid., 205

7 / Lost India

1. See discussion in chapter 1
2. See discussion in chapter 1
3. See discussion in chapter 2
4. See discussion in chapter 5
5. Georg Feuerstein, Subash Kak, David Frawley, *In Search of the Cradle of Civilization*, 52–9, Quest Books, Wheaton, Ill., USA, 1995
6. Ibid., 52
7. See discussion in Gregory L. Possehl, *Indus Age: The Beginnings*, 446ff, University of Pennsylvania Press, 1999
8. There is not space to review the literature here but there is little serious dispute amongst scholars that our ancestors were taking to the sea as much as 30,000 or more years ago. See, for example, Geoffrey Irwin, *The Prehistoric Exploration and Colonisation of the Pacific*, 3ff, Cambridge University Press, 1994
9. Frawley, in David Frawley, *Gods, Sages and Kings*, 45, Passage Press, Salt Lake City, 1991
10. Ibid., 45
11. Ibid., 45
12. Ibid., 45
13. S. P. Gupta, *The Indus-Sarasvati Civilization*, 145, Pratiba Prakashan, Delhi, 1996
14. E.g. see Werner Keller, *The Bible as History*, Bantam Books, New York, 1988
15. I described this effect at Easter Island in *Heaven's Mirror*, 244–5
16. Cited in chapter 4
17. Possehl, op. cit., 362
18. Ralph T. Griffith (trans.), *Hymns of the Rgveda*, vol. 1, 4, 12, footnote 99, Munisharam Manoharlal Publishers, Delhi, 1987 (first published 1889)
19. Ibid., vol. 2, 44, 6
20. Ibid. vol. 1, 676, 2, 7–9
21. Ibid., vol. 1, 677, note 2
22. Ibid., vol. 2, 510, 9
23. Possehl, op. cit., 363
24. Griffith, op. cit., vol. 2, 98–9, 1–2
25. Ramaswamy, Bakliwal and Verma, 1991, 'Remote Sensing and River Migration in Western India', *Remote Sensing*, 12 (12), 2597–2609, cited in Possehl, op. cit., 362
26. Ramaswamy et al., cited in Possehl, op. cit., and Possehl's own views, 362
27. Bhimal Ghose, Anil Kar and Zahrid Jussain, 'Comparative Role of the Aravali and Himalayan river systems in the fluvial sedimentation of the Rajasthan desert. Central Arid Zone Research

Institute, Jodhpur', cited in Frawley, op. cit., 75

28. B. Ghose et al., 1979, 'The lost courses of the Saraswati River in the Great Indian Desert, New Evidence from Landsat Imagery', cited in Gupta, op. cit., 15

29. B. P. Radhakrishna, 'Holocene Chronology and Indian Prehistory', in B. P. Radhakrishna and S. S. Merh (eds.) *Vedic Sarasvati: Evolutionary History of a Lost River of Northwestern India*, Geological Society of India, Bangalore, 1999

30. Possehl, op. cit., 372

31. Discussed in *Fingerprints of the Gods*, chapter 28

32. See *Heaven's Mirror, Fingerprints of the Gods*

33. Jacobi, *Indian Antiquary*, cited in Frawley, op. cit., 182

34. Lokamanya Bal Ganghadar Tilak, *The Orion or Researches into the Antiquity of the Vedas*, 220, 234, Tilak Bros, Poona, 1986

35. Ibid., 220

36. Ibid., 220

37. Frawley, op. cit., 189

38. Ibid., 198

39. Ibid., 198

40. Feuerstein et al., op. cit., 244–5

41. John E. Mitchiner, *Traditions of the Seven Rishis*, 134, Motilal Banarsidass, Delhi, 1982

42. Ibid., 139–41

43. Ibid., 158

44. Ibid., 158

45. Pliny, *Naturalis Historia*, 6, 59–60, cited in Mitchiner, op. cit., 158; Solinus, *Compendium*, 52.5, cited in Mitchiner, op. cit., 158–9

46. Mitchiner, op. cit., 158

47. Feuerstein et al., op. cit., 247

48. Mitchiner, op. cit., 160

8 / The Demon on the Mountain and the Rebirth of Civilization

1. Turania = Ancient Turkestan

2. S. P. Gupta, *The Indus-Sarasvati Civilization*, 16–17, Pratiba Prakashan, Delhi, 1996

3. Gregory L. Possehl, *Indus Age: The Beginnings*, 440, University of Pennsylvania Press, 1999

4. Ibid., 440

5. Jacques Cauvin, *The Birth of the Gods and the Origins of Agriculture*, 76, Cambridge University Press, 2000

6. Possehl, op. cit., 412

7. Ibid., 1

8. E.g. 'Childe 1936; Bradwood 1952; Binford 1968; Flannery 1968, 1986; Myers 1971; Harris 1972, 1977; Reed 1977; M. Cohen 1977; Rindos 1984; Henry 1989; McCorriston and Hole 1991; Belfer-Cohen 1991, to note a few', cited in ibid., 429

9. Ibid., 429

10. Bar-Yoseph and Meadow, etc., cited in ibid., 430–31

11. Bar-Yoseph and Meadow, cited in ibid., 430

12. Ibid., 430–31

13. Sauer, 1952, cited in ibid., 431

14. Ibid., 451

15. Ibid., 451, 465

16. Ibid., 453

17. Ibid., 453

18. Ibid., 453

19. Ibid., 453

20. Ibid., 453–7

21. Jarridge et al., cited in ibid., 455

22. Ibid., 453–7

23. Ibid., 238

24. Ibid., 238–9

25. Ibid., 460, 459

26. See discussion in chapter 6

27. Possehl, op. cit., 457–9

28. Ibid., 230

29. Ibid., 457–9

30. Ibid., 483–8

31. Ibid., 482
32. Ibid., 482
33. Jarridge, cited in ibid., 485
34. Ibid., 489. Emphasis added
35. Ibid., 489
36. Ibid., 489 and Jonathan Mark Kennoyer, *Ancient Cities of the Indus Valley Civilization*, 38, American Institute of Pakistan Studies, Oxford, 1998
37. Kennoyer, op. cit., 38
38. Possehl, op. cit., 460
39. Ibid., 21, 22, 491
40. Ibid., 447; Gupta, op. cit., 22
41. Possehl, op. cit., 447: 'Taken as a whole the dates from Mehrgarh I and II seem to indicate that there was a settlement there at the beginning of the eighth millennium BC'
42. David Frawley, *Gods, Sages and Kings*, 58 and 300, Passage Press, Salt Lake City, 1991
43. In Georg Feuerstein, Subash Kak, David Frawley, *In Search of the Cradle of Civilization*, 89, Quest Books, Wheaton, Ill., 1995
44. Ralph T. Griffith (trans.), *Hymns of the Rgveda*, vol. 1, 5, footnote 10, Munisharam Manoharlal Publishers, Delhi, 1987 (first published 1889)
45. Lokamanya Bal Ganghadar Tilak, *The Arctic Home in the Vedas*, 225, Tilak Bros, Poona, 1956
46. Wilson, cited in Griffith, op. cit., vol. 1, 47, note 1
47. See discussion in Griffith, op. cit., vol. 1, 48, note 12
48. Max Muller, cited in Griffith, op. cit., vol. 1, 48, note 12
49. Tilak, op. cit., 225
50. Ibid., 225
51. Nicholas Borozovic, Douglas W. Burbank, Andrew J. Meigs, 'Climatic Limits on Landscape Development in the Northwestern Himalaya', *Science*, vol. 276, no. 5312, 25 April 1997, 571–4
52. Edward Derbyshire, 'Quaternary glacial sediments, glaciation stule, climate and uplift in the Karakoram and northwest Himalaya: review and speculations', *in Palaeogeography, Palaeoclimatology, Palaeoecology*, 120, 1996, 147–57, see 151 and 153
53. S. K. Gupta, P. Sharma and S. K. Shah, *Journal of Quaternary Science*, 7 (4), 1992, 283–90, see 283
54. Jonathan A. Holmes, 'Present and Past Patterns of Glaciation in the Northwest Himalaya: Climate, Tectonic and Topographic Controls', in John F. Shroder Jr (ed.), *Himalaya to the Sea: Geology, Geomorphology and the Quaternary*, 72
55. Ibid., 84
56. Ibid., 86. Holmes' study suggests (p. 90) that the maximum extent of glaciation in the Himalayas and the Karakorams may even have been reached somewhat earlier than the maximum extent of glaciation elsewhere in the world and thus the Himalayan ice-cap may have endured a rather longer period of deep-freeze and 'slumber' than other ice-caps
57. Borozovic et al., op. cit.
58. Mohamed Amin, Duncan Willetts, Graham Hancock, *Journey Through Pakistan*, Camerapix Publishers International, Nairobi, 1982
59. Ibid., 112
60. Holmes in Shroder, op. cit., 73
61. Borozovic et al., op. cit., 573. P. J. Taylor has noted modern ELAs of between 4800 and 5500 metres in the Zanskar range, see *Glacial Geology and Geomorphology*, April 2001, *http://ggg.qub.c.uk/ggg/papers/full/2001/rp02/rp02.html*
62. Morner, cited in Derbyshire, op. cit., 154
63. Derbyshire, op. cit., 153
64. Ibid., 153
65. Some of the most useful papers are

gathered together in B. P. Radhakrishna and S. S. Merh (eds.), *Vedic Sarasvati: Evolutionary History of a Lost River of Northwestern India*, Geological Society of India, Bangalore, 1999.

66. Coxon et al., *Journal of Quaternary Science*, 11 (6), 1996, 495–510, see 495
67. Ibid., 495
68. Shroder, Owen, Derbyshire, 'Quaternary Glaciation', in Shroder, op. cit., 133
69. Coxon et al., op. cit., 498
70. Daniel Vuichard and Markus Zimerman, 'The 1985 Catastrophic Drainage, etc.', *Mountain Research and Development*, vol. 7, no. 2, 1987, 91–100, see 91.
71. Lewis A. Owen, 'Neotectonics and glacial, etc.', *Tectonophysics*, 163, 1989, 227–65, see 237
72. Butler, Owen and Prior, 'Flashfloods, earthquakes and uplift in the Pakistan Himalayas', *Geology Today*, 197, November–December 1998
73. Kenneth Hewitt, 'Natural dams and outburst floods of the Karakoram Himalaya', *Hydrological Aspects of Alpine and High Montane Areas* (Proceedings of the Exeter Symposium, July 1982), IAHS publ. no.138, see 259
74. Ibid., 259
75. Again the notable exception, where this matter is considered directly, is Radhakrishna and Merh, op. cit.
76. Elise Van Campo, *Quaternary Research*, 26, 1986, 376–8, see 376 and 384
77. Ibid., 385
78. Ibid., 384–5
79. Ibid., 385
80. See discussion in chapter 3
81. Thomas J. Crowley and Gerald R. North, *Palaeoclimatology*, 62, Oxford University Press, 1991
82. Lawrence Guy Straus et al., *Humans at the End of the Ice Age*, 66, 86, Plenum Press, New York and London, 1996. The Younger Dryas is explicitly a term for a European cold phase, although the phase itself was global. The same phase is thus sometimes referred to by different names in other places; but it is also a generic term and it is used as such here
83. Crowley and North, op. cit., 63
84. Adams and Otte give date of start of Younger Dryas cold period as 12,800 and the end as 11,400 calendar years ago, *Current Anthropology*, vol. 40, 1999, 73–7, see 73
85. Straus et al., op. cit., 86
86. Adams and Otte, op. cit., 73
87. Ibid., 73
88. M. A. J. Williams and M. F. Clarke, 'Late Quaternary environments in north-central India', in *Nature*, vol. 308, 12 April 1984, 633–5, see 633
89. See also B. P. Radhakrishna, 'Vedic Sarasvati and the Dawn of Indian Civilization', in Radhakrishna and Merh, op. cit., 7–8
90. Crowley and North, op. cit., 62

PART THREE: *India (2)*
9 / Fairytale Kingdom

1. Deo Prakash Sharma, *Harappan Seals, Sealings and Copper Tablets*, 20–21, National Museum, New Delhi, 2000
2. Jonathan Mark Kennoyer, *Ancient Cities of the Indus Valley Civilization*, 112, American Institute of Pakistan Studies, Oxford, 1998
3. *Encyclopaedia Britannica, Micropaedia*, vol. 12, 846
4. Personal communication from Hari Shankhar, Yoga teacher, Chennai, south India
5. Kennoyer, op. cit., 112
6. Sharma, op. cit., 20
7. For a superb inquiry into the nature of Siva see Stella Kramrisch, *The Pres-*

ence of Shiva, Motilal Banarsidass, Delhi, 1988

8. S. R. Rao, *Dawn and Devolution of the Indus Civilization*, 306, Aditya Prakashan, New Delhi, 1991

9. See for example discussion in Kennoyer, op. cit., 110

10. Gregory L. Possehl, *Indus Age: The Beginnings*, 80, University of Pennsylvania Press, 1999

11. Kennoyer, op. cit., 113

12. Full discussion in chapter 10

13. See discussion in chapter 8

14. Possehl, op. cit., 1

15. Captain M. W. Carr (ed.), *Descriptive and Historical Papers Relating to the Seven Pagodas of the Coromandel Coast* (first published 1869), reprinted Asian Educational Services, New Delhi, 1984

16. Shulman in Alan Dundes (ed.), *The Flood Myth*, 294, University of California Press, 1988

17. Ibid., 294–5

18. *Journal of Marine Archaeology* (JMA), vols. 5–6, 1995–6, 7ff, Society of Marine Archaeology, National Institute of Archaeology, Goa, India, 1997

19. Ibid., 7–8

20. Ibid., 7, 14ff

21. See discussion in chapter 1

22. Interview with Rao, 29 February 2000, cited in chapter 1

23. Ibid.

24. Ibid.

25. S. R. Rao, *The Lost City of Dvaraka*, fig. 55, Aditya Prakashan, New Delhi, 1999

26. Interview with Rao, 29 February 2000

27. JMA, op. cit., 64

28. Interview with Rao, 29 February 2000; and JMA, op. cit., 65

29. Many reproduced as plates to Rao's *The Lost City of Dvaraka*, op. cit.

30. Krishna as avatar of Vishnu – e.g. Danielou

31. Interview with Rao, 29 February 2000

32. Ibid.

33. As for example in 2001

34. Ananda K. Coomaraswamy, *Myths of the Hindus and Buddhists*, 393, Dover Publications, New York, 1967

10 / The Mystery of the Red Hill

1. Alain Danielou, *The Myths and Gods of India*, 221, Inner Traditions International, Rochester, 1991

2. See discussion in Ramana's *Arunachela: Ocean of Divine Grace*, 1, Sri Ramanasramam, Tiruvannamalai, India, 1998

3. M.C. Subramanian, *Glory of Arunachela*, 93, Sri Ramanasramam, Tiruvannamalai, India, 1999

4. Ibid., 100

5. Ibid., 103

6. Ibid., 104

7. Skandananda, *Arunachela Holy Hill*, xi, Sri Ramanasramam, Tiruvannamalai, India, 1995

8. Ibid., xi

9. Subramanian, op. cit., 104; Skandananda, op. cit., xix

10. Skandananda, op. cit., xl, note 20

11. John E. Mitchiner, *Traditions of the Seven Rishis*, 206, Motilal Banarsidass, Delhi, 1982

12. Ibid., 208

13. Ibid., 206

14. Subramanian, op. cit., 106

15. Skandananda, op. cit., xi

16. Subramanian, op. cit., 104

17. *Skanda Purana*, 12, chapter 2, verse 52

18. *Arunachela Mahatmyan*, 16

19. 'Thiruvannamalai and its surroundings a geological Paradise: a plea for preservation as a National Heritage site', paper by T. V. Viswanathan, former Deputy Director General, Geological Survey of India, handed to me at Sri Ramana Asram

20. E.g., see Jonathan Mark Kennoyer,

Ancient Cities of the Indus Valley Civilization, 110, American Institute of Pakistan Studies, Oxford, 1998

21. Sharp-eyed readers will have noticed an example of this in a passage quoted earlier in this chapter.

22. W. J. Wilkins, *Hindu Mythology*, 265–6, Heritage Publishers, New Delhi, 1991

23. Jan Knappert, *Indian Mythology*, 228 and 230, Diamond Books, London, 1995; *New Larousse Encyclopaedia of Mythology*, 341, Hamlyn, London, 1989

24. Stella Kamrisch, *The Presence of Shiva*, 7, Motilal Banarsidass, 1988

25. Alfred Hildebrandt, *Vedic Mythology*, vol. 2, 289, Motilal Banarsidass Publishers, Delhi, 1990; Wilkins, op. cit., 266

26. Hildebrandt, op. cit., vol. 2, 282; Wilkins, op. cit., 266

27. Danielou, op. cit., 192–4

28. Ralph T. Griffith (trans.), *Hymns of the Rgveda*, vol. 1, 318 and 319, Munisharam Manoharlal Publishers, Delhi, 1987 (first published 1889)

29. Ibid., vol. 1, 162–3

30. *Rig Veda*, 10.136.7, discussed in Hildebrandt, op. cit., vol. 2, 287; Kamrisch, op. cit., 20–21, 83–4; 424–6

31. Griffith, op. cit., vol. 2, 630; vol. 1, 162; Wilkins, op. cit., 265–6

32. Kamrisch, op. cit., 21

33. Ibid., 119 and 123

34. Griffith, op. cit., 8.13.20

35. *Larousse*, 376

36. Ibid., 337

37. Danielou, op. cit., 209

38. Wilkins, op. cit., 102–3

39. Kamrisch, op. cit., 425–6

40. Mitchiner, op. cit., 190–91

41. *Bhagvata Purana*, 6.15.11, vol. 8, 856, Motilal Banarsidas, Delhi, 1978

42. David Goodman (ed.), *Be As You Are: The Teachings of Sri Ramana*, 1, Penguin Books, New Delhi, 1992

43. Ibid., 1

44. Ibid., 1

45. Ibid., 3

46. *Skanda Purana* and *Arunachela Mahatmya*

47. For example, Inca myths regarding Tiahuanaco, Aztec myths regarding Teotihuacan and Ancient Egypt's Edfu cosmology

48. See discussion in *Keeper of Genesis/ Message of the Sphinx*, chapter 12

49. Cited above: Subramanian, op. cit., 104

50. See discussion in next chapter

51. C. Ramachandran Dikshitar, *Studies in Tamil Literature and History*, 7, The South India Sauiva Siddhanta Works Publishing Society, Madras, 1983

52. Saint Bernard of Clairvaux, *On Consideration*, cited in Robert Lawlor, *Sacred Geometry*, 6, Thames and Hudson, London, 1989

53. See discussions in *Fingerprints of the Gods* and *Heaven's Mirror*

54. Ibid.

55. See discussions in *Fingerprints of the Gods*

56. Giorgio de Santillana and Hertha von Dechend, *Hamlet's Mill*, Nonpareil, Boston, 1992

57. See discussion in *Fingerprints of the Gods*

58. See discussions in *Fingerprints of the Gods* and *Heaven's Mirror*

59. Santillana and von Dechend, op. cit.

60. See discussion in *Heaven's Mirror*

61. See discussion in *Heaven's Mirror*, where this grid is first proposed

62. Wilkins, op. cit., 353

63. Danielou, op. cit., 160

64. Ibid., 220–21

65. V. Naryanaswamy, *Thiruvannamalai*, 17, Manivasagar Pathippagam, Madras, 1992

66. S. Kamrisch, op. cit., 83–4, and see *Rig Veda*, 7.59.12

67. Donald A. McKenzie, *India: Myths and Legends*, 146–7, Mystic Press, London, 1987

68. M. Sundarraj, *Rig Vedic Studies*, 83, International Society for the Investigation of Ancient Civilization, Chennai, 1997

69. Ibid., 83

70. Griffith, op. cit., 1.154.1–3

71. Ibid., 1.155.4

72. Ibid., 1.155.6

73. Ibid., 6.49.13

74. Ibid., 1.164.11

75. See discussion in *Fingerprints of the Gods*, part 5

76. Richard L. Thompson, *Mysteries of the Sacred Universe: the Cosmology of the Bhagvata Purana*, Govhardan Hill Publishing, Florida, 2000

77. Ibid., 47

78. Ibid., 239

79. Ibid., 269

80. Ibid., 104

81. Ibid., 104

82. Ibid., 105–6

83. Schulman in Alan Dundes (ed.), *The Flood Myth*, 295, University of California Press, 1988

11 / The Quest for Kumari Kandam

1. See discussion in C. Ramachandran Dikshitar, *Studies in Tamil Literature and History*, chapter 1, The South India Sauiva Siddhanta Works Publishing Society, Madras, 1983

2. Discussed at length in K. N. Shivaraja Pillai, *The Chronology of the Early Tamils*, University of Madras, 1932

3. Ibid., 19

4. Ibid., 20

5. For example, it is claimed that Ethiopia's national epic, the *Kebra Nagast* ('Glory of Kings'), was concocted to legitimize the Solomonic dynasty

6. Pillai, op. cit., 19

7. Dikshitar, op. cit., 5

8. Ibid., 5

9. Ibid., 5–6, 13–14; P. Ramanathan, *A New Account of the History and Culture of the Tamils*, 8–10, Sauiva Siddhanta Works Publishing Society, Chennai, 1998; T. R. Sesha Iyenagar, *Dravidian India*, 154, Asian Educational Services, New Delhi, 1995

10. Dikshitar, op. cit., 6

11. N. Mahalingam, *Kumari Kandam – The Lost Continent*, 2, 59–60, Proceedings of the Fifth International Conference/Seminar of Tamil Studies, Madurai, Tamil Nadu, India, January 1981, International Association of Tamil Research, Madras

12. Personal communication from Dr T. N. P. Haran, American College, Madurai

13. Ibid.

14. Dikshitar, op. cit., 7

15. Ibid., 7

16. Ibid., 7

17. Ibid., 8

18. Ibid., 8

19. Dr D. Devakunjari, *Madurai Through the Ages*, 26, Society for Archaeological, Historical and Epigraphical Research, Madras

20. Dikshitar, op. cit., 8

21. Pillai, op. cit., 19

22. Dikshitar, op. cit., 4

23. Ibid., 4

24. Ibid., 5

25. Pillai, op. cit., 21. See also 19

26. V. Kanakasabhai, *The Tamils Eighteen Hundred Years Ago*, 21, Sauiva Siddhanta, Madras, 1966

27. Ibid., 21, note 3

28. Ibid., 21, note 3; Dikshitar, op. cit., 13–14

29. Ramanathan, op. cit., 8

30. Ibid., 8–9. Dikshitar, op. cit., 14. In his

commentary on the *Tolkappiyam*, Per-Asiyar calls this lost territory not Kumari Kandam but *Panainadu*

31. Ramanathan, op. cit., 32
32. Ibid., 32–3
33. Sesha Iyenagar, op. cit., 24, 25
34. Dr M. Sundaram, Chief Professor and Head of the Dept of Tamil, Presidency College, Madras, 'The Cultural Heritage of the Ancient Tamils', research paper
35. Shulman, in Alan Dundes (ed.), *The Flood Myth*, 301, University of California Press, 1988
36. William Geiger, *The Mahavamsa, or The Great Chronicle of Ceylon*, ix–x, Asian Educational Services, New Delhi, 1986 (first published 1912)
37. N. K. Mangalamurugesan, *Sangam Age*, Thendral Pathipakam, Madras, 1982, 47
38. Rajavali, vol. 2, 180, 190, cited in Kanakasabhai, op. cit., 21
39. R. Spence Hardy, *The Legends and Theories of the Buddhists*, 6, Sri Satguru Publications, Delhi, 1990 (first published 1866)
40. Ibid., 6
41. Ibid., 6
42. See chapters 7 and 11
43. Tennant's *Ceylon*, vol. 1, pages 6 and 7, cited in Kanakasabhai, op. cit., 21, footnote 4
44. Mahalingam, op. cit., 2–54
45. Ibid., 2, 59–60
46. Ibid., 2–54
47. Dikshitar, op. cit., 17, citing *Tiruvilaiyadal Puranam*
48. Dr D. Devakunjari, op. cit. To be precise, 'its two sides from north to south measure 720 feet and 729 feet, the two east to west sides measure 834 feet and 852 feet'
49. The dimensions of the Great Pyramid of Egypt are examined in detail in *Fingerprints of the Gods*, part 4
50. John Howley, Jada Bharata Dasa, *Holy Places and Temples in India*, 587, Spiritual Guides, 1996
51. V. Meena, *South India: A Travel Guide*, 35, Hari Kumari Arts, Kanyakumari
52. Howley and Dasa, op. cit., 589
53. Devakunjari, op. cit., 217
54. T. G. S. Balaram Iyer, *History and Description of Sri Meenakshu Temple*, 7, Sri Karthik Agency, Madurai, 1999; Devakunjari, op. cit., 214
55. Examples are to be seen in the Harappan Gallery of the National Archaeological Museum, New Delhi
56. Sesha Iyenagar, op. cit., 100
57. Cited in ibid., 100
58. Cited in David Frawley, *The Oracle of Rama*, 140, Motilal Banarsidass Publishers, Delhi, 1999
59. See chapter 7
60. Cited earlier in this chapter
61. Mahalingam, cited earlier in this chapter
62. Pillai, op. cit., 24
63. Cited earlier in this chapter
64. Dikshitar, op. cit., 9
65. Ibid., 9
66. This, at any rate, has been my experience
67. See chapter 1

12 / The Hidden Years

1. Gregory L. Possehl, *Indus Age: The Beginnings*, 431, University of Pennsylvania Press, 1999, is particularly critical of the climate stress theory
2. Ibid., 410–11, citing Sauer
3. *The Times*, London, 20 July 2001, citing a presentation by Dr Kevin Pope at the Society for American Archaeology in New Orleans

13 / Pyramid Islands

1. James Lyon, *Maldives*, 17, Lonely Planet, July 1997
2. *Encyclopaedia Britannica, Macropaedia*, 174; Lyon, op. cit., 17–18; *Hello Maldives*, 10, QR Publications, Maldives, 1999
3. Thor Heyerdahl, *The Maldives Mystery*, 197, Unwin Paperbacks, London, 1988
4. Discussed in Oppenheimer, *Eden in the East: The Drowned Continent of Southeast Asia*, 46–7, Weidenfeld and Nicholson, London, 1998
5. *Encyclopaedia Britannica, Macropaedia*, 174
6. Lyon, op. cit., 11
7. Heyerdahl, op. cit.
8. Kon Tiki Museum, 'Archaeological Test-Excavations on the Maldive Islands', Occasional Papers, vol. 2, 66, Oslo, 1991
9. Heyerdahl, op. cit., 197–8
10. Mohamed Amin, Duncan Willetts, Peter Marshall, *Journey Through the Maldives*, 16, Camerapix Publishers International, Nairobi, 1992
11. Ibid., 16–17
12. Kon Tiki Museum, op. cit.
13. Ibid., 66–73
14. *Divehi Writing Systems*, 5, National Centre for Linguistic and Historical Research, Maldives, 1999
15. Kon Tiki Museum, op. cit., 70
16. Ibid. 71–2
17. *Encyclopaedia Britannica, Micropaedia*, vol. 10, 837
18. Clarence Maloney, *People of the Maldive Islands*, Madras, 1980, cited in Kon Tiki Museum, op. cit., 70
19. Amin et al., op. cit., 12
20. T. R. Sesha Iyenagar, *Dravidian India*, 101, Asian Educational Services, New Delhi, 1995
21. Sesha Iyenagar, cited earlier
22. Interview with Naseema Mohamed by GH, Male, February 2001

23. Heyerdahl, op. cit., 169
24. Ibid., 220
25. Amin et al., op. cit., 20; Marshall does not mention the detail about the food being still warm by the time it was eaten; the source of this was Naseema Mohamed in her interview with me
26. Ralph T. Griffith (trans.), *Hymns of the Rgveda*, 1.116. 3–5, Munisharam Manoharlal Publishers, Delhi, 1987 (first published 1889)
27. Ibid., 1.182.5–6
28. Heyerdahl, op. cit., 159
29. Ibid., 312
30. *The Columbia Encyclopedia*, 6th edn (online), 2001

14 / Ghosts in the Water

1. See chapter 9
2. See chapter 9
3. Charles H. Hapgood, *Maps of Ancient Sea Kings*, 134–5, Adventures Unlimited Press (reprint), 1996
4. Ibid., 135

PART FOUR: *Malta*
15 / Smoke and Fire in Malta

1. See the 'Horizon Scandal' section on my website, www.grahamhancock.com
2. www.maltadiscovery.com
3. Archaeological Institute of America, 1999, http://www.archaeology.org/online/news/aliens.html
4. See discussion in David Trump, *Malta: An Archaeological Guide*, 67ff, Valletta, 1990
5. Chris Agius Sultana, personal communication
6. All three quotations are available to cite here thanks to original research into ancient texts concerning Malta carried out by Anton Mifsud, Simon Mifsud, Chris Agius Sultana and Charles Savona Ventura, and first cited in their book *Malta: Echoes of Plato's*

Island, 42, The Prehistoric Society of Malta, 2000

7. J. D. Evans, *The Prehistoric Antiquities of the Maltese Islands: A Survey*, 58, University of London, 1971; Trump, op. cit., 73; Colin Renfrew, *Before Civilization: The Radiocarbon Revolution and Prehistoric Europe*, 163, Pimlico, London, 1999

8. Personal communications with Museum officials

9. See discussion in Alastair Service and Jean Bradbury, *The Standing Stones of Europe*, 89, J. M. Dent, London, 1993

10. Ellul's annotation to map

11. From Ellul's annotated print of the photograph

12. See Mifsud et al., op. cit., 63, note 222

13. See chapter 16

14. Evans, op. cit., 44–5

15. Trump, op. cit., 19

16. Service and Bradbury, op. cit., 91–2

17. Renfrew, op. cit., 162

18. Ibid., 161

19. Anton Mifsud et al., op. cit., 58

20. Archaeological Institute of America, op. cit.

21. The implications of radiocarbon-dating for the Maltese temples are discussed in Renfrew, op. cit., 161ff

22. *www.grahamhancock.com*; see in particular 'Horizon Scandal'

23. This is the thesis of Karl Mayrhofer, *The Mystery of Hagar Qim*, Malta, 1996

24. See discussion in Mifsud et al., op. cit.

25. Ibid., 16

26. Most of the rest having been smoothed over by quarrying down the ages or simply covered up by modern developments – e.g. ironically beneath the National Archaeological Museum annexe and ticket office at Ghar Dhalam Cave. The remains of two very deep ruts can be seen protruding from beneath the rear of the building near the concrete steps that now lead down to the cave

27. Theories proposed include the use of carts or sleds to create the ruts for the transportation of agricultural produce, or of megaliths

28. See discussion in Parker, Rubenstein and Trump, *Malta's Ancient Temples and Ruts*, 45ff, Institute for Cultural Research, Tunbridge Wells, 1988. See also John Samut Tagliaferro, *Malta: Its Archaeology and History*, 36ff, Plurigraph, Italy, 2000. See Antony Bonanno, *Malta: An Archaeological Paradise*, 72, Valletta, 1997. And see Trump, op. cit., 107–8

29. Ibid.

30. Ibid.

31. Ibid.

32. Cited in Mifsud et al., op. cit., 24

33. Ibid., 24

34. Ibid., 24

35. Ibid., 24

36. Ibid., 24

37. Ibid., 24

38. Ibid., 24

39. Ibid.

40. Trump, op. cit., 28

41. Ibid., 29

42. Service and Bradbury, op. cit., 99

43. Trump, op. cit., 149

44. Ibid., 109

45. Ibid., 28

16 / Cave of Bones

1. J. D. Evans, *The Prehistoric Antiquities of the Maltese Islands: A Survey*, 45, University of London, 1971

2. Ibid., 59

3. Ibid., plans 14A and 14B

4. Ibid., 44–5

5. Ibid., 44

6. Ibid., 44–5, citing Sir Themistocles Zammit, 1910

7. Anton Mifsud, Simon Mifsud, Chris Agius Sultana and Charles Savona Ven-

tura, *Malta: Echoes of Plato's Island*, 38, The Prehistoric Society of Malta, 2000

8. Cited in ibid., 38

9. Evans, op. cit., citing Zammit, 45

10. Ibid., 45

11. Ibid., 45; David Trump, *Malta: An Archaeological Guide*, 67, Valletta, 1990

12. Trump, op. cit., 67

13. Zammit, 1910, cited in Mifsud et al., op. cit., 38

14. Ibid., 38

15. Bradley, Zammit, Pete, 1912, cited in ibid., 40

16. Trump, op. cit., 73

17. Cited by Mifsud in Anton Mifsud and Charles Savona Ventura (eds.), *Facets of Maltese Prehistory*, 155, The Prehistoric Society of Malta, 1999

18. Evans, op. cit., 58

19. Ibid., 58

20. Ibid., 57

21. Zammit, 1910, cited in ibid., 57

22. Ibid., 57

23. Ibid., 57–8

24. Ibid., 58

25. Trump, op. cit., 73

26. Colin Renfrew, *Before Civilization: The Radiocarbon Revolution and Prehistoric Europe*, 163, Pimlico, London, 1999

27. Marija Gimbutas, *The Civilization of the Goddess*, 286, Harper, San Francisco, 1991

28. See Paul G. Bahn and Jean Vertut, *Journey Through the Ice Age*, 13, 112–13, 161, Weidenfeld and Nicholson, London, 1997

29. Gimbutas, op. cit.

30. Ibid., 286–9

31. Evans, op. cit., 59. Evans thinks it is Room 22 and comments: 'Zammit and Singer (1924, p. 90) refer to these as having been found together in 1905 during Magri's directorship of the exca-

vations in a "deep pit of one of the painted rooms"'

32. Renfrew, op. cit., 164

33. Trump, op. cit., 77

34. Ibid., 77–8

35. Ibid., 52

36. Illustration and comments in ibid., 50–51

37. Illustration and comments, see Anthony Bonanno, *Malta: An Archaeological Paradise*, 25, Valletta, 1997

38. Trump, op. cit., 53

39. Ibid., 51

40. Although, in fact, nothing from the Hypogeum has been officially carbon-dated, according to Anthony Bonanno in a filmed interview with me, June 2001

41. Frendo, in Mifsud and Ventura, op. cit., 28

42. Ibid., 28

43. E.g. see Evans, op. cit., 208. Nobody is denying that such a Neolithic settlement of Malta did occur from Sicily. The point is that there may have been humans there before they arrived, and it may have been these earlier humans whose heritage led to the temples

44. Trump in Mifsud and Ventura, op. cit., 93

17 / The Thorn in the Flesh

1. E-mail from Anton Mifsud to GH, 15 July 2001

2. Anton Mifsud, Simon Mifsud, Chris Agius Sultana and Charles Savona Ventura, *Malta: Echoes of Plato's Island*, The Prehistoric Society of Malta, 2000

3. Ibid.

4. Ibid.

5. Ibid.

6. Ibid.

7. See discussions in *Fingerprints of the Gods* and in *Keeper of Genesis/Message of the Sphinx*

8. Fred Wendorf, Romuald Schild, 'Late Neolithic megalithic structures at Nabta Playa (Sahara) southwestern Egypt', http://www.comp-archaeology.org/WendorfSAA98.html

9. E.g. Minoan Crete, Troy, etc.

10. Also in the Andes and in Cuba in recent books

11. Anton Mifsud and Simon Mifsud, *Dossier Malta: Evidence for the Magdelenian*, 128, Malta, 1997

12. Ibid., 144

13. Bradley, Zammit, Pete, 1912, cited in Mifsud et al., op. cit., 40

14. Annual Report 1909/10, cited in Anton Mifsud and Charles Savona Ventura (eds.), *Facets of Maltese Prehistory*, 152, The Prehistoric Society of Malta, 1999

15. J. D. Evans, *The Prehistoric Antiquities of the Maltese Islands: A Survey*, 40, University of London, 1971

16. John Samut Tagliaferro, *Malta: Its Archaeology and History*, 30–31, Plurigraph, Italy, 2000

17. Zammit, 1910, cited in Mifsud et al., op. cit., 38

18. Zammit cited in ibid., 38

19. Mifsud in Mifsud and Ventura, op. cit., 163

20. See Malone, Stoddart et al., *Mortuary Ritual of the 4th Millennium BC*, Proceedings of the Prehistoric Society, 61, 1995, 303–45

21. Mifsud in Mifsud and Ventura, op. cit., 163

22. The Addolorata Cemetery, Mifsud et al., op. cit., 38

23. Ibid., 40, citing Museum of Archaeology Reports, 1973–4

24. Mifsud and Ventura, op. cit., 163

25. Mifsud in ibid., 153, citing Zammit, 1910

26. Zammit, 1910, 37, cited in ibid., 153

27. Mifsud and Mifsud, op. cit., 12

28. Marija Gimbutas, *The Civilization of the Goddess*, 286, Harper, San Francisco, 1991

29. Mifsud et al., op. cit., 47 and 58

30. Mifsud and Ventura, op. cit., 58

31. See for example David Trump, *Malta: An Archaeological Guide*, 75, Valletta, 1990

32. Ibid., 75

33. 'When the standard deviation of radiocarbon dates is taken into account, and after the exclusion of the very early dates with a wide range, the end of the Tarxien phase must have occurred between 2470 and 2140 BC,' Mifsud et al., op. cit., 47

34. Various model temples have survived from the megalithic period and illustrate roofs; a roof is also evident on a temple elevation carved into one of the megaliths at Mnajdra

35. Alastair Service and Jean Bradbury, *The Standing Stones of Europe*, 93, J. M. Dent, London, 1993

36. Ibid., 95

37. Ibid., 95

38. Ibid., 97

39. Ibid., 97

40. Trump, op. cit., 29

41. For example, in some of the more remote islands of Indonesia and in the Brazilian rainforest

42. I am assured the quote is reliably attributed to Picasso, but have been unable to find a published source

43. Discussed in Gregory L. Possehl, *Indus Age: The Beginnings*, 450ff, University of Pennsylvania Press, 1999

44. E.g. Trump, op. cit., 15

45. See chapter 16

46. Personal communication by e-mail, 15 July 2001

47. Trump, op. cit., 72

48. Mifsud and Mifsud, op. cit, 168

49. Ibid., 143

50. E-mail, 15 July 2001, personal communication with GH

51. E-mail, 15 July 2001, personal communication with GH (emphasis added)
52. Mifsud and Mifsud, op. cit., 150
53. Ibid., 139
54. Ibid., 142
55. Ibid., 144
56. Mifsud et al., op. cit., 61
57. Personal observation
58. Mifsud and Mifsud, op. cit., 144
59. Ibid., 143–4
60. Glyn Daniel, 1959, cited in Mifsud and Ventura, op. cit., 157
61. Mifsud in ibid., 157
62. Mifsud and Mifsud, op. cit., 127
63. Ibid., 128
64. Charles Savona Ventura and Anton Mifsud, *Hasan's Cave: Geology, Folklore and Antiquities*, Heritage Books, Valletta, 2000
65. E. Anati, 'Archaeological Exploration in Malta', *World Journal of Prehistoric and Primitive Art*, 28, 1995, 103–6
66. Mifsud and Mifsud, op. cit., 147
67. Ibid., 165–6, footnote 261
68. Ibid., 165–6, footnote 261
69. Ibid., 168
70. See next chapter
71. Tagliaferro, op. cit., 11
72. Ibid.

18 / The Masque of the Green Book

1. Evans, 1959, cited in Anton Mifsud and Simon Mifsud, *Dossier Malta: Evidence for the Magdelenian*, 100, Malta, 1997
2. Ibid., 68
3. Telephone interview with Sharif Sakr, 26 October 2001
4. Colin Renfrew, *Before Civilization: The Radiocarbon Revolution and Prehistoric Europe*, 161, Pimlico, London, 1999
5. Alastair Service and Jean Bradbury, *The Standing Stones of Europe*, 78–9, J. M. Dent, London, 1993
6. Veen and van der Blom, *The First Maltese*, 1992, 15–16; J. D. Evans, *The Prehistoric Antiquities of the Maltese Islands: A Survey*, 37, University of London, 1971; David Trump, *Malta: An Archaeological Guide*, 151–3, Valletta, 1990
7. Trump, op. cit., 153
8. Ibid., 153
9. Ibid., 153
10. Ibid., 47
11. Evans, op. cit., 37
12. Ibid., 166
13. Ibid., 166
14. John Samut Tagliaferro, *Malta: Its Archaeology and History*, 13, Plurigraph, Italy, 2000
15. Anthony Bonanno, *Malta: An Archaeological Paradise*, 44, Valletta, 1997
16. Trump, op. cit., 28
17. Tagliaferro, op. cit., 14
18. Ibid., 14
19. Trump, in Anton Mifsud and Charles Savona Ventura (eds.), *Facets of Maltese Prehistory*, 93, The Prehistoric Society of Malta, 1999
20. Trump, *Archaeological Guide*, 28
21. Evans, 1959, cited in Renfrew, op. cit., 165
22. Tagliaferro, op. cit., 11
23. Trump, op. cit., 91
24. Evans, op. cit., 20
25. Mifsud and Mifsud, op. cit., 37
26. Ibid., 36–7
27. Ibid., 37
28. Ibid., 38–9
29. Ibid., 39
30. Keith, 1924, cited by Mifsud and Mifsud, op. cit., 39
31. Ibid., 39–40
32. Ibid., 42
33. Ibid., 42
34. Ibid., 45
35. Ibid., 56
36. Ibid., 57
37. Trump, *Archaeological Guide*
38. Ibid., 91

39. Ibid., 91
40. Ibid., 92
41. Ibid., 19–20
42. Ibid., 92–3
43. Evans, op. cit., 18
44. Ibid., 19
45. Ibid., 19
46. Mifsud and Mifsud, 57
47. Ibid., 57
48. Ibid., 45–6
49. Caton-Thompson, 1925, 10, cited in Mifsud and Mifsud, op. cit., 44
50. See discussion in ibid., 44
51. Ibid., 58
52. *Encyclopaedia Britannica, Micropaedia*, vol. 9, 445
53. Trump, op. cit., 91
54. Mifsud and Mifsud, op. cit.
55. Evans, op. cit., 19
56. Mifsud and Mifsud, op. cit., 43
57. Ibid., 50
58. Ibid., 81
59. Ibid., 112, footnote 2
60. Ibid., 81–2
61. Ibid., 83
62. Ibid., 84–5
63. Ibid., 85
64. Ibid., 86
65. Ibid., 88
66. Personal communication to GH, e-mail, 15 July 2001
67. Mifsud and Mifsud, op. cit., 95 (emphasis added)
68. Ibid., 40–41
69. Discussed in Renfrew, op. cit.
70. Ibid., 165–6
71. Museum 1964 Report, cited in Mifsud and Mifsud, op. cit., 102
72. Evans, 1959, cited in ibid., 100
73. Ibid., 100–101
74. Ibid., 107
75. Ibid., 107–8
76. Frendo, in Mifsud and Ventura, op. cit., 28
77. Frendo, in ibid., 28
78. Frendo, in ibid., 28
79. Frendo, in ibid., 30
80. Frendo, in ibid., 30
81. For example, the nitrogen reading of 1.85 per cent for Ma.2 was published and taken as proof of a Neolithic date, while the nitrogen reading of 0.39 for Ma.1 was ignored. The deer reading of 0.17 per cent was taken as representative of the Cervus Layer such that even 0.39 per cent would have seemed later; whereas in fact deer and even hippo samples had yielded levels as high as 0.4 – showing that 0.17 is not a representative Cervus threshold. Moreover, the highly anomalous uranium oxide reading of 13 ppm for Ma.2 was ignored. Humphrey is ready to argue that even this result is ambiguous. But it is very difficult to see how so much uranium oxide could have filtered over just a few thousand years into a tooth that would have contained 0.1 ppm when its owner lived – especially since we now know that Ghar Dalam is an environment with low levels of uranium oxide in the percolating water
82. 24 October 2001
83. Mifsud and Mifsud, op. cit., plate 1; and see discussion on page 64
84. Ibid., 120, footnote 152
85. Ibid., 64, 105, 109
86. Frendo, in Mifsud and Ventura, op. cit., 30–31; and see discussion in Mifsud and Mifsud, op. cit., 65–6
87. Evans, op. cit., 19
88. Mifsud and Mifsud, op. cit., 105
89. Ibid., 64
90. Ibid., 120, footnote 152
91. Frendo, in Mifsud and Ventura, op. cit., 30

19 / Inundation

1. David Trump, *Malta: An Archaeological Guide*, 14–15, Valletta, 1990
2. Anton Mifsud and Simon Mifsud, *Dossier Malta: Evidence for the Magdelenian*, 12–13, Malta, 1997
3. Ibid., 27, 97
4. Ibid., 31, footnote 71
5. Ibid., 23
6. The Tyrrhenian Sea is the arm of the Mediterranean between Italy and the islands of Corsica, Sardinia and Sicily
7. 'Meltwater pulse 1-A' corresponds with the first of the three episodes of cataclysmic post-glacial flooding identified by Professor John Shaw of the University of Alberta (see chapter 3)
8. Van Andel, 'Late Quaternary Sea Level Changes and Archaeology', 737, *Antiquity*, 63, 1989, describes these Adriatic plains as 'one of the richest environments in the whole central northern Mediterranean'
9. *Malta: Echoes of Plato's Island*, 34, The Prehistoric Society of Malta, 2000
10. See chapter 17
11. Van Andel, op. cit., 737
12. See chapter 3

20 / The Morning of the World

1. See discussion by Trump, in Anton Mifsud and Charles Savona Ventura (eds.), *Facets of Maltese Prehistory*, 96, The Prehistoric Society of Malta, 1999; David Trump, *Malta: An Archaeological Guide*, 49, Valletta, 1990
2. See J. D. Evans, *The Prehistoric Antiquities of the Maltese Islands: A Survey*, University of London, 1971: compare plate 33 (11 and 12) with plate 47 (11, 9 and 10)
3. Trump, in Mifsud and Ventura, op. cit., 96; Trump, *Archaeological Guide*, 49
4. Paul I. Micallef, *Mnajdra Prehistoric*

Temple: A Calendar In Stone, Malta, 1992
5. Richard Walter, 'Wanderers Awheel in Malta', 253ff , *National Geographic*, March 1940
6. Micallef, op. cit., 35
7. Micallef makes these points in more detail in his unpublished paper 'Alignments along the main axes of Mnajdra', 5 June 2001, pages 6–7
8. Alexander Thom's proposed megalithic yard is discussed in Douglas C. Hegge, *Megalithic Science: Ancient Mathematics and Astronomy in Northwest Europe*, 55ff, Thames and Hudson, London, 1981
9. The boats appear in the form of roughly inscribed sketches or graffiti on slabs near the entrance to Tarxien
10. Use of obliquity errors for dating of other structures discussed in *Fingerprints of the Gods* and *Heaven's Mirror*
11. Micallef, op. cit., 32
12. Ibid., 32
13. See chapter 18
14. Evans, op. cit., 116–17
15. Personal communication from Anton Mifsud, and see Trump, op. cit., 176ff
16. Evans, op. cit., 95
17. Ibid., 80–81
18. Ibid., 172
19. Ibid., 172
20. Trump, *Journal of the Accordia Research Centre*, vols. 5–6, 173–7
21. Evans, *Antiquity*, vol. 35, no. 137, 143–4
22. Trump, *Journal of the Accordia Research Centre*, 173–7
23. Renfrew, *Antiquity*, vol. 46, 141–4
24. Trump, *Antiquity*, vol. 37, no. 148, 302–3
25. Anton Mifsud, Simon Mifsud, Chris Agius Sultana and Charles Savona Ventura, *Malta: Echoes of Plato's Island*,

34, The Prehistoric Society of Malta, 2000

26. Ibid., 34
27. Ibid., 34–5
28. Ibid.
29. Ibid., 18
30. Indeed, the idea of the Maltese archipelago as one larger single island survived until at least the late sixteenth century on some maps. For example, in *Tabula Europae VII*, one of a group of Ptolemaic maps published in Venice in 1598. See Mifsud et al., op. cit., 50
31. Ibid., 18–20
32. Ibid., 20–24, citing Ventura
33. Ibid., 24
34. Ibid., 22
35. Ibid., 49

PART FIVE: *Ancient Maps*
21 / *Terra Incognita*

1. Damiao Peres, *A History of the Portuguese Discoveries*, 56–72, Lisbon, 1960
2. Ibid., 87
3. Ibid., 87
4. Ibid., 87
5. Ibid., 87
6. Ibid., 87
7. Ibid., 113
8. Ibid., 88
9. Ibid., 89–90
10. Ibid., 93: 'This was the contemporary expression for shipwrecked ships from which there were no survivors or remains'
11. Ibid., 92
12. Ibid., 89–91
13. Ibid., 93
14. Ibid., 93
15. Ibid., 94
16. Ibid., 113
17. Ibid., 113
18. Ibid., 114
19. Ibid., 115
20. Ibid., 115

21. Thomas Suarez, *Early Mapping of Southeast Asia*, 64, 85, Periplus, Hong Kong, 1999
22. Peres, op. cit., 112
23. Notes on the 1502 Cantino, 1, see http://www.henry-davis.com/MAPS/Ren/Ren1/306mono.html
24. Peres, op. cit., 99
25. Notes on the 1502 Cantino, op. cit., 1; Peres, op. cit., 99; John Goss, *The Mapmaker's Art: A History of Cartography*, 64, Studio Editions, London, 1994
26. Notes on Cantino, op. cit., 1
27. Goss, op. cit., 64
28. For example, see ibid., 64; Peres, op. cit., 99ff
29. Goss, op. cit., 34
30. Ibid., 34
31. Ibid., 34
32. Ibid., 34
33. See Ibid., 35
34. Ibid., 35–40
35. Ibid., 35
36. *Encyclopaedia Britannica, Micropaedia*, vol. 9, 775
37. O. A. W, Dilke, *Greek and Roman Maps*, 75, Cornell University Press, 1985
38. Ibid., 75; Mostafa El-Abbadi, *Life and Fate of the Ancient Library of Alexandria*, 141, UNESCO, Paris, 1992
39. Dilke, op. cit., 75
40. *Encyclopaedia Britannica, Micropaedia*, vol. 9, 775
41. See discussion in Dilke, op. cit., 80–81
42. J. Oliver Thomson, *History of Ancient Geography*, 337, Biblo and Tannen, New York, 1965
43. Ibid., 336
44. Robert H. Fuson, *Legendary Islands of the Ocean Sea*, 11, Pineapple Press Inc., Florida, 1995
45. Ibid., 18–19
46. Ibid., 18

47. Ibid., 15–16
48. Ibid., 17. Curiously enough, Poseidonius had previously estimated the earth's circumference at 27,000 miles – much closer to the correct figure – then changed his mind. See Gregory C. McIntosh, *The Piri Reis Map of 1513*, 15, The University of Georgia Press, 2000
49. J. Lennart Berggren and Alexander Jones, *Ptolemy's Geography: An Annotated Translation of the Theoretical Chapters*, 22, Princeton University Press, 2000
50. Ibid., 22
51. Dilke, op. cit., 72
52. Cited in ibid., 73
53. See also ibid., 81: 'Research by E. Polashek makes it seem likely that different groups of manuscripts represent successive recensions of the coordinates in antiquity, at least the first of which may be attributred to Ptolemy himself'
54. See Fuson, op. cit., 18
55. Dilke, op. cit., 155–6
56. Ibid., 157
57. Goss, op. cit., 25
58. Ibid., 25
59. A. E. Nordenskiold, *Facsimile Atlas to the Early History of Cartography with Reproductions of the Most Important Maps Printed in the XV and XVI Centuries*, 45, Dover, New York, 1973 (reprint) (first published 1889)
60. Ibid., 45
61. Ibid., 45
62. Goss, op. cit., 41
63. Dilke, op. cit., 180
64. Goss, op. cit., 41
65. Dilke, op. cit., 180
66. Ibid., 180. Additionally, Barry Fell has reported strong evidence of use of magnetized needles by sailors in Spain in pre-Latin times: *Occasional Publications of the Epigraphic Society*, 3/57, Arlington, PA, USA
67. Nordenskiold, op. cit., 46
68. Charles H. Hapgood, *Maps of Ancient Sea Kings*, 116, Adventures Unlimited Press (reprint), 1996
69. *Fingerprints of the Gods*
70. Sharif Sakr, e-mail correspondence with G. McIntosh, 9 October 2000
71. Goss, op. cit., 41
72. Nordenskiold, op. cit., 48
73. Peter Whitfield, *The Charting of the Oceans: Ten Centuries of Maritime Maps*, 16, Pomegranate Art Books, California, 1996
74. Ibid., 17
75. Ibid., 19 (emphasis added)
76. Ibid., 19
77. Ibid., 19
78. Goss, op. cit., 41; Dilke, op. cit., 180–81
79. Goss, op. cit., 41
80. Nordenskiold, op. cit., 48
81. A. E. Nordenskiold, *Periplus: The Early History of Charts and Sailing Directions*, 45, Nart-Franklin, New York, 1967 (reprint)
82. Ibid., 45
83. Nordenskiold, *Facsimile Atlas*, 48
84. Ibid., 48
85. Ibid., 48
86. Ibid., 48
87. Ibid., 48
88. Svat Soucek, *Piri Reis and Turkish Mapmaking after Columbus*, 27, The Nour Foundation in association with Oxford University Press, 1996
89. Nordenskiold, *Facsimile Atlas*, 43
90. Frances Gibson, *The Seafarers: Pre-Columbian Voyages to America*, 253, Dorrance and Co., Philadelphia, 1974; McIntosh, op. cit., 31; Fuson, op. cit., 119
91. E.g. see Goss, op. cit., 54–5; Soucek, op. cit., 61–4; Fuson, op. cit., 119; Dilke, op. cit., 177

92. Nordenskiold, *Periplus*, 10
93. Fuson, op. cit., 9
94. Ibid., 119–20
95. E.g. Catalan Atlas, see http://www.bnf.fr/enluminures/manuscrits/aman6.htm
96. Christopher Columbus, 1484, quoted in *Historie*, 1571, cited in Fuson, op. cit., 185
97. Nordenskiold, *Periplus*, 15
98. Ibid., 14–15
99. Dilke, op. cit., 180
100. David Lewis, *We the Navigators: The Ancient Art of Landfinding in the Pacific*, 292, University of Hawaii Press, Honolulu, 1994
101. Ibid., 292
102. Ibid., 292
103. Ibid., 90
104. See chapter 14
105. Documented in Glenn Milne's inundation maps reproduced in chapters 7 and 11
106. Nordenskiold, *Periplus*, 47

22 / *The Secret Memories of Maps*

1. *Encyclopaedia Britannica, Micropaedia*, vol. 9, 571
2. Marco Polo, *The Travels of Marco Polo*, 205–11, Wordsworth, Classics, 1997; Marco Polo, *The Travels*, 243–9, Penguin, London, 1982; Robert H. Fuson, *Legendary Islands of the Ocean Sea*, 203, Pineapple Press Inc., Florida, 1995
3. *Encyclopaedia Britannica, Micropaedia*, vol. 9, 571–3
4. Polo, *Travels*, Penguin, 243; Polo, *Travels*, Wordsworth, 207
5. Polo, *Travels*, Wordsworth, 207
6. John Larner, *Marco Polo and the Discovery of the World*, 153, Yale University Press, 1999
7. Polo, *Travels*, Penguin, 258
8. Polo, *Travels*, Wordsworth, 224; Polo, *Travels*, Penguin, 258
9. Polo, *Travels*, Wordsworth, 224 (emphasis added)
10. Ibid., xv
11. Polo, *Travels*, Penguin, 258–9 (emphasis added)
12. Cited in Thomas Suarez, *Early Mapping of Southeast Asia*, 44, Periplus, Hong Kong, 1999 (emphasis added)
13. See in particular chapters 7 and 11
14. Chapters 7 and 11
15. Suarez, op. cit., 44
16. Ibid., 44
17. Ibid., 44
18. Ibid., 44
19. Ibid., 44: 'Ptolemy's *Geographia*, and maps constructed from it, were virtually unknown in Europe at this time, even among academics, and remained so until a century after Polo's return. Thus Polo clearly did not fabricate this key Ptolemaic error, which he himself did not understand. Ptolemy's *Geographia* was, however, known to Arab scholars, and had profoundly influenced the Arab conception of Southeast Asia. But the fact that the map seen by Polo retained such an incorrect dimension for Ceylon supports the view that native pilots guided their vessels by navigational texts and did not refer to the charts themselves.'
20. E.g. Taprobana identified with Sumatra on some maps
21. See chapter 21
22. See chapter 21
23. See discussion in chapter 21
24. Suarez, op. cit., 44
25. See chapter 21
26. Sharif Sakr, 'Was the World Mapped Before the End of the Ice Age?', research paper for GH, 5 February 2001
27. Luis Vaz de Camoes, *The Lusiads*, 221, Oxford University Press, 1997 World's Classics
28. Sharif Sakr, op. cit.

29. The most detailed description of the inundation of Sundaland is given in Oppenheimer, *Eden in the East: The Drowned Continent of Southeast Asia*, Weidenfeld and Nicholson, London, 1998

30. Armando Cortesao, *The Suma Oriental of Tome Pires and the Book of Francisco Rodrigues*, vol. 1, xi, Asian Educational Services, New Delhi, 1990

31. Cortesao, Introduction to *Suma Oriental*, ibid., vol. 1, xiii

32. Cortesao, Foreword to *Suma Oriental*, ibid., xi

33. Cortesao, Introduction to *Suma Oriental*, ibid., xiii–xviii

34. Ibid., xiii

35. Ibid., xiii

36. Ibid., lxxviii

37. Ibid., 128, footnote

38. Ibid., lxxviii (emphasis added)

39. Ibid., 45

40. Ibid., 45–6

41. 3 September 2001, 'Very interesting stuff from Pires' – e-mail from Sharif Sakr to GH

23 / Looking for the Lost on the Road to Nowhere

1. Cited in Robert H. Fuson, *Legendary Islands of the Ocean Sea*, 62, Pineapple Press Inc., Florida, 1995

2. Cited in Gregory C. McIntosh, *The Piri Reis Map of 1513*, 74, The University of Georgia Press, 2000

3. Fuson , op. cit., 43

4. Ibid., 62

5. Frances Gibson, *The Seafarers: Pre-Columbian Voyages to America*, 9ff, Dorrance and Co., Philadelphia, 1974

6. Ibid., 9ff

7. Ibid., 9–11

8. Cited in Fuson, op. cit., 23

9. Plato, *Timaeus and Critias*, 38, Penguin Books, London, 1977

10. Ibid., 37–8 (emphasis added)

11. Fuson, op. cit., 28

12. Ibid., 30

13. Svat Soucek, *Piri Reis and Turkish Mapmaking after Columbus*, 58, The Nour Foundation in association with Oxford University Press, 1996 (emphasis added)

14. McIntosh, op. cit., 140

15. Soucek, op. cit., 59; McIntosh, op. cit., 50–51

16. McIntosh, op. cit., 73

17. Ibid., 73

18. An extensive discussion of the Bahriye is found in Soucek, op. cit., 84ff

19. Ibid., 99

20. E.g. McIntosh, op. cit., 17, 19

21. Ibid., 17, 19

22. Fuson, op. cit., 186

23. The persistence of lost islands in nautical charts is discussed extensively in Fuson, op. cit.

24. Ibid.

25. E.g. see McIntosh, op. cit., 31 and 72; Dora Beale Polk, *The Island of California: The History of a Myth*, 24, University of Nebraska Press, 1991

26. George Firman, *Atlantis: A Definitive Study*, 33, California, 1985

27. Where the work of Vitaly Koudriatsvev, another Atlantis researcher, makes use of isostacy and the forebulge theory

28. Firman, op. cit., 33, 36–7

29. Ibid., 33

30. Ibid., 33

31. Ibid., 75

32. Plato, op. cit., 37–8

33. For some reason, never explained, the Soviet Union took a great interest in the underwater search for the remains of a lost civilization, principally in the Atlantic

34. Carried in National Geographic News (national geographic.com/news), 13 July 2001

35. Cited in Plato, op. cit., Appendix on Atlantis, 158

36. Please note that although not of significant size there were also a number of antediluvian islands on the eastern side of the Atlantic, one of them extremely close to the Straits of Gibraltar. Using inundation maps of this area, French scientist Jacques Collina-Girard of the University of the Mediterranean in Aix-en-Provence noticed an island (geologists call it Spartel) near the western end of the strait measuring 14 kilometres long by 5 kilometres wide that existed from the Last Glacial Maximum until it was flooded by rising sea-levels around 11,000 years ago. Despite its small size, and largely on account of its relationship to the Straits of Gibraltar mentioned by Plato, Collina-Girard had proposed this palaeo-island as a candidate for Atlantis. See NewScientist.com, 'Sea level study reveals Atlantis candidate', by Jon Copley, 18 September 2001. Note that there have been persistent reports by divers of significant unidentified underwater ruins off the coasts of Spain and Morocco. One long-running story, reported in the US in the *Orange County Register* in 1973, concerns an expedition by Maxine Asher of Pepperdine University to discover the ruins of Atlantis at a site about 20 kilometres off Cadiz in southern Spain. Before Spanish police stopped the expedition Asher's team did claim to have discovered 'pre-Roman and pre-Phoenician' underwater ancient ruins at the site (see *Orange County Register*, 27 March 1973, 17 July 1973, 22 July 1973, 26 July 1973, 22 August 1973). Presumably the problems with the Spanish police have ceased, because further expeditions have been mounted as recently as 2000 and Maxine Asher, reportedly, remains very much on the case. Good luck to her.

37. For details of the discovery see J. Manson Valentine, 'Underwater Archaeology in the Bahamas', *The Explorers Journal*, December 1976, 176–83

38. *National Geographic Research Reports*, vol. 12, 1980, 35

39. Ibid., 22–4

40. *Nature*, vol. 230, 2 April 1971, 287–8

41. *Nature*, vol. 287, 4 September 1980, 11–12

42. *The Explorers Journal*, December 1976, 177

43. Dr David Zink, *The Stones of Atlantis*, 50, Prentice-Hall, NJ, 1978

44. Mahlman and Zink, *1982 Conference on Underwater Archaeology*, 4, University of Pennsylvania, January 1982

24 / The Metamorphoses of Antilia

1. I was given this story in some detail in July 2001 in a filmed interview with an octogenarian resident of Bimini, Alvin Taylor, who said he used to watch 'Captain Webster's' barges from the US loading stone from the Bimini Road.

2. Mahlman and Zink, *1982 Conference on Underwater Archaeology*, 4, University of Pennsylvania, January 1982, 2–3

3. *Nature*, vol. 287, 4 September 1980, 12

4. Ibid., 12

5. Reuters, Monday 14 May 2001, 11:59 a.m. ET

6. Ibid.

7. Ibid.

8. Paul Weinzweg, co-founder of ADC, interviewed by Sharif Sakr, 21 May 2001

9. Al Hine, Marine Geologist, University of South Florida, interviewed by Sharif Sakr, 21 May 2001

10. Grenville Draper in e-mail exchange with Sharif Sakr, 24 May 2001
11. Christopher Columbus, 1484, quoted in *Historie*, 1571, cited in Fuson, op. cit., 185
12. Charles Duff, *The Truth About Columbus*, 28, Grayson and Grayson, London, 1936
13. Ibid., 116–17
14. Cited in ibid., 127
15. Cited in ibid., 123
16. Cited in ibid., 123
17. Cited in ibid., 128
18. Cited in ibid., 129
19. Ibid., 27ff
20. Robert H. Fuson, *Legendary Islands of the Ocean Sea*, 113 and 114, Pineapple Press Inc., Florida, 1995
21. Gregory C. McIntosh, *The Piri Reis Map of 1513*, 91, The University of Georgia Press, 2000
22. Duff, op. cit., 131
23. Ibid., 127
24. Cited in ibid., 141
25. Cited in ibid., 142
26. Cited in ibid., 142
27. Ibid., 222
28. Ibid., 222
29. Ibid., 225
30. McIntosh, op. cit., 113
31. Ibid., 91, 136
32. Ibid., discussion 135–7
33. Ibid., 91
34. Ibid., 91
35. Ibid., 137 and 136
36. Ibid., 88
37. Ibid., 113
38. Ibid., 91
39. Ibid., 115–16
40. Ibid., 115
41. Ibid., 115
42. E.g. Toscanelli
43. William Giles Nash, *America: the True History of its Discovery*, 37, Grant Richards Ltd, London, 1924
44. Ibid., 41–2
45. Cited in Duff, op. cit., 103–4
46. McIntosh, op. cit., 73–4
47. John Larner, *Marco Polo and the Discovery of the World*, 143–4, Yale University Press, 1999
48. Marco Polo, *The Travels of Marco Polo*, 207, Wordsworth Classics, 1997. Polo describes the palace of the ruler of Cipango: 'The entire roof is covered with a plating of gold, in the same manner as we cover houses or more properly churches, with lead. The ceilings of the halls are of the same precious metal.'
49. Fuson, op. cit., see in particular pages 185ff
50. Ibid., 193
51. Ibid., 195–6
52. Ibid., 196
53. Ibid., 198
54. Ibid., 199–205
55. Ibid., 204–5
56. Ibid., 191
57. Ibid., 191

PART SIX: *Japan, Taiwan, China*

25 / *The Land Beloved of the Gods*

1. Donald L. Philippi, *Norito: A Translation of the Ancient Japanese Ritual Prayers*, 53, Princeton University Press, 1990
2. Cited in Michael Czaja, *Gods of Myth and Stone*, 148, Weatherhill, New York, 1974
3. Robert H. Fuson, *Legendary Islands of the Ocean Sea*, Pineapple Press Inc., Florida, 1995
4. 'Akita pyramid-shaped hill built in Jomon era, experts say,' *Japan Times*, Tokyo, 16 November 1993
5. Ibid.
6. Ibid.
7. Ibid.
8. Ibid.
9. Ibid.

10. Irina Zhushchikkovskaya, 'On Early Pottery-Making in the Russian Far East', *Asian Perspectives*, vol. 36, no. 2, Fall 1997, 159–74

11. Douglas Moore Kenrick, *Jomon of Japan: The World's Oldest Pottery*, 5, Kegan Paul International, London, 1995

12. Matsuo Tsukuda, 'Vegetation in Prehistoric Japan: The Last 20,000 Years', in *Windows on the Japanese Past: Studies in Archaeology and Prehistory*, 12, Centre for Japanese Studies, University of Michigan, 1986

13. Information provided by Kiyoji Koita, Deputy Chairman, Prehistoric Cultural Research Council, Ena City Hall

14. Ibid.

15. Ibid.

16. Ibid.; observations and measurements by the Ena Prehistory Study Group

17. Information provided by Kiyoji Koita

18. Ibid.

19. *Omiwa Shrine*, 7, Moiwa Jinja, Miwamachi Sakuraishi Naraken, Japan

20. Ibid., 1

21. Ibid., 1

22. Personal observation

23. *Omiwa Shrine*, 7–8

24. Discussed by Steve Renshaw and Saori Ihara, 'Astronomy Amongst the Ancient Tombs and Relics in Asuka, Japan', March 1997 (unpublished)

25. *Guide to the Asuka Historical Museum*, 29, Asuka Historical Museum, 1978

26. Damaged summer 2000 when the central megalith was rolled off its platform; the official story is that exceptionally heavy typhoons were to blame.

27. PNAS, 31 July 2001, cited in Reuters report, Washington, 31 July 2001

28. *Washington Post*, 31 July 2001

29. Betty Meggers, Clifford Evans and Emilio Estrada, *Smithsonian Contributions to Anthropology*, vol. 1, 160ff

26 / Remembrance

1. Some Japanese scholars such as Yoshiro Saji and others have considered the possibility that the myths of the *Kojiki*, the *Fudoki* and the *Nihongi* may have originated in the Jomon period; however, this is very much a minority view. It has not received the support of mainstream academics who habitually maintain that the myths are of Yayoi origin.

2. See *New Larousse Encyclopaedia of Mythology*, 403ff, Hamlyn, London, 1989, and Post Wheeler, *The Sacred Scriptures of the Japanese*, 393–438, Henry Schuman Inc., 1952

3. Juliet Piggott, *Japanese Mythology*, 26, Paul Hamlyn, London, 1969

4. Wheeler, op. cit., xviii

5. *Larousse*, 403. Their function was to recite ancient legends during the great Shinto festivals

6. Wheeler, op. cit., xxii; *Larousse*, 404

7. Wheeler, op. cit., xxii. However, it is not clear that the reciter was male. *Larousse*, 404, makes her female – Hieda-no-Ara, an attendant lady at the court

8. Wheeler, op. cit., xxii; *Larousse*, 404

9. *The Kojiki: Records of Ancient Matters*, Basil Hall Chamberlain (trans.), inside front cover, Charles E. Tuttle Company, Tokyo, 1993

10. Wheeler, op. cit., xxii

11. Ibid. xii

12. *Nihongi: Chronicles of Japan from the Earliest Times to AD 697*, W. G. Aston (trans.), Charles E. Tuttle Company, Tokyo, 1998

13. Wheeler, op. cit., xxiv

14. Wheeler, op. cit., xi, xviii; *Larousse*, 404

15. *Larousse*, 404; Wheeler, op. cit., xi, xxiv–xxv

16. *Larousse*, 404; Wheeler, op. cit., xi, xxiv–xxvi

17. See discussion in *Larousse*, 404

18. Robert Graves, in his Introduction to *Larousse*, v
19. Ibid., v
20. Most famously Schliemann following mythical clues to discover Troy
21. The case of Immanuel Velikovsky, for example
22. Alan Dundes (ed.), *The Flood Myth*, 1, University of California Press, 1988
23. E.g. Matsuo Tsukuda, 'Vegetation in Prehistoric Japan: The Last 20,000 Years', in *Windows on the Japanese Past: Studies in Archaeology and Prehistory*, 12, Centre for Japanese Studies, University of Michigan, 1986, 11
24. 'Authentic history in Japan begins only in the fifth century. Whatever is earlier than that belongs to the age of tradition, which is supposed to maintain an unbroken record for ten thousand years', Romyn Hitchcock, *Shinto, Or the Mythology of the Japanese*, 489, Report of National Museum, 1891. See also 505: The imperial family claims officially to have ruled Japan 'for 2,550 years, tracing its ancestry for still 10,000 years back . . .'
25. Wheeler, op. cit., 21
26. *Nihongi*, 32; *Kojiki*, 50
27. *Kojiki*, 51; *Larousse*, 407
28. *Kojiki*, 51
29. *Nihongi*, 33; *Kojiki*, 51
30. *Nihongi*, 34
31. *Kojiki*, 52
32. *Kojiki*, 52–3; *Nihongi*, 34–5
33. *Nihongi*, 35
34. Ibid., 40–41
35. An alternative in the same vein is that the myth is a metaphor for an eclipse, or reflects 'primitive' fears of eclipses, etc.
36. In some translations 'myriad' is given but presumed to be a copyist's error for 'evil' – see *Kojiki*, 66, note 4

37. *Kojiki*, 63
38. The possible interaction between the increased volcanism that is known to have occurred at the end of the Ice Age and post-glacial sea-level rise is discussed in chapter 3
39. *Nihongi*, 49
40. Ibid., 50
41. T. E. G. Reynolds and S. C. Kanser, 'Japan', in O. Soffer and G. Gamble, *The World at 18,000 BP*, chapter 16, 227–41, Unwin Hyman, London, 1990; Y. Igarishi, 'A lateglacial climatic reversion in Hokkaido, northeast Asia, inferred from the *Larix* pollen record', *Quaternary Science Reviews*, vol. 15, 1996, 989–95; N. Ooi, 'Pollen spectra from around 20,000 years ago during the Last Glacial from the Nara Basin, Japan', *The Quaternary Research (Japan)*, vol. 31, 1992, 203–12; N. Ooi , M. Minaki and S. Noshiro, 'Vegetation changes around the Last Glacial Maximum and effects of the Aira-Tn Ash, at the Itai-Teragatani Site, Central Japan', *Ecological Research*, vol. 5, 1990, 81–91; N. Ooi and S. Tsuji, 'Palynological study of the Peat Sediments around the Last Glacial Maximum at Hikone, the east shore of Lake Biwa, Japan', *Journal of Phytogeography and Taxonomy*, vol. 37, 1989, 37–42.
42. Ibid.
43. *Nihongi*, 52–2; *Kojiki*, 71–3
44. *Nihongi*, 55
45. Ibid., 10–12
46. *Heaven's Mirror*
47. *Nihongi*, 15 and footnote 1
48. *Nihongi*, 15
49. *Larousse*, 58–60
50. *Nihongi*, 21; Wheeler, op. cit., 12
51. *Kojiki*, 32
52. See discussion of the Orpheus tale in W. K. C. Guthrie, *Orpheus and Greek Religion*, 29ff, Princeton University

Press, 1993; Persephone, *Encyclopae-dia Britannica*, 9, 307

53. *Nihongi*, 24; *Kojiki*, 38
54. *Kojiki*, 39; *Nihongi*, 24; *Encyclopaedia Brittanica, Micropaedia*, vol. 9, 307
55. *Nihongi*, 24, footnote 2
56. *Muir's Sanscrit Texts*, vol. 5, 329, cited in *Nihongi*, 24, footnote 2
57. *Encyclopaedia Britannica*, 8, 1012
58. *Nihongi*, 24
59. Ibid., 24; *Kojiki*, 39
60. *Nihongi*, 24–5
61. Ibid., 25
62. Wheeler, op. cit., 16
63. Ibid., 290–91
64. Ibid., 291
65. Ibid., 291
66. Ibid., 291
67. Ibid., 292–3
68. Ibid., 292
69. Juliet Piggott, op. cit., 123–4
70. *Nihongi*, 92
71. *Kojiki*, 145
72. Wheeler, op. cit., 425, on hunter/gatherer symbolism of Fire-Glow, Fire Fade. Archaeology confirms that fishing and the resources of the sea played a very important role for the Jomon
73. *Kojiki*, 145–6
74. Ibid., 146
75. *Kojiki*, 146
76. *Nihongi*, 92
77. Ibid., 92
78. Ibid., 92
79. *Kojiki*, 146
80. *Nihongi*, 92–3
81. Ibid., 93
82. Ibid., 93
83. Ibid., 93
84. Ibid., 93
85. Ibid., 94
86. Ibid., 94
87. Ibid., 94
88. Ibid., 95
89. *Kojiki*, 155
90. *Nihongi*, 94–5
91. Ibid., 95
92. Wheeler, op. cit., 89
93. *Kojiki*, 147
94. Ibid., 156–7
95. Wheeler, op. cit., 425

27 / Confronting Yonaguni

1. Points 1–8 cited verbatim from Professor Kimura, *Diving Survey Report for Submarine Ruins off Japan*, 178
2. Points 9–12, discussions with Professor Kimura, cited in *Heaven's Mirror*, 216–17
3. See his contribution to my 1998 television series, *Quest for the Lost Civilization*
4. See *Heaven's Mirror*, 215–16
5. See *Heaven's Mirror*, 217
6. *Horizon*, BBC2, 4 November 1999
7. Robert Schoch, *Voices of the Rocks*, 111–12, Harmony Books, New York, 1999
8. See ibid., 112–13; *Heaven's Mirror*, 217–21
9. Schoch, op. cit., 112
10. See discussion in *Heaven's Mirror*
11. *Der Spiegel*, 34/1999
12. *Der Spiegel*, 34/1999
13. www.grahamhancock.com, Articles
14. Interviewed by Tim Copestake for *Underworld* television series
15. TBS
16. TBS
17. Sundaresh report, see above
18. The boulder was rolled to the side, half on and half off the platform

28 / Maps of Japan and Taiwan 13,000 Years Ago?

1. In Lutz Walter (ed.), *Japan: A Cartographic Vision*, 2, Munich, NY, 1994
2. Robert H. Fuson, *Legendary Islands of the Ocean Sea*, 199, Pineapple Press Inc., Florida, 1995
3. See discussion in Joseph Needham, *Science and Civilization in China*, vol. 3,

497ff, Cambridge University Press, 1979 (first published 1959)

4. See chapter 24

5. Fuson, op. cit., 196

6. Discussed above, chapter 24

7. Fuson, op. cit., 196

29 / Confronting Kerama

1. *Collins English Dictionary*, 953, Collins, London, 1982

2. Two prominent Maltese sites contain a combination of rock-hewn structures and free-standing megaliths – the Hypogeum of Hal Saflieni and the Borchtorff Circle at Xaghra. The latter is semi-subterranean in form, rather similar to the Centre Circle complex at Kerama

30 / The Shark at the Gate

1. Janet B. Montgomery McGovern, *Among the Head Hunters of Formosa*, 39, SMC Publishing Inc., Taipei, 1997 (first published 1922)

2. Ibid., 39; Robert H. Fuson, *Legendary Islands of the Ocean Sea*, 193, Pineapple Press Inc., Florida, 1995

3. *Encyclopaedia Britannica, Micropaedia*, vol. 10, 272

4. Post Wheeler, *The Sacred Scriptures of the Japanese*, 425, Henry Schuman Inc., 1952

5. *Nihongi: Chronicles of Japan from the Earliest Times to AD 697*, W. G. Aston (trans.), 96, Charles E. Tuttle Company, Tokyo, 1998

6. *The Kojiki: Records of Ancient Matters*, Basil Hall Chamberlain (trans.), 147, Charles E. Tuttle Company, Tokyo, 1993; Wheeler, op. cit., 82

7. See discussions in chapter 26

8. *Shih Chi*, cited in Joseph Needham, *Science and Civilization in China*, vol. 4, part 3, 551, Cambridge University Press, 1979 (first published 1959)

9. Cited in ibid., vol. 4, part 3, 550

10. Ibid., vol. 4, part 3, 15

11. Ibid., vol. 4, part 3, 549

12. Ibid., vol. 4, part 3, 548

13. Betty Meggers, Clifford Evans and Emilio Estrada, *Smithsonian Contributions to Anthropology*, vol. 1

14. See discussion in chapter 25

15. See discussion, 'stone boat' in chapter 25

16. *Encyclopaedia Britannica*, vol. 7, 43–4

17. Needham, op. cit., vol. 4, part 3, 549. Needham would put it further east – perhaps even as far east as the Americas – though no one knows for sure, since its location is, after all, 'mythical'

18. Ibid., vol. 4, part 3, 549

19. Ibid., vol. 4, part 3, 551

20. E.g. see Gregory C. McIntosh, *The Piri Reis Map of 1513*, 72, 115, The University of Georgia Press, 2000; Svat Soucek, *Piri Reis and Turkish Mapmaking after Columbus*, 99, The Nour Foundation in association with Oxford University Press, 1996; Fuson, op. cit., 185

21. Cited in Needham, op. cit., vol. 4, part 3, 552

22. Cited in ibid., vol. 4, part 3, 553

23. Cited in ibid., vol. 4, part 3, 553

24. Cited in ibid., vol. 4, part 3, 551

25. Cited in ibid., vol. 4, part 3, 553

26. Ibid., vol. 4, part 3, 553

27. Ibid., vol. 4, part 3, 547

28. Ibid., vol. 4, part 3, 547–8

29. Ibid., vol. 4, part 3, 547–8

30. Ibid., vol. 4, part 3, 538. An alternative translation for Bird's-Eye map is 'Flying Bird Calendar'; I know which I prefer!

31. Cited in Needham, op. cit., vol. 4, part 3, 538

32. Ibid., vol. 4, part 3, 539

33. Date range approximate; source: Jacques Gernet, *A History of Chinese Civilization*, 39ff, Cambridge University Press, 1999

34. Cited in Needham, op. cit., vol. 4, part 3, 539
35. Ibid., vol. 4, part 3, 547
36. Translation in Wheeler, op. cit., 40–41
37. In ibid., 40
38. Ibid., 40–41
39. And collated by Sir James Frazer in *Folklore in the Old Testament*, vol. 1, 225–32, Macmillan, London
40. Ibid., 225–32
41. Ibid., 225–7
42. Ibid., 227
43. See *Fingerprints of the Gods*, chapter 24, for Noah-type flood myths from all around the world
44. These numbers are a focus of the discussion in Giorgio de Santillana and Hertha von Dechend, *Hamlet's Mill*, Nonpareil, Boston, 1992
45. Discussed at length in *Fingerprints of the Gods* and *Heaven's Mirror*
46. Thanks to Henry H. Y. Yuang for pointing this out

Index

Figures in italics refer to maps, those in bold type to illustrations.